VOLUME EIGHTY-ONE

VITAMINS AND HORMONES

Anandamide an Endogenous Cannabinoid

VITAMINS AND HORMONES

Editorial Board

TADHG P. BEGLEY
ANTHONY R. MEANS
BERT W. O'MALLEY
LYNN RIDDIFORD
ARMEN H. TASHJIAN, JR.

VOLUME EIGHTY-ONE

VITAMINS AND HORMONES

Anandamide an Endogenous Cannabinoid

Editor-in-Chief

GERALD LITWACK
*Chair, Department of Basic Sciences
The Commonwealth Medical College
Scranton, Pennsylvania*

AMSTERDAM • BOSTON • HEIDELBERG • LONDON
NEW YORK • OXFORD • PARIS • SAN DIEGO
SAN FRANCISCO • SINGAPORE • SYDNEY • TOKYO
Academic Press is an imprint of Elsevier

Cover photo credit: PDB 2vya
Mileni, M., Johnson, D. S., Wang, Z., Everdeen, D. S., Lilmatta, M., Pabst, B., Bhattacharya, K., Nugent, R. A., Kamtekar, S., Cravatt, B. F., Ahn, K., Stevens, R. C. Structure-guided inhibitor design for human FAAH by interspecies active site conversion. *Proc. Natl. Acad. Sci. USA* (2008) **105**, p. 12820.

Academic Press is an imprint of Elsevier
32 Jamestown Road, London, NW1 7BY, UK
Radarweg 29, PO Box 211, 1000 AE Amsterdam, The Netherlands
Linacre House, Jordan Hill, Oxford OX2 8DP, UK
30 Corporate Drive, Suite 400, Burlington, MA 01803, USA
525 B Street, Suite 1900, San Diego, CA 92101-4495, USA

First edition 2009

Copyright © 2009 Elsevier Inc. All rights reserved.

No part of this publication may be reproduced, stored in a retrieval system or transmitted in any form or by any means electronic, mechanical, photocopying, recording or otherwise without the prior written permission of the publisher

Permissions may be sought directly from Elsevier's Science & Technology Rights Department in Oxford, UK: phone (+44) (0) 1865 843830; fax (+44) (0) 1865 853333; email: permissions@elsevier.com. Alternatively you can submit your request online by visiting the Elsevier web site at http://elsevier.com/locate/permissions, and selecting *Obtaining permission to use Elsevier material*

Notice

No responsibility is assumed by the publisher for any injury and/or damage to persons or property as a matter of products liability, negligence or otherwise, or from any use or operation of any methods, products, instructions or ideas contained in the material herein. Because of rapid advances in the medical sciences, in particular, independent verification of diagnoses and drug dosages should be made

ISBN: 978-0-12-374782-2
ISSN: 0083-6729

For information on all Elsevier Academic Press publications
visit our website at elsevierdirect.com

Printed and bound in USA
09 10 11 12   10 9 8 7 6 5 4 3 2 1

Working together to grow
libraries in developing countries

www.elsevier.com | www.bookaid.org | www.sabre.org

Former Editors

ROBERT S. HARRIS
Newton, Massachusetts

JOHN A. LORRAINE
University of Edinburgh
Edinburgh, Scotland

PAUL L. MUNSON
University of North Carolina
Chapel Hill, North Carolina

JOHN GLOVER
University of Liverpool
Liverpool, England

GERALD D. AURBACH
Metabolic Diseases Branch
National Institute of
Diabetes and Digestive and
Kidney Diseases
National Institutes of Health
Bethesda, Maryland

KENNETH V. THIMANN
University of California
Santa Cruz, California

IRA G. WOOL
University of Chicago
Chicago, Illinois

EGON DICZFALUSY
Karolinska Sjukhuset
Stockholm, Sweden

ROBERT OLSEN
School of Medicine
State University of New York
at Stony Brook
Stony Brook, New York

DONALD B. MCCORMICK
Department of Biochemistry
Emory University School of
Medicine, Atlanta, Georgia

Contents

Contributors	xv
Preface	xxi

1. Enzymatic Formation of Anandamide — 1
Yasuo Okamoto, Kazuhito Tsuboi, and Natsuo Ueda

I.	The Transacylation–Phosphodiesterase Pathway for Anandamide Formation	2
II.	NAT	4
III.	NAPE-PLD	9
IV.	Alternative Pathways Forming NAEs from NAPEs	15
V.	Conclusions	18
	References	19

2. Organized Trafficking of Anandamide and Related Lipids — 25
Marla L. Yates and Eric L. Barker

I.	AEA and the Endocannabinoid System	26
II.	AEA Transport	31
	Acknowledgment	45
	References	45

3. Biosynthesis of Oleamide — 55
Gregory P. Mueller and William J. Driscoll

I.	Introduction	56
II.	Fatty Acid Amide Messengers: Structural Considerations	56
III.	Natural Occurrence of Oleamide	57
IV.	Biologic Actions of Oleamide	58
V.	Proposed Mechanisms for the Biosynthesis of Oleamide	59
VI.	Oleamide Biosynthesis by Peptidylglycine Alpha-amidating Monooxygenase	60
VII.	Discovery of Cytochrome *c* as an Oleamide Synthase	62
VIII.	Cytochrome *c* also Catalyzes the Formation of Oleoylglycine and Other Long-Chain Fatty Acylamino Acids	64
IX.	Proposal for an Oleamide Synthesome	66

X. Apoptosis: A Model for the Mechanism and Regulation
   of Oleamide Biosynthesis   67
XI. Considerations for the Investigation of Oleamide Biosynthesis   70
XII. Future Directions and Concluding Remarks   70
References   71

4. Anandamide Receptor Signal Transduction   79
Catherine E. Goodfellow and Michelle Glass

I. Introduction   80
II. Cannabinoid Receptor 1   81
III. Cannabinoid Receptor 2   89
IV. Transient Receptor Potential Vanilloid 1   90
V. Evidence for Additional Receptors   92
VI. Concluding Remarks   98
References   99

5. Is GPR55 an Anandamide Receptor?   111
Andrew J. Brown and C. Robin Hiley

I. Δ^9-Tetrahydrocannabinol, CB_1, and CB_2 Receptors   112
II. Functional Evidence for Novel Cannabinoid Receptors   113
III. Genomics of G Protein-Coupled Cannabinoid Receptors   116
IV. The Orphan Receptor GPR55   117
V. Endogenous Ligands for GPR55   121
VI. GPR55 Cellular Signaling Pathways   123
VII. Interactions Between GPR55 and CB_1 Receptors   126
VIII. Conclusion: GPR55 as an Anandamide Receptor   128
References   133

6. The Endocannabinoid System During Development:
   Emphasis on Perinatal Events and Delayed Effects   139
Ester Fride, Nikolai Gobshtis, Hodaya Dahan, Aron Weller,
Andrea Giuffrida, and Shimon Ben-Shabat

I. Introduction   140
II. Early Gestation   141
III. Neural Development   146
IV. Postnatal Development   147
V. Effects of Developmental Manipulation of the
   ECS System on the Offspring   150
VI. Conclusions   152
Acknowledgments   153
References   153

7. **Cannabinoid Receptor CB1 Antagonists: State of the Art and Challenges**   **159**

Maurizio Bifulco, Antonietta Santoro, Chiara Laezza, and Anna Maria Malfitano

    I. Introduction   160
    II. Endocannabinoid System: Control of Energy Balance   161
    III. Cannabinoid CB1 Receptors and CB1 Antagonists   162
    IV. CB1 Antagonists in the Treatment of Obesity and Related Comorbidities   169
    V. Other Emerging Effects of CB1 Antagonists   176
    VI. Therapeutic Prospects   179
    VII. Conclusions   180
    Acknowledgments   181
    References   181

8. **Novel Endogenous N-Acyl Glycines: Identification and Characterization**   **191**

Heather B. Bradshaw, Neta Rimmerman, Sherry S.-J. Hu, Sumner Burstein, and J. Michael Walker

    I. Historical View of Lipid Signaling Discoveries   192
    II. The Identification of Endogenous Signaling Lipids with Cannabimimetic Activity   192
    III. Identification of Additional N-Acyl Amides   193
    IV. N-Arachidonoyl Glycine Biological Activity   194
    V. N-Arachidonoyl Glycine Biosynthesis   194
    VI. N-Palmitoyl Glycine Biological Activity   195
    VII. N-Palmitoyl Glycine Biosynthesis   197
    VIII. PalGly Metabolism   197
    IX. Identification and Characterization of Additional Members of the N-Acyl Glycines   198
    X. Biological Activity of Novel N-Acyl Glycines   200
    XI. Conclusions   201
    References   203

9. **The Endocannabinoid Anandamide: From Immunomodulation to Neuroprotection. Implications for Multiple Sclerosis**   **207**

Fernando G. Correa, Leyre Mestre, Fabián Docagne, José Borrell, and Carmen Guaza

    I. Introduction   208
    II. AEA as a Neuroimmune Signal   212

III.	Anandamide and Multiple Sclerosis	219
IV.	Concluding Remarks	223
	Acknowledgment	224
	References	224

10. Modulation of the Endocannabinoid-Degrading Enzyme Fatty Acid Amide Hydrolase by Follicle-Stimulating Hormone   231

Paola Grimaldi, Gianna Rossi, Giuseppina Catanzaro, and Mauro Maccarrone

I.	Follicle-Stimulating Hormone: Signal Transduction and Molecular Targets	232
II.	Sertoli Cells: Activities and Biological Relevance	237
III.	Overview of the Endocannabinoid System	239
IV.	The ECS in Sertoli Cells	241
V.	Regulation of FAAH by FSH in Sertoli Cells	243
VI.	FAAH Is an Integrator of Fertility Signals	244
VII.	Conclusions	254
	Acknowledgments	255
	References	255

11. Glucocorticoid-Regulated Crosstalk Between Arachidonic Acid and Endocannabinoid Biochemical Pathways Coordinates Cognitive-, Neuroimmune-, and Energy Homeostasis-Related Adaptations to Stress   263

Renato Malcher-Lopes and Marcelo Buzzi

I.	Introduction	264
II.	The Arachidonic Acid Cascade	266
III.	Glucocorticoid-Mediated Inhibition of $cPLA_2$-Dependent AA Release from Membrane Phospholipids	269
IV.	Biosynthesis of the AA-Containing Endocannabinoids AEA and 2-AG	271
V.	Nongenomic Glucocorticoid-Induced Activation of Endocannabinoid Biosynthesis	274
VI.	Endocannabinoids Metabolization	278
VII.	Crosstalk Between GCs and COX_2 in the Control of Neuroinflammation and Neuroprotection	282
VIII.	Crosstalk Between GCs and COX_2 in the Control of Synaptic Plasticity and Learning Processes	285
IX.	Coordination of GC-Mediated Control of the Neuroimmune Response and Energy Homeostasis Control	288
	Acknowledgments	295
	References	295

12. Modulation of the Cys-Loop Ligand-Gated Ion Channels by Fatty Acid and Cannabinoids — 315

Li Zhang and Wei Xiong

    I. CB Receptor-Dependent and -Independent Effects of Endocannabinoids — 316
    II. Structure and Function of the Cys-Loop LGICs — 317
    III. Inhibition of 5-HT_3 Receptors by Cannabinoids — 318
    IV. Modulation of Gly Receptor Function by Cannabinoids — 324
    V. Inhibition of nACh Receptors by Endocannabinoids — 328
    VI. Modulation of $GABA_A$ Receptor Function by Fatty Acids — 329
    VII. Concluding Discussion — 330
    References — 331

13. Endogenous Cannabinoids and Neutrophil Chemotaxis — 337

Douglas McHugh and Ruth A. Ross

    I. Cellular Motility and Neutrophils — 338
    II. The Endogenous Cannabinoid System — 339
    III. Cannabinoids Modulate Cell Migration — 342
    IV. Endocannabinoid Effects on Basal Locomotion of Neutrophils — 343
    V. Endocannabinoid Effects on Induced Migration of Neutrophils — 343
    VI. Cannabinoid Receptor Expression in Neutrophils — 344
    VII. Inhibition of Induced Migration: Which Receptors are Involved? — 345
    VIII. Inhibitory Signal Transduction Mechanisms: Receptor Crosstalk — 350
    IX. Inhibitory Signal Transduction Mechanisms: Disruption of the Actin Cytoskeleton — 351
    X. Conclusion — 356
    References — 357

14. CB1 Activity in Male Reproduction: Mammalian and Nonmammalian Animal Models — 367

Riccardo Pierantoni, Gilda Cobellis, Rosaria Meccariello, Giovanna Cacciola, Rosanna Chianese, Teresa Chioccarelli, and Silvia Fasano

    I. Introduction — 368
    II. Receptor Properties — 368
    III. Brain–Pituitary Axis — 371
    IV. Testis — 374
    V. Excurrent Duct System — 379
    VI. Concluding Remarks — 382
    References — 382

15. Anandamide and the Vanilloid Receptor (TRPV1) — 389

Attila Tóth, Peter M. Blumberg, and Judit Boczán

I. Cannabinoid and Vanilloid Receptors — 390
II. Biochemistry of Anandamide — 392
III. Anandamide as Vanilloid Receptor (TRPV1) Ligand — 396
IV. Other Anandamide Receptors — 404
V. Physiological Actions of Anandamide on TRPV1 — 404
VI. Future Directions — 410
References — 411

16. Endocannabinoid System and Fear Conditioning — 421

Leonardo B. M. Resstel, Fabrício A. Moreira, and Francisco S. Guimarães

I. Introduction — 421
II. Fear Conditioning — 423
III. Influence of Endocannabinoids on Fear Conditioning — 426
IV. Brain Regions in which Endocannabinoids may Modulate Fear Conditioning — 429
V. Conclusion — 433
References — 433

17. Regulation of Gene Transcription and Keratinocyte Differentiation by Anandamide — 441

Nicoletta Pasquariello, Sergio Oddi, Marinella Malaponti, and Mauro Maccarrone

I. Introduction — 442
II. Epidermis — 445
III. Transcriptional Control of Skin Differentiation — 453
IV. Endocannabinoid System in Epidermis — 455
V. Modulation of the Endocannabinoid System in Differentiating Keratinocytes — 456
VI. Repression of Gene Transcription by Anandamide — 457
VII. Conclusions — 459
Acknowledgments — 460
References — 461

18. Changes in the Endocannabinoid System May Give Insight into new and Effective Treatments for Cancer — 469

Gianfranco Alpini and Sharon DeMorrow

I. Introduction — 470
II. Changes in the Endocannabinoid System in Cancer — 471

III.	Antiproliferative Effects of Anandamide	473
IV.	Effects of AEA on Migration, Invasion, and Angiogenesis	476
V.	Targeting Degradation Enzymes of Cannabinoids as an Anticancer Therapy	479
VI.	Tumor Promoting Effects of Anandamide	480
VII.	Conclusions	480
	Acknowledgments	481
	References	481

19. Use of Cannabinoids as a Novel Therapeutic Modality Against Autoimmune Hepatitis   487

Rupal Pandey, Venkatesh L. Hegde, Narendra P. Singh, Lorne Hofseth, Uday Singh, Swapan Ray, Mitzi Nagarkatti, and Prakash S. Nagarkatti

I.	Introduction	488
II.	The Endogenous Cannabinoid System	489
III.	The Biosynthesis of Endocannabinoids	490
IV.	Endocannabinoid System is Autoprotective	490
V.	Autoimmune Hepatitis	492
VI.	Treatment Drawbacks	494
VII.	Cannabinoid/Endocannabinoid System in Hepatitis	494
VIII.	Conclusions and Future Directions	499
	Acknowledgments	500
	References	500

Index   *505*

Contributors

Gianfranco Alpini
Division of Research, Central Texas Veterans Health Care System, Temple, Texas, USA, and Systems Biology and Translational Medicine, Texas A&M Health Science Center, College of Medicine, Temple, Texas, USA, and Department of Medicine, Texas A&M Health Science Center, College of Medicine, Temple, Texas, USA

Eric L. Barker
The Department of Medicinal Chemistry and Molecular Pharmacology, Purdue University School of Pharmacy and Pharmaceutical Sciences, West Lafayette, Indiana, USA

Shimon Ben-Shabat
Department of Pharmacology and School of Pharmacy, Ben-Gurion University of the Negev, Beer Sheva, Israel

Maurizio Bifulco
Dipartimento di Scienze Farmaceutiche, Università di Salerno, Fisciano (Salerno), Italy

Peter M. Blumberg
Molecular Mechanisms of Tumor Promotion Section, Laboratory of Cancer Biology and Genetics, National Cancer Institute, National Institutes of Health, Bethesda, Maryland, USA

Judit Boczán
Department of Neurology, University of Debrecen, Debrecen, Hungary

José Borrell
Neuroimmunology Group, Functional and Systems Neurobiology Department, Cajal Institute, CSIC, Avda Doctor Arce, Madrid, Spain

Heather B. Bradshaw
Department of Psychological and Brain Sciences, Indiana University, Bloomington, Indiana, USA

Andrew J. Brown
Department of Screening and Compound Profiling, Molecular Discovery Research, GlaxoSmithKline, New Frontiers Science Park, GlaxoSmithKline, Third Avenue, Harlow, Essex, United Kingdom

Sumner Burstein
Department of Biochemistry and Molecular Pharmacology, University of Massachusetts Medical School, Worcester, Massachusetts, USA

Marcelo Buzzi
Laboratory of Analytical Biochemistry, Molecular Pathology Division, SARAH Network of Rehabilitation Hospitals, Brasília-DF, Brazil

Giovanna Cacciola
Dipartimento di Medicina Sperimentale, Seconda Università di Napoli, Via Costantinopoli, Napoli, Italy

Giuseppina Catanzaro
European Center for Brain Research (CERC)/Santa Lucia Foundation, Rome, Italy, and Department of Biomedical Sciences, University of Teramo, Teramo, Italy

Rosanna Chianese
Dipartimento di Medicina Sperimentale, Seconda Università di Napoli, Via Costantinopoli, Napoli, Italy

Teresa Chioccarelli
Dipartimento di Medicina Sperimentale, Seconda Università di Napoli, Via Costantinopoli, Napoli, Italy

Gilda Cobellis
Dipartimento di Medicina Sperimentale, Seconda Università di Napoli, Via Costantinopoli, Napoli, Italy

Fernando G. Correa
Neuroimmunology Group, Functional and Systems Neurobiology Department, Cajal Institute, CSIC, Avda Doctor Arce, Madrid, Spain

Hodaya Dahan
Department of Psychology, Bar-Ilan University, Ramat Gan, Israel, and Departments of Behavioral Sciences and Molecular Biology, Ariel University Center of Samaria, Ariel, Israel

Sharon DeMorrow
Division of Research and Education, Scott & White Hospital, Temple, Texas, USA, and Department of Medicine, Texas A&M Health Science Center, College of Medicine, Temple, Texas, USA

Fabián Docagne
Neuroimmunology Group, Functional and Systems Neurobiology Department, Cajal Institute, CSIC, Avda Doctor Arce, Madrid, Spain

William J. Driscoll
Department of Anatomy, Physiology and Genetics, F. Edward Hebert School of Medicine, Uniformed Services University of the Health Sciences, Bethesda, Maryland, USA

Silvia Fasano
Dipartimento di Medicina Sperimentale, Seconda Università di Napoli, Via Costantinopoli, Napoli, Italy

Ester Fride
Departments of Behavioral Sciences and Molecular Biology, Ariel University Center of Samaria, Ariel, Israel

Andrea Giuffrida
Department of Pharmacology, University of Texas Health Science Center, San Antonio, Texas, USA

Michelle Glass
Department of Pharmacology and Clinical Pharmacology, University of Auckland, Auckland, New Zealand

Nikolai Gobshtis
Departments of Behavioral Sciences and Molecular Biology, Ariel University Center of Samaria, Ariel, Israel

Catherine E. Goodfellow
Department of Pharmacology and Clinical Pharmacology, University of Auckland, Auckland, New Zealand

Paola Grimaldi
Department of Public Health and Cell Biology, University of Rome Tor Vergata, Rome, Italy

Carmen Guaza
Neuroimmunology Group, Functional and Systems Neurobiology Department, Cajal Institute, CSIC, Avda Doctor Arce, Madrid, Spain

Francisco S. Guimarães
Department of Pharmacology, School of Medicine of Ribeirão Preto, University of São Paulo, Ribeirão Preto, São Paulo, Brazil

Venkatesh L. Hegde
Department of Pathology, Microbiology, and Immunology, University of South Carolina School of Medicine, Columbia, South Carolina, USA

C. Robin Hiley
Department of Pharmacology, University of Cambridge, Tennis Court Road, Cambridge, United Kingdom

Lorne Hofseth
Department of Pharmaceutical and Biomedical Sciences, College of Pharmacy, University of South Carolina, Columbia, South Carolina, USA

Sherry S.-J. Hu
Department of Psychological and Brain Sciences, Indiana University, Bloomington, Indiana, USA

Chiara Laezza
Istituto di Endocrinologia ed Oncologia Sperimentale, CNR, Naples, Italy

Mauro Maccarrone
Department of Biomedical Sciences, University of Teramo, Teramo, Italy, and European Center for Brain Research (CERC)/Santa Lucia Foundation, Rome, Italy

Marinella Malaponti
European Center for Brain Research (CERC)/Santa Lucia Foundation, Rome, Italy

Renato Malcher-Lopes
Laboratory of Mass Spectrometry, EMBRAPA-Center for Genetic Resources and Biotechnology, Brasília-DF, Brazil, and Genomic Sciences and Biotechnology Graduation Program of the Universidade Católica de Brasília

Anna Maria Malfitano
Dipartimento di Scienze Farmaceutiche, Università di Salerno, Fisciano (Salerno), Italy

Douglas McHugh
Department of Psychological and Brain Sciences, Indiana University, Bloomington, Indiana, USA

Rosaria Meccariello
Dipartimento di Studi delle Istituzioni e dei Sistemi Territoriali, Università Parthenope, Napoli, Italy

Leyre Mestre
Neuroimmunology Group, Functional and Systems Neurobiology Department, Cajal Institute, CSIC, Avda Doctor Arce, Madrid, Spain

Fabrício A. Moreira
Department of Pharmacology, Institute of Biological Sciences, Federal University of Minas Gerais, Belo Horizonte, Minas Gerais, Brazil

Gregory P. Mueller
Department of Anatomy, Physiology and Genetics, F. Edward Hebert School of Medicine, Uniformed Services University of the Health Sciences, Bethesda, Maryland, USA

Mitzi Nagarkatti
Department of Pathology, Microbiology, and Immunology, University of South Carolina School of Medicine, Columbia, South Carolina, USA

Prakash S. Nagarkatti
Department of Pathology, Microbiology, and Immunology, University of South Carolina School of Medicine, Columbia, South Carolina, USA

Sergio Oddi
European Center for Brain Research (CERC)/Santa Lucia Foundation, Rome, Italy, and Department of Biomedical Sciences, University of Teramo, Teramo, Italy

Yasuo Okamoto
The Department of Physiology, Graduate School of Medical Sciences, Kanazawa University, Kanazawa, Japan, and The Department of Biochemistry, Kagawa University School of Medicine, Kagawa, Japan

Rupal Pandey
Department of Pathology, Microbiology, and Immunology, University of South Carolina School of Medicine, Columbia, South Carolina, USA

Nicoletta Pasquariello
Department of Biomedical Sciences, University of Teramo, Teramo, Italy

Riccardo Pierantoni
Dipartimento di Medicina Sperimentale, Seconda Università di Napoli, Via Costantinopoli, Napoli, Italy

Swapan Ray
Department of Pathology, Microbiology, and Immunology, University of South Carolina School of Medicine, Columbia, South Carolina, USA

Leonardo B. M. Resstel
Department of Pharmacology, School of Medicine of Ribeirão Preto, University of São Paulo, Ribeirão Preto, São Paulo, Brazil

Neta Rimmerman
Department of Psychological and Brain Sciences, Indiana University, Bloomington, Indiana, USA

Ruth A. Ross
Institute of Medical Sciences, University of Aberdeen, Aberdeen, Scotland, United Kingdom

Gianna Rossi
Department of Health Sciences, University of L'Aquila, L'Aquila, Italy

Antonietta Santoro
Dipartimento di Scienze Farmaceutiche, Università di Salerno, Fisciano (Salerno), Italy

Narendra P. Singh
Department of Pathology, Microbiology, and Immunology, University of South Carolina School of Medicine, Columbia, South Carolina, USA

Uday Singh
Department of Pathology, Microbiology, and Immunology, University of South Carolina School of Medicine, Columbia, South Carolina, USA

Attila Tóth
Division of Clinical Physiology, Institute of Cardiology, University of Debrecen, Debrecen, Hungary

Kazuhito Tsuboi
The Department of Biochemistry, Kagawa University School of Medicine, Kagawa, Japan

Natsuo Ueda
The Department of Biochemistry, Kagawa University School of Medicine, Kagawa, Japan

J. Michael Walker
Department of Psychological and Brain Sciences, Indiana University, Bloomington, Indiana, USA

Aron Weller
Department of Psychology, Bar-Ilan University, Ramat Gan, Israel

Wei Xiong
Laboratory for Integrative Neuroscience, National Institute on Alcohol Abuse and Alcoholism, National Institutes of Health, Rockville, Maryland, USA

Marla L. Yates
The Department of Medicinal Chemistry and Molecular Pharmacology, Purdue University School of Pharmacy and Pharmaceutical Sciences, West Lafayette, Indiana, USA

Li Zhang
Laboratory for Integrative Neuroscience, National Institute on Alcohol Abuse and Alcoholism, National Institutes of Health, Rockville, Maryland, USA

Preface

It is interesting that the drug, morphine, used for 200 years to palliate pain was found in the last decade or so to mimic endogenous compounds made in the body. Now, we learn that the body also makes substances, anandamide for one, that act like cannabinoids found in plants that have been used recreationally and for relief of pain. The endocannabinoids are also involved in several systems in the body and in this volume are reviewed many aspects of anandamide and other endogenous cannabinoids.

The collection begins with more basic studies. The opening paper by Y. Okamoto, K. Tsuboi, and N. Ueda reports on the "Enzymatic formation of anandamide." M. L. Yates and E. L. Barker then review "Organized trafficking of anandamide and related lipids." G. P. Mueller and W. J. Driscoll report on "Biosynthesis of oleamide." C. E. Goodfellow and M. Glass follow with "Anandamide receptor signal transduction." A. J. Brown and C. R. Hiley ask "Is GPR55 an anandamide receptor?" Developmental aspects are the subject of a paper by E. Fride, N. Gobshtis, H. Dahan, A. Weller, G. Giuffrida, and S. Ben-Shabat: "The endocannabinoid system during development: emphasis on perinatal events and delayed effects." M. Bifulco, A. Santoro, C. Laezza, and A. M. Malfitano discuss the "Cannabinoid receptor CB1 antagonists: state of the art and challenges." Related compounds are addressed by H. B. Bradshaw, N. Rimmerman, S. S.-J. Hu, S. Burstein, and the late J. M. Walker in "Novel endogenous N-acyl glycines: identification and characterization." In relation to multiple sclerosis, F. G. Correa, L. Mestre, F. Docagne, J. Borrell, and C. Guaza write on "The endocannabinoid anandamide: from immunomodulation to neuroprotection, implications for multiple sclerosis." Referring to the enzyme that degrades anandamide, P. Grimaldi, G. Rossi, G. Catanzaro, and M. Maccarrone offer "Modulation of the endocannabinoid-degrading enzyme fatty acid amide hydrolase by follicle-stimulating hormone." "Glucocorticoid-regulated crosstalk between arachidonic acid and endocannabinoids biochemical pathways coordinates cognitive-, neuroimmune-, and energy homeostasis-related adaptations to stress" is the subject of R. Malcher-Lopes and M. Buzzi. L. Zhang and W. Xiong review "Modulation of the Cys-loop ligand-gated ion channels by fatty acid and cannabinoid."

This volume is rounded out by a group of papers that focus on biological and disease-related aspects of the endocannabinoids. D. McHugh and R. A. Ross discuss "Endogenous cannabinoids and neutrophil chemotaxis."

R. Pierantoni, G. Cobellis, R. Meccariello, G. Cacciola, R. Chianese, T. Chioccarelli, and S. Fasano report on "CB1 activity in male reproduction: mammalian and nonmammalian animal models." A. Toth, P. M. Blumberg, and J. Boczan study the relationship between anandamide and vanilloids in "Anandamide and the vanilloid receptor (TRPV1)." "Endocannabinoid system and fear conditioning" is the subject of L. B. M. Resstel, F. A. Moreira, and F. S. Guimaraes. N. Pasquariello, S. Oddi, M. Malaponti, and M. Maccarrone review "Regulation of gene transcription and keratinocyte differentiation by anandamide." Two disease-related papers close the volume: "Changes in the endocannabinoid system may give insight into new and effective treatments for cancer" by G. Alpini and S. DeMorrow and "Use of cannabinoids as a novel therapeutic modality against autoimmune hepatitis" by R. Panday, V. Hegde, N. Singh, L. Hofseth, U. Singh, S. Ray, M. Nagarkatti, and P. Nagarkatti.

The structure on the cover is a representation of the biological molecule of fatty acid amide hydrolase, the enzyme that inactivates anandamide. Its code is 2vya from the Protein Data Bank and the primary citation is M. Mileni, D. S. Johnson, Z. Wang, D. S. Everdeen, M. Lilmatta, B. Pabst, K. Bhattacharya, R. A. Nugent, S. Kamtekar, B. F. Cravatt, K. Ahn, and R. C. Stevens (2008). "Structure-guided inhibitor design for human FAAH by interspecies active site conversion." *Proc. Natl. Acad. Sci. USA* **105,** 12820.

Janice Hackenberg from Elsevier collaborated in the early phases of this volume and the bulk of processing of the manuscripts fell on Lisa Tickner and Narmada Thangavelu of Elsevier. Elsevier has made plans to decrease publication time. My thanks to all who participated in the production of this volume.

Gerald Litwack
March 9, 2009

CHAPTER ONE

Enzymatic Formation of Anandamide

Yasuo Okamoto,[*,†] Kazuhito Tsuboi,[*] and Natsuo Ueda[*]

Contents

I. The Transacylation–Phosphodiesterase Pathway for Anandamide Formation	2
II. NAT	4
A. Ca-NAT	4
B. iNAT	6
III. NAPE-PLD	9
A. Structure	9
B. Function	12
C. Tissue distribution	14
IV. Alternative Pathways Forming NAEs from NAPEs	15
V. Conclusions	18
References	19

Abstract

In animal tissues anandamide and other bioactive *N*-acylethanolamines are principally produced from glycerophospholipids through the transacylation–phosphodiesterase pathway consisting of two enzymatic reactions. The first reaction is the generation of *N*-acylphosphatidylethanolamine (NAPE) by transferring an acyl group esterified at *sn*-1 position of glycerophospholipid to the amino group of phosphatidylethanolamine. This reaction is catalyzed by Ca^{2+}-dependent *N*-acyltransferase. The discovery of Ca^{2+}-independent *N*-acyltransferase revealed the existence of plural enzymes which are capable of catalyzing this reaction. The second reaction is the release of *N*-acylethanolamine from NAPE catalyzed by NAPE-hydrolyzing phospholipase D (NAPE-PLD). The enzyme belongs to the metallo-β-lactamase family and specifically hydrolyzes NAPEs. Recent studies, including analysis of NAPE-PLD-deficient mice, led to the discovery of NAPE-PLD-independent pathways for the anandamide biosynthesis.

[*] The Department of Biochemistry, Kagawa University School of Medicine, Kagawa, Japan
[†] The Department of Physiology, Graduate School of Medical Sciences, Kanazawa University, Kanazawa, Japan

I. THE TRANSACYLATION–PHOSPHODIESTERASE PATHWAY FOR ANANDAMIDE FORMATION

Ethanolamides of long-chain fatty acids, referred to as N-acylethanolamines (NAEs), comprise several bioactive compounds such as N-arachidonoylethanolamine (anandamide), N-palmitoylethanolamine, and N-oleoylethanolamine. Anandamide was most extensively investigated (Di Marzo, 1998; Pacher et al., 2006) since the discovery in 1992 as the first endocannabinoid (endogenous ligand of cannabinoid receptors) (Devane et al., 1992). On the other hand, N-palmitoylethanolamine and N-oleoylethanolamine, which are insensitive to cannabinoid receptors, receive considerable attention due to anti-inflammatory and analgesic activities (Lambert et al., 2002) and anorexic activity (Rodríguez de Fonseca et al., 2001), respectively. Recent studies suggest involvement of the nuclear receptor PPAR-α (LoVerme et al., 2005) and G-protein-coupled receptor GPR-119 (Overton et al., 2006) in signal transduction by these NAEs.

It is widely accepted that NAEs are not stored in cellular vesicles but enzymatically formed from membrane phospholipid on demand. Thus, enzymes responsible for the biosynthetic pathways as well as degradation pathways are considered to be crucial in regulating the endogenous levels of NAEs. Earlier studies, mostly performed by Schmid and his colleagues, showed that in animal tissues NAEs are biosynthesized from glycerophospholipid through the transacylation–phosphodiesterase pathway (Fig. 1.1) (Schmid et al., 1990). Since at that time anandamide had not yet been isolated from animal tissues probably due to its relatively low content, the significance of this pathway was attributed to the biosyntheses of saturated and monounsaturated NAEs. However, after the discovery of anandamide, Di Marzo et al. (1994) and Sugiura et al. (1996a) suggested that anandamide is also formed through this pathway. On the other hand, anandamide was reported to be formed by the energy-independent condensation of free arachidonic acid with ethanolamine (Di Marzo et al., 1994; Sugiura et al., 1996a; Ueda et al., 1995). Later, by the use of recombinant fatty acid amide hydrolase (FAAH), this condensation reaction was demonstrated to be the reverse reaction of this enzyme (Arreaza et al., 1997; Kurahashi et al., 1997). However, several lines of evidence strongly suggested that the contribution of this reverse reaction to the *in vivo* formation of anandamide is, if any, minor (Katayama et al., 1999; Patricelli and Cravatt, 2001). The condensation by NAE-hydrolyzing acid amidase (NAAA), which is a lysosomal enzyme to hydrolyze NAEs (Tsuboi et al., 2007), is quite inefficient even with a high concentration of ethanolamine (Ueda et al., 2001b).

Although the presence of alternative pathways in animal tissues is known as discussed later, it is widely believed that the major route for the

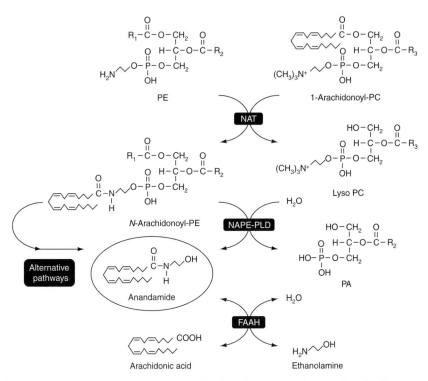

Figure 1.1 Biosynthesis of anandamide by the transacylation–phosphodiesterase pathway and the condensation reaction. PA, phosphatidic acid.

biosynthesis of anandamide as well as other NAEs is the transacylation–phosphodiesterase pathway. This pathway is initiated by transfer of a fatty acyl chain from *sn*-1 position of glycerophospholipid to the amino group of phosphatidylethanolamine (PE), resulting in the formation of *N*-acyl-PE (NAPE) and lysophospholipid. This reaction is catalyzed by an enzyme called as PE *N*-acyltransferase (NAT) or simply as NAT. NAPE is a phospholipid molecule with three fatty acyl chains (the chain at *sn*-1 position may be an alkenyl or alkyl group), and exists in biomembranes including plasma membrane (Cadas et al., 1996b). It should be noted that NAPE levels are highly increased upon tissue degeneration and inflammation (Hansen et al., 2000; Schmid et al., 1990; Schmid, 2000). Very recently, NAPE was reported to inhibit food intake (Gillum et al., 2008) and to inhibit macrophage phagocytosis through inhibition of Rac1 and Cdc42 (Shiratsuchi et al., 2009). The second step is the phospholipase (PL) D-type hydrolysis of NAPE to generate NAE and phosphatidic acid. The enzyme responsible for this reaction is highly specific for NAPE, and termed as NAPE-hydrolyzing PLD (NAPE-PLD). In this chapter, we will focus on the anandamide-forming enzymes NAT and NAPE-PLD and also refer to novel NAPE-PLD-independent pathways leading to the generation of anandamide.

II. NAT

A. Ca-NAT

NAT was first found in dog heart as an enzyme forming NAPE by PE N-acylation (Natarajan et al., 1982; Reddy et al., 1983a,b, 1984). Furthermore, NAT was isolated from other animal tissues such as dog brain (Natarajan et al., 1983), rat brain (Cadas et al., 1997; Natarajan et al., 1986; Sugiura et al., 1996a), and rat testis (Sugiura et al., 1996b) and crude or partially purified enzymes prepared from these tissues have been used for its characterization.

Earlier studies showed marked activation of NAT by Ca^{2+}. To distinguish this enzyme from recently discovered Ca^{2+}-independent NAT (iNAT) (Jin et al., 2007), we refer to the enzyme as Ca^{2+}-dependent N-acyltransferase (Ca-NAT) in this chapter. As acyl donors for the transacylation, Ca-NAT utilizes the fatty acyl chain esterified at sn-1 position of phosphatidylcholine (PC), 1-acyl-lyso PC, PE, and cardiolipin, but not free fatty acid or acyl-CoA. The abstracted acyl chain is then bound to the amino group of PE of the diacyl, alkylacyl, and plasmalogen types and lyso PE, resulting in the formation of their corresponding N-acylated ethanolamine phospholipids. The abstraction of an acyl chain from the sn-1 position, but not from sn-2 position, is highly specific (Jin et al., 2007; Sugiura et al., 1996a), while the enzyme does not seem to discriminate molecular species of the transferred acyl chain (Cadas et al., 1997; Sugiura et al., 1996a,b). Thus, the composition of the molecular species of N-acyl group in the generated NAPE should agree with those of the sn-1 acyl chain of the donor phospholipids. In fact, the anandamide precursor N-arachidonoyl-PE is a minor component among various NAPEs, including N-palmitoyl-PE, N-stearoyl-PE, and N-oleoyl-PE, present in animal tissues since the arachidonoyl chain is principally esterified at sn-2 position rather than sn-1 position of glycerophospholipids (Shetty et al., 1996). Recently, the acyl species present at sn-1 and sn-2 positions of N-arachidonoyl-PE were analyzed (Astarita et al., 2008). The major endogenous N-arachidonoyl-PE in rat brain contained alkenyl groups of C16:0, C18:0, and C:18:1 at sn-1 position and acyl groups of C18:1, C20:4, C22:4, and C22:6 at sn-2 position. Furthermore, Ca^{2+} stimulated the formation of N-arachidonoyl-PE containing O-polyunsaturated fatty acyl groups in the brain particulate fractions, suggesting a preference of Ca-NAT for O-polyunsaturated N-arachidonoyl-PE (Astarita et al., 2008).

Ca-NAT is a membrane-bound protein (Cadas et al., 1997; Natarajan et al., 1982), and nonionic detergent Nonidet P-40 was successfully used for its solubilization from the membrane (Cadas et al., 1997; Jin et al., 2007). As for subcellular distribution, NAT activity of canine brain was the highest in microsomes, followed by synaptosomes and mitochondria (Natarajan et al., 1983). The reported pH optima of Ca-NAT differ in the range of 7–10, depending on assay conditions and samples used (Jin et al., 2007). ATP or other high-energy compounds were not required. In the light of the marked stimulation of NAT by Ca^{2+} and the parallel increases in the endogenous levels of NAPEs and NAEs in response to various stimuli, NAT is considered to be the rate-limiting step of the transacylation–phosphodiesterase pathway. The activation of Ca-NAT by high concentrations of Ca^{2+} may trigger remarkable accumulation of NAPEs and NAEs in the lesions observed upon myocardial infarction (Epps et al., 1979, 1980), postdecapitative brain ischemia (Moesgaard et al., 2000; Schmid et al., 1995), $CdCl_2$-induced testicular inflammation (Kondo et al., 1998), glutamate-induced neuronal cytotoxicity (Hansen et al., 1995), and UV light irradiation to epidermal cells (Berdyshev et al., 2000). However, considering the requirement of millimolar concentrations of Ca^{2+} for full activity, it is unclear whether or not the enzyme is activated directly by Ca^{2+} in the cell. It was reported that agents which increase cyclic AMP levels potentiated the ionomycin-stimulated generation of NAPE (Cadas et al., 1996a). The stimulatory effect of Ca^{2+} could be replaced with Sr^{2+}, Mn^{2+}, and Ba^{2+} (Cadas et al., 1997; Jin et al., 2007; Reddy et al., 1984). Although specific inhibitors for Ca-NAT have not yet been developed, several nonspecific inhibitors were reported (Cadas et al., 1997). In addition to EDTA and EGTA, these inhibitors include alkylating reagents (dithionitrobenzoic acid and p-bromophenacylbromide) and cations (Be^{2+}, Cd^{2+}, Zn^{2+}, and Ag^+). (E)-6-(bromomethylene)-tetrahydro-3-(1-naphthalenyl)-2H-pyran-2-one, which was previously reported to be an inhibitor of Ca^{2+}-independent PLA_2, also inactivated Ca-NAT with an IC_{50} of about 2 μM (Cadas et al., 1997). On the other hand, millimolar concentrations of dithiothreitol were required for the full activity (Cadas et al., 1997; Jin et al., 2007).

Regarding the organ distribution in rats, the NAT activity was the highest in brain, followed by testis and muscle (Cadas et al., 1997). Other organs showed very low activities. In rat brain regions, the activity was the highest in brainstem, intermediate in cortex, striatum, hippocampus, medulla, and cerebellum, and the lowest in thalamus, hypothalamus, and olfactory bulb (Cadas et al., 1997). In rat brain the NAT activity decreased during development, and infant rats showed a several-fold higher activity than adult rats (Natarajan et al., 1986). High accumulation of NAPE in brain during postdecapitative ischemia was specifically observed with the youngest rats and was attributed to the age-dependent decrease in NAT activity concomitant with age-dependent increase in NAPE-PLD activity (Moesgaard et al., 2000). Although NAT isolated from dog heart was earlier characterized in details, the NAT activity in heart was largely different among animal species; the activity was high in dog and cat, but was very low or barely detectable in other mammals such as human, rat, and guinea-pig (Moesgaard et al., 2002). NAT solubilized from the particulate fractions of rat brain was partially purified by anion exchange column chromatography (Cadas et al., 1997; Jin et al., 2007). However, its further purification and cDNA cloning have not yet been achieved.

B. iNAT

Recently, our group discovered iNAT in rat, mouse, and human as an enzyme capable of catalyzing PE N-acylation to generate NAPE (Jin et al., 2007, 2009). cDNA of iNAT has been cloned as H-rev107-like protein 5 (HRLP5) or HRAS-like suppressor family member 5 (HRASLS5) and was noted to show homology to H-Rev107, a class II tumor suppressor. Although we referred to this protein as RLP-1 (rat LRAT-like protein 1) in our previous paper (Jin et al., 2007), the term iNAT will be used for the protein through the present chapter. The primary structures deduced from cDNA sequences show that rat, human, and mouse iNATs are composed of 287, 279, and 294 amino acid residues (Fig. 1.2A). Identity of human and mouse iNATs to rat iNAT was 65.8 and 87.1%, respectively, at protein level.

iNAT is a member of the LRAT protein family (Fig. 1.2B) (Anantharaman and Aravind, 2003). The representative protein in this family is lecithin retinol acyltransferase (LRAT), an enzyme to transfer a fatty acyl group esterified at sn-1 position of PC to all-*trans*-retinol (vitamin A), resulting in the formation of retinyl ester (Rando, 2002; Ruiz et al., 1999; Zolfaghari and Ross, 2000). This reaction is involved in intracellular storage of vitamin A and visual cycle of retina. It was interesting to us that LRAT specifically utilizes the fatty acyl group esterified at sn-1 position of PC as an acyl donor. Furthermore, retinylamine was also utilized as an acyl acceptor of LRAT, resulting in the formation of N-retinylamides (Golczak et al., 2005). Based on the functional similarity of LRAT to Ca-NAT, we wondered if Ca-NAT is a member of this protein family and characterized recombinant iNAT as a good candidate. However, iNAT differed from Ca-NAT on several points (Jin et al., 2007).

First, the activity of recombinant iNAT overexpressed in COS-7 cells was detected mainly in the cytosolic fraction, with a lower activity in the membrane fraction. Second, iNAT was almost independent of Ca^{2+}; the activity was only slightly increased or decreased in the presence of 5 mM

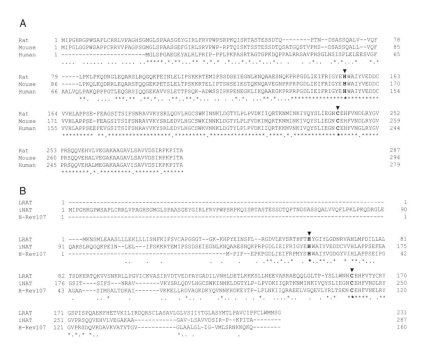

Figure 1.2 (A) The alignment of the amino acid sequences deduced from cDNAs for rat, mouse, and human iNATs. (B) The alignment of the amino acid sequences deduced from rat cDNAs for the members of the LRAT protein family, which show phospholipid-metabolizing activities. Asterisks and dots indicate identity shared by all three and any two sequences, respectively. Dashes denote lack of amino acid residues. Arrowheads show histidine and cysteine residues which are considered to constitute a catalytic dyad.

Ca^{2+} and 5 mM EDTA, respectively. Third, iNAT generated N-[^{14}C] palmitoyl-PE from both 1-[^{14}C]palmitoyl-PC and 2-[^{14}C]palmitoyl-PC in the presence of nonradioactive PE as an acyl acceptor, showing no selectivity in terms of the *sn*-1 and *sn*-2 positions of PC as an acyl donor. Moreover, N-[^{14}C]arachidonoyl-PE (the anandamide precursor) could be produced from 2-[^{14}C]arachidonoyl-PC and nonradiolabeled PE. Fourth, organ distribution of iNAT mRNA was considerably different from that of Ca-NAT activity. In rat and mouse, iNAT was expressed predominantly in testis, with much lower expression levels in other tissues including brain. In human, iNAT mRNA was abundant in testis and pancreas (Jin *et al.*, 2009). In agreement with these findings, a Ca^{2+}-independent PE N-acylation activity was detected with the cytosolic fraction of rat testis. A recent histochemical study showed that HRASLS5 (identical to iNAT) was mainly localized in spermatocytes of rat testis (Yamano *et al.*, 2008). Therefore, iNAT appeared to play a unique role in testis. It is still unclear whether or not iNAT contributes to the *in vivo* formation of NAPEs leading to the generation of bioactive NAEs.

His-60 and Cys-161 of rat LRAT are highly conserved within the LRAT family and considered to constitute a catalytic dyad in which the cysteine residue functions as the active site nucleophile (Fig. 1.2B) (Jahng *et al.*, 2003). The corresponding histidine and cysteine residues were also conserved in rat, human, and mouse iNATs (Fig. 1.2A), and the catalytic importance of the corresponding amino acids of rat iNAT (His-154 and Cys-241) was shown by site-directed mutagenesis (Jin *et al.*, 2009). Activation of iNAT by dithiothreitol and inhibition by iodoacetic acid also suggested that Cys-241 functions as the catalytic thiol residue. The N-terminal region (amino acid 1–138 of rat iNAT) shows low homology among rat, human, and mouse iNATs (Fig. 1.2A). The experiments using deletion mutants suggested that this region is involved in membrane association or protein–protein interaction, but not in catalytic activity.

When the purified iNAT was allowed to react with 1,2-[^{14}C]dipalmitoyl-PC in the presence of nonradioactive PE, not only *N*-[^{14}C]palmitoyl-PE and [^{14}C]lyso PC, but also several other radioactive compounds were produced. They contained free [^{14}C]palmitic acid, [^{14}C]lyso NAPE, and [^{14}C]PE (Jin *et al.*, 2007, 2009). The formation of free palmitic acid showed that the purified iNAT has PLA$_1$/A$_2$ activity. More interestingly, in the absence of nonradiolabeled PE, the formation of free [^{14}C]palmitic acid increased. [^{14}C] Lyso NAPE may be formed by transfer of a [^{14}C]palmitoyl group to the amino group of lyso PE which results from PE serving as acyl donor, or by abstraction of an acyl group from [^{14}C]NAPE serving as acyl donor. [^{14}C]PE may be formed by transfer of a [^{14}C]palmitoyl group to lyso PE. Thus, the formation of [^{14}C]PE suggests that iNAT also has *O*-acylation activity toward lysophospholipid. Since iNAT has a cysteine residue acting as catalytic nucleophile, iNAT appears to form an acyl-enzyme intermediate through the thioester bond, and then transfer the acyl group to the amino group of PE, water, or the hydroxyl group of lysophospholipid, leading to the formation of NAPE, free fatty acid, or phospholipid, respectively (Fig. 1.3). It is of interest to know if Ca-NAT is also such a multifunctional enzyme.

Moreover, we recently demonstrated that H-rev107, another member of the LRAT family (Fig. 1.2B), acts as Ca^{2+}-independent cytosolic PLA$_1$/A$_2$ of the thiol hydrolase-type toward various glycerophospholipids (Uyama *et al.*, 2009). PLA$_1$ activity was higher than PLA$_2$. Notably, the purified H-rev107 also showed a low PE *N*-acylation activity. Although H-rev107 has been analyzed as a tumor suppressor, it remains unclear whether this phospholipid-metabolizing activity is related to its tumor suppressing activity. Duncan *et al.* (2008) also reported that H-rev107 encodes white adipose tissue-specific PLA$_2$, and designated it AdPLA.

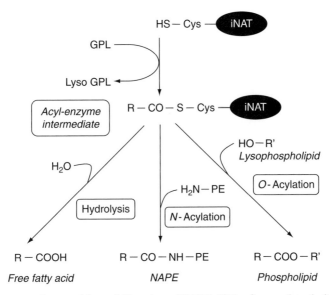

Figure 1.3 Presumable multifunction of iNAT. GPL, glycerophospholipid.

III. NAPE-PLD

A. Structure

Although NAPE-PLD, as an enzyme to release NAEs from NAPEs in animal tissues, has been recognized for more than 20 years (Liu *et al.*, 2002; Natarajan *et al.*, 1984; Petersen and Hansen, 1999; Schmid *et al.*, 1983; Sugiura *et al.*, 1996a,b; Ueda *et al.*, 2001a), information on this enzyme has been limited until recently. cDNA cloning of NAPE-PLD by our group, however, enabled molecular biological approach to this enzyme (Okamoto *et al.*, 2004).

We cloned NAPE-PLD from human, rat, and mouse. The deduced amino acid sequences are composed of 396 (mouse and rat) or 393 (human) residues (Fig. 1.4). The sequences of NAPE-PLD are highly conserved among these three animal species with more than 89% identity at the amino acid level (Okamoto *et al.*, 2004). The molecular masses based on the deduced sequences (45–46 kDa) were in good agreement with that of the purified rat heart NAPE-PLD estimated by SDS–PAGE and those of recombinant NAPE-PLDs detected by Western blotting using anti-NAPE-PLD antibody (Okamoto *et al.*, 2004). Obvious posttranslational modification was not observed. Currently the primary structures of putative NAPE-PLD orthologs of chimpanzee, rhesus monkey, orangutan, cow, and horse are also

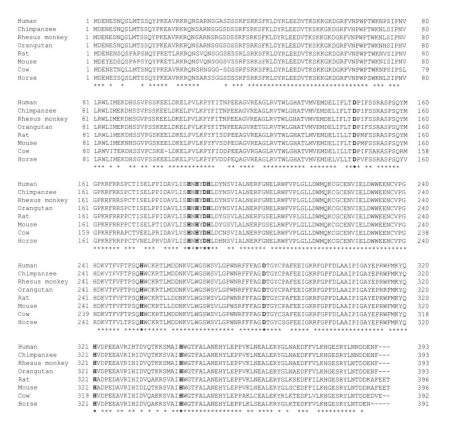

Figure 1.4 The alignment of the amino acid sequences deduced from cDNAs for putative NAPE-PLD orthologs of various mammals. Asterisks indicate identity shared by eight sequences. Dashes denote lack of amino acid residues. The catalytically important histidine and aspartic acid residues are shown in bold.

available from the NCBI database (Fig. 1.4), although it remains unclear whether these gene products have NAPE-PLD activities.

NAPE-PLD belongs to the metallo-β-lactamase family and is structurally unrelated to known PLD enzymes of the HKD/phosphatidyltransferase gene family (Liscovitch et al., 2000; Okamoto et al., 2004), such as mammalian PLD1 (Hammond et al., 1995) and PLD2 (Colley et al., 1997). Moreover, NAPE-PLD had no homology with *Streptomyces chromofuscus* PLD (Yang and Roberts, 2002), which was not a member of the HKD/phosphatidyltransferase gene family and has been noted because of its capability of hydrolyzing the phosphodiester bond of NAPE as well as those of other glycerophospholipids (Schmid et al., 1986). The metallo-β-lactamase family to which NAPE-PLD belongs is a superfamily, including a wide variety of hydrolases represented by class B β-lactamase (Aravind, 1999;

Daiyasu et al., 2001). NAPE-PLD has an highly conserved motif (H-x-[E/H]-x-D-[C/R/S/H]-x_{50-70}-H-x_{15-30}-[C/S/D]-x_{30-70}-H) which the family members share (Okamoto et al., 2004). Our site-directed mutagenesis addressed to the highly conserved aspartic acid and histidine residues of mouse NAPE-PLD suggested the catalytic importance of Asp-147, His-185, His-187, Asp-189, His-190, His-253, Asp-284, and His-321 (Wang et al., 2006b). In the members of this protein family, several aspartic acid and histidine residues are responsible for the binding of metal that is zinc in most cases (Aravind, 1999; Daiyasu et al., 2001; Wang et al., 1999). As expected, our metal analysis revealed the presence of catalytically essential zinc in NAPE-PLD (Wang et al., 2006b). Homology-based protein modeling of NAPE-PLD, using the crystal structure of Bacillus subtilis ribonuclease Z (PDB code 1y44) as template, predicted that His-185, His-187, and His-253 of NAPE-PLD bind to the first metal ion, Asp-189, His-190, and His-343 bind to the second metal ion, and Asp-284 forms a bridge between the two metal ions. This model also suggested that His-321 of NAPE-PLD is involved in the positioning of the phosphate moiety of substrate (Wang et al., 2006b). As a whole, these results support a hypothesis that the catalytic mechanism of NAPE-PLD is similar to those of the well-characterized members of the metallo-β-lactamase family.

In a mutagenesis study addressed to the conserved cysteine residues among NAPE-PLDs (Cys-170, Cys-222, Cys-224, Cys-237, Cys-255, and Cys-288), the mutant C224S showed a considerably reduced activity, while mutants of the other cysteine residues were as active as the wild type (Wang et al., 2006b). Cys-224 may be related to the inhibition of NAPE-PLD by sulfhydryl blocking reagents, which was previously reported (Morishita et al., 2005; Sasaki and Chang, 1997). Among four mutants (S152A, L207F, H380R, and D389N) discovered as single nucleotide polymorphism (SNP) of human NAPE-PLD, the activities of L207F and H380R were low (Wang et al., 2006b). However, according to the database, these mutants did not appear to link to any diseases. Although NAPE-PLD is tightly bound to the membrane, the region(s) responsible for the membrane association remains unclear. Judging from our results with deletion mutants, the N-terminal and C-terminal regions, located outside the conserved catalytic domain, appeared to be unrelated to membrane association (Wang et al., 2006b).

Further database search revealed that the two genes Y37E11AR.3c of Caenorhabditis elegans and YPL103c of Saccharomyces cerevisiae show considerable homology with NAPE-PLD (Ueda et al., 2005). These genes had 42.5 and 30.5% identity with mouse NAPE-PLD at amino acid level, respectively, when amino acid 52–361 of mouse NAPE-PLD was compared with the corresponding regions of these genes. The highly conserved aspartic acid and histidine residues in NAPE-PLD described above were also conserved in Y37E11AR.3c and YPL103c. Although the functions of the gene products remain unclear, Merkel et al. (2005) reported that the deletion of YPL103c gene in yeast caused a slight increase in the total amount of NAPEs and a concomitant decrease in that of NAEs by 60% compared to the wild type, suggesting partial contribution of YPL103c gene to the formation of NAEs from NAPEs. McPartland et al. (2006) also interpreted these genes of *S. cerevisiae* and *C. elegans* as functional NAPE-PLD orthologs based on the BLAST search.

The information on locus and intron–exon structure of the NAPE-PLD gene is available from the NCBI database. Human (GeneID: 222236) and mouse (GeneID: 242864) NAPE-PLD genes are assigned to chromosome 7q22 and chromosome 5 and presumably contain seven and five exons, respectively. The coding regions of human and mouse NAPE-PLD mRNAs appear to be derived from four exons, and the highly conserved motif within the metallo-β-lactamase family is localized in exons 3 and 4. Curtiss et al. (2005) analyzed chromosome 7q22 that had been defined as a commonly deleted segment in myeloid leukemia with a deletion of 7q, and suggested that the NAPE-PLD gene present in this locus is a candidate of myeloid tumor suppressor genes.

B. Function

A possible activation of NAPE-PLD by Ca^{2+} has been discussed with crude or partially purified enzyme preparations (Liu et al., 2002; Natarajan et al., 1984; Petersen and Hansen, 1999; Schmid et al., 1983; Sugiura et al., 1996a,b; Ueda et al., 2001a). With a highly purified recombinant enzyme, NAPE-PLD was demonstrated to be markedly stimulated by Ca^{2+} (Wang et al., 2006b). However, other divalent cations such as Mg^{2+}, Co^{2+}, Mn^{2+}, Ba^{2+}, and Sr^{2+} were similarly effective. In contrast, Fe^{2+}, Cu^{2+}, Hg^{2+}, and Zn^{2+} were inhibitory. Since millimolar concentrations of Ca^{2+} are required for the full activity, and since Ca^{2+} can be replaced with Mg^{2+} and other divalent cations, the physiological significance of Ca^{2+} as an endogenous activator of NAPE-PLD is doubtful. The stimulatory effect by Ca^{2+} becomes remarkable after solubilization of NAPE-PLD from the membrane (Ueda et al., 2001a). Consistent with this finding, we recently reported that the purified recombinant NAPE-PLD is activated by membrane fractions prepared from COS-7 cells or mouse brain tissue and this activation is followed by reduction of the stimulatory effects of Ca^{2+} (Wang et al., 2008a). PE, as one of major glycerophospholipids in biomembranes, also increased the NAPE-PLD activity. These results suggest that membrane-associated NAPE-PLD is constitutively kept as an active form by endogenous membrane components including PE, irrespective of intracellular Ca^{2+} concentrations. The nonionic detergent Triton X-100 has been used as an activator of NAPE-PLD and is effective for both membrane-associated and solubilized NAPE-PLDs (Natarajan et al., 1984; Schmid et al., 1983; Sugiura et al., 1996b; Ueda et al., 2001a). Triton X-100 is therefore a useful activator to detect NAPE-PLD activity in crude preparations. The organic cations spermine, spermidine, and putrescine also stimulated NAPE-PLD (Liu et al., 2002). However, a synergistic effect of Ca^{2+}, Triton X-100, and spermine was not observed (Liu et al., 2002). Octyl glucoside (0.1%, w/v) weakly activated the purified recombinant enzyme and showed a synergistic effect with Mg^{2+} (Wang et al., 2006b).

Petersen and Hansen (1999) showed that NAPE-PLD is incapable of catalyzing transphosphatidylation, which generates phosphatidylalcohol in the presence of primary alcohol and is observed with most of the known PLDs. We confirmed this finding with recombinant NAPE-PLD, indicating that the enzyme is not only structurally but also functionally distinct from the known PLDs (Okamoto et al., 2004). Earlier studies with crude preparations also suggested high substrate specificity of NAPE-PLD toward NAPEs (Schmid et al., 1983). The purified recombinant enzyme was almost inactive with major membrane glycerophospholipids such as PC, PE, phosphatidylserine, and phosphatidylinositol (Wang et al., 2006b). As a constitutively active enzyme, NAPE-PLD may require this strict substrate specificity to minimize damage of the membrane components by undesirable side reactions. As discussed below, NAPEs can be hydrolyzed by phospholipases other than NAPE-PLD. Thus, N-palmitoyl-lyso-PE, glycerophospho(N-palmitoyl)ethanolamine, and N-palmitoylethanolamine phosphate can be generated from N-palmitoyl-PE. The purified NAPE-PLD, however, showed low reactivities or was almost inactive with these compounds, suggesting the importance of the sn-1 and sn-2 acyl chains and glycerol backbone of NAPE in the substrate recognition (Wang et al., 2006b). With respect to the length of N-acyl groups of NAPEs, NAPE-PLD did not have selectivity for medium- and long-chain N-acyl species (C_4–C_{20}) (Wang et al., 2006b). N-Acetyl-PE and N-formyl-PE were poor substrates. These results clearly show that NAPE-PLD is capable of forming all the bioactive NAEs, including anandamide, N-palmitoylethanolamine, and N-oleoylethanolamine, from their corresponding NAPEs. In spite of the discovery of alternative pathways, so far there is no evidence showing the presence of isozymes of NAPE-PLD in animal tissues.

The development of specific inhibitors of NAPE-PLD and generation of NAPE-PLD-deficient animals should be of grate use to elucidate physiological significance of this enzyme. Specific inhibitors have not yet been reported, while NAPE-PLD knockout mice (NAPE-PLD$^{-/-}$) were generated by Leung et al. (2006). NAPE-PLD$^{-/-}$ mice were born at the expected Mendelian frequency, were viable and healthy, and did not show any phenotypes. As expected, disruption of NAPE-PLD caused a remarkable increase in the endogenous brain levels of saturated and monounsaturated NAPEs with a concomitant decrease in the levels of their corresponding NAEs. However, the levels of polyunsaturated NAEs and

NAPEs, including anandamide and N-arachidonoyl-PE, remained unchanged. Furthermore, although the N-palmitoylethanolamine-forming activity from N-palmitoyl-PE in various organs of NAPE-PLD$^{-/-}$ mice was decreased in cell-free assay systems, the anandamide-forming activity from N-arachidonoyl-PE was unchanged. These results suggested that NAPE-PLD is actually responsible for the conversion of NAPEs to NAEs *in vivo*, but other enzyme(s) or pathway(s) also participate in the syntheses of NAEs, especially anandamide and other polyunsaturated NAEs.

C. Tissue distribution

NAPE-PLD is widely distributed in mammalian tissues with obvious species difference (Moesgaard et al., 2002). NAPE-PLD is expressed in almost all organs of mouse, with higher specific activities in brain, kidney, and testis (Okamoto et al., 2004). Among rat organs, the highest specific activity was detected in heart, followed by testis and brain, and lower activities were seen in many other organs including kidney, spleen, liver, and lung (Liu et al., 2002; Schmid et al., 1983). In bovine, the brain exhibited the highest specific activity, followed by kidney, spleen, lung, heart, and liver (Petersen et al., 2000). In all animal species examined, the brain consistently displays a high NAPE-PLD activity (Morishita et al., 2005).

Considering possible physiological roles of anandamide and other NAEs in brain, distribution of NAPE-PLD in this organ must be useful information. We showed the wide distribution of NAPE-PLD in various brain regions of rats, with the highest expression level in thalamus (Morishita et al., 2005). Notably, the predominant expression of NAPE-PLD in thalamus was not observed with mouse brain (Egertová et al., 2008). Age-dependent increase in the NAPE-PLD activity in rat and mouse brains was in sharp contrast with age-dependent decrease in NAT activity in the same organ as described above (Moesgaard et al., 2000, 2003; Morishita et al., 2005; Natarajan et al., 1986). Such an age-dependent increase in the NAPE-PLD activity was well correlated with the increase at protein and mRNA levels (shown by Western blotting and real-time PCR, respectively), suggesting the regulation of this enzyme at transcriptional level (Morishita et al., 2005). Recent studies using *in situ* hybridization or immunohistochemical and immunocytochemical analyses identified NAPE-PLD-expressing cells in mouse brain (Cristino et al., 2008; Egertová et al., 2008; Nyilas et al., 2008). The NAPE-PLD positive regions contained the dentate gyrus and Ammon's horn of hippocampus, the cerebellar cortex, and other regions such as cortex, thalamus, and hypothalamus. In the dentate gyrus, intense NAPE-PLD immunoreactivity was detected in the axons of granule cells (Egertová et al., 2008; Nyilas et al., 2008) and in many neurons of the hilus region (Cristino et al., 2008; Nyilas et al., 2008). Collectively, these findings indicate that NAPE-PLD is expressed by specific populations of neurons in the brain and targeted to axonal processes. In contrast to the other endocannabinoid 2-arachidonoylglycerol confirmed as a retrograde synaptic molecule (Kano et al., 2009), anandamide and other NAEs generated by NAPE-PLD in axons were suggested to act as anterograde synaptic signaling molecules that regulate the activity of postsynaptic neurons (Egertová et al., 2008).

The involvement of the endocannabinoid system in the regulation of female and male fertility in mammals also attracts much attention (Guo et al., 2005; Maccarrone et al., 2005; Schuel et al., 2002; Wang et al., 2006a). NAPE-PLD was expressed in mouse oviduct on days 1–4 of pregnancy and preimplantation embryos (Wang et al., 2006a). NAPE-PLD was detected in the oviduct epithelium at the isthmus region with much lower levels in the ampullary region by *in situ* hybridization, and in embryos at all stages from fertilized 1-cell embryos to blastocysts by reverse transcriptase-PCR and Western blotting (Wang et al., 2006a). NAPE-PLD was also reported to be principally localized in the luminal and glandular epithelium with lower levels of accumulation in mouse uterus prior to implantation (on days 1–4 of pregnancy) by *in situ* hybridization and immunohistochemical analysis. Its accumulation in the luminal epithelium at the interimplantation sites was increased significantly after the initiation of implantation (on days 5–7) (Guo et al., 2005; Wang et al., 2006a). The endogenous anandamide level in uterus was also changed in parallel with the NAPE-PLD activity during implantation. Furthermore, it was shown that ovarian steroid hormones including progesterone and estradiol-17β downregulated the uterine NAPE-PLD expression via their nuclear receptors. These results suggest that NAPE-PLD is a major player in regulating the anandamide level in the uterus during early pregnancy (Guo et al., 2005).

Moreover, boar sperm cells possess NAPE-PLD as well as other endocannabinoid-related proteins including the cannabinoid receptor CB1, the transient receptor potential vanilloid type I (TRPV1), anandamide membrane transporter, and FAAH (Maccarrone et al., 2005). Endogenous anandamide was also detected in the sperm cells as well as seminal plasma (Maccarrone et al., 2005; Schuel et al., 2002). We recently found expression of NAPE-PLD mRNA in human prostate epithelial cells and several prostate cancer cells (Wang et al., 2008b). These findings suggest the physiological roles of the endocannabinoid anandamide and its signaling system in male fertility in mammals.

IV. Alternative Pathways Forming NAEs from NAPEs

Apart from the physiological importance of NAPE-PLD in the formation of bioactive NAEs, recent analysis of NAPE-PLD$^{-/-}$ mice revealed presence of other enzyme(s) or pathway(s) responsible for the formation of

NAEs from NAPEs (Leung et al., 2006). Previously, Natarajan et al. (1984) discussed possible pathway(s) for the biosynthesis of NAEs and suggested a possibility that O-acyl chain esterified at sn-1 or sn-2 position of NAPE or both acyl chains were first hydrolyzed to generate N-acyl-lyso-PE (lyso NAPE) or glycerophospho (GP)-NAE, respectively, and subsequent cleavage of the phosphodiester bonds of lyso NAPE and GP-NAE would generate NAE. In dog and rat brains the PLD-type hydrolyzing activities toward lyso NAPE and GP-NAE were actually observed (Natarajan et al., 1984; Wang et al., 2006b). Considering that NAPE-PLD has poor or no activities for these compounds as described above, it was suggested that phosphodiesterases other than NAPE-PLD are mostly responsible for the PLD-type hydrolysis of these compounds (Wang et al., 2006b).

Our group reported that consecutive reactions of PLA_2 and a lyso PLD found in rat tissues lead to the formation of NAE from NAPE via lyso NAPE (Fig. 1.5) (Sun et al., 2004). We showed a PLA_1/A_2 activity to hydrolyze N-palmitoyl-PE to N-palmitoyl-lyso-PE in various rat tissues with by far the highest activity in stomach. The stomach enzyme was purified and identified as group IB secretory PLA_2 ($sPLA_2$-IB) (Sun et al., 2004).

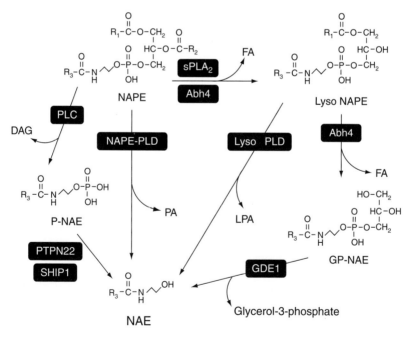

Figure 1.5 Multiple pathways for the biosynthesis of anandamide and other NAEs in animal tissues. DAG, diacylglycerol; FA, fatty acid; LPA, lysophosphatidic acid; PA, phosphatidic acid; P-NAE, phospho-NAE.

Recombinant preparations of group IB, IIA, and V of sPLA$_2$ were also active with N-palmitoyl-PE, while group X sPLA$_2$ and cytosolic PLA$_2$ α were much less active. We also detected a lyso PLD-like activity releasing N-palmitoylethanolamine from N-palmitoyl-lyso-PE in various rat tissues, with higher activities in brain and testis. The lyso PLD-like enzyme was membrane-associated and was different from NAPE-PLD in catalytic properties. N-Arachidonoyl-PE and N-arachidonoyl-lyso-PE also served as active substrates for sPLA$_2$-IB and the lyso PLD-like enzyme, showing a possibility that this pathway contributes to the biosynthesis of anandamide. However, further characterization of the lyso PLD-like enzyme has not been performed.

Following the analysis of NAPE-PLD$^{-/-}$ mice, Simon and Cravatt (2006) proposed a new pathway for producing NAEs that involves the double-O-deacylation of NAPEs by a phospholipase to form GP-NAEs, followed by the hydrolysis of the resultant GP-NAEs to NAEs and glycerol-3-phosphate by a phosphodiesterase (Fig. 1.5). Using the functional proteomic isolation, they identified α/β-hydrolase 4 (Abh4) as a lipase to deacylate both lyso NAPE and NAPE. The lyso NAPE-lipase activity was widely distributed in mouse organs with the highest activity in brain, spinal cord, and testis, followed by liver, kidney, and heart. The mRNA of Abh4 was similarly distributed, suggesting that Abh4 is responsible for this activity in various mouse organs. Abh4 showed a higher reactivity with lyso NAPE as compared with other lysophospholipids, but did not discriminate different N-acyl species of lyso NAPE including N-arachidonoyl-lyso-PE. Furthermore, mouse brain exhibited the enzyme activity to rapidly cleave the phosphodiester bond of GP-NAE to generate NAE. Recently, the same group investigated the ability of five proteins classified as glycerophosphodiesterases (GDEs) to covert GP-NAE to NAE and identified GDE1, also known as MIR16, as a GP-NAE phosphodiesterase (Simon and Cravatt, 2008). This enzyme is membrane bound and stimulated by Mg^{2+}. The addition of EDTA potently inhibits GDE1. Distribution of the GP-NAE phosphodiesterase activity in mouse organs showed the highest activity in brain, spinal cord, testis, and kidney, followed by liver, spleen, and heart. The mRNA level of GDE1 was similarly distributed. GDE1 hydrolyzed GP-NAEs with various N-acyl chains including GP-N-arachidonoylethanolamine. Furthermore, they detected GP-N-arachidonoylethanolamine and other GP-NAEs as endogenous constituents of mouse brain tissue. Thus, the combination of Abh4 and GDE1 may participate in the formation of NAEs from NAPEs as an NAPE-PLD-independent pathway. However, it remains unclear whether this pathway is responsible for the selective formation of polyunsaturated NAEs suggested in the analysis of NAPE-PLD$^{-/-}$ mice.

Another anandamide biosynthetic pathway was proposed with mouse brain and bacterial lipopolysaccharide (LPS)-induced macrophages, which involves PLC-like enzyme-dependent conversion of N-arachidonoyl-PE to anandamide phosphate, followed by phosphatase-mediated hydrolysis of anandamide phosphate to anandamide (Fig. 1.5) (Liu et al., 2006). LPS downregulated the expression of NAPE-PLD mRNA in RAW264.7 macrophage cells, and the PLC inhibitor (neomycin) or phosphatase inhibitor (sodium orthovanadate) reduced the anandamide formation in macrophages and mouse brain. Furthermore, knockdown of NAPE-PLD expression by RNA interference in RAW264.7 cells did not affect the increase of anandamide level by LPS. Thus, this pathway appeared to contribute to the anandamide formation stimulated by LPS. LPS instead induced the expression of PTPN22, a nonreceptor protein tyrosine phosphatase, which was shown to convert anandamide phosphate to anandamide (Liu et al., 2006). Consistent with the role of PTPN22 in anandamide formation, overexpression of PTPN22 in RAW264.7 cells resulted in a twofold increase in anandamide levels. Recently, with NAPE-PLD$^{-/-}$ mice it was demonstrated that conversion of synthetic N-arachidonoyl-PE to anandamide by brain homogenates proceeds through both Abh4-mediated pathway and PLC-mediated pathway with the former being dominant at shorter (<10 min) and the latter at longer (60 min) incubations (Liu et al., 2008). It was also reported that SHIP1, an inositol 5'-phosphatase, contributes to the generation of anandamide from anandamide phosphate. However, the PLC-like enzyme remains unidentified.

V. Conclusions

cDNA cloning of NAPE-PLD and subsequent characterization of the recombinant enzyme demonstrated that NAPE-PLD is a novel enzyme exclusively responsible for the transacylation–phosphodiesterase pathway and enabled molecular biological studies on this pathway for the first time. However, further studies, including analysis of NAPE-PLD$^{-/-}$ mice, revealed that the biosynthesis of anandamide and other NAEs is more complex than presumed before. Several enzymes involved in the NAPE-PLD-independent anandamide biosynthesis pathways were already identified and characterized. Such advances are greatly contributing to comprehensive understanding of not only the anandamide synthesis system, but also the phospholipid turnover in biomembrane. Discovery and characterization of iNAT by our group showed the existence of plural enzymes that are capable of forming NAPE. However, Ca-NAT is still considered to play a central role in the formation of anandamide and other bioactive NAEs, and its cDNA cloning is eagerly anticipated.

REFERENCES

Anantharaman, V., and Aravind, L. (2003). Evolutionary history, structural features and biochemical diversity of the NlpC/P60 superfamily of enzymes. *Genome Biol.* **4,** R11.

Aravind, L. (1999). An evolutionary classification of the metallo-beta-lactamase fold proteins. *In Silico Biol.* **1,** 69–91.

Arreaza, G., Devane, W. A., Omeir, R. L., Sajnani, G., Kunz, J., Cravatt, B. F., and Deutsch, D. G. (1997). The cloned rat hydrolytic enzyme responsible for the breakdown of anandamide also catalyzes its formation via the condensation of arachidonic acid and ethanolamine. *Neurosci. Lett.* **234,** 59–62.

Astarita, G., Ahmed, F., and Piomelli, D. (2008). Identification of biosynthetic precursors for the endocannabinoid anandamide in the rat brain. *J. Lipid Res.* **49,** 48–57.

Berdyshev, E. V., Schmid, P. C., Dong, Z., and Schmid, H. H. O. (2000). Stress-induced generation of N-acylethanolamines in mouse epidermal JB6 P^+ cells. *Biochem. J.* **346,** 369–374.

Cadas, H., di Tomaso, E., and Piomelli, D. (1997). Occurrence and biosynthesis of endogenous cannabinoid precursor, N-arachidonoyl phosphatidylethanolamine, in rat brain. *J. Neurosci.* **17,** 1226–1242.

Cadas, H., Gaillet, S., Beltramo, M., Venance, L., and Piomelli, D. (1996a). Biosynthesis of an endogenous cannabinoid precursor in neurons and its control by calcium and cAMP. *J. Neurosci.* **16,** 3934–3942.

Cadas, H., Schinelli, S., and Piomelli, D. (1996b). Membrane localization of N-acylphosphatidylethanolamine in central neurons: Studies with exogenous phospholipases. *J. Lipid Mediat. Cell Signal.* **14,** 63–70.

Colley, W. C., Sung, T.-C., Roll, R., Jenco, J., Hammond, S. M., Altshuller, Y., Bar-Sagi, D., Morris, A. J., and Frohman, M. A. (1997). Phospholipase D2, a distinct phospholipase D isoform with novel regulatory properties that provokes cytoskeletal reorganization. *Curr. Biol.* **7,** 191–201.

Cristino, L., Starowicz, K., De Petrocellis, L., Morishita, J., Ueda, N., Guglielmotti, V., and Di Marzo, V. (2008). Immunohistochemical localization of anabolic and catabolic enzymes for anandamide and other putative endovanilloids in the hippocampus and cerebellar cortex of the mouse brain. *Neuroscience* **151,** 955–968.

Curtiss, N. P., Bonifas, J. M., Lauchle, J. O., Balkman, J. D., Kratz, C. P., Emerling, B. M., Green, E. D., Le Beau, M. M., and Shannon, K. M. (2005). Isolation and analysis of candidate myeloid tumor suppressor genes from a commonly deleted segment of 7q22. *Genomics* **85,** 600–607.

Daiyasu, H., Osaka, K., Ishino, Y., and Toh, H. (2001). Expansion of the zinc metallohydrolase family of the β-lactamase fold. *FEBS Lett.* **503,** 1–6.

Devane, W. A., Hanus, L., Breuer, A., Pertwee, R. G., Stevenson, L. A., Griffin, G., Gibson, D., Mandelbaum, A., Etinger, A., and Mechoulam, R. (1992). Isolation and structure of a brain constituent that binds to the cannabinoid receptor. *Science* **258,** 1946–1949.

Di Marzo, V. (1998). "Endocannabinoids" and other fatty acid derivatives with cannabimimetic properties: Biochemistry and possible physiopathological relevance. *Biochim. Biophys. Acta* **1392,** 153–175.

Di Marzo, V., Fontana, A., Cadas, H., Schinelli, S., Cimino, G., Schwartz, J.-C., and Piomelli, D. (1994). Formation and inactivation of endogenous cannabinoid anandamide in central neurons. *Nature* **372,** 686–691.

Duncan, R. E., Sarkadi-Nagy, E., Jaworski, K., Ahmadian, M., and Sul, H. S. (2008). Identification and functional characterization of adipose-specific phospholipase A_2 (AdPLA). *J. Biol. Chem.* **283,** 25428–25436.

Egertová, M., Simon, G. M., Cravatt, B. F., and Elphick, M. R. (2008). Localization of N-acyl phosphatidylethanolamine phospholipase D (NAPE-PLD) expression in mouse brain: A new perspective on N-acylethanolamines as neural signaling molecules. *J. Comp. Neurol.* **506,** 604–615.

Epps, D. E., Schmid, P. C., Natarajan, V., and Schmid, H. H. O. (1979). N-Acylethanolamine accumulation in infarcted myocardium. *Biochem. Biophys. Res. Commun.* **90,** 628–633.

Epps, D. E., Natarajan, V., Schmid, P. C., and Schmid, H. H. O. (1980). Accumulation of N-acylethanolamine glycerophospholipids in infarcted myocardium. *Biochim. Biophys. Acta* **618,** 420–430.

Gillum, M. P., Zhang, D., Zhang, X.-M., Erion, D. M., Jamison, R. A., Choi, C., Dong, J., Shanabrough, M., Duenas, H. R., Frederick, D. W., Hsiao, J. J., Horvath, T. L., *et al.* (2008). N-acylphosphatidylethanolamine, a gut-derived circulating factor induced by fat ingestion, inhibits food intake. *Cell* **135,** 813–824.

Golczak, M., Imanishi, Y., Kuksa, V., Maeda, T., Kubota, R., and Palczewski, K. (2005). Lecithin:retinol acyltransferase is responsible for amidation of retinylamine, a potent inhibitor of the retinoid cycle. *J. Biol. Chem.* **280,** 42263–42273.

Guo, Y., Wang, H., Okamoto, Y., Ueda, N., Kingsley, P. J., Marnett, L. J., Schmid, H. H. O., Das, S. K., and Dey, S. K. (2005). N-acylphosphatidylethanolamine-hydrolyzing phospholipase D is an important determinant of uterine anandamide levels during implantation. *J. Biol. Chem.* **280,** 23429–23432.

Hammond, S. M., Altshuller, Y. M., Sung, T.-C., Rudge, S. A., Rose, K., Engebrecht, J., Morris, A. J., and Frohman, M. A. (1995). Human ADP-ribosylation factor-activated phosphatidylcholine-specific phospholipase D defines a new and highly conserved gene family. *J. Biol. Chem.* **270,** 29640–29643.

Hansen, H. S., Lauritzen, L., Strand, A. M., Moesgaard, B., and Frandsen, A. (1995). Glutamate stimulates the formation of N-acylphosphatidylethanolamine and N-acylethanolamine in cortical neurons in culture. *Biochim. Biophys. Acta* **1258,** 303–308.

Hansen, H. S., Moesgaard, B., Hansen, H. H., and Petersen, G. (2000). N-Acylethanolamines and precursor phospholipids – Relation to cell injury. *Chem. Phys. Lipids* **108,** 135–150.

Jahng, W. J., Xue, L., and Rando, R. R. (2003). Lecithin retinol acyltransferase is a founder member of a novel family of enzymes. *Biochemistry* **42,** 12805–12812.

Jin, X.-H., Okamoto, Y., Morishita, J., Tsuboi, K., Tonai, T., and Ueda, N. (2007). Discovery and characterization of a Ca^{2+}-independent phosphatidylethanolamine N-acyltransferase generating the anandamide precursor and its congeners. *J. Biol. Chem.* **282,** 3614–3623.

Jin, X.-H., Uyama, T., Wang, J., Okamoto, Y., Tonai, T., and Ueda, N. (2009). cDNA cloning and characterization of human and mouse Ca^{2+}-independent phosphatidylethanolamine N-acyltransferases. *Biochim. Biophys. Acta* **1791,** 32–38.

Kano, M., Ohno-Shosaku, T., Hashimotodani, Y., Uchigashima, M., and Watanabe, M. (2009). Endocannabinoid-mediated control of synaptic transmission. *Physiol. Rev.* **89,** 309–380.

Katayama, K., Ueda, N., Katoh, I., and Yamamoto, S. (1999). Equilibrium in the hydrolysis and synthesis of cannabimimetic anandamide demonstrated by a purified enzyme. *Biochim. Biophys. Acta* **1440,** 205–214.

Kondo, S., Sugiura, T., Kodaka, T., Kudo, N., Waku, K., and Tokumura, A. (1998). Accumulation of various N-acylethanolamines including N-arachidonoylethanolamine (anandamide) in cadmium chloride-administered rat testis. *Arch. Biochem. Biophys.* **354,** 303–310.

Kurahashi, Y., Ueda, N., Suzuki, H., Suzuki, M., and Yamamoto, S. (1997). Reversible hydrolysis and synthesis of anandamide demonstrated by recombinant rat fatty-acid amide hydrolase. *Biochem. Biophys. Res. Commun.* **237,** 512–515.

Lambert, D. M., Vandevoorde, S., Jonsson, K.-O., and Fowler, C. J. (2002). The palmitoylethanolamide family: A new class of anti-inflammatory agents? *Curr. Med. Chem.* **9,** 663–674.

Leung, D., Saghatelian, A., Simon, G. M., and Cravatt, B. F. (2006). Inactivation of N-acyl phosphatidylethanolamine phospholipase D reveals multiple mechanisms for the biosynthesis of endocannabinoids. *Biochemistry* **45,** 4720–4726.

Liscovitch, M., Czarny, M., Fiucci, G., and Tang, X. (2000). Phospholipase D: Molecular and cell biology of a novel gene family. *Biochem. J.* **345,** 401–415.

Liu, Q., Tonai, T., and Ueda, N. (2002). Activation of N-acylethanolamine-releasing phospholipase D by polyamines. *Chem. Phys. Lipids* **115,** 77–84.

Liu, J., Wang, L., Harvey-White, J., Osei-Hyiaman, D., Razdan, R., Gong, Q., Chan, A. C., Zhou, Z., Huang, B. X., Kim, H.-Y., and Kunos, G. (2006). A biosynthetic pathway for anandamide. *Proc. Natl. Acad. Sci. USA* **103,** 13345–13350.

Liu, J., Wang, L., Harvey-White, J., Huang, B. X., Kim, H.-Y., Luquet, S., Palmiter, R. D., Krystal, G., Rai, R., Mahadevan, A., Razdan, R. K., and Kunos, G. (2008). Multiple pathways involved in the biosynthesis of anandamide. *Neuropharmacology* **54,** 1–7.

LoVerme, J., La Rana, G., Russo, R., Calignano, A., and Piomelli, D. (2005). The search for the palmitoylethanolamide receptor. *Life Sci.* **77,** 1685–1698.

Maccarrone, M., Barboni, B., Paradisi, A., Bernabò, N., Gasperi, V., Pistilli, M. G., Fezza, F., Lucidi, P., and Mattioli, M. (2005). Characterization of the endocannabinoid system in boar spermatozoa and implications for sperm capacitation and acrosome reaction. *J. Cell Sci.* **118,** 4393–4404.

McPartland, J. M., Matias, I., Di Marzo, V., and Glass, M. (2006). Evolutionary origins of the endocannabinoid system. *Gene* **370,** 64–74.

Merkel, O., Schmid, P. C., Paltauf, F., and Schmid, H. H. O. (2005). Presence and potential signaling function of N-acylethanolamines and their phospholipid precursors in the yeast *Saccharomyces cerevisiae*. *Biochim. Biophys. Acta* **1734,** 215–219.

Moesgaard, B., Petersen, G., Jaroszewski, J. W., and Hansen, H. S. (2000). Age dependent accumulation of N-acyl-ethanolamine phospholipids in ischemic rat brain: A ^{31}P NMR and enzyme activity study. *J. Lipid Res.* **41,** 985–990.

Moesgaard, B., Petersen, G., Mortensen, S. A., and Hansen, H. S. (2002). Substantial species differences in relation to formation and degradation of N-acyl-ethanolamine phospholipids in heart tissue: An enzyme activity study. *Comp. Biochem. Physiol. B Biochem. Mol. Biol.* **131,** 475–482.

Moesgaard, B., Hansen, H. H., Hansen, S. L., Hansen, S. H., Petersen, G., and Hansen, H. S. (2003). Brain levels of N-acylethanolamine phospholipids in mice during pentylenetetrazol-induced seizure. *Lipids* **38,** 387–390.

Morishita, J., Okamoto, Y., Tsuboi, K., Ueno, M., Sakamoto, H., Maekawa, N., and Ueda, N. (2005). Regional distribution and age-dependent expression of N-acylphosphatidylethanolamine-hydrolyzing phospholipase D in rat brain. *J. Neurochem.* **94,** 753–762.

Natarajan, V., Reddy, P. V., Schmid, P. C., and Schmid, H. H. O. (1982). N-Acylation of ethanolamine phospholipids in canine myocardium. *Biochim. Biophys. Acta* **712,** 342–355.

Natarajan, V., Schmid, P. C., Reddy, P. V., Zuzarte-Augustin, M. L., and Schmid, H. H. O. (1983). Biosynthesis of N-acylethanolamine phospholipids by dog brain preparations. *J. Neurochem.* **41,** 1303–1312.

Natarajan, V., Schmid, P. C., Reddy, P. V., and Schmid, H. H. O. (1984). Catabolism of N-acylethanolamine phospholipids by dog brain preparations. *J. Neurochem.* **42,** 1613–1619.

Natarajan, V., Schmid, P. C., and Schmid, H. H. O. (1986). N-Acylethanolamine phospholipid metabolism in normal and ischemic rat brain. *Biochim. Biophys. Acta* **878,** 32–41.

Nyilas, R., Dudok, B., Urban, G. M., Mackie, K., Watanabe, M., Cravatt, B. F., Freund, T. F., and Katona, I. (2008). Enzymatic machinery for endocannabinoid biosynthesis associated with calcium stores in glutamatergic axon terminals. *J. Neurosci.* **28,** 1058–1063.

Okamoto, Y., Morishita, J., Tsuboi, K., Tonai, T., and Ueda, N. (2004). Molecular characterization of a phospholipase D generating anandamide and its congeners. *J. Biol. Chem.* **279,** 5298–5305.

Overton, H. A., Babbs, A. J., Doel, S. M., Fyfe, M. C. T., Gardner, L. S., Griffin, G., Jackson, H. C., Procter, M. J., Rasamison, C. M., Tang-Christensen, M., Widdowson, P. S., Williams, G. M., *et al.* (2006). Deorphanization of a G protein-coupled receptor for oleoylethanolamide and its use in the discovery of small-molecule hypophagic agents. *Cell Metab.* **3,** 167–175.

Pacher, P., Bátkai, S., and Kunos, G. (2006). The endocannabinoid system as an emerging target of pharmacotherapy. *Pharmacol. Rev.* **58,** 389–462.

Patricelli, M. P., and Cravatt, B. F. (2001). Proteins regulating the biosynthesis and inactivation of neuromodulatory fatty acid amides. *Vitam. Horm.* **62,** 95–131.

Petersen, G., and Hansen, H. S. (1999). N-acylphosphatidylethanolamine-hydrolysing phospholipase D lacks the ability to transphosphatidylate. *FEBS Lett.* **455,** 41–44.

Petersen, G., Chapman, K. D., and Hansen, H. S. (2000). A rapid phospholipase D assay using zirconium precipitation of anionic substrate phospholipids: Application to N-acylethanolamine formation *in vitro*. *J. Lipid Res.* **41,** 1532–1538.

Rando, R. R. (2002). Membrane-bound lecithin-retinol acyltransferase. *Biochem. Biophys. Res. Commun.* **292,** 1243–1250.

Reddy, P. V., Natarajan, V., Schmid, P. C., and Schmid, H. H. O. (1983a). N-Acylation of dog heart ethanolamine phospholipids by transacylase activity. *Biochim. Biophys. Acta* **750,** 472–480.

Reddy, P. V., Schmid, P. C., Natarajan, V., and Schmid, H. H. O. (1983b). The role of cardiolipin as an acyl donor in dog heart N-acylethanolamine phospholipid biosynthesis. *Biochim. Biophys. Acta* **751,** 241–246.

Reddy, P. V., Schmid, P. C., Natarajan, V., Muramatsu, T., and Schmid, H. H. O. (1984). Properties of canine myocardial phosphatidylethanolamine N-acyltransferase. *Biochim. Biophys. Acta* **795,** 130–136.

Rodríguez de Fonseca, F., Navarro, M., Gómez, R., Escuredo, L., Nava, F., Fu, J., Murillo-Rodríguez, E., Giuffrida, A., LoVerme, J., Gaetani, S., Kathuria, S., Gall, C., *et al.* (2001). An anorexic lipid mediator regulated by feeding. *Nature* **414,** 209–212.

Ruiz, A., Winston, A., Lim, Y.-H., Gilbert, B. A., Rando, R. R., and Bok, D. (1999). Molecular and biochemical characterization of lecithin retinol acyltransferase. *J. Biol. Chem.* **274,** 3834–3841.

Sasaki, T., and Chang, M. C. (1997). N-arachidonylethanolamine (anandamide) formation from N-arachidonylphosphatidylethanolamine in rat brain membranes. *Life Sci.* **61,** 1803–1810.

Schmid, H. H. O. (2000). Pathways and mechanisms of N-acylethanolamine biosynthesis: Can anandamide be generated selectively? *Chem. Phys. Lipids* **108,** 71–87.

Schmid, P. C., Natarajan, V., Weis, B. K., and Schmid, H. H. O. (1986). Hydrolysis of N-acylated glycerophospholipids by phospholipases A_2 and D: A method of identification and analysis. *Chem. Phys. Lipids* **41,** 195–207.

Schmid, P. C., Reddy, P. V., Natarajan, V., and Schmid, H. H. O. (1983). Metabolism of N-acylethanolamine phospholipids by a mammalian phosphodiesterase of the phospholipase D type. *J. Biol. Chem.* **258,** 9302–9306.

Schmid, H. H. O., Schmid, P. C., and Natarajan, V. (1990). N-Acylated glycerophospholipids and their derivatives. *Prog. Lipid Res.* **29,** 1–43.

Schmid, P. C., Krebsbach, R. J., Perry, S. R., Dettmer, T. M., Maasson, J. L., and Schmid, H. H. O. (1995). Occurrence and postmortem generation of anandamide and other long-chain N-acylethanolamines in mammalian brain. *FEBS Lett.* **375,** 117–120.

Schuel, H., Burkman, L. J., Lippes, J., Crickard, K., Forester, E., Piomelli, D., and Giuffrida, A. (2002). N-Acylethanolamines in human reproductive fluids. *Chem. Phys. Lipids* **121,** 211–227.

Shetty, H. U., Smith, Q. R., Washizaki, K., Rapoport, S. I., and Purdon, A. D. (1996). Identification of two molecular species of rat brain phosphatidylcholine that rapidly incorporate and turn over arachidonic acid *in vivo. J. Neurochem.* **67,** 1702–1710.

Shiratsuchi, A., Ichiki, M., Okamoto, Y., Ueda, N., Sugimoto, N., Takuwa, Y., and Nakanishi, Y. (2009). Inhibitory effect of N-palmitoylphosphatidylethanolamine on macrophage phagocytosis through inhibition of Rac1 and Cdc42. *J. Biochem.* **145,** 43–50.

Simon, G. M., and Cravatt, B. F. (2006). Endocannabinoid biosynthesis proceeding through glycerophospho-N-acyl ethanolamine and a role for α/β-hydrolase 4 in this pathway. *J. Biol. Chem.* **281,** 26465–26472.

Simon, G. M., and Cravatt, B. F. (2008). Anandamide biosynthesis catalyzed by the phosphodiesterase GDE1 and detection of glycerophospho-N-acyl ethanolamine precursors in mouse brain. *J. Biol. Chem.* **283,** 9341–9349.

Sugiura, T., Kondo, S., Sukagawa, A., Tonegawa, T., Nakane, S., Yamashita, A., Ishima, Y., and Waku, K. (1996a). Transacylase-mediated and phosphodiesterase-mediated synthesis of N-arachidonoylethanolamine, an endogenous cannabinoid-receptor ligand, in rat brain microsomes. Comparison with synthesis from free arachidonic acid and ethanolamine. *Eur. J. Biochem.* **240,** 53–62.

Sugiura, T., Kondo, S., Sukagawa, A., Tonegawa, T., Nakane, S., Yamashita, A., and Waku, K. (1996b). Enzymatic synthesis of anandamide, an endogenous cannabinoid receptor ligand, through N-acylphosphatidylethanolamine pathway in testis: Involvement of Ca^{2+}-dependent transacylase and phosphodiesterase activities. *Biochem. Biophys. Res. Commun.* **218,** 113–117.

Sun, Y.-X., Tsuboi, K., Okamoto, Y., Tonai, T., Murakami, M., Kudo, I., and Ueda, N. (2004). Biosynthesis of anandamide and N-palmitoylethanolamine by sequential actions of phospholipase A_2 and lysophospholipase D. *Biochem. J.* **380,** 749–756.

Tsuboi, K., Takezaki, N., and Ueda, N. (2007). The N-acylethanolamine-hydrolyzing acid amidase (NAAA). *Chem. Biodivers.* **4,** 1914–1925.

Ueda, N., Kurahashi, Y., Yamamoto, S., and Tokunaga, T. (1995). Partial purification and characterization of the porcine brain enzyme hydrolyzing and synthesizing anandamide. *J. Biol. Chem.* **270,** 23823–23827.

Ueda, N., Liu, Q., and Yamanaka, K. (2001a). Marked activation of the N-acylphosphatidylethanolamine-hydrolyzing phosphodiesterase by divalent cations. *Biochim. Biophys. Acta* **1532,** 121–127.

Ueda, N., Yamanaka, K., and Yamamoto, S. (2001b). Purification and characterization of an acid amidase selective for N-palmitoylethanolamine, a putative endogenous anti-inflammatory substance. *J. Biol. Chem.* **276,** 35552–35557.

Ueda, N., Okamoto, Y., and Morishita, J. (2005). N-acylphosphatidylethanolamine-hydrolyzing phospholipase D: A novel enzyme of the β-lactamase fold family releasing anandamide and other N-acylethanolamines. *Life Sci.* **77,** 1750–1758.

Uyama, T., Morishita, J., Jin, X.-H., Okamoto, Y., Tsuboi, K., and Ueda, N. (2009). The tumor suppressor gene H-rev107 functions as a novel Ca^{2+}-independent cytosolic phospholipase $A_{1/2}$ of the thiol hydrolase type. *J. Lipid Res.* **50,** 685–693.

Wang, H., Xie, H., Guo, Y., Zhang, H., Takahashi, T., Kingsley, P. J., Marnett, L. J., Das, S. K., Cravatt, B. F., and Dey, S. K. (2006a). Fatty acid amide hydrolase deficiency limits early pregnancy events. *J. Clin. Invest.* **116,** 2122–2131.

Wang, J., Okamoto, Y., Morishita, J., Tsuboi, K., Miyatake, A., and Ueda, N. (2006b). Functional analysis of the purified anandamide-generating phospholipase D as a member of the metallo-β-lactamase family. *J. Biol. Chem.* **281,** 12325–12335.

Wang, J., Okamoto, Y., Tsuboi, K., and Ueda, N. (2008a). The stimulatory effect of phosphatidylethanolamine on N-acylphosphatidylethanolamine-hydrolyzing phospholipase D (NAPE-PLD). *Neuropharmacology* **54,** 8–15.

Wang, J., Zhao, L.-Y., Uyama, T., Tsuboi, K., Wu, X.-X., Kakehi, Y., and Ueda, N. (2008b). Expression and secretion of N-acylethanolamine-hydrolysing acid amidase in human prostate cancer cells. *J. Biochem.* **144,** 685–690.

Wang, Z., Fast, W., Valentine, A. M., and Benkovic, S. J. (1999). Metallo-β-lactamase: Structure and mechanism. *Curr. Opin. Chem. Biol.* **3,** 614–622.

Yamano, Y., Asano, A., Ohyama, K., Ohta, M., Nishio, R., and Morishima, I. (2008). Expression of the Ha-ras suppressor family member 5 gene in the maturing rat testis. *Biosci. Biotechnol. Biochem.* **72,** 1360–1363.

Yang, H., and Roberts, M. F. (2002). Cloning, overexpression, and characterization of a bacterial Ca^{2+}-dependent phospholipase D. *Protein Sci.* **11,** 2958–2968.

Zolfaghari, R., and Ross, A. C. (2000). Lecithin:retinol acyltransferase from mouse and rat liver: cDNA cloning and liver-specific regulation by dietary vitamin A and retinoic acid. *J. Lipid Res.* **41,** 2024–2034.

CHAPTER TWO

ORGANIZED TRAFFICKING OF ANANDAMIDE AND RELATED LIPIDS

Marla L. Yates *and* Eric L. Barker

Contents

I. AEA and the Endocannabinoid System	26
A. Discovery of endocannabinoids	26
B. AEA synthesis	26
C. AEA signaling	27
D. AEA degradation	28
E. Lipid rafts and the fate of AEA metabolites	29
II. AEA Transport	31
A. Fatty acid transporters	32
B. Lipid transfer proteins	35
C. Characteristics of AEA transport	36
D. Proposed models for cellular AEA accumulation	37
E. Implications of pharmacologically altered AEA signaling	44
Acknowledgment	45
References	45

Abstract

N-arachidonylethanolamide (anandamide or AEA) is an endogenous long-chain fatty acid ethanolamide with activity at both the cannabinoid 1 (CB_1) and cannabinoid 2 (CB_2) receptors, as well as the transient receptor potential vanilloid 1 (TRPV1) receptor. Whereas the mechanisms for both AEA biosynthesis and metabolism are fairly well established, the manner by which AEA is accumulated into cells remains controversial. The overwhelming majority of scientific reports indicate that this lipid neuromodulator is taken into cells via a facilitated process. Some reports have suggested that AEA uptake occurs by facilitated diffusion. Recent evidence indicates that AEA uptake may occur via endocytosis, contesting the premise that passive diffusion is the mechanism by which AEA transverses the plasma membrane. This chapter serves as an

---

The Department of Medicinal Chemistry and Molecular Pharmacology, Purdue University School of Pharmacy and Pharmaceutical Sciences, West Lafayette, Indiana, USA

introduction to the endocannabinoid field with an emphasis on the various proposed mechanisms for the cellular uptake of endocannabinoids and other related hydrophobic molecules.

I. AEA AND THE ENDOCANNABINOID SYSTEM

A. Discovery of endocannabinoids

The psychotropic properties and medicinal uses of marijuana have been long-known. However, it was not until the 1970s that the plant's active ingredient, Δ^9-tetrahydrocannabinol (THC), was discovered (Gaoni and Mechoulam, 1971). For years, the mechanism by which THC exerts its physiological and psychoactive effects escaped scientists, until the discovery of the cannabinoid receptors, CB_1 and CB_2, in the early-1990s (Matsuda et al., 1990; Munro et al., 1993). These groundbreaking discoveries led to another—the identification of an endogenous CB receptor lipid ligand termed AEA (Devane et al., 1992). Almost 20 years later, AEA remains under intense scientific investigation. In addition, several other endocannabinoids have also been identified, including 2-arachidonylglycerol (2-AG) (Mechoulam et al., 1995; Sugiura et al., 1995), virodhamine (Porter et al., 2002), and noladin ether (Hanus et al., 2001) (Fig. 2.1). Below, we will focus on the endocannabinoid AEA, broadly covering its metabolism and signaling, and more specifically, the mechanism by which it is accumulated in cells.

B. AEA synthesis

AEA is the most well-characterized endocannabinoid. Unlike typical neuromodulators, AEA is not presynthesized and stored in synaptic vesicles. Rather, the most generally accepted hypothesis is that AEA is synthesized

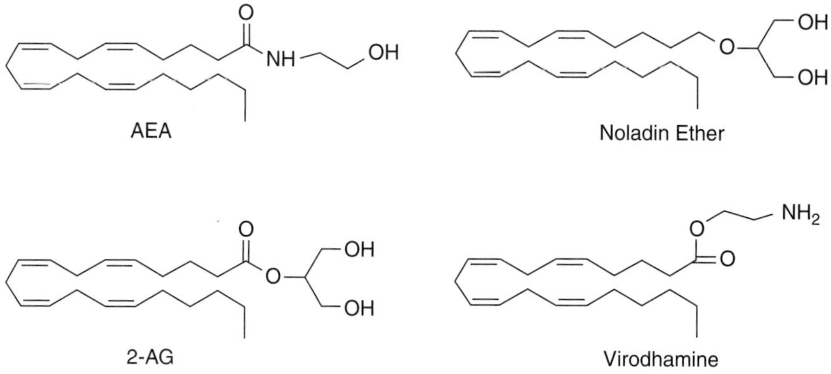

Figure 2.1 Endogenous cannabinoids.

on-demand in a Ca^{2+}-dependent manner via N-archidonylphosphatidylethanolamine (NAPE) phospholipase D (PLD) (Cadas et al., 1997; Di Marzo et al., 1994). First, NAPE is formed by the transfer of an N-archidonyl group to a phosphatidylethanolamine moiety, although the precise enzymes involved in this process are currently under debate (Jin et al., 2007). Then, upon demand, AEA is synthesized from its precursor NAPE. Originally, the cleavage of NAPE to AEA was thought to occur via the Ca^{2+}-dependent phospholipase D activity of NAPE-PLD. Yet, even though the recently cloned NAPE-PLD (Okamoto et al., 2004) is capable of generating AEA from NAPE, studies using NAPE-PLD-null transgenic mice indicate that this may not be the dominant enzyme responsible for AEA biosynthesis (Leung et al., 2006) (for a recent review of this topic, see Okamoto et al. (2007)). In addition, other calcium-independent mechanisms for AEA biosynthesis have been proposed (Liu et al., 2006; Simon and Cravatt, 2008; Vellani et al., 2008).

C. AEA signaling

Regardless of the precise mechanisms responsible for its biosynthesis, AEA is released into the extracellular space, where it is known to act on the cell-surface CB_1 and CB_2 receptors (Felder et al., 1993; Munro et al., 1993), among other potentially distinct targets (Mackie and Stella, 2006). AEA also acts as an endogenous agonist for the transient receptor potential vanilloid type-1 (TRPV1) receptor (De Petrocellis et al., 2001; Ralevic et al., 2001; Ross, 2003; Smart et al., 2000; Zygmunt et al., 1999).

CB1 and CB2 receptor signaling. The cannabinoid receptors (CB_1 and CB_2) are seven-transmembrane (7TM) G protein-coupled receptors (GPCRs) that are predominantly coupled to $G_{\alpha i/o}$ (Matsuda et al., 1990), the inhibitory G protein subunit, but have also been shown to couple with $G_{\alpha s}$ (Jarrahian et al., 2004) and $G_{\alpha q}$ (Lauckner et al., 2005). Activation of these receptors leads to dissociation of the $G_{\alpha i/o}$ subunit from the $G_{\beta \gamma}$ subunits, subsequent inhibition of adenylyl cyclase (AC) via $G_{\alpha i/o}$, and downregulation of cAMP production.

CB_1 receptors are predominantly localized to the central nervous system (CNS) where AEA is released from postsynaptic neurons and signals in a retrograde fashion to bind CB_1 receptors located on the presynaptic axon termini (Leung et al., 2006). CB_1 receptor activation leads to the opening of inwardly rectifying K^+ channels (Mu et al., 1999), closing of T-, N-, and P/Q-type voltage-gated Ca^{2+} channels (Caulfield and Brown, 1992; Lovinger, 2008; Mackie and Hille, 1992; Twitchell et al., 1997), and the activation of various protein kinases (Howlett, 2005). Such anandamide-stimulated CB_1 receptor signaling controls the presynaptic release of neurotransmitters such as gamma-aminobutyric acid (GABA), glutamate, and acetylcholine [reviewed by Freund et al. (2003) and Lovinger (2008)].

Unlike CB_1 receptors, the majority of CB_2 receptors are expressed in the periphery (Klein et al., 2000; Munro et al., 1993). Their activation does not modulate ion channel activity, but does lead to activation of protein kinases (Guindon and Hohmann, 2007; Howlett, 2005). Such CB_2 receptor signaling regulates the immune response, as well as fertility (Huffman and Marriott, 2008; Klein et al., 2000; Maccarrone, 2007). Interestingly, a role for CB_2 receptors in the CNS has also recently been proposed (Guindon and Hohmann, 2007). However, the exact location and signaling events associated with these apparent nonperipherally localized CB_2 receptors is still under investigation.

TRPV1 receptor signaling. AEA also has activity at the TRPV1 receptor, a member of the TRP family of ion channels (Caterina et al., 1997; Ralevic et al., 2001; Ross, 2003; Smart et al., 2000; Zygmunt et al., 1999). The TRPV1 receptor is a ligand-gated Ca^{2+}-permeable nonselective cation channel that is activated by multiple noxious stimuli, including heat, protons, and capsaicin (Caterina et al., 1997). AEA interacts with an intracellular binding site on TRPV1, stimulating channel activation and Ca^{2+} influx (De Petrocellis et al., 2001). AEA-stimulated TRPV1 activation regulates the pain response, circulatory effects, and locomotor skills (Ralevic et al., 2002; Starowicz et al., 2007). Currently, the rationale for AEA activation of both TRPV1 and CB1 receptors, which seemingly elicit opposing signaling pathways, is unknown.

D. AEA degradation

FAAH-1 and FAAH-2. Following uptake into cells, AEA is rapidly broken down into the metabolites arachidonic acid and ethanolamine via fatty acid amide hydrolase (FAAH). Two human FAAH isoforms, which share approximately 20% sequence identity, have been identified and are referred to as FAAH-1 and FAAH-2, respectively (Cravatt et al., 1996; Wei et al., 2006).

Both FAAH-1 and FAAH-2 are members of the large and varied amidase signature (AS) family of proteins (Cravatt et al., 1996; McKinney and Cravatt, 2005; Wei et al., 2006). Members of this family share a long stretch of sequence similarity known as the amidase signature. However, despite the conserved AS domain, a very diverse group of molecules is substrates for AS family member proteins (McKinney and Cravatt, 2005). When comparing the FAAH isoforms, both degrade primary fatty acid amides with equal activity (Wei et al., 2006). Yet, FAAH-1 has greater activity with regard to fatty acid taurines and fatty acid ethanolamides (Wei et al., 2006).

Although most AS family members are soluble proteins, both FAAH enzymes are integral membrane proteins that contain a single transmembrane domain on their N-termini (Cravatt et al., 1996; Wei et al., 2006). We and others have used immunocytochemistry to localize FAAH to an intracellular location in, or around, the endoplasmic reticulum (ER) (Giang and Cravatt, 1997; McFarland et al., 2006). However, the proposed orientation of these two enzymes within the ER membrane is different (Wei et al., 2006). The C-terminal catalytic domain of FAAH-1 has been reported to have a cytoplasmic orientation as compared to FAAH-2, which exposes its C-terminus to the ER lumen (Wei et al., 2006). Such altered arrangements of the FAAH enzymes within the membrane could be indicative of their respective substrate pools within a cell, however, this has yet to be determined (Wei et al., 2006).

Further comparisons reveal that the expression pattern of the two FAAH isoforms varies among tissues and across species (Wei et al., 2006). FAAH-1 has been cloned from various species, including humans, rats, and mice, whereas FAAH-2 is apparently not expressed in murids (McKinney and Cravatt, 2005; Wei et al., 2006). Throughout human tissues, FAAH-1 and FAAH-2 have similar expression levels overall, except that FAAH-2 is the isoform predominantly expressed in the heart, whereas FAAH-1 is preferentially expressed in the brain, testis, and small intestine (Wei et al., 2006).

E. Lipid rafts and the fate of AEA metabolites

We have found that following AEA uptake and metabolism, the AEA metabolites arachidonic acid and ethanolamine are trafficked to caveolae or caveolin-rich membrane (CRM) domains (McFarland et al., 2004) (Fig. 2.2). Caveolae/CRM domains are specialized microdomains within the plasma membrane that are enriched in cholesterol, sphingomyelin, arachidonic acid, plasmenylethanolamine, and the protein caveolin and are considered a subset of lipid rafts (Brown and London, 2000; Pike et al., 2002). Because of their ordered structure relative to the rest of the plasma membrane (Brown and London, 2000), these microdomains are continually implicated in various cellular roles, such as the organization of proteins involved in various signal transduction pathways, including CB_1 receptor signaling (Dainese et al., 2007).

AEA metabolites can theoretically be used to form NAPE, the precursor to AEA, and thus, these CRM microdomains might serve as "warehouses" for on-demand AEA biosynthesis. Indeed, our group has previously reported that caveolae/CRM domains may serve as platforms for the biosynthesis of AEA, and potentially other endocannabinoids (McFarland et al., 2006). When caveolae/CRM domain organization is disrupted, Ca^{2+} ionophore-stimulated AEA synthesis is decreased, indicating the potential importance of these microdomains in the biosynthesis of AEA (Placzek et al., 2008). Additionally, NAPE-PLD and diacylglycerol-lipase, the respective major biosynthetic enzymes responsible for AEA and 2-AG production, have been localized to CRM fractions (Rimmerman et al., 2007).

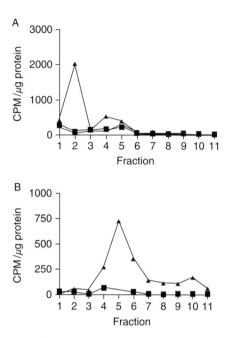

Figure 2.2 Following uptake and catabolism by FAAH, the AEA metabolites arachidonic acid and ethanolamine are trafficked to lipid rafts/caveolae. RBL-2H3 cells were treated with [3H]AEA for 5 min and subsequently lysed and subjected to sucrose density gradient ultracentrifugation to isolate 11 one-mL fractions. The second fraction contains the detergent-resistant CRM domains/caveolae (▲). The effect of the uptake inhibitor AM404 (100 μM) (■) and the FAAH inhibitor methyl arachidonyl fluorophosphonate (250 nM) (▼) on the trafficking of AEA metabolites is also shown. Figure taken from McFarland et al. (2004).

Taking into account the above data, one can propose the following model for AEA biosynthesis and degradation (Fig. 2.3): Following uptake, AEA is shuttled to the intracellularly localized FAAH, where it is degraded. The resulting metabolites arachidonic acid and ethanolamine are re-esterified and then shuttled to caveolae/CRM domains, where they are transformed into NAPE. Upon demand, NAPE is cleaved, and AEA is released into the extracellular space. Such a model is further supported by the localization of NAPE-PLD and diacylglycerol-lipase, the respective major biosynthetic enzymes responsible for AEA and 2-AG production, to CRM fractions (Rimmerman et al., 2007).

In addition to potentially serving as the platform for endocannabinoid biosynthesis, our lab recently demonstrated a role for caveolae/CRM domains in the protein-mediated endocytic uptake of AEA (McFarland et al., 2004, 2008). This topic will be covered in more detail below. First, however, we will address why a lipophilic molecule, such as AEA, might

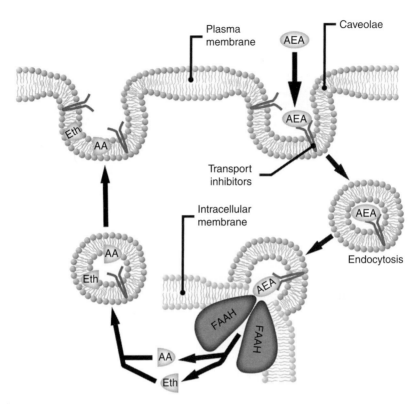

Figure 2.3 The caveolae-related endocytic uptake and trafficking of AEA metabolites to plasma membrane CRM domains. AEA binds a putative carrier protein located within plasma membrane caveolae. Following caveolae-related endocytosis, AEA is trafficked to the intracellularly localized FAAH enzyme. There, AEA is broken down into the metabolites arachidonic acid (AA) and ethanolamine (Eth). The AEA metabolites are then trafficked back to the caveolae/CRM domains via an unknown mechanism, where they are possibly stored for the on-demand biosynthesis of new AEA molecules.

require a specific protein-mediated uptake mechanism for efficient passage across the cell plasma membrane.

II. AEA Transport

Although the mechanisms for AEA biosynthesis, signaling, and degradation are relatively well defined, there is a great deal of controversy surrounding the process(es) controlling AEA transport across the cell membrane. Several hypotheses have been proposed over the years, including a model of passive diffusion (Fasia et al., 2003; Glaser et al., 2003; Kaczocha *et al.*, 2006). However, the majority of data suggest that AEA uptake is a protein-facilitated process (Di Marzo *et al.*, 1994; Fegley *et al.*, 2004; Hillard *et al.*, 1997; Ligresti *et al.*, 2004; Maccarrone *et al.*, 2000; McFarland *et al.*, 2004, 2008; Ortega-Gutierrez *et al.*, 2004; Rakhshan *et al.*, 2000). We will focus on the evidence supporting the protein-facilitated uptake of AEA here including our own data that suggest a caveolae-related endocytic process for the cellular accumulation of AEA (McFarland *et al.*, 2004, 2008). First, though, we will briefly explore the protein-mediated mechanisms responsible for transporting other extracellular lipids and long-chain fatty acid molecules, similar to AEA, across the plasma membrane.

A. Fatty acid transporters

Long-chain fatty acids are used by all cell types in various physiological processes and are emerging as potential modulators of specific diseases, including cardiovascular disease, diabetes, and obesity (Bonen *et al.*, 2004; Coort *et al.*, 2007; Hamilton, 2007; Schwenk *et al.*, 2008). However, besides lipogenic tissues (liver, adipose tissue, lactating mammary gland), all other cell types must obtain their fatty acids from exogenous sources, including cardiac tissue and muscle (Jia *et al.*, 2007). Hence, extracellular fatty acids must be capable of crossing the plasma membrane. After the fatty acid enters the cell, it is either activated by thioesterification to acetyl Co-A immediately at the membrane (vectorial acylation) (Kampf and Kleinfeld, 2007; Watkins, 1997) or is shuttled to intracellular organelles and activated there (Hamilton, 2007; Kampf and Kleinfeld, 2007; Zimmerman and Veerkamp, 2002). Once activated, the cell can then use the fatty acid molecule in various metabolic pathways (Jia *et al.*, 2007).

Previously, there has been debate as to whether long-chain fatty acids cross the cell membrane via a specific mechanism or by passive diffusion (Bonen *et al.*, 2007; Doege and Stahl, 2006). However, it is now believed that specific protein-mediated transport mechanisms do exist for the cellular accumulation of long-chain fatty acid molecules (Abumrad *et al.*, 1998). Two individual proteins and an entire protein family have emerged as potential key regulators of fatty acid transport across the plasma membrane: the plasma membrane fatty acid binding protein (FABPpm), the fatty acid translocase (FAT/CD36), and the fatty acid transport protein (FATP) family (Bonen *et al.*, 2007; Chabowski *et al.*, 2007; Doege and Stahl, 2006). Whether these proteins work independently or in concert to transport fatty acids is unclear and controversial (Bonen *et al.*, 2007; Chabowski *et al.*, 2007; Doege and Stahl, 2006). Here, we will briefly review these established fatty acid transporters to set a precedent for the premise that AEA, a long-chain fatty acid ethanolamide, crosses the membrane via a protein-mediated process, rather than by simple diffusion.

Plasma membrane fatty acid binding protein. FABPpm was the original reported candidate for the fatty acid transporters. This approximately 40 kDa protein, originally discovered by Stremmel and colleagues (1985), is a peripheral membrane protein that has since been found to be identical to the mitochondrial aspartate aminotransferase (mAspAt) (Berk *et al.*, 1990; Isola *et al.*, 1995). Apparently, this protein has different functions depending on its cellular localization to either the plasma membrane or mitochondria (Holloway *et al.*, 2007).

Although the precise mechanism is unclear, there is a great deal of evidence to suggest that FABPpm plays a role in fatty acid transport (Berk *et al.*, 1990, 1997; Heather *et al.*, 2006; Isola *et al.*, 1995; Stump *et al.*, 1993; Turcotte *et al.*, 1999; Zhou *et al.*, 1995). The assignment of a protein as a fatty acid transporter is based on several criteria (Hui and Bernlohr, 1997), and FABPpm has been observed to meet these criteria (Berk *et al.*, 1990, 1997; Heather *et al.*, 2006; Isola *et al.*, 1995; Stump *et al.*, 1993; Turcotte *et al.*, 1999; Zhou *et al.*, 1995): (1) localization to the plasma membrane, (2) a selective and reversible high affinity for fatty acids, (3) an increase in saturable fatty acid transport during overexpression of the protein, and (4) a decrease in fatty acid transport after treatment with antibodies or siRNA targeting the protein. FABPpm localizes to the outer leaflet of the plasma membrane and is not a member of the cytosolic fatty acid binding protein (FABP) family, which at first, may be implied due to the name (Stump *et al.*, 1993; Zhou *et al.*, 1995). Overexpression of FABPpm in *Xenopus laevis* oocytes and in NIH-3T3 fibroblasts in both cases corresponded to an increase in saturable fatty acid transport (Isola *et al.*, 1995; Zhou *et al.*, 1992). Likewise, alterations in the expression levels of FABPpm seem to generally correlate with the levels of fatty acid uptake required by different cell types (Berk *et al.*, 1997; Heather *et al.*, 2006; Turcotte *et al.*, 1999) although changes in protein expression are not necessarily always proportionate to fatty acid transport (Chabowski *et al.*, 2007). In addition, treatment of cells with FABPpm-recognizing antibodies inhibits long-chain fatty acid uptake selectively (Berk *et al.*, 1990; Zhou *et al.*, 1995).

Fatty acid translocase (FAT/CD36). The heavily glycosylated fatty acid translocase (FAT) was cloned by Abumrad and colleagues in 1993, and was determined to have 85% sequence similarity with the class B scavenger receptor protein CD36 (Abumrad *et al.*, 1993). Whereas FAT/CD36 has various cellular functions (Febbraio *et al.*, 2001; Holloway *et al.*, 2008), the protein has repeatedly been shown to also play a role in the translocation of fatty acids across the plasma membrane (Abumrad *et al.*, 1999; Bonen *et al.*, 2007; Febbraio *et al.*, 2002; Rac *et al.*, 2007). We will focus on that function here.

Similar to FABPpm, FAT/CD36 meets the criteria defining a fatty acid transporter (Abumrad *et al.*, 1999; Bonen *et al.*, 2007; Febbraio *et al.*, 2002; Hui and Bernlohr, 1997; Rac *et al.*, 2007). FAT/CD36 has been isolated to the plasma membrane of various cell types and is believed to contain two transmembrane domains (Rac et al., 2007). Recently, the cytoplasmic C-terminus was implicated in the localization of FAT/CD36 to the plasma membrane as well as the protein's FAT activity (Eyre et al., 2007). Both *in vitro* and *in vivo* studies show a positive correlation between expression levels of the 88 kDa FAT/CD36 and saturable and selective long-chain fatty acid uptake in adipocytes and myocytes [reviewed by Abumrad et al. (1999), Bonen et al. (2007), Febrraio et al. (2002)].

Treatment of cells with long-chain fatty acids increased FAT/CD36 expression, possibly via PPARγ-mediated transcriptional events (Sfeir et al., 1997; Yang et al., 2007). Interestingly, treatment of human peripheral mononuclear blood cells with AEA, a proposed PPARγ agonist (Bouaboula et al., 2005; O'Sullivan, 2007), was recently shown to increase CD36 mRNA and protein expression (Malfitano et al., 2007). Such findings may indicate a role for FAT/CD36 in the uptake of AEA although to our knowledge, no detailed reports investigating this premise exist at this time.

Fatty acid transport protein (FATP) family. Within the solute carrier family 27 (SLC27) of proteins, there are six mammalian isoforms of fatty acid transport proteins (FATPs1–6), which are apparently identical to the six members of the very long-chain acyl-CoA synthetase (VLACS) protein family (Jia et al., 2007). While FATPs are thought to participate in fatty acid transport, their roles are controversial and not well defined.

FATPs are integral membrane proteins that contain one or more aminoterminal transmembrane domains (Lewis et al., 2001; Schaap et al., 1997). In addition, these family members contain two cytoplasmic-facing motifs found within the approximately 300 amino acid-long FATP signature sequence: an adenosine triphosphate/adenosine monophosphate (ATP/AMP) motif and an FATP/VLACS motif (Stahl, 2004; Stahl et al., 2001). The ATP/AMP motif is important for both ATP binding and fatty acid transport whereas the FATP/VLACS motif is believed to be involved solely in fatty acid transport (DiRusso et al., 2008; Stahl, 2004; Stahl et al., 2001; Stuhlsatz-Krouper et al., 1998, 1999). The individual FATPs exhibit differential tissue- and species-specific expression patterns, and this topic has been covered in more detail in several reviews Bonen et al. (2007), Doege and Stahl (2006), Ehehalt et al. (2006), Stahl (2004), Stahl et al. (2001).

FATPs have been characterized as FATPs based upon the observations that fatty acid transport increases with FATP expression levels and that these overexpressed proteins are localized at the plasma membrane (DiRusso et al., 2005; Ehehalt et al., 2006; Stahl, 2004). However, it was recently shown that overexpressed FATPs display cellular localizations distinct from their endogenous counterparts (Jia et al., 2007). Thus, there is now debate as to whether these proteins are associated with the plasma membrane or the membrane of intracellular organelles (Jia et al., 2007; Pei et al., 2004). Obviously, if FATPs truly act as fatty acid "transporters," they must localize to the plasma membrane in native cells. The lack of such evidence has prompted some researchers to question what role, if any, that FATPs play in fatty acid transport (Doege and Stahl, 2006; Jia *et al.*, 2007).

The fatty acyl CoA synthetase activity of FATP family members has stirred additional debate regarding the mechanism by which FATPs promote fatty acid transport. As mentioned earlier, FATPs are apparently identical to the VLACS protein family members (Jia *et al.*, 2007) and have long- and very long-chain (i.e., fatty acid chains >18 carbons) fatty acyl CoA synthetase activity (Coe *et al.*, 1999; DiRusso *et al.*, 2005; Pei *et al.*, 2004). This intrinsic enzyme activity, rather than serving an actual transporter function, has been proposed to be responsible for the FATP-driven translocation of fatty acids across the plasma membrane (Jia *et al.*, 2007; Pohl *et al.*, 2004; Stahl, 2004). The acylation of intracellular fatty acids prevents them from crossing the plasma membrane and reentering the extracellular space (Watkins, 1997). Therefore, FATPs could be promoting fatty acid transport, not by physically transporting the molecules across the membrane, but rather by essentially altering the concentration gradient along which the molecules move (Ehehalt *et al.*, 2006). The FATPs may "remove" the activated fatty acids from the intracellular pool of inactivated fatty acids that defines the concentration gradient driving the influx of inactivated fatty acids (Ehehalt *et al.*, 2006). Yet, in contrast to this hypothesis, several studies have recently used yeast overexpression studies to demonstrate that the acyl CoA synthetase activity of FATPs is distinguishable from their transport function (DiRusso *et al.*, 2005, 2008; Pei *et al.*, 2004).

B. Lipid transfer proteins

StAR-related lipid transfer (START) protein family. The steroidogenic acute regulatory protein (StAR)-related lipid transfer (START) domain protein family is a class of proteins involved in the intracellular transfer of hydrophobic ligands between membranes (Soccio and Breslow, 2003; Wirtz, 2006). In humans, there are 15 START domain protein family members (StARD1–15) comprising six START domain subfamilies.

Every START protein contains one C-terminal START domain consisting of approximately 210 amino acid residues (Ponting and Aravind, 1999). The structure of the START domain consists of a "helix-grip fold" comprised of a central nine-stranded antiparallel incomplete β-barrel surrounded by long N-terminal and C-terminal α-helices ($\alpha1$ and $\alpha4$) in addition to two shorter α-helices ($\alpha2$ and $\alpha3$) (Iyer *et al.*, 2001; Tsujishita and Hurley, 2000). The ligand-binding site lies within a hydrophobic tunnel composed of the β-sheet and α-helices 2, 3, and 4 (Alpy and Tomasetto, 2005). The C-terminal portion of the START domain ($\alpha4$) may act as a lid atop this hydrophobic tunnel and appears to play an essential role in START domain function (Alpy and Tomasetto, 2005; Arakane *et al.*, 1996; Feng *et al.*, 2000). In addition to this START domain, different subfamily members contain other functional domains, including the Rho-GAP or thioesterase domains (Alpy and Tomasetto, 2005; Soccio and Breslow, 2003). While not verified, it has been suggested that the function of START proteins containing such additional domains is regulated in a lipid-dependent manner using the START domain as a "lipid sensor" (Alpy and Tomasetto, 2005).

There are many proposed models for how START proteins transfer lipids from one membrane to another (Alpy and Tomasetto, 2005; Miller, 2007). However, most data regarding the original START protein, StAR, support a model whereby this protein transports cholesterol to the mitochondria by acting exclusively at the outer mitochondrial membrane (Alpy and Tomasetto, 2005; Miller, 2007). Alpy and Tomasetto (2005) have thus applied this mechanism of action to the entire START protein family. In their model, a START protein accepts a lipid ligand by opening its C-terminal lid and dipping into a membrane source (plasma, organelle, etc.) (Alpy and Tomasetto, 2005). Once the lid closes, the START protein releases from the donor membrane, travels to another membrane, and then transfers the lipid to that acceptor membrane via the reverse of the above process (Alpy and Tomasetto, 2005). In addition, they note that while START proteins have only been reported to donate their lipid ligands to membrane acceptors, it is possible that they may also transfer lipids to protein acceptors, as well (Alpy and Tomasetto, 2005).

C. Characteristics of AEA transport

AEA accumulation has been observed in a wide variety of neuronal and nonneuronal cell types, including primary cortical neurons (Di Marzo *et al.*, 1994; Ortega-Gutierrez *et al.*, 2004), mouse Cath.a-differentiated (CAD) neuronal cells (McFarland *et al.*, 2008), primary rat cerebellar granule cells (Hillard *et al.*, 1997), rat basophilic leukemia (RBL-2H3) cells (Ligresti *et al.*, 2004; Rakhshan *et al.*, 2000), rat C6 glioma cells (Ligresti *et al.*, 2004), HeLa cells (Day *et al.*, 2001), 3T3-L1 preadipocytes (Thors *et al.*, 2007), human endothelial cells (HUVECs) (Maccarrone *et al.*, 2000), two prostate cancer cell lines (PC-3 and R3327 AT-1) (Thors *et al.*, 2007), and the intact retina (Glaser *et al.*, 2005a). Many of these studies have characterized AEA uptake as a rapid and saturable process that is temperature-dependent and independent of energy in the form of either ATP or ion gradients (Di Marzo *et al.*, 1994; Fegley *et al.*, 2004; Hillard *et al.*, 1997; Ligresti *et al.*, 2004; Maccarrone *et al.*, 2000; McFarland *et al.*, 2008; Ortega-Gutierrez *et al.*, 2004; Rakhshan *et al.*, 2000). Furthermore, there appears to be a structural requisite for AEA uptake (reviewed by Reggio and Traore (2000)), and AEA accumulation can be pharmacologically inhibited, although the precise protein target(s) of these inhibitory compounds is unclear (Fowler et al., 2004). Overall, such characteristics correspond well with a process of facilitated diffusion. However, despite there being a general consensus that AEA uptake occurs via a facilitated process (Di Marzo et al., 1994; Fegley et al., 2004; Hillard et al., 1997; Ligresti et al., 2004; Maccarrone et al., 2000; McFarland et al., 2004, 2008; Ortega-Gutierrez et al., 2004; Rakhshan et al., 2000), there is debate regarding the exact proteins involved (Hillard and Jarrahian, 2003). In addition, other researchers have offered an alternative, passive diffusion-based model for AEA uptake (Fasia et al., 2003; Glaser et al., 2003; Kaczocha et al., 2006). Below, we will summarize the various proposed models for the cellular accumulation of AEA.

D. Proposed models for cellular AEA accumulation

Passive diffusion model for AEA uptake. While the majority of studies characterize AEA uptake as a facilitated process (Di Marzo et al., 1994; Fegley et al., 2004; Hillard et al., 1997; Ligresti et al., 2004; Maccarrone et al., 2000; McFarland et al., 2004, 2008; Ortega-Gutierrez et al., 2004; Rakhshan et al., 2000), a few studies have concluded that AEA crosses the plasma membrane via passive diffusion (Fasia et al., 2003; Glaser et al., 2003; Kaczocha et al., 2006). These researchers suggest a model where no membrane protein carrier exists, but rather movement of AEA across the membrane is merely driven via its metabolism by FAAH and/or its sequestration by other intracellular binding proteins (Fasia et al., 2003; Glaser et al., 2003; Kaczocha et al., 2006).

Hillard and colleagues have suggested a model of intracellular AEA sequestration that simultaneously supports the passive diffusion of AEA across the cell membrane while explaining the characteristics of AEA uptake that are consistent with a process of facilitated diffusion (Hillard and Jarrahian, 2003). In this model, extracellular AEA can cross the plasma membrane along the concentration gradient un-aided by any protein carrier (Hillard and Jarrahian, 2003). Once across, AEA is believed to bind intracellular binding proteins and is thus removed from the pool of "free AEA," allowing for maintenance of the concentration gradient (Hillard and Jarrahian, 2003). In this hypothesis, AEA uptake would be saturated only once there are no available intracellular binding proteins. This model for the sequestration of intracellular AEA has been supported by observations that, at equilibrium, the intracellular concentration of AEA is actually higher than that in the extracellular space (Deutsch et al., 2001; Hillard and Jarrahian, 2000; Rakhshan et al., 2000). However, to our knowledge, there are no reports identifying a specific intracellular binding protein that plays a role in the AEA uptake process. Furthermore, although this model describes AEA uptake as a passive diffusion process, Hillard and colleagues do not refute the possible existence of a plasma membrane-localized protein carrier in certain cell types (Hillard and Jarrahian, 2003).

Kaczocha *et al.* (2006) have proposed an alternative passive diffusion model for AEA uptake. They suggest that any saturating effects observed during initial AEA uptake are the result of the permeation of AEA through an unstirred water layer surrounding cells. These authors propose that the movement of AEA through this water layer around cells can mimic characteristics of facilitated transport when measured via standard kinetic assays (Bojesen and Hansen, 2003; Kaczocha *et al.*, 2006).

As was discussed by Glaser *et al.* (2005b) much of the contention between the facilitated and passive diffusion models for AEA uptake stem from the varying protocols used to study the process. The time points at which AEA uptake is measured seem to influence the characterization of the transport process as either a facilitated or passive diffusion event. Recently, a few reports have shown that initial AEA uptake kinetics are linear, and thus, suggest passive diffusion of AEA across the plasma membrane (Glaser *et al.*, 2003; Kaczocha *et al.*, 2006; Sandberg and Fowler, 2005). In these studies, researchers measured AEA uptake at time points less than 1 min (Glaser *et al.*, 2003; Kaczocha *et al.*, 2006; Sandberg and Fowler, 2005), as opposed to experiments determining steady-state uptake kinetics at time points ranging anywhere from 5 min to 40 h (Dickason-Chesterfield *et al.*, 2006; Fegley *et al.*, 2004; Maccarrone *et al.*, 2000; McFarland *et al.*, 2008; Moore *et al.*, 2005; Rakhshan *et al.*, 2000). It was argued that at early time points under 1 min, FAAH has not yet begun to metabolize AEA and therefore cannot influence transport or interfere with the analysis of AEA uptake kinetics (Glaser *et al.*, 2003; Kaczocha *et al.*, 2006). Thus, these initial measurements may better represent the mechanism by which extracellular AEA crosses the plasma membrane. However, in contrast to the results of these studies, not all measurements of initial AEA uptake (time points <5 min) correspond to a model of passive diffusion (Ligresti *et al.*, 2004). Ligresti and colleagues (2004) found that at low concentrations, AEA uptake is saturable within 90 s in both RBL-2H3 and C6 glioma cell lines, as well as in mouse brain synaptosomes from both wild-type and $FAAH^{-/-}$ mice. Such discrepant results suggest that either AEA uptake is mediated by cell type-specific mechanisms or that the experimental variability within the assays has led to potentially artifactual results in some, or all, studies.

Another source of experimental variability that has created debate regarding the mechanism for AEA uptake is the presence or absence of bovine serum albumin (BSA) in the assay buffer (Glaser *et al.*, 2005b). Due to its lipophilic nature, AEA has a tendency to adhere to the extracellular cell surface and also to the plastic tissue culture plates most often used for the transport assays (Karlsson *et al.*, 2004; Ligresti *et al.*, 2004; Ortega-Gutierrez *et al.*, 2004). BSA is known to bind AEA (Bojesen and Hansen, 2003), but its use in uptake assays to prevent the nonspecific binding of AEA to plasticware and cells has been widely debated. Although the use of BSA in assay buffer does seem to affect AEA uptake (Di Marzo *et al.*, 1994; Karlsson *et al.*, 2004; Ligresti *et al.*, 2004; Ortega-Gutierrez *et al.*, 2004), Glaser *et al.* (2003) eloquently noted in a recent review that these results are interpreted very differently. Some researchers believe that BSA interferes with and prevents AEA uptake (Di Marzo *et al.*, 1994; Ligresti *et al.*, 2004) and thus, choose to eliminate it from assay buffers. Yet, others insist that assays conducted in the absence of BSA result in artifactual data that support the mistaken characterization of AEA uptake as facilitated process (Karlsson *et al.*, 2004; Ortega-Gutierrez *et al.*, 2004). Karlsson *et al.* (2004) and Ortega-Gutierrez *et al.* (2004) demonstrated that not only does AEA bind to the wells of plastic tissue culture plates, but also this binding is temperature-dependent and blocked by supposed AEA uptake inhibitors in a dose-dependent manner, thus potentially mimicking a protein-facilitated uptake event. Today, there appears to be no clear resolution to the dilemma regarding the various protocols used to measure AEA uptake, whether concerning the inclusion of BSA in assay buffers or at what time points to measure AEA transport. Therefore, these issues remain as points of both internal and external conflict among many researchers in the field.

Facilitated processes for AEA uptake. Whereas, as described above, there is some evidence to suggest a model of passive diffusion for AEA uptake, the vast majority of data suggest that AEA crosses the plasma membrane via a process of facilitated diffusion mediated by a specific, but unidentified protein (Di Marzo *et al.*, 1994; Fegley *et al.*, 2004; Hillard *et al.*, 1997; Ligresti *et al.*, 2004; Maccarrone *et al.*, 2000; McFarland *et al.*, 2004, 2008; Ortega-Gutierrez *et al.*, 2004; Rakhshan *et al.*, 2000). At first, it may seem as though a lipophilic molecule such as AEA should not require additional assistance to cross the plasma lipid bilayer. However, as was summarized in the previous section, it is generally accepted that many lipids and fatty acid molecules are transported across the cell membrane via a protein-mediated event (Abumrad *et al.*, 1998). Yet, even those in agreement about a model of facilitated diffusion for AEA uptake debate the exact mechanism by which the protein-mediated process occurs. Some have argued for the existence of a plasma membrane-localized carrier protein that helps AEA transverse the plasma membrane (Day *et al.*, 2001; Deutsch *et al.*, 2001; Hillard and Jarrahian, 2000, 2003; Hillard *et al.*, 1997; Ligresti *et al.*, 2004; Rakhshan *et al.*, 2000), whereas others have suggested an intracellular AEA-binding protein distinct from FAAH (Fegley *et al.*, 2004; Hillard and Jarrahian, 2003; Hillard *et al.*, 2007). Alternatively, our lab has demonstrated that AEA uptake can be mediated by a caveolae-related endocytic event (McFarland *et al.*, 2004, 2008). Below, we will discuss these variations on the facilitated diffusion model proposed to explain AEA uptake.

There is evidence to suggest that the movement of AEA across the plasma membrane is bidirectional, indicating the existence of a plasma membrane-localized carrier protein (Gerdeman et al., 2002; Hillard and Jarrahian, 2000; Hillard et al., 1997; Ligresti et al., 2004; Maccarrone et al., 2002). Several different cell lines, including cerebellar granule cells (Hillard et al., 1997), human umbilical vein endothelial cells (HUVECs) (Maccarrone et al., 2002), and striatal neurons (Gerdeman et al., 2002), are capable of AEA efflux. In HEK-293 cells, the release of de novo biosynthesized AEA is significantly decreased by the AEA uptake inhibitor VDM11 (Ligresti et al., 2004). Additionally, AEA transport has been shown to exhibit the *trans* effect of flux coupling (Hillard and Jarrahian, 2000). During this phenomenon, carrier proteins accumulate and orient themselves toward the extracellular space in response to accumulating intracellular AEA (Hillard and Jarrahian, 2000). Thus, AEA can be transported against the concentration gradient without requiring energy (Hillard and Jarrahian, 2000). However, the intracellular sequestration of AEA by a binding protein (Hillard and Jarrahian, 2003), a model described earlier, could also account for the observation that intracellular AEA concentrations can exceed those in the extracellular space (Hillard and Jarrahian, 2003).

Initially, the synthesis and characterization of specific AEA uptake inhibitors further supported the model whereby AEA uptake occurred via a membrane-localized protein carrier (Fig. 2.4). However, it has since been determined that nearly all of these "specific" AEA uptake inhibitors also inhibit FAAH activity, to one extent or another (Alexander and Cravatt, 2006; Dickason-Chesterfield et al., 2006; Hillard et al., 2007). This revelation has created much debate as to whether such inhibitory compounds exert their effects via targeting of the putative transporter, FAAH, or both [for a review of this topic, see Fowler et al. (2004)]. Hence, such questions have led researchers to investigate the role that FAAH plays in AEA uptake.

AEA uptake is driven by, but not necessarily dependent upon, FAAH activity (Day et al., 2001; Deutsch et al., 2001; Fegley et al., 2004; Hillard et al., 2007; Ligresti et al., 2004). It is clear that FAAH activity stimulates AEA transport, presumably by maintaining the concentration gradient between the extracellular and intracellular AEA pools (Day et al., 2001; Ligresti et al., 2004). Overexpression of FAAH in mammalian cells enhances AEA uptake (Day et al., 2001). These results are consistent with those from studies comparing AEA uptake in cells from transgenic $FAAH^{-/-}$ versus wild-type mice (Ligresti et al., 2004). However, FAAH does not appear to be the only protein driving AEA uptake. For instance, although AEA uptake is heightened in wild-type mice, cells from $FAAH^{-/-}$ mice still display saturable and pharmacologically inhibitable AEA accumulation (Fegley et al., 2004; Ligresti et al., 2004). Likewise, our lab has found that HeLa cells, which are devoid of endogenous FAAH activity, still accumulate AEA in a manner consistent with facilitated diffusion (Day et al., 2001).

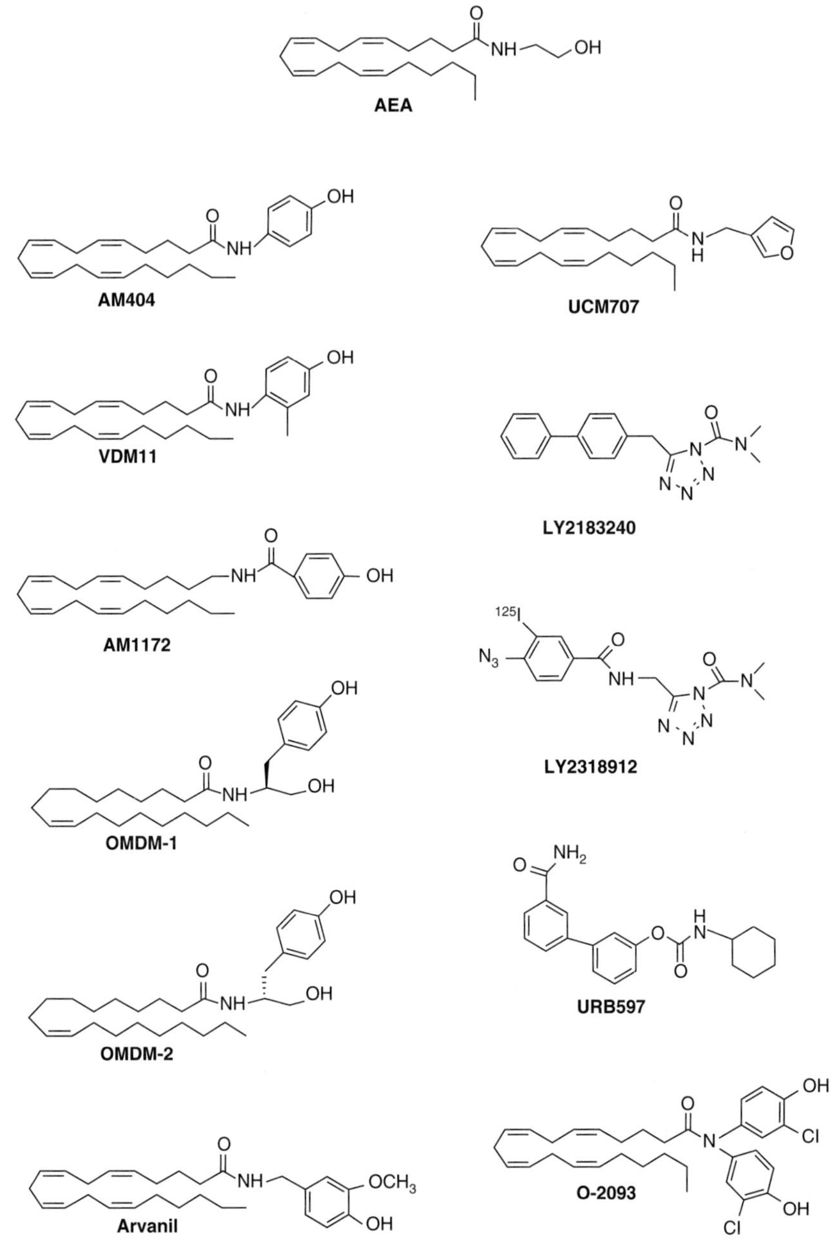

Figure 2.4 AEA uptake inhibitors.

Furthermore, Hillard and colleagues (2007) have reported that while "specific" AEA uptake inhibitors inhibit FAAH activity, the associated reduction in AEA accumulation is not solely due to their prevention of AEA hydrolysis. Such data suggest that another protein, or group of proteins, is also responsible for mediating AEA uptake.

Our lab has proposed a protein-mediated caveolae-related endocytic model for the cellular accumulation of AEA (McFarland *et al.*, 2004, 2008). This model is consistent with the aforementioned hypotheses concerning the facilitated uptake of AEA as well as the intracellular sequestration model put forth by Hillard and colleagues (Hillard and Jarrahian, 2003; McFarland *et al.*, 2004, 2008). We have shown that the disruption of caveolae/CRM domain formation by the cholesterol-depleting compounds nystatin and progesterone inhibited AEA uptake by approximately 50% in the RBL-2H3 cell line (McFarland *et al.*, 2004) (Fig. 2.5). More recently,

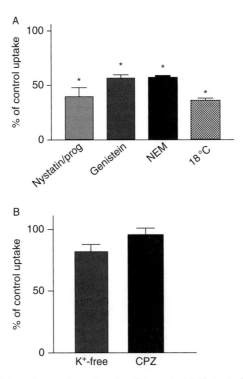

Figure 2.5 Inhibition of caveolae-related endocytosis (clathrin-independent) reduces cellular AEA accumulation. AEA uptake was assessed in RBL-2H3 cells treated to inhibit caveolae-related endocytosis (25 μg/mL nystatin and 10 μg/mL progesterone, 200 μM genistein, 500 μM N-ethylmaleimide (NEM), or 18 °C temperature hold (A). (B) Inhibition of clathrin-dependent endocytosis by removal of potassium from the assay buffer or treatment with 100 μM chlorpromazine (CPZ) did not affect AEA uptake. Figure taken from McFarland *et al.* (2004).

we showed that caveolae also appear to mediate AEA uptake in neuronal cells, as well (McFarland et al., 2008). Small inhibitory RNA (siRNA)-mediated knockdown of dynamin 2 (Dyn2), a protein involved in both the distinct caveolae-related and clathrin-dependent endocytosis events, inhibited the uptake of the fluorescent AEA analog SKM 4-45-1 in mouse neuronal CAD cells (McFarland et al., 2008). However, treatment of these cells with siRNA targeting the $\beta 2$ adaptin subunit of the clathrin-associated activator protein 2 complex (AP2), which is involved only in clathrin-dependent endocytosis, had no effect (McFarland et al., 2008). The decrease in Dyn2 expression had no observable effect on either FAAH expression or activity suggesting that the disruption of caveolae-related endocytosis alone led to the decrease in AEA uptake (McFarland et al., 2008).

The caveolae-mediated endocytic uptake of AEA has been further supported by other published reports (Bari et al., 2005a,b, 2006; Rimmerman et al., 2007). Recently, Rimmerman et al. (2007) showed that exogenously applied AEA localizes to lipid rafts. In 2005, Bari and colleagues (2005a,b) reported that following lipid raft/CRM domain disruption with the cholesterol-depleting agent methyl-β cyclodextrin, CB_1 receptor signaling increased by twofold in C6 glioma cells. These results indicate that inhibition of AEA uptake via the disruption of lipid rafts/CRM domains led to an increased availability of extracellular AEA for CB_1 receptor activation (Bari et al., 2005a,b). Additionally, other endocannabinoids may utilize CRM domains to enter the cell as disruption of lipid raft/CRM domain formation was also shown to decrease uptake of the endocannabinoid 2-AG (Bari et al., 2006).

Interestingly, these specialized plasma membrane microdomains have been previously implicated in the endocytic uptake of the long-chain fatty acid oleate (Pohl et al., 2002; Ring et al., 2002) and the fatty acid- and AEA-binding protein albumin (Vogel et al., 2001). In vascular endothelial cells, the caveolae-mediated transcytosis of albumin is apparently triggered by the binding of albumin to its receptor, gp60 (Vogel et al., 2001). As AEA has an affinity for albumin (Bojesen and Hansen, 2003) and has been shown to be transported in HUVECs (Maccarrone et al., 2000), one could question whether AEA uptake into these cells is also dependent upon a caveolae-related endocytic process, as appears to be the case in other cell lines (McFarland et al., 2004). Furthermore, one could also ask whether the caveolae-related endocytosis of AEA is dependent upon albumin binding its apparent caveolae-localized receptor. To our knowledge, there are no reports outlining such studies.

In addition to providing evidence of a role for caveolae in AEA uptake, our data support the existence of a plasma membrane-localized AEA binding protein (McFarland et al., 2008). Knockdown of Dyn2 in CAD cells only affected the uptake of the fluorescent AEA analog SKM 4-45-1 and did not alter the apparent accumulation of radiolabeled AEA (McFarland et al., 2008). SKM 4-45-1, which is transported via the same mechanism as AEA, only fluoresces once inside the cell where the fluorescein moiety can be cleaved from anandamide by intracellular esterases (Muthian et al., 2000). Therefore, we interpreted these results to suggest that using the radioactive uptake assay, one cannot discern between the association of AEA with the plasma membrane, possibly via a binding protein, and the actual internalization of AEA into the cell (McFarland et al., 2008).

Recently Moore and colleagues (2005) reported the development of the cell-impermeable high-affinity AEA uptake inhibitor LY2318912. Although the parent compound, LY2183240, has since been found to inhibit FAAH activity (Alexander and Cravatt, 2006), Moore et al. (2005) report that LY2318912 binds in a pharmacologically similar manner to cells both expressing and devoid of FAAH. These data suggest that this compound interacts with a target that is distinct from FAAH (Moore et al., 2005). Perhaps the putative binding protein inferred from our data and discussed above is, in fact, the target of this unique AEA uptake inhibitor.

As has been reviewed above, there is still much debate concerning the mechanism(s) by which AEA is accumulated in cells. However, the majority of data support a model for a protein-facilitated process (Di Marzo et al., 1994; Fegley et al., 2004; Hillard et al., 1997; Ligresti et al., 2004; Maccarrone et al., 2000; McFarland et al., 2004, 2008; Ortega-Gutierrez et al., 2004; Rakhshan et al., 2000). New developments, including the synthesis of biotin-tagged AEA (Fezza et al., 2008) and several photoaffinity ligands for the putative AEA carrier protein (Balas et al., 2005; Moore et al., 2005; Moriello et al., 2006), keep the debate surrounding AEA uptake alive and indicate that we may soon have a better understanding of this controversial process.

E. Implications of pharmacologically altered AEA signaling

There are an overwhelming number of implications for AEA and other endocannabinoids in various physiologically and pathophysiologically relevant human states including pain (Hohmann and Suplita, 2006), appetite (Kirkham and Tucci, 2006), reproduction (Wang et al., 2006), mood and anxiety (Witkin et al., 2005), the cardiovascular system (Ashton and Smith, 2007), alcoholism (Rodriguez de Fonseca et al., 2005), spasticity (Pertwee, 2002), neurodegeneration (Battista et al., 2006), and inflammation (Ashton, 2007). In fact, these cited references are merely examples of the numerous reviews recently written on each representative topic.

One way to alter AEA signaling, to achieve desirable health benefits, is to modulate its uptake into cells. As was noted by Moore et al. (2005) modulating AEA signaling in this way would be superior to modulation of AEA receptor signaling directly because inhibition of AEA uptake would allow for increased signaling only in areas where AEA was already present and physiologically active. Hence, the putative AEA-transport protein should prove to be a significant pharmacological target. To truly understand and pharmacologically exploit AEA signaling, we must continue to focus our efforts on better understanding the mechanism(s) governing AEA uptake in various cell types, as well as continuing the search for the putative and elusive AEA carrier protein(s).

ACKNOWLEDGMENT

We thank David Allen for illustrations.

REFERENCES

Abumrad, N. A., el Maghrabi, M. R., Amri, E. Z., Lopez, E., and Grimaldi, P. A. (1993). Cloning of a rat adipocyte membrane protein implicated in binding or transport of long-chain fatty acids that is induced during preadipocyte differentiation. Homology with human CD36. *J. Biol. Chem.* **268**, 17665–17668.

Abumrad, N., Harmon, C., and Ibrahimi, A. (1998). Membrane transport of long-chain fatty acids: Evidence for a facilitated process. *J. Lipid Res.* **39**, 2309–2318.

Abumrad, N., Coburn, C., and Ibrahimi, A. (1999). Membrane proteins implicated in long-chain fatty acid uptake by mammalian cells: CD36, FATP and FABPm. *Biochim. Biophys. Acta Mol. Cell Biol. Lipids* **1441**, 4–13.

Alexander, J. P., and Cravatt, B. F. (2006). The putative endocannabinoid transport blocker LY2183240 is a potent inhibitor of FAAH and several other brain serine Hydrolases. *J. Am. Chem. Soc.* **128**, 9699–9704.

Alpy, F., and Tomasetto, C. (2005). Give Lipids a START: The StAR-related lipid transfer (START) domain in mammals. *J. Cell Sci.* **118**, 2791–2801.

Arakane, F., Sugawara, T., Nishino, H., Liu, Z., Holt, J., Pain, D., Stocco, D., Miller, W., and Strauss, J. (1996). Steroidogenic acute regulatory protein (StAR) retains activity in the absence of its mitochondrial import sequence: Implications for the mechanism of StAR action. *Proc. Natl. Acad. Sci. USA* **93**, 13731–13736.

Ashton, J. C. (2007). Cannabinoids for the treatment of inflammation. *Curr. Opin. Invest. Drugs* **8**, 373–384.

Ashton, J. C., and Smith, P. F. (2007). Cannabinoids and cardiovascular disease: The outlook for clinical treatments. *Curr. Vasc. Pharmacol.* **5**, 175–185.

Balas, L., Hellal, M., Rossi, J. C., and Durand, T. (2005). Synthesis of a photoactivatable probe of the anandamide re-uptake. *Nat. Prod. Res.* **19**, 419–423.

Bari, M., Paradisi, A., Pasquariello, N., and Maccarrone, M. (2005a). Cholesterol-dependent modulation of type 1 cannabinoid receptors in nerve cells. *J. Neurosci. Res.* **81**, 275–283.

Bari, M., Battista, N., Fezza, F., Finazzi-Agro, A., and Maccarrone, M. (2005b). Lipid rafts control signaling of type-1 cannabinoid receptors in neuronal cells: Implications for anandamide-induced apoptosis. *J. Biol. Chem.* **280**, 12212–12220.

Bari, M., Spagnuolo, P., Fezza, F., Oddi, S., Pasquariello, N., Finazzi-Agro, A., and Maccarrone, M. (2006). Effect of lipid rafts on CB2 receptor signaling and 2-arachidonoyl-glycerol metabolism in human immune cells. *J. Immunol.* **177**, 4971–4980.

Battista, N., Fezza, F., Finazzi-Agro, A., and Maccarrone, M. (2006). The endocannabinoid system in neurodegeneration. *Ital. J. Biochem.* **55**, 283–289.

Berk, P. D., Wada, H., Horio, Y., Potter, B. J., Sorrentino, D., Zhou, S., Isola, L. M., Stump, D., Kiang, C., and Thung, S. (1990). Plasma membrane fatty acid-binding protein and mitochondrial glutamic-oxaloacetic transaminase of rat liver are related. *Proc. Natl. Acad. Sci. USA* **87**, 3484–3488.

Berk, P. D., Zhou, S. L., Kiang, C. L., Stump, D., Bradbury, M., and Isola, L. M. (1997). Uptake of long chain free fatty acids is selectively up-regulated in adipocytes of zucker rats with genetic obesity and non-insulin-dependent diabetes mellitus. *J. Biol. Chem.* **272**, 8830–8835.

Bojesen, I. N., and Hansen, H. S. (2003). Binding of anandamide to bovine serum albumin. *J. Lipid Res.* **44**, 1790–1794.

Bonen, A., Parolin, M. L., Steinberg, G. R., Calles-Escandon, J., Tandon, N. N., Glatz, J. F. C., Luiken, J. J. F. P., Heigenhauser, G., and Dyck, D. J. (2004). Triacylglycerol Accumulation in human obesity and type 2 diabetes is associated with increased rates of skeletal muscle fatty acid transport and increased sarcolemmal FAT/CD36. *FASEB J.* **18**, 1144–1146.

Bonen, A., Chabowski, A., Luiken, J. J. F. P., and Glatz, J. F. C. (2007). Mechanisms and regulation of protein-mediated cellular fatty acid uptake: Molecular, biochemical, and physiological evidence. *Physiology* **22**, 15–28.

Bouaboula, M., Hilairet, S., Marchand, J., Fajas, L., Fur, G. L., and Casellas, P. (2005). Anandamide induced PPARγ transcriptional activation and 3T3-L1 preadipocyte differentiation. *Eur. J. Pharmacol.* **517**, 174–181.

Brown, D. A., and London, E. (2000). Structure and function of sphingolipid- and cholesterol-rich membrane rafts. *J. Biol. Chem.* **275**, 17221–17224.

Cadas, H., di Tomaso, E., and Piomelli, D. (1997). Occurrence and biosynthesis of endogenous cannabinoid precursor, N-arachidonoyl phosphatidylethanolamine, in rat brain. *J. Neurosci.* **17**, 1226–1242.

Caterina, M. J., Schumacher, M. A., Tominaga, M., Rosen, T. A., Levine, J. D., and Julius, D. (1997). The capsaicin receptor: A heat-activated ion channel in the pain pathway. *Nature* **389**, 816–824.

Caulfield, M. P., and Brown, D. A. (1992). Cannabinoid receptor agonists inhibit Ca current in NG108-15 neuroblastoma cells via a pertussis toxin-sensitive mechanism. *Br. J. Pharmacol.* **106**, 231–232.

Chabowski, A., Gorski, J., Luiken, J. J. F. P., Glatz, J. F. C., and Bonen, A. (2007). Evidence for concerted action of FAT/CD36 and FABPpm to increase fatty acid transport across the plasma membrane. *Prostaglandins Leukot. Essent. Fatty Acids* **77**, 345–353.

Coe, N. R., Smith, A. J., Frohnert, B. I., Watkins, P. A., and Bernlohr, D. A. (1999). The fatty acid transport protein (FATP1) is a very long chain Acyl-CoA synthetase. *J. Biol. Chem.* **274**, 36300–36304.

Coort, S., Bonen, A., van der Vusse, G., Glatz, J., and Luiken, J. (2007). Cardiac substrate uptake and metabolism in obesity and type-2 diabetes: Role of sarcolemmal substrate transporters. *Mol. Cell. Biochem.* **299**, 5–18.

Cravatt, B. F., Giang, D. K., Mayfield, S. P., Boger, D. L., Lerner, R. A., and Gilula, N. B. (1996). Molecular characterization of an enzyme that degrades neuromodulatory fatty-acid amides. *Nature* **384**, 83–87.

Dainese, E., Oddi, S., Bari, M., and Maccarrone, M. (2007). Modulation of the endocannabinoid system by lipid rafts. *Curr. Med. Chem.* **14**, 2702–2715.

Day, T. A., Rakhshan, F., Deutsch, D. G., and Barker, E. L. (2001). Role of fatty acid amide hydrolase in the transport of the endogenous cannabinoid anandamide. *Mol. Pharmacol.* **59**, 1369–1375.

De Petrocellis, L., Bisogno, T., Maccarrone, M., Davis, J. B., Finazzi-Agro, A., and Di Marzo, V. (2001). The activity of anandamide at vanilloid VR1 receptors requires facilitated transport across the cell membrane and is limited by intracellular metabolism. *J. Biol. Chem.* **276**, 12856–12863.

Deutsch, D. G., Glaser, S. T., Howell, J. M., Kunz, J. S., Puffenbarger, R. A., Hillard, C. J., and Anbumrad, N. (2001). The cellular uptake of anandamide is coupled to its breakdown by fatty-acid amide hydrolase. *J. Biol. Chem.* **276,** 6967–6973.

Devane, W. A., Hanus, L., Breuer, A., Pertwee, R. G., Stevenson, L. A., Griffin, G., Gibson, D., Mandelbaum, A., Etinger, A., and Mechoulam, R. (1992). Isolation and structure of a brain constituent that binds to the cannabinoid receptor. *Science* **258,** 1946–1949.

Dickason-Chesterfield, A., Kidd, S., Moore, S., Schaus, J., Liu, B., Nomikos, G., and Felder, C. (2006). Pharmacological characterization of endocannabinoid transport and fatty acid amide hydrolase inhibitors. *Cell. Mol. Neurobiol.* **26,** 1–17.

Di Marzo, V., Fontana, A., Cadas, H., Schinelli, S., Cimino, G., Schwartz, J. C., and Piomelli, D. (1994). Formation and inactivation of endogenous cannabinoid anandanide in central neurons. *Nature* **372,** 686–691.

DiRusso, C. C., Li, H., Darwis, D., Watkins, P. A., Berger, J., and Black, P. N. (2005). Comparative biochemical studies of the murine fatty acid transport proteins (FATP) expressed in yeast. *J. Biol. Chem.* **280,** 16829–16837.

DiRusso, C. C., Darwis, D., Obermeyer, T., and Black, P. N. (2008). Functional domains of the fatty acid transport proteins: Studies using protein chimeras. *Biochim. Biophys. Acta Mol. Cell Biol. Lipids* **1781,** 135–143.

Doege, H., and Stahl, A. (2006). Protein-mediated fatty acid uptake: Novel insights from in vivo models. *Physiology* **21,** 259–268.

Ehehalt, R., Fullekrug, J., Pohl, J., Ring, A., Herrmann, T., and Stremmel, W. (2006). Translocation of long chain fatty acids across the plasma membrane—Lipid rafts and fatty acid transport proteins. *Mol. Cell. Biochem.* **284,** 135–140.

Eyre, N. S., Cleland, L. G., Tandon, N. N., and Mayrhofer, G. (2007). Importance of the carboxyl terminus of FAT/CD36 for plasma membrane localization and function in long-chain fatty acid uptake. *J. Lipid Res.* **48,** 528–542.

Fasia, L., Karava, V., and Siafaka-Kapadai, A. (2003). Uptake and metabolism of [3H] anandamide by rabbit platelets—Lack of transporter? *Eur. J. Biochem.* **270,** 3498–3506.

Febbraio, M., Hajjar, D. P., and Silverstein, R. L. (2001). CD36: A class B scavenger receptor involved in angiogenesis, atherosclerosis, inflammation, and lipid metabolism. *J. Clin. Invest.* **108,** 785–791.

Febbraio, M., Guy, E., Coburn, C., Knapp, F. F., Beets, A. L., Abumrad, N. A., and Silverstein, R. L. (2002). The impact of overexpression and deficiency of fatty acid translocase (FAT)/CD36. *Mol. Cell. Biochem.* **239,** 193–197.

Fegley, D., Kathuria, S., Mercier, R., Li, C., Goutopoulos, A., Makriyannis, A., and Piomelli, D. (2004). Anandamide transport is independent of fatty-acid amide hydrolase activity and is blocked by the hydrolysis-resistant inhibitor AM1172. *Proc. Natl. Acad. Sci. USA* **101,** 8756–8761.

Felder, C. C., Briley, E. M., Axelrod, J., Simpson, J. T., Mackie, K., and Devane, W. A. (1993). Anandamide, an endogenous cannabimimetic eicosanoid, binds to the cloned human cannabinoid receptor and stimulates receptor-mediated signal transduction. *Proc. Natl. Acad. Sci. USA* **90,** 7656–7660.

Feng, L., Chan, W. W., Roderick, S. L., and Cohen, D. E. (2000). High-level expression and mutagenesis of recombinant human phosphatidylcholine transfer protein using a synthetic gene: Evidence for a C-terminal membrane binding domain. *Biochemistry* **39,** 15399–15409.

Fezza, F., Oddi, S., Di Tommaso, M., De Simone, C., Rapino, C., Pasquariello, N., Dainese, E., Finazzi-Agro, A., and Maccarrone, M. (2008). Characterization of biotin-anandamide, a novel tool for the visualization of anandamide accumulation. *J. Lipid Res.* **49,** 1216–1223.

Fowler, C. J., Tiger, G., Ligresti, A., Lopez-Rodriguez, M. L., and Di Marzo, V. (2004). Selective inhibition of anandamide cellular uptake versus enzymatic hydrolysis—A difficult issue to handle. *Eur. J. Pharmacol.* **492**, 1–11.

Freund, T. F., Katona, I., and Piomelli, D. (2003). Role of endogenous cannabinoids in synaptic signaling. *Physiol. Rev.* **83**, 1017–1066.

Gaoni, Y., and Mechoulam, R. (1971). The isolation and structure of delta-1-tetrahydrocannabinol and other neutral cannabinoids from hashish. *J. Am. Chem. Soc.* **93**, 217–224.

Gerdeman, G. L., Ronesi, J., and Lovinger, D. M. (2002). Postsynaptic endocannabinoid release is critical to long-term depression in the striatum. *Nat. Neurosci.* **5**, 446–451.

Giang, D. K., and Cravatt, B. F. (1997). Molecular characterization of human and mouse fatty acid amide hydrolases. *Proc. Natl. Acad. Sci. USA* **94**, 2238–2242.

Glaser, S. T., Abumrad, N. A., Fatade, F., Kaczocha, M., Studholme, K. M., and Deutsch, D. G. (2003). Evidence against the presence of an anandamide transporter. *Proc. Natl. Acad. Sci. USA* **100**, 4269–4274.

Glaser, S. T., Deutsch, D. G., Studholme, K. M., Zimov, S., and Yazulla, S. (2005a). Endocannabinoids in the intact retina: [3H]Anandamide uptake, fatty acid amide hydrolase immunoreactivity and hydrolysis of anandamide. *Visual Neurosci.* **22**, 693–705.

Glaser, S. T., Kaczocha, M., and Deutsch, D. G. (2005b). Anandamide transport: A critical review. *Life Sci.* **77**, 1584–1604.

Guindon, J., and Hohmann, A. G. (2007). Cannabinoid CB2 receptors: A therapeutic target for the treatment of inflammatory and neuropathic pain. *Br. J. Pharmacol.* **153**, 319–334.

Hamilton, J. A. (2007). New insights into the roles of proteins and lipids in membrane transport of fatty acids. *Prostaglandins Leukot. Essent. Fatty Acids* **77**, 355–361.

Hanus, L., Abu-Lafi, S., Fride, E., Breuer, A., Vogel, Z., Shalev, D. E., Kustanovich, I., and Mechoulam, R. (2001). 2-arachidonyl glyceryl ether, an endogenous agonist of the cannabinoid CB1 receptor. *Proc. Natl. Acad. Sci. USA* **98**, 3662–3665.

Heather, L. C., Cole, M. A., Lygate, C. A., Evans, R. D., Stuckey, D. J., Murray, A. J., Neubauer, S., and Clarke, K. (2006). Fatty acid transporter levels and palmitate oxidation rate correlate with ejection fraction in the infarcted rat heart. *Cardiovasc. Res.* **72**, 430–437.

Hillard, C. J., and Jarrahian, A. (2000). The movement of n-arachidonoylethanolamine (anandamide) across cellular membranes. *Chem. Phys. Lipids* **108**, 123–134.

Hillard, C. J., and Jarrahian, A. (2003). Cellular accumulation of anandamide: Consensus and controversy. *Br. J. Pharmacol.* **140**, 802–808.

Hillard, C. J., Edgemond, W. S., Jarrahian, A., and Campbell, W. B. (1997). Accumulation of n-arachidonoylethanolamine (anandamide) into cerebellar granule cells occurs via facilitated diffusion. *J. Neurochem.* **69**, 631–638.

Hillard, C. J., Shi, L., Tuniki, V. R., Falck, J. R., and Campbell, W. B. (2007). Studies of anandamide accumulation inhibitors in cerebellar granule neurons: Comparison to inhibition of fatty acid amide hydrolase. *J. Mol. Neurosci.* **33**, 18–24.

Hohmann, A. G., and Suplita, R. L. (2006). Endocannabinoid mechanisms of pain modulation. *AAPS J.* **8**, E693–E708.

Holloway, G. P., Lally, J., Nickerson, J. G., Alkhateeb, H., Snook, L. A., Heigenhauser, G. J. F., Calles-Escandon, J., Glatz, J. F. C., Luiken, J. J. F. P., Spriet, L. L., and Bonen, A. (2007). Fatty acid binding protein facilitates sarcolemmal fatty acid transport but not mitochondrial oxidation in rat and human skeletal muscle. *J. Physiol. (Oxford, UK)* **582**, 393–405.

Holloway, G. P., Luiken, J. J. F. P., Glatz, J. F. C., Spriet, L. L., and Bonen, A. (2008). Contribution of FAT/CD36 to the regulation of skeletal muscle fatty acid oxidation: An overview. *Acta Physiol.* **194**, 293–309.

Howlett, A. C. (2005). Cannabinoid receptor signaling. *In* "Cannabinoids" (R. G. Pertwee, Ed.), pp. 53–79. Springer, Berlin-Heidelberg.

Huffman, J. W., and Marriott, K. S. C. (2008). Recent advances in the development of selective ligands for the cannabinoid CB2 receptor. *Curr. Top. Med. Chem. (Sharjah, United Arab Emirates)* **8,** 187–204.

Hui, T. Y., and Bernlohr, D. A. (1997). Fatty acid transporters in animal cells. *Front. Biosci.* **2,** 222–231.

Isola, L. M., Zhou, S., Kiang, C., Stump, D. D., Bradbury, M. W., and Berk, P. D. (1995). 3T3 Fibroblasts transfected with a cDNA for mitochondrial aspartate aminotransferase express plasma membrane fatty acid-binding protein and saturable fatty acid uptake. *Proc. Natl. Acad. Sci. USA* **92,** 9866–9870.

Iyer, L. M., Koonin, E. V., and Aravind, L. (2001). Adaptations of the helix-grip fold for ligand binding and catalysis in the START domain superfamily. *Proteins* **43,** 134–144.

Jarrahian, A., Watts, V. J., and Barker, E. L. (2004). D2 dopamine receptors modulate Gα-subunit coupling of the CB1 cannabinoid receptor. *J. Pharmacol. Exp. Ther.* **308,** 880–886.

Jia, Z., Pei, Z., Maiguel, D., Toomer, C., and Watkins, P. (2007). The fatty acid transport protein (FATP) family: Very long chain acyl-CoA synthetases or solute carriers? *J. Mol. Neurosci.* **33,** 25–31.

Jin, X. H., Okamoto, Y., Morishita, J., Tsuboi, K., Tonai, T., and Ueda, N. (2007). Discovery and characterization of Ca^{2+}-independent phosphatidylethanolamine n-acyltransferase generating the anandamide precursor and its congeners. *J. Biol. Chem.* **282,** 3614–3623.

Kaczocha, M., Hermann, A., Glaser, S. T., Bojesen, I. N., and Deutsch, D. G. (2006). Anandamide uptake is consistent with rate-limited diffusion and is regulated by the degree of its hydrolysis by fatty acid amide hydrolase. *J. Biol. Chem.* **281,** 9066–9075.

Kampf, J. P., and Kleinfeld, A. M. (2007). Is membrane transport of FFA mediated by lipid, protein, or both? *Physiology* **22,** 7–14.

Karlsson, M., Pahlsson, C., and Fowler, C. J. (2004). Reversible, temperature-dependent, and AM404-inhibitable adsorption of anandamide to cell culture wells as a confounding factor in release experiments. *Eur. J. Pharm. Sci.* **22,** 181–189.

Kirkham, T. C., and Tucci, S. A. (2006). Endocannabinoids in appetite control and the treatment of obesity. *CNS Neurol. Disord. Drug Targets* **5,** 272–292.

Klein, T. W., Lane, B., Newton, C. A., and Friedman, H. (2000). The cannabinoid system and cytokine network. *Exp. Biol. Med.* **225,** 1–8.

Lauckner, J. E., Hille, B., and Mackie, K. (2005). The cannabinoid agonist WIN55,212-2 increases intracellular calcium via CB1 receptor coupling to $G_{Q/11}$ G proteins. *Proc. Natl. Acad. Sci. USA* **102,** 19144–19149.

Leung, D., Saghatelian, A., Simon, G. M., and Cravatt, B. F. (2006). Inactivation of n-acyl phosphatidylethanolamine phospholipase D reveals multiple mechanisms for the biosynthesis of endocannabinoids. *Biochemistry* **45,** 4720–4726.

Lewis, S. E., Listenberger, L. L., Ory, D. S., and Schaffer, J. E. (2001). Membrane topology of the murine fatty acid transport protein 1. *J. Biol. Chem.* **276,** 37042–37050.

Ligresti, A., Morera, E., Van Der Stelt, M., Monory, K., Lutz, B., Ortar, G., and Di Marzo, V. (2004). Further evidence for the existence of a specific process for the membrane transport of anandamide. *Biochem. J.* **380,** 265–272.

Liu, J., Wang, L., Harvey-White, J., Osei-Hyiaman, D., Razdan, R., Gong, Q., Chan, A. C., Zhou, Z., Huang, B. X., Kim, H. Y., and Kunos, G. (2006). A biosynthetic pathway for anandamide. *Proc. Natl. Acad. Sci. USA* **103,** 13345–13350.

Lovinger, D. (2008). Presynaptic modulation by endocannabinoids. *In* "Pharmacology of Neurotransmitter Release" (T. C. Sudholf and K. Starke, Eds.), pp. 435–477. Springer, Berlin.

Maccarrone, M. (2007). CB2 receptors in reproduction. *Br. J. Pharmacol.* **153,** 189–198.

Maccarrone, M., Bari, M., Lorenzon, T., Bisogno, T., Di Marzo, V., and Finazzi-Agro, A. (2000). Anandamide uptake by human endothelial cells and its regulation by nitric oxide. *J. Biol. Chem.* **275,** 13484–13492.

Maccarrone, M., Bari, M., Battista, N., and Finazzi-Agro, A. (2002). Estrogen stimulates arachidonoylethanolamide release from human endothelial cells and platelet activation. *Blood* **100,** 4040–4048.

Mackie, K., and Hille, B. (1992). Cannabinoids inhibit n-type calcium channels in neuroblastoma-glioma cells. *Proc. Natl. Acad. Sci. USA* **89,** 3825–3829.

Mackie, K., and Stella, N. (2006). Cannabinoid receptors and endocannabinoids: Evidence for new players. *AAPS J.* **8,** E298–E306.

Malfitano, A. M., Toruner, G. A., Gazzerro, P., Laezza, C., Husain, S., Eletto, D., Orlando, P., De Petrocellis, L., Terskiy, A., Schwalb, M., Vitale, E., and Bifulco, M. (2007). Arvanil and anandamide up-regulate CD36 expression in human peripheral blood mononuclear cells. *Immunol. Lett.* **109,** 145–154.

Matsuda, L. A., Lolait, S. J., Brownstein, M. J., Young, A. C., and Bonner, T. I. (1990). Structure of a cannabinoid receptor and functional expression of the cloned cDNA. *Nature* **346,** 561–564.

McFarland, M. J., and Barker, E. L. (2004). Anandamide transport. *Pharmacol. Ther.* **104,** 117–135.

McFarland, M. J., Porter, A. C., Rakhshan, F. R., Rawat, D. S., Gibbs, R. A., and Barker, E. L. (2004). A role for caveolae/lipid rafts in the uptake and recycling of the endogenous cannabinoid anandamide. *J. Biol. Chem.* **279,** 41991–41997.

McFarland, M. J., Terebova, E. A., and Barker, E. L. (2006). Detergent-resistant membrane microdomains in the disposition of the lipid signaling molecule anandamide. *AAPS J.* **8,** E95–E100.

McFarland, M. J., Bardell, T. K., Yates, M. L., Placzek, E. A., and Barker, E. L. (2008). RNA interference-mediated knockdown of dynamin 2 reduces endocannabinoid uptake into neuronal dCAD cells. *Mol. Pharmacol.* **74,** 101–108.

McKinney, M. K., and Cravatt, B. F. (2005). Structure and function of fatty acid amide hydrolase. *Annu. Rev. Biochem.* **74,** 411–432.

Mechoulam, R., Ben Shabat, S., Hanus, L., Ligumsky, M., Kaminski, N. E., Schatz, A. R., Gopher, A., Almog, S., Martin, B. R., Compton, D. R., Pertwee, R. G., Griffin, G., et al. (1995). Identification of an endogenous 2-monoglyceride, present in canine gut, that binds to cannabinoid receptors. *Biochem. Pharmacol.* **50,** 83–90.

Miller, W. L. (2007). StAR search—What we know about how the Steroidogenic acute regulatory protein mediates mitochondrial cholesterol import. *Mol. Endocrinol.* **21,** 589–601.

Moore, S. A., Nomikos, G. G., Dickason-Chesterfield, A. K., Schober, D. A., Schaus, J. M., Ying, B. P., Xu, Y. C., Phebus, L., Simmons, R. M. A., Li, D., Iyengar, S., and Felder, C. C. (2005). Identification of a high-affinity binding site involved in the transport of endocannabinoids. *Proc. Natl. Acad. Sci. USA* **102,** 17852–17857.

Moriello, A. S., Balas, L., Ligresti, A., Cascio, M. G., Durand, T., Morera, E., Ortar, G., and Di Marzo, V. (2006). Development of the first potential covalent inhibitors of anandamide cellular uptake. *J. Med. Chem.* **49,** 2320–2332.

Mu, J., Zhuang, S. Y., Kirby, M. T., Hampson, R. E., and Deadwyler, S. A. (1999). Cannabinoid receptors differentially modulate potassium A and D currents in hippocampal neurons in culture. *J. Pharmacol. Exp. Ther.* **291,** 893–902.

Munro, S., Thomas, K. L., and Abu-Shaar, M. (1993). Molecular characterization of a peripheral receptor for cannabinoids. *Nature* **365,** 61–65.

Muthian, S., Nithipatikom, K., Campbell, W. B., and Hillard, C. J. (2000). Synthesis and Characterization of a fluorescent substrate for the n-arachidonoylethanolamine (anandamide) transmembrane carrier. *J. Pharmacol. Exp. Ther.* **293,** 289–295.

Okamoto, Y., Morishita, J., Tsuboi, K., Tonai, T., and Ueda, N. (2004). Molecular characterization of a phospholipase D generating anandamide and its congeners. *J. Biol. Chem.* **279,** 5298–5305.

Okamoto, Y., Wang, J., Morishita, J., and Ueda, N. (2007). Biosynthetic pathways of the endocannabinoid anandamide. *Chem. Biodiversity* **4,** 1842–1857.

Ortega-Gutierrez, S., Hawkins, E. G., Viso, A., Lopez-Rodriguez, M. L., and Cravatt, B. F. (2004). Comparison of anandamide transport in FAAH wild-type and knockout neurons: Evidence for contributions by both FAAH and the CB1 receptor to anandamide uptake. *Biochemistry* **43,** 8184–8190.

O'Sullivan, S. E. (2007). Cannabinoids go nuclear: Evidence for activation of peroxisome proliferator-activated receptors. *Br. J. Pharmacol.* **152,** 576–582.

Pei, Z., Fraisl, P., Berger, J., Jia, Z., Forss-Petter, S., and Watkins, P. A. (2004). Mouse very long-chain acyl-CoA synthetase 3/fatty acid transport protein 3 catalyzes fatty acid activation but not fatty acid transport in MA-10 Cells. *J. Biol. Chem.* **279,** 54454–54462.

Pertwee, R. G. (2002). Cannabinoids and multiple sclerosis. *Pharmacol. Ther.* **95,** 165–174.

Pike, L. J., Han, X., Chung, K. N., and Gross, R. W. (2002). Lipid rafts are enriched in arachidonic acid and plasmenylethanolamine and their composition is independent of caveolin-1 expression: A quantitative electrospray ionization/mass spectrometric analysis. *Biochemistry* **41,** 2075–2088.

Placzek, E. A., Okamoto, Y., Ueda, N., and Barker, E. L. (2008). Membrane microdomains and metabolic pathways that define anandamide and 2-arachidonyl glycerol biosynthesis and breakdown. *Neuropharmacology* **55,** 1095–1104.

Pohl, J., Ring, A., and Stremmel, W. (2002). Uptake of long-chain fatty acids in HepG2 cells involves caveolae: Analysis of a novel pathway. *J. Lipid Res.* **43,** 1390–1399.

Pohl, J., Ring, A., Ehehalt, R., Herrmann, T., and Stremmel, W. (2004). New concepts of cellular fatty acid uptake: Role of fatty acid transport proteins and of caveolae. *Proc. Nutr. Soc.* **63,** 259–262.

Ponting, C. P., and Aravind, L. (1999). START: A lipid-binding domain in StAR, HD-ZIP and signaling proteins. *Trends Biochem. Sci.* **24,** 130–132.

Porter, A. C., Sauer, J. M., Knierman, M. D., Becker, G. W., Berna, M. J., Bao, J., Nomikos, G. G., Carter, P., Bymaster, F. P., Leese, A. B., and Felder, C. C. (2002). Characterization of a novel endocannabinoid, virodhamine, with antagonist activity at the CB1 receptor. *J. Pharmacol. Exp. Ther.* **301,** 1020–1024.

Rac, M. E., Safranow, K., and Poncyljusz, W. (2007). Molecular basis of human CD36 gene mutations. *Mol. Med.* **13,** 288–296.

Rakhshan, F., Day, T. A., Blakely, R. D., and Barker, E. L. (2000). Carrier-mediated uptake of the endogenous cannabinoid anandamide in RBL-2H3 cells. *J. Pharmacol. Exp. Ther.* **292,** 960–967.

Ralevic, V., Kendall, D. A., Jerman, J. C., Middlemiss, D. N., and Smart, D. (2001). Cannabinoid activation of recombinant and endogenous vanilloid receptors. *Eur. J. Pharmacol.* **424,** 211–219.

Ralevic, V., Kendall, D. A., Randall, M. D., and Smart, D. (2002). Cannabinoid modulation of sensory neurotransmission via cannabinoid and vanilloid receptors: Roles in regulation of cardiovascular function. *Life Sci.* **71,** 2577–2594.

Reggio, P. H., and Traore, H. (2000). Conformational requirements for endocannabinoid interaction with the cannabinoid receptors, the anandamide transporter and fatty acid amidohydrolase. *Chem. Phys. Lipids* **108,** 15–35.

Rimmerman, N., Hughes, H. V., Bradshaw, H. B., Pazos, M. X., Mackie, K., Prieto, A. L., and Walker, J. M. (2007). Compartmentalization of endocannabinoids into lipid rafts in a dorsal root ganglion cell line. *Br. J. Pharmacol.* **153,** 380–389.

Ring, A., Pohl, J., Volkl, A., and Stremmel, W. (2002). Evidence for vesicles that mediate long-chain fatty acid uptake by human microvascular endothelial cells. *J. Lipid Res.* **43,** 2095–2104.

Rodriguez de Fonseca, F., Del Arco, I., Bermudez-Silva, F. J., Bilbao, A., Cippitelli, A., and Navarro, M. (2005). The endocannabinoid system: Physiology and pharmacology. *Alcohol Alcohol* **40**, 2–14.

Ross, R. A. (2003). Anandamide and vanilloid TRPV1 receptors. *Br. J. Pharmacol.* **140**, 790–801.

Sandberg, A., and Fowler, C. J. (2005). Measurement of saturable and non-saturable components of anandamide uptake into P19 embryonic carcinoma cells in the presence of fatty acid-free bovine serum albumin. *Chem. Phys. Lipids* **134**, 131–139.

Schaap, F. G., Hamers, L., Van der Vusse, G. J., and Glatz, J. F. C. (1997). Molecular cloning of fatty acid-transport protein cDNA from rat. *Biochim. Biophys. Acta (BBA) Gene Struct. Expr.* **1354**, 29–34.

Schwenk, R. W., Luiken, J. J. F. P., Bonen, A., and Glatz, J. F. C. (2008). Regulation of sarcolemmal glucose and fatty acid transporters in cardiac disease. *Cardiovasc. Res.* **79**, 249–258.

Sfeir, Z., Ibrahimi, A., Amri, E., Grimaldi, P., and Abumrad, N. (1997). Regulation of FAT/CD36 gene expression: Further evidence in support of a role of the protein in fatty acid binding/transport. *Prostaglandins Leukot. Essent. Fatty Acids* **57**, 17–21.

Simon, G. M., and Cravatt, B. F. (2008). Anandamide biosynthesis catalyzed by the phosphodiesterase GDE1 and detection of glycerophospho-n-acyl ethanolamine precursors in mouse brain. *J. Biol. Chem.* **283**, 9341–9349.

Smart, D., Gunthorpe, M. J., Jerman, J. C., Nasir, S., Gray, J., Muir, A. I., Chambers, J. K., Randall, A. D., and Davis, J. B. (2000). The endogenous lipid anandamide is a full agonist at the human vanilloid receptor (HVR1). *Br. J. Pharmacol.* **129**, 227–230.

Soccio, R. E., and Breslow, J. L. (2003). StAR-related lipid transfer (START) proteins: Mediators of intracellular lipid metabolism. *J. Biol. Chem.* **278**, 22183–22186.

Stahl, A. (2004). A current review of fatty acid transport proteins (SLC27). *Pflugers Arch.* **447**, 722–727.

Stahl, A., Gimeno, R. E., Tartaglia, L. A., and Lodish, H. F. (2001). Fatty acid transport proteins: A current view of a growing family. *Trends Endocrinol. Metab.* **12**, 266–273.

Starowicz, K., Nigam, S., and Di Marzo, V. (2007). Biochemistry and pharmacology of endovanilloids. *Pharmacol. Ther.* **114**, 13–33.

Stremmel, W., Strohmeyer, G., Borchard, F., Kochwa, S., and Berk, P. D. (1985). Isolation and partial characterization of a fatty acid binding protein in rat liver plasma membranes. *Proc. Natl. Acad. Sci. USA* **82**, 4–8.

Stuhlsatz-Krouper, S. M., Bennett, N. E., and Schaffer, J. E. (1998). Substitution of alanine for serine 250 in the murine fatty acid transport protein inhibits long chain fatty acid transport. *J. Biol. Chem.* **273**, 28642–28650.

Stuhlsatz-Krouper, S. M., Bennett, N. E., and Schaffer, J. E. (1999). Molecular aspects of fatty acid transport: Mutations in the IYTSGTTGXPK motif impair fatty acid transport protein function. *Prostaglandins Leukot. Essent. Fatty Acids* **60**, 285–289.

Stump, D. D., Zhou, S. L., and Berk, P. D. (1993). Comparison of plasma membrane FABP and mitochondrial isoform of aspartate aminotransferase from rat liver. *Am. J. Physiol. Gastrointest. Liver Physiol.* **265**, G894–G902.

Sugiura, T., Kondo, S., Sukagawa, A., Nakane, S., Shinoda, A., Itoh, K., Yamashita, A., and Waku, K. (1995). 2-arachidonoylgylcerol: A possible endogenous cannabinoid receptor ligand in brain. *Biochem. Biophys. Res. Comm.* **215**, 89–97.

Thors, L., Eriksson, J., and Fowler, C. J. (2007). Inhibition of the cellular uptake of anandamide by genistein and its analogue daidzein in cells with different levels of fatty acid amide hydrolase-driven uptake. *Br. J. Pharmacol.* **152**, 744–750.

Tsujishita, Y., and Hurley, J. H. (2000). Structure and lipid transport mechanism of a StAR-related domain. *Nat. Struct. Mol. Biol.* **7**, 408–414.

Turcotte, L. P., Swenberger, J. R., Tucker, M. Z., and Yee, A. J. (1999). Training-induced elevation in FABPPM Is associated with increased palmitate use in contracting muscle. *J. Appl. Physiol.* **87,** 285–293.

Twitchell, W., Brown, S., and Mackie, K. (1997). Cannabinoids inhibit N- and P/Q-type calcium channels in cultured rat hippocampal neurons. *J. Neurophysiol.* **78,** 43–50.

Vellani, V., Petrosino, S., De Petrocellis, L., Valenti, M., Prandini, M., Magherini, P. C., McNaughton, P. A., and Di Marzo, V. (2008). Functional lipidomics. Calcium-independent activation of endocannabinoid/endovanilloid lipid signalling in sensory neurons by protein kinases C and A and thrombin. *Neuropharmacology* **55,** 1274–1279.

Vogel, S. M., Minshall, R. D., Pilipovic, M., Tiruppathi, C., and Malik, A. B. (2001). Albumin uptake and transcytosis in endothelial cells *in vivo* induced by albumin-binding protein. *Am. J. Physiol. Lung Cell. Mol. Physiol.* **281,** L1512–L1522.

Wang, H., Dey, S. K., and Maccarrone, M. (2006). Jekyll and Hyde: Two faces of cannabinoid signaling in male and female fertility. *Endocrinol. Rev.* **27,** 427–448.

Watkins, P. A. (1997). Fatty acid activation. *Prog. Lipid Res.* **36,** 55–83.

Wei, B. Q., Mikkelsen, T. S., McKinney, M. K., Lander, E. S., and Cravatt, B. F. (2006). A second fatty acid amide hydrolase with variable distribution among placental mammals. *J. Biol. Chem.* **281,** 36569–36578.

Wirtz, K. W. A. (2006). Phospholipid transfer proteins in perspective. *FEBS Lett.* **580,** 5436–5441.

Witkin, J. M., Tzavara, E. T., and Nomikos, G. G. (2005). A role for cannabinoid CB1 receptors in mood and anxiety disorders. *Behav. Pharmacol.* **16,** 315–331.

Yang, Y., Chen, M., Loux, T., and Harmon, C. (2007). Regulation of FAT/CD36 mRNA gene expression by long chain fatty acids in the differentiated 3T3-L1 cells. *Pediatr. Surg. Int.* **23,** 675–683.

Zhou, S. L., Stump, D., Sorrentino, D., Potter, B. J., and Berk, P. D. (1992). Adipocyte differentiation of 3T3-L1 cells involves augmented expression of a 43-kDa plasma membrane fatty acid-binding protein. *J. Biol. Chem.* **267,** 14456–14461.

Zhou, S. L., Stump, D. D., Kinag, C. L., Isola, L. M., and Berk, P. D. (1995). Mitochondrial aspartate aminotransferase expressed on the surface of 3T3-L1 adipocytes mediates saturable fatty acid uptake. *Proc. Soc. Exp. Biol. Med.* **208,** 263–270.

Zimmerman, A. W., and Veerkamp, J. H. (2002). New insights into the structure and function of fatty acid-binding proteins. *Cell. Mol. Life Sci.* **59,** 1096–1116.

Zygmunt, P. M., Petersson, J., Andersson, D. A., Chuang, H. H., Sorgard, M., Di Marzo, V., Julius, D., and Hogestatt, E. D. (1999). Vanilloid receptors on sensory nerves mediate the vasodilator action of anandamide. *Nature* **400,** 452–457.

CHAPTER THREE

Biosynthesis of Oleamide

Gregory P. Mueller *and* William J. Driscoll

Contents

I. Introduction	56
II. Fatty Acid Amide Messengers: Structural Considerations	56
III. Natural Occurrence of Oleamide	57
IV. Biologic Actions of Oleamide	58
V. Proposed Mechanisms for the Biosynthesis of Oleamide	59
VI. Oleamide Biosynthesis by Peptidylglycine Alpha-amidating Monooxygenase	60
VII. Discovery of Cytochrome *c* as an Oleamide Synthase	62
VIII. Cytochrome *c* also Catalyzes the Formation of Oleoylglycine and Other Long-Chain Fatty Acylamino Acids	64
IX. Proposal for an Oleamide Synthesome	66
X. Apoptosis: A Model for the Mechanism and Regulation of Oleamide Biosynthesis	67
XI. Considerations for the Investigation of Oleamide Biosynthesis	70
XII. Future Directions and Concluding Remarks	70
References	71

Abstract

Oleamide (cis-9-octadecenamide) is the prototype long chain primary fatty acid amide lipid messenger. The natural occurrence of oleamide was first reported in human serum in 1989. Subsequently oleamide was shown to accumulate in the cerebrospinal fluid of sleep-deprived cats and to induce sleep when administered to experimental animals. Accordingly, oleamide first became known for its potential role in the mechanisms that mediate the drive to sleep. Oleamide also has profound effects on thermoregulation and acts as an analgesic in several models of experimental pain. Although these important pharmacologic effects are well establish, the biochemical mechanism for the synthesis of oleamide has not yet been defined. This chapter reviews the biosynthetic pathways that have been proposed and highlights two mechanisms which are most supported by experimental evidence: the generation of oleamide from oleoylglycine by the

Department of Anatomy, Physiology and Genetics, F. Edward Hebert School of Medicine, Uniformed Services University of the Health Sciences, Bethesda, Maryland, USA

neuropeptide processing enzyme, peptidylglycine alpha-amidating monooxygenase (PAM), and alternatively, the direct amidation of oleic acid via oleoyl coenzyme A by cytochrome c using ammonia as the nitrogen source. The latter mechanism is discussed in the context of apoptosis where oleamide may play a role in regulating gap junction communication. Lastly, several considerations and caveats pertinent to the future study oleamide biosynthesis are discussed.

I. Introduction

The diverse chemical nature and widespread biologic importance of lipid signaling molecules became appreciated with the discovery of the prostaglandins, leukotrienes, and thromboxanes. These molecules joined the already established cholesterol-based steroid hormones as mediators of intercellular communication. More recently, long-chain fatty acid amides have emerged as important additions to the superfamily of lipid messengers. While attention initially focused on anandamide, a secondary fatty acid amide and the endogenous ligand for the CB_1 cannabinoid receptor, oleamide, the prototype primary fatty acid amide, has become recognized for its potential roles in sleep and thermoregulation. While significant inroads have been made in understanding the biology of oleamide, the precise mechanism of its biosynthesis still remains undefined. Here we discuss current understanding of the potential mechanisms for oleamide biosynthesis and how these may be linked to cellular responses.

The natural occurrence of oleamide was first reported in human serum in 1989 (Arafat *et al.*, 1989). Subsequently oleamide was shown to accumulate in the cerebrospinal fluid of sleep-deprived cats (Cravatt *et al.*, 1995) and to produce sleep when administered to experimental animals (Cravatt *et al.*, 1995; Mendelson and Basile, 2001). Accordingly, oleamide first became known for its potential role in the mechanisms that mediate the drive to sleep. Oleamide also has profound effects on thermoregulation and locomotion (Basile *et al.*, 1999; Chaturvedi *et al.*, 2006; Fedorova *et al.*, 2001; Huitron-Resendiz *et al.*, 2001; Lichtman *et al.*, 2002; Martinez-Gonzalez *et al.*, 2004; Murillo-Rodriguez *et al.*, 2001) and produces antinociception, or analgesia, in models of experimental pain (Fedorova *et al.*, 2001). Although these pharmacologic effects are well established, the underlying mechanisms involved are not understood.

II. Fatty Acid Amide Messengers: Structural Considerations

Collectively, the long-chain fatty acid amides constitute a diverse assortment of lipid messengers, however, they share the common characteristics of containing an alkyl chain of 16 or more carbon atoms and a single

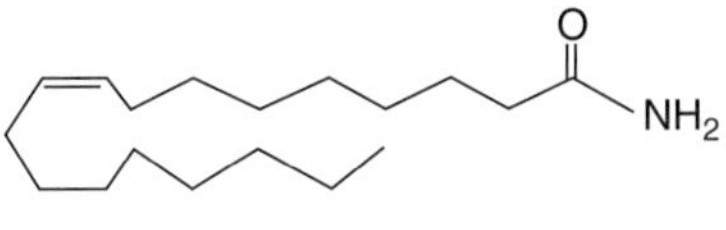

Oleamide

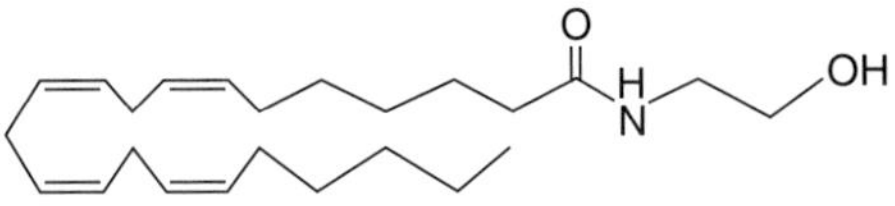

Anandamide

Figure 3.1 Structures of the prototype long-chain primary and secondary fatty acid amide classes of lipid messengers. Oleamide represents the long-chain primary fatty acid amides. This group is characterized by an alkyl chain of 16 or more carbons and a terminal primary amide, R_1–CO–NH_2. Long-chain secondary amides, represented by anandamide, are characterized by the structure R_1–CO–NH–R_2, where R_1 is a fatty acyl group and –NH–R_2 is a defining moiety such as ethanolamine, an amino acid, dopamine, serotonin, etc.

amide moiety. These molecules may be divided into two major categories consisting of the primary fatty acid amides, represented by oleamide, and secondary fatty acid amides, most often represented by anandamide (Fig. 3.1). Oleamide is the best known member of the class of long-chain primary fatty acid amides which includes palmitamide, stearamide, erucamide, and related long-chain primary fatty acid amides (Farrell and Merkler, 2008; Maccarrone and Finazzi-Agro, 2002; Nichols et al., 2007; Patricelli and Cravatt, 2001; Wakamatsu et al., 1990). While structurally similar, primary fatty acid amide messengers achieve distinct biochemical identities by varied carbon chain length, and by degree, position and stereochemistry of unsaturation. In contrast, secondary amides constitute a much more heterogeneous group of lipid messengers which may be further subdivided into the N-acylethanolamides, N-acylamino acids, and N-acyl conjugates of dopamine, serotonin and gamma amino butyric acid (Farrell and Merkler, 2008). While the structural distinction between the primary and secondary fatty acid amides is undoubtedly a critical feature to their respective biologic actions, recent evidence suggests that primary fatty acid amides and certain of the secondary fatty acid amides may share a common pathway for biosynthesis (Driscoll et al., 2007; McCue et al., 2008; Mueller and Driscoll, 2007).

III. Natural Occurrence of Oleamide

The discoveries of oleamide in human serum in 1989 (Arafat et al., 1989) and rat cerebral spinal fluid in 1995 (Cravatt et al., 1995) were followed by the studies of Basile and coworkers in 1999 who were the first to quantitatively report on oleamide levels in rat plasma (10 ng/mL) and cerebral spinal fluid (44 ng/mL) (Hanus *et al.*, 1999). Oleamide has also been reported to be present in three human breast cancer lines (Bisogno *et al.*, 1998) and in human tear film, although there is disagreement over the relative contribution of oleamide to the total content of nonpolar lipids in this secretion (Butovich *et al.*, 2007; Nichols *et al.*, 2007). Reports on the natural occurrence of oleamide must be considered with caution, due to its widespread use as a slip agent in the production of polyethylene products commonly used in laboratory research. This complication is avoided with a biosynthetic labeling approach using either radioactive or stable mass isotopes. Biosynthetic labeling has been used to demonstrate the biosynthesis of oleamide by different cell types. Initial studies were carried out in the mouse neuroblastoma $N_{18}TG_2$ cell line (Bisogno *et al.*, 1997; Merkler *et al.*, 2004). As will be discussed below, these cells were of particular interest because they express the peptide amidating enzyme, peptidylglycine alpha-amidating monooxygenase (PAM) and produce amidated peptide messengers (Ford and Mueller, 1993; Ritenour-Rodgers *et al.*, 2000). Recently, we have observed oleamide synthesis by HeLa cells, mouse fibroblasts, and human mast cells (personal observations, unpublished). These findings indicate that the synthesis of oleamide is not restricted to a single cell type and may be a process shared by many, if not all, cells.

IV. Biologic Actions of Oleamide

To date, a specific "oleamide receptor" has not been identified. However, there are examples where oleamide was shown to have effects on well-established receptor systems, most likely through allosteric interactions with the receptor proteins themselves. These systems include the $GABA_A$ (Laposky *et al.*, 2001; Lees *et al.*, 1998; Verdon *et al.*, 2000), serotonin$_{2A}$ (Boger *et al.*, 1998a; Huidobro-Toro and Harris, 1996; Thomas *et al.*, 1997, 1998), serotonin$_{2C}$ (Huidobro-Toro and Harris, 1996), and serotonin$_7$ (Alberts *et al.*, 2001; Hedlund *et al.*, 1999, 2003; Thomas *et al.*, 1997, 1998) receptors. In general, the effects of oleamide on these systems are remarkably sensitive to specific structural features of oleamide whereby, subtle changes in acyl chain length, presence/position and stereochemistry of the double bond or alterations in the amide terminus all diminished the oleamide effects. The fact that oleamide modulates 5-HT receptor activity suggests that, in addition to sleep regulation, it also may be an important factor in depression, anxiety, and analgesia, all of which involved altered serotonergic neurotransmission (Jann and Slade, 2007; Moltzen and Bang-Andersen, 2006; Suzuki *et al.*, 2004; Zohar and Westenberg, 2000).

Intriguingly, the first reported cellular action of oleamide was inhibition of gap junction communication (Boger *et al.*, 1998b; Guan *et al.*, 1997; Huang *et al.*, 1998; Lerner, 1997). Gap junctions are protein channels that form between adjacent cells and allow for direct electrical coupling and the exchange of small molecules (<1500 Da) (Bernstein and Morley, 2006; Sohl *et al.*, 2005; Theis *et al.*, 2005; van Veen *et al.*, 2006). The gap junction structure consists of two hemichannels, or connexons, in apposed plasma membranes. Connexons dock with each other to form the gap junction channel. Each hemichannel is composed of six connexin or pannexin proteins (Sohl *et al.*, 2005). The ability of oleamide to inhibit gap junction communication is well established in a variety of cellular models. However, it still remains unclear whether this cellular effect is mediated by specific oleamide receptors or via allosteric interactions with the gap junction proteins themselves. Nevertheless, oleamide is widely used as an experimental tool for studying gap junction function. The earliest studies focused on gap junctions in glial cells (Boger *et al.*, 1998b; Guan *et al.*, 1997). Subsequently, evidence that oleamide inhibits gap junction function has been reported in midbrain dopaminergic neurons (SiuYi Leung *et al.*, 2001), motor neurons (Coleman and Sengelaub, 2002), migrating rat neural crest cells (Bannerman *et al.*, 2000), and in organotypic preparations of dentate gyrus (Schweitzer *et al.*, 2000). Similar demonstrations were made with epithelial cells (Boitano and Evans, 2000), endothelial cells (Nagasawa *et al.*, 2006), osteoblasts (Schiller *et al.*, 2001a,b), osteoclasts (Ransjo *et al.*, 2003), sertoli cells (Decrouy *et al.*, 2004; Gilleron *et al.*, 2006), and Chinese hamster ovary (CHO) cells (Subauste *et al.*, 2001). The diversity of cells expressing oleamide-sensitive gap junctions indicates that the biosynthesis of oleamide may be widespread across many cell types and tissues.

V. Proposed Mechanisms for the Biosynthesis of Oleamide

The biosynthetic pathway for oleamide has been a focus of investigation since the mid-1990s. Considerable effort has been invested in this area, and several mechanisms have emerged as intriguing possibilities. Table 3.1 lists the pathways that have been considered. The most compelling evidence supports the mechanism involving the conversion of oleoylglycine to oleamide by the neuropeptide processing enzyme, PAM. Recently, however, a second mechanism involving the condensation of oleoyl coenzyme A (oleoyl-CoA) and ammonia by cytochrome *c* has emerged as an alternative pathway for the biosynthesis of oleamide. Based upon what is known about the regional distribution of PAM and the more generalized production of oleamide, it is reasonable to predict that both mechanisms may contribute to the biosynthesis of oleamide in a cell and tissue specific manner.

Table 3.1 Potential mechanisms for oleamide biosynthesis

From	Amidation via
1. Oleoyl CoA and NH_3	Cytochrome c
2. Oleoylglycine	PAM
3. Oleic acid and NH_3	FAAH: operating in reverse
4. Oleic acid containing phospholipids	Aminolysis
5. Sphingomyelin	Catabolism
6. Oleoyl CoA and NH_3	Serum synthase

Other possible mechanisms for oleamide biosynthesis have been proposed which include (i) the hydrolytic enzyme, fatty acid amide hydrolase "working in reverse" (Bisogno et al., 1997; Sugiura et al., 1996), (ii) aminolysis of oleic acid containing phospholipids (Bisogno et al., 1997), and (iii) catabolism of sphingomyelin (Lerner et al., 1994). While these alternatives are conceptually consistent, experimental evidence supporting their biologic relevance is lacking. For example, pharmacologic inhibition of fatty acid amide hydrolase was shown to actually *increase* oleamide levels in neuroblastoma cells (Bisogno et al., 1997), indicating that the hydrolase does not mediate oleamide biosynthesis *in vivo*. Additionally, attempts to demonstrate oleamide release via the aminolysis of oleate-containing phospholipids were unsuccessful (Bisogno et al., 1997) and there is no direct evidence for a pathway involving the catabolism of sphingomyelin (Lerner et al., 1994). Finally, we have recently discovered the existence of a circulating oleamide synthase that is biochemically and functionally distinct from both cytochrome c and PAM. This activity is enriched in fetal bovine serum used to supplement cell culture media. The serum synthase utilizes oleoyl-CoA and ammonia as substrates for the formation of oleamide, and glycine can substitute for ammonia resulting in the formation of oleoylglycine. The serum synthase activity is inactivated by heat, hydrogen peroxide, limited proteolytic digestion by trypsin and HPLC solvent conditions (acetonitrile/trifluoroacetic acid), all in contrast to the synthase activity of cytochrome c. Ion exchange chromatography and molecular sizing reveal the serum synthase activity to be negatively charged and approximately 65,000 Da in mass.

VI. Oleamide Biosynthesis by Peptidylglycine Alpha-amidating Monooxygenase

A leading hypothesis for oleamide biosynthesis holds that the neuropeptide processing enzyme, PAM, mediates the generation of oleamide *in vivo* (Merkler et al., 1996, 2004; Ritenour-Rodgers et al., 2000). PAM is a bifunctional enzyme that is expressed in neuroendocrine tissues where it is known to catalyze the formation of alpha-amidated peptide messengers from their glycine-extended precursors (Eipper et al., 1992). PAM is also capable of experimentally catalyzing the amidation of a variety of nonpeptide compounds that contain a reactive glycine (DeBlassio et al., 2000; King et al., 2000; McIntyre et al., 2006; Merkler et al., 1996; Miller et al., 2003; Shonsey et al., 2005; Wilcox et al., 1999). Notably, this group includes long-chain fatty acylglycines (Merkler et al., 1996; Wilcox et al., 1999). Although a role for PAM in generating primary fatty acid amides *in vivo* is certainly feasible in those cells that express PAM, evidence supporting this possibility is quite limited. The most compelling evidence was reported by Merkler and co-workers in 2004 who used a pharmacologic approach to demonstrate a PAM-based pathway for oleamide synthesis in the $N_{18}TG_2$ neuroblastoma cell line (Merkler et al., 2004). The $N_{18}TG_2$ line expresses PAM (Ritenour-Rodgers et al., 2000) and had been shown previously to synthesize oleamide (Bisogno et al., 1997; Merkler et al., 2004). Merkler et al. observed that treatment with a PAM inhibitor reduced the incorporation of radiolabeled oleic acid into oleamide and further that this reduction was associated with a detectable increment in cellular concentrations of radiolabeled material that co-chromtographed with oleoylglycine in high-pressure liquid chromatography (Merkler et al., 2004). While these experimental findings support a role for PAM in the biosynthesis of oleamide, this mechanism has not been studied in other cellular systems, and to date, there have been no further reports to extend the original findings of 2004.

A persistent challenge to the PAM-based mechanism for oleamide biosynthesis has been the lack of a defined pathway for the formation of oleoylglycine, the obligatory substrate for the generation of oleamide by PAM. While glycination is essential in the synthesis of bile acids (Kase and Bjorkhem, 1989; Kase et al., 1986) and metabolism of short chain fatty acids (Bonafe et al., 2000) and small-molecule xenobiotics (Kelley and Vessey, 1994; Mawal and Qureshi, 1994), the enzymes that mediate these conventional conjugations are highly specific and are essentially unreactive with unsaturated long-chain fatty acyl-CoAs (O'Byrne et al., 2003). There are, however, two other potential mechanisms for the biosynthesis of oleoylglycine. One mechanism involves the oxidative metabolism of oleoylethanolamide via the sequential actions of fatty alcohol and fatty aldehyde dehydrogenases (Burstein et al., 2000). This putative pathway is supported by the existence of established dehydrogenases which catalyze reactions analogous to those proposed for the generation of oleoylglycine from oleoylethanolamide and the relative abundance of oleoyethanolamide in brain which exceed that of anadamide by eightfold (Koga et al., 1997). To date, however, there are no *in vivo* data to either support or refute this pathway. A second potential mechanism has emerged with the intriguing discovery that cytochrome *c* can catalyze the formation of oleoylglycine from oleoyl-CoA and glycine in the presence of hydrogen peroxide (H_2O_2) (Mueller and Driscoll, 2007). Accordingly, present evidence points to two possible mechanisms for the generation of the PAM substrate, oleoylglycine.

VII. Discovery of Cytochrome *c* as an Oleamide Synthase

Recently, a cellular activity capable of synthesizing oleamide was isolated from rat kidney and subsequently identified as cytochrome *c* (Driscoll et al., 2007). This surprising discovery was the result of a proteomics-based investigation designed to define the biosynthetic mechanism for oleamide. Because fatty acids are universally activated as their coenzyme A derivatives (Watkins, 1997), oleoyl-CoA was used as a substrate to screen for oleamide synthesizing activity in rat tissues. The oleamide synthetase activity was observed in ammonium sulfate fractionations of rat brain, liver, and kidney extracts. Rat kidney, the most abundant source, was used for detailed characterization (Driscoll et al., 2007). The activity was found to be stable to heat denaturation and limited exposure to proteolytic enzymes. Ammonium ion was required for oleamide synthesis, and the reaction was significantly enhanced by H_2O_2. The activity displayed physiologic temperature and pH optima, specificity for long-chain acyl-CoA substrates and a K_m for oleoyl-CoA of 21 μM. Proteomic, biochemical, and immunologic analyses were used to definitively identify the source of the oleamide synthase activity as cytochrome *c* (Fig. 3.2) (Driscoll et al., 2007). Figure 3.3 shows the profound effects of H_2O_2 on the efficiency and kinetics of the reaction as catalyzed by commercially obtained cytochrome *c*.

While incomplete, our working model for the biosynthesis of oleamide by cytochrome *c* predicts that oleoyl-CoA binds to cytochrome *c* with a geometry that places the thioester adjacent to the heme center. In this position the iron can withdraw an electron from the thioester group to promote nucleophilic attack of ammonia at the thioester bond. In this mechanism, Fe^{3+} is reduced to Fe^{2+} and subsequently reoxidized by H_2O_2. This model is supported by evidence that cytochrome *c* exhibits (i) high binding affinity for long-chain fatty acyl-CoA molecules (Stewart et al., 2000), (ii) peroxidase enzymatic activity (Barros et al., 2003; Wang et al., 2003; Zhao and Xu, 2004; Zhao et al., 2003), (iii) inhibition of the reaction by the iron chelator, deferoxamine (Driscoll et al., 2007), and (iv) opening of the protein structure surrounding the heme center following binding of long-chain fatty acyl compounds (Droghetti et al., 2006; Pinheiro et al., 1997; Sakono et al., 2000; Sanghera and Pinheiro, 2000).

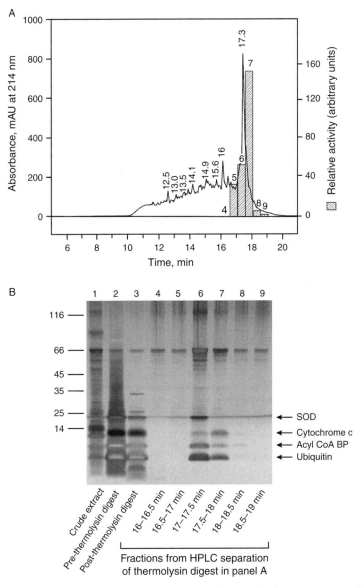

Figure 3.2 Oleamide synthesizing activity correlates with the occurrence of cytochrome *c* protein. (A) Rat kidney oleamide synthesizing activity was prepared by limited thermolysin digestion and subjected to reverse phase HPLC. Fractions were collected and assayed for oleamide synthase activity (cross-hatched bars). (B) Silver-stained polyacrylamide gel showing the crude kidney extract (lane 1), the partially purified activity prior to thermolysin digestion (lane 2), the same sample following digestion (lane 3) and HPLC fractions collected in (A) (lanes 4–9). The identities of the protein bands taken for peptide mass fingerprinting are listed to the right of the gel. Note that the only protein band that coordinately increases in staining intensity with increased activity between fractions 6 and 7 is identified as cytochrome *c*.

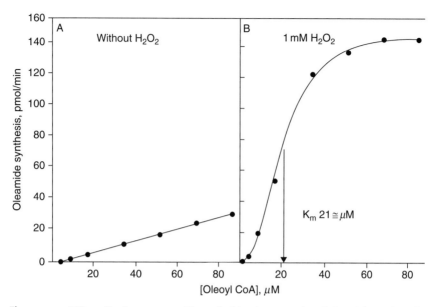

Figure 3.3 Effect of hydrogen peroxide on the kinetic properties of oleamide synthesis by cytochrome c. Duplicate assays were performed using 400 ng/reaction of commercially obtained rat cytochrome c in 100 μL of 50 mM Tris, pH 7.4 containing 125 mM NH_4Cl and increasing concentrations of $[^{14}C]$oleoyl-CoA as indicated. Left panel: The analysis was performed without the addition of H_2O_2 (maximal conversion of substrate to product = 3.4%). Right panel: The analysis was performed with the addition of 1 mM H_2O_2 (maximal conversion of substrate to product = 39.5%). Oleamide synthesis was calculated on the basis of the specific activity of the $[^{14}C]$oleoyl-CoA substrate (54 mCi/mmol).

VIII. Cytochrome *c* Also Catalyzes the Formation of Oleoylglycine and Other Long-Chain Fatty Acylamino Acids

The ability of cytochrome c to utilize ammonia for the generation of oleamide from oleoyl-CoA prompted us to explore the possibility that other primary amines might substitute for ammonia as nitrogen donors. In exploring this possibility, we found that glycine effectively supported the formation of oleoylglycine by cytochrome c (Fig. 3.4) (Mueller and Driscoll, 2007). The reaction exhibits similar reactant concentration optima, preference for long-chain acyl-CoAs and Michaelis–Menten kinetics, with respect to oleoyl-CoA, as was observed for the generation oleamide (Mueller and Driscoll, 2007). These findings have suggested the intriguing possibility that cytochrome c mediates the biosynthesis of oleamide through two mechanisms: (i) direct synthesis of oleamide from oleoyl-CoA and ammonia and (ii) generation of oleoylglycine for subsequent amidation

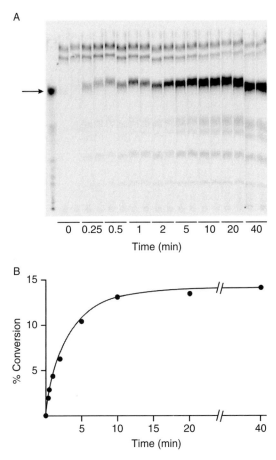

Figure 3.4 Time course for the formation of oleoylglycine by rat cytochrome c. Rat cytochrome c (500 ng/reaction) was incubated in 50 mM Tris, pH 7.4 (100 μL) containing 10 μM [^{14}C]oleoyl CoA, 150 mM glycine, and 1 mM H_2O_2 at 37 °C for the times indicated. Reaction products were extracted into ethyl acetate and separated by TLC. (A) Phosphorimage of the TLC plate (*arrow* and *first lane* show oleoylglycine standard). (B) Graphic representation of the TLC image data. Data plotted are the mean values for each set of duplicates expressed as pmol oleoylglycine product formed.

by PAM. A composite of currently proposed biosynthetic pathways for oleamide is presented in Fig. 3.5. Recent investigations have revealed that cytochrome c will also catalyze the formation of several other, but not all, oleoylamino acid conjugates (personal observation, unpublished). Lastly, we have found that cytochrome c is capable of catalyzing the synthesis of arachidonoylglycine from glycine and arachidonoyl-CoA in the presence of H_2O_2 (McCue *et al.*, 2008). Collectively, these findings suggest that cytochrome c could play a central role in a novel pathway for the generation of several classes of lipid signaling molecules.

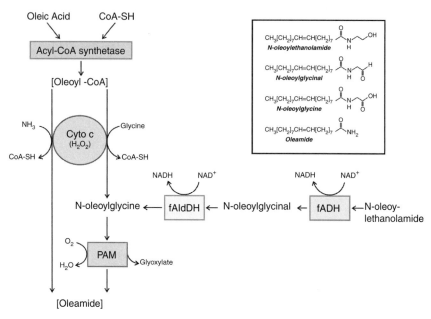

Figure 3.5 Schematic overview of proposed biosynthetic pathways for oleamide. Catalytic enzymes are boxed or circled. Reference structures for fatty amides are shown in the inset. Cyto c, cytochrome c; PAM, peptidylglycine α-amidating monooxygenase; fAldDH, fatty aldehyde dehydrogenase; fADH, fatty alcohol dehydrogenase.

IX. Proposal for an Oleamide Synthesome

The findings that the oleamide synthetic reaction catalyzed by cytochrome c requires high concentrations of ammonium ion and H_2O_2 and that cytochrome c co-purifies with superoxide dismutase and fatty acyl-CoA binding protein (Fig. 3.2) (Driscoll et al., 2007), all support the proposal that oleamide biosynthesis may be mediated by a functional complex of proteins, or a synthesome. We hypothesize that such a complex (Fig. 3.6) would minimally contain cytochrome c as the catalytic enzyme, superoxide dismutase for the local production of H_2O_2, long-chain fatty acyl-CoA binding protein for the presentation of oleoyl-CoA, and an amino transferase for the generation of the amide nitrogen, possibly from arginine, glutamine, or other similar source. The fact that superoxide dismutase and fatty acyl-CoA binding protein co-purified with cytochrome c though a rigorous procedure of heat denaturation, proteolysis with thermolysin and reverse phase high-pressure liquid chromatography (Driscoll et al., 2007) indicates that the three proteins are tightly associated in a manner that confers a remarkable physicochemical stability to the complex. The

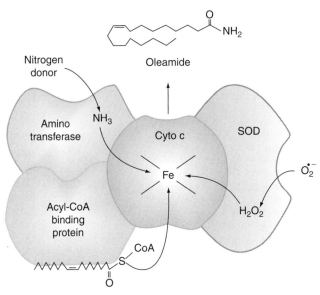

Figure 3.6 Conceptual diagram of the hypothetical oleamide "synthesome." It is proposed that a putative oleamide synthetic complex would minimally consist of cytochrome *c* (Cyto *c*) as the catalytic enzyme, superoxide dismutase (SOD) for the local production of H_2O_2 from superoxide anion, acyl CoA binding protein for the presentation of oleoyl-CoA and an amino transferase activity for the generation of the amide nitrogen. While incomplete, a working model for the catalytic mechanism predicts that oleoyl-CoA binds to cytochrome *c* with the thioester adjacent to the oxidized heme center such that the iron could accept an electron, thus reducing Fe^{3+} to Fe^{2+}. The amino transferase would then provide the amide nitrogen (possibly in the form of ammonia) for nucleophilic attack at the thioester bond to produce oleamide. Finally, the role of H_2O_2, generated by SOD, would be to reoxidize the heme iron of cytochrome *c*.

possibility that an amino transferase enzyme also functions as part of the oleamide synthesome has not been revealed by the proteomic data to date.

X. APOPTOSIS: A MODEL FOR THE MECHANISM AND REGULATION OF OLEAMIDE BIOSYNTHESIS

The fundamental importance of apoptosis in normal embryogenesis and adult tissue remodeling is well recognized (Danial and Korsmeyer, 2004; Rathmell and Thompson, 2002; Thompson, 1995; Yuan and Yankner, 2000). The ability of cytochrome *c* to catalyze the synthesis of oleamide has intriguing implications in this process. Central to this apoptosis is the release of cytochrome *c* from mitochondria which triggers the

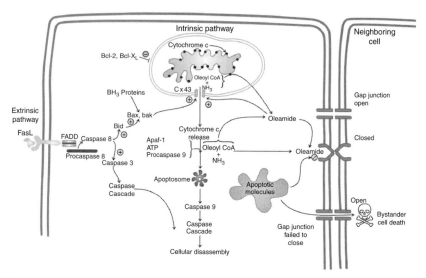

Figure 3.7 Mechanistic model for the roles of cytochrome c, oleamide, and gap junctions in programmed cell death. The intrinsic and extrinsic apoptotic pathways for programmed cell death converge on the release of mitochondrial cytochrome c and increased oleamide production. Activation of the intrinsic pathway is mediated by BH_3/Bcl proteins that monitor shifts in intracellular homeostasis, aging, and accumulations of cytotoxic molecules including Ca^{2+}, H_2O_2, and proteins modified by free radical damage, aggregation, or misfolding. Prosurvival proteins (Bcl-2 and Bcl-X_L) and pro-apoptotic molecules (Bax and Bak) interact directly, and indirectly, to maintain survival or trigger apoptosis. Once activated, Bax and Bak induce a permeability transition pore in the mitochondrial outer membrane which leads to the release of cytochrome c and several other pro-apoptotic molecules not depicted (PUMA, SMAC/DIABLO, AIP, Endo G, HtrA2/Omi). These events are accompanied by an increase in oleamide production and the closure of gap junctions. Oleamide is also reported to act back on mitochondria, via the gap junction protein connexin 43 (Cx43), to increase the release of cytochrome c and other apoptotic factors. This would serve to accelerate programmed cell death in a positive feedback fashion. Cytochrome c released into the cytosol is known to complex with Apaf1, ATP, and procaspase 9 to form the apoptosome. The apoptosome activates caspase 9 and thereby initiates the intrinsic apoptotic caspase cascade, resulting in cellular disassembly. The extrinsic pathway is triggered by ligand activation of cell surface death receptors (e.g., FAS, TNFR1, DR4/5) and subsequent cleavage of procaspase 8. Activated caspase 8 initiates the extrinsic apoptotic caspase cascade and also activates Bid. Bid, in turn, activates Bax and Bak which bring about the release of mitochondrial cytochrome c. Accordingly, Bax and Bak function in both the intrinsic and extrinsic pathways. This linkage may provide a mechanism whereby both pathways trigger cytochrome c release, induction of oleamide biosynthesis and protection against bystander killing. This condensed overview of apoptosis is reflective of the more than 900 review articles that have been published on this topic since 2004.

apoptotic pathway (Kluck *et al.*, 1997; Liu *et al.*, 1996; Yang *et al.*, 1997) via formation of the apoptosome (Schafer and Kornbluth, 2006; Zou *et al.*, 1999) and the resulting activation of caspase 9 (Hu *et al.*, 1999; Kuida *et al.*, 1998; Zou *et al.*, 1999). Figure 3.7 summarizes some of the key events involved and outlines a proposed role for cytochrome *c* in oleamide production for the regulation of gap junction communication during apoptosis. This role for oleamide signaling in apoptosis is based upon the findings that (i) cytochrome *c* catalyzes the synthesis of oleamide (Driscoll et al., 2007), (ii) oleamide production is upregulated early in apoptosis (preliminary personal observations, unpublished), and (iii) oleamide inhibits gap junction communication (Boger et al., 1998b; Guan et al., 1997; Huang et al., 1998; Lerner, 1997; Quist et al., 2000; Schiller et al., 2001b; Subauste et al., 2001; Unger et al., 1999).

Physiologically, gap junctions promote cell survival by creating an electrical and chemical syncytium. This syncytium provides for the delivery of nutrients and growth factors, removal of excess metabolites, and buffering of potentially toxic molecules which may accumulate as a result of cellular metabolism, neuronal depolarization, and other stressors (Bernstein and Morley, 2006; Sohl et al., 2005; Theis et al., 2005; van Veen et al., 2006). Thus, the mitochondrial localization of cytochrome *c*, coupled with its potential role as an oleamide synthase, leads to the proposal that oleamide may serve as an intracellular messenger that links mitochondrial metabolic status with gap junction communication.

Under pathological conditions, however, the beneficial functions of gap junctions can be subverted, resulting in the transmission of pro-apoptotic molecules which precipitate a wave of apoptosis that can spread across networks of otherwise healthy cells (Andrade-Rozental et al., 2000; Bernstein and Morley, 2006; Elshami et al., 1996; Lin et al., 1998; Mesnil et al., 1996; Reznikov et al., 2000; Theiss et al., 2007; van Veen et al., 2006). This propagation of apoptosis, referred to as bystander killing, exacerbates cellular loss in cerebral ischemia and stroke (Cotrina et al., 1998; Lin et al., 1998; Perez Velazquez et al., 2006; Rawanduzy et al., 1997) and myocardial infarction (Miura et al., 2004) and appears to also play a role in neurodegeneration (Nakase and Naus, 2004; Yuan and Yankner, 2000). While the specific molecules that mediate bystander killing have yet to be defined, oleamide and other treatments that block gap junctions have been shown to reduce the magnitude of bystander killing. Specifically, oleamide reduced the spread of apoptosis in response to metabolic depression in primary hippocampal cultures (Nodin et al., 2005) and K^+ deprivation in cerebellar granule cells (Yang et al., 2002). Further, alpha-glycyrrhetinic acid and two other gap junction blockers were reported to reduce infarct size following myocardial infarction (Miura et al., 2004), stroke (Rawanduzy et al., 1997), and global transient ischemia (Rami et al., 2001). Related studies indicate that gap junctions also have a central role in ischemia preconditioning (Boengler et al., 2005, 2006, 2007; Halestrap, 2006). Finally, in marked contrast to situations where reductions in gap junction communication and bystander killing are desired, strengthening gap junction communication may increase the effectiveness of cancer chemotherapies. This is because cancer cells are generally resistant to bystander killing due to reduced gap junction communication (Leithe *et al.*, 2006; Matono *et al.*, 2003; Trosko and Ruch, 2002). Accordingly, inhibitors of oleamide synthesis or action may offer benefits in the treatment of cancer.

XI. Considerations for the Investigation of Oleamide Biosynthesis

Several considerations must be taken into account when using [^{14}C] oleic acid as a biosynthetic precursor for oleamide in cell culture systems. First, commercial sources of [^{14}C]oleic acid are often contaminated with trace amounts [^{14}C]oleamide. While very small, this contamination can profoundly compromise experimental findings and must be controlled for. Cultured cells readily take up and concentrate oleamide from media in a time-, temperature-, and cell type-dependent manner. This uptake results in experimental findings that may be mistakenly interpreted as evidence for cellular biosynthesis of oleamide. Second, and equally important, is that consideration must be given to the composition of the growth medium used in biosynthetic labeling experiments. As described in Section V, we have recently determined that the fetal bovine serum used to supplement cell culture media contains an oleamide synthase activity that is biochemically and functionally distinct from both cytochrome *c* and PAM. Accordingly, the possibility exists that oleamide could be synthesized extracellularly and then taken up into cultured cells. Finally, the levels radiolabeled oleamide recovered from cells cultured in the presence of [^{14}C]oleic acid are modest. Experimental protocols, therefore, need to be optimized for both labeling and recovery of [^{14}C]oleamide.

XII. Future Directions and Concluding Remarks

The biosynthesis of oleamide has been under investigation since the mid-1990s. Evidence to date points to two potential mechanisms for the generation of oleamide *in vivo*: one mechanism is mediated by the well-established neuropeptide amidating enzyme, PAM, and the other by the mitochondrial protein, cytochrome *c*. While there are key experimental observations supporting each of these pathways, the findings are far from complete. Accordingly, the next important step for this area of research is to establish if PAM, cytochrome *c*, or both of these possibilities constitute the *in vivo* pathway for oleamide biosynthesis. Alternatively, further study may reveal a completely unexpected mechanism for the generation of oleamide. Such a mechanism may involve the serum synthase activity discussed here.

Strategies for defining the biosynthetic pathway for oleamide will include both pharmacologic- and genetic-based strategies although substantial challenges face both approaches. For example, while several drugs target PAM *in vitro* (Driscoll et al., 2000; Mains et al., 1986) these agents lack specificity and high efficacy *in vivo* (Mueller et al., 1993, 1999). Moreover, pharmacologic agents against cytochrome c have not been developed, although, immune-neutralization with monoclonal antibodies may prove useful for this purpose. It appears, therefore, that genetic strategies offer the most effective means for defining the mechanism(s) for oleamide biosynthesis. These strategies are particularly powerful in cell culture where targeted gene disruption and RNA inhibition (RNAi) can be used to selectively eliminate the expression of PAM or cytochrome c in specific cell types. In this regard, we have observed in preliminary studies that cells lacking both somatic and testicular forms of cytochrome c (Vempati et al., 2007) fail to upregulate oleamide biosynthesis during apoptosis (personal observations, unpublished). In considering this type of data, it is important to emphasize that ablation of cytochrome c causes dramatic adaptive shifts in whole cell physiology such that the loss of oleamide biosynthesis may be an indirect consequence of the absence of cytochrome c. Similar knockout studies have yet to be performed with PAM. Genetic studies in whole organisms are problematic due to embryonic lethality of the PAM and cytochrome c null conditions (Czyzyk et al., 2005; Vempati et al., 2007). However, experimental approaches employing conditional and tissue-specific knockout strategies hold the potential of providing important insights into the *in vivo* roles of PAM and cytochrome c in oleamide biosynthesis.

In summary, although the natural occurrence and pharmacologic actions of oleamide have been appreciated for more than a decade, the actual mechanism(s) that mediates oleamide biosynthesis has yet to be defined. Nevertheless, recent evidence supporting two intriguing biosynthetic pathways now provides the necessary foundation to guide hypothesis-driven research to elucidate the precise reactions involved. Ultimately, knowing how oleamide is made should open our understanding of the biochemical processes that govern the synthesis of all long-chain primary fatty acid amide messengers.

REFERENCES

Alberts, G. L., Chio, C. L., and Im, W. B. (2001). Allosteric modulation of the human 5-HT(7A) receptor by lipidic amphipathic compounds. *Mol. Pharmacol.* **60,** 1349–1355.

Andrade-Rozental, A. F., Rozental, R., Hopperstad, M. G., Wu, J. K., Vrionis, F. D., and Spray, D. C. (2000). Gap junctions: The "kiss of death" and the "kiss of life". *Brain Res. Brain Res. Rev.* **32,** 308–315.

Arafat, E. S., Trimble, J. W., Andersen, R. N., Dass, C., and Desiderio, D. M. (1989). Identification of fatty acid amides in human plasma. *Life Sci.* **45,** 1679–1687.

Bannerman, P., Nichols, W., Puhalla, S., Oliver, T., Berman, M., and Pleasure, D. (2000). Early migratory rat neural crest cells express functional gap junctions: Evidence that neural crest cell survival requires gap junction function. *J. Neurosci. Res.* **61,** 605–615.

Barros, M. H., Netto, L. E., and Kowaltowski, A. J. (2003). H(2)O(2) generation in Saccharomyces cerevisiae respiratory pet mutants: Effect of cytochrome c. *Free Radic. Biol. Med.* **35,** 179–188.

Basile, A. S., Hanus, L., and Mendelson, W. B. (1999). Characterization of the hypnotic properties of oleamide. *Neuroreport* **10,** 947–951.

Bernstein, S. A., and Morley, G. E. (2006). Gap junctions and propagation of the cardiac action potential. *Adv. Cardiol.* **42,** 71–85.

Bisogno, T., Sepe, N., De Petrocellis, L., Mechoulam, R., and Di Marzo, V. (1997). The sleep inducing factor oleamide is produced by mouse neuroblastoma cells. *Biochem. Biophys. Res. Commun.* **239,** 473–479.

Bisogno, T., Katayama, K., Melck, D., Ueda, N., De Petrocellis, L., et al. (1998). Biosynthesis and degradation of bioactive fatty acid amides in human breast cancer and rat pheochromocytoma cells—Implications for cell proliferation and differentiation. *Eur. J. Biochem.* **254,** 634–642.

Boengler, K., Dodoni, G., Rodriguez-Sinovas, A., Cabestrero, A., Ruiz-Meana, M., et al. (2005). Connexin 43 in cardiomyocyte mitochondria and its increase by ischemic preconditioning. *Cardiovasc. Res.* **67,** 234–244.

Boengler, K., Schulz, R., and Heusch, G. (2006). Connexin 43 signalling and cardioprotection. *Heart* **92,** 1724–1727.

Boengler, K., Konietzka, I., Buechert, A., Heinen, Y., Garcia-Dorado, D., et al. (2007). Loss of ischemic preconditioning's cardioprotection in aged mouse hearts is associated with reduced gap junctional and mitochondrial levels of connexin 43. *Am. J. Physiol. Heart Circ. Physiol.* **292,** H1764–H1769.

Boger, D. L., Patterson, J. E., and Jin, Q. (1998a). Structural requirements for 5-HT2A and 5-HT1A serotonin receptor potentiation by the biologically active lipid oleamide. *Proc. Natl. Acad. Sci. USA* **95,** 4102–4107.

Boger, D. L., Patterson, J. E., Guan, X., Cravatt, B. F., Lerner, R. A., and Gilula, N. B. (1998b). Chemical requirements for inhibition of gap junction communication by the biologically active lipid oleamide. *Proc. Natl. Acad. Sci. USA* **95,** 4810–4815.

Boitano, S., and Evans, W. H. (2000). Connexin mimetic peptides reversibly inhibit Ca(2+) signaling through gap junctions in airway cells. *Am. J. Physiol. Lung Cell Mol. Physiol.* **279,** L623–L630.

Bonafe, L., Troxler, H., Kuster, T., Heizmann, C. W., Chamoles, N. A., et al. (2000). Evaluation of urinary acylglycines by electrospray tandem mass spectrometry in mitochondrial energy metabolism defects and organic acidurias. *Mol. Genet. Metab.* **69,** 302–311.

Burstein, S. H., Rossetti, R. G., Yagen, B., and Zurier, R. B. (2000). Oxidative metabolism of anandamide. *Prostaglandins Other Lipid Mediat.* **61,** 29–41.

Butovich, I. A., Uchiyama, E., and McCulley, J. P. (2007). Lipids of human meibum: Mass-spectrometric analysis and structural elucidation. *J. Lipid Res.* **48,** 2220–2235.

Chaturvedi, S., Driscoll, W. J., Elliot, B. M., Faraday, M. M., Grunberg, N. E., and Mueller, G. P. (2006). In vivo evidence that N-oleoylglycine acts independently of its conversion to oleamide. *Prostaglandins Other Lipid Mediat.* **81,** 136–149.

Coleman, A. M., and Sengelaub, D. R. (2002). Patterns of dye coupling in lumbar motor nuclei of the rat. *J. Comp. Neurol.* **454,** 34–41.

Cotrina, M. L., Kang, J., Lin, J. H., Bueno, E., Hansen, T. W., et al. (1998). Astrocytic gap junctions remain open during ischemic conditions. *J. Neurosci.* **18,** 2520–2537.

Cravatt, B. F., Prospero-Garcia, O., Siuzdak, G., Gilula, N. B., Henriksen, S. J., et al. (1995). Chemical characterization of a family of brain lipids that induce sleep. *Science* **268,** 1506–1509.

Czyzyk, T. A., Ning, Y., Hsu, M. S., Peng, B., Mains, R. E., et al. (2005). Deletion of peptide amidation enzymatic activity leads to edema and embryonic lethality in the mouse. *Dev. Biol.* **287,** 301–313.
Danial, N. N., and Korsmeyer, S. J. (2004). Cell death: Critical control points. *Cell* **116,** 205–219.
DeBlassio, J. L., deLong, M. A., Glufke, U., Kulathila, R., Merkler, K. A., et al. (2000). Amidation of salicyluric acid and gentisuric acid: A possible role for peptidylglycine alpha-amidating monooxygenase in the metabolism of aspirin. *Arch. Biochem. Biophys.* **383,** 46–55.
Decrouy, X., Gasc, J. M., Pointis, G., and Segretain, D. (2004). Functional characterization of Cx43 based gap junctions during spermatogenesis. *J. Cell Physiol.* **200,** 146–154.
Driscoll, W. J., Konig, S., Fales, H. M., Pannell, L. K., Eipper, B. A., and Mueller, G. P. (2000). Peptidylglycine-alpha-hydroxylating monooxygenase generates two hydroxylated products from its mechanism-based suicide substrate, 4-phenyl-3-butenoic acid. *Biochemistry* **39,** 8007–8016.
Driscoll, W. J., Chaturvedi, S., and Mueller, G. P. (2007). Oleamide synthesizing activity from rat kidney: Identification as cytochrome c. *J. Biol. Chem.* **282,** 22353–22363.
Droghetti, E., Oellerich, S., Hildebrandt, P., and Smulevich, G. (2006). Heme coordination states of unfolded ferrous cytochrome C. *Biophys. J.* **91,** 3022–3031.
Eipper, B. A., Stoffers, D. A., and Mains, R. E. (1992). The biosynthesis of neuropeptides: Peptide alpha-amidation. *Annu. Rev. Neurosci.* **15,** 57–85.
Elshami, A. A., Saavedra, A., Zhang, H., Kucharczuk, J. C., Spray, D. C., Fishman, G. I., Amin, K. M., Kaiser, L. R., and Albelda, S. M. (1996). Gap junctions play a role in the 'bystander effect' of the herpes simplex virus thymidine kinase/ganciclovir system *in vitro*. *Gene Ther.* **3,** 85–92.
Farrell, E. K., and Merkler, D. J. (2008). Biosynthesis, degradation and pharmacological importance of the fatty acid amides. *Drug Discov. Today* **13,** 558–568.
Fedorova, I., Hashimoto, A., Fecik, R. A., Hedrick, M. P., Hanus, L. O., Boger, D. L., Rice, K. C., and Basile, A. S. (2001). Behavioral evidence for the interaction of oleamide with multiple neurotransmitter systems. *J. Pharmacol. Exp. Ther.* **299,** 332–342.
Ford, T. A., and Mueller, G. P. (1993). Induction of peptidylglycine alpha-hydroxylating monooxygenase activity by nerve growth factor in PC12 cells. *J. Mol. Neurosci.* **4,** 97–105.
Gilleron, J., Nebout, M., Scarabelli, L., Senegas-Balas, F., Palmero, S., Segretain, D., and Pointis, G. (2006). A potential novel mechanism involving connexin 43 gap junction for control of sertoli cell proliferation by thyroid hormones. *J. Cell Physiol.* **209,** 153–161.
Guan, X., Cravatt, B. F., Ehring, G. R., Hall, J. E., Boger, D. L., Lerner, R. A., and Gilula, N. B. (1997). The sleep-inducing lipid oleamide deconvolutes gap junction communication and calcium wave transmission in glial cells. *J. Cell Biol.* **139,** 1785–1792.
Halestrap, A. P. (2006). Mitochondria and preconditioning: A connexin connection? *Circ. Res.* **99,** 10–12.
Hanus, L. O., Fales, H. M., Spande, T. F., and Basile, A. S. (1999). A gas chromatographic-mass spectral assay for the quantitative determination of oleamide in biological fluids. *Anal. Biochem.* **270,** 159–166.
Hedlund, P. B., Carson, M. J., Sutcliffe, J. G., and Thomas, E. A. (1999). Allosteric regulation by oleamide of the binding properties of 5-hydroxytryptamine7 receptors. *Biochem. Pharmacol.* **58,** 1807–1813.
Hedlund, P. B., Danielson, P. E., Thomas, E. A., Slanina, K., Carson, M. J., and Sutcliffe, J. G. (2003). No hypothermic response to serotonin in 5-HT7 receptor knockout mice. *Proc. Natl. Acad. Sci. USA* **100,** 1375–1380.
Huang, G. Y., Cooper, E. S., Waldo, K., Kirby, M. L., Gilula, N. B., and Lo, C. W. (1998). Gap junction-mediated cell-cell communication modulates mouse neural crest migration. *J. Cell Biol.* **143,** 1725–1734.

Huidobro-Toro, J. P., and Harris, R. A. (1996). Brain lipids that induce sleep are novel modulators of 5-hydroxytrypamine receptors. *Proc. Natl. Acad. Sci. USA* **93**, 8078–8082.

Huitron-Resendiz, S., Gombart, L., Cravatt, B. F., and Henriksen, S. J. (2001). Effect of oleamide on sleep and its relationship to blood pressure, body temperature, and locomotor activity in rats. *Exp. Neurol.* **172**, 235–243.

Hu, Y., Benedict, M. A., Ding, L., and Nunez, G. (1999). Role of cytochrome c and dATP/ATP hydrolysis in Apaf-1-mediated caspase-9 activation and apoptosis. *EMBO J.* **18**, 3586–3595.

Jann, M. W., and Slade, J. H. (2007). Antidepressant agents for the treatment of chronic pain and depression. *Pharmacotherapy* **27**, 1571–1587.

Kase, B. F., and Bjorkhem, I. (1989). Peroxisomal bile acid-CoA:amino-acid N-acyltransferase in rat liver. *J. Biol. Chem.* **264**, 9220–9223.

Kase, B. F., Prydz, K., Bjorkhem, I., and Pedersen, J. I. (1986). Conjugation of cholic acid with taurine and glycine by rat liver peroxisomes. *Biochem. Biophys. Res. Commun.* **138**, 167–173.

Kelley, M., and Vessey, D. A. (1994). Characterization of the acyl-CoA:amino acid N-acyltransferases from primate liver mitochondria. *J. Biochem. Toxicol.* **9**, 153–158.

King, L., 3rd, Barnes, S., Glufke, U., Henz, M. E., Kirk, M., Merkler, K. A., Vederas, J. C., Wilcox, B. J., and Merkler, D. J. (2000). The enzymatic formation of novel bile acid primary amides. *Arch. Biochem. Biophys.* **374**, 107–117.

Kluck, R. M., Bossy-Wetzel, E., Green, D. R., and Newmeyer, D. D. (1997). The release of cytochrome c from mitochondria: A primary site for Bcl-2 regulation of apoptosis. *Science* **275**, 1132–1136.

Koga, D., Santa, T., Fukushima, T., Homma, H., and Imai, K. (1997). Liquid chromatographic-atmospheric pressure chemical ionization mass spectrometric determination of anandamide and its analogs in rat brain and peripheral tissues. *J. Chromatogr. B Biomed. Sci. Appl.* **690**, 7–13.

Kuida, K., Haydar, T. F., Kuan, C. Y., Gu, Y., Taya, C., Karasuyama, H., Su, M. S., Rakic, P., and Flavell, R. A. (1998). Reduced apoptosis and cytochrome c-mediated caspase activation in mice lacking caspase 9. *Cell* **94**, 325–337.

Laposky, A. D., Homanics, G. E., Basile, A., and Mendelson, W. B. (2001). Deletion of the GABA(A) receptor beta 3 subunit eliminates the hypnotic actions of oleamide in mice. *Neuroreport* **12**, 4143–4147.

Lees, G., Edwards, M. D., Hassoni, A. A., Ganellin, C. R., and Galanakis, D. (1998). Modulation of GABA(A) receptors and inhibitory synaptic currents by the endogenous CNS sleep regulator cis-9,10-octadecenoamide (cOA). *Br. J. Pharmacol.* **124**, 873–882.

Leithe, E., Sirnes, S., Omori, Y., and Rivedal, E. (2006). Downregulation of gap junctions in cancer cells. *Crit. Rev. Oncog.* **12**, 225–256.

Lerner, R. A. (1997). A hypothesis about the endogenous analogue of general anesthesia. *Proc. Natl. Acad. Sci. USA* **94**, 13375–13377.

Lerner, R. A., Siuzdak, G., Prospero-Garcia, O., Henriksen, S. J., Boger, D. L., and Cravatt, B. F. (1994). Cerebrodiene: A brain lipid isolated from sleep-deprived cats. *Proc. Natl. Acad. Sci. USA* **91**, 9505–9508.

Lichtman, A. H., Hawkins, E. G., Griffin, G., and Cravatt, B. F. (2002). Pharmacological activity of fatty acid amides is regulated, but not mediated, by fatty acid amide hydrolase in vivo. *J. Pharmacol. Exp. Ther.* **302**, 73–79.

Lin, J. H., Weigel, H., Cotrina, M. L., Liu, S., Bueno, E., Hansen, A. J., Hansen, T. W., Goldman, S., and Nedergaard, M. (1998). Gap-junction-mediated propagation and amplification of cell injury. *Nat. Neurosci.* **1**, 494–500.

Liu, X., Kim, C. N., Yang, J., Jemmerson, R., and Wang, X. (1996). Induction of apoptotic program in cell-free extracts: Requirement for dATP and cytochrome c. *Cell* **86**, 147–157.

Maccarrone, M., and Finazzi-Agro, A. (2002). Endocannabinoids and their actions. *Vitam. Horm.* **65**, 225–255.

Mains, R. E., Park, L. P., and Eipper, B. A. (1986). Inhibition of peptide amidation by disulfiram and diethyldithiocarbamate. *J. Biol. Chem.* **261,** 11938–11941.

Martinez-Gonzalez, D., Bonilla-Jaime, H., Morales-Otal, A., Henriksen, S. J., Velazquez-Moctezuma, J., and Prospero-Garcia, O. (2004). Oleamide and anandamide effects on food intake and sexual behavior of rats. *Neurosci. Lett.* **364,** 1–6.

Matono, S., Tanaka, T., Sueyoshi, S., Yamana, H., Fujita, H., and Shirouzu, K. (2003). Bystander effect in suicide gene therapy is directly proportional to the degree of gap junctional intercellular communication in esophageal cancer. *Int. J. Oncol.* **23,** 1309–1315.

Mawal, Y. R., and Qureshi, I. A. (1994). Purification to homogeneity of mitochondrial acyl coa:glycine n-acyltransferase from human liver. *Biochem. Biophys. Res. Commun.* **205,** 1373–1379.

McCue, J. M., Driscoll, W. J., and Mueller, G. P. (2008). Cytochrome c catalyzes the in vitro synthesis of arachidonoyl glycine. *Biochem. Biophys. Res. Commun.* **365,** 322–327.

McIntyre, N. R., Lowe, E. W., Jr., Chew, G. H., Owen, T. C., and Merkler, D. J. (2006). Thiorphan, tiopronin, and related analogs as substrates and inhibitors of peptidylglycine alpha-amidating monooxygenase (PAM). *FEBS Lett.* **580,** 521–532.

Mendelson, W. B., and Basile, A. S. (2001). The hypnotic actions of the fatty acid amide, oleamide. *Neuropsychopharmacology* **25,** S36–S39.

Merkler, D. J., Merkler, K. A., Stern, W., and Fleming, F. F. (1996). Fatty acid amide biosynthesis: A possible new role for peptidylglycine alpha-amidating enzyme and acyl-coenzyme A:glycine N-acyltransferase. *Arch. Biochem. Biophys.* **330,** 430–434.

Merkler, D. J., Chew, G. H., Gee, A. J., Merkler, K. A., Sorondo, J. P., and Johnson, M. E. (2004). Oleic acid derived metabolites in mouse neuroblastoma N18TG2 cells. *Biochemistry* **43,** 12667–12674.

Mesnil, M., Piccoli, C., Tiraby, G., Willecke, K., and Yamasaki, H. (1996). Bystander killing of cancer cells by herpes simplex virus thymidine kinase gene is mediated by connexins. *Proc. Natl. Acad. Sci. USA* **93,** 1831–1835.

Miller, L. A., Baumgart, L. E., Chew, G. H., deLong, M. A., Galloway, L. C., Jung, K. W., Merkler, K. A., Nagle, A. S., Poore, D. D., Yoon, C. H., and Merkler, D. J. (2003). Glutathione, S-substituted glutathiones, and leukotriene C4 as substrates for peptidylglycine alpha-amidating monooxygenase. *Arch. Biochem. Biophys.* **412,** 3–12.

Miura, T., Ohnuma, Y., Kuno, A., Tanno, M., Ichikawa, Y., Nakamura, Y., Yano, T., Miki, T., Sakamoto, J., and Shimamoto, K. (2004). Protective role of gap junctions in preconditioning against myocardial infarction. *Am. J. Physiol. Heart Circ. Physiol.* **286,** H214–H221.

Moltzen, E. K., and Bang-Andersen, B. (2006). Serotonin reuptake inhibitors: The corner stone in treatment of depression for half a century—A medicinal chemistry survey. *Curr. Top. Med. Chem.* **6,** 1801–1823.

Mueller, G. P., and Driscoll, W. J. (2007). In vitro synthesis of oleoylglycine by cytochrome c points to a novel pathway for the production of lipid signaling molecules. *J. Biol. Chem.* **282**(31), 22364–22369.

Mueller, G. P., Husten, E. J., Mains, R. E., and Eipper, B. A. (1993). Peptide alpha-amidation and peptidylglycine alpha-hydroxylating monooxygenase: Control by disulfiram. *Mol. Pharmacol.* **44,** 972–980.

Mueller, G. P., Driscoll, W. J., and Eipper, B. A. (1999). In vivo inhibition of peptidylglycine-alpha-hydroxylating monooxygenase by 4-phenyl-3-butenoic acid. *J. Pharmacol. Exp. Ther.* **290,** 1331–1336.

Murillo-Rodriguez, E., Giordano, M., Cabeza, R., Henriksen, S. J., Mendez Diaz, M., Navarro, L., and Prospero-Garcia, O. (2001). Oleamide modulates memory in rats. *Neurosci. Lett.* **313,** 61–64.

Nagasawa, K., Chiba, H., Fujita, H., Kojima, T., Saito, T., Endo, T., and Sawada, N. (2006). Possible involvement of gap junctions in the barrier function of tight junctions of brain and lung endothelial cells. *J. Cell Physiol.* **208**, 123–132.

Nakase, T., and Naus, C. C. (2004). Gap junctions and neurological disorders of the central nervous system. *Biochim. Biophys. Acta* **1662**, 149–158.

Nichols, K. K., Ham, B. M., Nichols, J. J., Ziegler, C., and Green-Church, K. B. (2007). Identification of fatty acids and fatty acid amides in human meibomian gland secretions. *Invest Ophthalmol. Vis. Sci.* **48**, 34–39.

Nodin, C., Nilsson, M., and Blomstrand, F. (2005). Gap junction blockage limits intercellular spreading of astrocytic apoptosis induced by metabolic depression. *J. Neurochem.* **94**, 1111–1123.

O'Byrne, J., Hunt, M. C., Rai, D. K., Saeki, M., and Alexson, S. E. (2003). The human bile acid-CoA:amino acid N-acyltransferase functions in the conjugation of fatty acids to glycine. *J. Biol. Chem.* **278**, 34237–34244.

Patricelli, M. P., and Cravatt, B. F. (2001). Proteins regulating the biosynthesis and inactivation of neuromodulatory fatty acid amides. *Vitam. Horm.* **62**, 95–131.

Perez Velazquez, J. L., Kokarovtseva, L., Sarbaziha, R., Jeyapalan, Z., and Leshchenko, Y. (2006). Role of gap junctional coupling in astrocytic networks in the determination of global ischaemia-induced oxidative stress and hippocampal damage. *Eur. J. Neurosci.* **23**, 1–10.

Pinheiro, T. J., Elove, G. A., Watts, A., and Roder, H. (1997). Structural and kinetic description of cytochrome c unfolding induced by the interaction with lipid vesicles. *Biochemistry* **36**, 13122–13132.

Quist, A. P., Rhee, S. K., Lin, H., and Lal, R. (2000). Physiological role of gap-junctional hemichannels. Extracellular calcium-dependent isosmotic volume regulation. *J. Cell Biol.* **148**, 1063–1074.

Rami, A., Volkmann, T., and Winckler, J. (2001). Effective reduction of neuronal death by inhibiting gap junctional intercellular communication in a rodent model of global transient cerebral ischemia. *Exp. Neurol.* **170**, 297–304.

Ransjo, M., Sahli, J., and Lie, A. (2003). Expression of connexin 43 mRNA in microisolated murine osteoclasts and regulation of bone resorption *in vitro* by gap junction inhibitors. *Biochem. Biophys. Res. Commun.* **303**, 1179–1185.

Rathmell, J. C., and Thompson, C. B. (2002). Pathways of apoptosis in lymphocyte development, homeostasis, and disease. *Cell* **109**(Suppl), S97–S107.

Rawanduzy, A., Hansen, A., Hansen, T. W., and Nedergaard, M. (1997). Effective reduction of infarct volume by gap junction blockade in a rodent model of stroke. *J. Neurosurg.* **87**, 916–920.

Reznikov, K., Kolesnikova, L., Pramanik, A., Tan-No, K., Gileva, I., Yakovleva, T., Rigler, R., Terenius, L., and Bakalkin, G. (2000). Clustering of apoptotic cells via bystander killing by peroxides. *FASEB J.* **14**, 1754–1764.

Ritenour-Rodgers, K. J., Driscoll, W. J., Merkler, K. A., Merkler, D. J., and Mueller, G. P. (2000). Induction of peptidylglycine alpha-amidating monooxygenase in N(18)TG(2) cells: A model for studying oleamide biosynthesis. *Biochem. Biophys. Res. Commun.* **267**, 521–526.

Sakono, M., Goto, M., and Furusaki, S. (2000). Refolding of cytochrome c using reversed micelles. *J. Biosci. Bioeng.* **89**, 458–462.

Sanghera, N., and Pinheiro, T. J. (2000). Unfolding and refolding of cytochrome c driven by the interaction with lipid micelles. *Protein Sci.* **9**, 1194–1202.

Schafer, Z. T., and Kornbluth, S. (2006). The apoptosome: Physiological, developmental, and pathological modes of regulation. *Dev. Cell* **10**, 549–561.

Schiller, P. C., D'Ippolito, G., Balkan, W., Roos, B. A., and Howard, G. A. (2001a). Gap-junctional communication is required for the maturation process of osteoblastic cells in culture. *Bone* **28**, 362–369.

Schiller, P. C., D'Ippolito, G., Brambilla, R., Roos, B. A., and Howard, G. A. (2001b). Inhibition of gap-junctional communication induces the trans-differentiation of osteoblasts to an adipocytic phenotype *in vitro*. *J. Biol. Chem.* **276,** 14133–14138.
Schweitzer, J. S., Wang, H., Xiong, Z. Q., and Stringer, J. L. (2000). pH Sensitivity of nonsynaptic field bursts in the dentate gyrus. *J. Neurophysiol.* **84,** 927–933.
Shonsey, E. M., Sfakianos, M., Johnson, M., He, D., Falany, C. N., Falany, J., Merkler, D. J., and Barnes, S. (2005). Bile acid coenzyme A:amino acid N-acyltransferase in the amino acid conjugation of bile acids. *Methods Enzymol.* **400,** 374–394.
SiuYi Leung, D., Unsicker, K., and Reuss, B. (2001). Gap junctions modulate survival-promoting effects of fibroblast growth factor-2 on cultured midbrain dopaminergic neurons. *Mol. Cell Neurosci.* **18,** 44–55.
Sohl, G., Maxeiner, S., and Willecke, K. (2005). Expression and functions of neuronal gap junctions. *Nat. Rev. Neurosci.* **6,** 191–200.
Stewart, J. M., Blakely, J. A., and Johnson, M. D. (2000). The interaction of ferrocytochrome c with long-chain fatty acids and their CoA and carnitine esters. *Biochem. Cell Biol.* **78,** 675–681.
Subauste, M. C., List, B., Guan, X., Hahn, K. M., Lerner, R., and Gilula, N. B. (2001). A catalytic antibody produces fluorescent tracers of gap junction communication in living cells. *J. Biol. Chem.* **276,** 49164–49168.
Sugiura, T., Kondo, S., Kodaka, T., Tonegawa, T., Nakane, S., Yamashita, A., Ishima, Y., and Waku, K. (1996). Enzymatic synthesis of oleamide (cis-9, 10-octadecenoamide), an endogenous sleep-inducing lipid, by rat brain microsomes. *Biochem. Mol. Biol. Int.* **40,** 931–938.
Suzuki, R., Rygh, L. J., and Dickenson, A. H. (2004). Bad news from the brain: Descending 5-HT pathways that control spinal pain processing. *Trends Pharmacol. Sci.* **25,** 613–617.
Theis, M., Sohl, G., Eiberger, J., and Willecke, K. (2005). Emerging complexities in identity and function of glial connexins. *Trends Neurosci.* **28,** 188–195.
Theiss, C., Mazur, A., Meller, K., and Mannherz, H. G. (2007). Changes in gap junction organization and decreased coupling during induced apoptosis in lens epithelial and NIH-3T3 cells. *Exp. Cell Res.* **313,** 38–52.
Thomas, E. A., Carson, M. J., Neal, M. J., and Sutcliffe, J. G. (1997). Unique allosteric regulation of 5-hydroxytryptamine receptor-mediated signal transduction by oleamide. *Proc. Natl. Acad. Sci. USA* **94,** 14115–14119.
Thomas, E. A., Carson, M. J., and Sutcliffe, J. G. (1998). Oleamide-induced modulation of 5-hydroxytryptamine receptor-mediated signaling. *Ann. NY Acad. Sci.* **861,** 183–189.
Thompson, C. B. (1995). Apoptosis in the pathogenesis and treatment of disease. *Science* **267,** 1456–1462.
Trosko, J. E., and Ruch, R. J. (2002). Gap junctions as targets for cancer chemoprevention and chemotherapy. *Curr. Drug Targets* **3,** 465–482.
Unger, V. M., Kumar, N. M., Gilula, N. B., and Yeager, M. (1999). Electron cryocrystallography of a recombinant cardiac gap junction channel. *Novartis Found Symp.* **219,** 22–30; discussion 1–43.
van Veen, T. A., van Rijen, H. V., and Jongsma, H. J. (2006). Physiology of cardiovascular gap junctions. *Adv. Cardiol.* **42,** 18–40.
Vempati, U. D., Diaz, F., Barrientos, A., Narisawa, S., Mian, A. M., Millan, J. L., Boise, L. H., and Moraes, C. T. (2007). Role of cytochrome C in apoptosis: Increased sensitivity to tumor necrosis factor alpha is associated with respiratory defects but not with lack of cytochrome C release. *Mol. Cell Biol.* **27,** 1771–1783.
Verdon, B., Zheng, J., Nicholson, R. A., Ganelli, C. R., and Lees, G. (2000). Stereoselective modulatory actions of oleamide on GABA(A) receptors and voltage-gated Na(+) channels *in vitro*: A putative endogenous ligand for depressant drug sites in CNS. *Br. J. Pharmacol.* **129,** 283–290.

Wakamatsu, K., Masaki, T., Itoh, F., Kondo, K., and Sudo, K. (1990). Isolation of fatty acid amide as an angiogenic principle from bovine mesentery. *Biochem. Biophys. Res. Commun.* **168,** 423–429.

Wang, Z. B., Li, M., Zhao, Y., and Xu, J. X. (2003). Cytochrome C is a hydrogen peroxide scavenger in mitochondria. *Protein Pept. Lett.* **10,** 247–253.

Watkins, P. A. (1997). Fatty acid activation. *Prog. Lipid Res.* **36,** 55–83.

Wilcox, B. J., Ritenour-Rodgers, K. J., Asser, A. S., Baumgart, L. E., Baumgart, M. A., Boger, D. L., DeBlassio, J. L., deLong, M. A., Glufke, U., Henz, M. E., King, L., 3rd, Merkler, K. A., *et al.* (1999). N-acylglycine amidation: Implications for the biosynthesis of fatty acid primary amides. *Biochemistry* **38,** 3235–3245.

Yang, J., Liu, X., Bhalla, K., Kim, C. N., Ibrado, A. M., Cai, J., Peng, T. I., Jones, D. P., and Wang, X. (1997). Prevention of apoptosis by Bcl-2: Release of cytochrome c from mitochondria blocked. *Science* **275,** 1129–1132.

Yang, J. Y., Abe, K., Xu, N. J., Matsuki, N., and Wu, C. F. (2002). Oleamide attenuates apoptotic death in cultured rat cerebellar granule neurons. *Neurosci. Lett.* **328,** 165–169.

Yuan, J., and Yankner, B. A. (2000). Apoptosis in the nervous system. *Nature* **407,** 802–809.

Zhao, Y., and Xu, J. X. (2004). The operation of the alternative electron-leak pathways mediated by cytochrome c in mitochondria. *Biochem. Biophys. Res. Commun.* **317,** 980–987.

Zhao, Y., Wang, Z. B., and Xu, J. X. (2003). Effect of cytochrome c on the generation and elimination of O2*- and H2O2 in mitochondria. *J. Biol. Chem.* **278,** 2356–2360.

Zohar, J., and Westenberg, H. G. (2000). Anxiety disorders: A review of tricyclic antidepressants and selective serotonin reuptake inhibitors. *Acta Psychiatr. Scand. Suppl.* **403,** 39–49.

Zou, H., Li, Y., Liu, X., and Wang, X. (1999). An APAF-1.cytochrome c multimeric complex is a functional apoptosome that activates procaspase-9. *J. Biol. Chem.* **274,** 11549–11556.

CHAPTER FOUR

Anandamide Receptor Signal Transduction

Catherine E. Goodfellow *and* Michelle Glass

Contents

I. Introduction	80
II. Cannabinoid Receptor 1	81
A. Regulation of cyclic AMP	83
B. Nuclear signaling pathways	83
C. CB1 Receptor-mediated regulation of ion channels	85
III. Cannabinoid Receptor 2	89
A. Signaling pathways	89
IV. Transient Receptor Potential Vanilloid 1	90
A. In the nervous system	91
B. In vasodilation and bronchoconstriction	92
V. Evidence for Additional Receptors	92
A. Direct activation of ion channels	92
B. 5-HT_{3A} receptors	94
C. Nicotinic acetylcholine receptors	95
D. Glycine receptors	95
E. NMDA receptors	96
F. Peroxisome proliferator-activated receptors	96
G. Other GPCRs	97
VI. Concluding Remarks	98
References	99

Abstract

In the 15 years since its discovery anandamide has been implicated in many physiological processes. The signaling pathways mediating many of these processes are now coming to light, particularly in the CNS. The complexity of the cannabinoid system and the identification of many potential other receptors for anandamide have made conclusive evidence of molecular pathways stimulated by this molecule significantly more difficult to achieve. It is becoming obvious that anandamide receptor signal transduction is not a simple process

Department of Pharmacology and Clinical Pharmacology, University of Auckland, Auckland, New Zealand

and that many different cascades can be activated depending on a range of both experimental and physiological variables. This chapter explores the signaling pathways activated by anandamide both through the cannabinoid receptors and through other cellular targets.

I. Introduction

It has been long established that compounds produced by the *Cannabis sativa* plant cause a range of psychological effects. The isolation of delta-9-tetrahydrocannabinol (Δ^9THC) as the main psychoactive constituent (Gaoni and Mechoulam, 1964) led to many studies investigating its *in vivo* effects. However, it was not until 1990, when the first cannabinoid receptor was cloned (Matsuda *et al.*, 1990), followed by the cloning of cannabinoid receptor 2 in 1993 (Munro *et al.*, 1993) that research into the function of cannabinoids escalated.

Anandamide was the first endogenous cannabinoid to be isolated and described (Devane *et al.*, 1992) but a period of almost 10 years remained until evidence emerged of its physiological relevance. In 2001 the endocannabinoids were identified as the signal messengers responsible for the phenomenon of retrograde signaling, providing a physiological rationale for their presence in the central nervous system (CNS) (Ohno-Shosaku *et al.*, 2001; Wilson and Nicoll, 2001). Shortly after the discovery of anandamide, a second, now well-studied endocannabinoid, 2-arachidonoylglycerol (2-AG) was identified (Mechoulam *et al.*, 1995).

The cannabinoid receptors belong to a family of G protein-coupled receptors (GPCRs) which exhibit a canonical seven transmembrane structure and the ability to couple to G proteins intracellularly. Anandamide exerts its actions by binding and activating cannabinoid receptors and other targets. Currently only two cannabinoid receptors have been cloned and extensively investigated; CB1 and CB2 receptors (cannabinoid receptors 1 and 2, respectively). However, it should not be overlooked that there is substantial evidence for additional receptors and targets through which anandamide may act. This chapter will encompass the current knowledge of the signaling pathways activated by the action of anandamide on known and putative receptors.

Upon its discovery anandamide was shown to inhibit the cyclic adenosine monophosphate (cAMP) generating enzyme adenylate cyclase (Vogel *et al.*, 1993) and N-type calcium currents (Mackie *et al.*, 1993) in CB1 receptor expressing cell lines, consistent with the effects of synthetic- and plant-derived cannabinoids. Rapid hydrolysis of anandamide can lead to an under estimation of the potency of the ligand and complications of active breakdown products exerting cellular effects. This can be circumvented by employing the use of a metabolically stable analog of anandamide (methanandamide) or inhibiting fatty acid amide hydrolase (FAAH), the enzyme primarily responsible for anandamide hydrolysis (Abadji *et al.*, 1994; Pertwee, 1997).

To date there is a lack of signal transduction information pertaining to anandamide actions *in vivo*, largely due to experimental difficulties working with this lipophilic and rapidly hydrolyzed molecule as well as the multiple targets it is able to interact with to elicit responses.

II. Cannabinoid Receptor 1

There are three known mRNA splice variants of the human CB1 receptor known as the full-length CB1, CB1a, and CB1b receptors (Ryberg *et al.*, 2005). Rodents appear to lack the CB1b receptor variant (Ryberg *et al.*, 2005). Of crucial importance, however, is the relative lack of affinity that anandamide has for the CB1a and CB1b receptor variants with IC_{50} values reported at ~50, 10967, and 12467 nM respectively (Ryberg *et al.*, 2005), although these figures have recently been challenged by Xiao *et al.* (2008). In a large range of tissues the full-length CB1 receptor mRNA is generally the highest expressed with much lower levels of the two splice variants. The CB1 splice variants have only been detected at the mRNA level to date and little is known of their protein expression, or their cellular role, however, CB1a does exhibit a high level of conservation across species (Shire *et al.*, 1995).

Many studies confirm the ability of anandamide to bind to the CB1 receptor but variation in reported affinities is common. The K_i of anandamide binding to the CB1 receptor ranges from 37 to 781 nM between cell lines transfected with CB1 and various mammalian brain preparations (Pertwee, 1997). This is likely to be a result of a combination of factors such as species differences, disparity in the expression profile of the cell type and inconsistency in experimental protocols and conditions. Similarly, the reported efficacy of anandamide in its ability to stimulate CB1 receptor coupling to GTPγS varies between species and expression systems (Burkey *et al.*, 1997; Hillard *et al.*, 1999).

It has taken some time and effort to convince many researchers of anandamide's classification as a partial agonist as this seems an unusual and uncommon property of an endogenous ligand. For many years, reports of this nature were largely brushed aside as being a result of rapid metabolism. These concerns, although valid, were countered by the use of the metabolically stable derivative, methanandamide, and inhibitors of anandamide degradation coupled with more stringent controls in many publications (Abadji et al., 1994; Breivogel et al., 1998; Childers et al., 1994; Deutsch, 2005; Pinto et al., 1994).

In GTPγS assays performed with both rat brain membranes and membranes from CHO cells transfected with human CB1 receptor anandamide acts as a partial agonist (Breivogel et al., 1998; Gonsiorek et al., 2000). Anandamide exhibited the same property in cAMP assays, whether coupling to $G_{i/o}$ or G_s (Bonhaus et al., 1998; Childers et al., 1994; Hillard et al., 1999). Furthermore, in NG108-15 cells anandamide stimulation resulted in transient elevations of free intracellular Ca^{2+} but the maximal response was significantly lower then that elicited by 2-AG and other full agonists (Sugiura et al., 1996, 1999). Inhibition of N-type calcium currents by anandamide also exhibits properties of partial agonism (Mackie et al., 1993).

On the converse side, anandamide has been reported as a full agonist in some CB1-mediated signaling pathways. In rat hippocampal cultures, anandamide showed the same inhibition of induced calcium spikes as the full agonist WIN55,212-2, however, the concentration–response curve was particularly shallow raising concerns of additional targets being involved (Shen et al., 1996). Perhaps more convincing was the ability of anandamide to maximally activate inwardly rectifying potassium currents in AtT20 cells expressing the rat CB1 receptor (Mackie et al., 1995). However, in a *Xenopus* oocyte model expressing specific subtypes of inwardly rectifying potassium channels and the human CB1 receptor anandamide exhibited typical partial agonist properties (McAllister et al., 1999).

The partial agonistic properties of anandamide should not be misconstrued as redundancy, several hypotheses of interactions utilizing this property have been put forward (Childers, 2006; Gonsiorek et al., 2000; Ross, 2003; Sim et al., 1996). It is plausible that anandamide need not be a full agonist in defined regions of the brain where complete receptor occupancy and activation may not be necessary to elicit a maximal response as the CB1 receptor levels are so high.

Interestingly, at human CB2 receptors expressed by CHO cells anandamide is a partial agonist and antagonizes the full agonist 2-AG activity in GTPγS assays (Gonsiorek et al., 2000). The same may be true of CB1 receptors as 2-AG has been reported as a full agonist here as well (Sugiura et al., 1999). This indicates that *in vivo* the role of anandamide may be much more complicated then initially thought. In scenarios of both 2-AG and anandamide synthesis, the relative concentrations of each would alter the activation of cannabinoid receptors, from strong activation in the presence of 2-AG alone which may be antagonized by varying levels of anandamide (Gonsiorek et al., 2000). The ability of the latter to act as a partial agonist at the CB1 receptor would enable it to mediate a wider range of responses determined by competition with other ligands, receptor levels and downstream signaling amplification.

A. Regulation of cyclic AMP

The CB1 receptor primarily couples with $G_{i/o}$ proteins which upon activation bind and inhibit adenylate cyclase and prevent accumulation of cAMP. There is a robust response upon activation of CB1 receptors with anandamide in various cell models and brain membranes (Childers *et al.*, 1994; Felder *et al.*, 1993, 1995; Hillard *et al.*, 1999; Pinto *et al.*, 1994; Vogel *et al.*, 1993). *In vitro* CB1 receptors show preferential binding to the G_i subtype of G proteins and lower affinity coupling with G_o (Glass and Northup, 1999).

Under some conditions, anandamide is also capable of increasing cAMP levels by coupling with G_s instead of $G_{i/o}$ (Bonhaus *et al.*, 1998; Felder *et al.*, 1993; Glass and Felder, 1997). Determination of which G protein will couple to the CB1 receptor and hence which downstream pathways will be stimulated is influenced by several factors. For example, ligand binding can modulate the capacity of these receptors to couple to different G proteins (Glass and Northup, 1999; Mukhopadhyay and Howlett, 2005). Furthermore, the expression and activation of other GPCRs at the cell membrane may alter the availability of certain G protein subtypes (Jarrahian *et al.*, 2004) which is also complicated by the differential distribution of the G_i, G_o, and G_s subtypes in the brain (Okuhara *et al.*, 1996).

One downstream consequence of cAMP inhibition by anandamide may be the modulation of neuronal structure. In PC12 cells expressing, the CB1 receptor anandamide inhibits neuronal differentiation, and in neural progenitor cells anandamide promotes the retraction of neurites. These effects were mediated via the inhibition of cAMP accumulation and PKA activation (Rueda *et al.*, 2002). The modulation of neuronal structure may be at least partially attributable to anandamide-induced phosphorylation of focal adhesion kinase (pp125FAK) via cAMP/PKA as this tyrosine kinase is important in interactions between integrins and the cytoskeleton (Derkinderen *et al.*, 1996).

B. Nuclear signaling pathways

A range of physiological functions have been suggested to be modulated by CB1 receptor-mediated activation of nuclear signaling pathways related in particular to differentiation, cell proliferation, and cell metabolism.

The differentiation of neural precursors into mature neurons is inhibited by stimulation of CB1 receptors with anandamide (Rueda *et al.*, 2002). In PC12 cells expressing human CB1 receptors and induced to differentiate with nerve growth factor (NGF), anandamide inhibited differentiation by blocking phosphorylation of the NGF receptor, TrkA. This leads to attenuation of Rap1/B-Raf activation and subsequent inhibition of the p42/44 MAPK pathway. Activation of p42/44 is important for phosphorylation of the Elk transcription factor crucial for neuronal differentiation

(Rueda et al., 2002). Activation of the p38 MAPKs, and to a lesser extent p42/44, by anandamide stimulation of CB1 receptors is responsible for reduced transcription and differentiation of keratinocytes (Paradisi et al., 2008). Anandamide stimulation of CB1 receptor transfected and endogenous cell lines also exhibit activation of p42/44 MAPK through CB1 receptors (Bouaboula et al., 1995), which is likely responsible for induction of the gene Krox-24 (Graham et al., 2006).

In various tumor models, anandamide exhibits both anti- and pro-tumorigenic properties, as reviewed by Flygare and Sander (2008). Little is known about the signaling pathways involved in these events but depending on the tumor type and location, these may include activation of MAPKs and transcriptional regulation of various genes. In various cancer-derived cell lines, anandamide leads to epidermal growth factor receptor phosphorylation and subsequent activation of p42/44 MAPK and Akt signaling pathways although it is not clear which cannabinoid receptor is responsible for these effects (Hart et al., 2004). In human breast cancer cells, the cell division induced by incubation with NGF is inhibited by anandamide acting through the CB1 receptor. In contrast to the pathways described in PC12 cells (Rueda et al., 2002), subsequent inhibition of cAMP accumulation and activation of Raf-1 in these cells lead to MAPK activation and reduced expression of the NGF receptor, Trk (Melck et al., 1999, 2000). Inhibition of expression of both vascular endothelial growth factor and its receptor, and p27 induction has also been reported in tumor models following anandamide activation of CB1 receptors (Portella et al., 2003). In MCL cells anandamide stimulation of both CB1 and CB2 receptor signaling was required to initiate cell death through activation of p38 MAPK leading to caspase activation (Gustafsson et al., 2006). Anandamide stimulation of a human-derived cell line from umbilical endothelial cells induced a robust coupling to the p42/44 MAPK pathway via PKC and tyrosine kinases. This coupling was only partially inhibited by CB1 receptor antagonism or knockdown indicating the presence of additional receptors also capable of stimulating this pathway when activated by anandamide (Liu et al., 2000). In the same model anandamide also activates p38 kinase and c-Jun kinase and may be involved in regulating cell proliferation (Liu et al., 2000).

Astrocytes are intimately involved in regulating energy metabolism in the CNS (Magistretti and Pellerin, 1999) and it has been suggested that the endogenous cannabinoid system plays a role in regulating astrocyte metabolic activity (reviewed by Guzmán and Sánchez, 1999). Cannabinoid stimulation has certainly been linked to changes in metabolic activity in astrocytes but the majority of work to date has utilized plant-derived and synthetic cannabinoids which may not be indicative of the actions of the endocannabinoids in *in vivo* systems. In various astrocyte-related cell lines activation of the MAPK pathway has been described upon stimulation with anandamide and other cannabinoids (Bouaboula et al., 1995; Wartmann et al., 1995). This pathway of activation was investigated in more depth in primary rat astrocytes where it was found that both Δ^9THC and the potent synthetic cannabinoid, HU210, induced p42/44 MAPK induction via Raf-1 to increase glucose metabolism and glycogen synthesis through the CB1 receptor. This effect was prevented by addition of PTX suggesting this pathway requires activation of $G_{i/o}$ linked proteins (Sanchez et al., 1998).

C. CB1 Receptor-mediated regulation of ion channels

1. Calcium channels

Anandamide has been demonstrated to modulate calcium channels through a range of pathways. Calcium spikes typical of cultured rat hippocampal neurons stimulated with glutamate are inhibited by anandamide with an IC_{50} of 71 nM (Shen et al., 1996). In cerebellar granule cells methanandamide increased Ca^{2+} elevations induced by stimulation with glutamate and high K^+. This effect was CB1 receptor dependent and occurred through activation of phospholipase C (PLC) and inositol triphosphate (IP_3) pathways (Netzeband et al., 1999). Coupling of the CB1 receptor to the IP_3 pathway is likely to be a region and cell specific event as in transfected CHO cells this mechanism is not seen (Felder et al., 1993, 1995) and activation of CB1 receptors in hippocampal cultures produced a slight inhibition of IP_3 which is pertussis toxin (PTX) insensitive (Nah et al., 1993).

The voltage-gated Ca^{2+} channels are classified into five categories, L-, N-, P/Q-, and R-type. L-type is expressed widely through the skeletal and cardiac system as well as in cortical neurons; they induce long lasting currents but require potent stimulation to become activated. P/Q and N currents are also activated by high voltages (strong membrane depolarization) with localization limited to the brain and specifically Purkinje and cerebellar granule cells for P/Q-Ca^{2+} channels. Transient currents (T-type) are characterized by activation with moderate membrane depolarization, slow inactivation and inactivity at resting membrane potential, this current type is limited to neurons, bone cells, and various cells with pacemaker activity. The residual Ca^{2+} currents with inhibition of each of the other subtypes are termed the R-type Ca^{2+} current and may be comprised of more subtypes that cannot be distinguished from each other with current tools (Catterall et al., 2005). Anandamide inhibits N-type calcium channels in neuronal cell types (Felder et al., 1993; Mackie et al., 1993; Sugiura et al., 1997), mouse blastocysts (Wang et al., 2003) and primary cultured cerebellar (Brown et al., 2004), striatal (Huang et al., 2001), and brainstem neurons (Liang et al., 2004) through the CB1 receptor and is likely to be able to inhibit Q-type currents via the rat CB1 receptor (Mackie et al., 1995). Anandamide also blocks P/Q-type currents in rat cortical and cerebellar slices (Hampson et al., 1998). Similarly, inhibition of opening of L-type calcium channels in cat cerebral arterial smooth muscle cells leading to vasodilation occurs maximally at nanomolar concentrations of anandamide (Gebremedhin et al., 1999).

A more poorly characterized response to anandamide is the stimulation of increases in intracellular calcium levels. Anandamide acts as a partial agonist in NG108-15 cells where micromolar concentrations are required to increase free intracellular Ca^{2+} through CB1 receptor stimulation (Sugiura et al., 1996, 1999). Evidence suggests increases in intracellular calcium may be induced by three mechanisms coupled to the CB1 receptor: (1) through G_s coupling leading to cAMP production and protein kinase A (PKA) activation which in turn generates IP_3-mediated release of Ca^{2+} from internal stores (Bash et al., 2003), (2) activation of a tyrosine kinase through $G_{i/o}$ coupling and subsequent protein kinase C (PKC) activation and stimulation of a mitogen-activated protein kinase (MAPK) pathway (Rubovitch et al., 2004), or (3) by coupling to $G_{q/11}$ activating PLC and subsequent IP_3 generation and intracellular Ca^{2+} release (Lauckner et al., 2005).

In addition to neuronal cells, anandamide may also produce responses in glial cells in the brain. In a glial cell model application of anandamide or elevation of endogenous levels by inhibiting FAAH activity reduced intracellular Ca^{2+} and S100B protein levels as well as limiting cell proliferation through the CB1 receptor (Iuvone et al., 2007). In this model anandamide was also shown to act as a pro-survival factor to cocultured neuronal cells. Confirmation of glial responses to anandamide was published by Navarrete and Araque (2008) who showed CB1 receptor stimulation with cannabinoids caused increased intracellular Ca^{2+} levels in primary cultured rodent astrocytes. This was probably mediated through coupling to $G_{q/11}$ G proteins in hippocampal mouse slices as the effect was PLC dependent. However, coupling of the CB1 receptor to $G_{q/11}$ G proteins may be a cell and agonist specific phenomenon and requires further investigation. In HEK 293 cells transfected with the rat CB1 receptor (Lauckner et al., 2005) and in a human ocular cell model (McIntosh et al., 2007) WIN55,212-2 induces increases in intracellular Ca^{2+} via coupling to $G_{q/11}$, however, other cannabinoids, including methanandamide, could not induce this response in these cells.

2. Potassium channels

Anandamide, like other cannabinoids, activates inwardly rectifying potassium currents through CB1 receptor activation (Mackie et al., 1995; McAllister et al., 1999). Of the inwardly rectifying potassium channel family one class, the G protein-coupled inwardly rectifying potassium channels or GIRKs are capable of activation by interaction with the $\beta\gamma$ subunit of G proteins and are likely to be the targets of CB1 receptor signaling. In AtT20 cells transfected with rat CB1 receptor anandamide elicits a similar enhancement of potassium currents to the full agonist WIN55,212-2 (Mackie et al., 1995). In a similar fashion, *Xenopus* oocytes expressing human CB1 receptor and GIRK1/4 also respond to anandamide with increases in potassium current but anandamide acts only as a partial agonist (McAllister et al., 1999). This differential response to anandamide is not unexpected given AtT20 cells express GIRK subtypes 1 and 2 (Kuzhikandathil et al., 1998) and suggests different GIRK subtype composition affects potential responses to anandamide stimulation.

It is also probable that metabolic products of anandamide, such as arachidonic acid, mediate their own effects, as has been demonstrated in the rat dorsal root ganglion (DRG) (Evans et al., 2008). Here, the inhibition of potassium currents was shown to be lifted when anandamide hydrolysis was prevented, indicating it was the metabolites of anandamide inducing the response independent of the CB1 receptor.

3. Neuronal plasticity

As described in the above sections anandamide acting at CB1 results in the modulation of a wide range of ion channels. In neuronal cells this is likely the basis of its role in neuronal plasticity. Endocannabinoids have been identified as retrograde messengers involved in modulating processes involved in neuronal plasticity such as DSI, DSE, and LTP. While it is clear that neuronal depolarization stimulates the release of endocannabinoids from the postsynaptic terminal it is likely that the concentrations of endocannabinoids released differs between cell types and brain regions (Alger, 2002). At least in the hippocampus it appears 2-AG is likely to be the predominant retrograde messenger as significant increases in its levels were detected in stimulated hippocampal slices and cultured cortical neurons with no change in anandamide levels (Stella et al., 1997). Also, inhibition of monoacylglycerol lipase, the enzyme that degrades 2-AG, but not other endocannabinoid inactivating enzymes enhanced the cannabinoid receptor-mediated inhibition of inhibitory postsynaptic currents (IPSCs) in hippocampal neurons (Hashimotodani et al., 2007). However, in the striatum neural activity only increases anandamide release, with no change in 2-AG, implicating anandamide as an important retrograde messenger in this region (Giuffrida et al., 1999). Stimulation of NMDA receptors on cortical rat neurons results in an increase in 2-AG, however, when both NMDA and acetylcholine receptors are stimulated the levels of both endocannabinoids increase (Stella and Piomelli, 2001). Thus, cell type, the receptor complement, as well as the stimulatory inputs received, are likely to determine which endocannabinoid is produced. Most work to date does not distinguish between endocannabinoids but as the quality of pharmacological tools improves, the specific role of each endocannabinoid will no doubt be teased out allowing greater insight into many aspects of brain functioning.

It is now well accepted that endocannabinoid-mediated depolarization-induced suppression of inhibition (DSI) occurs in CA1 pyramidal neurons as the inhibitory neurotransmitter GABA is prevented from release from the presynaptic terminal (Kreitzer and Regehr, 2002; Wilson and Nicoll, 2002). Subsequent to this DSI has been shown to take place in a variety of other brain regions including between Purkinje cells and GABAergic cerebellar interneurons (Kreitzer and Regehr, 2001a), neocortical pyramidal cells (Trettel and Levine, 2003), and substantia nigra neurons (Yoshida et al., 2002). However, in hippocampal pyramidal neurons DSI is not affected by inhibition of FAAH indicating it is not mediated by anandamide (Kim and Alger, 2004) and there is accumulating evidence indicating 2-AG is the primary mediator of this response (reviewed by Alger, 2005). However, this does not eliminate the potential for additional signaling molecules as in another study, in rat hippocampal slices, DSI was not able to be blocked by inhibition of 2-AG synthesis (Chevaleyre and Castillo, 2003).

In addition to DSI cannabinoids have also been suggested to be involved in depolarization-induced suppression of excitation (DSE) with endocannabinoids acting at the CB1 receptors of excitatory neurotransmitter releasing neurons. This form of suppression is observed in glutamatergic synaptic inputs onto cerebellar Purkinje cells (Kreitzer and Regehr, 2001b; Maejima et al., 2001) as well as in dopaminergic neurons within the ventral tegmental region (Melis et al., 2004). No doubt further examples will emerge as research continues.

Several mechanisms are suggested for CB1 receptor mediation of DSI and DSE, outlined by Marco and Alain (2004). CB1 receptor-mediated inhibition of voltage-gated Ca^{2+} channels would lead to reduced Ca^{2+} entry and therefore diminished neurotransmitter release. Activation of potassium currents can further inhibit calcium influx as well as increasing the threshold for action potential firing. Studies with WIN55,212-2 indicate neurotransmitter release is inhibited by indirect inhibition of Ca^{2+} influx into the presynaptic terminal through activation of both presynaptic voltage-gated potassium channels and inwardly rectifying potassium channels. Inhibition of glutamate release showed no dependence on the cAMP/PKA or MAPK pathways (Azad et al., 2003; Daniel and Crepel, 2001; Daniel et al., 2004). Additional mechanisms are, however, likely as these effects on ion channels do not explain the commonly observed alterations in the frequency of synaptic currents.

Establishment of long-term potentiation (LTP) within the hippocampus is a crucial step for memory formation and is generated by repetitive stimulation of synapses leading to increased and long lasting synaptic plasticity. This enables more sensitive and rapid signaling through the synapse and is thought to be the underlying mechanism of memory formation. LTP in hippocampal slices is inhibited by anandamide acting at CB1 receptors (Collin et al., 1995; Terranova et al., 1995). This inhibition of LTP formation in hippocampal cultures is likely mediated through reducing the excitability of presynaptic neurons and hence reduced glutamate release to stimulate the postsynaptic terminal (Shen et al., 1996).

Anandamide is also capable of inducing reduced plasticity in the form of long-term depression in the adult striatum by acting at the presynaptic CB1 receptor (Ade and Lovinger, 2007; Gerdeman et al., 2002). Thus it is clear that anandamide plays a key role in the establishment of synaptic plasticity, however, the signaling pathways involved remain to be fully elucidated.

III. Cannabinoid Receptor 2

The second cannabinoid receptor was originally cloned from the human-derived leukemia cell line HL-60 (Munro et al., 1993) and has now been cloned from several diverse species including mouse (Shire et al., 1996), rat (Brown et al., 2002), Japanese pufferfish (Elphick, 2002), and zebrafish (Rodriguez-Martin et al., 2007).

In general, the CB2 receptor has a lower affinity for anandamide compared to the CB1 receptor. Human CB2 receptors expressed in AtT20, or COS, cells exhibit K_i values of 1940 and 1600 nM, respectively (Felder et al., 1995; Munro et al., 1993). This low affinity has led some investigators to suggest that 2-AG is the endogenous ligand for the CB2 receptor (Sugiura et al., 1999), however, numerous effects of anandamide on the CB2 receptor have been reported and the physiological relevance of this interaction remains to be established. In light of this low affinity, it is perhaps not surprising that there are considerably more data available on the activation of the CB2 receptor by 2-AG than by anandamide. However, in human CB2 receptor transfected CHO cells anandamide antagonizes the actions of 2-AG (Gonsiorek et al., 2000), thus *in vivo* anandamide may act to modulate the effects of 2-AG at this receptor.

A. Signaling pathways

The CB2 receptor, like CB1, couples with $G_{i/o}$ proteins when stimulated with anandamide, leading to inhibition of adenylate cyclase and reduced cAMP levels in transfected CHO cells and rat salivary glands (Felder et al., 1995; Gonsiorek et al., 2000; Hillard et al., 1999; Prestifilippo et al., 2006). These effects may be mediated entirely by G_i as the association with G_o is particularly low affinity in recombinant expression systems (Glass and Northup, 1999). As for the CB1 receptor, a downstream consequence of CB2 activation is the stimulation of p42/44 MAPK pathways and activation of Krox 24 (Bouaboula et al., 1996). Unlike CB1, CB2 receptors appear unable to modulate the activity of ion channels (Felder et al., 1995).

Some of the signaling pathways activated by the CB2 receptor may be cell, or species, specific. For example, anandamide stimulation of CB2 receptors expressed by calf pulmonary endothelial cells caused a transient increase in cytoplasmic calcium via PLC generation of IP_3 (Zoratti et al., 2003), however, these pathways are not activated by anandamide in AtT20 cells transfected with human CB2 receptor (Felder et al., 1995).

IV. Transient Receptor Potential Vanilloid 1

The transient receptor potential (TRP) family of proteins detects transient changes in the environment such as temperature and acidity and responds by altering their permeability to specific ions thereby modifying the membrane potential. Transient receptor potential vanilloid 1 (TRPV1) was the first of the temperature-activated TRP family to be cloned (Caterina et al., 1997) and is known to respond to capsaicin (Szallasi and Blumberg, 1999), various plant toxins (Szallasi and Blumberg, 1999), high temperatures (Caterina et al., 1997), low pH (Tominaga et al., 1998), and several inflammatory factors (De Petrocellis and Di Marzo, 2005; Tominaga and Caterina, 2004).

Capsaicin, the first identified TRPV1 ligand has a comparable structure to anandamide which led to investigation and subsequent reporting of anandamide as an additional ligand (Smart et al., 2000; Zygmunt et al., 1999). There has been some contention as to the labeling of anandamide as a TRPV1 agonist due to inconsistencies in the responses elicited by the ligand (reviewed in Starowicz et al., 2007). It seems likely however that this is an artifact of different experimental protocols and in particular sensitivity states of TRPV1. The phosphorylation state of the receptor alters the sensitivity threshold and hence the effective concentrations of anandamide or other stimulants required to elicit a response (De Petrocellis et al., 2001; Lee et al., 2005).

Many reports have described TRPV1 expression in varied tissues and cell types including neuronal cells of the central and peripheral nervous system (Birder et al., 2001; Caterina et al., 1997; Cristino et al., 2006; Mezey et al., 2000; Roberts et al., 2002; Toth et al., 2005), various subtypes of white blood cells (Chen et al., 2003; Heiner et al., 2003; Saunders et al., 2007), vascular endothelial cells (Golech et al., 2004), keratinocytes of the epidermis (Southall et al., 2003), beta-pancreatic cells (Akiba et al., 2004), and in cultured human lung and liver cells (Reilly et al., 2003). Anandamide is also synthesized by many peripheral and central cell types suggesting anandamide action on TRPV1 channels may have many widespread effects.

Unfortunately the specific pathways activated by TRPV1-mediated influx of Ca^{2+} are not yet known and are likely to vary depending on the individual intracellular proteome of the cells expressing TRPV1. So far most research into the functionality of the receptor has focused on a physiological response or detection of inward currents and increases in intracellular calcium, but no reports of the pathways linking these events. This work has, however, been crucial in assigning TRPV1 importance in several physiological processes. Certainly, signal transduction pathways generated by TRPV1 activation by heat are proving difficult to decipher and likely to vary considerably with regional and cellular differences (Rau et al., 2007).

A. In the nervous system

Anandamide activation of TRPV1 receptors has been implicated in pain detection and transmission (Starowicz et al., 2007). Similar to its actions at CB1 receptors, anandamide can exhibit both full agonist and partial agonist properties at TRPV1 receptors, depending on the expression system, species and models used (reviewed by Ross, 2003). Inward currents and accumulation of intracellular Ca^{2+} via anandamide stimulation of TRPV1 channels has been reported in DRG neurons (Jerman et al., 2002; Olah et al., 2002) and in mouse trigeminal neurons (Roberts et al., 2002). The DRG is a nodule of sensory neurons near the spinal cord which relay sensory signals to the CNS and activation of TRPV1 or CB1 receptors in this region produces analgesia. Stimulation of anandamide synthesis in the DRG and transfected cells was sufficient to stimulate TRPV1-mediated Ca^{2+} influx and was potentiated by inhibition of anandamide efflux from cells (van der Stelt et al., 2005). This work indicates that anandamide binds to an intracellular site on TRPV1 receptors where it may act prior to stimulation of CB1/CB2 receptors at extracellular binding sites.

An emerging role of TRPV1 in regulating movement has recently attracted attention. It appears that in the basal ganglia (a region important for controlling movement) anandamide activation of TRPV1 resulted in decreased dopamine release and reduced movement of rodents (de Lago et al., 2004; Lee et al., 2006; Tzavara et al., 2006). Conversely, in the substantia nigra anandamide increased spontaneous excitatory postsynaptic currents through presynaptic TRPV1 receptors and release of glutamate onto dopaminergic neurons (Marinelli et al., 2003). In agreement with this, anandamide increased NMDA-induced excitatory postsynaptic currents observed in dopaminergic synapses of the ventral tegmental region in rat slices are mediated by TRPV1 activation. Interestingly, when TRPV1 receptor activity was blocked the canonical inhibitory effect of anandamide acting through CB1 receptors was observed suggesting these pathways may act in competition (Melis et al., 2004).

B. In vasodilation and bronchoconstriction

In the human periphery anandamide most likely acts in an auto- and/or paracrine manner on TRPV1 receptors to cause vasodilatation of dermal microvasculature rather than as a circulating hormone (Movahed et al., 2005). Anandamide activates TRPV1 receptors to cause vasodilation at sub-micromolar doses (Andersson et al., 2002; Domenicali et al., 2005; Ho and Hiley, 2003; Ralevic et al., 2001; Zygmunt et al., 1999) and bronchoconstriction at higher concentrations (Andersson et al., 2002; Craib et al., 2001; Tucker et al., 2001). It is suggested that TRPV1-induced vasorelaxation is mediated through receptors at perivascular nerve endings inducing the release of calcitonin gene-related peptides via nitric oxide (NO) production. This in turn provokes hyperpolarization of smooth muscle and subsequent relaxation (Breyne et al., 2006; Mukhopadhyay et al., 2002; Zygmunt et al., 1999). In contrast to isolated tissue preparations, in vivo measurement of vasorelaxation via TRPV1 receptors requires high doses of anandamide and is short lived (Pacher et al., 2004), however, this may be due to rapid hydrolysis of anandamide reducing its effective concentration.

V. Evidence for Additional Receptors

Cannabinoids are generally of a highly lipophilic nature; this raises complications in their mechanisms of action as cannabinoid interactions with integral membrane proteins are not limited to their exposed surface. Anandamide may be able to pass through the plasma membrane by passive diffusion or active transport to access both the intracellular and extracellular domains of proteins as well as gaining access to their membrane spanning regions. Many of the effects of anandamide, which cannot be attributed to the cannabinoid receptors or TRPV1, appear to be mediated through direct interactions with a variety of proteins and modulation of their activity. The apparent promiscuity of anandamide may be attributed, at least in part, to its incorporation into and subtle alteration of the lipid bilayer. This would also impact the proteins within the modified region of bilayer, potentially modifying their activity state (Morris and Juranka, 2007; Oz, 2006).

A. Direct activation of ion channels

1. Voltage-gated Ca^{2+} channels

The voltage-gated Ca^{2+} channels are found in excitable cell types and generally become permeable to Ca^{2+} when the cell membrane is depolarized. The resultant influx of calcium leads to activation of specialized functions defined by the cell type; neurons fire action potentials and muscle cells contract. A growing body of evidence indicates that anandamide inhibition of voltage-gated Ca^{2+} channels can be mediated through direct interaction rather than via classical cannabinoid receptors. The use of inhibitors of anandamide uptake and release from the intracellular site of synthesis, as well as clamping of inside out membrane patches, indicates that anandamide's site of action may be intracellular (Chemin et al., 2001, 2007; Oz et al., 2000). This suggests the exciting possibility that anandamide elicits effects within the postsynaptic terminal where it is synthesized before traveling in a retrograde direction across the synapse to activate presynaptic CB1 receptors.

Anandamide appears to directly inhibit T-type Ca^{2+} channels in a variety of transfected and endogenous cell lines (Chemin et al., 2001) and P-type Ca^{2+} channels in rat Purkinje cells (Fisyunov et al., 2006) at nanomolar and low micromolar levels. At higher concentrations, anandamide causes inhibition of L-type Ca^{2+} channels and vasodilation in rat gastric arteries (Breyne et al., 2006). This corresponds well to the micromolar concentrations required to compete for binding to L-type Ca^{2+} channels (Johnson et al., 1993; Oz et al., 2000; Shimasue et al., 1996). It is likely that the CB1/CB2 and TRPV1-receptor independent component of vasorelaxation seen in small mesenteric rat arteries is mediated by direct inhibition of L-type calcium channels in a similar mechanism to that seen in the gastric arteries (Ho and Hiley, 2003).

In cells expressing the CB1 receptor, in addition to voltage-gated Ca^{2+} channels, the effects of direct interaction of anandamide with these channels can be masked by the action of anandamide at the CB1 receptor (Guo and Ikeda, 2004). It is unclear exactly how this additional inhibition of Ca^{2+} currents is physiologically significant but it is not difficult to appreciate that the ability of anandamide to modulate the activity of several types of calcium channel may enable this one signaling molecule to produce varying effects by acting on cell types exhibiting different proportions of these channels.

2. Voltage-gated sodium channels

Voltage-gated sodium channels are intricately involved in the initial phase of action potentials as they open rapidly to allow Na^+ influx and fast membrane depolarization (Rogers et al., 2006). Anandamide has been demonstrated to compete for direct binding to sodium channels in mouse brain synaptosomes (Nicholson et al., 2003) as well as inhibiting membrane depolarization and neurotransmitter release through voltage dependent sodium channel inhibition in the DRG (Kim et al., 2005), cortical neurons (Duan et al., 2007), and brain synaptosomes (Nicholson et al., 2003). Cannabinoid and TRPV1 receptor antagonists do not alter this inhibition (Duan et al., 2007) suggesting interaction with an unknown receptor or directly on the channels themselves.

3. Voltage-gated potassium channels

There are four main types of potassium channels: calcium activated, inwardly rectifying, tandem pore domain, and voltage gated. Anandamide inhibits activation of voltage-gated potassium channels at low micromolar concentrations in transfected and primary cell models (Oliver *et al.*, 2004; Poling *et al.*, 1996; Van den Bossche and Vanheel, 2000). Action at the potassium channels is generally rapid onset, PTX insensitive and independent of SR141716A treatment, indicating it is not mediated through the CB1/CB2 receptors (Poling *et al.*, 1996; Van den Bossche and Vanheel, 2000). It is possible that the inhibition of potassium channels is mediated by products of anandamide hydrolysis, as in cultured DRG neurons inhibition of FAAH inhibited the ability of anandamide to reduce potassium currents (Evans *et al.*, 2008). However, it is also possible that different subtypes of potassium channels have differential abilities to respond to different cannabinoids. Certainly, anandamide, methanandamide, Δ^9THC, arachidonic acid, and WIN55,212-2 are all capable of inhibiting potassium currents in various models, despite their structural differences (Oliver *et al.*, 2004; Poling *et al.*, 1996; Van den Bossche and Vanheel, 2000).

4. Task-1

The TASK-1 channel belongs to the tandem pore domain group of potassium channels and plays an important role in setting the neuronal resting membrane potential; they are constantly active at negative membrane potentials and allow passage of K^+ to the extracellular space driving the resting membrane potential to a more negative state. Inhibition of TASK-1 channels results in depolarization of the plasma membrane. Sub-micromolar concentrations of anandamide effectively blocked TASK-1 activity in transfected COS cells, rat cerebellar granule neurons, and rat pulmonary arteries (Gonczi *et al.*, 2006; Maingret *et al.*, 2001). The signaling pathway leading to this inhibition was not via the classical cannabinoid receptors and hence suggested to be a result of direct interaction of anandamide with the channel (Maingret *et al.*, 2001).

B. 5-HT$_{3A}$ receptors

The 5-HT$_3$ (serotonin type three receptor) is a neuronally expressed ligand-gated ion channel permeable to sodium (influx), potassium (efflux), and calcium (influx), generating a fast inward current when activated and resulting in neuronal excitation (Costall and Naylor, 2004). Anandamide is a noncompetitive inhibitor of 5-HT$_3$ receptors in rat nodose ganglion neurons, *Xenopus* oocytes, and HEK 293 cells expressing human 5-HT$_{3A}$ subunits (Barann *et al.*, 2002; Fan, 1995; Oz *et al.*, 2002). *In vivo* both WIN55,212-2 and CP55,940 appear to act through the same mechanism to elicit equivalent responses (Godlewski et al., 2003). It seems unlikely that direct inhibition is occurring given the length of the incubation required to achieve maximal inhibition in these reports, but the effect is not mediated by cannabinoid receptors, opioid receptors, PTX sensitive pathways, cAMP signaling, or changes in intracellular calcium. Discrepancies in the exceptionally high levels of anandamide required to inhibit 5-HT$_3$ receptors expressed in *Xenopus* oocytes and the more moderate concentrations required in rat neurons may be explained by the fact that the ability of anandamide to inhibit serotonin-induced currents through 5-HT$_{3A}$ receptors is highly dependent on the abundance of the receptor at the cell surface (Xiong et al., 2008).

C. Nicotinic acetylcholine receptors

There is a high degree of homology between the α7-nicotinic acetylcholine receptors (α7-nACh) and 5-HT$_3$ receptors. Activation of α7-nACh stimulates opening of an ion channel with selectivity for Ca^{2+} resulting in influx of this ion and increased neuronal excitability. In transfected *Xenopus* oocytes methanandamide inhibits α7-nACh current with an IC$_{50}$ in the nanomolar range (Oz et al., 2003). The same effect is seen in SH-EP1 cells transfected with α4β2-nACh receptors where anandamide exhibits a slightly lower efficacy (Spivak et al., 2007). Although this evidence may suggest the direct action of anandamide at nicotinic acetylcholine receptors, as with 5-HT$_{3A}$ the time frame for maximal inhibition is much longer then would be expected for a direct interaction, however, inhibition of α7-nACh current does not occur via the cannabinoid or TRPV1 receptors (Oz et al., 2003; Spivak et al., 2007), indicating an as yet undetermined mechanism or these receptors are more susceptible to lipid alterations induced by the lipophilic qualities of anandamide (Morris and Juranka, 2007; Oz, 2006).

D. Glycine receptors

Anandamide alters the properties of the inhibitory ionotropic glycine receptor, which allows influx of Cl$^-$ and generation of inhibitory postsynaptic potentials. Contradictory reports as to whether anandamide potentiates or inhibits glycine receptor activity are likely to be a function of the different models used. Hejazi and colleagues showed that in both rat ventral tegmental neurons and *Xenopus* oocytes expressing glycine receptors, anandamide concentration-dependently enhanced glycine-mediated currents (Hejazi et al., 2006), whereas in hippocampal pyramidal and Purkinje cerebellar neurons, anandamide inhibited glycine-induced currents (Lozovaya et al., 2005). The effects of anandamide on glycine signaling may not be limited to modifications of the glycine receptor as it has also been demonstrated to increase the activity of the GLYT1A glycine transporter at high concentrations (Pearlman et al., 2003). If this occurs physiologically its effects would be complex. Increased clearance of glycine from the synapse could cause a reduction in excitatory signals via reduced coactivation of the NMDA receptor or alternatively reduce inhibitory signals via the glycine receptor.

E. NMDA receptors

A well-established effect of anandamide acting through the CB1 receptor in the brain is inhibition of NMDA-induced calcium influx. A more poorly defined effect of anandamide is the enhancement of NMDA-induced calcium influx when the CB1 receptor is blocked, which possibly involves a direct interaction of anandamide with the NMDA receptor (Hampson et al., 1998). The concentrations of anandamide required for augmentation of NMDA-induced calcium influx varies in different brain regions, with nanomolar concentrations required in cortical slices but low micromolar concentrations in cerebellar slices (Hampson et al., 1998). In the hippocampus, although functional CB1 receptors are present, only the stimulatory effect of anandamide on NMDA receptors was observed (Hampson et al., 1998). However, in subsequent reports of anandamide enhancing NMDA-mediated Ca^{2+} influx, no direct action at NMDA receptors was indicated and the effects were consistently shown to be initiated by action at the CB1 receptor (Mendiguren and Pineda, 2004; Netzeband et al., 1999).

F. Peroxisome proliferator-activated receptors

Anandamide inhibits transcription of various genes independently of CB1, CB2, and TRPV1 receptors at concentrations above 5 μM in primary activated microglial cultures (Correa et al., 2008). This activity may be mediated by direct binding to transcriptional regulators as anandamide has also been shown to bind peroxisome proliferator-activated receptors (PPARs). Anandamide binds directly to and activates both PPARα (Sun et al., 2006, 2007) and PPARγ (Bouaboula et al., 2005) in the high micromolar range in vitro. At micromolar concentrations, anandamide induced expression of several PPARγ sensitive genes in 3T3-L1 fibroblasts (Bouaboula et al., 2005) and caused inhibition of interleukin-2 expression via PPARγ activation in primary splenocytes (Rockwell and Kaminski, 2004). The high concentrations necessary for anandamide binding and activation are similar to the required concentrations of fatty acids reported for binding to PPARs (Kliewer et al., 1997) and suggest it is more likely anandamide would act on PPARs near its site of synthesis where such high concentrations may be attainable.

G. Other GPCRs

The GPR55 receptor is a GPCR with little homology to the classical cannabinoid receptors. It was cloned almost a decade ago and has been detected in both the brain and peripheral tissues although a comprehensive investigation of its distribution is lacking (Sawzdargo et al., 1999).

Anandamide binds to GPR55 and activates coupling to GTPγS with an EC$_{50}$ of 18 nM (Drmota et al., 2004; Ryberg et al., 2007). Coupling of this GPCR to G$_{13}$ as well as G$_{12}$ and G$_q$ subtypes of G proteins has been indicated (Baker et al., 2006; Brown and Wise, 2001; Lauckner et al., 2008; Ryberg et al., 2007). At micromolar concentrations, anandamide stimulation of GPR55 increases intracellular Ca^{2+} in transiently transfected HEK 293 cells (Lauckner et al., 2008). The same response, when elicited by alternative cannabinoids, was dependant on G$_q$, G$_{12}$, PLC and RhoA activation and subsequent release of Ca^{2+} from internal stores by IP$_3$ (Lauckner et al., 2008) and similarly, in GPR55 stably transfected HEK 293 cells, anandamide stimulation resulted in activation of cdc42, rac1, and RhoA (Ryberg et al., 2007).

Recently, further insights into the signaling pathways involved in GPR55 stimulation by anandamide have been elucidated. Waldeck-Weiermair et al. (2008) describe a role for integrin clustering in CB1 and GPR55 receptor signaling; when integrins are unclustered, signaling via the CB1 receptor activates spleen tyrosine kinase (Syk), an inhibitor of PI3K and as GPR55 signaling is dependent on PI3K activity, when the CB1 receptor is activated GPR55 signaling is inhibited. The inhibition of GPR55 signaling is lifted when integrins are induced to form clusters and dissociate from the CB1 receptor (by addition of Mn^{2+} or removal of Ca^{2+} from the extracellular space). The authors suggest that this involves the physical movement of the CB1 receptor complex, including Syk, away from integrin-associated GPR55, allowing activation of PI3K which proceeds to activate PLC via a tyrosine kinase Bmx/Etk. PLC is responsible for the generation of IP$_3$ which stimulates release of Ca^{2+} from intracellular stores, subsequently activating the transcription factor NFAT. Upregulation of p42/44 MAPK occurs under both integrin clustering conditions but it is not clear if this is due to continued CB1 receptor signaling independent of integrin association, through GPR55 signaling pathways, or a combination of both (Waldeck-Weiermair et al., 2008). This work highlights that careful characterization of cell lines and models as well as experimental protocols are important in determining cell signaling pathways. In this case, the expression profile of other membrane proteins strongly influences the ability of GPR55 to respond to stimuli and is further modulated by the presence or absence of various ions during stimulation (Waldeck-Weiermair et al., 2008).

Anandamide has been suggested to mediate a non-CB1/CB2 or TRPV1 receptor-mediated vasorelaxation of blood vessels (White et al., 2001).

GPR55 has been suggested as a candidate for the "endothelial anandamide receptor" involved in vasodilatory effects (Hiley and Kaup, 2007; Offertaler *et al.*, 2003). Although few publications focusing on GPR55-mediated vasodilation are available, the results are controversial. Many of the vasodilatory effects mediated by the unidentified endothelial receptor are dependent on coupling to $G_{i/o}$ which has not been observed with GPR55. In addition, GPR55 knockout mice fails to exhibit any changes in basal blood pressure or heart rate, and isolated knockout blood vessels continue to exhibit vasodilatory responses to agonists of the receptor (Johns *et al.*, 2007). On the other hand Waldeck-Weiermair *et al.* (2008) saw an increase in the anandamide-induced intracellular Ca^{2+} response when GPR55 was over-expressed in endothelial cells and a corresponding decrease in response when GPR55 expression was downregulated with siRNA. This indicates that in this model, at least, GPR55 is the likely endothelial anandamide receptor. Although GPR55 may be involved in the vasorelaxation effects of anandamide there appears to still be an unidentified receptor or receptors that also mediate this effect.

Although only briefly noted by Overton and colleagues, anandamide, but not methanandamide, induced a mild response in a yeast reporter assay with GPR119, another orphan GPCR (Overton *et al.*, 2006). Additional evidence and characterization of this interaction may provide an alternative candidate for the endothelial receptor.

It is apparent that anandamide modulates the activity of many targets besides the better characterized interactions with CB1, CB2, and TRPV1 receptors. Many of these appear to be specific to the nervous system and functioning of neurons. As research progresses in this area, it would be expected that some of the currently suggested direct targets for anandamide action will be confirmed and others placed further down the signal transduction pathways stimulated by this molecule. The role of anandamide in altering lipid bilayer character should not be overlooked as a mechanism to achieve structural constraints and changes in the activity of integral membrane proteins, particularly given that the majority of putative anandamide receptors are ion channels.

VI. Concluding Remarks

A large proportion of work investigating the cell signaling pathways of cannabinoid receptors has been carried out in transfected cell models. This allows large numbers of cells to be generated and can help control for stimulation of other targets by comparisons with untransfected control cells as well as removing complications of multiple cell types found within tissues *in vivo*. However, while of great value these models have limitations.

Species differences in cell signaling are possible as well as the model used not expressing key factors important for signaling. It has been demonstrated that anandamide acts as a full agonist on the insect cell line Sf9 transfected with the human CB2 receptor in GTPγS experiments. However, the same experiment using CHO cells transfected with hCB2 indicated that anandamide acts only as a partial agonist (Gonsiorek *et al.*, 2000). Although cell models can suggest to us how anandamide may work *in vivo* the reality of its activity must be investigated directly. Furthermore, it cannot be taken for granted that an activity seen by anandamide within the brain can be directly transposed to the same activity on peripheral cells, or even from one region of the brain to another. Every different cell type has a specific expression profile which will influence its ability to respond to stimulations and hence the response detected by experimental methods.

There are several factors which may contribute to the varied responses of anandamide at its receptors. These include, but are probably not limited to, species differences, expression levels of the receptor and other potential receptors, the ability of the receptor to multiply the signal intracellularly and the presence of other competing ligands. Together this has caused much confusion over the role of anandamide *in vivo* and particular care must be observed when working with this ligand due to its inherent promiscuity and tendency to be broken down into further bioactive products.

Although the receptor signaling profile of anandamide remains complicated, and in many cases conflicting, considerable progress has been made in this field upon which to build. As we gain a better understanding of the widespread targets of anandamide we can proceed to view previous results more critically and design future protocols with this in mind. There is still much to be elucidated, confirmation of the putative anandamide receptors and identification of the structural binding sites of anandamide in particular as well as determining the interplay between anandamide and the CB2 receptor and if this is of physiological relevance.

REFERENCES

Abadji, V., Lin, S., Taha, G., Griffin, G., Stevenson, L. A., Pertwee, R. G., and Makriyannis, A. (1994). (R)-Methanandamide: A chiral novel anandamide possessing higher potency and metabolic atability. *J. Med. Chem.* **37,** 1889–1893.

Ade, K. K., and Lovinger, D. M. (2007). Anandamide regulates postnatal development of long-term synaptic plasticity in the rat dorsolateral striatum. *J. Neurosci.* **27,** 2403–2409.

Akiba, Y., Kato, S., Katsube, K., Nakamura, M., Takeuchi, K., Ishii, H., and Hibi, T. (2004). Transient receptor potential vanilloid subfamily 1 expressed in pancreatic islet beta cells modulates insulin secretion in rats. *Biochem. Biophys. Res. Commun.* **321,** 219–225.

Alger, B. E. (2002). Retrograde signaling in the regulation of synaptic transmission: Focus on endocannabinoids. *Prog. Neurobiol.* **68,** 247–286.

Alger, B. E. (2005). Endocannabinoid identification in the brain: Studies of breakdown lead to breakthrough, and there may be no hope. *Sci. Signal.* **2005**, pe51.

Andersson, D. A., Adner, M., Hogestatt, E. D., and Zygmunt, P. M. (2002). Mechanisms underlying tissue selectivity of anandamide and other vanilloid receptor agonists. *Mol. Pharmacol.* **62**, 705–713.

Azad, S. C., Eder, M., Marsicano, G., Lutz, B., Zieglgansberger, W., and Rammes, G. (2003). Activation of the cannabinoid receptor type 1 decreases glutamatergic and GABAergic synaptic transmission in the lateral amygdala of the mouse. *Learn. Mem.* **10**, 116–128.

Baker, D., Pryce, G., Davies, W. L., and Hiley, C. R. (2006). *In silico* patent searching reveals a new cannabinoid receptor. *Trends Pharmacol. Sci.* **27**, 1–4.

Barann, M., Molderings, G., Bruss, M., Bonisch, H., Urban, B. W., and Gothert, M. (2002). Direct inhibition by cannabinoids of human 5-HT3A receptors: Probable involvement of an allosteric modulatory site. *Br. J. Pharmacol.* **137**, 589–596.

Bash, R., Rubovitch, V., Gafni, M., and Sarne, Y. (2003). The stimulatory effect of cannabinoids on calcium uptake is mediated by Gs GTP-binding proteins and cAMP formation. *Neurosignals* **12**, 39–44.

Birder, L. A., Kanai, A. J., de Groat, W. C., Kiss, S., Nealen, M. L., Burke, N. E., Dineley, K. E., Watkins, S., Reynolds, I. J., and Caterina, M. J. (2001). Vanilloid receptor expression suggests a sensory role for urinary bladder epithelial cells. *Proc. Natl. Acad. Sci. USA* **98**, 13396–13401.

Bonhaus, D. W., Chang, L. K., Kwan, J., and Martin, G. R. (1998). Dual activation and inhibition of adenylyl cyclase by cannabinoid receptor agonists: Evidence for agonist-specific trafficking of intracellular responses. *J. Pharmacol. Exp. Ther.* **287**, 884–888.

Bouaboula, M., Bourrié, B., Rinaldi-Carmona, M., Shire, D., Fur, G. L., and Casellas, P. (1995). Stimulation of cannabinoid receptor CB1 induces krox-24 expression in human astrocytoma cells. *J. Biol. Chem.* **270**, 13973–13980.

Bouaboula, M., Poinot-Chazel, C., Marchand, J., Canat, X., Bourrie, B., Rinaldi-Carmona, M., Calandra, B., Fur, G., and Casellas, P. (1996). Signaling pathway associated with stimulation of CB2 peripheral cannabinoid receptor. Involvement of both mitogen-activated protein kinase and induction of Krox-24 expression. *Euro. J. Biochem.* **237**, 704–711. doi:10.1111/j.1432-1033.1996.0704p.x.

Bouaboula, M., Hilairet, S., Marchand, J., Fajas, L., Fur, G. L., and Casellas, P. (2005). Anandamide induced PPARg transcriptional activation and 3T3-L1 preadipocyte differentiation. *Eur. J. Pharmacol.* **517**, 174–181.

Breivogel, C. S., Selley, D. E., and Childers, S. R. (1998). Cannabinoid receptor agonist efficacy for stimulating [35S]GTPgamma S binding to rat cerebellar membranes correlates with agonist-induced decreases in GDP affinity. *J. Biol. Chem.* **273**, 16865–16873.

Breyne, J., Van De Voorde, J., and Vanheel, B. (2006). Characterization of the vasorelaxation to methanandamide in rat gastric arteries. *Can. J. Physiol. Pharmacol.* **84**, 1121–1132.

Brown, A., and Wise, A. (2001). "Identification of Modulators of GPR55 Activity." Vol. WO200186305. GlaxoSmithKline, USA.

Brown, S. M., Wager-Miller, J., and Mackie, K. (2002). Cloning and molecular characterization of the rat CB2 cannabinoid receptor. *Biochim. Biophys. Acta* **1576**, 255–264.

Brown, S. P., Safo, P. K., and Regehr, W. G. (2004). Endocannabinoids inhibit transmission at granule cell to Purkinje cell synapses by modulating three types of presynaptic calcium channels. *J. Neurosci.* **24**, 5623–5631.

Burkey, T. H., Quock, R. M., Consroe, P., Ehlert, F. J., Hosohata, Y., Roeske, W. R., and Yamamura, H. I. (1997). Relative efficacies of cannabinoid CB1 receptor agonists in the mouse brain. *Eur. J. Pharmacol.* **336**, 295–298.

Caterina, M. J., Schumacher, M. A., Tominaga, M., Rosen, T. A., Levine, J. D., and Julius, D. (1997). The capsaicin receptor: A heat-activated ion channel in the pain pathway. *Nature* **389,** 816–824.
Catterall, W. A., Perez-Reyes, E., Snutch, T. P., and Striessnig, J. (2005). International union of pharmacology. XLVIII. Nomenclature and structure-function relationships of voltage-gated calcium channels. *Pharmacol. Rev.* **57,** 411–425.
Chemin, J., Monteil, A., Perez-Reyes, E., Nargeot, J., and Lory, P. (2001). Direct inhibition of T-type calcium channels by the endogenous cannabinoid anandamide. *EMBO J.* **20,** 7033–7040.
Chemin, J., Nargeot, J., and Lory, P. (2007). Chemical determinants involved in anandamide-induced inhibition of T-type calcium channels. *J. Biol. Chem.* **282,** 2314–2323.
Chen, C., Lee, S. T., Wu, W. T., Fu, W., Ho, F., and Lin, W. W. (2003). Signal transduction for inhibition of inducible nitric oxide synthase and cyclooxygenase-2 induction by capsaicin and related analogs in macrophages. *Br. J. Pharmacol.* **140,** 1077–1087.
Chevaleyre, V., and Castillo, P. E. (2003). Heterosynaptic LTD of hippocampal GABAergic synapses: A novel role of endocannabinoids in regulating excitability. *Neuron* **38,** 461–472.
Childers, S. R. (2006). Activation of G-proteins in brain by endogenous and exogenous cannabinoids. *AAPS J.* **8,** E112–E117.
Childers, S. R., Sexton, T., and Roy, M. B. (1994). Effects of anandamide on cannabinoid receptors in rat brain membranes. *Biochem. Pharmacol.* **47,** 711–715.
Collin, C., Devane, W. A., Dahl, D., Lee, C. J., Axelrod, J., and Alkon, D. L. (1995). Long-term synaptic transformation of hippocampal CA1 gamma-aminobutyric acid synapses and the effect of anandamide. *Proc. Natl. Acad. Sci. USA* **92,** 10167–10171.
Correa, F., Docagne, F., Clemente, D., Mestre, L., Becker, C., and Guaza, C. (2008). Anandamide inhibits IL-12p40 production by acting on the promoter repressor element GA-12: Possible involvement of the COX-2 metabolite prostamide E2. *Biochem. J.* **409,** 761–770.
Costall, B., and Naylor, R. J. (2004). 5-HT3 receptors. *Curr. Drug Targ.* **3,** 27–37.
Craib, S. J., Ellington, H. C., Pertwee, R. G., and Ross, R. A. (2001). A possible role of lipoxygenase in the activation of vanilloid receptors by anandamide in the guinea-pig bronchus. *Br. J. Pharmacol.* **134,** 30–37.
Cristino, L., de Petrocellis, L., Pryce, G., Baker, D., Guglielmotti, V., and Di Marzo, V. (2006). Immunohistochemical localization of cannabinoid type 1 and vanilloid transient receptor potential vanilloid type 1 receptors in the mouse brain. *Neuroscience* **139,** 1405–1415.
Daniel, H., and Crepel, F. (2001). Control of Ca^{2+} influx by cannabinoid and metabotropic glutamate receptors in rat cerebellar cortex requires K^+ channels. *J. Physiol.* **537,** 793–800.
Daniel, H., Rancillac, A., and Crepel, F. (2004). Mechanisms underlying cannabinoid inhibition of presynaptic Ca^{2+} influx at parallel fibre synapses of the rat cerebellum. *J. Physiol.* **557,** 159–174.
de Lago, E., de Miguel, R., Lastres-Becker, I., Ramos, J. A., and Fernández-Ruiz, J. (2004). Involvement of vanilloid-like receptors in the effects of anandamide on motor behavior and nigrostriatal dopaminergic activity: *In vivo* and *in vitro* evidence. *Brain Res.* **1007,** 152–159.
De Petrocellis, L., and Di Marzo, V. (2005). Lipids as regulators of the activity of transient receptor potential type V1 (TRPV1) channels. *Life Sci.* **77,** 1651–1666.
De Petrocellis, L., Harrison, S., Bisogno, T., Tognetto, M., Brandi, I., Smith, G. D., Creminon, C., Davis, J. B., Geppetti, P., and Di Marzo, V. (2001). The vanilloid

receptor (VR1)-mediated effects of anandamide are potently enhanced by the cAMP-dependent protein kinase. *J. Neurochem.* **77,** 1660–1663.

Derkinderen, P., Toutant, M., Burgaya, F., Le Bert, M., Siciliano, J. C., de Franciscis, V., Gelman, M., and Girault, J. A. (1996). Regulation of a neuronal form of focal adhesion kinase by anandamide. *Science* **273,** 1719–1722.

Deutsch, D. G. (2005). Design of on-target FAAH inhibitors. *Chem. Biol.* **12,** 1157–1158.

Devane, W. A., Hanus, L., Breuer, A., Pertwee, R. G., Stevenson, L. A., Griffin, G., Gibson, D., Mandelbaum, A., Etinger, A., and Mechoulam, R. (1992). Isolation and structure of a brain constituent that binds to the cannabinoid receptor. *Science* **258,** 1946–1949.

Domenicali, M., Ros, J., Fernandez-Varo, G., Cejudo-Martin, P., Crespo, M., Morales-Ruiz, M., Briones, A. M., Campistol, J. M., Arroyo, V., Vila, E., Rodes, J., and Jimenez, W. (2005). Increased anandamide induced relaxation in mesenteric arteries of cirrhotic rats: Role of cannabinoid and vanilloid receptors. *Gut* **54,** 522–527.

Drmota, E., Greasley, P., and Groblewski, T. (2004). "Screening Assays for Cannabinoid-Ligand Type Modulators." Vol. WO2005074844. AstraZeneca, USA.

Duan, Y., Zheng, J., and Nicholson, R. A. (2007). Vanilloid (subtype 1) receptor-modulatory drugs inhibit [3H]batrachotoxinin-A 20-a-benzoate binding to Na^+ channels. *Basic Clin. Pharmacol. Toxicol.* **100,** 91–95.

Elphick, M. R. (2002). Evolution of cannabinoid receptors in vertebrates: Identification of a CB2 gene in the puffer fish Fugu rubripes. *Biol. Bull.* **202,** 104–107.

Evans, R. M., Wease, K. N., MacDonald, C. J., Khairy, H. A., Ross, R. A., and Scott, R. H. (2008). Modulation of sensory neuron potassium conductances by anandamide indicates roles for metabolites. *Br. J Pharmacol.* **154,** 480–492.

Fan, P. (1995). Cannabinoid agonists inhibit the activation of 5-HT3 receptors in rat nodose ganglion neurons. *J. Neurophysiol.* **73,** 907–910.

Felder, C. C., Briley, E. M., Axelrod, J., Simpson, J. T., Mackie, K., and Devane, W. A. (1993). Anandamide, an endogenous cannabimimetic eicosanoid, binds to the cloned human cannabinoid receptor and stimulates receptor-mediated signal transduction. *Proc. Natl. Acad. Sci. USA* **90,** 7656–7660.

Felder, C. C., Joyce, K. E., Briley, E. M., Mansouri, J., Mackie, K., Blond, O., Lai, Y., Ma, A. L., and Mitchell, R. L. (1995). Comparison of the pharmacology and signal transduction of the human cannabinoid CB1 and CB2 receptors. *Mol. Pharmacol.* **48,** 443–450.

Fisyunov, A., Tsintsadze, V., Min, R., Burnashev, N., and Lozovaya, N. (2006). Cannabinoids modulate the P-type high-voltage-activated calcium currents in Purkinje neurons. *J. Neurophysiol.* **96,** 1267–1277.

Flygare, J., and Sander, B. (2008). The endocannabinoid system in cancer—Potential therapeutic target? *Semin. Cancer Biol.* **18,** 176–189.

Gaoni, Y., and Mechoulam, R. (1964). Isolation, structure, and partial synthesis of an active constituent of Hashish. *J. Am. Chem. Soc.* **86,** 1646–1647.

Gebremedhin, D., Lange, A. R., Campbell, W. B., Hillard, C. J., and Harder, D. R. (1999). Cannabinoid CB1 receptor of cat cerebral arterial muscle functions to inhibit L-type Ca^{2+} channel current. *Am. J. Physiol.* **276**(6 Pt 2), H2085–H2093.

Gerdeman, G. L., Ronesi, J., and Lovinger, D. M. (2002). Postsynaptic endocannabinoid release is critical to long-term depression in the striatum. *Nat. Neurosci.* **5,** 446–451.

Giuffrida, A., Parsons, L. H., Kerr, T. M., de Fonseca, F. R., Navarro, M., and Piomelli, D. (1999). Dopamine activation of endogenous cannabinoid signaling in dorsal striatum. *Nat. Neurosci.* **2,** 358–363.

Glass, M., and Felder, C. C. (1997). Concurrent stimulation of cannabinoid CB1 and dopamine D2 receptors augments cAMP accumulation in striatal neurons: Evidence for a Gs linkage to the CB1 receptor. *J. Neurosci.* **17,** 5327–5333.

Glass, M., and Northup, J. K. (1999). Agonist selective regulation of G proteins by cannabinoid CB1 and CB2 receptors. *Mol. Pharmacol.* **56,** 1362–1369.

Godlewski, G., Gothert, M., and Malinowska, B. (2003). Cannabinoid receptor-independent inhibition by cannabinoid agonists of the peripheral 5-HT3 receptor-mediatedvon Bezold-Jarisch reflex. *Br. J. Pharmacol.* **138,** 767–774.

Golech, S. A., McCarron, R. M., Chen, Y., Bembry, J., Lenz, F., Mechoulam, R., Shohami, E., and Spatz, M. (2004). Human brain endothelium: Coexpression and function of vanilloid and endocannabinoid receptors. *Brain Res. Mol. Brain Res.* **132,** 87–92.

Gonczi, M., Szentandrassy, N., Johnson, I. T., Heagerty, A. M., and Weston, A. H. (2006). Investigation of the role of TASK-2 channels in rat pulmonary arteries; pharmacological and functional studies following RNA interference procedures. *Br. J. Pharmacol.* **147,** 496–505.

Gonsiorek, W., Lunn, C., Fan, X., Narula, S., Lundell, D., and Hipkin, R. W. (2000). Endocannabinoid 2-arachidonyl glycerol is a full agonist through human type 2 cannabinoid receptor: Antagonism by anandamide. *Mol. Pharmacol.* **57,** 1045–1050.

Graham, E. S., Ball, N., Scotter, E. L., Narayan, P., Dragunow, M., and Glass, M. (2006). Induction of Krox-24 by endogenous cannabinoid type 1 receptors in Neuro2A cells is mediated by the MEK-ERK MAPK pathway and is suppressed by the phosphatidylinositol 3-kinase pathway. *J. Biol. Chem.* **281,** 29085–29095.

Guo, J., and Ikeda, S. R. (2004). Endocannabinoids modulate N-type calcium channels and G-protein-coupled inwardly rectifying potassium channels via CB1 cannabinoid receptors heterologously expressed in mammalian neurons. *Mol. Pharmacol.* **65,** 665–674.

Gustafsson, K., Christensson, B., Sander, B., and Flygare, J. (2006). Cannabinoid receptor-mediated apoptosis induced by R(+)-methanandamide and Win55,212-2 is associated with ceramide accumulation and p38 activation in mantle cell lymphoma. *Mol. Pharmacol.* **70,** 1612–1620.

Guzmán, M., and Sánchez, C. (1999). Effects of cannabinoids on energy metabolism. *Life Sci.* **65,** 657–664.

Hampson, A. J., Bornheim, L. M., Scanziani, M., Yost, C. S., Gray, A. T., Hansen, B. M., Leonoudakis, D. J., and Bickler, P. E. (1998). Dual effects of anandamide on NMDA receptor-mediated responses and neurotransmission. *J. Neurochem.* **70,** 671–676.

Hart, S., Fischer, O. M., and Ullrich, A. (2004). Cannabinoids induce cancer cell proliferation via tumor necrosis factor a-converting enzyme (TACE/ADAM17)-mediated transactivation of the epidermal growth factor receptor. *Cancer Res.* **64,** 1943–1950.

Hashimotodani, Y., Ohno-Shosaku, T., and Kano, M. (2007). Presynaptic monoacylglycerol lipase activity determines basal endocannabinoid tone and terminates retrograde endocannabinoid. *J. Neurosci.* **27,** 1211–1219.

Heiner, I., Eisfeld, J., Halaszovich, C. R., Wehage, E., Jungling, E., Zitt, C., and Luckhoff, A. (2003). Expression profile of the transient receptor potential (TRP) family in neutrophil granulocytes: Evidence for currents through long TRP channel 2 induced by ADP-ribose and NAD. *Biochem. J.* **371,** 1045–1053.

Hejazi, N., Zhou, C., Oz, M., Sun, H., Ye, J. H., and Zhang, L. (2006). Delta9-tetrahydrocannabinol and endogenous cannabinoid anandamide directly potentiate the function of glycine receptors. *Mol. Pharmacol.* **69,** 991–997.

Hiley, C. R., and Kaup, S. S. (2007). GPR55 and the vascular receptors for cannabinoids. *Br. J. Pharmacol.* **152,** 559–561.

Hillard, C. J., Manna, S., Greenberg, M. J., DiCamelli, R., Ross, R. A., Stevenson, L. A., Murphy, V., Pertwee, R. G., and Campbell, W. B. (1999). Synthesis and characterization of potent and selective agonists of the neuronal cannabinoid receptor (CB1). *J. Pharmacol. Exp. Ther.* **289,** 1427–1433.

Ho, V. W. S., and Hiley, R. C. (2003). Endothelium-independent relaxation to cannabinoids in rat-isolated mesenteric artery and role of Ca^{2+} influx. *Br. J. Pharmacol.* **139**, 585–597.

Huang, C. C., Lo, S. W., and Hsu, K. S. (2001). Presynaptic mechanisms underlying cannabinoid inhibition of excitatory synaptic transmission in rat striatal neurons. *J. Physiol.* **532**, 731–748.

Iuvone, T., Esposito, G., De Filippis, D., Bisogno, T., Petrosino, S., Scuderi, C., Di Marzo, V., Steardo, L., and Endocannabinoid Research Group. (2007). Cannabinoid CB1 receptor stimulation affords neuroprotection in MPTP-induced neurotoxicity by attenuating S100B up-regulation *in vitro. J. Mol. Med.* **85**, 1379–1392.

Jarrahian, A., Watts, V. J., and Barker, E. L. (2004). D2 Dopamine receptors modulate Ga-subunit coupling of the CB1 cannabinoid receptor. *J. Pharmacol. Exp. Ther.* **308**, 880–886.

Jerman, J. C., Gray, J., Brough, S. J., Ooi, L., Owen, D., Davis, J. B., and Smart, D. (2002). Comparison of effects of anandamide at recombinant and endogenous rat vanilloid receptors. *Br. J. Anaesth.* **89**, 882–887.

Johns, D. G., Behm, D. J., Walker, D. J., Ao, Z., Shapland, E. M., Daniels, D. A., Riddick, M., Dowell, S., Staton, P. C., Green, P., Shabon, U., Bao, W., et al. (2007). The novel endocannabinoid receptor GPR55 is activated by atypical cannabinoids but does not mediate their vasodilator effects. *Brit. J. Pharmacol.* **152**, 825–831.

Johnson, D. E., Heald, S. L., Dally, R. D., and Janis, R. A. (1993). Isolation, identification and synthesis of an endogenous arachidonic amide that inhibits calcium channel antagonist 1,4-dihydropyridine binding. *Prostaglandins Leukot. Essent. Fatty Acids* **48**, 429–437.

Kim, J., and Alger, B. E. (2004). Inhibition of cyclooxygenase-2 potentiates retrograde endocannabinoid effects in hippocampus. *Nat. Neurosci.* **7**, 697–698.

Kim, H. I., Kim, T. H., Shin, Y. K., Lee, C. S., Park, M., and Song, J. H. (2005). Anandamide suppression of Na^+ currents in rat dorsal root ganglion neurons. *Brain Res.* **1062**, 39–47.

Kliewer, S. A., Sundseth, S. S., Jones, S. A., Brown, P. J., Wisely, G. B., Koble, C. S., Devchand, P., Wahli, W., Willson, T. M., Lenhard, J. M., and Lehmann, J. M. (1997). Fatty acids and eicosanoids regulate gene expression through direct interactions with peroxisome proliferator-activated receptors alpha and gamma. *Proc. Natl. Acad. Sci. USA* **94**, 4318–4323.

Kreitzer, A. C., and Regehr, W. G. (2001a). Cerebellar depolarization-induced suppression of inhibition is mediated by endogenous cannabinoids. *J. Neurosci.* **21**, 174RC.

Kreitzer, A. C., and Regehr, W. G. (2001b). Retrograde inhibition of presynaptic calcium influx by endogenous cannabinoids at excitatory synapses onto Purkinje cells. *Neuron* **29**, 717–727.

Kreitzer, A. C., and Regehr, W. G. (2002). Retrograde signaling by endocannabinoids. *Curr. Opin. Neurobiol.* **12**, 324–330.

Kuzhikandathil, E. V., Yu, W., and Oxford, G. S. (1998). Human Dopamine D3 and D2L receptors couple to inward rectifier potassium channels in mammalian cell lines. *Mol. Cell. Neurosci.* **12**, 390–402.

Lauckner, J. E., Hille, B., and Mackie, K. (2005). The cannabinoid agonist WIN55,212-2 increases intracellular calcium via CB1 receptor coupling to Gq/11 G proteins. *Proc. Natl. Acad. Sci. USA* **102**, 19144–19149.

Lauckner, J. E., Jensen, J. B., Chen, H.-Y., Lu, H.-C., Hille, B., and Mackie, K. (2008). GPR55 is a cannabinoid receptor that increases intracellular calcium and inhibits M current. *Proc. Natl. Acad. Sci. USA* **105**, 2699–2704.

Lee, S. Y., Lee, J. H., Kang, K. K., Hwang, S. Y., Choi, K. D., and Oh, U. (2005). Sensitization of vanilloid receptor involves an increase in the phosphorylated form of the channel. *Arch. Pharm. Res.* **28**, 405–412.

Lee, J., Di Marzo, V., and Brotchie, J. M. (2006). A role for vanilloid receptor 1 (TRPV1) and endocannabinnoid signalling in the regulation of spontaneous and L-DOPA induced locomotion in normal and reserpine-treated rats. *Neuropharmacology* **51**, 557–565.

Liang, Y. C., Huang, C. C., Hsu, K. S., and Takahashi, T. (2004). Cannabinoid-induced presynaptic inhibition at the primary afferent trigeminal synapse of juvenile rat brainstem slices. *J. Physiol.* **555**, 85–96.

Liu, J., Gao, B., Mirshahi, F., Sanyal, A. J., Khanolkar, A. D., Makriyannis, A., and Kunos, G. (2000). Functional CB1 cannabinoid receptors in human vascular endothelial cells. *Biochem. J.* **346**, 835–840.

Lozovaya, N., Yatsenko, N., Beketov, A., Tsintsadze, T., and Burnashev, N. (2005). Glycine receptors in CNS neurons as a target for nonretrograde action of cannabinoids. *J. Neurosci.* **25**, 7499–7506.

Mackie, K., Devane, W. A., and Hille, B. (1993). Anandamide, an endogenous cannabinoid, inhibits calcium currents as a partial agonist in N18 neuroblastoma cells. *Mol. Pharmacol.* **44**, 498–503.

Mackie, K., Lai, Y., Westenbroek, R., and Mitchell, R. (1995). Cannabinoids activate an inwardly rectifying potassium conductance and inhibit Q-type calcium currents in AtT20 cells transfected with rat brain cannabinoid receptor. *J. Neurosci.* **15**, 6552–6561.

Maejima, T., Hashimoto, K., Yoshida, T., Aiba, A., and Kano, M. (2001). Presynaptic inhibition caused by retrograde signal from metabotropic glutamate to cannabinoid receptors. *Neuron* **31**, 463–475.

Magistretti, P. J., and Pellerin, L. (1999). Astrocytes couple synaptic activity to glucose utilization in the brain. *News. Physiol. Sci.* **14**, 177–182.

Maingret, F., Patel, A. J., Lazdunski, M., and Honore, E. (2001). The endocannabinoid anandamide is a direct and selective blocker of the background K^+ channel TASK-1. *EMBO J.* **20**, 47–54.

Marco, D. A., and Alain, M. (2004). Endocannabinoid-mediated short-term synaptic plasticity: Depolarization-induced suppression of inhibition (DSI) and depolarization-induced suppression of excitation (DSE). *Br. J. Pharmacol.* **142**, 9–19.

Marinelli, S., Di Marzo, V., Berretta, N., Matias, I., Maccarrone, M., Bernardi, G., and Mercuri, N. B. (2003). Presynaptic facilitation of glutamatergic synapses to dopaminergic neurons of the rat substantia nigra by endogenous stimulation of vanilloid receptors. *J. Neurosci.* **23**, 3136–3144.

Matsuda, L. A., Lolait, S. J., Brownstein, M. J., Young, A. C., and Bonner, T. I. (1990). Structure of a cannabinoid receptor and functional expression of the cloned cDNA. *Nature* **346**, 561–564.

McAllister, S. D., Griffin, G., Satin, L. S., and Abood, M. E. (1999). Cannabinoid receptors can activate and inhibit G protein-coupled inwardly rectifying potassium channels in a *Xenopus* oocyte expression system. *J. Pharmacol. Exp. Ther.* **291**, 618–626.

McIntosh, B. T., Hudson, B., Yegorova, S., Jollimore, C. A. B., and Kelly, M. E. M. (2007). Agonist-dependent cannabinoid receptor signalling in human trabecular meshwork cells. *Br. J. Pharmacol.* **152**, 1111–1120.

Mechoulam, R., Ben-Shabat, S., Hanus, L., Ligumsky, M., Kaminski, N. E., Schatz, A. R., Gopher, A., Almog, S., Martin, B. R., Compton, D. R., Pertwee, R. G., Griffin, G., *et al.* (1995). Identification of an endogenous 2-monoglyceride, present in canine gut, that binds to cannabinoid receptors. *Biochem. Pharmacol.* **50**, 83–90.

Melck, D., Rueda, D., Galve-Roperh, I., De Petrocellis, L., Guzman, M., and Di Marzo, V. (1999). Involvement of the cAMP/protein kinase A pathway and of mitogen-activated protein kinase in the anti-proliferative effects of anandamide in human breast cancer cells. *FEBS Lett.* **463**, 235–240.

Melck, D., De Petrocellis, L., Orlando, P., Bisogno, T., Laezza, C., Bifulco, M., and Di Marzo, V. (2000). Suppression of nerve growth factor Trk receptors and prolactin receptors by endocannabinoids leads to inhibition of human breast and prostate cancer cell proliferation. *Endocrinology* **141,** 118–126.

Melis, M., Pistis, M., Perra, S., Muntoni, A. L., Pillolla, G., and Gessa, G. L. (2004). Endocannabinoids mediate presynaptic inhibition of glutamatergic transmission in rat ventral tegmental area Dopamine neurons through activation of CB1 receptors. *J. Neurosci.* **24,** 53–62.

Mendiguren, A., and Pineda, J. (2004). Cannabinoids enhance N-methyl-aspartate-induced excitation of locus coeruleus neurons by CB1 receptors in rat brain slices. *Neurosci. Lett.* **363,** 1–5.

Mezey, E., Toth, Z. E., Cortright, D. N., Arzubi, M. K., Krause, J. E., Elde, R., Guo, A., Blumberg, P. M., and Szallasi, A. (2000). Distribution of mRNA for vanilloid receptor subtype 1 (VR1), and VR1-like immunoreactivity, in the central nervous system of the rat and human. *Proc. Natl. Acad. Sci. USA* **97,** 3655–3660.

Morris, C. E., and Juranka, P. F. (2007). Lipid stress at play: Mechanosensitivity of voltage-gated channels. *Curr. Top. Membr.* **59,** 297–338.

Movahed, P., Evilevitch, V., Andersson, T. L. G., Jonsson, B. A. G., Wollmer, P., Zygmunt, P. M., and Hogestatt, E. D. (2005). Vascular effects of anandamide and N-acylvanillylamines in the human forearm and skin microcirculation. *Br. J. Pharmacol.* **146,** 171–179.

Mukhopadhyay, S., and Howlett, A. C. (2005). Chemically distinct ligands promote differential CB1 cannabinoid receptor-Gi protein interactions. *Mol. Pharmacol.* **67,** 2016–2024.

Mukhopadhyay, S., Chapnick, B. M., and Howlett, A. C. (2002). Anandamide-induced vasorelaxation in rabbit aortic rings has two components: G protein dependent and independent. *Am. J. Physiol. Heart Circ. Physiol.* **282,** 2046–2054.

Munro, S., Thomas, K. L., and Abu-Shaar, M. (1993). Molecular characterization of a peripheral receptor for cannabinoids. *Nature* **365,** 61–65.

Nah, S.-Y., Saya, D., and Vogel, Z. (1993). Cannabinoids inhibit agonist-stimulated formation of inositol phosphates in rat hippocampal cultures. *Eur. J. Pharmacol. (Mol. Pharmacol.)* **246,** 19–24.

Navarrete, M., and Araque, A. (2008). Endocannabinoids mediate neuron-astrocyte communication. *Neuron* **57,** 883–893.

Netzeband, J. G., Conroy, S. M., Parsons, K. L., and Gruol, D. L. (1999). Cannabinoids enhance NMDA-elicited Ca^{2+} signals in cerebellar granule neurons in culture. *J. Neurosci.* **19,** 8765–8777.

Nicholson, R. A., Liao, C., Zheng, J., David, L. S., Coyne, L., Errington, A. C., Singh, G., and Lees, G. (2003). Sodium channel inhibition by anandamide and synthetic cannabimimetics in brain. *Brain Res.* **978,** 194–204.

Offertaler, L., Mo, F. M., Batkai, S., Liu, J., Begg, M., Razdan, R. K., Martin, B. R., Bukoski, R. D., and Kunos, G. (2003). Selective ligands and cellular effectors of a G protein-coupled endothelial cannabinoid receptor. *Mol. Pharmacol.* **63,** 699–705.

Ohno-Shosaku, T., Maejima, T., and Kano, M. (2001). Endogenous cannabinoids mediate retrograde signals from depolarized postsynaptic neurons to presynaptic terminals. *Neuron* **29,** 729–738.

Okuhara, D. Y., Lee, J. M., Beck, S. G., and Muma, N. A. (1996). Differential immunohistochemical labeling of G_s, $G_{i1\ and\ 2}$, and G_o-subunits in rat forebrain. *Synapse* **23,** 246–257.

Olah, Z., Karai, L., and Iadarola, M. J. (2002). Protein kinase Cα is required for vanilloid receptor 1 activation. Evidence for multiple signaling pathways. *J. Biol. Chem.* **277,** 35752–35759.

Oliver, D., Lien, C. C., Soom, M., Baukrowitz, T., Jonas, P., and Fakler, B. (2004). Functional conversion between A-type and delayed rectifier K^+ channels by membrane lipids. *Science* **304,** 265–270.

Overton, H. A., Babbs, A. J., Doel, S. M., Fyfe, M. C. T., Gardner, L. S., Griffin, G., Jackson, H. C., Procter, M. J., Rasamison, C. M., Tang-Christensen, M., Widdowson, P. S., Williams, G. M., et al. (2006). Deorphanization of a G protein-coupled receptor for oleoylethanolamide and its use in the discovery of small-molecule hypophagic agents. *Cell Metab.* **3,** 167–175.

Oz, M. (2006). Receptor-independent actions of cannabinoids on cell membranes: Focus on endocannabinoids. *Pharmacol. Ther.* **111,** 114–144.

Oz, M., Tchugunova, Y. B., and Dunn, S. M. (2000). Endogenous cannabinoid anandamide directly inhibits voltage-dependent Ca(2+) fluxes in rabbit T-tubule membranes. *Eur. J. Pharmacol.* **404,** 13–20.

Oz, M., Zhang, L., and Morales, M. (2002). Endogenous cannabinoid, anandamide, acts as a noncompetitive inhibitor on 5-HT3 receptor-mediated responses in *Xenopus* oocytes. *Synapse* **46,** 150–156.

Oz, M., Ravindran, A., Diaz-Ruiz, O., Zhang, L., and Morales, M. (2003). The endogenous cannabinoid anandamide Inhibits a7 nicotinic acetylcholine receptor-mediated responses in *Xenopus* oocytes. *J. Pharmacol. Exp. Ther.* **306,** 1003–1010.

Pacher, P., Batkai, S., and Kunos, G. (2004). Haemodynamic profile and responsiveness to anandamide of TRPV1 receptor knock-out mice. *J. Physiol.* **558,** 647–657.

Paradisi, A., Pasquariello, N., Barcaroli, D., and Maccarrone, M. (2008). Anandamide regulates keratinocyte differentiation by inducing DNA methylation in a CB1 receptor-dependent manner. *J. Biol. Chem.* **283,** 6005–6012.

Pearlman, R. J., Aubrey, K. R., and Vandenberg, R. J. (2003). Arachidonic acid and anandamide have opposite modulatory actions at the glycine transporter, GLYT1a. *J. Neurochem.* **84,** 592–601.

Pertwee, R. G. (1997). Pharmacology of cannabinoid CB1 and CB2 receptors. *Pharmacol. Ther.* **74,** 129–180.

Pinto, J. C., Potie, F., Rice, K. C., Boring, D., Johnson, M. R., Evans, D. M., Wilken, G. H., Cantrell, C. H., and Howlett, A. C. (1994). Cannabinoid receptor binding and agonist activity of amides and esters of arachidonic acid. *Mol. Pharmacol.* **46,** 516–522.

Poling, J. S., Rogawski, M. A., Salem, N. Jr, and Vicini, S. (1996). Anandamide, an endogenous cannabinoid, inhibits Shaker-related voltage-gated K^+ channels. *Neuropharmacol.* **35,** 983–991.

Portella, G., Laezza, C., Laccetti, P., De Petrocellis, L., Di Marzo, V., and Bifulco, M. (2003). Inhibitory effects of cannabinoid CB1 receptor stimulation on tumor growth and metastatic spreading: Actions on signals involved in angiogenesis and metastasis. *FASEB J.* **17**(12), 1771–1773.

Prestifilippo, J. P., Fernandez-Solari, J., de la Cal, C., Iribarne, M., Suburo, A. M., Rettori, V., McCann, S. M., and Elverdin, J. C. (2006). Inhibition of salivary secretion by activation of cannabinoid receptors. *Exp. Biol. Med.* **231,** 1421–1429.

Ralevic, V., Kendall, D. A., Jerman, J. C., Middlemiss, D. N., and Smart, D. (2001). Cannabinoid activation of recombinant and endogenous vanilloid receptors. *Eur. J. Pharmacol.* **424,** 211–219.

Rau, K. K., Jiang, N., Johnson, R. D., and Cooper, B. Y. (2007). Heat sensitization in skin and muscle nociceptors expressing distinct combinations of TRPV1 and TRPV2 protein. *J. Neurophysiol.* **97,** 2651–2662.

Reilly, C. A., Taylor, J. L., Lanza, D. L., Carr, B. A., Crouch, D. J., and Yost, G. S. (2003). Capsaicinoids cause inflammation and epithelial cell death through activation of vanilloid receptors. *Toxicol. Sci.* **73,** 170–181.

Roberts, L. A., Christie, M. J., and Connor, M. (2002). Anandamide is a partial agonist at native vanilloid receptors in acutely isolated mouse trigeminal sensory neurons. *Br. J. Pharmacol.* **137,** 421–428.

Rockwell, C. E., and Kaminski, N. E. (2004). A cyclooxygenase metabolite of anandamide causes inhibition of Interleukin-2 secretion in murine splenocytes. *J. Pharmacol. Exp. Ther.* **311,** 683–690.

Rodriguez-Martin, I., Herrero-Turrion, M. J., Marron Fdez de Velasco, E., Gonzalez-Sarmiento, R., and Rodriguez, R. E. (2007). Characterization of two duplicate zebrafish Cb2-like cannabinoid receptors. *Gene* **389,** 36–44.

Rogers, M., Tang, L., Madge, D. J., and Stevens, E. B. (2006). The role of sodium channels in neuropathic pain. *Semin. Cell Dev. Biol.* **17,** 571–581.

Ross, R. A. (2003). Anandamide and vanilloid TRPVI receptors. *Br. J. Pharmacol.* **140,** 790–801.

Rubovitch, V., Gafni, M., and Sarne, Y. (2004). The involvement of VEGF receptors and MAPK in the cannabinoid potentiation of Ca^{2+} flux into N18TG2 neuroblastoma cells. *Brain Res.* **120,** 138–144.

Rueda, D., Navarro, B., Martinez-Serrano, A., Guzman, M., and Galve-Roperh, I. (2002). The endocannabinoid anandamide inhibits neuronal progenitor cell differentiation through attenuation of the Rap1/B-Raf/ERK pathway. *J. Biol. Chem.* **277,** 46645–46650.

Ryberg, E., Vu, H. K., Larsson, N., Groblewski, T., Hjorth, S., Elebring, T., Sjogren, S., and Greasley, P. J. (2005). Identification and characterisation of a novel splice variant of the human CB1 receptor. *FEBS Lett.* **579,** 259–264.

Ryberg, E., Larsson, N., Sjogren, S., Hjorth, S., Hermansson, N. O., Leonova, J., Elebring, T., Nilsson, K., Drmota, T., and Greasley, P. J. (2007). The orphan receptor GPR55 is a novel cannabinoid receptor. *Br. J. Pharmacol.* **152,** 1092–1101.

Sanchez, C., Galve-Roperh, I., Rueda, D., and Guzman, M. (1998). Involvement of sphingomyelin hydrolysis and the mitogen-activated protein kinase cascade in the delta 9-tetrahydrocannabinol-induced stimulation of glucose metabolism in primary astrocytes. *Mol. Pharmacol.* **54,** 834–843.

Saunders, C. I., Kunde, D. A., Crawford, A., and Geraghty, D. P. (2007). Expression of transient receptor potential vanilloid 1 (TRPV1) and 2 (TRPV2) in human peripheral blood. *Mol. Immunol.* **44,** 1429–1435.

Sawzdargo, M., Nguyen, T., Lee, D. K., Lynch, K. R., Cheng, R., Heng, H. H. Q., George, S. R., and O'Dowd, B. F. (1999). Identification and cloning of three novel human G protein-coupled receptor genes GPR52, YGPR53 and GPR55: GPR55 is extensively expressed in human brain. *Mol. Brain Res.* **64,** 193–198.

Shen, M., Piser, T. M., Seybold, V. S., and Thayer, S. A. (1996). Cannabinoid receptor agonists inhibit glutamatergic synaptic transmission in rat hippocampal cultures. *J. Neurosci.* **16,** 4322–4334.

Shimasue, K., Urushidani, T., Hagiwara, M., and Nagao, T. (1996). Effects of anandamide and arachidonic acid on specific binding of (+)-PN200-110, diltiazem and (-)-desmethoxyverapamil to L-type Ca^{2+} channel. *Eur. J. Pharmacol.* **296,** 347–350.

Shire, D., Carillon, C., Kaghad, M., Calandra, B., Rinaldi-Carmona, M., Le Fur, G., Caput, D., and Ferrara, P. (1995). An amino-terminal variant of the central cannabinoid receptor resulting from alternative splicing. *J. Biol. Chem.* **270,** 3726–3731.

Shire, D., Calandra, B., Rinaldi-Carmona, M., Oustric, D., Pessegue, B., Bonnin-Cabanne, O., Fur, G. L., Caput, D., and Ferrara, P. (1996). Molecular cloning, expression and function of the murine CB2 peripheral cannabinoid receptor. *Biochim. Biophys. Acta* **1307,** 132–136.

Sim, L. J., Selley, D. E., Xiao, R., and Childers, S. R. (1996). Differences in G-protein activation by mu- and delta-opioid, and cannabinoid, receptors in rat striatum. *Eur. J. Pharmacol.* **307,** 97–105.
Smart, D., Gunthorpe, M. J., Jerman, J. C., Nasir, S., Gray, J., Muir, A. I., Chambers, J. K., Randall, A. D., and Davis, J. B. (2000). The endogenous lipid anandamide is a full agonist at the human vanilloid receptor (hVR1). *Br. J. Pharmacol.* **129,** 227–230.
Southall, M. D., Li, T., Gharibova, L. S., Pei, Y., Nicol, G. D., and Travers, J. B. (2003). Activation of epidermal vanilloid receptor-1 induces release of proinflammatory mediators in human keratinocytes. *J. Pharmacol. Exp. Ther.* **304,** 217–222.
Spivak, C. E., Lupica, C. R., and Oz, M. (2007). The endocannabinoid anandamide inhibits the function of a4b2 nicotinic acetylcholine receptors. *Mol. Pharmacol.* **72,** 1024–1032.
Starowicz, K., Nigam, S., and Di Marzo, V. (2007). Biochemistry and pharmacology of endovanilloids. *Pharmacol. Ther.* **114,** 13–33.
Stella, N., and Piomelli, D. (2001). Receptor-dependent formation of endogenous cannabinoids in cortical neurons. *Eur. J. Pharmacol.* **425,** 189–196.
Stella, N., Schweitzer, P., and Plomelli, D. (1997). A second endogenous cannabinoid that modulates long-term potentiation. *Nature* **388,** 773–778.
Sugiura, T., Kodaka, T., Kondo, S., Tonegawa, T., Nakane, S., Kishimoto, S., Yamashita, A., and Waku, K. (1996). 2-Arachidonoylglycerol, a putative endogenous cannabinoid receptor ligand, induces rapid, transient elevation of intracellular free Ca^{2+} in neuroblastoma $\times$ glioma hybrid NG108-15 cells. *Biochem. Biophys. Res. Commun.* **229,** 58–64.
Sugiura, T., Kodaka, T., Kondo, S., Tonegawa, T., Nakane, S., Kishimoto, S., Yamashita, A., and Waku, K. (1997). Inhibition by 2-Arachidonoylglycerol, a novel type of possible neuromodulator, of the depolarization-induced increase in intracellular free calcium in neuroblastoma $\times$ glioma hybrid NG108-15 cells. *Biochem. Biophys. Res. Commun.* **233,** 207–210.
Sugiura, T., Kodaka, T., Nakane, S., Miyashita, T., Kondo, S., Suhara, Y., Takayama, H., Waku, K., Seki, C., Baba, N., and Ishima, Y. (1999). Evidence that the cannabinoid CB1 receptor is a 2-arachidonoylglycerol receptor. Structure-activity relationship of 2-arachidonylglycerol, ether-linked analogues, and related compounds. *J. Biol. Chem.* **274,** 2794–2801.
Sun, Y., Alexander, S. P. H., Kendall, D. A., and Bennett, A. J. (2006). Cannabinoids and PPARalpha signalling. *Biochem. Soc. Trans.* **34,** 1095–1097.
Sun, Y., Alexander, S. P. H., Garle, M. J., Gibson, C. L., Hewitt, K., Murphy, S. P., Kendall, D. A., and Bennett, A. J. (2007). Cannabinoid activation of PPARa; a novel neuroprotective mechanism. *Br. J. Pharmacol.* **152,** 734–743.
Szallasi, A., and Blumberg, P. M. (1999). Vanilloid (Capsaicin) receptors and mechanisms. *Pharmacol. Rev.* **51,** 159–212.
Terranova, J. P., Michaud, J. C., Le Fur, G., and Soubrie, P. (1995). Inhibition of long-term potentiation in rat hippocampal slices by anandamide and WIN55212–2: Reversal by SR141716 A, a selective antagonist of CB1 cannabinoid receptors. *Naunyn Schmiedebergs Arch Pharmacol.* **352,** 576–579.
Tominaga, M., and Caterina, M. J. (2004). Thermosensation and pain. *J. Neurobiol.* **61,** 3–12.
Tominaga, M., Caterina, M. J., Malmberg, A. B., Rosen, T. A., Gilbert, H., Skinner, K., Raumann, B. E., Basbaum, A. I., and Julius, D. (1998). The cloned capsaicin receptor integrates multiple pain-producing stimuli. *Neuron* **21,** 531–543.
Toth, A., Boczan, J., Kedei, N., Lizanecz, E., Bagi, Z., Papp, Z., Edes, I., Csiba, L., and Blumberg, P. M. (2005). Expression and distribution of vanilloid receptor 1 (TRPV1) in the adult rat brain. *Brain Res.* **135,** 162–168.

Trettel, J., and Levine, E. S. (2003). Endocannabinoids mediate rapid retrograde signaling at interneuron right-arrow pyramidal neuron synapses of the neocortex. *J. Neurophysiol.* **89**, 2334–2338.

Tucker, R. C., Kagaya, M., Page, C. P., and Spina, D. (2001). The endogenous cannabinoid agonist, anandamide stimulates sensory nerves in guinea-pig airways. *Br. J. Pharmacol.* **132**, 1127–1135.

Tzavara, E. T., Li, D. L., Moutsimilli, L., Bisogno, T., Di Marzo, V., Phebus, L. A., Nomikos, G. G., and Giros, B. (2006). Endocannabinoids activate transient receptor potential Vanilloid 1 receptors to reduce hyperdopaminergia-related hyperactivity: Therapeutic implications. *Biol. Psychiatry* **59**, 508–515.

Van den Bossche, I., and Vanheel, B. (2000). Influence of cannabinoids on the delayed rectifier in freshly dissociated smooth muscle cells of the rat aorta. *Br. J. Pharmacol.* **131**, 85–93.

van der Stelt, M., Trevisani, M., Vellani, V., De Petrocellis, L., Schiano Moriello, A., Campi, B., McNaughton, P., Geppetti, P., and Di Marzo, V. (2005). Anandamide acts as an intracellular messenger amplifying Ca^{2+} influx via TRPV1 channels. *EMBO J.* **24**, 3026–3037.

Vogel, Z., Barg, J., Levy, R., Saya, D., Heldman, E., and Mechoulam, R. (1993). Anandamide, a brain endogenous compound, interacts specifically with cannabinoid receptors and inhibits adenylate cyclase. *J. Neurochem.* **61**, 352–355.

Waldeck-Weiermair, M., Zoratti, C., Osibow, K., Balenga, N., Goessnitzer, E., Waldhoer, M., Malli, R., and Graier, W. F. (2008). Integrin clustering enables anandamide-induced Ca^{2+} signaling in endothelial cells via GPR55 by protection against CB1-receptor-triggered repression. *J. Cell. Sci.* **121**, 1704–1717.

Wang, H., Matsumoto, H., Guo, Y., Paria, B. C., Roberts, R. L., and Dey, S. K. (2003). Differential G protein-coupled cannabinoid receptor signaling by anandamide directs blastocyst activation for implantation. *Proc. Natl. Acad. Sci. USA* **100**, 14914–14919.

Wartmann, M., Campbell, D., Subramanian, A., Burstein, S. H., and Davis, R. J. (1995). The MAP kinase signal transduction pathway is activated by the endogenous cannabinoid anandamide. *FEBS Lett.* **359**, 133–136.

White, R., Ho, W. S., Bottrill, F. E., Ford, W. R., and Hiley, C. R. (2001). Mechanisms of anandamide-induced vasorelaxation in rat isolated coronary arteries. *Br. J. Pharmacol.* **134**, 921–929.

Wilson, R. I., and Nicoll, R. A. (2001). Endogenous cannabinoids mediate retrograde signalling at hippocampal synapses. *Nature* **410**, 588–592.

Wilson, R. I., and Nicoll, R. A. (2002). Endocannabinoid signaling in the brain. *Science* **296**, 678–682.

Xiao, J. C., Jewell, J. P., Lin, L. S., Hagmann, W. K., Fong, T. M., and Shen, C.-P. (2008). Similar *in vitro* pharmacology of human cannabinoid CB1 receptor variants expressed in CHO cells. *Brain Res.* **1238**, 36–43.

Xiong, W., Hosoi, M., Koo, B. N., and Zhang, L. (2008). Anandamide inhibition of 5-HT3A receptors varies with receptor density and desensitization. *Mol. Pharmacol.* **73**, 314–322.

Yoshida, T., Hashimoto, K., Zimmer, A., Maejima, T., Araishi, K., and Kano, M. (2002). The cannabinoid CB1 receptor mediates retrograde signals for depolarization-induced suppression of inhibition in cerebellar Purkinje cells. *J. Neurosci.* **22**, 1690–1697.

Zoratti, C., Kipmen-Korgun, D., Osibow, K., Malli, R., and Graier, W. F. (2003). Anandamide initiates Ca^{2+} signaling via CB2 receptor linked to phospholipase C in calf pulmonary endothelial cells. *Br. J. Pharmacol.* **140**, 1351–1362.

Zygmunt, P. M., Petersson, J., Andersson, D. A., Chuang, H., Sorgard, M., Di Marzo, V., Julius, D., and Hogestatt, E. D. (1999). Vanilloid receptors on sensory nerves mediate the vasodilator action of anandamide. *Nature* **400**, 452–457.

CHAPTER FIVE

Is GPR55 an Anandamide Receptor?

Andrew J. Brown* and C. Robin Hiley[†]

Contents

I. Δ^9-Tetrahydrocannabinol, CB_1, and CB_2 Receptors	112
II. Functional Evidence for Novel Cannabinoid Receptors	113
III. Genomics of G Protein-Coupled Cannabinoid Receptors	116
IV. The Orphan Receptor GPR55	117
A. Patent reports	118
B. Pharmacology of GPR55	119
V. Endogenous Ligands for GPR55	121
A. Lysophosphatidylinositol	121
B. LPI as a signaling molecule	122
C. LPI and cannabinoids at GPR55: Biased agonism	122
VI. GPR55 Cellular Signaling Pathways	123
A. $G_{\alpha 12}$ and $G_{\alpha 13}$	123
B. Signaling downstream of $G_{\alpha 12}$ and $G_{\alpha 13}$	123
C. GPR55 and Ca^{2+} signals	124
VII. Interactions Between GPR55 and CB_1 Receptors	126
VIII. Conclusion: GPR55 as an Anandamide Receptor	128
A. Multiple signaling modalities for GPR55	128
B. Receptor dimerization in GPCRs	130
C. Allosteric effects and biased agonism	131
D. Pharmacology in recombinant and natural systems	131
E. Anandamide and GPR55	132
References	133

Abstract

Anandamide activates CB_1 cannabinoid receptors but also has effects, particularly in the vasculature, that cannot be explained by actions at either this or the other cloned cannabinoid receptor, the CB_2 receptor. These effects are probably mediated by a novel G protein-coupled receptor, but genome searching has not

---

* Department of Screening and Compound Profiling, Molecular Discovery Research, GlaxoSmithKline, New Frontiers Science Park, GlaxoSmithKline, Third Avenue, Harlow, Essex, United Kingdom
[†] Department of Pharmacology, University of Cambridge, Tennis Court Road, Cambridge, United Kingdom

Vitamins and Hormones, Volume 81   © 2009 Elsevier Inc.
ISSN 0083-6729, DOI: 10.1016/S0083-6729(09)81005-4   All rights reserved.

revealed a strong candidate. Several approaches have suggested that an orphan receptor, GPR55, is a target for anandamide, but the pharmacology of this receptor is such that it cannot be categorically identified as a cannabinoid receptor. GPR55 appears primarily to be a receptor for lysophosphatidylinositol which may exhibit biased agonism, leading to it also responding to anandamide. GPR55 activates $G_{\alpha 12}$ and $G_{\alpha 13}$ and thence RhoA, leading to an oscillatory intracellular Ca^{2+} signal. Further complexity arises from possible interactions between the anandamide-sensitive CB_1 receptor and GPR55. Overall, it appears that GPR55 has several signaling modalities and that, while anandamide can activate systems containing this receptor, GPR55 cannot yet be primarily designated a receptor for this endocannabinoid.

I. Δ^9-Tetrahydrocannabinol, CB_1, and CB_2 Receptors

The hemp plant, *Cannabis sativa*, has been exploited by humankind for over 5000 years. According to the archeological record, the first use was apparently for its fibers and later it was found to be a highly nutritious foodstuff. Its medicinal properties were being exploited in Chinese culture about 4000 years ago and it was prescribed for treatment of malarial fever, constipation, and arthritic pains. Much later, Indo-Aryan cultures exploited its psychoactive properties as well as its benefits in fevers and gastrointestinal conditions. Despite this long history, it is less than 50 years since the groundbreaking demonstration by Gaoni and Mechoulam (1964) that Δ^9-tetrahydrocannabinol was the primary active ingredient of cannabis. For some years after this discovery there was a debate as to whether or not the drug acted on specific receptors. One hypothesis was that its primary interaction with cells took the form of a disturbance of the function of the cell membrane arising from its highly lipophilic character (Lawrence and Gill, 1975). However, it was also observed that the difference in potency between the optical isomers of Δ^9-tetrahydrocannabinol affected its activity in a way that suggested that there was a stereoselective site mediating some of its actions in the central nervous system (Jones *et al.*, 1974).

The debate was very effectively ended when Matsuda *et al.* (1990) cloned a G protein-coupled receptor (GPCR) for Δ^9-tetrahydrocannabinol and other cannabinoids from a cDNA library of rat cerebral cortex; this was serendipitous since the search had not initially been for a cannabinoid receptor but for homologues of the bovine substance K receptor. The new cannabinoid receptor coupled to $G_{i/o}$ since agonists decreased adenylyl cyclase activity and the response was sensitive to pertussis toxin (PTX). This linked the clone to previous functional studies since PTX sensitivity was known to be a characteristic of cannabinoid signaling in the brain (Howlett *et al.*, 1986). Shortly after the cloning of the cannabinoid receptor, two major advances were made. Firstly, N-arachidonoylethanolamine (anandamide; Di Marzo et al., 1994) and 2-arachidonoylglycerol (Mechoulam et al., 1995) were identified as endogenous ligands for the new receptor. Secondly, a peripheral cannabinoid receptor was identified by using degenerate primers to search a cDNA library from the HL60 human leukemia cell line (Munro et al., 1993). This second receptor, which also couples negatively to adenylyl cyclase in a PTX-sensitive manner (Bayewitch et al., 1995), is designated the CB_2 cannabinoid receptor to distinguish it from the first sequence which became the CB_1 receptor (Howlett et al., 2002). Photoaffinity labeling with [^{32}P]azidoaniline GTP revealed CB_1 receptors activated the G_α subunits $G_{i\alpha 2}$, $G_{i\alpha 3}$, $G_{o\alpha 1}$, and $G_{o\alpha 2}$ (Ho et al., 2000). The CB_1 receptor can also stimulate adenylyl cyclase activity in primary cultured neurons that have been pretreated with PTX (Glass and Felder, 1997) suggesting it can also couple through G_s. The effect on the cyclase might be related to the isoform present since both CB_1 and CB_2 receptors inhibited types I, V, VI, and VIII of adenylyl cyclase while types II, IV, and VII were stimulated (Rhee et al., 1998).

Activation of G proteins by CB_1 cannabinoid receptors is now known to result in a diversity of signaling outcomes. These include inhibition of voltage-gated N- and Q-types of Ca^{2+} channel and activation of p42/p44 mitogen-activated protein kinase (MAPK) (Mukhopadhyay et al., 2002). The CB_1 receptor also activates the extracellular signal-regulated kinase (ERK) cascade, via phosphatidylinositol 3′-kinase (PI3K) and protein kinase B/Akt, probably by $\beta\gamma$ subunit activation, and this is able to oppose ceramide-induced apoptosis in U373 MG human astrocytoma cells (Galve-Roperh et al., 2002). CB_2 receptors, on the other hand, have been shown to increase apoptosis through ceramide signaling, a process that is associated with internalization of the receptor (Galve-Roperh et al., 2002; Sanchez et al., 2001). These now "classical" cannabinoid receptors are therefore capable of regulation of cellular activity not only on a minute-to-minute basis but also in the longer term by means of effects on gene activity. This confers upon cannabinoid agonists, especially endocannabinoids, great potential in regulating the physiological processes of the body. However, there is growing evidence for the existence of other GPCRs, also PTX-sensitive, which mediate the responses of cannabinoids, especially anandamide.

II. Functional Evidence for Novel Cannabinoid Receptors

Investigations of the vascular effects of anandamide in particular have revealed that it induces relaxation by means that are independent of either CB_1 or CB_2 receptors. It can act independently of GPCRs by activation of

TRPV1 channels on sensory nerve endings in the vascular adventitia, which causes the release of the vasodilator calcitonin gene-related peptide (CGRP) (Zygmunt et al., 1999). But not all the vasodilator activity of anandamide (see Fig. 5.1) can be explained by this mechanism since, for example, relaxation of rat coronary arteries by the endocannabinoid is not sensitive to TRPV1 desensitization with capsaicin (White et al., 2001) and relaxation of rat mesenteric arteries is sensitive to PTX (White and Hiley, 1997a).

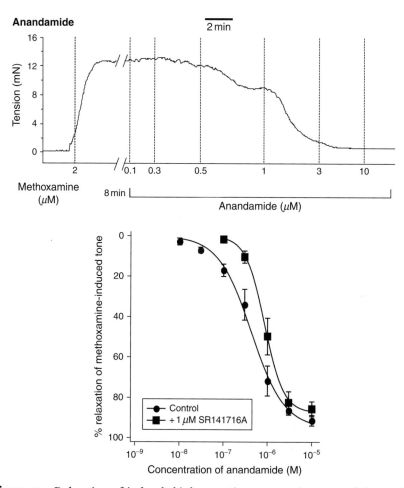

Figure 5.1 Relaxation of isolated third generation mesenteric artery of the rat by anandamide and its inhibition by 1 μM rimonabant (SR141716A). The upper panel shows a trace from a typical experiment showing the induction of contractile tone with the α-adrenoceptor agonist methoxamine and then relaxation of the vessel by cumulative addition of anandamide. The lower panel shows the concentration–response curves obtained from 10 (control) and 8 (+ rimonabant) experiments. Data replotted from White (1999).

Although rimonabant, a CB_1 receptor antagonist, antagonizes these responses, it does so to a limited extent and then only at very high concentrations (≥ 1 μM). The small degree of antagonism observed is incompatible with actions at CB_1 receptors since it suggests a K_d well above that reported for CB_1 receptor-mediated actions (White and Hiley, 1997a). In view of the pharmacology of GPR55 which will be discussed later, it should be noted here that mesenteric vasodilatation cannot be elicited by potent cannabinoid agonists such as WIN55,212-2 or HU210, nor by the endocannabinoid 2-arachidonoylglycerol (Wagner et al., 1999). Abnormal cannabidiol, a behaviorally inactive cannabinoid, can also relax rat mesenteric vessels in a rimonabant-sensitive manner (Járai et al., 1999). Furthermore, it evokes hypotension and mesenteric vasodilatation in CB_1 and CB_1/CB_2 receptor knockout mice (Járai et al., 1999). Like the mesenteric response to anandamide, that to abnormal cannabidiol is PTX-sensitive and, although rimonabant blocks the vasorelaxation, neither another CB_1 receptor antagonist, AM251, nor CB_2 receptor antagonists are effective blockers (Ho and Hiley, 2003b). Pharmacological investigation of the novel receptor mediating these responses has been facilitated by the development of O-1918, an analogue of abnormal cannabidiol, which blocks the vasodilatation to both the parent compound and anandamide (Offertáler et al., 2003). Interestingly, cannabidiol itself antagonizes the effects of abnormal cannabidiol in rat mesentery, but not those of anandamide (Járai et al., 1999).

Other non-CB_1/non-CB_2 receptor-mediated responses can show different pharmacology, in particular they are not always sensitive to rimonabant. For example, noladin ether, a putative endocannabinoid, inhibits the vasorelaxation mediated by CGRP released by electrical stimulation of the nerves to the rat mesenteric bed. This effect is PTX-sensitive but insensitive not only to the CB_2 antagonist SR144528, but also to both rimonabant (at 1 μM) and another CB_1 receptor antagonist LY320135 (Duncan et al., 2004).

The central nervous system also displays responses that are not mediated by either of the two "classical" cannabinoid GPCRs. Breivogel et al. (2001) showed that anandamide and the synthetic cannabinoid WIN55,212-2 initiated [^{35}S]GTPγS binding to brain membranes from both wild-type and CB_1 receptor-knockout mice, although other cannabinoids like CP55940 and HU210 were inactive. Anandamide and WIN55,212-2 were probably acting through the same receptor since their effects were not additive. In contrast, the actions of anandamide summated with those of either a μ opioid agonist [D-Ala, N-Me-Phe^4, Gly^5-ol]-enkephalin or an adenosine A_1 receptor agonist N^6-(2-phenylisopropyl)adenosine). Interestingly, like the results obtained in the vasculature, the responses to anandamide in the $CB_1^{-/-}$ mice were antagonized by a high concentration of rimonabant. Radioligand binding revealed sites for [^{3}H]WIN55,212-2, but not for [^{3}H]CP55940, in the brains of the CB_1 receptor-null mice and these sites showed a distinct regional distribution. It has also been found that rimonabant has effects on suckling in CB_1 receptor-knockout mice which are significant, even if they are of lesser magnitude than in the wild-type controls. This has been interpreted as showing that there is another rimonabant-sensitive receptor, possibly for cannabinoids, in the mammalian central nervous system (Fride et al., 2003).

Yet other studies in the brain on CB_1 receptor-null mice showed that both WIN55,212-2 and CP55940 inhibit glutamate-induced excitatory postsynaptic currents in the CA1 cells of the hippocampus (Hajos et al., 2001). Although these responses were also inhibited by PTX and by high concentrations of rimonabant, the agonist activity of CP55940 gave these responses a slightly different pharmacology to that found in the GTPγS binding studies of Breivogel et al. (2001). The pharmacology is also different from the vascular studies since abnormal cannabidiol had no effect on the glutamatergic postsynaptic currents in the hippocampus.

In summary, at least one novel PTX-sensitive, and therefore presumably G protein-coupled, receptor for anandamide occurs in the brain and vasculature. Responses mediated by this receptor are sensitive to rimonabant, but only when it is used at concentrations (≤ 1 μM) well in excess of those necessary to block CB_1 receptors (where its K_i is 2 nM; Rinaldi-Carmona et al., 1995). The receptor might also be that which mediates the vascular responses to abnormal cannabidiol as both are antagonized by O-1918. However, there are many variations in the pharmacology of these non-CB_1, non-CB_2 sites with regard to the abilities of other cannabinoid agonists to bring about the observed responses.

III. Genomics of G Protein-Coupled Cannabinoid Receptors

The molecular search for other cannabinoid GPCRs has been complicated by the fact that even the two receptors identified in the 1990s show remarkably little homology in their amino acid sequences. The CB_1/CB_2 receptor sequence comparison shows only 44% overall homology which rises to a still modest 68% when only the transmembrane helices are considered (Munro et al., 1993). This suggests that they arose some considerable time apart in the evolutionary process though the urochordate *Ciona intestinalis*, an invertebrate, has only a single genomic sequence which resembles those of the mammalian cannabinoid receptors (Elphick and Egertová, 2005). The gene designated *CiCBR* shows 28 and 24% homology, respectively, with human CB_1 and CB_2 sequences (Elphick et al., 2003). It was postulated that perhaps the CB_1 and CB_2 receptors, like many other sets of two-or-more related genes in vertebrate genomes, diverged at an evolutionary stage which occurred after a duplication event of the invertebrate gene. The phylogenetic tree for *CiCBR* shows it clades with the human cannabinoid receptors rather than with those other human GPCRs which most closely resemble the cannabinoid receptors. These latter receptors, for lysophosphatidic acid (the LPA_1 and LPA_2 receptors), for sphingosine-1-phosphate (the $S1P_1$ and $S1P_3$ receptors), and for melanocortin (MC_1–MC_5 receptors), form separate clades.

The relative lack of homology between the human CB_1 and CB_2 receptors has made it difficult to identify other possible cannabinoid receptors in the human genome using sequence analysis searching. A series of searches (using Ensembl; W.L. Davies and C.R. Hiley, unpublished results) against the transmembrane domains (TMDs) of the CB_1 receptor reveals homology primarily only with the CB_2 receptor, and that occurred in TMDs 1, 2, 3, 4, and 7. Focus on orphan receptors reveals that GPR3 (TMD2, TMD4, TMD7), GPR6 (TMD4, TMD7), and GPR12 (TMD2, TMD4, TMD7) share identity over multiple TMDs and it is noteworthy that the binding sites for rimonabant (TMD3–6; McAllister *et al.*, 2003) and anandamide (TMD2, TMD3, TMD7; Barnett-Norris *et al.*, 2002) show a limited overlap with these TMDs. However, the few studies that have described effects of ligands on GPR3, GPR6, and GPR12 suggest sphingosine-1-phosphate and related lipids are agonists, not endocannabinoids (Uhlenbrock *et al.*, 2002). This and other studies have detected high levels of constitutive activity simply by inducing cellular expression of these receptors, so they may function by being active whenever expressed (analogous to virally encoded GPCRs) rather than being active conditionally on receiving a ligand signal.

A further challenge to sequence searches intended to identify novel GPCRs is the precedent for receptor pairs activated by the same endogenous agonist ligand (particularly lipids) that lack any sequence similarity beyond the basic features common across the receptor superfamily. Several such pairs, for example, the prostanoid DP_1 and DP_2 receptors, occur on opposite sides of the receptor phylogenetic tree and may have arisen by convergent evolution (Brown, 2007).

An alternative method to identifying new receptors is not to use a direct genomic route but rather a functional approach, such as ligand-fishing. Screening compound banks allows identification of ligands for those putative receptor-coding sequences seen in the genome which are collectively referred to as "orphan receptors." This has led to the GPCR, GPR55, being proposed as a possible cannabinoid receptor (Brown, 2007).

IV. THE ORPHAN RECEPTOR GPR55

GPR55 was first cloned from a commercial human genomic DNA library. This was screened for the full-length open reading frame which contained the sequence for a putative TMD which had been identified by *in silico* searching of an expressed sequence tag database (Sawzdargo *et al.*, 1999). GPR55 is a member of the purine receptor subset of the GPCR superfamily, having highest identity to GPR35 (37% identity), P2Y$_5$ (30%), GPR92 (30%), and GPR23 (29%) (Baker *et al.*, 2006). Three of these related receptors are reportedly activated by lysophosphatidic acid (P2Y$_5$, GPR92, and GPR23; Lee *et al.*, 2006; Noguchi *et al.*, 2003; Pasternack *et al.*, 2008). GPR35 is weakly activated by kynurenic acid (Wang *et al.*, 2006), but conceivably may also be activated by as yet unidentified endogenous lipid mediators. GPR55 mRNA was detected by Northern blotting in human brain (in the caudate nucleus and putamen, but not in the hippocampus, thalamus, and mid-brain). *In situ* hybridization in the rat brain showed expression in the hippocampus, parts of the thalamus and midbrain. In peripheral tissues, expression was found in the spleen and intestine and there was also mRNA in fetal tissues (Sawzdargo *et al.*, 1999).

A. Patent reports

After the initial report by Sawzdargo *et al.* (1999), no more appeared in the academic literature about this orphan receptor until 2006, when attention was drawn to two patents which suggested GPR55 could bind to, and be acted upon, cannabinoid ligands (Baker *et al.*, 2006). A GlaxoSmithKline patent described the use of yeast host strains for the coexpression of human GPR55 and chimeric yeast/human G proteins to detect ligands for the GPCR (Brown and Wise, 2001). The yeast cells are designed such that they will only propagate if there is activation of the receptor of interest and as such they can reveal both spontaneous receptor activity and, more importantly in the present context, agonist activation of the receptor (Brown *et al.*, 2000). This system is well suited to ligand detection for orphan receptors since the cells do not have any of their own GPCRs (Dowell and Brown, 2002). Its use showed that both rimonabant and the CB$_1$ antagonist AM251 could activate GPR55 (Brown and Wise, 2001), thereby establishing a link with cannabinoid ligands.

A group from AstraZeneca (Drmota *et al.*, 2004) reported the use of [^{3}H]CP55940 and [^{3}H]rimonabant as radiolabels for detecting ligands for GPR55 expressed in HEK293 cells. The cell membranes did not bind another cannabinoid ligand, [^{3}H]WIN55,212-2, and therefore showed some similarity with the functional studies which suggest that WIN55,212-2 is usually not active at the rimonabant-sensitive cardiovascular sites for cannabinoids which mediate non-CB$_1$/non-CB$_2$ activity (White and Hiley, 1998). Surprisingly, rimonabant, which shows antagonist or inverse agonist activity in functional assays of CB$_1$ receptors or against the non-CB$_1$ cannabinoid effects in the cardiovascular system, was an agonist in GTPγS-binding assays using membranes containing GPR55. Other agonists in this assay included other ligands at CB$_1$ and CB$_2$ receptors, not only classical agonists (e.g., anandamide, 2-arachidonoylglycerol, Δ^9-tetrahydrocannabinol, and virodhamine) but also other the CB_2 cannabinoid receptor antagonist SR144528.

It also has to be remembered that rimonabant is a relatively weak antagonist of the vascular relaxation evoked by anandamide in the rat ($K_d \approx$ 0.7 μM; White and Hiley, 1997b) whereas successful binding with [^{3}H]-rimonabant took place with 1.6 nM ligand; this implies either a very high level of receptor expression (as this concentration of rimonabant would occupy $\sim$0.25% of the sites at the estimated K_d) or a much higher affinity for the ligand at the cloned GPR55 than at the vascular receptor. In the HEK293 cell assay, anandamide was a potent agonist (EC_{50} = 18.4 nM) as was Δ^9-tetrahydrocannabinol (EC_{50} = 9.4 nM), while abnormal cannabidiol was over 100 times less potent (EC_{50} = 2.78 μM). Therefore it might be that GPR55 is *a* receptor for anandamide but not *the* one mediating the responses to abnormal cannabidiol, which in the vascular assay is of similar potency (EC_{50} = 630 nM; Ho and Hiley, 2003b) to anandamide (EC_{50} = 650 nM; White and Hiley, 1997a).

B. Pharmacology of GPR55

The two patents clearly showed that GPR55 binds to, and is activated by, a variety of cannabinoid ligands but aspects of its pharmacology are not consistent with it being the receptor which mediates all, if any, of the functional effects ascribed to actions of anandamide and other cannabinoids at receptors other than the classical CB_1 and CB_2 types.

More recently, Johns *et al.* (2007), from GlaxoSmithKline, confirmed that abnormal cannabidiol and O-1602 activated GTPγS binding in HEK293T cells expressing GPR55, while the CB_1/CB_2 cannabinoid ligand WIN55,212-2 was inactive. In one of the inconsistencies of the literature on GPR55, they found abnormal cannabidiol to have an EC_{50} of 2.5 nM in their GTPγS assay, which was similar to that for O-1602, but very different from the value given by (Drmota *et al.*, 2004). Abnormal cannabidiol also relaxed mesenteric blood vessels from mice, in an O-1918-sensitive manner, even when the vessels were taken from $GPR55^{-/-}$ animals. Therefore, although GPR55 is activated by abnormal cannabidiol, it is not the receptor responsible for the vasodilator actions of this cannabinoid. Unfortunately, Johns *et al.* (2007) did not test the effects of anandamide due to limited availability of tissue from $GPR55^{-/-}$ animals (D.G. Johns, personal communication), so it is not possible to conclude with certainty that GPR55 is not the receptor mediating relaxations to the endocannabinoid (Hiley and Kaup, 2007).

Shortly afterward, the AstraZeneca group reported on the pharmacology of GPR55 using a different range of ligands (Ryberg *et al.*, 2007). In a GTPγS binding assay, Δ^9-tetrahydrocannabinol and the synthetic cannabinoid,

CP55940, activated the receptor as did cannabidiol, abnormal cannabidiol, O-1602 and the endocannabinoids, anandamide and virodhamine. Since there is agreement between the groups that WIN55,212-2 is inactive, GPR55 is not the receptor mediating the brain responses reported by Breivogel et al. (2001) in CB_1 receptor-knockout mice. Ryberg et al. (2007) carried out binding studies with [^{3}H]rimonabant which showed a low level of binding to transfected HEK293 cells (about twice that seen in nontransfected cells, but <10% of that seen with [^{3}H]CP55940). As has been noted, rimonabant is a weak antagonist of the vascular response, requiring concentrations around 1 μM to produce measurable antagonism. Therefore, if the radioligand were binding to a site similar to that in the vasculature, it would bind to <10% of the receptors (using the estimated K_d range from the functional experiments by White and Hiley (1997a) at the concentration of 50 nM used. Unfortunately, the affinities of neither [^{3}H]rimonabant nor [^{3}H]-CP55940 were reported, although CP55940 was shown to be a potent agonist, and so it is not possible to explain the great different in binding seen with these two ligands on the data available in this chapter. Another intriguing observation is that the CB_1 receptor antagonist AM251, which is closely related to rimonabant, was an agonist in the GTPγS assay. This corresponds with the previous observation for rimonabant and the report that AM251 increased reporter gene activity in a yeast assay for GPR55 activation (Brown and Wise, 2001). So far, AM251 has not been reported to be a vasodilator compound in its own right, although it might not have been looked for in any of the currently reported studies.

Lauckner et al. (2008) investigated the pharmacology of GPR55 not only in HEK293 cells expressing human and mouse GPR55 but also in large (diameter >35 μm) dorsal root ganglion cells of the mouse. In HEK293 cells transiently expressing the human sequence of the receptor, intracellular Ca^{2+} concentration was increased by Δ^9-tetrahydrocannabinol and JWH015 (an agonist at CB_2 cannabinoid receptors with an aminoalkylindole structure like WIN55,212-2, which itself was inactive) as well as by anandamide and its metabolically stable analogue, methanandamide (at concentrations of 5 μM). Perhaps of more interest are the observations that there were no Ca^{2+} responses to 2-arachidonoylglycerol, CP55940, abnormal cannabidiol, virodhamine, or palmitoylethanolamide (all used at 5 μM). Similar pharmacology was seen in the ganglionic neurons with both Δ^9-tetrahydrocannabinol and JWH015 being active. Rimonabant was found to be an antagonist at GPR55 (contrasting with agonist activity reported in the AstraZeneca patent; Drmota et al., 2004), with a potency that could be consistent with GPR55 being the receptor for anandamide in the blood vessels as the concentration needed for blockade (2 μM) was similar to that in the vascular experiments.

More recently, Henstridge et al. (2009) also published results from a study in which changes in intracellular Ca^{2+} concentration were measured in HEK293 cells stably expressing recombinant GPR55; several of their observations are not consistent with those of Lauckner et al. (2008). They observed characteristic oscillatory increases in cytoplasmic Ca^{2+} concentration evoked by receptor activation (primarily using lysophosphatidylinositol (LPI) as the agonist—see below). Crucially, of the cannabinoid ligands described, neither 2-arachidonoylglycerol nor, and more importantly, anandamide (30 μM) activated Ca^{2+} oscillations in GPR55-expressing cells, whereas AM251 behaved as a weak agonist with an EC_{50} of 612 nM. CP55940 (3 μM) had no agonist activity but acted as an antagonist, causing a rightward shift in agonist concentration–response curves. This closely recapitulates the pattern of ligand activity previously observed in yeast (Brown, 2007). Hence we can conclude that two strikingly different patterns of ligand specificity are manifested on recombinant expression of GPR55. The first is revealed by a direct measure of G protein activation and typified by activation of GTPγS-binding on GPR55-expressing membranes by a wide range of cannabinoids including anandamide. The second is revealed by measuring downstream signal transduction events (oscillatory Ca^{2+} signals in HEK293 cells or gene induction in yeast) and typified by AM251 having agonist activity, CP55,940 having antagonist activity, and anandamide and other cannabinoids being inactive.

V. Endogenous Ligands for GPR55

A. Lysophosphatidylinositol

These contradictory findings with cannabinoid ligands may become easier to reconcile following recently published data describing the likely endogenous ligand for GPR55 as being LPI. Oka et al. (2007) described experiments in which activation of ERK1/ERK2 in GPR55-transfected HEK293 cells was used as the assay. GPR55-specific phosphorylation of the kinases occurred in response to LPI, whereas anandamide, 2-arachidonoylglycerol, virodhamine, oleoylethanolamide, and palmitoylethanolamide were ineffective as were the synthetic cannabinoids CP55,940 and WIN55,212-2. Lauckner et al. (2008) also found that LPI was an agonist in the large dorsal root ganglion neurons supporting the concept that the lipid might be an endogenous ligand for GPR55 when the receptor is naturally expressed. In corroboration of the suggestion that LPI is the natural ligand, the study of Henstridge et al. (2009) also showed it evoked oscillatory Ca^{2+} signals in GPR55-HEK293 cells. Furthermore, the identification of GPR55 as an LPI receptor is consistent with its phylogeny since two of its closest homologues GPR23 and GPR92, formerly both orphan receptors, have been shown to respond to a related lipid, lysophosphatidic acid (Lee et al., 2006; Noguchi et al., 2003). The extent to which LPI would behave as an agonist in the

GTPγS-binding assays of Ryberg et al. (2007) and Johns et al. (2007) is not yet clear. Oka et al. (2007) showed LPI-mediated activation of GTPγS binding in GPR55-containing membranes, but did not test cannabinoid ligands in parallel.

B. LPI as a signaling molecule

LPI has a wide range of biological activities including, for example, increases in intracellular Ca^{2+} concentration in rat liver cells (Baran and Kelly, 1988) and opening of two pore-domain K^+ channels, such as TREK-1 (Maingret et al., 2000). As a derivative of a long-chain fatty acid, some of its actions have been ascribed to its capacity to interact physically with membranes (its critical micellar concentration is around 75 μM), and this seems to be the case with respect to its actions, and those of other lysophospholipids, on K^+ channels which can be correlated with the shape of the molecules (Patel et al., 2001). However, there are many systems where the effects of LPI are more easily explained by it interacting with a specific receptor, for example, in its mitogenic action in a differentiated thyroid cell line (Falasca et al., 1995). Although this action was not sensitive to PTX, it was inhibited by tyrphostins AG18 and AG561 and so is dependent on tyrosine kinase activity.

LPI is synthesized by the phospholipase A (PLA)-mediated removal of one of the acyl moieties of phosphatidylinositol (for a more full description of LPI metabolism, see the review by Corda et al., 2002). There may be specific enzymes involved in the generation and breakdown of LPI, which would strengthen its claim as a signaling molecule in its own right. Both Ca^{2+}-independent PLA_1 and Ca^{2+}-dependent PLA_2 activity are found in platelets and, in the rat brain there are both a Ca^{2+}-independent, phosphatidylinositol-preferring, phospholipase A_1 and an LPI-specific PLC (Kobayashi et al., 1996). With the identification of an LPI-preferring GPCR in GPR55, the lysophospholipid gains credibility as a biological signaling molecule. The problem now becomes how to explain the apparent appearance and disappearance of anandamide sensitivity in studies with GPR55.

C. LPI and cannabinoids at GPR55: Biased agonism

One mechanism that could provide the basis for disparate behavior between agonists in different systems which involve the same receptor is that of "biased agonism" (Kenakin, 1995). This phenomenon, also called "functional selectivity," "stimulus trafficking," "agonist trafficking," and "collateral efficacy" (Bosier and Hermans, 2007; Kenakin, 2007b), was reviewed in the August 2007 edition of *Trends in Pharmacological Sciences* (see Kenakin, 2007a). In this paradigm, the existence of multiple conformations of a receptor, which can activate different G proteins to modify intracellular function, can lead to different orders of efficacy being observed for the same set of agonists according to the cellular end-effect being monitored. In the case of the α_{2A}-adrenoceptor this has been shown as the basis of response differences with agonists classed as catecholamine like, imidazoline, and noncatecholamine–nonimidazoline types (Kukkonen et al., 2001). Another example, this time involving the cannabinoids, is the opposite changes in tyrosine hydroxylase gene promoter activity (assayed using a luciferase reporter assay in mouse neuroblastoma cells) related to CB_1 receptor activation which can occur with the cannabinoid agonists HU210 and CP55,940 (Bosier et al., 2007). Thus, if GPR55 can generate a variety of G protein-activating conformations, this might produce greater or lesser efficacy for a given ligand depending on the G protein and signaling profile of the cellular system involved. It is therefore worthwhile exploring the signaling pathways thought to be associated with GPR55.

VI. GPR55 Cellular Signaling Pathways

A. $G_{\alpha 12}$ and $G_{\alpha 13}$

Ryberg et al. (2007) reported that the responses evoked by the cannabinoids in GPR55-expressing cells were insensitive to PTX. Therefore, to identify properly the G proteins involved, they performed a series of experiments in which C-terminal fragments of, or antibodies against, G_α proteins were used to assess their effects on the GTPγS binding induced by O-1602; this showed that coupling was through $G_{\alpha 13}$ and not by means of $G_{\alpha i1/2}$ or $G_{\alpha s}$. This confirmed results reported both in the yeast assay (Brown and Wise, 2001) and in GPR55-HEK293 cells, where expression of a dominant-inhibitory mutant of $G_{\alpha 13}$ blocked the LPI-evoked oscillatory Ca^{2+} signals (Henstridge et al., 2009).

B. Signaling downstream of $G_{\alpha 12}$ and $G_{\alpha 13}$

Once the G protein is activated, it appears that later stages in the signaling pathway involve RhoA, cdc42, and rac1. In the cell-line studied by Ryberg et al. (2007), anandamide and O-1602 activated these small G proteins and cannabidiol blocked these effects. On the other hand, in the cell line of Henstridge et al. (2009), activation of RhoA occurred in response to LPI; moreover the oscillatory Ca^{2+} signal was blocked by a dominant-inhibitory mutant of RhoA or by a small-molecule inhibitor of Rho-kinase. This pattern of activation of small G proteins is typical of $G_{\alpha 13}$-mediated responses. For example, CHO cells endogenously express PAR-1 thrombin receptors and $P2Y_2$ purinoceptors which can activate Ca^{2+}-dependent PLA_2. This can be through G_q, but overexpression of $G_{\alpha 13}$ reveals that both thrombin and ATP can activate cytosolic PLA_2 through the activation of RhoA and, subsequently, ERK1/ERK2 (Mariggio et al., 2006). Also, lysophosphatidic acid regulates β_2-adrenoceptor trafficking (decreasing β_2-adrenoceptors on the cell surface) through a $G_{\alpha13}$ signaling pathway which involves p115 Rho guanine nucleotide exchange factor (p115RhoGEF), RhoA and the MAPK cascade which ultimately activates c-Jun N-terminal kinase (JNK) (Shumay et al., 2007). The JNK then acts to reduce the ability of β_2-adrenoceptors to activate the cells.

In the context of the cardiovascular system, angiotensin II can act through $G_{\alpha13}$, p115RhoGEF, and RhoA in human right atrial appendage (Kilts et al., 2007). Photoaffinity labeling with [^{32}P]azidoaniline GTP showed that angiotensin II activated $G_{\alpha13}$, while another vasoconstrictor peptide, endothelin-1, activated $G_{\alpha12}$ although α_1-adrencoceptor activation with phenylephrine had no effect on either of these homologous G proteins. Activation of $G_{\alpha13}$ with angiotensin II, but not of $G_{\alpha12}$ with endothelin-1, recruited p115RhoGEF from the cytosol to the membrane of the atrial cells.

In the vasculature, and perhaps more importantly in view of the discovery of GPR55 in human umbilical vein endothelial cells, $G_{\alpha13}$ signaling regulates endothelial cell migration as shown by its involvement in the effects of activation of the cell membrane-bound estrogen receptor-α (Simoncini et al., 2006). In this case, Rho activates Rho-associated kinase to phosphorylate Thr^{558} of moesin, one of the proteins that regulates membrane/cytoskeleton interactions. This modifies the binding of moesin to actin and leads to the remodeling of the actin cytoskeleton and cellular shape changes. Thus GPR55 may regulate endothelial cell migration after stimulation with an exogenous cannabinoid, an endocannabinoid, or other endogenous ligand. There is also evidence that the vasodilator-associated protein (VASP) pathway is activated by $G_{\alpha13}$ and VASP also regulates cell shape and activation in platelets and endothelial cells (Profirovic et al., 2005). However, $G_{\alpha13}$ and RhoA signaling is usually associated with smooth muscle contraction rather than the relaxation of vascular smooth muscle that is induced by anandamide and abnormal cannabidiol. Thus the sustained phase of endothelin-1-induced contraction of intestinal smooth muscle is dependent on this pathway while the initial contraction is G_q-mediated (Hersch et al., 2004).

C. GPR55 and Ca^{2+} signals

The signaling pathway proposed by Henstridge et al. (2009), shown in Fig. 5.2), follows activation of GPR55, $G_{\alpha13}$, RhoA, and Rho-kinase. U73122, an inhibitor of phospholipase C(PLC), also blocked oscillatory Ca^{2+} signals, implying activation of PLC downstream of Rho-kinase, and hence inositol trisphosphate production (though this was not measured), and activation of the inositol trisphosphate receptor of the endoplasmic

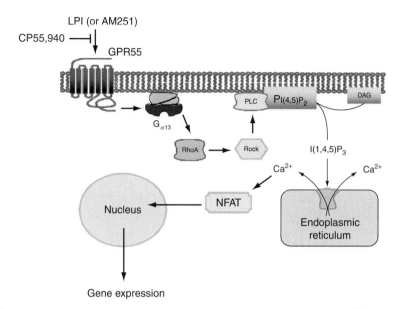

Figure 5.2 Cellular signaling pathways for the induction of oscillatory Ca^{2+} signals in GPR55-expressing HEK293 cells as proposed by Henstridge et al. (2009). Lysophosphatidylinositol (LPI) and the CB_1 cannabinoid receptor agonist AM251 act as agonists on GPR55 and, through $G_{\alpha 13}$, these activate the small G protein RhoA. In turn this leads to generation of inositol 1,4,5-trisphosphate through activation of RhoA kinase (ROCK) and phospholipase C (PLC). This process also leads to an increase in activity of a transcription factor, the nuclear factor of activated T-cells (NFAT). Reproduced with permission.

reticulum and causing Ca^{2+} release. The feature of this pathway responsible for the oscillatory nature of the Ca^{2+} response, which is highly unusual for a GPCR and which appears to persist long after wash-out of agonist, is not clear. Lysophosphatidic acid, tested as a control in the same cells, and presumably acting by activating endogenous $G_{\alpha q}$-coupled receptors, evokes transient increases in cytoplasmic Ca^{2+}, but no subsequent oscillations, and presumably this signal also is mediated by inositol trisphosphate acting on the endoplasmic reticulum. Henstridge et al. (2009) also found GPR55-dependent activation of the transcription factor, NFAT, by nuclear translocation which presumably arose as a consequence of the Ca^{2+} oscillations.

At first sight, these effects on cellular Ca^{2+} do not seem to fit easily with the observed vascular responses to anandamide, which are of vasodilatation, and so might be expected to involve a decrease in the cytoplasmic concentration of Ca^{2+} in the vascular smooth muscle. However, endothelium-dependent vasodilatation is dependent on increased cytoplasmic Ca^{2+} in the endothelial cells. The increased Ca^{2+} concentration in turn leads to enhanced activity of nitric oxide synthase, which causes smooth muscle cell relaxation by increasing intracellular cyclic GMP, or of Ca^{2+}-activated K^+ channels, which bring about hyperpolarization of the smooth muscle cells and hence decrease developed tension (Lagaud *et al.*, 1999). Therefore, if the "endothelial anandamide receptor" were GPR55, then increases in endothelial Ca^{2+} could result in blood vessel relaxation, but it should be noted that vasodilatation induced by cannabinoids shows varying degrees of dependence on the endothelium and it is by no means clear that endothelial mechanisms are dominant in cannabinoid-induced vasorelaxation (see, e.g., Ho and Hiley, 2003a,b, 2004).

VII. Interactions Between GPR55 and CB_1 Receptors

It has been argued that some of the unusual pharmacology of GPR55 might arise as a result of crosstalk between the signaling pathways which it activates. This has been most clearly shown in cultured human umbilical endothelial cells which have been suggested to express constitutively both GPR55 and the CB_1 cannabinoid receptor. Waldeck-Weiermair *et al.* (2008) found that the presence or absence of extracellular Ca^{2+} appeared to change the dominant receptor pathway activated by anandamide (Fig. 5.3).

When Ca^{2+} is present in the bathing fluid, anandamide only evokes a small increase in intracellular Ca^{2+} whereas, in nominally Ca^{2+}-free solution, there is a large rise in the cytosolic Ca^{2+} concentration which approaches the magnitude of the signal evoked by histamine. The response to anandamide in the absence of Ca^{2+} is sensitive to rimonabant but not to the CB_2 receptor antagonist SR144528. Furthermore, in these conditions the synthetic cannabinoid agonist HU210 had no effect on the intracellular Ca^{2+} signal despite the fact that CB_1 receptor message could be detected by RT-PCR. When these CB_1 receptors were blocked by AM251 (the close relative of rimonabant that does not block the responses to anandamide in blood vessels; White *et al.*, 2001) there was a marked response to anandamide, even in the presence of Ca^{2+}. This shows that CB_1 receptor activation can inhibit responses evoked by anandamide acting through another receptor. Since PTX also resulted in a large cytosolic Ca^{2+} signal to anandamide, even in the presence of Ca^{2+} in the cellular bathing medium, it is likely that the inhibitory effect of CB_1 receptor stimulation on the cellular Ca^{2+} responses to anandamide is mediated through $G_{i/o}$ proteins.

The receptor, which had its actions modified by CB_1 receptor activation, was proposed to be GPR55. The presence of its message was shown at the molecular level by RT-PCR and, functionally, by the ability of O-1602 to evoke a strong Ca^{2+} signal, but no definitive demonstration of receptor

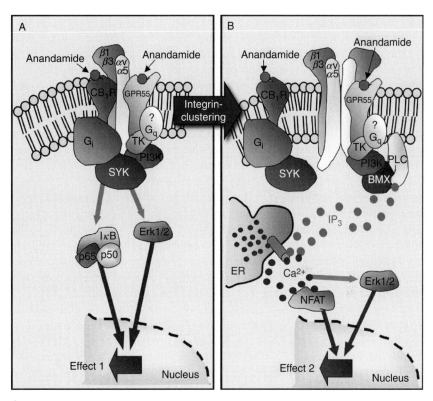

Figure 5.3 The signaling pathway for the interaction between GPR55 and the CB_1 receptor proposed by Waldeck-Weiermair et al. (2008). In (A) integrins ($\beta 1$, $1\beta 3$, αv, $\alpha 5$) are not clustered and anandamide activates the cells through the cannabinoid CB_1 receptor. The response is evoked through $G_{i/o}$-mediated activation of spleen tyrosine kinase (Syk) which, in turn, leads to enhanced activity of the extracellular signal-regulated kinases 1 and 2 (Erk1/2) and nuclear factor κB as a result of decreased IκB activity. In this condition, phosphatidylinositol 3-kinase (PI3K) is inhibited by Syk and so there is no response evoked by anandamide binding to GPR55. (B) shows the responses generated when the integrins are clustered. The CB_1 cannabinoid receptor is no longer associated with the $\beta 1$ integrin and the inhibition of PI3K is thereby released. Anandamide binding to GPR55 now activates bone marrow kinase, X-linked/epithelial and endothelial tyrosine kinase (Bmx/Etk) and PI3K signaling. The consequential generation of inositol trisphosphate causes release of Ca^{2+} from intracellular stores. In this case, as in Fig. 5.2, the intracellular Ca^{2+} activates NFAT and the kinases Erk1/2. Reproduced with permission.

protein was made. Waldeck-Weiermair et al. (2008) also showed that the responses to O-1602 were antagonized by O-1918, which blocks the vascular responses to anandamide and abnormal cannabidiol (Begg et al., 2003; Offertáler et al., 2003). siRNA directed against GPR55 decreased both the message, and the response to anandamide, but the effects of GPR55-siRNA on control Ca^{2+} responses mediated by unrelated receptors (such as those for histamine or acetylcholine) were not investigated and it therefore remains possible that the observed effect was unrelated to GPR55.

After having demonstrated responses to anandamide that were mediated by both CB_1 receptors and another receptor, Waldeck-Weiermair et al. (2008) went on to determine the signaling pathways associated with, and regulating the effects of, the two receptors. Firstly, they showed that the effect of Ca^{2+} appeared to be due to integrin clustering. Sr^{2+} or Ba^{2+} had similar effects to those of Ca^{2+}, and the IC_{50} for Ca^{2+} on the response to anandamide was similar to that determined for Ca^{2+} inhibition of integrin clustering. On the other hand, induction of clustering with 70 μM Mn^{2+} allowed anandamide to cause a large enhancement of cytosolic Ca^{2+}, even in the presence of Ca^{2+}; that is, it removed the inhibition of response that was presumably due to actions of anandamide at the CB_1 receptor. Inhibition of RhoA-associated kinases 1 and 2 (ROCK1 and ROCK2) with Y27632, which would be expected to decrease integrin clustering, decreased the intracellular Ca^{2+} response to anandamide that was observed in the absence of extracellular Ca^{2+}. The particular integrins involved appear to be $\alpha_v\beta_3$ (the vitronectin receptor) and $\alpha_5\beta_1$ (the fibronectin receptor).

It was hypothesized that, in the absence of integrin clustering, anandamide activates both CB_1 receptors and GPR55. The CB_1 receptor in these conditions is associated with the β_1 integrin chain, and this allows activation of spleen tyrosine kinase (Syk) which leads to activation of nuclear factor κB (NFκB) as well as ERK1 and ERK2. Syk involvement results in inhibition of PI3K. However, when integrins are clustered, PI3K inhibition is removed and anandamide can activate this enzyme through GPR55. In turn this leads to activation of PLC, bone marrow kinase, X-linked/epithelial and endothelial tyrosine kinase (Bmx/Etk) and, then, through generation of inositol trisphosphate, release of intracellular Ca^{2+}. The G protein, or proteins, involved in this cascade remains unknown. This signaling cascade contrasts with that elucidated by Henstridge et al. (2009), and further studies will be required to define if the LPI-evoked signal in GPR55-HEK293 cells involves tyrosine kinases and PI3K, and if the anandamide-evoked signal in endothelial cells is mediated by $G_{\alpha 13}$.

VIII. Conclusion: GPR55 as an Anandamide Receptor

A. Multiple signaling modalities for GPR55

In concluding, the major issue that arises is why do systems thought to express GPR55 give varying patterns of ligand specificity and signaling when studied in different ways? The different patterns of GPR55 ligand

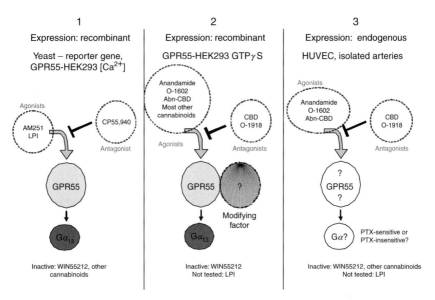

Figure 5.4 Different ligand specificities for GPR55 showing the three signaling modalities that have been reported. See text for explanation.

specificity that have been described are summarized in Fig. 5.4 as three signaling modalities.

In the first modality, AM251 and LPI behave as agonists which are antagonized by CP55,940, and signaling downstream of GPR55 is mediated via $G_{\alpha 13}$. This pattern of ligand specificity has been observed in yeast and in mammalian HEK293 cells expressing recombinant GPR55, but has not yet been observed in cells endogenously expressing GPR55. In both of these recombinant systems, a signaling end-point distal to the receptor was measured (Brown, 2007; Henstridge et al., 2009). The second modality shows many different cannabinoids and lipids behaving as agonists, including anandamide, O-1602, abnormal cannabidiol, AM251, and CP55,940. The activity of LPI has not yet been reported. Cannabidiol and O-1918 behave as antagonists, and signaling downstream of GPR55 is mediated by $G_{\alpha 13}$. This pattern of ligand specificity was revealed by measuring a signaling end-point near to the receptor, that is, binding of GTPγS to membranes from HEK293 cells expressing recombinant GPR55 (Johns et al., 2007; Ryberg et al., 2007). Again, this pattern has not yet been observed at GPR55 expressed in its natural physiological context. In the third modality, anandamide, O-1602, and abnormal cannabidiol behave as agonists at a receptor expressed endogenously in vascular endothelial cells; their effects are antagonized by cannabidiol and O-1918. In vascular preparations, these agonists cause vasodilatation.

In the endothelium-derived immortalized cell-line studied by Waldeck-Weiermair et al. (2008), activation of a receptor with this ligand specificity results in mobilization of intracellular Ca^{2+}. RNA interference directed against GPR55 reduced the anandamide-evoked Ca^{2+} signal, identifying the receptor as GPR55, although, as already noted, this specificity has not yet been tested in systems expressing recombinant GPR55. From these three observed modalities, the question arises "How can a single molecular species manifest these multiple pharmacological phenotypes?"

B. Receptor dimerization in GPCRs

The contrast between modalities 1 and 2 is particularly striking because both have been seen with recombinant expression of GPR55 in HEK293 cells. There are now multiple examples in the superfamily of GPCRs where ligand specificity is modified by interaction with other proteins. For example, Family B GPCRs can associate with single transmembrane proteins known as receptor activity modifying proteins (RAMPs). In the case of the calcitonin receptor, association with RAMP1 confers the ability to be activated by amylin (Hay et al., 2006). Within Family A, several pairs of GPCRs have been shown to heterodimerize with resultant effects on ligand specificity (Milligan, 2007). Thus the Δ-opioid receptor and the sensory neuron-specific receptor-4 (SNSR-4), both of which are expressed in dorsal root ganglion cells, appear to heterodimerize when recombinants are coexpressed. In coexpression, SNSR-4 signaling is retained, but that to Δ-opioid receptors are masked, and selective agonists for the latter receptor fail to activate the dimer (Breit et al., 2006). Perhaps GPR55 in a similar fashion interacts with a secondary factor that modifies its ligand-binding profile.

The correlation between ligand specificity in yeast, where GPR55 is isolated from other mammalian proteins, with that described by Henstridge et al. (2009) in GPR55-HEK293 cells suggests interaction with AM251, LPI, and CP55,940 (Modality 1) is an intrinsic property of GPR55. By inference, interaction with the wider array of cannabinoids (Modality 2) could result from the interaction of GPR55 with a secondary modifying factor. The defining property of this factor would be that it confers on GPR55 affinity for diverse cannabinoid ligands. Given the propensity of GPCRs to heterodimerize we should consider the possibility that cannabinoid receptors combine with GPR55 to yield the specificity observed by Ryberg et al. (2007). This is attractive because several of the ligands differing between modalities 1 and 2 are of course validated ligands at CB_1 receptors, for example, anandamide. Moreover, this hypothesis is testable by combinatorial expression of CB_1 with GPR55. Signaling of CB_1 in this notional $GPR55/CB_1$ heterodimer is presumably masked, similar to the Δ-opioid receptor/SNSR-4 heterodimer, since Ryberg et al. (2007) observed no activation of $G_{\alpha i}$, and WIN55,212-2 was inactive. HEK293 cells, used to express GPR55 by both Ryberg et al. (2007) and Johns et al. (2007), are not generally regarded as having endogenous cannabinoid pharmacology, so contingent on this hypothesis would be the ability of GPR55 to upregulate expression of CB$_1$ receptors. A clue about GPR55-interacting factors may be contained in the fascinating observation that antibodies directed against β1 or β3 integrins, or the integrin ligand fibronectin, can block the anandamide-evoked Ca^{2+} response in endothelial cells (Waldeck-Weiermair et al., 2008). If this Ca^{2+} response is indeed mediated by GPR55, this raises immediate questions of whether the GPR55-integrin interaction involves direct protein–protein contact and, if so, is it sufficient to modify the ligand specificity of GPR55.

C. Allosteric effects and biased agonism

Alternatively, the apparent agonist effects in GTPγS-binding may be due to allosteric modulation. If, for example, low concentrations of LPI were present in the membrane preparations used by Ryberg et al. (2007) and Johns et al. (2007) cannabinoids such as anandamide might act at as modulators, potentiating effects at an orthosteric LPI-binding site but without having being pure agonists in their own right. This seems unlikely, because CP55,940 would be predicted to behave as an antagonist, but in fact it activates GTPγS binding. Furthermore, O-1602 and abnormal cannabidiol have not been observed to potentiate GPR55 agonists in the yeast assay (A. J. Brown, unpublished observations). A third property of GPCRs that can underlie differing patterns of ligand specificity is biased agonism. This is not fully satisfactory for GPR55, however, primarily because most examples of biased agonism involve changes in orders of efficacy among panels of ligands when comparing different signaling cascades (generally, different G proteins), whereas the Ca^{2+} and GTPγS-binding assays of GPR55 both involve signaling via G$_{\alpha 13}$. To probe these models, it will be useful to establish both Ca^{2+} and GTPγS-binding assays in a single, GPR55-expressing, host cell line.

D. Pharmacology in recombinant and natural systems

Screening activities to identify novel small-molecule ligands and drug leads for GPCRs take place almost wholly in recombinant systems. Therefore, a critical consideration is the extent to which recombinant pharmacology parallels receptor specificity at the physiological site of action. At first sight, there are clear similarities between the behavior of GPR55 in GTPγS-binding and the previously characterized endothelial vasodilator site. In both cases, anandamide, O-1602, and abnormal cannabidiol activate, and cannabidiol and O-1918 antagonize, the target (Fig. 5.4). However, there are several important differences. As has been noted, in GTPγS-binding, rimonabant behaves as an agonist (Brown and Wise, 2001), whereas antagonism by rimonabant is a defining property of the vascular endothelial anandamide site (Wagner *et al.*, 1999; White and Hiley, 1997a). Rimonabant also acts as an antagonist at the endothelial Ca^{2+}-coupled site characterized by Waldeck-Weiermair *et al.* (2008), the overall pharmacology of which is closely reminiscent of the anandamide site in vascular tissue. Likewise, HU210 and AM251 (Ryberg *et al.*, 2007) behave as agonists in GTPγS-binding but are inactive at endothelial sites in both tissue- and cell-based experiments. It is thus not yet clear which assay, if any, best predicts whether newly identified ligands are likely to have GPR55-mediated effects *in vivo*.

There are two opposing interpretations of the data of Waldeck-Weiermair *et al.* (2008). If their identification of the Ca^{2+}-coupled anandamide receptor as GPR55 is correct, the third signaling modality illustrated in Fig. 5.4 must represent yet another phenotype manifested by GPR55. This would mean that the recombinant assay systems do not fully align with *in vivo* ligand specificity, emphasizing our need to understand the mechanisms involved, and with implications for progression of GPR55 as a therapeutic target. The opposing interpretation is that the identification of GPR55 by Waldeck-Weiermair *et al.* (2008) is not accurate. If, for example, the knockdown or overexpression of GPR55 in EA.hy926 cells had pleiotropic effects, for example, on cell health, the observed effects on Ca^{2+} responses might not be the specific result of modulating GPR55 protein levels. This could account for the paradox between the observations of Waldeck-Weiermair *et al.* (2008) and the lack of effect on vasodilator responses following genetic ablation of GPR55 (Johns *et al.*, 2007).

To resolve these two viewpoints, a key area for future enquiry will be to gain greater validation for GPR55 function in endothelial cells. A fuller investigation of the G protein involved in the anandamide-evoked Ca^{2+} response, and in particular whether $G_{\alpha 13}$ is involved, could support GPR55, as could specific anti-GPR55 antibodies. However, full confirmation is likely to require new and selective pharmacological tools. Even if GPR55 does not mediate the anandamide-evoked Ca^{2+} response, EA.hy926 cells nonetheless offer a unique insight into the signaling mechanisms downstream of this receptor and an excellent opportunity to identify them at the molecular level.

E. Anandamide and GPR55

In summary, anandamide does appear capable of activating GPR55, though under restricted circumstances. The effect is not consistently observed across all assay systems suggesting the effect is not simple or classical agonism. Various hypotheses to explain the data have been explored here, but none offer a fully satisfactory explanation. In particular it is not clear whether anandamide binds directly to GPR55 or that the binding site arises from interaction between GPR55 and partner proteins. Lastly, there is currently insufficient clarity to allow classification of GPR55 alongside CB_1 and CB_2 as a third cannabinoid receptor. LPI is apparently a more likely endogenous ligand for GPR55 (following IUPHAR guidelines, GPR55 would be classified as LPI_1), although we would recommend characterization of the molecule as GPR55 should continue at least until the mechanisms of receptor activation underlying the paradoxical data reviewed here are better understood.

REFERENCES

Baker, D., Pryce, G., Davies, W. L., and Hiley, C. R. (2006). In silico patent searching reveals a new cannabinoid receptor. *Trends Pharmacol. Sci.* **27,** 1–4.

Baran, D. T., and Kelly, A. M. (1988). Lysophosphatidylinositol: A potential mediator of 1,25-dihydroxyvitamin D-induced increments in hepatocyte cytosolic calcium. *Endocrinology* **122,** 930–934.

Barnett-Norris, J., Hurst, D. P., Lynch, D. L., Guarnieri, F., Makriyannis, A., and Reggio, P. H. (2002). Conformational memories and the endocannabinoid binding site at the cannabinoid CB1 receptor. *J. Med. Chem.* **45,** 3649–3659.

Bayewitch, M., Avidor-Reiss, T., Levy, R., Barg, J., Mechoulam, R., and Vogel, Z. (1995). The peripheral cannabinoid receptor: Adenylate cyclase inhibition and G protein coupling. *FEBS Lett.* **375,** 143–147.

Begg, M., Mo, F. M., Offertaler, L., Batkai, S., Pacher, P., Razdan, R. K., Lovinger, D. M., and Kunos, G. (2003). G protein-coupled endothelial receptor for atypical cannabinoid ligands modulates a Ca^{2+}-dependent K^+ current. *J. Biol. Chem.* **278,** 46188–46194.

Bosier, B., and Hermans, E. (2007). Versatility of GPCR recognition by drugs: From biological implications to therapeutic relevance. *Trends Pharmacol. Sci.* **28,** 438–446.

Bosier, B., Tilleux, S., Najimi, M., Lambert, D. M., and Hermans, E. (2007). Agonist selective modulation of tyrosine hydroxylase expression by cannabinoid ligands in a murine neuroblastoma cell line. *J. Neurochem.* **102,** 1996–2007.

Breit, A., Gagnidze, K., Devi, L. A., Lagace, M., and Bouvier, M. (2006). Simultaneous activation of the delta opioid receptor (deltaOR)/sensory neuron-specific receptor-4 (SNSR-4) hetero-oligomer by the mixed bivalent agonist bovine adrenal medulla peptide 22 activates SNSR-4 but inhibits deltaOR signaling. *Mol. Pharmacol.* **70,** 686–696.

Breivogel, C. S., Griffin, G., Di Marzo, V., and Martin, B. R. (2001). Evidence for a new G protein-coupled cannabinoid receptor in mouse brain. *Mol. Pharmacol.* **60,** 155–163.

Brown, A. J. (2007). Novel cannabinoid receptors. *Br. J. Pharmacol.* **152,** 567–575.

Brown, A. J., Dyos, S. L., Whiteway, M. S., White, J. H., Watson, M. A., Marzioch, M., Clare, J. J., Cousens, D. J., Paddon, C., Plumpton, C., Romanos, M. A., and Dowell, S. J. (2000). Functional coupling of mammalian receptors to the yeast mating pathway using novel yeast/mammalian G protein alpha-subunit chimeras. *Yeast* **16,** 11–22.

Brown, A. J., and Wise, A. (2001). Identification of modulators of GPR55 activity, GlaxoSmithKline. Patent WO0186305.

Corda, D., Iurisci, C., and Berrie, C. P. (2002). Biological activities and metabolism of the lysophoinositides and glycerophoinositols. *Biochim. Biophys. Acta* **1582,** 52–69.

Di Marzo, V., Fontana, A., Cadas, H., Schinelli, S., Cimino, G., Schwartz, J. C., and Piomelli, D. (1994). Formation and inactivation of endogenous cannabinoid anandamide in central neurons. *Nature* **372**, 686–691.

Dowell, S. J., and Brown, A. J. (2002). Yeast assays for G-protein-coupled receptors. *Recept. Channels* **8**, 343–352.

Drmota, E., Greasley, P., and Groblewski, T. (2004). Screening assays for cannabinoid-ligand type modulators, AstraZeneca. Patent WO2004074844.

Duncan, M., Millns, P., Smart, D., Wright, J. E., Kendall, D. A., and Ralevic, V. (2004). Noladin ether, a putative endocannabinoid, attenuates sensory neurotransmission in the rat isolated mesenteric arterial bed via a non-CB1/CB2 $G_{i/o}$ linked receptor. *Br. J. Pharmacol.* **142**, 509–518.

Elphick, M. R., and Egertová, M. (2005). The phylogenetic distribution and evolutionary origins of endocannabinoid signalling. *Handbook Exp. Pharmacol.* **168**, 283–297.

Elphick, M. R., Satou, Y., and Satoh, N. (2003). The invertebrate ancestry of endocannabinoid signalling: An orthologue of vertebrate cannabinoid receptors in the urochordate Ciona intestinalis. *Gene* **302**, 95–101.

Falasca, M., Silletta, M. G., Carvelli, A., Di Francesco, A. L., Fusco, A., Ramakrishna, V., and Corda, D. (1995). Signalling pathways involved in the mitogenic action of lysophosphatidylinositol. *Oncogene* **10**, 2113–2124.

Fride, E., Foox, A., Rosenberg, E., Faigenboim, M., Cohen, V., Barda, L., Blau, H., and Mechoulam, R. (2003). Milk intake and survival in newborn cannabinoid CB1 receptor knockout mice: Evidence for a "CB3" receptor. *Eur. J. Pharmacol.* **461**, 27–34.

Galve-Roperh, I., Rueda, D., Gomez del Pulgar, T., Velasco, G., and Guzman, M. (2002). Mechanism of extracellular signal-regulated kinase activation by the CB_1 cannabinoid receptor. *Mol. Pharmacol.* **62**, 1385–1392.

Gaoni, Y., and Mechoulam, R. (1964). Isolation, structure and partial synthesis of an active constituent of haschish. *J. Am. Chem. Soc.* **86**, 1646–1648.

Glass, M., and Felder, C. C. (1997). Concurrent stimulation of cannabinoid CB_1 and dopamine D_2 receptors augments cAMP accumulation in striatal neurons: Evidence for a Gs linkage to the CB_1 receptor. *J. Neurosci.* **17**, 5327–5333.

Hajos, N., Ledent, C., and Freund, T. F. (2001). Novel cannabinoid-sensitive receptor mediates inhibition of glutamatergic synaptic transmission in the hippocampus. *Neuroscience* **106**, 1–4.

Hay, D. L., Poyner, D. R., and Sexton, P. M. (2006). GPCR modulation by RAMPs. *Pharmacol. Ther.* **109**, 173–197.

Henstridge, C. M., Balenga, N. A., Ford, L. A., Ross, R. A., Waldhoer, M., and Irving, A. J. (2009). The GPR55 ligand L-α-lysophosphatidylinositol promotes RhoA-dependent Ca^{2+} signaling and NFAT activation. *FASEB J.* **23**(1), 183–193.

Hersch, E., Huang, J., Grider, J. R., and Murthy, K. S. (2004). Gq/G13 signaling by ET-1 in smooth muscle: MYPT1 phosphorylation via ET_A and CPI-17 dephosphorylation via ETB. *Am. J. Physiol.* **287**, C1209–1218.

Hiley, C. R., and Kaup, S. S. (2007). GPR55 and the vascular receptors for cannabinoids. *Br. J. Pharmacol.* **152**, 559–561.

Ho, B. Y., Stadnicka, A., Prather, P. L., Buckley, A. R., Current, L. L., Bosnjak, Z. J., and Kwok, W. M. (2000). Cannabinoid CB_1 receptor-mediated inhibition of prolactin release and signaling mechanisms in GH4C1 cells. *Endocrinology* **141**, 1675–1685.

Ho, W. S., and Hiley, C. R. (2003a). Endothelium-independent relaxation to cannabinoids in rat-isolated mesenteric artery and role of Ca^{2+} influx. *Br. J. Pharmacol.* **139**, 585–597.

Ho, W. S., and Hiley, C. R. (2003b). Vasodilator actions of abnormal-cannabidiol in rat isolated small mesenteric artery. *Br. J. Pharmacol.* **138**, 1320–1332.

Ho, W. S., and Hiley, C. R. (2004). Vasorelaxant activities of the putative endocannabinoid virodhamine in rat isolated small mesenteric artery. *J. Pharm. Pharmacol.* **56**, 869–875.

Howlett, A. C., Barth, F., Bonner, T. I., Cabral, G., Casellas, P., Devane, W. A., Felder, C. C., Herkenham, M., Mackie, K., Martin, B. R., Mechoulam, R., and Pertwee, R. G. (2002). International Union of Pharmacology. XXVII. Classification of cannabinoid receptors. *Pharmacol. Rev.* **54,** 161–202.

Howlett, A. C., Qualy, J. M., and Khachatrian, L. L. (1986). Involvement of G_i in the inhibition of adenylate cyclase by cannabimimetic drugs. *Mol. Pharmacol.* **29,** 307–313.

Járai, Z., Wagner, J. A., Varga, K., Lake, K. D., Compton, D. R., Martin, B. R., Zimmer, A. M., Bonner, T. I., Buckley, N. E., Mezey, E., Razdan, R. K., Zimmer, A., et al. (1999). Cannabinoid-induced mesenteric vasodilation through an endothelial site distinct from CB_1 or CB_2 receptors. *Proc. Natl. Acad. Sci. USA* **96,** 14136–14141.

Johns, D. G., Behm, D. J., Walker, D. J., Ao, Z., Shapland, E. M., Daniels, D. A., Riddick, M., Dowell, S., Staton, P. C., Green, P., Shabon, U., Bao, W., et al. (2007). The novel endocannabinoid receptor GPR55 is activated by atypical cannabinoids but does not mediate their vasodilator effects. *Br. J. Pharmacol.* **152,** 825–831.

Jones, G., Pertwee, R. G., Gill, E. W., Paton, W. D., Nilsson, I. M., Widman, M., and Agurell, S. (1974). Relative pharmacological potency in mice of optical isomers of delta 1-tetrahydrocannabinol. *Biochem. Pharmacol.* **23,** 439–446.

Kenakin, T. (1995). Agonist-receptor efficacy. II. Agonist trafficking of receptor signals. *Trends Pharmacol. Sci.* **16,** 232–238.

Kenakin, T. (2007a). Collateral efficacy in drug discovery: Taking advantage of the good (allosteric) nature of 7TM receptors. *Trends Pharmacol. Sci.* **28,** 407–415.

Kenakin, T. (2007b). Functional selectivity through protean and biased agonism: Who steers the ship? *Mol. Pharmacol.* **72,** 1393–1401.

Kilts, J. D., Lin, S. S., Lowe, J. E., and Kwatra, M. M. (2007). Selective activation of human atrial Galpha12 and Galpha13 by $G_{\alpha q}$-coupled angiotensin and endothelin receptors. *J. Cardiovasc. Pharmacol.* **50,** 299–303.

Kobayashi, T., Kishimoto, M., and Okuyama, H. (1996). Phospholipases involved in lysophosphatidylinositol metabolism in rat brain. *J. Lipid Mediat. Cell Signal.* **14,** 33–37.

Kukkonen, J. P., Jansson, C. C., and Akerman, K. E. (2001). Agonist trafficking of $G_{i/o}$-mediated alpha(2A)-adrenoceptor responses in HEL 92.1.7 cells. *Br. J. Pharmacol.* **132,** 1477–1484.

Lagaud, G. J., Skarsgard, P. L., Laher, I., and van Breemen, C. (1999). Heterogeneity of endothelium-dependent vasodilation in pressurized cerebral and small mesenteric resistance arteries of the rat. *J. Pharmacol. Exp. Ther.* **290,** 832–839.

Lauckner, J. E., Jensen, J. B., Chen, H. Y., Lu, H. C., Hille, B., and Mackie, K. (2008). GPR55 is a cannabinoid receptor that increases intracellular calcium and inhibits M current. *Proc. Natl. Acad. Sci. USA* **105,** 2699–2704.

Lawrence, D. K., and Gill, E. W. (1975). The effects of delta1-tetrahydrocannabinol and other cannabinoids on spin-labeled liposomes and their relationship to mechanisms of general anesthesia. *Mol. Pharmacol.* **11,** 595–602.

Lee, C. W., Rivera, R., Gardell, S., Dubin, A. E., and Chun, J. (2006). GPR92 as a new $G_{12/13}$- and G_q-coupled lysophosphatidic acid receptor that increases cAMP, LPA5. *J. Biol. Chem.* **281,** 23589–23597.

Maingret, F., Patel, A. J., Lesage, F., Lazdunski, M., and Honore, E. (2000). Lysophospholipids open the two-pore domain mechano-gated K^+ channels TREK-1 and TRAAK. *J. Biol. Chem.* **275,** 10128–10133.

Mariggio, S., Bavec, A., Natale, E., Zizza, P., Salmona, M., Corda, D., and Di Girolamo, M. (2006). $G_{\alpha 13}$ mediates activation of the cytosolic phospholipase A2α through fine regulation of ERK phosphorylation. *Cell. Signal.* **18,** 2200–2208.

Matsuda, L. A., Lolait, S. J., Brownstein, M. J., Young, A. C., and Bonner, T. I. (1990). Structure of a cannabinoid receptor and functional expression of the cloned cDNA. *Nature* **346,** 561–564.

McAllister, S. D., Rizvi, G., Anavi-Goffer, S., Hurst, D. P., Barnett-Norris, J., Lynch, D. L., Reggio, P. H., and Abood, M. E. (2003). An aromatic microdomain at the cannabinoid CB(1) receptor constitutes an agonist/inverse agonist binding region. *J. Med. Chem.* **46,** 5139–5152.

Mechoulam, R., Ben-Shabat, S., Hanus, L., Ligumsky, M., Kaminski, N. E., Schatz, A. R., Gopher, A., Almog, S., Martin, B. R., Compton, D. R., Pertwee, R., Griffin, G., et al. (1995). Identification of an endogenous 2-monoglyceride, present in canine gut, that binds to cannabinoid receptors. *Biochem. Pharmacol.* **50,** 83–90.

Milligan, G. (2007). G protein-coupled receptor dimerisation: Molecular basis and relevance to function. *Biochim. Biophys. Acta* **1768,** 825–835.

Mukhopadhyay, S., Shim, J. Y., Assi, A. A., Norford, D., and Howlett, A. C. (2002). CB_1 cannabinoid receptor-G protein association: A possible mechanism for differential signaling. *Chem. Phys. Lipids* **121,** 91–109.

Munro, S., Thomas, K. L., and Abu-Shaar, M. (1993). Molecular characterization of a peripheral receptor for cannabinoids. *Nature* **365,** 61–65.

Noguchi, K., Ishii, S., and Shimizu, T. (2003). Identification of p2y9/GPR23 as a novel G protein-coupled receptor for lysophosphatidic acid, structurally distant from the Edg family. *J. Biol. Chem.* **278,** 25600–25606.

Offertáler, L., Mo, F. M., Bátkai, S., Liu, J., Begg, M., Razdan, R. K., Martin, B. R., Bukoski, R. D., and Kunos, G. (2003). Selective ligands and cellular effectors of a G protein-coupled endothelial cannabinoid receptor. *Mol. Pharmacol.* **63,** 699–705.

Oka, S., Nakajima, K., Yamashita, A., Kishimoto, S., and Sugiura, T. (2007). Identification of GPR55 as a lysophosphatidylinositol receptor. *Biochem. Biophys. Res. Commun.* **362,** 928–934.

Pasternack, S. M., von Kügelgen, I., Aboud, K. A., Lee, Y. A., Rüschendorf, F., Voss, K., Hillmer, A. M., Molderings, G. J., Franz, T., Ramirez, A., Nürnberg, P., Nöthen, M. M., et al. (2008). G protein-coupled receptor P2Y5 and its ligand LPA are involved in maintenance of human hair growth. *Nat. Genet.* **40,** 329–334.

Patel, A. J., Lazdunski, M., and Honore, E. (2001). Lipid and mechano-gated 2P domain K^+ channels. *Curr. Opin. Cell Biol.* **13,** 422–428.

Profirovic, J., Gorovoy, M., Niu, J., Pavlovic, S., and Voyno-Yasenetskaya, T. (2005). A novel mechanism of G protein-dependent phosphorylation of vasodilator-stimulated phosphoprotein. *J. Biol. Chem.* **280,** 32866–32876.

Rhee, M. H., Bayewitch, M., Avidor-Reiss, T., Levy, R., and Vogel, Z. (1998). Cannabinoid receptor activation differentially regulates the various adenylyl cyclase isozymes. *J. Neurochem.* **71,** 1525–1534.

Rinaldi-Carmona, M., Barth, F., Heaulme, M., Alonso, R., Shire, D., Congy, C., Soubrie, P., Breliere, J. C., and Le Fur, G. (1995). Biochemical and pharmacological characterisation of SR141716A, the first potent and selective brain cannabinoid receptor antagonist. *Life Sci.* **56,** 1941–1947.

Ryberg, E., Larsson, N., Sjögren, S., Hjorth, S., Hermansson, N. O., Leonova, J., Elebring, T., Nilsson, K., Drmota, T., and Greasley, P. J. (2007). The orphan receptor GPR55 is a novel cannabinoid receptor. *Br. J. Pharmacol.* **152,** 1092–1101.

Sanchez, C., de Ceballos, M. L., del Pulgar, T. G., Rueda, D., Corbacho, C., Velasco, G., Galve-Roperh, I., Huffman, J. W., Ramon y Cajal, S., and Guzman, M. (2001). Inhibition of glioma growth *in vivo* by selective activation of the CB_2 cannabinoid receptor. *Cancer Res.* **61,** 5784–5789.

Sawzdargo, M., Nguyen, T., Lee, D. K., Lynch, K. R., Cheng, R., Heng, H. H., George, S. R., and O'Dowd, B. F. (1999). Identification and cloning of three novel human G protein-coupled receptor genes GPR52, PsiGPR53 and GPR55: GPR55 is extensively expressed in human brain. *Brain Res. Mol. Brain Res.* **64,** 193–198.

Shumay, E., Tao, J., Wang, H. Y., and Malbon, C. C. (2007). Lysophosphatidic acid regulates trafficking of β_2-adrenergic receptors: The $G_{\alpha 13}$/p115RhoGEF/JNK pathway stimulates receptor internalization. *J. Biol. Chem.* **282,** 21529–21541.

Simoncini, T., Scorticati, C., Mannella, P., Fadiel, A., Giretti, M. S., Fu, X. D., Baldacci, C., Garibaldi, S., Caruso, A., Fornari, L., Naftolin, F., and Genazzani, A. R. (2006). Estrogen receptor alpha interacts with Galpha13 to drive actin remodeling and endothelial cell migration via the RhoA/Rho kinase/moesin pathway. *Mol. Endocrinol.* **20,** 1756–1771.

Uhlenbrock, K., Gassenhuber, H., and Kostenis, E. (2002). Sphingosine 1-phosphate is a ligand of the human gpr3, gpr6 and gpr12 family of constitutively active G protein-coupled receptors. *Cell. Signal.* **14,** 941–953.

Wagner, J. A., Varga, K., Jarai, Z., and Kunos, G. (1999). Mesenteric vasodilation mediated by endothelial anandamide receptors. *Hypertension* **33,** 429–434.

Waldeck-Weiermair, M., Zoratti, C., Osibow, K., Balenga, N., Goessnitzer, E., Waldhoer, M., Malli, R., and Graier, W. F. (2008). Integrin clustering enables anandamide-induced Ca^{2+} signaling in endothelial cells via GPR55 by protection against CB1-receptor-triggered repression. *J. Cell Sci.* **121,** 1704–1717.

Wang, J., Simonavicius, N., Wu, X., Swaminath, G., Reagan, J., Tian, H., and Ling, L. (2006). Kynurenic acid as a ligand for orphan G protein-coupled receptor GPR35. *J. Biol. Chem.* **281,** 22021–22028.

White, R. (1999). Mechanisms of vasodilatation in the rat mesenteric artery. PhD Thesis, University of Cambridge.

White, R., and Hiley, C. R. (1997a). A comparison of EDHF-mediated and anandamide-induced relaxations in the rat isolated mesenteric artery. *Br. J. Pharmacol.* **122,** 1573–1584.

White, R., and Hiley, C. R. (1997b). Endothelium and cannabinoid receptor involvement in levcromakalim vasorelaxation. *Eur. J. Pharmacol.* **339,** 157–160.

White, R., and Hiley, C. R. (1998). The actions of some cannabinoid receptor ligands in the rat isolated mesenteric artery. *Br. J. Pharmacol.* **125,** 533–541.

White, R., Ho, W. S., Bottrill, F. E., Ford, W. R., and Hiley, C. R. (2001). Mechanisms of anandamide-induced vasorelaxation in rat isolated coronary arteries. *Br. J. Pharmacol.* **134,** 921–929.

Zygmunt, P. M., Petersson, J., Andersson, D. A., Chuang, H., Sorgard, M., Di Marzo, V., Julius, D., and Hogestatt, E. D. (1999). Vanilloid receptors on sensory nerves mediate the vasodilator action of anandamide. *Nature* **400,** 452–457.

CHAPTER SIX

The Endocannabinoid System During Development: Emphasis on Perinatal Events and Delayed Effects

Ester Fride,[*] Nikolai Gobshtis,[*] Hodaya Dahan,[*,†] Aron Weller,[†] Andrea Giuffrida,[‡] *and* Shimon Ben-Shabat[§]

Contents

I. Introduction	140
II. Early Gestation	141
A. The preimplantation embryo and implantation—Critical role for the ECS system	141
B. CB_1 receptors and preterm birth: Corticosterone and prenatal stress	144
III. Neural Development	146
IV. Postnatal Development	147
V. Effects of Developmental Manipulation of the ECS System on the Offspring	150
VI. Conclusions	152
Acknowledgments	153
References	153

Abstract

The endocannabinoid system (ECS) including its receptors, endogenous ligands ("endocannabinoids"), synthesizing and degradating enzymes, and transporter molecules has been detected from the earliest embryonal stages and throughout pre- and postnatal development; endocannabinoids, notably 2-arachidonoylglycerol, are also present in maternal milk.

---

[*] Departments of Behavioral Sciences and Molecular Biology, Ariel University Center of Samaria, Ariel, Israel
[†] Department of Psychology, Bar-Ilan University, Ramat Gan, Israel
[‡] Department of Pharmacology, University of Texas Health Science Center, San Antonio, Texas, USA
[§] Department of Pharmacology and School of Pharmacy, Ben-Gurion University of the Negev, Beer Sheva, Israel

During three developmental stages, (1) early embryonal, (2) prenatal brain development, and (3) postnatal suckling, the ECS plays an essential role for development and survival. During early gestation, successful embryonal passage through the oviduct and implantation into the uterus require critical enzymatic control of the endocannabinoids. During fetal life, endocannabinoids and the cannabinoid CB_1 receptor are important for brain development, regulating neural progenitor differentiation and guiding axonal migration and synaptogenesis. Postnatally, CB_1 receptor activation by 2-arachidonoylglycerol appears to play a critical role in the initiation of milk suckling in mouse pups, possibly by enabling innervation and/or activation of the tongue muscles.

Perinatal manipulation of the ECS, by administering cannabinoids or by maternal marijuana consumption, alters neurotransmitter and behavioral functions in the offspring. Interestingly, the sequelae of prenatal cannabinoids are similar to many effects of prenatal stress, which may suggest that prenatal stress impacts on the ECS and that vice versa prenatal cannabinoid exposure may interfere with the ability of the fetus to cope with the stress.

Future studies should further clarify the mechanisms involved in the developmental roles of the ECS and understand better the adverse effects of prenatal exposure, to design strategies for the treatment of conditions including infertility, addiction, and failure-to-thrive.

I. Introduction

The endocannabinoid system (ECS) including its well-established receptors CB_1 and CB_2, the putative CB_3 (Fride *et al.*, 2003; Howlett *et al.*, 2002), the orphan receptor GPR55 (which not only shares some pharmacological properties with CB_1 and CB_2, but also differs from CB_1 and CB_2 in cannabinoid responsivity; Baker *et al.*, 2006), endogenous ligands ("endocannabinoids"), synthesizing and degradating enzymes, as well as transporter molecules, have been observed in almost every brain structure and organ system and have been shown to participate in a large number of physiological functions (Fride, 2002, 2006).

The ECS system is actively present from the earliest stages of ontogenetic development. In this chapter, the essential role of the ECS system at three crucial stages of development will be described: (1) embryonal (pre) implantation, (2) prenatal neuronal development, and (3) newborn suckling (see also Fride, 2004; Harkany *et al.*, 2007). In view of these areas of involvement in the developing brain, we will review some of the consequences of early life manipulations of the ECS.

II. Early Gestation

A. The preimplantation embryo and implantation—Critical role for the ECS system

The ECS system has been detected in virtually all components of the reproductive system and at virtually all stages of fertilization and development (Maccarrone and Finazzi-Agro, 2004; Schuel and Burkman, 2005). Thus for example, cannabinoid receptors and anandamide have been detected in sperm cells from sea urchins as well as humans and other species. Anandamide may be released from sea urchin egg cells, when it is postulated to modulate the rate of their fertilization by the sperm. This modulation is accomplished by inhibition of the acrosome reaction (Schuel and Burkman, 2005). Anandamide is also synthesized in rodent testis, uterus, and oviduct and has been found in human uterus and seminal fluid, while the anandamide degrading enzyme FAAH (fatty acid amide hydrolase; see Fride, 2002) has been detected in the uterus (see Schuel and Burkman, 2005), oviduct, and preimplantation (blastocyst) embryo (Sun and Dey, 2008). Furthermore, CB_1 and CB_2 receptors have been described in the preimplantation mouse embryo (Yang et al., 1996), the CB_1 receptor at higher concentrations than those in the brain (Yang et al., 1996). CB_1 receptors are also present in the oviduct and uterus (Maccarrone, 2008; Maccarrone and Finazzi-Agro, 2004). These observations led to the discovery that the ECS, primarily through the activation of CB_1 receptors, tightly regulates both the oviductal transport of the embryo into the uterus, and its implantation into the uterine wall (Maccarrone, 2008; Wang et al., 2006b). Thus, cannabinoids and endocannabinoids arrest the development of two-cell embryos into blastocysts, whereas anandamide levels in the uterus are very high and anandamide binding to CB_1 receptors on blastocytes conversely would lead to cell death (Maccarrone and Finazzi-Agro, 2004). Therefore, for implantation to take place, uterine anandamide levels have to be reduced on the day and at the site of implantation. Failing to do so prevents implantation of the embryo (Maccarrone and Finazzi-Agro, 2004; Paria and Dey, 2000). Thus, reducing anandamide levels at the implantation site is a critical condition for the implantation and hence survival of the embryo. On the other hand, anandamide binding to CB_1 receptors at the interimplantations sites is required to inhibit gap junctions and thus facilitates the necessary uterine changes for normal gestation (Maccarrone and Finazzi-Agro, 2004).

As to the importance of the ECS for oviductal transport, either CB_1 receptor blockade with SR141716 or pharmacological activation with THC or anandamide resulted in pregnancy loss due to oviductal retention, that is, failure to reach the uterus for subsequent implantation. Similarly, CB_1 receptor knockout mice (but not CB_2 receptor knockouts) displayed 40% pregnancy loss and dysfunctional oviductal transport (Wang et al., 2006a,b). These observations imply that an optimal endocannabinoid tone is required for oviductal transport of the embryo (Sun and Dey, 2008).

Anandamide levels in turn appear to be fine-tuned by varying levels of FAAH. We further know that $mRNA^{FAAH}$ is present in preimplantation and implanting embryos as well as at the implantation site of the uterus (Paria and Dey, 2000; Paria et al., 1999), where it inversely correlates with anandamide levels. Maccarrone and colleagues reported FAAH downregulation in the uterus of pregnant and pseudopregnant mice during the implantation period (Maccarrone et al., 2000a), and in clinical studies this group showed that FAAH concentrations in lymphocytes of women who miscarried were lower than lymphocyte FAAH concentrations from women who gave birth (Maccarrone et al., 2000b). In addition, plasma anandamide levels were significantly higher in a subsample of pregnant women with a threatened miscarriage than in those women who gave birth at term (Habayeb et al., 2008). Recently, in placental tissue samples from pregnant women, an association was found between the expression of FAAH in the nuclei of trophoblasts, suggesting that low anandamide levels at this locality may be an additional factor potentially compromising successful implantation (Chamley et al., 2008).

An additional locale for anandamide's control of successful gestation has also been demonstrated. Increasing levels of the anandamide synthesizing enzyme NAPE–PLD (N-acylphosphatidylethanolamine–selective phospholipase D) together with decreasing FAAH levels along the oviduct was shown to be essential for successful transport of the preimplantation embryo and unimpaired pregnancy outcome (Wang et al., 2006b). Taken together, it appears that the control of anandamide levels by its enzymes and possibly a specific anandamide membrane transporter (AMT) is responsible for creating optimal anandamide concentrations at the localities involved in embryonal implantation and development (Maccarrone and Finazzi-Agro, 2004).

In view of the multiple and critical roles of FAAH in pregnancy outcome, FAAH-like compounds could be developed, as suggested previously, as a novel approach to infertility therapy (Maccarrone and Finazzi-Agro, 2004; Maccarrone et al., 2005). Recently, the endocannabinoid 2-arachidonoylglycerol (2-AG) was identified in uterus and blastocytes and seems to have a similar role in embryonal implantation to anandamide (Wang et al., 2007). Interestingly, uterine 2-AG levels, similarly to those measured in milk (Fride et al., 2001) and in the developing brain (Fernandez-Ruiz et al., 2004), occur at about 1000-fold higher levels compared to those of anandamide (Wang et al., 2007).

Studies with selective receptor antagonists indicated that cannabinoid-induced embryonal growth arrest is mediated by CB_1 and not by CB_2

receptors (Paria and Dey, 2000). However, observations on CB_1- and CB_2-deficient mice as well as CB_1–CB_2 double knockouts have provided further evidence for a role of both receptor subtypes in the synchronization of early gestational events necessary for embryonal implantation (Paria et al., 2001).

The adipocyte-secreted hormone leptin is negatively balanced with endocannabinoids in regulating energy homeostasis (Di Marzo et al., 2001). Leptin serves additional functions including fertility, by stimulating embryonal development (Mechoulam and Fride, 2001). Leptin defective ob/ob mice are infertile. Investigating the cause of this infertility, we have reported previously that levels of both anandamide and 2-AG were elevated in the uterus of ob/ob mice with respect to wild-type littermates, due to reduced FAAH activity in the case of anandamide, and to reduced monoacylglycerol lipase and enhanced diacylglycerol lipase activity in the case of 2-AG. The process mediating endocannabinoid cellular uptake was also impaired in ob/ob mice, whereas the levels of cannabinoid and anandamide receptors were not modified. Leptin reversed these effects in the ob/ob mice (Maccarrone et al., 2005). Despite the negative interaction between the endocannabinoid and leptin systems at the biochemical level, in vivo infertility of ob/ob mice was not restored by chronic administration of the CB_1 receptor antagonist SR141716 (rimonabant), nor did the endocannabinoid reuptake inhibitor OMDM-1 interfere with fertility in wild-type females (Maccarrone et al., 2005). Thus, the interaction between the leptin and ECS systems for in vivo regulation of fertility should be studied more closely.

Taken together, the complex yet essential role of the ECS system in fertilization, oviductal transport and implantation, may explain the reported association between early miscarriage and marijuana smoking (see Maccarrone and Finazzi-Agro, 2004) and perhaps also the observation that maternal and/or paternal marijuana consumption, before conception and/or birth has been shown to negatively affect pregnancy outcomes in couples undergoing in vitro fertilization, including fewer oocyte retrieval and lower birth weights (Klonoff-Cohen et al., 2006). Adverse effects, primarily lower birth weights, and greater degrees of premature births have been associated previously with pregnancies of women who had regularly consumed marijuana during pregnancy (Hingson et al., 1982; Sherwood et al., 1999; Zuckerman et al., 1989). However, other studies have yielded negative (Shiono et al., 1995) or inconclusive outcomes (Day et al., 1991; Fergusson et al., 2002; Hatch and Bracken, 1986; Kline et al., 1987; Linn et al., 1983). A meta-analysis which included 10 previously published studies controlling for cigarette smoking found no evidence that prenatal exposure to cannabis through maternal consumption resulted in low birth weight (English et al., 1997).

In view of the multilevel involvement of the ECS in early gestational processes, the rather modest and controversial effects of prenatal cannabinoid exposure on fertility and birth outcomes seems surprising. However, it is possible that the specific requirements at different localities and points in time during pregnancy produce variable effects of gestational cannabinoid exposure on fertility and birth. Therefore, more specific manipulations of these events, in place and time, may yield more consistent outcomes and clarify the mechanisms involved. Moreover, manipulating multiple components of the ECS system simultaneously based on the complex involvement of the ECS in gestational events should be explored as a therapeutic approach to infertility.

B. CB_1 receptors and preterm birth: Corticosterone and prenatal stress

Cannabinoid CB_1 receptor blockade with rimonabant (SR141716) administered during late pregnancy in mice resulted in a modest (less than 1 day in mice) but highly significant shortening of pregnancy. Similarly, CB_1 receptor knockout mice displayed a shorter gestational period than *wild-type* mice. Analogous silencing of the CB_2 receptor did not affect gestational length (Wang *et al.*, 2008). Interestingly, a slight but significant reduction on gestational length was reported in a population of heavy (>6 times a week) marijuana smoking women (Fried *et al.*, 1984).

The study by Dey and colleagues (Wang *et al.*, 2008) also reported a shift in peak levels of corticosterone (CCS) to an earlier gestational age in CB_1 receptor *knockout* mice (from day 17–18 in *wild type* to day 15–16 CB_1 in *knockouts*). Strikingly, a similar shift of the maternal CCS peak values was observed in prenatally stressed rats (day 17 compared to day 18 of gestation in controls) (Fride and Weinstock, 1985; Weinstock *et al.*, 1988). These observations support our earlier contention that the sequelae of prenatal manipulation of the ECS, such as seen after marijuana consumption during pregnancy, overlap with the consequences of prenatal stress (Fride and Mechoulam, 1996). Therefore, we speculate that the impairments in emotional regulation and cognitive performance induced by prenatal exposure to stress (Fride and Weinstock, 1988; Fride *et al.*, 1985, 1986, 2004), which have also been observed as a consequence of pre- and perinatal cannabinoid exposure (Campolongo *et al.*, 2007; Mereu *et al.*, 2003; Newsom and Kelly, 2008; Trezza *et al.*, 2008), may be explained by maternal stress–induced changes in fetal ECS development (Fride and Mechoulam, 1996). Indeed, recently, we have shown that "ultramild" prenatal stress (Meek *et al.*, 2000) results in an alteration of the "prepulse inhibition (PPI) of the startle reflex" (Fig. 6.1A), a behavioral assay for sensorimotor gating, used as an assessment of adaptability and schizophrenia-like symptoms (Crawley, 2000; Varty *et al.*, 2001). Thus, we observed that the PPI in prenatally stressed mice was not adversely affected; rather a more robust PPI response was observed than in controls. Similarly, a lack of adverse effects on the PPI response was reported both in mice which were exposed to repeated postnatal stress

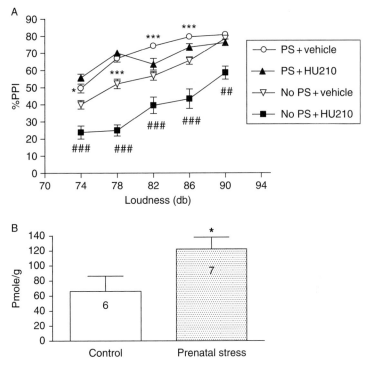

Figure 6.1 Prenatal stress enhances hippocampal levels of anandamide and sensorimotor gating (prepulse inhibition of the startle reflex) in mice and decreases the response to a cannabinoid CB receptor agonist. Male mice which had been stressed prenatally (PS) using an "ultramild stress regime" (Meek et al., 2000) in utero was (A) tested for sensorimotor gating by using the "prepulse inhibition of the startle response" assay (PPI, Kinder|Scientific, San Diego). Half the mice were injected 24 h previously with vehicle, the other half with the CB agonist HU210 (0.1 mg/kg). The PPI (expressed in % as a function of the acoustic startle response) was significantly elevated in PS animals. However, PS did not respond to the administration of HU210 (two-way ANOVAs were followed by Bonferroni post-hoc analyses). (B) The hippocampal anandamide levels of the PS animals were significantly higher compared to those in controls (two-tailed t-test, $p < 0.05$). Anandamide was extracted from hippocampal tissue punches in chloroform/methanol/water (2:1:0.5 v/v/v), purified by solid-phase extraction, derivitized with BSTFA (30 μl) and quantified by gas chromatography/mass spectrometry using an isotope dilution assay as previously described (Hardison et al., 2006). *$p < 0.05$ significantly different from nonstressed controls; ***$p < 0.001$ significantly different from nonstressed controls; ##$p < 0.01$ significantly different from nonstressed, vehicle controls; ###$p < 0.001$ significantly different from nonstressed, vehicle control.

(maternal separation) (Millstein et al., 2006) and in rats which had been exposed prenatally to a cannabinoid receptor agonist (WIN 55,212-2) (Bortolato et al., 2006). Importantly, when we acutely administered the potent synthetic cannabinoid receptor agonist HU210 (0.1 mg/kg), PPI was again not affected in the prenatally stressed males, as opposed to the severe impairment seen in controls (Fig. 6.1A). The latter observation is compatible with the previously reported cannabinoid-induced reduction in the PPI response in naive animals (Schneider and Koch, 2002). Our finding of a reduced response to a CB_1 receptor agonist after prenatal stress suggests that the ECS had become less sensitive after prenatal stress. Moreover, this altered response pattern was accompanied by significantly higher levels of anandamide in the hippocampi of prenatally stressed mice (Fig. 6.1B). The role of the hippocampus in PPI has been described previously (Bast and Feldon, 2003). Although we did not assess CB_1 receptor expression in this study, we may speculate that as a result of elevated anandamide levels, these receptors may have been downregulated thus explaining the hyporesponsiveness to HU210. An alternative scenario would suggest a compensatory elevation of anandamide levels in response to downregulated CB_1 receptors. Although the precise mechanism awaits further clarification, our data taken together demonstrate that prenatal stress alters the development of the ECS which is expressed at adulthood, at least in male mice subjected to ultramild stress.

The clinical importance of this observation is that the consumption of marijuana or the administration of cannabinoid-based medicinal drugs during pregnancy may interfere with ECS-mediated protection of the developing fetus against physiological perturbations due to the maternal stress.

III. Neural Development

Cannabinoid CB_1 and CB_2 receptor mRNA has been described around day 11 of gestation in embryonic rat brain (Buckley et al., 1998). Postnatally, a gradual increase in CB_1 receptor mRNA (McLaughlin and Abood, 1993), and in the density of CB_1 receptors has been measured (Rodriguez de Fonseca et al., 1993) in whole brain. Similar developmental patterns of CB_1 receptors were found during human pre- and postnatal development. Thus, CB_1 receptors were detected at week 14 of gestation in the human embryo (Biegon and Kerman, 2001). A progressive increase in the concentrations of CB_1 receptors was found in the frontal cortex, hippocampus basal ganglia, and cerebellum between the fetal period and adulthood (Mato et al., 2003). Interestingly, during the 20th week of gestation, a selective expression of CB_1 receptors was recorded in the limbic area CA of the hippocampus and the basal nuclear group of the amygdala, as compared to a more homogeneous expression of CB_1 mRNA in the adult human brain (Wang et al., 2003), suggesting a role for CB_1 receptors in the development of emotional processing.

Endocannabinoids are also present from early development. During the fetal period, anandamide is present at much lower (almost 1000-fold) concentrations than 2-AG (Fernandez-Ruiz et al., 2000). Moreover, the developmental pattern differs between the two endocannabinoids. Thus, whereas concentrations of anandamide gradually increase throughout development until adult levels are reached (Berrendero et al., 1999) (Fig. 6.1A), fetal levels of 2-AG are similar to those in young and in adult brains, with a peak value however, on the first day after birth in rats (Berrendero et al., 1999) (Fig. 6.1B).

A remarkable aspect of CB_1 receptor presence in the developing brain is its transient appearance on neural fiber tracts (white matter areas). Thus in human fetal brain, the highest cannabinoid receptor-binding densities (measured by [^{3}H]CP55940 autoradiography) appear in the pyramidal tract, brachium conjunctivum and subventricular germinative (proliferative) zones (Mato et al., 2003) as well as in the corpus callosum, stria terminalis, anterior commissure, midbrain, and cerebral cortex (Fernandez-Ruiz et al., 2000, 2004; Romero et al., 1997). In these same areas, WIN 55,212-2 (a cannabinoid receptor agonist)-stimulated [^{35}S] GTPγS binding (a measure of receptor activation) was also at its highest, indicating that these CB_1 receptors were functional (Mato et al., 2003).

The abundance of the endocannabinoids and their receptors in the developing nervous system and in these "atypical" regions renders the ECS a likely candidate for the regulation of the structural and functional maturation of the nervous system. Indeed, a series of recent studies has demonstrated a fundamental role for the endocannabinoids and the CB_1 receptor in various aspects of neural development including neurogenesis, glia formation, neuronal migration, axonal elongation, and fasciculation (Berghuis et al., 2007; Fernandez-Ruiz et al., 2004; Harkany et al., 2007; Mulder et al., 2008). A neurogenic effect of the synthetic cannabinoid HU210 was reported in the embryonic and adult hippocampus. The cellular effect in adult rats was accompanied by an antianxiety and antidepressant activity of chronic treatment with HU210. These results were interpreted to mean that HU210 promotes neurogenesis which in turn produces behavioral alterations (Jiang et al., 2005).

IV. Postnatal Development

Weanling offspring of undernourished dams displayed lower body weights and levels of anandamide compared to controls, while 2-AG concentrations were not influenced (Matias et al., 2003). Pups depend on maternal fatty acid precursor supply for the production of long chain polyunsaturated fatty acids. Therefore, given that dietary supplementation with essential fatty acids increased concentrations of anandamide but not of 2-AG in piglets (Berger et al., 2001), Matias et al. (2003) estimated that maternal undernutrition-induced decrease in hypothalamic anandamide concentrations in the offspring may have resulted from a disruption in essential fatty acids supplies from the maternal blood and/or from milk.

Several years ago, we reported the presence of endocannabinoids in bovine as well as human milk; 2-AG was present in at least 100–1000-fold higher concentrations than anandamide (Fride et al., 2001). This observation, together with the high levels of CB_1 receptor mRNA and 2-AG which have been observed on the first day of life in structures including the hypothalamic ventromedial nucleus (Berrendero et al., 1999) (which is associated with feeding behavior), suggested that pup brain-derived 2-AG comprises a major stimulus for the newborn to initiate milk ingestion immediately after birth. Indeed, in a series of studies performed in neonatal mice, we have demonstrated that CB_1 receptor activation is critically important for the initiation of the suckling response. Thus, when the CB_1 receptor antagonist SR141716 is injected in newborn mice, milk ingestion and subsequent growth are dramatically compromised in most pups (75–100%) and death follows within days after antagonist administration (Fride et al., 2001). The antagonist must be administered within 24 h of birth to obtain the full effect: injections on day 2 result in a 50% death rate; SR141716 administration on day 5 has no effect at all on pup growth and survival (Fride et al., 2003). To determine whether the proximity to birth, rather than the developmental stage of the pup is critical for the impaired suckling induced by SR141716, we recently injected the antagonist into newborn precocial mice (Egyptian Spiny mice, *Acomys cahirinus*), which are born with open eyes and the ability to walk, run, and ingest solid food. Our data show that SR141716 also significantly delayed physical growth in these pups (E. Fride et al., unpublished results).

Furthermore, we investigated the effects of VCHSR1, a CB_1 receptor antagonist, which unlike the inverse agonist activity of SR141716, presumably only causes a neutral receptor blockade (Hurst et al., 2002). The data demonstrated that VCHSR has growth arresting effects, similarly to those of SR141716 (Fride et al., 2007). Thus, the impaired suckling induced by neonatally administered SR141716 is not limited to this specific compound, but is apparently due to the cessation of an endocannabinoid "tone" or the inhibition of constitutive CB_1 receptor activity in the newborn.

Additional studies indicated that the dramatic effect of CB_1 receptor blockade is dose dependent and further supported a specific CB_1 receptor-mediated effect. Thus, coapplication of Δ^9-THC with SR141716 almost completely reversed the SR141716-induced growth failure (Fride et al., 2001). Further, CB_1 receptor-deficient mice displayed deficient milk suckling on the first days of life, while by day 3 of life they had developed normal suckling behavior. Their weight gain though, remained significantly lower than the C57BL/6 background strain. Further, as expected, the growth curve of CB_1 receptor knockout mice was not affected by neonatal injections of the CB_1 antagonist. On the other hand, survival rate and the initiation of the suckling response were significantly inhibited by the CB_1 receptor blocker, suggesting the existence of an additional "CB_3" receptor, possibly upregulated in the $CB_1^{-/-}$ knockout mice (Fride et al., 2003). The phenomenon appears to have a genetic component since its severity varied between the three strains of mice studied (Sabra, C57BL/6, and ICR, unpublished results).

In a further set of experiments, 2–11-day-old pups which had been injected with SR141716, or with vehicle on day 1 of life, were exposed to anesthetized nursing dams. While vehicle-injected pups all located the nipples and nursed from the dam on every testing day, the SR141716-injected pups approached the nipple but could not suckle, thus lacking the oral-motor strength to ingest milk through the nipple. However, we observed recently that when exposed to a dish with a milk/cream mixture, which can be ingested by licking ("lapping") without the need for sucking, the SR141716-treated pups were able to ingest the same amount of milk as controls (Fig. 6.3A) and also derive significant benefit from this ingestion in terms of body weight gain (Fig. 6.3B and C). This series of experiments suggests that the SR141716-treated pups have severe oral-motor impairment. Interestingly, as described above, anandamide plays a fundamental role in axon guidance and synaptogenesis (Berghuis et al., 2007), while 2-AG has been shown to be required for axonal growth (Williams et al., 2003). Moreover, blockade of CB_1 receptors inhibited the axonal targeting of CB_1 receptors by causing a sequestration of CB_1 receptors on somatodendritic membranes in cultured hippocampal neurons (Leterrier et al., 2006). It has also been shown that endocannabinoid signaling is required for elongation and fasciculation of cerebral cortical neurons (Mulder et al., 2008). Therefore, it is possible that neonatal CB_1 receptor blockade interferes with CB_1 trafficking to the synaptic region. If similar processes occur in peripheral nerves, we hypothesize that the CB_1 receptor antagonist at birth may interfere with fasciculation of axonal tract formation of suckling-relevant nerves, thus compromising milk suckling. A highly relevant finding is that CB_1 receptor activation participates in the modulation of glycinergic synaptic currents in hypoglossal motoneurons of postnatal rats (Mukhtarov et al., 2005), while resection of the hypoglossal nerve in rat pups compromised milk suckling, resulting in 100% mortality (Fujita et al., 2006). Thus, we speculate that when pups are treated with SR141716 at birth, incomplete synaptogenesis of the hypoglossal nerve may fail to adequately activate tongue movement (Fujita et al., 2006) which is critical for sucking (Mukhtarov et al., 2005).

Based on the critical importance of the timing of CB_1 receptor blockade and the abundance of 2-AG in maternal milk and in the postnatal brain as described above, we designed a working model (Fride, 2008), which suggests that pup-derived 2-AG release at birth enables the first milk sucking session (via CB_1 receptor activation). In the normal situation, 2-AG will be supplemented with milk-derived 2-AG, thus enabling CB_1 receptor activation during the next nursing bout. With blocked CB_1 receptors soon after birth, we have postulated that the sucking apparatus cannot be activated by the pup's 2-AG; milk is not ingested and brain-derived 2-AG, being now the sole source of 2-AG, reverts to low levels, insufficient to activate enough CB_1 receptors required for sucking during the next nursing session. As a result, the neonate does not ingest sufficient milk for growth and survival (see Fride, 2008 for a diagrammatic model). This functional deficit may complement the hypothesized structural deficit in hypoglossal nerve synapses described in the previous paragraph.

V. Effects of Developmental Manipulation of the ECS System on the Offspring

In this chapter, three developmental stages have been described in which the ECS system plays a crucial role. It is therefore not surprising that maternal exposure to external cannabinoids during pregnancy or the nursing period (by marijuana smoking or administration of cannabinoids in animal experiments) produces long-term effects on the fetus and postnatal or adult offspring (Campolongo et al., 2007; Fride, 2004; Fride and Mechoulam, 1996; Ramos et al., 2002). For example, at midgestation, human fetuses of marijuana smoking mothers had lower body weights than controls (Hurd et al., 2005) while in another study on midgestational fetuses of marijuana smoking mothers, a selective decrease in dopamine D2 receptors was observed in the amygdala, but no changes in CB_1 mRNA and in dopamine D1 receptors were found in this structure. Neither were there any changes in the other structures studied, the striatum and hippocampus (Wang et al., 2004). Data from animal studies have shown that prenatal and postnatal exposure to Δ^9-THC interfered with normal dopamine-dependent motor functions and the hypothalamic–pituitary–adrenal axis in the adult offspring (Ramos et al., 2002). Further, prenatal Δ^9-THC given to rats, facilitated morphine and heroin self administration in the female offspring (Spano et al., 2007; Vela et al., 1998), while a number of brain areas including the prefrontal cortex, amygdala, and hippocampus displayed altered concentrations of μ-opioid receptors (Vela et al., 1998); another study reported changes in preproenkephalin mRNA expression in the nucleus accumbens and amygdala of the offspring (Spano et al., 2007). Memory retention in the adult offspring in a passive avoidance task was disrupted by prenatal exposure to the synthetic cannabinoid WIN 55,212.

The memory impairment was correlated with a shortening of long-term potentiation and a reduction in extracellular glutamate in the hippocampus (Mereu et al., 2003).

In an elegantly designed prospective study of the children of marijuana smoking mothers, Fried and colleagues have specifically pointed at a subtle but significant impairment of higher cognitive ("executive") functioning, which is ascribed to the prefrontal cortex (Fried et al., 2003) and only becomes apparent from the age of 4 (Fried and Smith, 2001). Importantly, in a study using functional MRI recording, the same offspring, but now at young adulthood, displayed a bilateral increase in neural activity in the prefrontal cortex and elevated activity in the right premotor cortex (Smith et al., 2004). In a different prospective study, on children of low socioeconomic status, marijuana-consuming pregnant women, cognitive and attention impairments were detected at young ages and significantly higher rates of depressive symptoms compared to the control population (Gray et al., 2005; Trezza et al., 2008).

We have performed experiments on the adult offspring of mice injected daily with Δ^9-THC during the last week of gestation, with the aim to detect changes in the ECS system *per se* (Fride and Mechoulam, 1996). Thus, we found both behavioral (Fride and Mechoulam, 1996) and, recently, biochemical evidence (Fig. 6.2) that the CB_1 receptors were upregulated in the

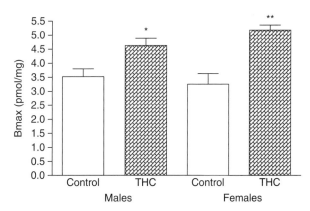

Figure 6.2 Cannabinoid CB_1 receptors' density (B_{max}) was significantly elevated in male and female offspring of mothers which had been treated with daily Δ^9-THC (20 mg/kg) over the last week of gestation. Scatchard analysis was performed to determine affinity (K_i) and density (B_{max}) of CB_1 receptors. CB_1 receptor affinity and density in the forebrains of the adult offspring were measured by competition-binding assay using [3-H]HU-243 as the labeled ligand, as described in detail (Ben-Shabat et al., 1998; Rhee et al., 1997). Δ^9-THC-treated male and female animals displayed a highly significant elevation in CB_1 receptors (two-way ANOVA, drug effect: $F_{(1,14)} = 28.3$, $p = 0.0001$), which was especially evident in the females. *$p < 0.05$, Bonferroni post-hoc analysis, cf. control; ***$p < 0.001$, Bonferroni post-hoc analysis, cf. control.

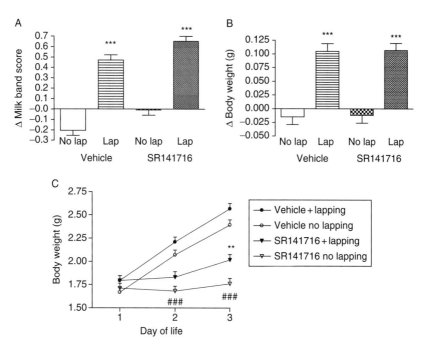

Figure 6.3 Cannabinoid CB_1 receptor blockade at birth allows pups to lick ("lap") food and gain weight. Pups (ICR strain) which had been treated with the CB_1 receptor antagonist SR141716 (rimonabant 20 mg/kg on day 1 of life) after 3 h of maternal deprivation. Preliminary data indicate clearly that both vehicle- and rimonabant-treated pups were able to effectively ingest a mixture of milk (50%) and cream (50%), as measured by (A) the presence of milk into the stomach ("milkbands") and (B) body weight increase after the "meal." (C) Over the course of the first 3 days of life, the rimonabant-treated pups which were allowed to lap gained significantly more weight than those that were not ($p < 0.001$, two-way ANOVA), although the growth delay compared to vehicle-treated pups remained ($p < 0.001$, two-way ANOVA).

brains of prenatally exposed offspring. These data are consistent with an overactive ECS resulting perhaps, in a greater vulnerability to the addictive potential of cannabis or other drugs (Vela et al., 1998). In support of this suggestion, offspring of marijuana smoking mothers were found to be more frequent marijuana users at the age of 14 (Day et al., 2006).

VI. Conclusions

Endocannabinoids and their receptors are highly abundant components of the developing organism as has been demonstrated in a number of species. This ECS system is not only "concerned" with its own preparation for proper functioning in the adult organism, but it also critically impacts on general development of the organism. We have highlighted here its essential role in embryonal implantation and the events preceding and resulting from it; in the development of the nervous system, regulating processes including neurogenesis, axon guidance, and synaptogenesis. Immediately after birth, cannabinoid CB_1 receptor activation appears to play an essential role in the initiation of suckling, which is necessary for successful postnatal growth, development, and thriving.

Much more knowledge still needs to be accumulated to understand the precise mechanisms by which the ECS system controls development and to utilize this knowledge toward the alleviation of conditions such as infertility, mental retardation, and failure-to-thrive.

ACKNOWLEDGMENTS

EF was supported by National Institute for Psychobiology in Israel. HD was supported by the Israel foundation for research and educational support for doctoral students. Some of the data presented here (Fig. 6.3) are part of the requirements toward the PhD dissertation of HD.

REFERENCES

Baker, D., Pryce, G., Davies, W. L., and Hiley, C. R. (2006). In silico patent searching reveals a new cannabinoid receptor. *Trends Pharmacol. Sci.* **27,** 1–4.

Bast, T., and Feldon, J. (2003). Hippocampal modulation of sensorimotor processes. *Prog. Neurobiol.* **70,** 319–345.

Ben-Shabat, S., Fride, E., Sheskin, T., Tamiri, T., Rhee, M. H., Vogel, Z., Bisogno, T., De Petrocellis, L., Di Marzo, V., and Mechoulam, R. (1998). An entourage effect: Inactive endogenous fatty acid glycerol esters enhance 2-arachidonoyl-glycerol cannabinoid activity. *Eur. J. Pharmacol.* **353,** 23–31.

Berger, A., Crozier, G., Bisogno, T., Cavaliere, P., Innis, S., and Di Marzo, V. (2001). Anandamide and diet: Inclusion of dietary arachidonate and docosahexaenoate leads to increased brain levels of the corresponding N-acylethanolamines in piglets. *Proc. Natl. Acad. Sci. USA* **98,** 6402–6406.

Berghuis, P., Rajnicek, A. M., Morozov, Y. M., Ross, R. A., Mulder, J., Urban, G. M., Monory, K., Marsicano, G., Matteoli, M., Canty, A., Irving, A. J., Katona, I., et al. (2007). Hardwiring the brain: Endocannabinoids shape neuronal connectivity. *Science* **316,** 1212–1216.

Berrendero, F., Sepe, N., Ramos, J. A., Di Marzo, V., and Fernandez-Ruiz, J. J. (1999). Analysis of cannabinoid receptor binding and mRNA expression and endogenous cannabinoid contents in the developing rat brain during late gestation and early postnatal period. *Synapse* **33,** 181–191.

Biegon, A., and Kerman, I. A. (2001). Autoradiographic study of pre- and postnatal distribution of cannabinoid receptors in human brain. *Neuroimage* **14,** 1463–1468.

Bortolato, M., Frau, R., Orru, M., Casti, A., Aru, G. N., Fa, M., Manunta, M., Usai, A., Mereu, G., and Gessa, G. L. (2006). Prenatal exposure to a cannabinoid receptor agonist does not affect sensorimotor gating in rats. *Eur. J. Pharmacol.* **531,** 166–170.

Buckley, N. E., Hansson, S., Harta, G., and Mezey, E. (1998). Expression of the CB1 and CB2 receptor messenger RNAs during embryonic development in the rat. *Neuroscience* **82,** 1131–1149.

Campolongo, P., Trezza, V., Cassano, T., Gaetani, S., Morgese, M. G., Ubaldi, M., Soverchia, L., Antonelli, T., Ferraro, L., Massi, M., Ciccocioppo, R., and Cuomo, V. (2007). Perinatal exposure to delta-9-tetrahydrocannabinol causes enduring cognitive deficits associated with alteration of cortical gene expression and neurotransmission in rats. *Addict. Biol.* **12,** 485–495.

Chamley, L. W., Bhalla, A., Stone, P. R., Liddell, H., O'Carroll, S., Kearn, C., and Glass, M. (2008). Nuclear localisation of the endocannabinoid metabolizing enzyme fatty acid amide hydrolase (FAAH) in invasive trophoblasts and an association with recurrent miscarriage. *Placenta* **29,** 970–975.

Crawley, J. N. (2000). "What's Wrong with My Mouse? Behavioral Phenotyping of Transgenic and Knockout Mice." Wiley-Liss, New York.

Day, N., Sambamoorthi, U., Taylor, P., Richardson, G., Robles, N., Jhon, Y., Scher, M., Stoffer, D., Cornelius, M., and Jasperse, D. (1991). Prenatal marijuana use and neonatal outcome. *Neurotoxicol. Teratol.* **13,** 329–334.

Day, N. L., Goldschmidt, L., and Thomas, C. A. (2006). Prenatal marijuana exposure contributes to the prediction of marijuana use at age 14. *Addiction* **101,** 1313–1322.

Di Marzo, V., Goparaju, S. K., Wang, L., Liu, J., Batkai, S., Jarai, Z., Fezza, F., Miura, G. I., Palmiter, R. D., Sugiura, T., and Kunos, G. (2001). Leptin-regulated endocannabinoids are involved in maintaining food intake. *Nature* **410,** 822–825.

English, D. R., Hulse, G. K., Milne, E., Holman, C. D., and Bower, C. I. (1997). Maternal cannabis use and birth weight: A meta-analysis. *Addiction* **92,** 1553–1560.

Fergusson, D. M., Horwood, L. J., and Northstone, K. (2002). Maternal use of cannabis and pregnancy outcome. *BJOG* **109,** 21–27.

Fernandez-Ruiz, J., Berrendero, F., Hernandez, M. L., and Ramos, J. A. (2000). The endogenous cannabinoid system and brain development. *Trends Neurosci.* **23,** 14–20.

Fernandez-Ruiz, J., Gomez, M., Hernandez, M., de Miguel, R., and Ramos, J. A. (2004). Cannabinoids and gene expression during brain development. *Neurotox. Res.* **6,** 389–401.

Fride, E. (2002). Endocannabinoids in the central nervous system—An overview. *Prostaglandins Leukot. Essent. Fatty Acids* **66,** 221–233.

Fride, E. (2004). The endocannabinoid-CB1 receptor system during gestation and postnatal development. *Eur. J. Pharmacol.* **500,** 289–297.

Fride, E. (2006). The endocannabinoid-CB receptor system: A new player in the brain-gut-adipose field. *Biomed. Rev.* **17,** 1–20.

Fride, E. (2008). Multiple roles for the endocannabinoid system during the earliest stages of life: Pre- and postnatal development. *J. Neuroendocrinol.* **20**(Suppl. 1), 75–81.

Fride, E., and Mechoulam, R. (1996). Developmental aspects of anandamide: Ontogeny of response and prenatal exposure. *Psychoneuroendocrinology* **21,** 157–172.

Fride, E., and Weinstock, M. (1985). The role of maternal corticosterone in the mediation of prenatal predictable and unpredictable stress. *J. Neurochem.* **44,** S70C.

Fride, E., and Weinstock, M. (1988). Prenatal stress increases anxiety related behavior and alters cerebral lateralization of dopamine activity. *Life Sci.* **42,** 1059–1065.

Fride, E., Dan, Y., Gavish, M., and Weinstock, M. (1985). Prenatal stress impairs maternal behavior in a conflict situation and reduces hippocampal benzodiazepine receptors. *Life Sci.* **36,** 2103–2109.

Fride, E., Dan, Y., Feldon, J., Halevy, G., and Weinstock, M. (1986). Effects of prenatal stress on vulnerability to stress in prepubertal and adult rats. *Physiol. Behav.* **37,** 681–687.

Fride, E., Ginzburg, Y., Breuer, A., Bisogno, T., Di Marzo, V., and Mechoulam, R. (2001). Critical role of the endogenous cannabinoid system in mouse pup suckling and growth. *Eur. J. Pharmacol.* **419,** 207–214.

Fride, E., Foox, A., Rosenberg, E., Faigenboim, M., Cohen, V., Barda, L., Blau, H., and Mechoulam, R. (2003). Milk intake and survival in newborn cannabinoid CB1 receptor knockout mice: Evidence for a "CB3" receptor. *Eur. J. Pharmacol.* **461**, 27–34.

Fride, E., Ben-Shabat, S., and Gobshtis, N. (2004). Prenatal stress, the endogenous cannabinoid system, schizophrenia and depression. *In* "32nd Annual Meeting of the Society for Neuroscience", San Diego, USA.

Fride, E., Braun, H., Matan, H., Steinberg, S., Reggio, P. H., and Seltzman, H. H. (2007). Inhibition of milk ingestion and growth after administration of a neutral cannabinoid CB1 receptor antagonist on the first postnatal day in the mouse. *Pediatr. Res.* **62**, 533–536.

Fried, P. A., and Smith, A. M. (2001). A literature review of the consequences of prenatal marihuana exposure. An emerging theme of a deficiency in aspects of executive function. *Neurotoxicol. Teratol.* **23**, 1–11.

Fried, P. A., Watkinson, B., and Willan, A. (1984). Marijuana use during pregnancy and decreased length of gestation. *Am. J. Obstet. Gynecol.* **150**, 23–27.

Fried, P. A., Watkinson, B., and Gray, R. (2003). Differential effects on cognitive functioning in 13- to 16-year-olds prenatally exposed to cigarettes and marihuana. *Neurotoxicol. Teratol.* **25**, 427–436.

Fujita, K., Yokouchi, K., Fukuyama, T., Fukushima, N., Kawagishi, K., and Moriizumi, T. (2006). Effects of hypoglossal and facial nerve injuries on milk-suckling. *Int. J. Dev. Neurosci.* **24**, 29–34.

Gray, K. A., Day, N. L., Leech, S., and Richardson, G. A. (2005). Prenatal marijuana exposure: Effect on child depressive symptoms at ten years of age. *Neurotoxicol. Teratol.* **27**, 439–448.

Habayeb, O. M., Taylor, A. H., Finney, M., Evans, M. D., and Konje, J. C. (2008). Plasma anandamide concentration and pregnancy outcome in women with threatened miscarriage. *JAMA* **299**, 1135–1136.

Hardison, S., Weintraub, S. T., and Giuffrida, A. (2006). Quantification of endocannabinoids in rat biological samples by GC/MS: Technical and theoretical considerations. *Prostaglandins Other Lipid Mediat.* **81**, 106–112.

Harkany, T., Guzman, M., Galve-Roperh, I., Berghuis, P., Devi, L. A., and Mackie, K. (2007). The emerging functions of endocannabinoid signaling during CNS development. *Trends Pharmacol. Sci.* **28**, 83–92.

Hatch, E. E., and Bracken, M. B. (1986). Effect of marijuana use in pregnancy on fetal growth. *Am. J. Epidemiol.* **124**, 986–993.

Hingson, R., Alpert, J. J., Day, N., Dooling, E., Kayne, H., Morelock, S., Oppenheimer, E., and Zuckerman, B. (1982). Effects of maternal drinking and marijuana use on fetal growth and development. *Pediatrics* **70**, 539–546.

Howlett, A. C., Barth, F., Bonner, T. I., Cabral, G., Casellas, P., Devane, W. A., Felder, C. C., Herkenham, M., Mackie, K., Martin, B. R., Mechoulam, R., and Pertwee, R. G. (2002). International Union of Pharmacology. XXVII. Classification of cannabinoid receptors. *Pharmacol. Rev.* **54**, 161–202.

Hurd, Y. L., Wang, X., Anderson, V., Beck, O., Minkoff, H., and Dow-Edwards, D. (2005). Marijuana impairs growth in mid-gestation fetuses. *Neurotoxicol. Teratol.* **27**, 221–229.

Hurst, D. P., Lynch, D. L., Barnett-Norris, J., Hyatt, S. M., Seltzman, H. H., Zhong, M., Song, Z. H., Nie, J., Lewis, D., and Reggio, P. H. (2002). N-(piperidin-1-yl)-5-(4-chlorophenyl)-1-(2,4-dichlorophenyl)-4-methyl-1H-pyrazole-3-carboxamide (SR141716A) interaction with LYS 3.28(192) is crucial for its inverse agonism at the cannabinoid CB1 receptor. *Mol. Pharmacol.* **62**, 1274–1287.

Jiang, W., Zhang, Y., Xiao, L., Van Cleemput, J., Ji, S. P., Bai, G., and Zhang, X. (2005). Cannabinoids promote embryonic and adult hippocampus neurogenesis and produce anxiolytic- and antidepressant-like effects. *J. Clin. Invest.* **115**, 3104–3116.

Kline, J., Stein, Z., and Hutzler, M. (1987). Cigarettes, alcohol and marijuana: Varying associations with birth weight. *Int. J. Epidemiol.* **16,** 44–51.
Klonoff-Cohen, H. S., Natarajan, L., and Chen, R. V. (2006). A prospective study of the effects of female and male marijuana use on *in vitro* fertilization (IVF) and gamete intrafallopian transfer (GIFT) outcomes. *Am. J. Obstet. Gynecol.* **194,** 369–376.
Leterrier, C., Laine, J., Darmon, M., Boudin, H., Rossier, J., and Lenkei, Z. (2006). Constitutive activation drives compartment-selective endocytosis and axonal targeting of type 1 cannabinoid receptors. *J. Neurosci.* **26,** 3141–3153.
Linn, S., Schoenbaum, S. C., Monson, R. R., Rosner, R., Stubblefield, P. C., and Ryan, K. J. (1983). The association of marijuana use with outcome of pregnancy. *Am. J. Public Health* **73,** 1161–1164.
Maccarrone, M. (2008). CB2 receptors in reproduction. *Br. J. Pharmacol.* **153,** 189–198.
Maccarrone, M., and Finazzi-Agro, A. (2004). Anandamide hydrolase: A guardian angel of human reproduction? *Trends Pharmacol. Sci.* **25,** 353–357.
Maccarrone, M., De Felici, M., Bari, M., Klinger, F., Siracusa, G., and Finazzi-Agro, A. (2000a). Down-regulation of anandamide hydrolase in mouse uterus by sex hormones. *Eur. J. Biochem.* **267,** 2991–2997.
Maccarrone, M., Valensise, H., Bari, M., Lazzarin, N., Romanini, C., and Finazzi-Agro, A. (2000b). Relation between decreased anandamide hydrolase concentrations in human lymphocytes and miscarriage. *Lancet* **355,** 1326–1329.
Maccarrone, M., Fride, E., Bisogno, T., Bari, M., Cascio, M. G., Battista, N., Finazzi Agro, A., Suris, R., Mechoulam, R., and Di Marzo, V. (2005). Up-regulation of the endocannabinoid system in the uterus of leptin knockout (ob/ob) mice and implications for fertility. *Mol. Hum. Reprod.* **11,** 21–28.
Matias, I., Leonhardt, M., Lesage, J., De Petrocellis, L., Dupouy, J. P., Vieau, D., and Di Marzo, V. (2003). Effect of maternal under-nutrition on pup body weight and hypothalamic endocannabinoid levels. *Cell. Mol. Life Sci.* **60,** 382–389.
Mato, S., Del Olmo, E., and Pazos, A. (2003). Ontogenetic development of cannabinoid receptor expression and signal transduction functionality in the human brain. *Eur. J. Neurosci.* **17,** 1747–1754.
McLaughlin, C. R., and Abood, M. E. (1993). Developmental expression of cannabinoid receptor mRNA. *Brain Res. Dev. Brain Res.* **76,** 75–78.
Mechoulam, R., and Fride, E. (2001). Physiology. A hunger for cannabinoids. *Nature* **410,** 763–765.
Meek, L. R., Burda, K. M., and Paster, E. (2000). Effects of prenatal stress on development in mice: Maturation and learning. *Physiol. Behav.* **71,** 543–549.
Mereu, G., Fa, M., Ferraro, L., Cagiano, R., Antonelli, T., Tattoli, M., Ghiglieri, V., Tanganelli, S., Gessa, G. L., and Cuomo, V. (2003). Prenatal exposure to a cannabinoid agonist produces memory deficits linked to dysfunction in hippocampal long-term potentiation and glutamate release. *Proc. Natl. Acad. Sci. USA* **100,** 4915–4920.
Millstein, R. A., Ralph, R. J., Yang, R. J., and Holmes, A. (2006). Effects of repeated maternal separation on prepulse inhibition of startle across inbred mouse strains. *Genes Brain Behav.* **5,** 346–354.
Mukhtarov, M., Ragozzino, D., and Bregestovski, P. (2005). Dual Ca^{2+} modulation of glycinergic synaptic currents in rodent hypoglossal motoneurones. *J. Physiol.* **569,** 817–831.
Mulder, J., Aguado, T., Keimpema, E., Barabas, K., Ballester Rosado, C. J., Nguyen, L., Monory, K., Marsicano, G., Di Marzo, V., Hurd, Y. L., Guillemot, F., Mackie, K., *et al.* (2008). Endocannabinoid signaling controls pyramidal cell specification and long-range axon patterning. *Proc. Natl. Acad. Sci. USA* **105,** 8760–8765.
Newsom, R. J., and Kelly, S. J. (2008). Perinatal delta-9-tetrahydrocannabinol exposure disrupts social and open field behavior in adult male rats. *Neurotoxicol. Teratol.* **30,** 213–219.

Paria, B. C., and Dey, S. K. (2000). Ligand-receptor signaling with endocannabinoids in preimplantation embryo development and implantation. *Chem. Phys. Lipids* **108**, 211–220.

Paria, B. C., Zhao, X., Wang, J., Das, S. K., and Dey, S. K. (1999). Fatty-acid amide hydrolase is expressed in the mouse uterus and embryo during the periimplantation period. *Biol. Reprod.* **60**, 1151–1157.

Paria, B. C., Song, H., Wang, X., Schmid, P. C., Krebsbach, R. J., Schmid, H. H., Bonner, T. I., Zimmer, A., and Dey, S. K. (2001). Dysregulated cannabinoid signaling disrupts uterine receptivity for embryo implantation. *J. Biol. Chem.* **276**, 20523–20528.

Ramos, J. A., De Miguel, R., Cebeira, M., Hernandez, M., and Fernandez-Ruiz, J. (2002). Exposure to cannabinoids in the development of endogenous cannabinoid system. *Neurotox. Res.* **4**, 363–372.

Rhee, M. H., Vogel, Z., Barg, J., Bayewitch, M., Levy, R., Hanus, L., Breuer, A., and Mechoulam, R. (1997). Cannabinol derivatives: Binding to cannabinoid receptors and inhibition of adenylylcyclase. *J. Med. Chem.* **40**, 3228–3233.

Rodriguez de Fonseca, F., Ramos, J. A., Bonnin, A., and Fernandez-Ruiz, J. J. (1993). Presence of cannabinoid binding sites in the brain from early postnatal ages. *Neuroreport* **4**, 135–138.

Romero, J., Garcia-Palomero, E., Berrendero, F., Garcia-Gil, L., Hernandez, M. L., Ramos, J. A., and Fernandez-Ruiz, J. J. (1997). Atypical location of cannabinoid receptors in white matter areas during rat brain development. *Synapse* **26**, 317–323.

Schneider, M., and Koch, M. (2002). The cannabinoid agonist WIN 55,212-2 reduces sensorimotor gating and recognition memory in rats. *Behav. Pharmacol.* **13**, 29–37.

Schuel, H., and Burkman, L. J. (2005). A tale of two cells: Endocannabinoid-signaling regulates functions of neurons and sperm. *Biol. Reprod.* **73**, 1078–1086.

Sherwood, R. A., Keating, J., Kavvadia, V., Greenough, A., and Peters, T. J. (1999). Substance misuse in early pregnancy and relationship to fetal outcome. *Eur. J. Pediatr.* **158**, 488–492.

Shiono, P. H., Klebanoff, M. A., Nugent, R. P., Cotch, M. F., Wilkins, D. G., Rollins, D. E., Carey, J. C., and Behrman, R. E. (1995). The impact of cocaine and marijuana use on low birth weight and preterm birth: A multicenter study. *Am. J. Obstet. Gynecol.* **172**, 19–27.

Smith, A. M., Fried, P. A., Hogan, M. J., and Cameron, I. (2004). Effects of prenatal marijuana on response inhibition: An FMRI study of young adults. *Neurotoxicol. Teratol.* **26**, 533–542.

Spano, M. S., Ellgren, M., Wang, X., and Hurd, Y. L. (2007). Prenatal cannabis exposure increases heroin seeking with allostatic changes in limbic enkephalin systems in adulthood. *Biol. Psychiatry* **61**, 554–563.

Sun, X., and Dey, S. K. (2008). Aspects of endocannabinoid signaling in periimplantation biology. *Mol. Cell. Endocrinol.* **286**, S3–S11.

Trezza, V., Campolongo, P., Cassano, T., Macheda, T., Dipasquale, P., Carratu, M. R., Gaetani, S., and Cuomo, V. (2008). Effects of perinatal exposure to delta-9-tetrahydrocannabinol on the emotional reactivity of the offspring: A longitudinal behavioral study in Wistar rats. *Psychopharmacology (Berlin)* **198**, 529–537.

Varty, G. B., Walters, N., Cohen-Williams, M., and Carey, G. J. (2001). Comparison of apomorphine, amphetamine and dizocilpine disruptions of prepulse inhibition in inbred and outbred mice strains. *Eur. J. Pharmacol.* **424**, 27–36.

Vela, G., Martin, S., Garcia-Gil, L., Crespo, J. A., Ruiz-Gayo, M., Javier Fernandez-Ruiz, J., Garcia-Lecumberri, C., Pelaprat, D., Fuentes, J. A., Ramos, J. A., and Ambrosio, E. (1998). Maternal exposure to delta9-tetrahydrocannabinol facilitates morphine self-administration behavior and changes regional binding to central mu opioid receptors in adult offspring female rats. *Brain Res.* **807**, 101–109.

Wang, X., Dow-Edwards, D., Keller, E., and Hurd, Y. L. (2003). Preferential limbic expression of the cannabinoid receptor mRNA in the human fetal brain. *Neuroscience* **118,** 681–694.

Wang, X., Dow-Edwards, D., Anderson, V., Minkoff, H., and Hurd, Y. L. (2004). In utero marijuana exposure associated with abnormal amygdala dopamine D2 gene expression in the human fetus. *Biol. Psychiatry* **56,** 909–915.

Wang, H., Dey, S. K., and Maccarrone, M. (2006a). Jekyll and hyde: Two faces of cannabinoid signaling in male and female fertility. *Endocr. Rev.* **27,** 427–448.

Wang, H., Xie, H., Guo, Y., Zhang, H., Takahashi, T., Kingsley, P. J., Marnett, L. J., Das, S. K., Cravatt, B. F., and Dey, S. K. (2006b). Fatty acid amide hydrolase deficiency limits early pregnancy events. *J. Clin. Invest.* **116,** 2122–2131.

Wang, H., Xie, H., Sun, X., Kingsley, P. J., Marnett, L. J., Cravatt, B. F., and Dey, S. K. (2007). Differential regulation of endocannabinoid synthesis and degradation in the uterus during embryo implantation. *Prostaglandins Other Lipid Mediat.* **83,** 62–74.

Wang, H., Xie, H., and Dey, S. K. (2008). Loss of cannabinoid receptor CB1 induces preterm birth. *PLoS ONE* **3,** e3320.

Weinstock, M., Fride, E., and Hertzberg, R. (1988). Prenatal stress effects on functional development of the offspring. *Prog. Brain Res.* **73,** 319–331.

Williams, E. J., Walsh, F. S., and Doherty, P. (2003). The FGF receptor uses the endocannabinoid signaling system to couple to an axonal growth response. *J. Cell Biol.* **160,** 481–486.

Yang, Z. M., Paria, B. C., and Dey, S. K. (1996). Activation of brain-type cannabinoid receptors interferes with preimplantation mouse embryo development. *Biol. Reprod.* **55,** 756–761.

Zuckerman, B., Frank, D. A., Hingson, R., Amaro, H., Levenson, S. M., Kayne, H., Parker, S., Vinci, R., Aboagye, K., Fried, L. E., *et al.* (1989). Effects of maternal marijuana and cocaine use on fetal growth. *N. Engl. J. Med.* **320,** 762–768.

CHAPTER SEVEN

Cannabinoid Receptor CB1 Antagonists: State of the Art and Challenges

Maurizio Bifulco,* Antonietta Santoro,* Chiara Laezza,[†] and Anna Maria Malfitano*

Contents

I. Introduction	160
II. Endocannabinoid System: Control of Energy Balance	161
III. Cannabinoid CB1 Receptors and CB1 Antagonists	162
IV. CB1 Antagonists in the Treatment of Obesity and Related Comorbidities	169
A. Clinical trials	169
B. Studies *in vitro* and in animal models	172
C. Safety and adverse effects	174
V. Other Emerging Effects of CB1 Antagonists	176
VI. Therapeutic Prospects	179
VII. Conclusions	180
Acknowledgments	181
References	181

Abstract

The discovery of cannabinoid receptors led to the development of several compounds targeted against these receptors. In particular, CB1 receptor antagonists have been described to possess key functions in the treatment of obesity and obesity-related pathologies. Numerous clinical trials revealed the advantage of strategies designed to block CB1 receptor but also evidenced the limitations due to side effects exerted by these substances. Recent studies have highlighted that CB1 antagonists could have other effects and find applications even in other pathologies like hepatic fibrosis, chronic inflammatory conditions, diabetes, and cancer. Since the suspending sales of the lead compound, rimonabant, and the discontinuation of all ongoing clinical trials of

---

* Dipartimento di Scienze Farmaceutiche, Università di Salerno, Fisciano (Salerno), Italy
[†] Istituto di Endocrinologia ed Oncologia Sperimentale, CNR, Naples, Italy

CB1 blockers, alternative strategies could emerge and lead to the development of further basic research studies to redirect these compounds.

I. INTRODUCTION

Cannabinoids are the main constituents of the marijuana plant (*Cannabis sativa*), and it is well known that Δ-9-tetrahydrocannabinol (THC) is the primary psychoactive substance in marijuana. Two receptor sites for cannabinoids have been cloned: the most abundant CB1 and the more restricted CB2. Since CB receptors discovery, two major endogenous cannabinoids (endocannabinoids) have been identified: anandamide (AEA) and 2-arachidonoylglycerol (2-AG) (Devane *et al.*, 1992; Matsuda *et al.*, 1990; Munro *et al.*, 1993). Both eicosanoids are derived from lipids and chemically characterized by an arachidonoic acid moiety. A set of enzymes catalyze their biogenesis and degradation (Blankman *et al.*, 2007; Liu *et al.*, 2008; Wang and Ueda, 2008) and together with CB receptors and their endogenous ligands comprise the endocannabinoid system (EC).

Numerous evidences showed that the EC system plays a significant role not only in appetite drive and associated behaviors, but also in endocrine and metabolic regulation and energy balance (Pagotto *et al.*, 2006). Indeed, CB receptors, especially CB1 receptors, are present not only in the brain, but also in peripheral organs, that is, adipose tissue, gut, liver, skeletal muscle, and pancreas (Cota, 2007; Matias and Di Marzo 2007; Pagotto *et al.*, 2006). They participate in the physiological modulation of many central and peripheral functions (Cota, 2007; Matias and Di Marzo 2007; Pagotto *et al.*, 2006). Many studies pointed out that altered endocannabinoid signaling and CB1 receptor expression are involved in several pathophysiological situations, ranging from neurological and psychiatric diseases to eating, cardiovascular, and reproductive disorders. In particular, EC system overactivity has been demonstrated in human obesity, especially in the visceral adipose tissue (Cote *et al.*, 2007; Deedwania, 2009; Matias *et al.*, 2006), which is closely related to high risk of type 2 diabetes and cardiovascular diseases (CVD) (Després and Lemieux, 2006; Van Gaal *et al.*, 2006). The presence of the EC system and CB1 receptors in several organs that play an important role in metabolic disturbances offers a great opportunity for new pharmacological approaches (Di Marzo *et al.*, 2004; Pacher *et al.*, 2006). With the growing worldwide epidemic of both obesity and diabetes mellitus, efforts have intensified to find novel therapies to help patients to lose weight and either to prevent or to control diabetes and its associated cardiovascular diseases. Weight loss is associated with favorable changes in lipid profiles and C-reactive protein, as well as with improved glycemic control and decreased mortality. However, success with behavior modification and currently available medications in achieving and sustaining even mild weight loss is limited. Thus, the EC system represents a new target for pharmacological modulation (Di Marzo *et al.*, 2004; Pacher *et al.*, 2006) and its manipulation via blockade of the CB1 receptor is intriguing particularly in the treatment of obesity and related pathologies.

The aim of this chapter is to evidence the recent emerging knowledge about the modulation of CB1 receptors by their antagonists highlighting the promising therapeutic potential applications and the related difficulties of these compounds.

II. Endocannabinoid System: Control of Energy Balance

Based on the abundance of CB1 receptors in the brain and the tightly regulated expression of endocannabinoids in tegmental area where modulation of rewarding properties of food occurs, and in the hypothalamus where food intake is centrally regulated (Herkenham *et al.*, 1990; Melis *et al.*, 2004), the EC system is considered a major homeostatic system controlling energy balance.

In the central nervous system (CNS) endocannabinoids are synthesized by neurons in response to depolarization (Freund *et al.*, 2003), and act as neurotransmitters mediating a retrograde signal to the presynaptic site where they inhibit GABA and glutamate release in a CB1 receptor-dependent mechanism (Wilson and Nicoll, 2002). Interestingly, endocannabinoids are likely released immediately after biosynthesis and still now no evidence exists for their storage in secretory vesicles. Endocannabinoid production increases between meals and rapidly decreases following access to food suggesting a direct modulation of feeding behavior (Kirkham *et al.*, 2002). Endocannabinoids, by interacting with CB1 receptors, induce a dose-dependent orexigenic effect. The direct administration of AEA into the ventromedial nucleus of the hypothalamus causes a general hyperphagic effect blocked by administration of CB1 antagonists (Jamshidi and Taylor, 2001). In mouse models of obesity such as ob/ob and db/db mice, characterized by an impairment of leptinergic signaling, increased levels of endocannabinoids have been found (Di Marzo *et al.*, 2001). Moreover, it has been demonstrated a functional colocalization of CB1 receptor either with orexin 1 receptor or with leptin receptor in orexigenic neurons of the hypothalamus (Cota *et al.*, 2003; Yo *et al.*, 2005). Recently, numerous evidences suggested that EC system is also involved in the neuron release of peptides that regulates food intake. AEA increases depolarization-induced neuropeptide Y (NPY) release (Gamber *et al.*, 2005). CB1 activation inhibits the cocaine- and amphetamine-related transcript (CART) and plays a key role in the changes of α-melanocyte-stimulating hormone (MSH) levels occurring in feeding behavior (Osei-Hyaman et al., 2005a; Verty et al., 2004). The central role of EC system in energy balance is also corroborated by studies showing that the administration of the CB1 antagonist rimonabant reduces ghrelin-mediated hyperphagic effect and decreases plasma ghrelin levels in rats (Cani et al., 2004; Kola et al., 2008; Tucci et al., 2004). Furthermore, the long-term weight loss associated with rimonabant treatment is due at least in part to a higher energy expenditure, as demonstrated in a recent study by elevated temperature recorded in the brown adipose tissue (BAT), which is mediated primarily by the central EC system (Verty et al., 2009).

As well-known energy balance is controlled not only at central but also at peripheral level. CB1 receptor activation mediates lipogenetic activity in primary murine adipocytes (Cota et al., 2003). In obese Zucker (fa/fa) rats, CB1 blockage by rimonabant stimulates Acrp30 mRNA, an adipocytokine exclusively expressed by adipocytes, and regulates fatty acid oxidation. This stimulation has been also found to occur in cultured mouse 3T3 F442A preadipocyte cells treated with rimonabant (Bensaid et al., 2003). The Acrp30 mRNA inhibition was accompanied by a reduced adipocyte proliferation and enhanced cell maturation (Gary-Bobo et al., 2006). The effect was mediated by the inhibition of MAPK phosphorylation. Endocannabinoid tonic activation of CB1 receptor in liver induces mRNA expression of the lipogenic transcription factor SREBP-1c, its target enzymes acetyl coenzyme-A carboxylase (ACC1) and fatty acid synthase (FAS), also increasing the rate of fatty acid synthesis (Osei-Hyiaman et al., 2005b). Recent data indicated that endocannabinoids decrease fatty acid oxidation and oxygen consumption in skeletal muscle (Cavuoto et al., 2007; Liu et al., 2005). CB1 receptor blockage selectively increased 2-deoxyglucose uptake in L6 myotubes by regulating the expression of phosphatidylinositol-3-kinase (Esposito et al., 2008). Moreover, endocannabinoids are able to regulate directly glucagon and insulin secretion by pancreatic cells, therefore confirming the importance of EC system in the control of energy homeostasis both at central and peripheral levels.

III. Cannabinoid CB1 Receptors and CB1 Antagonists

The CB1 receptor is expressed in the cortex, striatum, hippocampus, and cerebellum (Howlett, 2005), found predominantly at central and peripheral nerve terminals, mediates inhibition of transmitter release (Howlett et al., 2002). It is also expressed at lower levels by certain non-neuronal cells and tissues, for example the pituitary gland, immune cells, and reproductive tissues. The CB1 receptor belongs to the G protein-coupled receptor (GPCR) type and is coupled to G proteins and inhibits certain adenylyl cyclase isozymes, leading to decreased cAMP production, decreased Ca^{2+} conductance, increased K^+ conductance, and increased mitogen-activated protein kinase activity (Di Marzo et al., 2004). The distribution pattern of CB1 receptors within the CNS accounts for several prominent effects of THC. Examples include its ability to decrease motor activity as indicated in rodents by hypokinesia and catalepsy, to induce signs of analgesia in animals and man and to stimulate food intake (Berry and Mechoulam, 2002; Howlett et al., 2002; Pertwee, 2001).

The discovery of CB1 and CB2 receptors was followed by the development of CB1-selective cannabinoid receptor antagonists (Howlett et al., 2002). The compound SR141716, also known as rimonabant (Acomplia), was the first cannabinoid receptor blocker showing high potency and selectivity for the CB1 receptor (Rinaldi-Carmona et al., 1994). Developed at Sanofi-Recherche, and described in the scientific literature as a CB1 antagonist (Rinaldi-Carmona et al., 1994, 1995), the pharmacology and medicinal chemistry of rimonabant have been extensively studied. Rimonabant is also described as an inverse agonist at the CB1 cannabinoid receptor, displaying negative intrinsic activity in both heterologous and constitutive systems (for a review, see Pertwee, 2005). These inverse cannabimimetic effects are opposite in direction from those produced by agonists for these receptors. Thus, for example, in vivo inverse effects of rimonabant in rats or mice include signs of hyperalgesia in models of inflammatory and neuropathic pain, stimulation of intestinal motility and suppression of food consumption. In vitro inverse effects of rimonabant include enhancement of ongoing release of acetylcholine, noradrenaline, and γ-aminobutyric acid in hippocampal slices and enhancement of the amplitude of electrically evoked contractions of the mouse isolated vas deferens (Pertwee, 2005). The affinity of rimonabant for the CB1 receptor lies in the range 1–20 nM, depending on the studies, its affinity for the CB2 receptor being in the micromolar range. Rimonabant potency, selectivity, and oral bioavailability made it a good pharmacological tool, as well as a promising drug candidate. Other analogues of rimonabant have been developed and compared for their ability to antagonize CB1 receptors with the lead compound.

Numerous studies were conducted with CP-55940 (AM251), a close analogue of rimonabant only differing by the exchange of an I-atom for a Cl-atom on the 5-phenyl ring (see Table 7.1) (Gatley et al., 1996; Lan et al., 1999). However, although this compound is very similar to rimonabant, blocks CB1 receptors, and produces inverse cannabimimetic effects, several pharmacological differences between rimonabant and AM251 have been detected in vitro for example in experiments with cardiovascular tissue (reviewed in Pertwee, 2005) and in studies on GABAergic and glutamatergic

Table 7.1 Chemical structures and effects on energy metabolism and human obesity of the main CB1 receptor antagonists

Compound	Chemical structure	Clinical studies	Preclinical studies
Rimonabant		Weight loss and reduced waist circumference; increased HDL levels and plasma adiponectin; reduction of triglycerides, and insulin resistance (Després et al., 2005; Pi-Sunyer et al., 2006; Van Gaal et al., 2005)	Stimulation of Acrp30 mRNA; inhibition of cell proliferation and MAPK activity in mouse preadipocytes; induction of adiponectin in Zucker rats (Gary-Bobo et al., 2006)
		Improvement of glucose control and metabolic parameters in type 2 diabetes (Iranmnesh et al., 2006; Rosenstock et al., 2008)	Inhibition of starvation-induced hyperphagia in NPY-deficient mice (Di Marzo et al., 2001)
		Reduced progression of coronary disease in obese patients with metabolic syndrome and improvement of atherogenic factors (Després et al., 2008; Nissen et al., 2008)	Reduction of *de novo* fatty acid synthesis in mice (Osei-Hyiaman et al., 2005a)
		Reduction of hypertension in obese patients (Ruilope et al., 2008)	Reversion of diet-induced obesity and regulation of lipolysis in DIO mice (Jbilo et al., 2005)

Surinabant	[chemical structure]	Reduction of plasma ghrelin levels in rats (Kola et al., 2008; Tucci et al., 2004) Increased mitochondrial biogenesis and eNOS in mouse adipocytes (Tedesco et al., 2008) Reduced food intake and ethanol or sucrose consumption in rats (Rinaldi-Carmona et al., 2004)
AM251	[chemical structure]	Suppression of rat food intake and food-reinforced behavior and inhibition of basal G protein activity in rat cerebellar membranes (Savinainen et al., 2003)

(continued)

Table 7.1 (continued)

Compound	Chemical structure	Clinical studies	Preclinical studies
Taranabant		Reduction of body weight, decreased triglyceride levels, and increased HDL levels. Increase in fat acid oxidation (Addy et al., 2008)	
Otenabant			Reduced food intake and increased energy expenditure and fat oxidation in rodents (Griffith et al., 2008)

AVE1625 — Induced lipolysis in fat tissue and glycogenolysis from the liver in Wister rats (Herling et al., 2007)

LH-21 — Improved lipid biochemistry in Zucker rats. No amelioration of cholesterol and triglyceride levels in Zucker rats (Pavon et al., 2006, 2008)

transmission in the hypothalami of both wild-type and CB1$^{-/-}$ knockout mice. These results suggested that AM251 could be more selective compared to rimonabant, which kept some activity in the knockout tissues (Hajos and Freund, 2002; Hajos et al., 2001; Haller et al., 2002, 2004).

Another pyrazole derivative closely related to rimonabant, SR147778 (surinabant), was described as a potent, orally active antagonist/inverse agonist of the CB1 receptor (Table 7.1). It possesses very similar pharmacological properties in terms of affinity and functionality (Rinaldi-Carmona et al., 2004) as the lead compound rimonabant, and to date, no differences were reported when used in in vivo models (Gessa et al., 2005; Lallemand and De Witte, 2006; Pamplona and Takahashi, 2006). Two other CB1 receptor antagonists/inverse agonists are CP-945,598 (otenabant, Table 7.1) produced by Pfizer and Ave1625 from Sanofi-Aventis, and likewise the above-mentioned CB1 antagonists, they have been considered promising drug candidates for therapeutic applications.

There are antagonists known as "neutral" antagonists that share the ability of rimonabant to block responses to CB1 receptor agonists but lack its apparent ability to produce inverse cannabimimetic effects in CB1-containing systems in the absence of any endogenously released or exogenously added CB1 receptor agonist. Examples of "neutral" CB1 receptor antagonists are 600-azidohex-200-yne-cannabidiol (O-2654), O-2050, a sulfonamide analogue of D8-THC with an acetylenic side chain, and two SR141716 analogues, VCHSR and NESS (Pertwee, 2005). Other antagonists not based on a cyclic central moiety were also developed (Muccioli et al., 2005), among these compounds there is Merck's MK-0364 (taranabant) (Hagmann, 2008; Lin et al., 2006). Its chemical structure is given in Table 7.1. Recently, the design of new CB1 receptor antagonists has been published, in particular a new series of tetrazole–biarylpyrazole analogues were efficiently prepared and bioassayed for binding to cannabinoid CB1 receptor and some of them demonstrated good binding affinity and selectivity (Kang et al., 2008). A series of diarylimidazolyl oxadiazole and thiadiazole derivatives as antagonists to the cannabinoid CB1 and CB2 receptors have been investigated (Lange and Kruse, 2008). Several of the compounds in these series exceeded or maintained the potency of known CB1 antagonists also displaying good selectivity for CB1 over CB2. Thus, the diarylimidazolyl oxadiazole or thiadiazole class of compounds possesses promising therapeutic possibility as a CB1 receptor antagonist for the treatment of obesity (Kim et al., 2009). Among "neutral antagonists," it is emerging a new class of CB1 antagonists derived from 1,2,4-triazole which represent a novel entry in cannabinoid chemistry exhibiting particular properties. LH-21 (Table 7.1) is an in vivo CB1 antagonist with a paradoxical low affinity in vitro for CB1 receptors and devoid of inverse agonistic properties (Jagerovich et al., 2004). LH-21 has a poor penetration into the brain, therefore acting as a peripheral antagonist of the CB1 receptor (Pavon et al., 2008).

IV. CB1 ANTAGONISTS IN THE TREATMENT OF OBESITY AND RELATED COMORBIDITIES

A. Clinical trials

The interest to develop cannabinoid antagonists derives from the multiple functions in which the endogenous cannabinoid system is involved mainly the control of appetite and lipid and glucose metabolism (for a review, see Di Marzo, 2008a; Piomelli *et al.*, 2000). We will focus more in detail on the effects of rimonabant as this lead compound has been widely used and described in literature (for a review, see Bifulco *et al.*, 2007a). The pharmacokinetic/pharmacodynamic profile of rimonabant, as expected by both preclinical and clinical studies showed that rimonabant is distributed widely in brown fat, it could reduce total energy intake and body weight gain in obese rats and the most effective dose in reducing body weight in obese human subjects was 20 mg/day.

The assessment of the clinical efficacy of rimonabant as an antiobesity drug was carried out in multinational, randomized, and placebo-controlled trials on overweight (with a body mass index—BMI—higher than 27 kg/m^2) or obese (BMI $\geq$ 30 kg/m^2) patients (Pi-Sunyer *et al.*, 2006; Van Gaal *et al.*, 2005). A cumulative weight loss and a significant change of waist circumference from the baseline were observed in patients receiving 20 mg/day of rimonabant for 1 year. It appears that the weight loss achieved an average 5% of decrease compared to placebo in the RIO studies showing a reduction of body weight not larger than other drugs commercially available (Padwal and Majumdar, 2007; Rucker *et al.*, 2007). However, rimonabant seems to produce an amelioration of lipid and glucose metabolism not found with other drugs (Deedwania, 2009). Rimonabant caused an increase of high-density lipoprotein (HDL) cholesterol levels, a reduction of triglycerides, fasting insulin, and insulin resistance that were twice those attributable to the concurrent weight loss alone as assessed by analysis of covariance (Pi-Sunyer *et al.*, 2006). In addition, in patients who completed the study in the second year and received 20 mg of rimonabant, levels of triglycerides and fasting insulin declined rapidly from baseline, suggesting a direct pharmacological effect of rimonabant on glucose and lipid metabolism outside the weight loss achieved (Di Marzo, 2008b; Pi-Sunyer *et al.*, 2006; Scheen and Paquot, 2008; Van Gaal *et al.*, 2008b). In another recent trial aimed to assess the effects of rimonabant in overweight patients with dyslipidemia (Déspres *et al.*, 2005), in the majority of patients completing 12 months of study and receiving 20 mg/day of rimonabant, the levels of triglycerides significantly decreased, whereas HDL cholesterol increased compared with placebo group. Furthermore, the prevalence of metabolic syndrome, a collection of factors (abdominal adiposity, hypertriglyceridemia, low HDL cholesterol, hypertension, and fasting hyperglycemia) increasing the risk of type 2 diabetes and cardiovascular disease (Nesto, 2005), was reduced in the same subset of patients (Pi-Sunyer et al., 2006). Indeed, the population of patients matching the National Cholesterol Education Program's Adult Treatment Panel III (NCEP-ATPIII) criteria for the metabolic syndrome (54% of the total), fell to the half of baseline after they received 20 mg rimonabant (Déspres et al., 2005).

The overall observations strongly suggest that rimonabant could modulate positively risk factors for a number of obesity-related comorbidities through weight loss-dependent and -independent pathways. Of note, the treatment with 20 mg rimonabant for 1 year increased significantly the levels of plasma adiponectin compared to the placebo group (Després et al., 2005). Indeed, serum adiponectin levels are inversely correlated to obesity, metabolic syndrome and type 2 diabetes (Gable et al., 2006), therefore the observation that rimonabant improves the levels of either adiponectin or fasting insulin and induces favorable changes in insulin resistance has suggested a possible pharmacological application of rimonabant in diabetes. Recently, the RIO diabetes and the SERENADE study demonstrated that rimonabant improves multiple cardiovascular and metabolic risk factors in patients with type 2 diabetes (Iranmanesh et al., 2006; Rosenstock et al., 2008; Scheen et al., 2006b). Although all the efficacy data currently available demonstrate considerable benefit in a number of cardiometabolic parameters (Wright et al., 2008), determining the true magnitude of reduction in cardiometabolic risk will require long-term clinical trials examining the effect on more clinically relevant endpoints than lipid parameters and levels of neurohormones. A recent work evaluated the impact of rimonabant on the blood pressure in patients recruited for the RIO studies, being hypertension a key risk factor for CVDs. In the individual RIO trials and the pooled data, small but consistent decrease in both systolic blood pressure (SBP) and diastolic blood pressure (DBP) was observed following 1 year of treatment with rimonabant 20 mg/day (Ruilope et al., 2008). The blood pressure-lowering effect of rimonabant was entirely attributable to the concomitant body weight loss; indeed, there was no evidence of blood pressure increase with rimonabant treatment. In patients with high blood pressure at baseline, improvement in SBP and DBP was more pronounced than in the overall pooled population. Taken collectively, the effects on blood pressure, plus the demonstrated improvements in lipid profile, glycemic control, and abdominal obesity support the use of selective CB1 blockade with rimonabant as a novel approach to the management of multiple cardiometabolic risk factors (Ruilope et al., 2008). Very recently, the results of the Strategy to Reduce Atherosclerosis Development Involving Administration of Rimonabant—The Intravascular Ultrasound Study (STRADIVARIUS) have been published. In this trial, the effects of rimonabant on progression of atherosclerosis in abdominally obese patients with the metabolic syndrome and preexisting coronary disease were reported (Nissen et al., 2008). The STRADIVARIUS trial was a randomized, double-blinded, placebo-controlled, two-group, parallel-group trial (enrolment December 2004–December 2005) comparing rimonabant to placebo in 839 patients, at 112 centers in North America, Europe, and Australia. Patients received dietary counseling were randomized to receive rimonabant (20 mg daily) or matching placebo, and underwent coronary intravascular ultrasonography at baseline ($n = 839$) and study completion ($n = 676$). The primary efficacy parameter was change in percent atheroma volume (PAV) and the secondary efficacy parameter was the change in normalized total atheroma volume (TAV). In the rimonabant versus placebo groups, PAV increased 0.25% versus 0.51%, respectively, and TAV decreased 2.2 mm^3 versus an increase of 0.88 mm^3. In the rimonabant versus placebo groups, imputing results based on baseline characteristics for patients not completing the trial, PAV increased 0.25% versus 0.57%, and TAV decreased 1.95 mm^3 versus an increase of 1.19 mm^3. Rimonabant-treated patients had a larger reduction in body weight and greater decrease in waist circumference. A general improvement of metabolic profile was also reported in this study. HDL cholesterol levels increased and median triglyceride levels decreased. Rimonabant-treated patients had significant decreases in high-sensitivity C-reactive protein and less increase in glycated hemoglobin levels. After 18 months of treatment, the study failed to show an effect for rimonabant on disease progression for the primary endpoint (PAV) but showed a favorable effect on the secondary endpoint (TAV) (Nissen et al., 2008). Very recently, the publication of the results of the ADAGIO-lipids study (Després et al., 2008), carried out on 803 obese patients with atherogenic dyslipidemia followed up for 1 year, confirmed the favorable changes of rimonabant on cardiometabolic risk factors. The study reported that in 20 mg rimonabant-treated patients, HDL cholesterol levels increased by 7.4% and triglycerides decreased by 18%. Interestingly, rimonabant also produced a significant change in the size of HDL and low-density lipoprotein (LDL) particles. In particular, it induced a significant reduction in the proportion of small, atherogenic LDL and a concomitant increase in the concentration of large LDL. Moreover, an increase of HDL particle size and apolipoprotein A1 and B were observed in rimonabant-treated patients compared with placebo (Després et al., 2008). The outcomes of the ADAGIO-lipids study also reported that rimonabant decreased abdominal subcutaneous adipose tissue and liver fat content thus suggesting an improvement in cardiometabolic risk factors and the reduction in both intra-abdominal and liver fat. The completed clinical trials, from RIO studies to STRADIVARIUS and ADAGIO, reporting a reduction of HDL, triglycerides, and fasting insulin and the increase of adiponectin, suggest a direct pharmacological effect of rimonabant on glucose and lipid metabolism (Di Marzo, 2008a). However, determining whether rimonabant is useful in the management of coronary disease and amelioration of long-term risk of cardiovascular events requires additional outcomes trials, which are currently under way. Another rimonabant-strictly related compound, surinabant, is still in phase II clinical trial for the treatment of obesity but results are not yet disclosed (Di Marzo, 2008a). A general view of rimonabant-induced effects on human obesity and energy metabolism as well as its related compounds is showed in Table 7.1.

In 2008, the results of a clinical trial with another CB1 receptor antagonist have been reported. It concerns the use of taranabant in the treatment of obese patients with a BMI of 30–43 kg/m^2 (Addy et al., 2008). After 1 year, the population who completed the study showed a reduction in body weight that reached 6.3 kg in patients receiving a daily dose of 6 mg taranabant (the reduction in body weight of subjects treated with placebo was of 1.4 kg). Interestingly, during the second year of treatment, there was a significant decrease in triglyceride levels and prevalence of metabolic syndrome and an increase in HDL as well as in LDL levels. However, comparing results of rimonabant (20 mg/day) and taranabant (4 mg/day) no significant differences in weight reduction and metabolic parameters seem to emerge. Furthermore, by using positron emission tomography with a CB1 receptor-selective ligand, authors reported in the same study that a daily dose of 4–6 mg of taranabant is able to occupy 25–30% of CB1 receptors in the brain (Addy et al., 2008). Resting energy expenditure was also measured 5 and 24 h after treatment with 4 and 12 mg of taranabant using indirect calorimetry. The authors demonstrated a significant increase in resting energy expenditure after a treatment of 5 h with 12 mg of taranabant thus suggesting a direct increase in fat oxidation.

Additional CB1 antagonists are now in clinical development including otenabant and AVE1625 for the treatment of obesity. AVE1625 is also in clinical trials for the treatment of nicotine dependence and cognitive impairment (Di Marzo, 2008a).

B. Studies *in vitro* and in animal models

Although rimonabant is a powerful agent for the treatment of obesity and its metabolic complications, as assessed by the mentioned clinical trials, its biological mechanism of action is not clear yet.

In nongenetic-induced obese mice, body weight loss seems to consist of an early phase that depends on regulation of food intake and in a second phase, food intake regulation independent, in which weight loss is maintained probably through a sustained reduction of adiposity. The treatment with rimonabant of mice lacking the cocaine–amphetamine-regulated transcript (CART), an anorexigenic peptide, does not affect feeding behavior (Ravinet Trillou et al., 2003), whereas this compound inhibits starvation-induced hyperphagia in NPY-deficient mice and reduces body weight in leptin knockout mice (Di Marzo et al., 2001). On the other hand, the finding that the CB1 receptor-mediated lipogenetic activity in primary murine adipocytes can be blocked by rimonabant (Cota et al., 2003), intriguingly supports the idea that this compound also functions at peripheral level by decreasing lipoprotein lipase activity. Recently, a study showed that in mouse white adipocytes rimonabant increases mitochondrial biogenesis by inducing the expression of endothelial NO synthase (eNOS), the production of nitric oxide (NO) induces mitochondrial biogenesis and function in adipocytes. This effect is linked to the prevention of high-fat diet-induced fat accumulation, without concomitant changes in food intake (Tedesco et al., 2008).

In line with these observations, studies with rimonabant in obese Zucker (fa/fa) rats showed that the compound stimulates Acrp30 mRNA (Gary-Bobo et al., 2006), and rimonabant working as an inverse agonist, switches off MAPK activation from the insulin receptor–tyrosine kinase and insulin-like growth factor (IGF1) receptors (Bouaboula et al., 1997). Blockage of CB1 receptor stimulation by rimonabant significantly reduces de novo fatty acid synthesis in mice, thus providing the evidence for the involvement of fatty acid biosynthetic pathway in the sustained reduction of body weight. Moreover, rimonabant reverses the diet-induced obesity (DIO) phenotype through the regulation of lipolysis and energy expenditure (Jbilo et al., 2005). The gene expression profile of DIO mice treated with rimonabant (10 mg/kg for 40 days) was analyzed to compare with respect to nontreated mice the transcriptional effect of rimonabant. Results demonstrated that rimonabant induces the expression of genes directly involved in fatty acid oxidation and modulates several enzymes regulating the activation of lipolysis whereas the high-fat regimen regulated the same genes in an opposite way (Jbilo et al., 2005). A recent study assessed and compared the respective potential relevance of CNS versus peripheral CB1 receptors in the regulation of energy homeostasis and lipid and glucose metabolism in DIO rats (Nogueiras et al., 2008). It was observed that specific CNS–CB1 blockade decreases body weight and food intake but, independent of those effects, had no beneficial influence on peripheral lipid and glucose metabolism. A reduction of food intake and body weight was observed by peripheral treatment with rimonabant but, in addition, it independently triggers lipid mobilization pathways in white adipose tissue and cellular glucose uptake. Furthermore, during peripheral infusion of rimonabant, insulin sensitivity and skeletal muscle glucose uptake were enhanced, while hepatic glucose production was decreased. However, these effects depend on the antagonist-elicited reduction of food intake. These data suggest that several relevant metabolic processes appear to benefit independently from peripheral blockade of CB1, while CNS–CB1 blockade alone predominantly affects food intake and body weight (Nogueiras et al., 2008). Recently, the body weight loss by rimonabant has been associated with amelioration of insulin resistance via adiponectin-dependent and adiponectin-independent pathways (Watanabe et al., 2009).

Despite clinical trials, more data reported the effects of CB1 antagonists in reducing food intake and increasing energy expenditure *in vitro* and in animal models. For example, AM251 induced suppression of rat food intake and food-reinforced behavior and inhibition of basal G protein activity in rat cerebellar membranes (Savinainen et al., 2003). Few studies on surinabant, AVE1625, and otenabant reported an effect on feeding behavior in animal models. Surinabant was able to reduce both food intake in rats and ethanol and sucrose consumption in mice and rats (Herling et al., 2007; Rinaldi-Carmona et al., 2004). Moreover, AVE1625 has been reported to increase energy expenditure in Wistar rats associated to an enhancement of fat and glucose oxidation (Herling et al., 2007). A recent report on otenabant demonstrated that the compound reduced food intake and increased energy expenditure and fat oxidation in rodents (Griffith et al., 2009). Finally, recent reports have investigated the pharmacology of LH-21 and its effects on feeding behavior and weight gain. It has been reported that LH-21, at doses unable to penetrate in the brain, is able to inhibit feeding (Pavon et al., 2008). Encouraging results demonstrate that LH-21 enhances the efficacy of oleoylethanolamide (OEA), a lipid mediator acting in coordination with anandamide in regulating energy metabolism but with opposite effects (Rodriguez de Fonseca et al., 2001), and improves lipid biochemistry in Zucker rats (Pavon et al., 2006). Notably, LH-21 was unable to improve serum levels of cholesterol, triglycerides, and glucose in the same animal model, thus suggesting that these parameters are centrally regulated (Pavon et al., 2008). A summary reported in Table 7.1 shows the *in vitro* and *in vivo* experiments carried out with rimonabant and the other main CB1 antagonists.

C. Safety and adverse effects

Although CB1 antagonists have been developed and tested in clinical trials, for most of them data concerning adverse effects are still lacking and/or incomplete. Among the CB1 antagonists, rimonabant was the most investigated compound and four RIO studies have been completed. Safety profile and adverse events (AEs) observed with rimonabant showed, from all RIO studies, most frequently gastrointestinal, neurological, and psychiatric side effects (Curioni and Andre, 2006; Henness et al., 2006; Scheen et al., 2006a). The most commonly reported AEs were depressive disorders, anxiety, and nausea. Most AEs occurred during the first few weeks to months of rimonabant treatment. During the second year, AEs were low and similar with rimonabant and placebo in a pooled analysis of RIO Europe and RIO North America trials. The overall safety of rimonabant in the RIO program has been extensively reviewed (Van Gaal et al., 2008a,b), a recently published independent meta-analysis confirmed that rimonabant caused significantly more AEs than did placebo and more serious adverse events during the first year of treatment (Christensen et al., 2007). In particular, because of anxiety and depressive mood disorders, patients treated with rimonabant 20 mg were 2.5 times more likely to discontinue the treatment compared with placebo. The Food and Drug Administration (FDA) assessed the overall safety profile of rimonabant providing a more extensive additional safety set of data (Food and Drug Administration, 2007). The main FDA concern was a higher incidence of suicidal ideation in rimonabant-treated patients compared with placebo-treated overweight/ obese patients. In the entire rimonabant clinical trials database, there have been two completed suicides. Most importantly, the development of depression was more frequent among patients with a previous history of depressive disorders than in patients with no previous history. Thus, it is worth noting that the risk of depressive disorders was considerably lower in patients without previous history of depression but receiving rimonabant 20 mg combined with diet and exercise advice than in patients with a previous history of depression and receiving placebo and diet and exercise counseling. It is likely that anxiety and depression result from the pharmacological CB1 antagonist activity of rimonabant in the brain. Indeed, although reported data are conflicting, it is known that the pharmacological enhancement of the EC system activity at the CB1 receptor level appears to exert an antidepressant-like effect in some animal models of depression. By contrast, a reduced activity of the EC system, in particular an alteration of EC serum levels seems to be associated with an animal model of depression (Serra and Fratta, 2007). Anyway, CB1 antagonists have been suggested to interfere with altered endocannabinoid levels consequent to new stressful conditions, thus partly explaining the anxiogenic and prodepressant effects observed in obese patients under treatment (Di Marzo, 2008c). Therefore, it is highly probable that the psychological adverse effects of rimonabant might be observed also with other CB1 receptor antagonists (Scheen and Paquot, 2008). In a recent study, also administration of taranabant was associated with a dose-related increased incidence of mild to moderate psychiatric AEs (Addy et al., 2008).

The most common treatment-emergent AEs reported from the STRADIVARIUS trial confirmed psychiatric disorders observed from the RIO studies, which occurred in 28.4% of placebo-treated patients and 43.4% of rimonabant-treated patients (Nissen et al., 2008). These AEs consisted primarily of an increase in anxiety and depression (Doggrell, 2008). Severe psychiatric AEs, defined as major depression, suicidal ideation, or attempted or successful suicide occurred with similar frequency in the placebo- and rimonabant-treated patients. An increase of gastrointestinal tract adverse effects, especially nausea was also observed. These adverse effects resulted in an increase in drug discontinuations, 73% of patients were able to successfully complete 18 months of rimonabant therapy, compared with 84% in the placebo group. The study had some limitations: (a) a narrow population that may not be representative of the broader population with coronary artery disease; (b) the relatively short duration of exposure to rimonabant (18 months), which may have been inadequate to observe a treatment effect on the rate of progression of coronary atherosclerosis; and (c) the psychiatric and gastrointestinal tract adverse effects observed. Although these restrictions, we have to point out that AEs were usually mild or moderate in severity and, unlike previous studies with rimonabant, patients receiving treatment for depression were not excluded from the trial. Concerning the ADAGIO study and the other clinical trials conducted in obese patients with type 2 diabetes, no substantial differences in AEs were observed compared to RIO studies; therefore, it remains still unproved whether rimonabant induces severe side effects in patients with a past history of depression or in obese/overweight patients not having psychiatric disorders or both.

V. Other Emerging Effects of CB1 Antagonists

Many studies conducted with rimonabant showed its pleiotropic effects from obesity to drug dependence and memory impairment (Bifulco et al., 2007a). In this section, we aim to mention recent findings on antitumor properties of rimonabant (and other antagonists) as these studies suggest that targeting the EC system, via modulation of the CB1 receptor, could be a promising therapeutic strategy for cancer management.

The studies conducted from the late 1990s on the endocannabinoid system have provided strong evidence for a key role of the endocannabinoid in the control of cancer cell growth, invasion, and metastasis development in a way dependent on CB receptor activation (for a review, see Bifulco and Di Marzo, 2002; Bifulco et al., 2006, 2007b).

It has been demonstrated that rimonabant attenuates the antitumor effects of anandamide-related compounds or other cannabinoid agonists in thyroid, breast, and prostate cancers (Bifulco et al., 2001; Grimaldi et al., 2006; Portella et al., 2003; Sarfaraz et al., 2005), the effects being dependent on CB1 receptor activation. In other tumor types, like glioma, rimonabant failed to revert the antiproliferative action of cannabinoid agonists whereas the selective CB2 antagonist, SR144528 (Sanchez et al., 2001) or a combination of the CB1/CB2 antagonists can partially prevent this effect (Jacobsson et al., 2001). However, a 48-h preincubation with these antagonists seems to enhance the AEA-mediated cell death of glioma cells, suggesting a more complex mechanism of action (Maccarrone et al., 2000).

Considering the antitumor properties of the cannabinoid receptor agonists, it could be expected that cannabinoid receptor antagonists, like rimonabant, if used alone, would enhance proliferation of normal and malignant cells leading to cancer. Some data excluded this possibility, rather reporting that not only agonists to cannabinoid receptors but also antagonists, used alone, are able to inhibit cancer growth (Bifulco et al., 2004) or induce apoptosis in cancer cells (Derocq et al., 1998; Powles et al., 2005). The first observation of a rimonabant potential antitumor action was provided by our group in rat thyroid cancer cells (KiMol) *in vitro* and in thyroid tumor xenografts induced by KiMol injection in athymic mice. In this model, rimonabant was able to prevent partially the antitumor effect of the inhibitors of endocannabinoid degradation, and the anandamide metabolically stable analogue (Met-F-AEA). However, rimonabant, when used alone, in the same model and at the dose previously shown to counteract the Met-F-AEA effect did not enhance tumor growth exerting *per se* a significant antitumor effect both *in vitro* and *in vivo* (Bifulco et al., 2004; Fig. 7.1). Interestingly, micromolar concentrations of rimonabant decreased viability of primary mantle lymphoma cells isolated from tumor biopsies of two patients (Flygare et al., 2005; Fig. 7.1). Moreover, rimonabant showed an additive negative effect also on the viability of the mantle cell lymphoma cell line Rec-1 when combined with equipotent doses of AEA. Both Bifulco and Flygare supported the evidence of the antitumor action of rimonabant but they did not investigate or provide a molecular mechanism of action. They proposed that the observed effects could be ascribed to a tonic antiproliferative action mediated by the local endocannabinoids and the inverse agonist properties of rimonabant on the receptor. These possibilities could explain the paradox whereby both CB1 agonists and antagonists display antitumor activity. Recently, we reported that rimonabant exerts antitumor effects on breast cancer *in vitro* and in a

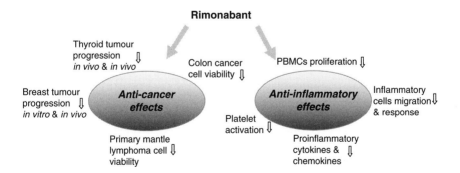

Figure 7.1 The anticancer and anti-inflammatory effects of rimonabant are described in the figure.

mouse model *in vivo*, providing for the first time a new mechanism of action of this drug (Sarnataro *et al.*, 2006; Fig. 7.1). Rimonabant, at nanomolar concentrations, inhibits human breast cancer cell proliferation being more effective in highly invasive metastatic MDA-MB-231 cells, than in less invasive T47D and MCF-7 cells, depending on both the presence and the different expression levels of the CB1 receptor. The antiproliferative effect is characterized by a G1/S phase cell cycle arrest, without induction of apoptosis. *In vivo* studies confirmed the effect observed *in vitro* in particular, rimonabant reduces the volume of xenografts tumors induced by MDA-MB-231 injection in mice, after 2 weeks of treatment. The molecular mechanism at the basis of rimonabant function involves inhibition of p42/44 MAPK phosphorylation and needs lipid rafts/caveolae integrity to occur. This suggests that rimonabant effects on cell proliferation and signaling requires the presence of CB1 receptor in lipid rafts (Sarnataro *et al.*, 2006). In addition, we recently reported that rimonabant could possess antitumor properties also in colon cancer, as it significantly decreased human adenocarcinoma DLD1 cell viability (Santoro *et al.*, 2008; Fig. 7.1). Contrarily to rimonabant, its analogue AM251 does not exhibit any antitumor action but it was shown to increase the number of polyps and tumors in small intestine and colon of Apc Min/+ mice after a treatment of 7 weeks with 10 mg/kg body (Wang *et al.*, 2008). Furthermore according to its antagonistic properties, AM251 was able to prevent the proapoptotic activity of the CB1 agonist ACEA (arachidonoyl-2′-chloroethylamide) in human colon cancer cells (Cianchi *et al.*, 2008).

Other effects of rimonabant have been reported, a recent study by our group demonstrated that rimonabant inhibits the proliferative response of mitogen-activated peripheral blood mononuclear cells and probably acting as an inverse agonist, enhances the antiproliferative effect of anandamide in the same *in vitro* model (Malfitano *et al.*, 2008; Fig. 7.1). These findings suggest that rimonabant exerts immunomodulatory and anti-inflammatory effects as previously observed by our group using CB1 agonists (Malfitano *et al.*, 2006, 2007). A recent study showed the involvement of CB1 receptors in immune functions, over the CB2 receptors. In this study, patients with coronary artery disease demonstrated the activation of the EC system with elevated levels of blood endocannabinoids and increased expression of CB1 receptor in coronary atheroma. The block of CB1 receptor exhibited anti-inflammatory effects on macrophages, which might provide beneficial effects on atherogenesis as result of CB1 inhibition (Sugamura *et al.*, 2009). Furthermore, a recent finding reported the antiatherosclerotic effects of rimonabant in $LDLR^{-/-}$ mice. In particular, rimonabant reduces food intake, weight gain, serum total cholesterol, and atherosclerotic lesion development in the aorta and aortic sinus, indeed rimonabant decreases plasma levels of proinflammatory cytokines (Fig. 7.1), thus suggesting a relation between the antiatherosclerotic effects and its anti-inflammatory properties (Dol Gleizes et al., 2009). Accordingly, rimonabant in addition to improving plasma lipid alterations and decreasing inflammatory cell migration and inflammatory response (Fig. 7.1) may exert beneficial effects in atherosclerosis and restenosis by decreasing vascular smooth muscle proliferation and migration (Rajesh et al., 2008). The anti-inflammatory effect of rimonabant has been investigated also in type 2 diabetic Zucker rats; administration of rimonabant positive modulates circulating neutrophil and monocyte numbers, reduces platelet activation, and lowers proinflammatory chemokines levels (Fig. 7.1). These effects may potentially contribute to a reduction of cardiovascular risk (Di Marzo and Szallasi, 2008; Schäfer et al., 2008).

VI. Therapeutic Prospects

The CB1 receptors have been considered for years main targets in the treatment of obesity and obesity-related pathologies. Although the CB1 antagonists have been widely used in clinical trials, it is recent news that the European Medicines Agency recommended suspending sales of the lead compound, rimonabant, which has been approved in Europe for 2 years but has not been approved in the US. The reason provided by the agency is that the experience with rimonabant since approval has indicated that serious psychiatric disorders may be more common than in the clinical trials used in the initial assessment of the medicine. Following this recommendation, several companies have decided to discontinue all ongoing clinical trials of CB1 blockers, in particular, Sanofi-Aventis' rimonabant has been withdrawn from the market, Merck's taranabant, which had reached Phase III trials and Pfizer's otenabant, which had also been in Phase III trials for obesity have been discontinued. However, from the recent trials of CB1 antagonists as antiobesity drugs resulted that the incidence of depression due to rimonabant treatment is more pronounced in patients with a previous history of depression (Nissen et al., 2008). According to these findings, an alternative procedure might be the selection of patients before the treatments; furthermore, variants in the codifying or promoting regions of *CNR1* gene, encoding the CB1 receptor, have been associated with mood disorders and predisposition to depression. It could be hypothesized that a lack or reduction of psychiatric side effects of rimonabant or other CB1 antagonists could be determined by polymorphic variants; thus, a subselection of patients' eligibility based on the level or polymorphic CB1 expression could improve the choice of the treatment initially addressed by the clinical profiles. These therapeutic procedures could reduce the incidence of adverse or toxic side effects in selected patients. Further research are needed as it is not completely clear what underlies the psychiatric side effects of CB1 antagonists; however to overcome the difficulty of such limiting effects, the development of novel CB1 blockers not penetrating in the brain is highly required. LH-21 (Table 7.1) is an example of neutral CB1 antagonist that exhibits some anorexigenic effect in rats. Indeed, it has been reported that LH-21 failed to ameliorate hepatic damage or to improve insulin secretion whereas rimonabant did it (Pavon *et al.*, 2006, 2008). However, blocking peripheral CB1 receptors could play a role in pathologies not directly related to obesity, such as hepatic fibrosis, chronic inflammatory conditions, diabetes, and cancer. Alternative applications of novel high selective peripheral CB1 antagonists and lower doses of CB1 antagonists should be taken into account to avoid central psychotropic side effects, or to overcome psychotropic side effects limitations. Both efficacy and side effects increase with dose, but the 2-year RIO-EUROPE study suggests that doses lower than that normally prescribed can still have beneficial effects on metabolic factors that lead to diabetes and cardiovascular disease, even though these doses do not induce such a pronounced effect on body weight loss.

The potential role of CB1 receptors in cancer therapy is also an interesting target to develop. CB1 antagonists used in selective regimens in combination with chemotherapeutic drugs could represent a relevant strategy to reduce doses, to avoid resistance, and to exert a more potent clinical effect; further studies are required to assess and ensure the efficacy of CB1 antagonists as anticancer drugs, as well as their potential application in other peripheral pathologies. These considerations suggest the relevance of basic research studies to clarify the role of the EC system and its regulation that is still highly attractive and represents a promising target to better understand and treat diseases in which it is dysregulated.

VII. Conclusions

In light of the public health implications of the obesity pandemic, CB1 blockade strategy aimed to treat obesity and related disorders has encouraged several pharmaceutical companies in these years to develop new and more selective CB1 antagonists.

The literature currently available on the effects of CB1 antagonists, particularly rimonabant, demonstrates that the use of these drugs is controversial due to the double aspect concerning their potential as antiobesity molecules and their side effects. Indeed, side effects could be controlled by using lower doses in obese/overweight patients with a previous history of depression. Furthermore, the development of peripheral CB1 antagonists incapable to penetrate brain barrier could be of particular importance to limit side effects. However, it is undeniable that, even though rimonabant does not induce a higher weight loss compared to other drugs available on the market, it exerts beneficial effects on metabolic profile and cardiovascular risk factors outside the weight loss achieved. Therefore, it would be necessary to perform more long-term clinical trials and to investigate in detail its safety and efficacy on the basis of an appropriate selection of patients. In our opinion, rimonabant does not represent the "panacea" for obesity, but taking into account its pleiotropic effects, it should not be "thrown away." Rimonabant and other CB1 antagonists could have alternative applications in pathologies not directly related to obesity such as hepatic fibrosis, chronic inflammatory conditions, diabetes, and cancer. The potentiality of targeting CB1 receptors in cancer therapy is particularly intriguing. This issue should be investigated to ascertain and ensure the efficacy of CB1 antagonists as anticancer drugs, as well as their potential application in other peripheral pathologies reinforcing the basic research on the EC system and its role in these pathological conditions.

ACKNOWLEDGMENTS

We thank the Associazione Educazione e Ricerca Medica Salernitana (ERMES) for supporting our studies. A.M. Malfitano was supported by a fellowship from AIRC.

REFERENCES

Addy, C., Wright, H., Van Laere, K., Gantz, I., Erondu, N., Musser, B. J., Lu, K., Yuan, J., Sanabria-Bohórquez, S. M., Stoch, A., Stevens, C., Fong, T. M., et al. (2008). The acyclic CB1R inverse agonist taranabant mediates weight loss by increasing energy expenditure and decreasing caloric intake. *Cell Metab.* **7,** 68–78.

Bensaid, M., Gary-Bobo, M., Esclangon, A., Maffrand, J. P., Le Fur, G., Oury-Donat, F., and Soubrié, P. (2003). The cannabinoid CB_1 receptor antagonist SR141716 increases Acrp30 mRNA expression in Adipose tissue of obese fa/fa rats and in cultured adipocyte cells. *Mol. Pharmacol.* **63,** 908–914.

Berry, E. M., and Mechoulam, R. (2002). Tetrahydrocannabinol and endocannabinoids in feeding and appetite. *Pharmacol. Ther.* **95,** 185–190.

Bifulco, M., and Di Marzo, V. (2002). Targeting the endocannabinoid system in cancer therapy: A call for further research. *Nat. Med.* **8,** 547–550.

Bifulco, M., Laezza, C., Portella, G., Vitale, M., Orlando, P., De Petrocellis, L., and Di Marzo, V. (2001). Control by the endogenous cannabinoid system of ras oncogene-dependent tumor growth. *FASEB J.* **15,** 2745–2747.

Bifulco, M., Laezza, C., Valenti, M., Ligresti, A., Portella, G., and Di Marzo, V. (2004). A new strategy to block tumor growth by inhibiting endocannabinoid inactivation. *FASEB J.* **18,** 1606–1608.

Bifulco, M., Laezza, C., Pisanti, S., and Gazzerro, P. (2006). Cannabinoids and cancer: Pros and cons of an antitumour strategy. *Br. J. Pharmacol.* **148,** 123–135.

Bifulco, M., Grimaldi, C., Gazzerro, P., Pisanti, S., and Santoro, A. (2007a). Rimonabant: Just an antiobesity drug? Current evidence on its pleiotropic effects. *Mol. Pharmacol.* **71,** 1445–1456.

Bifulco, M., Gazzerro, P., Laezza, C., and Pentimalli, F. (2007b). Endocannabinoids as emerging suppressors of angiogenesis and tumor invasion. *Oncol. Rep.* **17,** 813–816.

Blankman, J. L., Simon, G. M., and Cravatt, B. F. (2007). A comprehensive profile of brain enzyme that hydrolize the endocannabinoid 2-arachidonoylglycerol. *Chem. Biol.* **14,** 1347–1356.

Bouaboula, M., Perrachon, S., Milligan, L., Canat, X., Rinaldi-Carmona, M., Portier, M., Barth, F., Calandra, B., Pecceu, F., Lupker, J., Maffrand, J. P., Le Fur, G., et al. (1997). A selective inverse agonist for central cannabinoid receptor inhibits mitogen-activated protein kinase activation stimulated by insulin or insulin-like growth factor 1. Evidence for a new model of receptor/ligand interactions. *J. Biol. Chem.* **272,** 22330–22339.

Cani, P. D., Montoya, M. L., Neyrinck, A. M., Delzenne, N. M., and Lambert, D. M. (2004). Potential modulation of plasma ghrelin and glucagons-like peptide 1 by anorexigenic cannabinoids compound, SR141716 (rimonabant) and oleoylethanolamide. *Br. J. Nutr.* **92,** 757–761.

Cavuoto, P., McAincha, A. J., Hatzinikolas, G., Cameron-Smith, D., and Wittert, G. A. (2007). Effects of cannabinoid receptors on skeletal muscle oxidative pathways. *Mol. Cell. Endocrinol.* **267,** 63–69.

Christensen, R., Kristensen, P. K., Bartels, E. M., Bliddal, H., and Astrup, A. (2007). Efficacy and safety of the weight-loss drug rimonabant: A meta-analysis of randomised trials. *Lancet* **370,** 1706–1713.

Cianchi, F., Papucci, L., Schiavone, N., Lulli, M., Magnelli, L., Vinci, M. C., Messerini, L., Manera, C., Ronconi, E., Romagnani, P., Donnini, M., Perigli, G., et al. (2008). Cannabinoid receptor activation induces apoptosis through tumor necrosis factor α-mediated de novo synthesis in colon cancer cells. *Clin. Cancer Res.* **14,** 7691–7700.

Cota, D. (2007). CB1 receptors: Emerging evidence for central and peripheral mechanisms that regulate energy balance, metabolism, and cardiovascular health. *Diabetes Metab. Res. Rev.* **23,** 507–517.

Cota, D., Marsicano, G., Tschoep, M., Gruebler, Y., Flachskamm, C., Schubert, M., Auer, D., Yassouridis, A., Thone-Reineke, C., Ortmann, S., Tomassoni, F., Cervino, C., et al. (2003). The endogenous cannabinoid system affects energy balance via central orexigenic drive and peripheral lipogenesis. *J. Clin. Invest.* **112,** 423–431.

Cote, M., Matias, I., Lemieux, I., Petrosino, S., Almeras, N., Després, J. P., and Di Marzo, V. (2007). Circulating endocannabinoid levels, abdominal adiposity and related cardiometabolic risk factors in obese men. *Int. J. Obesity* **31,** 692–699.

Curioni, C., and Andre, C. (2006). Rimonabant for overweight or obesity. *Cochrane Database Syst. Rev.* **4,** CD006162.

Deedwania, P. (2009). The endocannabinoid system and cardiometabolic risk: Effects of CB (1) receptor blockade on lipid metabolism. *Int. J. Cardiol.* **131,** 305–312.

Derocq, J. M., Bouaboula, M., Marchand, J., Rinaldi-Carmona, M., Segui, M., and Casellas, P. (1998). The endogenous cannabinoid anandamide is a lipid messenger activating cell growth via a cannabinoid receptor-independent pathway in hematopoietic cell lines. *FEBS Lett.* **425,** 419–425.

Després, J. P., and Lemieux, I. (2006). Abdominal obesity and metabolic syndrome. *Nature* **444,** 881–887.

Després, J. P., Golay, A., and Sjöström, L., and Rimonabant in Obesity-Lipids Study Group (2005). Effects of rimonabant on metabolic risk factors in overweight patients with dyslipidemia. *N. Engl. J. Med.* **353,** 2121–2134.

Després, J. P., Ross, R., Boka, G., Alméras, N., Lemieux, I., and ADAGIO-LIPIDS Investigators (2008). Effect of rimonabant on the high-triglyceride/low-HDL-cholesterol dyslipidemia intraabdominal adiposity and liver fat. The ADAGIO-lipids trial. *Arterioscler. Thromb. Vasc. Biol.* **29,** 416–423.

Devane, W. A., Hanus, L., Breuer, A., Pertwee, R. G., Stevenson, L. A., Griffin, G., Gibson, D., Mandelbaum, A., Etinger, A., and Mechoulam, R. (1992). Isolation and structure of a brain constituent that binds to the cannabinoid receptor. *Science* **258,** 1946–1949.
Di Marzo, V. (2008a). Targeting the endocannabinoid system: To enhance or reduce? *Nat. Rev.Drug Discov.* **7,** 438–453.
Di Marzo, V. (2008b). CB(1) receptor antagonism: Biological basis for metabolic effects. *Drug Discov. Today* **13,** 1026–1041.
Di Marzo, V. (2008c). Play an ADAGIO with a STRADIVARIUS: The right patient for CB1 receptor antagonists? *Nat. Clin. Pract. Cardiovasc. Med.* **5**(10), 610–612.
Di Marzo, V., and Szallasi, A. (2008). Rimonabant in rats with a metabolic syndrome: Good news after the depression. *Br. J. Pharmacol.* **154,** 915–917.
Di Marzo, V., Goparaju, S. K., Wang, L., Liu, J., Batkai, S., Jarai, Z., Fezza, F., Miura, G. I., Palmiter, R. D., Sugiura, T., and Kunos, G. (2001). Leptin-regulated endocannabinoids are involved in maintaining food intake. *Nature* **410,** 822–825.
Di Marzo, V., Bifulco, M., and De Petrocellis, L. (2004). The endocannabinoid system and its therapeutic exploitation. *Nat. Rev. Drug Discov.* **3,** 771–784.
Doggrell, S. A. (2008). Is rimonabant efficacious and safe in the treatment of obesity? *Expert Opin. Pharmacother.* **9,** 2727–2731.
Dol-Gleizes, F., Paumelle, R., Visentin, V., Marés, A. M., Desitter, P., Hennuyer, N., Gilde, A., Staels, B., Schaeffer, P., and Bono, F. (2009). Rimonabant, a selective cannabinoid CB1 receptor antagonist, inhibits atherosclerosis in LDL receptor-deficient mice. *Arterioscler. Thromb. Vasc. Biol.* **29,** 12–18.
Esposito, I., Proto, M. C., Gazzerro, P., Laezza, C., Miele, C., Alberobello, A. T., D'Esposito, V., Beguinot, F., Formisano, P., and Bifulco, M. (2008). The cannabinoid CB1 receptor antagonist rimonabant stimulates 2-deoxyglucose uptake in skeletal muscle cells by regulating the expression of phosphatidylinositol-3-kinase. *Mol. Pharmacol.* **74,** 1678–1686.
Flygare, J., Gustafsson, K., Kimby, E., Christensson, B., and Sander, B. (2005). Cannabinoid receptor ligands mediate growth inhibition and cell death in mantle cell lymphoma. *FEBS Lett.* **579,** 6885–6889.
Food and Drug Administration http://www.fda.gov/ohrms/dockets/ac/07/briefing/2007-4306b1-00-index.htm.
Freund, T. F., Katona, I., and Piomelli, D. (2003). Role of endogenous cannabinoids in synaptic signaling. *Physiol. Rev.* **83,** 1017–1066.
Gable, D. R., Hurel, S. J., and Humphries, S. E. (2006). Adiponectin and its gene variants as risk factors for insulin resistance, the metabolic syndrome and cardiovascular disease. *Atherosclerosis* **188,** 231–244.
Gamber, K. M., Macartur, H., and Westfall, T. C. (2005). Cannabinoids augment the release of neuropeptide Y in the rat hypothalamus. *Neuropharmacology* **143,** 520–523.
Gary-Bobo, M., Elachouri, G., Scatton, B., Le Fur, G., Oury-Donat, F., and Bensaid, M. (2006). The cannabinoid CB1 receptor antagonist rimonabant (SR141716) inhibits cell proliferation and increases markers of adipocyte maturation in cultured mouse 3T3 F442A preadipocytes. *Mol. Pharmacol.* **69,** 471–478.
Gatley, S. J., Gifford, A. N., Volkow, N. D., Lan, R., and Makriyannis, A. (1996). 123I-labeled AM251: A radioiodinated ligand which binds *in vivo* to mouse brain cannabinoid CB1 receptors. *Eur. J. Pharmacol.* **307,** 331–338.
Gessa, G. L., Serra, S., Vacca, G., Carai, M. A. M., and Colombo, G. (2005). Suppressing effect of the cannabinoid CB1 receptor antagonist, SR147778, on alcohol intake and motivational properties of alcohol in alcohol-preferring sP rats. *Alcohol Alcoholism* **40,** 46–53.

Griffith, D. A., Hadcock, J. R., Black, S. C., Iredale, P. A., Carpino, P. A., Da Silva-Jardine, P., Day, R., DiBrino, J., Dow, R. L., Landis, M. S., O'Connor, R. E., *et al.* (2009). Discovery of 1-[9-(4-chlorophenyl)-8-(2-chlorophenyl)-9*H*-purin-6-yl]-4-ethylaminopiperidine-4-carboxylic acid amide hydrochloride (CP-945,598), a novel, potent, and selective cannabinoid type 1 receptor antagonist. *J. Med. Chem.* **52,** 234–237.

Grimaldi, C., Pisanti, C., Laezza, C., Malfitano, A. M., Santoro, A., Vitale, M., Caruso, M., Notarnicola, M., Iacuzzo, I., Portella, G., Di Marzo, V., *et al.* (2006). Anandamide inhibits adhesion and migration of breast cancer cells. *Exp. Cell Res.* **312,** 363–373.

Hagmann, W. K. (2008). The discovery of taranabant, a selective cannabinoid-1 receptor inverse agonist for the treatment of obesity. *Arch. Pharm. (Weinheim)* **341,** 405–411.

Hajos, N., and Freund, T. F. (2002). Pharmacological separation of cannabinoid sensitive receptors on hippocampal excitatory and inhibitory fibers. *Neuropharmacology* **43,** 503–510.

Hajos, N., Ledent, C., and Freund, T. F. (2001). Novel cannabinoid-sensitive receptor mediates inhibition of glutamatergic synaptic transmission in the hippocampus. *Neuroscience* **106,** 1–4.

Haller, J., Bakos, N., Szirmay, M., Ledent, C., and Freund, T. F. (2002). The effects of genetic and pharmacological blockade of the CB1 cannabinoid receptor on anxiety. *Eur. J. Neurosci.* **16,** 1395–1398.

Haller, J., Varga, B., Ledent, C., and Freund, T. F. (2004). CB1 cannabinoid receptors mediate anxiolytic effects: Convergent genetic and pharmacological evidence with CB1-specific agents. *Behav. Pharmacol.* **15,** 299–304.

Henness, S., Robinson, D. M., and Lyseng-Williamson, K. A. (2006). Rimonabant. *Drugs* **66,** 2109–2119discussion 2120–2121.

Herkenham, M., Lynn, A. B., Little, M. D., Johnson, M. R., Melvin, L. S., de Costa, B. R., and Rice, K. C. (1990). Cannabinoid receptor localization in brain. *Proc. Natl. Acad. Sci. USA* **87,** 1932–1936.

Herling, A. W., Gossel, M., Haschke, G., Stengelin, S., Kuhlmann, J., Muller, G., Schmoll, D., and Kramer, W. (2007). CB1 receptor antagonist AVE1625 affects primarily metabolic parameters independently of reduced food intake in Wistar rats. *Am. J. Physiol. Endocrinol. Metab.* **293,** E826–E832.

Howlett, A. C. (2005). Cannabinoid receptor signaling. *Handbook Exp. Pharmacol.* **168,** 53–79.

Howlett, A. C., Barth, F., Bonner, T. I., Cabral, G., Casellas, P., Devane, W. A., Felder, C. C., Herkenham, M., Mackie, K., Martin, B. R., Mechoulam, R., *et al.* (2002). International Union of Pharmacology. XXVII. Classification of cannabinoid receptors. *Pharmacol. Rev.* **54,** 161–202.

Iranmanesh, A., Rosenstock, J., Hollander, P., and SERENADE Study-Group (2006). SERENADE: Rimonabant monotherapy for treatment of multiple cardiometabolic risk factors in treatment-naive patients with type 2 diabetes. *Diabet. Med.* **23**(Suppl. 4), 200–410.

Jacobsson, S. O., Wallin, T., and Fowler, C. J. (2001). Inhibition of rat C6 glioma cell proliferation by endogenous and synthetic cannabinoids. Relative involvement of cannabinoid and vanilloid receptors. *J. Pharmacol. Exp. Ther.* **299,** 951–959.

Jagerovic, N., Hernandez-Folgado, L., Alkorta, I., Goya, P., Navarro, M., Serrano, A., Rodriguez de Fonseca, F., Dannert, M. T., Alsasua, A., Suardiaz, M., Pascual, D., *et al.* (2004). Discovery of 5-(4-chlorophenyl)-1-(2,4-dichlorophenyl)-3-hexyl-1*H*-1,2,4-triazole, a novel *in vivo* cannabinoid antagonist containing a 1,2,4-triazole motif. *J. Med. Chem.* **47,** 2939–2942.

Jamshidi, N., and Taylor, D. A. (2001). Anandamide administration into the ventromedial hypothalamus stimulates appetite in rats. *Br. J. Pharmacol.* **134,** 1151–1154.

Jbilo, O., Ravinet-Trillou, C., Arnone, M., Buisson, I., Bribes, E., Peleraux, A., Pénarier, G., Soubrié, P., Le Fur, G., Galiègue, S., and Casellas, P. (2005). The CB1 receptor antagonist rimonabant reverses the diet-induced obesity phenotype through the regulation of lipolysis and energy balance. *FASEB J.* **19,** 1567–1569.

Kang, S. Y., Lee, S., Seo, H. J., Jung, M. E., Ahn, K., Kim, J., and Lee, J. (2008). Tetrazole–biarylpyrazole derivatives as cannabinoid CB1 receptor antagonists. *Bioorg. Med. Chem. Lett.* **18**, 2385–2389.

Kim, J. Y., Seo, H. J., Lee, S. H., Jung, M. E., Ahn, K., Kim, J., and Lee, J. (2009). Diarylimidazolyl oxadiazole and thiadiazole derivatives as cannabinoid CB1 receptor antagonists. *Bioorg. Med. Chem. Lett.* **19**, 142–145.

Kirkham, T. C., Williams, C. M., Fezza, F., and Di Marzo, V. (2002). Endocannabinoid levels in rat limbic forebrain and hypothalamus in relation to fasting, feeding and satiation: Stimulation on eating by 2-arachidonoyl glycerol. *Br. J. Pharmacol.* **136**, 550–557.

Kola, B., Farkas, I., Christ-Crain, M., Wittmann, G., Lolli, F., Amin, F., Harvey-White, J., Liposits, Z., Kunos, G., Grossman, A. B., Fekete, C., et al. (2008). The orexigenic effect of ghrelin is mediated through central activation of the endogenous cannabinoid system. *PLoS ONE* **12**, e1797.

Lallemand, F., and De Witte, P. (2006). SR147778, a CB1 cannabinoid receptor antagonist, suppresses ethanol preference in chronically alcoholized Wistar rats. *Alcohol* **39**, 125–134.

Lan, R., Liu, Q., Fan, P., Lin, S., Fernando, S. R., McCallion, D., Pertwee, R., and Makriyannis, A. (1999). Structure–activity relationships of pyrazole derivatives as cannabinoid receptor antagonists. *J. Med. Chem.* **42**, 769–776.

Lange, J. H., and Kruse, C. G. (2008). Cannabinoid CB1 receptor antagonists in therapeutic and structural perspectives. *Chem. Rec.* **8**, 156–168.

Lin, L. S., Lanza, T. J., Jewell, J. P., Liu, P., Shah, S. K., Qi, H., Tong, X., Wang, J., Xu, S. S., Fong, T. M., Shen, C. P., Lao, J., et al. (2006). Discovery of N-[(1S,2S)-3-(4-chlorophenyl)-2-(3-cyanophenyl)-1-methylpropyl]-2-methyl-2-{[5-(trifluoromethyl)pyridin-2yl]oxy}propanamide (MK-0364), a novel, acyclic cannabinoid-1 receptor inverse agonist for the treatment of obesity. *J. Med. Chem.* **49**, 7584–7587.

Liu, Y. L., Connoley, I. P., Wilson, C. A., and Stock, M. J. (2005). Effects of the cannabinoid CB1 receptor antagonist SR141716 on oxygen consumption and soleus muscle glucose uptake in Lep(ob)/Lep(ob) mice. *Int. J. Obesity* **29**, 183–187.

Liu, J., Wang, L., Harvey-White, J., Huang, B. X., Kim, H. Y., Luquet, S., Palmiter, R. D., Krystal, G., Rai, R., Mahadevan, A., Razdan, R. K., et al. (2008). Multiple pathways involved in the biosynthesis of anandamide. *Neuropharmacology* **54**, 1–7.

Maccarrone, M., Lorenzon, T., Bari, M., Melino, G., and Finazzi-Agrò, A. (2000). Anandamide induces apoptosis in human cells via vanilloid receptors. Evidence for a protective role of cannabinoid receptors. *J. Biol. Chem.* **275**, 31938–31945.

Malfitano, A. M., Matarese, G., Pisanti, S., Grimaldi, C., Laezza, C., Bisogno, T., Di Marzo, V., Lechler, R. I., and Bifulco, M. (2006). Arvanil inhibits T lymphocyte activation and ameliorates autoimmune encephalomyelitis. *J. Neuroimmunol.* **171**, 110–119.

Malfitano, A. M., Toruner, G. A., Gazzerro, P., Laezza, C., Husain, S., Eletto, D., Orlando, P., De Petrocellis, L., Terskiy, A., Schwalb, M., Vitale, E., et al. (2007). Arvanil and anandamide up-regulate CD36 expression in human peripheral blood mononuclear cells. *Immunol. Lett.* **109**, 145–154.

Malfitano, A. M., Laezza, C., Pisanti, S., Gazzerro, P., and Bifulco, M. (2008). Rimonabant (SR141716) exerts anti-proliferative and immunomodulatory effects in human peripheral blood mononuclear cells. *Br. J. Pharmacol.* **153**, 1003–1010.

Matias, I., and Di Marzo, V. (2007). Endocannabinoids and the control of energy balance. *Trends Endocrinol. Metab.* **18**, 27–37.

Matias, I., Gonthier, M. P., Orlando, P., Martiadis, V., De Petrocellis, L., Cervino, C., Petrosino, S., Hoareau, L., Festy, F., Pasquali, R., Roche, R., Maj, M., et al. (2006). Regulation, function, and dysregulation of endocannabinoids in models of adipose and b-pancreatic cells and in obesity and hyperglycemia. *J. Clin. Endocrinol. Metab.* **91**, 3171–3180.

Matsuda, L. A., Lolait, S. J., Brownstein, M. J., Young, A. C., and Bonner, T. I. (1990). Structure of a cannabinoid receptor and functional expression of the cloned cDNA. *Nature* **346**, 561–564.

Melis, M., Pistis, M., Perra, S., Muntoni, A. L., Pillolla, G., and Gessa, G. L. (2004). Endocannabinoids mediate presynaptic inhibition of glutamatergic transmission in rat ventral tegmental area dopamine neurons through activation of CB1 receptors. *J. Neurosci.* **24**, 53–62.

Muccioli, G. G., Wouters, J., Scriba, G. K. E., Poppitz, W., Poupaert, J. H., and Lambert, D. M. (2005). 1-Benzhydryl-3-phenylurea and 1-benzhydryl-3-phenylthiourea derivatives: New templates among the CB1 cannabinoid receptor inverse agonists. *J. Med. Chem.* **48**, 7486–7490.

Munro, S., Thomas, K. L., and Abu-Shaar, M. (1993). Molecular characterization of a peripheral receptor for cannabinoids. *Nature* **365**, 61–65.

Nesto, R. W. (2005). Managing cardiovascular risk in patients with metabolic syndrome. *Clin. Cornerstone* **7**, 46–51.

Nissen, S. E., Nicholls, S. J., Wolski, K., Rodés-Cabau, J., Cannon, C. P., Deanfield, J. E., Després, J. P., Kastelein, J. J., Steinhubl, S. R., Kapadia, S., Yasin, M., Ruzyllo, W., et al. (2008). Effect of rimonabant on progression of atherosclerosis in patients with abdominal obesity and coronary artery disease The STRADIVARIUS randomized controlled trial. *JAMA* **299**(13), 1547–1560.

Nogueiras, R., Veyrat-Durebex, C., Suchanek, P. M., Klein, M., Tschöp, J., Caldwell, C., Woods, S. C., Wittmann, G., Watanabe, M., Liposits, Z., Fekete, C., Reizes, O., et al. (2008). Peripheral, but not central, CB1 antagonism provides food intake-independent metabolic benefits in diet-induced obese rats. *Diabetes* **57**, 2977–2991.

Osei-Hyiaman, D., Depetrillo, M., Harvey-White, J., Bannon, A. W., Cravatt, B. F., Kuhar, M. J., Mackie, K., Palkovits, M., and Kunos, G. (2005a). Cocaine- and amphetamine-related transcript is involved in the orexigenic effect of anandamide. *Neuroendocrinology* **81**, 273–282.

Osei-Hyiaman, D., DePetrillo, M., Pacher, P., Liu, J., Radaeva, S., Batkai, S., Harvey-White, J., Mackie, K., Offertaler, L., Wang, L., and Kunos, G. (2005b). Endocannabinoids activation at hepatic CB1 receptors stimulates fatty acid synthesis and contributes to diet induced obesity. *J. Clin. Invest.* **115**, 1298–1305.

Pacher, P., Batkai, S., and Kunos, G. (2006). The endocannabinoid system as an emerging target of pharmacotherapy. *Pharmacol. Rev.* **58**, 389–462.

Padwal, R. S., and Majumdar, S. R. (2007). Drug treatments for obesity: Orlistat, sibutramine, and rimonabant. *Lancet* **369**, 71–77.

Pagotto, U., Marsicano, G., Cota, D., Lutz, B., and Pasquali, R. (2006). The emerging role of the endocannabinoid system in endocrine regulation and energy balance. *Endocr. Rev.* **27**, 73–100.

Pamplona, F. A., and Takahashi, R. N. (2006). WIN 55212-2 impairs contextual fear conditioning through the activation of CB1 cannabinoid receptors. *Neurosci. Lett.* **397**, 88–92.

Pavon, F. J., Bilbao, A., Hernandez-Folgado, L., Cippitelli, A., Jagerovic, N., Abellan, G., Rodriguez-Franco, M. A., Serrano, A., Macias, M., Gomez, R., Navarro, M., Goya, P., et al. (2006). Antiobesity effects of the novel *in vivo* neutral cannabinoid receptor antagonist 5-(4-chlorophenyl)-1-(2,4-dichlorophenyl)-3-hexyl-1H-1,2,4-triazole-LH-21. *Neuropharmacology* **51**, 358–366.

Pavon, F. J., Serrano, A., Pérez-Valero, V., Jagerovic, N., Hernandez-Folgado, L., Bermudez-Silva, F. J., Macias, M., Goya, M., and Rodriguez de Fonseca, F. (2008). Central versus peripheral antagonism of cannabinoid CB1 receptor in obesity: Effects of LH-21, a peripherally acting neutral cannabinoid receptor antagonist, in Zucker rats. *J. Neuroendocrinol.* **20**, 116–123.

Pertwee, R. G. (2001). Cannabinoid receptors and pain. *Prog. Neurobiol.* **63**, 569–611.

Pertwee, R. G. (2005). Inverse agonism and neutral antagonism at cannabinoid CB1 receptors. *Life Sci.* **76**, 1307–1324.

Piomelli, D., Giuffrida, A., Calignano, A., and de Fonseca, F. R. (2000). The endocannabinoid system as a target for therapeutic drugs. *Trends Pharmacol. Sci.* **21**, 218–224.

Pi-Sunyer, X. F., Aronne, L. J., Heshmati, H. M., Devin, J., Rosenstock, J., and RIO-North America Study Group (2006). Effect of Rimonabant, a cannabinoid-1 receptor blocker, on weight and cardiometabolic risk factors in overweight or obese patients: A randomized controlled trial. *JAMA* **295**, 761–775.

Portella, G., Laezza, C., Laccetti, P., De Petrocellis, L., Di Marzo, V., and Bifulco, M. (2003). Inhibitory effects of cannabinoid CB1 receptor stimulation on tumor growth and metastatic spreading: Actions on signals involved in angiogenesis and metastasis. *FASEB J.* **17**, 1771–1773.

Powles, T., Poele, R., Shamash, J., Chaplin, T., Propper, D., Joel, S., Oliver, T., and Liu, W. M. (2005). Cannabis-induced cytotoxicity in leukemic cell lines: The role of the cannabinoid receptors and the MAPK pathway. *Blood* **105**, 1214–1221.

Rajesh, M., Mukhopadhyay, P., Haskó, G., and Pacher, P. (2008). Cannabinoid CB1 receptor inhibition decreases vascular smooth muscle migration and proliferation. *Biochem. Biophys. Res. Commun.* **377**, 1248–1252.

Ravinet Trillou, C., Arnone, M., Delgorge, C., Gonalons, N., Keane, P., Maffrand, J. P., and Soubrié, P. (2003). Anti-obesity effect of SR141716, a CB1 receptor antagonist, in diet-induced obese mice. *Am. J. Regul. Integr. Comp. Physiol.* **284**, R345–R353.

Rinaldi-Carmona, M., Barth, F., Healume, M., Shire, D., Calandra, B., Congy, C., Martinez, S., Maruani, J., Neliat, G., Caput, D., Ferrara, P., et al. (1994). SR141716A, a potent and selective antagonist of the brain cannabinoid receptor. *FEBS Lett.* **350**, 240–244.

Rinaldi-Carmona, M., Barth, F., Heaulme, M., Alonso, R., Shire, D., Congy, C., Soubrie, P., Breliere, J. C., and Le Fur, G. (1995). Biochemical and pharmacological characterisation of SR141716A, the first potent and selective brain cannabinoid receptor antagonist. *Life Sci.* **56**, 1941–1947.

Rinaldi-Carmona, M., Barth, F., Congy, C., Martinez, S., Oustric, D., Perio, A., Poncelet, M., Maruani, J., Arnone, M., Finance, O., Soubrie, P., et al. (2004). SR147778 [5-(4-bromophenyl)-1-(2,4-dichlorophenyl)-4-ethyl-N-(1-piperidinyl)-1H-pyrazole-3-carboxamide], a new potent and selective antagonist of the CB1 cannabinoid receptor: Biochemical and pharmacological characterization. *J. Pharmacol. Exp. Ther.* **310**, 905–914.

Rodriguez de Fonseca, F., Navarro, M., Gomez, R., Escuredo, L., Nava, F., Fu, J., Murillo-Rodriguez, E., Giuffrida, A., LoVerme, J., Gaetani, S., Kathuria, S., Gall, C., et al. (2001). An anorexic lipid mediator regulated by feeding. *Nature* **414**, 209–212.

Rosenstock, J., Hollander, P., Chevalier, S., Iranmanesh, A., and SERENADE Study Group (2008). SERENADE: The study evaluating rimonabant efficacy in drug-naive diabetic patients: Effects of monotherapy with rimonabant, the first selective CB1 receptor antagonist, on glycemic control, body weight, and lipid profile in drug-naive type 2 diabetes. *Diabetes Care* **31**, 2169–2176.

Rucker, D., Padwal, R., Li, S. K., Curioni, C., and Lau, D. C. (2007). Long term pharmacotherapy for obesity and overweight: Updated meta-analysis. *BMJ* **335**, 1194–1199.

Ruilope, L. M., Després, J. P., Scheen, A., Pi-Sunyer, X., Mancia, G., Zanchetti, A., and Van Gaal, L. (2008). Effect of rimonabant on blood pressure in overweight/obese patients with/without co-morbidities: Analysis of pooled RIO study results. *J. Hypertens.* **26**, 357–367.

Sanchez, C., de Ceballos, M. L., del Pulgar, T. G., Rueda, D., Corbacho, C., Velasco, G., Galve-Roperh, I., Huffman, J. W., Ramon y Cajal, S., and Guzman, M. (2001). Inhibition of glioma growth *in vivo* by selective activation of the CB(2) cannabinoid receptor. *Cancer Res.* **61**, 5784–5789.

Santoro, A., Gazzerro, P., Malfitano, A. M., Pisanti, S., Laezza, C., and Bifulco, M. (2008). Reply to the letter to the editor "Long-term cannabinoid receptor (CB1) blockade in obesity: Implications for the development of colorectal cancer" *Int. J. Cancer* **122**, 243–244.

Sarfaraz, S., Afaq, F., Adhami, V. M., and Mukhtar, H. (2005). Cannabinoid receptor as a novel target for the treatment of prostate cancer. *Cancer Res.* **65**, 1635–1641.

Sarnataro, D., Pisanti, S., Santoro, A., Gazzerro, P., Malfitano, A. M., Laezza, C., and Bifulco, M. (2006). The cannabinoid CB1 receptor antagonist rimonabant (SR141716) inhibits human breast cancer cell proliferation through a lipid rafts mediated mechanism. *Mol. Pharmacol.* **70**, 1298–1306.

Savinainen, J. R., Saario, S. M., Niemi, R., Järvinen, T., and Laitinen, J. T. (2003). An optimized approach to study endocannabinoid signaling: Evidence against constitutive activity of rat brain adenosine A1 and cannabinoid CB1 receptors. *Br. J. Pharmacol.* **140**, 1451–1459.

Schäfer, A., Pfrang, J., Neumüller, J., Fiedler, S., Ertl, G., and Bauersachs, J. (2008). The cannabinoid receptor-1 antagonist rimonabant inhibits platelet activation and reduces pro-inflammatory chemokines and leukocytes in Zucker rats. *Br. J. Pharmacol.* **154**, 1047–1054.

Scheen, A. J., and Paquot, N. (2008). Inhibitors of cannabinoid receptors and glucose metabolism. *Curr. Opin. Clin. Nutr. Metab. Care* **11**, 505–511.

Scheen, A. J., Fine, N., Hollander, P., Jensen, M. D., Van Gaal, L. F., and RIO-Diabetes Study Group. (2006a). Efficacy and tolerability of rimonabant in overweight or obese patients with type 2 diabetes: A randomized controlled study. *Lancet* **368**, 1660–1672.

Scheen, A. J., Van Gaal, L. F., Després, J. P., Pi-Sunyer, X., Golay, A., and Hanotin, C. (2006b). Le rimonabant améliore le profil de risque cardio-métabolique chez le sujet obe'se ou en surpoids: Synthe'se des études "RIO". *Rev. Med. Suisse* **2**, 1916–1923.

Serra, G., and Fratta, W. (2007). A possible role for the endocannabinoid system in the neurobiology of depression. *Clin. Pract. Epidemiol. Mental Health* **3**, 25doi:10.1186/1745-0179-3-25.

Sugamura, K., Sugiyama, S., Nozaki, T., Matsuzawa, Y., Izumiya, Y., Miyata, K., Nakayama, M., Kaikita, K., Obata, T., Takeya, M., and Ogawa, H. (2009). Activated endocannabinoid system in coronary artery disease and antiinflammatory effects of cannabinoid 1 receptor blockade on macrophages. *Circulation* **119**, 28–36.

Tedesco, L., Valerio, A., Cervino, C., Cardile, A., Pagano, C., Vettor, R., Pasquali, R., Carruba, M. O., Marsicano, G., Lutz, B., Pagotto, U., et al. (2008). Cannabinoid type 1 receptor blockade promotes mitochondrial biogenesis through endothelial nitric oxide synthase expression in white adipocytes. *Diabetes* **57**, 2028–2036.

Tucci, S. A., Rogers, E. K., Korbonits, M., and Kirlham, T. C. (2004). The cannabinoid receptor antagonist SR141716 blocks the orexigenic effect of intrahypothalamic ghrelin. *Br. J. Pharmacol.* **143**, 520–523.

Van Gaal, L. F., Rissanen, A. M., Scheen, A., Ziegler, O., Rössner, S., and RIO-Europe Study Group (2005). Effects of the cannabinoid-1 receptor blocker rimonabant on weight reduction and cardiovascular risk factors in overweight patients: 1-year experience from the RIO-Europe study. *Lancet* **365**, 1389–1397.

Van Gaal, L. F., Mertens, I. L., and De Block, C. E. (2006). Mechanisms linking obesity with cardiovascular disease. *Nature* **444**, 875–880.

Van Gaal, L. F., Pi-Sunyer, X., Després, J. P., Mc Carthy, C., and Scheen, A. J. (2008a). Efficacy and safety of rimonabant for improvement of multiple cardiometabolic risk factors in overweight/obese patients: Pooled 1-year data from the RIO program. *Diabetes Care* **31**(Suppl. 2), S229–S240.

Van Gaal, L. F., Scheen, A. J., Rissanen, A. M., Rössner, S., Hanotin, C., Ziegler, O., and RIO-Europe Study Group (2008b). Long-term effect of CB1 blockade with rimonabant on cardiometabolic risk factors: Two year results from the RIO-Europe Study. *Eur. Heart J.* **29,** 1761–1771.

Verty, A. N., McFarlane, J. R., McGregor, I. S., and Mallet, P. E. (2004). Evidence for an interaction between CB1 cannabinoid and melanocortin MC4 receptors in regulating food intake. *Endocrinology* **145,** 3224–3231.

Verty, A. N., Allen, A. M., and Oldfield, B. J. (2009). The effects of rimonabant on brown adipose tissue in rat: Implications for energy expenditure. *Obesity (Silver Spring)* **17,** 254–261.

Wang, J., and Ueda, N. (2008). Biology of endocannabinoid synthesis system. *Prostaglandins Other Lipid Mediat.* (in press).

Wang, D., Wang, H., Ning, W., Backlun, M. G., Dey, S., and DuBois, R. (2008). Loss of cannabinoid receptor 1 accelerates intestinal tumor growth. *Cancer Res.* **68,** 6468–6476.

Watanabe, T., Kubota, N., Ohsugi, M., Kubota, T., Takamoto, I., Iwabu, M., Awazawa, M., Katsuyama, H., Hasegawa, C., Tokuyama, K., Moroi, M., Sugi, K., *et al.* (2009). Rimonabant ameliorates insulin resistance via both adiponectin-dependent and adiponectin-independent pathways. *J. Biol. Chem.* **284,** 1803–1812.

Wilson, R. I., and Nicoll, R. A. (2002). Endocannabinoids signaling in the brain. *Science* **296,** 678–682.

Wright, S. M., Dikkers, C., and Aronne, L. J. (2008). Rimonabant: New data and emerging experience. *Curr. Atheroscler. Rep.* **10,** 71–78.

Yo, J. H., Chen, Y. J., Talmage, D. A., and Role, L. W. (2005). Integration of leptin signalling in an appetite-related neural circuit. *Neuron* **48,** 1055–1066.

CHAPTER EIGHT

Novel Endogenous N-Acyl Glycines: Identification and Characterization

Heather B. Bradshaw,* Neta Rimmerman,* Sherry S.-J. Hu,* Sumner Burstein,[†] *and* J. Michael Walker*

Contents

I. Historical View of Lipid Signaling Discoveries	192
II. The Identification of Endogenous Signaling Lipids with Cannabimimetic Activity	192
III. Identification of Additional N-Acyl Amides	193
IV. N-Arachidonoyl Glycine Biological Activity	194
V. N-Arachidonoyl Glycine Biosynthesis	194
VI. N-Palmitoyl Glycine Biological Activity	195
VII. N-Palmitoyl Glycine Biosynthesis	197
VIII. PalGly Metabolism	197
IX. Identification and Characterization of Additional Members of the N-Acyl Glycines	198
X. Biological Activity of Novel N-Acyl Glycines	200
XI. Conclusions	201
References	203

Abstract

Discovery of the endogenous cannabinoid and N-acyl amide, anandamide (N-arachidonoyl ethanolamine), paved the way for lipidomics discoveries in the growing family of N-acyl amides. Lipidomics is a field that is broadening our view of the molecular world to include a wide variety of endogenous lipid signaling molecules. Many of these lipids will undoubtedly provide new insights into old questions while others will provide broad platforms for new questions. J Michael Walker's last 8 years were dedicated to this search and he lived long enough to see 54 novel lipids isolated from biological tissues in his laboratory. Here, we summarize the biosynthesis, metabolism and biological activity of two of the family of N-acyl glycines, N-arachidonoyl glycine and

* Department of Psychological and Brain Sciences, Indiana University, Bloomington, Indiana, USA
† Department of Biochemistry and Molecular Pharmacology, University of Massachusetts Medical School, Worcester, Massachusetts, USA

N-palmitoyl glycine, and introduce four additional members: N-stearoyl glycine, N-linoleoyl glycine, N-oleoyl glycine, and N-docosahexaenoyl glycine. Each of these compounds is found throughout the body at differing levels suggesting region-specific functionality and at least four of the N-acyl glycines are regulated by the enzyme fatty acid amide hydrolase. The family of N-acyl glycines presented here is merely a sampling of what is to come in the continuing discovery of novel endogenous lipids.

I. Historical View of Lipid Signaling Discoveries

The use of natural products for a variety of ailments in folklore medicine attracted scientists to explore their mechanisms of bioactivity (Caterina et al., 1997; Mechoulam and Gaoni, 1965; Naef et al., 2003; Urca et al., 1977). The finding that plant-derived active ingredients such as the opium-derived morphine and cannabis-derived Δ^9-hydrotetracannabinol (Δ^9-THC; cannabinoids) that had a wide range of biological activity promoted the identification of novel endogenous cannabinoid-like signaling molecules and pathways in the mammalian nervous system that would likewise drive these actions. The discovery of these signaling molecules has opened new avenues into research that continue to expand our knowledge of basic biochemical functioning.

II. The Identification of Endogenous Signaling Lipids with Cannabimimetic Activity

The first mammalian counterpart of the analgesic phytocannabinoid Δ^9-THC was identified in porcine brain and named anandamide after the Sanskrit word for bliss, *ananda* (also, N-arachidonoyl ethanolamine—AEA; Fig. 8.1A) [5]. Both exogenous and endogenous compounds were shown to activate the G protein-coupled receptor (GPCR) cannabinoid receptor 1 (CB_1) and induced several physiological and behavioral outcomes including hypothermia, analgesia, hypoactivity, and catalepsy (Devane et al., 1992; Mechoulam and Gaoni, 1965). Additional characteristics of AEA included binding to the GPCR cannabinoid receptor 2 (CB_2) and activating the transient receptor potential vanilloid type-1 channel (Felder et al., 1995; Ross, 2003). Furthermore, unique retrograde signaling properties affecting release of classical neurotransmitters were revealed (Elphick and Egertova, 2001; Ueda et al., 2005). A second endogenous cannabinoid 2-arachidonoyl glycerol (2-AG) binding to both cannabinoid receptors was later identified in rat brain and canine gut (Mechoulam et al., 1995; Sugiura et al., 1995).

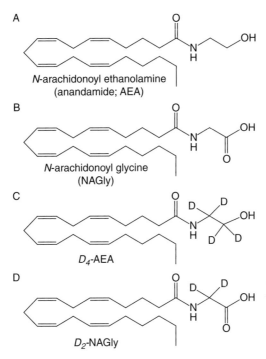

Figure 8.1 Molecular structures of N-arachidonoyl ethanolamine (AEA; A) and N-arachidonoyl glycine (NAGly; B). (C) Deuterium-labeled AEA in which the deuteriums are located on the ethanolamine moiety. (D) Deuterium-labeled NAGly in which the deuteriums are on the glycine moiety from the theoretical conversion of AEA to NAGly via the initial step of alcohol dehydrogenase.

Together these two endogenous lipids paved the way for the identification of a family of lipids often referred to as endocannabinoids.

III. IDENTIFICATION OF ADDITIONAL N-ACYL AMIDES

Burstein et al. (1997) suggested that N-arachidonoyl glycine (NAGly; Fig. 8.1B) was a putative endogenous compound in 1997. The methodologies used in the isolation and measurements of AEA in biological samples (lipid extractions and HPLC/MS/MS; Walker et al., 1999) enabled Walker and colleagues to search for other N-acyl amides of similar structure, which were hypothesized to have similar function. Huang et al. (2001) were able to isolate three novel N-acyl amide molecules in the brain and periphery: NAGly, N-arachidonoyl GABA, and N-arachidonoyl alanine. Further work from the Walker laboratory identified N-oleoyl dopamine, N-stearoyl dopamine, N-palmitoyl dopamine, and N-arachidonoyl dopamine (Chu et al., 2003; Huang et al., 2002). Other research groups identified and characterized N-arachidonoyl serine (Milman et al., 2006) and the N-acyl taurines (McKinney and Cravatt, 2006) adding support to the hypothesis that there is a potentially large family of N-acyl amides that are putative signaling molecules with a wide range of biological activity (for review, see Bradshaw and Walker, 2005).

IV. N-Arachidonoyl Glycine Biological Activity

NAGly was originally synthesized as part of a structure activity relationship study of the endocannabinoid AEA (Burstein et al., 1997; Sheskin et al., 1997). NAGly differs from AEA by the oxidation state of the carbon β to the amido nitrogen (Fig. 8.1B); a modification that drastically reduces its activity at both cannabinoid receptors (Sheskin et al., 1997). NAGly produces antinociceptive and anti-inflammatory effects in a variety of pain models (Burstein et al., 2000; Huang et al., 2001; Succar et al., 2007; Vuong et al., 2008). Studies by Kohno et al. (2006) found that low concentrations ($EC_{50} \sim 20$ nM) of NAGly activate GPR18, an orphan GPCR. Consistent with the anti-inflammatory effects of NAGly, GPR18 is highly expressed in peripheral blood leukocytes and several hematopoietic cell lines. Burstein et al. (2000) showed that at low concentrations NAGly induces proliferation of T cells, but suppresses production of IL-1β. NAGly also reduces proliferation of the Caco-2, human rectal carcinoma cell line (Gustafsson et al., 2009). In pancreatic β-cells, NAGly caused intracellular calcium mobilization and insulin release (Ikeda et al., 2005). This effect was blocked by the L-type voltage-gated channel nitredipine (1 μM) or in the absence of extracellular calcium. Additionally, NAGly inhibited the glycine transporter, GLYT2a through direct, noncompetitive interactions (Wiles et al., 2006). These data support the hypothesis that NAGly is an endogenous signaling molecule with multiple biological activities.

V. N-Arachidonoyl Glycine Biosynthesis

Unlike the acyl glycerols and N-acyl ethanolamines, the biosynthesis of N-acyl glycines cannot logically be derived from phospholipid biochemistry. Two primary pathways for the biosynthesis of the NAGly have been proposed: (1) conjugation of arachidonyl CoA and glycine (Burstein et al.,

2002; Huang *et al.*, 2001; McCue *et al.*, 2008) and (2) oxygenation of *N*-arachidonoyl ethanolamine via the sequential enzymatic reaction of alcohol dehydrogenase (ADH) and aldehyde dehydrogenase (Burstein *et al.*, 2000).

Huang *et al.* (2001) proposed that NAGly is synthesized by the condensation of arachidonic acid (AA) with glycine based upon the formation of deuterated NAGly following incubations of brain membranes with deuterated AA and deuterated glycine. McCue *et al.* (2008) demonstrated that NAGly is formed via cytochrome *c* acting on arachidonoyl CoA and glycine in support of this conjugation pathway. We have additional data to support this hypothesis; however, the AA that conjugates with glycine appears to be a result of the hydrolysis of AEA (Bradshaw *et al.*, 2009). This evidenced by results that show deuterium-labeled AEA (labeled on that arachidonoyl chain) incubated in C6 glioma cells is converted to deuterium-labeled NAGly (Bradshaw *et al.*, 2009). This pathway is blocked by the fatty acid amide hydrolase (FAAH) inhibitor, URB597 *in vivo* and *in vitro* (Bradshaw *et al.*, 2009). The paper by Mueller in this special edition suggests a mitochondrial synthesome for the biosynthesis of oleamide, an important lipid signaling acyl amine (Huitron-Resendiz *et al.*, 2001; Martinez-Gonzalez *et al.*, 2004). It is possible that NAGly biosynthesis is part of this mitochondrial enzyme complex and that the hydrolysis of AEA is a rate-limiting step.

An alternative pathway was proposed by Burstein *et al.* (2000) who speculated that NAGly is produced by the oxidation of the ethanolamine in AEA, presumably through ADH. We have evidence to support this pathway as well in that deuterium-labeled AEA (deuterium on the ethanolamine moiety; Fig. 8.1C) incubated in RAW264.7 cells is converted to deuterium-labeled NAGly (Fig. 8.1D; Bradshaw *et al.*, 2009). This pathway is illustrated by Mueller and Driscoll in this volume to demonstrate the potential biosynthesis of *N*-oleoyl glycine from *N*-oleoyl ethanolamine. Therefore, there is evidence for both proposed pathways and they both have an *N*-acyl ethanolamine as the precursor molecule.

VI. *N*-Palmitoyl Glycine Biological Activity

An additional member of the *N*-acyl glycine family was identified in the Walker group. Rimmerman *et al.* (2008) showed that *N*-palmitoyl glycine (PalGly; Fig. 8.2A) is produced throughout the body and plays a role in sensory neuronal signaling. The authors showed that PalGly is produced following cellular stimulation and occurs in high levels in rat skin and spinal cord. PalGly was upregulated in FAAH knockout (KO)

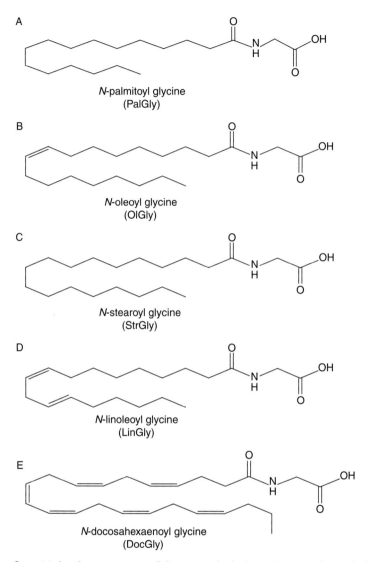

Figure 8.2 Molecular structures of five N-acyl glycines: (A) N-palmitoyl glycine (PalGly), which was recently identified (Rimmerman); four other N-acyl glycines identified here: (B) N-oleoyl glycine (OlGly), (C) N-stearoyl glycine (StrGly), (D) N-linoleoyl glycine (LinGly), and (E) N-docosahexaenoyl glycine (DocGly).

mice suggesting a pathway for enzymatic regulation. PalGly potently inhibited heat-evoked firing of nociceptive neurons in rat dorsal horn. In addition, PalGly induced transient calcium influx in native adult DRG cells and in a DRG-like cell line (F-11). The effect of PalGly on the latter was characterized by strict structural requirements, PTX sensitivity, and dependence on the presence of extracellular calcium. PalGly-induced calcium influx was blocked by the nonselective calcium channel blockers ruthenium red, SKF96365, and La^{3+}. Furthermore, PalGly contributed to the production of nitric oxide (NO) through calcium-sensitive nitric oxide synthase (NOS) enzymes present in F-11 cells, and was inhibited by the NOS inhibitor 7-NI.

VII. N-PALMITOYL GLYCINE BIOSYNTHESIS

PalGly is comprised of an 18-carbon saturated fatty acid that is amide-linked to glycine (Fig. 8.2A) and, therefore, has structural similarities to the phospholipid-derived N-acyl ethanolamines. Several biosynthetic pathways for the production of PalGly are possible. Historically glycine conjugation was investigated in the context of glycine N-acylase (Schachter and Taggart, 1954). The enzyme was purified from the mitochondria of bovine liver, was glycine specific, and active with aliphatic short and medium carbon chains (2–10 carbons), and aromatic acyl thioesters, yielding only short and medium chain N-acyl glycines (Schachter and Taggart, 1954). Another glycine-conjugating enzyme, bile acid CoA:amino acid N-acyl transferase (BACAT), was found in microsomes and peroxisomes and shown to conjugate bile acids mainly to glycine and taurine amino acids (O'Byrne et al., 2003). Furthermore, it was shown that human BACAT can conjugate saturated 16–20 carbon fatty acids to glycine in vitro. The enzyme was found in liver and gallbladder mucosa with lower expression in skin and lung. However, the activity of BACAT with fatty acid CoAs was only 20% of the activity reported for bile acids. Recently, Mueller and Driscoll (2007) demonstrated the production of several acyl glycines (including PalGly) via the enzyme cytochrome c acting in the presence of hydrogen peroxide, glycine, and acyl CoAs in vitro.

Based on its structural similarity to NAGly, the alcohol dehydrogenase biosynthetic scheme may also be a viable biosynthetic route. Further investigations are needed to determine the production of PalGly via this pathway.

VIII. PALGLY METABOLISM

We recently observed that both the FAAH inhibitor URB597 treated rats and FAAH KO mice have increased brain levels of PalGly (Rimmerman et al., 2008). These findings suggest that FAAH may be a major metabolic pathway for the hydrolysis of saturated fatty acid glycine molecules. *In vitro* metabolism of PalGly was demonstrated using a bacterial cytochrome P450 (CYPBM-3). PalGly bound the enzyme with higher affinity than any other tested compound. The products of the enzymatic oxidation were ω-1-, ω-2-, and ω-3-monohydroxylated metabolites of PalGly (Haines *et al.*, 2001).

IX. Identification and Characterization of Additional Members of the *N*-Acyl Glycines

The recognition of the growing family of *N*-acyl amides (Farrell and Merkler, 2008; Tan *et al.*, 2006) has been made possible by the developing field of lipidomics. Lipidomics is the lipid corollary to proteomics for proteins with the exception that there is no genomic template from which to predict lipids in the way that novel proteins are extrapolated from the genome. The lipidomics of the *N*-acyl amides is based on the knowledge that there are a certain number of predominate fatty acids (acyl chains) and amines (e.g., amino acids) that could possibly form conjugations. Here, using lipidomics approaches similar to those used to identify PalGly (Rimmerman *et al.*, 2008), we continue with the identification of the endogenous *N*-acyl glycines with the isolation and characterization of *N*-oleoyl glycine (OlGly; Fig. 8.2B), *N*-stearoyl glycine (StrGly; Fig. 8.2C), *N*-linoleoyl glycine (LinGly; Fig. 8.2D), and *N*-docosahexaenoyl glycine (DocGly; Fig. 8.2E).

Comparisons of a partially purified brain matrix and the product ion scan of a synthesized OlGly standard demonstrate that there is an identical molecular species of lipid compound in the mammalian brain (Fig. 8.3). Equivalent analyses were performed for each of the additional *N*-acyl glycines listed and the molecular ion and fragment ions in parts per million (ppm) for each is shown in Table 8.1.

To determine the distribution of each of these *N*-acyl glycines throughout the body, extraction, purification, and HPLC/MS/MS methods were optimized for each compound using the synthetic standards with a methodology previously described (Rimmerman *et al.*, 2008). Figure 8.4 shows an example of a chromatogram of an OlGly standard and of this HPLC/MS/MS method used on purified spinal cord extracts. Twelve tissue types (skin, lung, spinal cord, ovary, kidney, liver, spleen, brain, small intestine, uterus, testes, and heart) were subjected to lipid extraction and partial purification as previously outlined for PalGly and the amounts of

Novel Endogenous N-Acyl Glycines

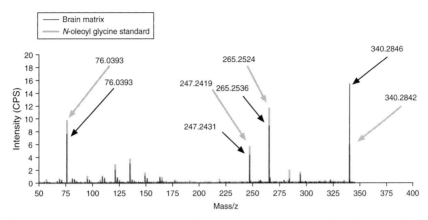

Figure 8.3 Mass spectrums of brain lipid extract match those of synthetic N-oleoyl glycine (OlGly). Product ion scan mass spectrum of the OlGly synthetic standard (gray peaks) positively charged molecular ion in which the arrows are directed to the peaks of the proposed molecular and fragment ions and labeled with the calculated exact mass. The black peaks indicated the product ion scan mass spectrum of the brain lipid extract tuned to the positively charged OlGly molecular ion in which the arrows are directed to the peaks of the proposed molecular and fragment ions and labeled with the calculated exact mass.

each novel lipid quantified (Fig. 8.5A–D). StrGly had the highest levels of the four novel N-acyl glycines reported here, with levels that were similar to PalGly reported earlier (Rimmerman *et al.*, 2008). Additionally, like PalGly, but unlike NAGly, the levels in the skin, lung, and spinal cord were the highest in StrGly, OlGly, and DocGly. LinGly levels differed in that the relative levels in the spinal cord were much less than the other N-acyl glycines measured here. The fairly ubiquitous nature of these molecules suggests that they have potential roles either as signaling molecules or as precursors to other bioactive lipids or both. The identification of OlGly as an endogenous molecule that is measured throughout the body lends evidence to the hypothesis that it is produced endogenously and, therefore, available to be the precursor for oleamide as summarized in this volume by Mueller and Driscoll.

Finally, the metabolism of these novel N-acyl glycines was examined after the systemic injection of the FAAH inhibitor, URB597 (Piomelli *et al.*, 2006). Levels of each N-acyl glycine were measured from rat striatum after systemic vehicle (DMSO) or URB597 i.p. injection (3 mg/kg). We found that both StrGly and OlGly were significantly increased in striatum after the FAAH blocker was present for 30 min (Fig. 8.6). LinGly levels were not

Table 8.1 Molecular and fragmentation ions of N-acyl glycines

Molecular and fragment ions (ppm difference from theoretical exact mass)	Proposed formulae	Proposed fragmentation
N-oleoyl glycine		
340.2846 (−0.061)	$C_{20}H_{38}NO_3$	MH^+
294.2801 (−9.9)	$C_{19}H_{36}NO$	$MH^+ - CH_2O_2$
265.2536 (−10.2)	$C_{18}H_{33}O$	$MH^+ - C_2H_5NO_2$
247.2431 (−11.0)	$C_{18}H_{31}$	$MH^+ - C_2H_7NO_3$
76.0393 (−0.065)	$C_2H_6NO_2$	$MH^+ - C_{18}H_{32}O$
N-stearoyl glycine		
342.3001 (2.1020)	$C_{20}H_{40}NO_3$	MH^+
324.2970 (20.7994)	$C_{20}H_{38}NO_2$	$MH^+ - H_2O$
296.2993 (13.3642)	$C_{19}H_{38}NO$	$MH^+ - CH_2O_2$
267.2636 (19.4229)	$C_{18}H_{35}O$	$MH^+ - C_2H_5NO_2$
76.0391 (9.9097)	$C_2H_6NO_2$	$MH^+ - C_{18}H_{34}O$
N-linoleoyl glycine		
338.2703 (3.9294)	$C_{20}H_{36}NO_3$	MH^+
245.2203 (−24.7841)	$C_{18}H_{29}$	$MH^+ - C_2H_7NO_3$
76.0401 (10.4556)	$C_2H_6NO_2$	$M^+ - C_{18}H_{30}O$
N-docosahexaenoyl glycine		
336.2681 (−2.2543)	$C_{24}H_{36}NO_3$	MH^+
76.0396 (3.38802)	$C_2H_6NO_2$	$MH^+ - C_{22}H_{30}O$

Mass measurements of the molecular and fragment ions as measured using nano-HPLC coupled to a C18 capillary column with a QSTAR pulsar Q-TOF detector. Proposed formulae of these ions were extrapolated by the fragmentation patterns and determinations for four putative endogenous N-acyl glycines were found in lipid extracts of murine brain and liver. The parts per million (ppm) differences from the theoretical exact masses of each molecular ion and fragment ion are represented in parenthesis.

detected in striatum and DocGly levels were just at detection limits and no significant differences were measured. This demonstrates that these endogenous N-acyl glycines are metabolically regulated by FAAH. The biosynthesis of these molecules is still under investigation.

X. Biological Activity of Novel N-Acyl Glycines

There is developing evidence for biological activity of these novel N-acyl glycines. Chaturvedi et al. (2006) suggest that OlGly possess biologic activity that is independent of its conversion to oleamide. Recent data by

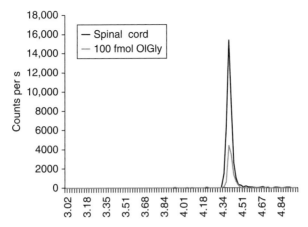

Figure 8.4 Chromatograms of N-oleoyl glycine (OlGly) and spinal cord matrix. A 100 fmol standard of OlGly was analyzed using the API 3000 LC/MS/MS MRM method that optimizes on the molecular ion 338 and the fragment ion 74 in negative ion mode and a chromatographic gradient optimized for C18 column retention and release of the standard (gray line) and an overlay of a chromatogram from the injection of a methanolic extract partially purified on C18 solid phase extraction columns using the same MRM method (black line).

Burstein et al. (2007) demonstrate that NAGly, DocGly, and LinGly suppress proliferation of the murine macrophage cell line, RAW264.7, whereas OlGly and PalGly had no effect. Additionally, his work shows that NAGly, DocGly, and LinGly increase PGJ2 immunoreactivity, whereas again OlGly and PalGly had no effect.

XI. Conclusions

Lipidomics is a field that is broadening our view of the molecular world to include growing numbers of lipid signaling molecules. Many of these lipids will undoubtedly provide new insights into old questions while others will provide broad platforms for new questions. The family of N-acyl glycines presented here is merely a sampling of what is to come in the discovery of novel lipids. Basic combinatorial math of as few as 7 fatty acid chains and 20 amino acids yields 140 novel N-acyl amino acids. Michael Walker's last 8 years were dedicated to this search and he lived long enough to see 54 novel lipids elucidated in biological tissue in his laboratory. His dream lives on in all those he trained and in the larger community of scientists who continue to isolate and characterize novel endogenous lipids.

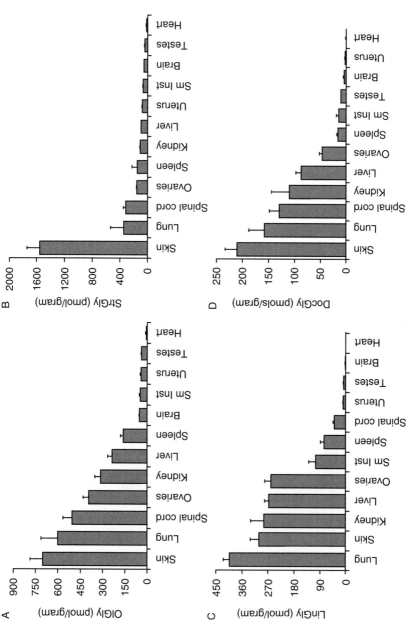

Figure 8.5 *N*-acyl glycines are found throughout the CNS and body. Constitutive production of *N*-oleoyl glycine (OlGly), *N*-stearoyl glycine (StrGly), *N*-linoleoyl glycine (LinGly), and *N*-docosahexaenoyl glycine (DocGly) was measured in partially purified lipid extracts using LC/MS/MS optimized to each of the synthetic standards. As a matter of weight standardization, dry tissue weights were estimated from the ratio of lyophilized to wet weight of each type of tissue that underwent methanolic lipid extraction and partial purification on C18 solid phase extraction columns directly after dissection. The 12 types of tissue are listed on the *x*-axis in relative order of production per gram dry weight.

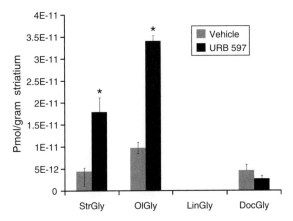

Figure 8.6 FAAH metabolism of acyl glycines. Comparison of production levels of N-stearoyl glycine (StrGly), N-oleoyl glycine (OlGly), N-linoleoyl glycine (LinGly), and N-docosahexaenoyl glycine (DocGly) in rat striatum: Vehicle (gray bars) and 3 mg/kg URB597 i.p. (black bars). Brains were removed and striatum dissected 30 min after injection. Tissue underwent methanolic extraction and partial purification on C18 solid phase extraction columns before eluants were measured with LC/MS/MS as outlined in Figs. 8.3–8.5.

REFERENCES

Bradshaw, H. B., and Walker, J. M. (2005). The expanding field of cannabimimetic and related lipid mediators. *Br. J. Pharmacol.* **144**(4), 459–465.

Bradshaw, H. B., Rimmerman, N., Hu, S. S-J., Benton, V. M., Stuart, J. M., Masuda, K., Cravatt, B. F., O'Dell, D. K., and Walker, J. M. (2009). The endocannabinoid anandamide is a precursor for the signaling lipid, N-arachidonoyl glycine through two distinct pathways. *BMC Biochemistry* **10**, 14.

Burstein, S., Monaghan, A., Pearson, W., Rooney, T., Yagen, B., Zipkin, R., and Zurier, A. (1997). Studies with analogs of anandamide and indomethacin. In "Symposium on Cannabinoids" p. 31. International Cannabinoid Research Society, Burlington, VT.

Burstein, S. H., Rossetti, R. G., Yagen, B., and Zurier, R. B. (2000). Oxidative metabolism of anandamide. *Prostaglandins Other Lipid Mediat.* **61**(1–2), 29–41.

Burstein, S. H., Huang, S. M., Petros, T. J., Rossetti, R. G., Walker, J. M., and Zurier, R. B. (2002). Regulation of anandamide tissue levels by N-arachidonylglycine. *Biochem. Pharmacol.* **64**(7), 1147–1150.

Burstein, S. H., Adams, J. K., Bradshaw, H. B., Fraioli, C., Rossetti, R. G., Salmonsen, R. A., Shaw, J. W., Walker, J. M., Zipkin, R. E., and Zurier, R. B. (2007). Potential anti-inflammatory actions of the elmiric (lipoamino) acids. *Bioorg. Med. Chem.* **15**(10), 3345–3355.

Caterina, M. J., Schumacher, M. A., Tominaga, M., Rosen, T. A., Levine, J. D., and Julius, D. (1997). The capsaicin receptor: A heat-activated ion channel in the pain pathway. *Nature* **389**(6653), 816–824.

Chaturvedi, S., Driscoll, W. J., Elliot, B. M., Faraday, M. M., Grunberg, N. E., and Mueller, G. P. (2006). *In vivo* evidence that N-oleoylglycine acts independently of its conversion to oleamide. *Prostaglandins Other Lipid Mediat.* **81**(3–4), 136–149.

Chu, C. J., Huang, S. M., De Petrocellis, L., Bisogno, T., Ewing, S. A., Miller, J. D., Zipkin, R. E., Daddario, N., Appendino, G., Di Marzo, V., and Walker, J. M. (2003). N-oleoyldopamine, a novel endogenous capsaicin-like lipid that produces hyperalgesia. *J. Biol. Chem.* **278**(16), 13633–13639.

Devane, W. A., Hanus, L., Breuer, A., Pertwee, R. G., Stevenson, L. A., Griffin, G., Gibson, D., Mandelbaum, A., Etinger, A., and Mechoulam, R. (1992). Isolation and structure of a brain constituent that binds to the cannabinoid receptor. *Science* **258**(5090), 1946–1949.

Elphick, M. R., and Egertova, M. (2001). The neurobiology and evolution of cannabinoid signalling. *Philos. Trans. R. Soc. Lond. B Biol. Sci.* **356**(1407), 381–408.

Farrell, E. K., and Merkler, D. J. (2008). Biosynthesis, degradation and pharmacological importance of the fatty acid amides. *Drug Discov. Today* **13**(13–14), 558–568.

Felder, C. C., Joyce, K. E., Briley, E. M., Mansouri, J., Mackie, K., Blond, O., Lai, Y., Ma, A. L., and Mitchell, R. L. (1995). Comparison of the pharmacology and signal transduction of the human cannabinoid CB1 and CB2 receptors. *Mol. Pharmacol.* **48**(3), 443–450.

Gustafsson, S. B., Lindgren, T., Jonsson, M., and Jacobsson, S. O. (2009). Cannabinoid receptor-independent cytotoxic effects of cannabinoids in human colorectal carcinoma cells: Synergism with 5-fluorouracil. *Cancer Chemother. Pharmacol.* **63**(4), 691–701.

Haines, D. C., Tomchick, D. R., Machius, M., and Peterson, J. A. (2001). Pivotal role of water in the mechanism of P450BM-3. *Biochemistry* **40**(45), 13456–13465.

Huang, S. M., Bisogno, T., Petros, T. J., Chang, S. Y., Zavitsanos, P. A., Zipkin, R. E., Sivakumar, R., Coop, A., Maeda, D. Y., De Petrocellis, L., Burstein, S., Di Marzo, V., et al. (2001). Identification of a new class of molecules, the arachidonyl amino acids, and characterization of one member that inhibits pain. *J. Biol. Chem.* **276**(46), 42639–42644.

Huang, S. M., Bisogno, T., Trevisani, M., Al-Hayani, A., De Petrocellis, L., Fezza, F., Tognetto, M., Petros, T. J., Krey, J. F., Chu, C. J., Miller, J. D., Davies, S. N., et al. (2002). An endogenous capsaicin-like substance with high potency at recombinant and native vanilloid VR1 receptors. *Proc. Natl. Acad. Sci. USA* **99**(12), 8400–8405.

Huitron-Resendiz, S., Gombart, L., Cravatt, B. F., and Henriksen, S. J. (2001). Effect of oleamide on sleep and its relationship to blood pressure, body temperature, and locomotor activity in rats. *Exp. Neurol.* **172**(1), 235–243.

Ikeda, Y., Iguchi, H., Nakata, M., Ioka, R. X., Tanaka, T., Iwasaki, S., Magoori, K., Takayasu, S., Yamamoto, T. T., Kodama, T., Yada, T., Sakurai, T., et al. (2005). Identification of N-arachidonylglycine, U18666A, and 4-androstene-3,17-dione as novel insulin Secretagogues. *Biochem. Biophys. Res. Commun.* **333**(3), 778–786.

Kohno, M., Hasegawa, H., Inoue, A., Muraoka, M., Miyazaki, T., Oka, K., and Yasukawa, M. (2006). Identification of N-arachidonylglycine as the endogenous ligand for orphan G-protein-coupled receptor GPR18. *Biochem. Biophys. Res. Commun.* **347**(3), 827–832.

Martinez-Gonzalez, D., Bonilla-Jaime, H., Morales-Otal, A., Henriksen, S. J., Velazquez-Moctezuma, J., and Prospero-Garcia, O. (2004). Oleamide and anandamide effects on food intake and sexual behavior of rats. *Neurosci. Lett.* **364**(1), 1–6.

McCue, J. M., Driscoll, W. J., and Mueller, G. P. (2008). Cytochrome *c* catalyzes the *in vitro* synthesis of arachidonoyl glycine. *Biochem. Biophys. Res. Commun.* **365**(2), 322–327.

McKinney, M. K., and Cravatt, B. F. (2006). Structure-based design of a FAAH variant that discriminates between the N-acyl ethanolamine and taurine families of signaling lipids. *Biochemistry* **45**(30), 9016–9022.

Mechoulam, R., and Gaoni, Y. (1965). A total synthesis of dl-delta-1-tetrahydrocannabinol, the active constituent of hashish. *J. Am. Chem. Soc.* **87**, 3273–3275.

Mechoulam, R., Ben-Shabat, S., Hanus, L., Ligumsky, M., Kaminski, N. E., Schatz, A. R., Gopher, A., Almog, S., Martin, B. R., Compton, D. R., et al. (1995). Identification of an endogenous 2-monoglyceride, present in canine gut, that binds to cannabinoid receptors. *Biochem. Pharmacol.* **50**(1), 83–90.

Milman, G., Maor, Y., Abu-Lafi, S., Horowitz, M., Gallily, R., Batkai, S., Mo, F. M., Offertaler, L., Pacher, P., Kunos, G., and Mechoulam, R. (2006). N-arachidonoyl L-serine, an endocannabinoid-like brain constituent with vasodilatory properties. *Proc. Natl. Acad. Sci. USA* **103**(7), 2428–2433.

Mueller, G. P., and Driscoll, W. J. (2007). In vitro synthesis of oleoylglycine by cytochrome c points to a novel pathway for the production of lipid signaling molecules. *J. Biol. Chem.* **282**(31), 22364–22369.

Naef, M., Curatolo, M., Petersen-Felix, S., Arendt-Nielsen, L., Zbinden, A., and Brenneisen, R. (2003). The analgesic effect of oral delta-9-tetrahydrocannabinol (THC), morphine, and a THC-morphine combination in healthy subjects under experimental pain conditions. *Pain* **105**(1–2), 79–88.

O'Byrne, J., Hunt, M. C., Rai, D. K., Saeki, M., and Alexson, S. E. (2003). The human bile acid-CoA:amino acid N-acyltransferase functions in the conjugation of fatty acids to glycine. *J. Biol. Chem.* **278**(36), 34237–34244.

Piomelli, D., Tarzia, G., Duranti, A., Tontini, A., Mor, M., Compton, T. R., Dasse, O., Monaghan, E. P., Parrott, J. A., and Putman, D. (2006). Pharmacological profile of the selective FAAH inhibitor KDS-4103 (URB597). *CNS Drug Rev.* **12**(1), 21–38.

Rimmerman, N., Bradshaw, H. B., Hughes, H. V., Chen, J. S., Hu, S. S., McHugh, D., Vefring, E., Jahnsen, J. A., Thompson, E. L., Masuda, K., Cravatt, B. F., Burstein, S., et al. (2008). N-palmitoyl glycine, a novel endogenous lipid that acts as a modulator of calcium influx and nitric oxide production in sensory neurons. *Mol. Pharmacol.* **74**(1), 213–224.

Ross, R. A. (2003). Anandamide and vanilloid TRPV1 receptors. *Br. J. Pharmacol.* **140**(5), 790–801.

Schachter, D., and Taggart, J. V. (1954). Glycine N-acylase: Purification and properties. *J. Biol. Chem.* **208**(1), 263–275.

Sheskin, T., Hanus, L., Slager, J., Vogel, Z., and Mechoulam, R. (1997). Structural requirements for binding of anandamide-type compounds to the brain cannabinoid receptor. *J. Med. Chem.* **40**(5), 659–667.

Succar, R., Mitchell, V. A., and Vaughan, C. W. (2007). Actions of N-arachidonyl-glycine in a rat inflammatory pain model. *Mol. Pain* **3,** 24.

Sugiura, T., Kondo, S., Sukagawa, A., Nakane, S., Shinoda, A., Itoh, K., Yamashita, A., and Waku, K. (1995). 2-Arachidonoylglycerol: A possible endogenous cannabinoid receptor ligand in brain. *Biochem. Biophys. Res. Commun.* **215**(1), 89–97.

Tan, B., Bradshaw, H. B., Rimmerman, N., Srinivasan, H., Yu, Y. W., Krey, J. F., Monn, M. F., Chen, J. S., Hu, S. S., Pickens, S. R., and Walker, J. M. (2006). Targeted lipidomics: Discovery of new fatty acyl amides. *AAPS J.* **8**(3), E461–E465.

Ueda, N., Okamoto, Y., and Tsuboi, K. (2005). Endocannabinoid-related enzymes as drug targets with special reference to N-acylphosphatidylethanolamine-hydrolyzing phospholipase D. *Curr. Med. Chem.* **12**(12), 1413–1422.

Urca, G., Frenk, H., Liebeskind, J. C., and Taylor, A. N. (1977). Morphine and enkephalin: Analgesic and epileptic properties. *Science* **197**(4298), 83–86.

Vuong, L. A., Mitchell, V. A., and Vaughan, C. W. (2008). Actions of N-arachidonyl-glycine in a rat neuropathic pain model. *Neuropharmacology* **54**(1), 189–193.

Walker, J. M., Huang, S. M., Strangman, N. M., Tsou, K., and Sanudo-Pena, M. C. (1999). Pain modulation by release of the endogenous cannabinoid anandamide. *Proc. Natl. Acad. Sci. USA* **96**(21), 12198–12203.

Wiles, A. L., Pearlman, R. J., Rosvall, M., Aubrey, K. R., and Vandenberg, R. J. (2006). N-Arachidonyl-glycine inhibits the glycine transporter GLYT2a. *J. Neurochem.* **99**(3), 781–786.

CHAPTER NINE

The Endocannabinoid Anandamide: From Immunomodulation to Neuroprotection. Implications for Multiple Sclerosis

Fernando G. Correa, Leyre Mestre, Fabián Docagne, José Borrell, *and* Carmen Guaza

Contents

I. Introduction	208
II. AEA as a Neuroimmune Signal	212
A. AEA as modulator of immune function	212
B. AEA as modulator of cytokine network in glial cells	214
C. Regulatory role of anandamide on the production of cytokines of the IL-12 family	215
III. Anandamide and Multiple Sclerosis	219
A. Inhibitors of AEA uptake and metabolism	219
B. Anandamide upregulation under neuroinflammatory condition	220
C. AEA as a neuroprotective agent	222
IV. Concluding Remarks	223
Acknowledgment	224
References	224

Abstract

Over the last decade, the endocannabinoid system (ECS) has emerged as a potential target for multiple sclerosis (MS) management. A growing amount of evidence suggests that cannabinoids may be neuroprotective during CNS inflammation. Advances in the understanding of the physiology and pharmacology of the ECS have potentiated the interest of several components of this system as useful biological targets for disease management. Alterations of the ECS have been recently implicated in a number of neuroinflammatory and neurodegenerative conditions, so that the pharmacological modulation of cannabinoid (CB) receptors and/or of the enzymes controlling synthesis, transport,

Neuroimmunology Group, Functional and Systems Neurobiology Department, Cajal Institute, CSIC, Avda Doctor Arce, Madrid, Spain

and degradation of these lipid mediators is considered an option to treat several neurological diseases. This chapter focuses on our current understanding of the function of anandamide (AEA), its biological and therapeutic implications, as well as a description of its effects on neuroimmune modulation.

I. INTRODUCTION

During the last decade, our knowledge about the role played by the endocannabinoid system (ECS), in physiological and pathological situations, has increased in an exponential manner. *Cannabis sativa* derivatives have been used in medicine and as recreational drugs for many centuries and currently, there is a great interest in the study of their therapeutic effects in a variety of disorders. This process started with the identification and chemical synthesis of Δ-9-tetrahydrocannabinol (THC) (Gaoni and Mechoulan, 1964), the study of its pharmacological properties, and the identification of its receptors in animal organisms and of their endogenous ligands, and it continued with the progressive understanding of the physiological and pathological roles of this newly discovered signaling system. The ECS consisting of cannabinoid receptors, the endocannabinoids and the synthetic and metabolic enzymatic machinery, is an important intercellular signaling system involved in a wide variety of physiological processes (Fig. 9.1). The purpose of this chapter is to highlight the role of the endocannabinoid, N-arachidonoylethanolamine (anandamide—AEA), a small lipid molecule similar to eicosanoids, in the regulation of neuroinflammation and its contribution to balance brain cytokine network and to discuss its implication in multiple sclerosis (MS) and other CNS inflammation-related damage situations. It will also provide essential details of the basic components of the ECS, namely concerning AEA synthesis, actions, and inactivation mechanisms.

The endocannabinoids, arachidonoylethanolamide named anandamide (AEA), and 2-arachidonoylglycerol (2-AG) are the most studied members of this family (Fig. 9.2) and are generally considered to be released from cells immediately after biosynthesis, as no evidence exists for the storage in secretory vesicles (Lambert and Fowler, 2005). AEA was identified in 1992 by Raphael Mechoulam and coworkers (Devane *et al.*,1992) as the first endocannabinoid, and in 1995 was identified a second endogenous cannabinoid ligand, 2-AG by two laboratories (Mechoulam *et al.*, 1995; Sugiura *et al.*, 1995). Although multiple pathways for endocannabinoid synthesis have been described, the main routes are showed in Fig. 9.3. Nevertheless, the prevalent pathway for their *in vivo* biosynthesis has yet to be determined. This is an important issue since the diversity of the signaling pathways in which AEA, 2-AG, and perhaps, other endogenous cannabinoids

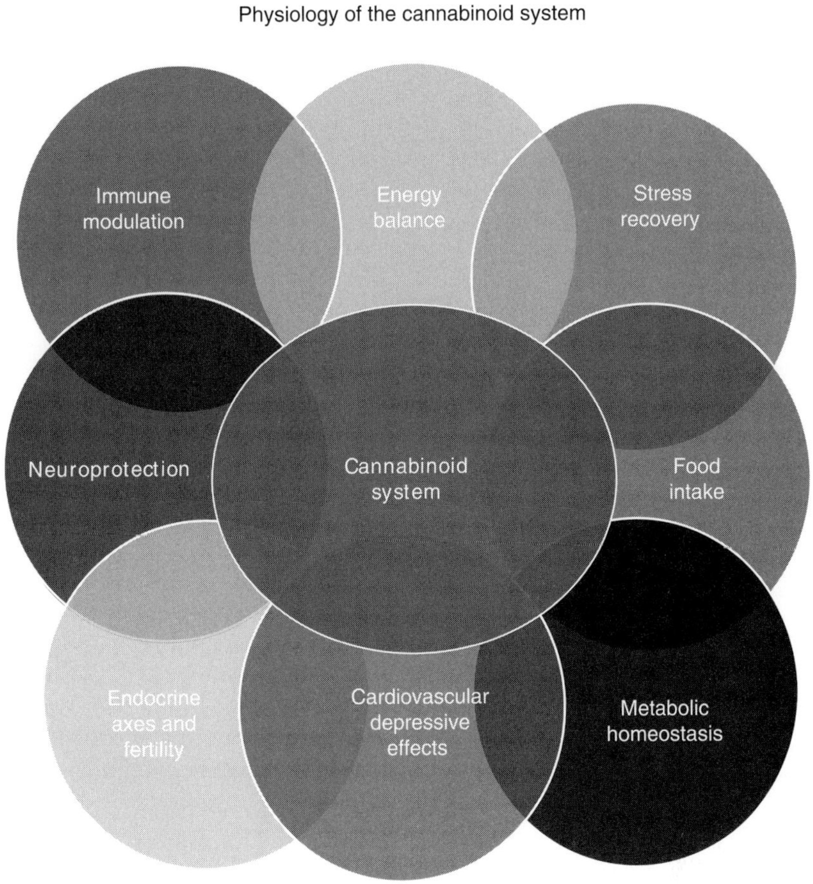

Physiology of the cannabinoid system

Figure 9.1 Endocannabinoid signaling contributes to the regulation of essential physiological processes.

participate may be determined by their biosynthetic route affecting the role(s) that these molecules play in various physiological and pathological conditions. In the case of AEA, it mainly originates from a phospholipid precursor, N-arachidonoyl-phosphatidylethanolamine (NArPE) which is formed from the N-arachidoylation of phosphatidylethanolamine via N-acyltransferases (NATs) (Di Marzo et al.,1994). NArPE is then transformed into AEA by several possible pathways, but the direct conversion is catalyzed by N-acyl-phosphatidylethanolamine-selective phosphodiesterase (NAPE-PLD). Biological inactivation of AEA occurs by rapid cellular uptake followed by intracellular hydrolysis via fatty acid amide hydrolase (FAAH) (Cravatt et al., 2001). The cellular uptake mechanism of AEA is yet to be characterized and it also appears to mediate the release of the *de novo*

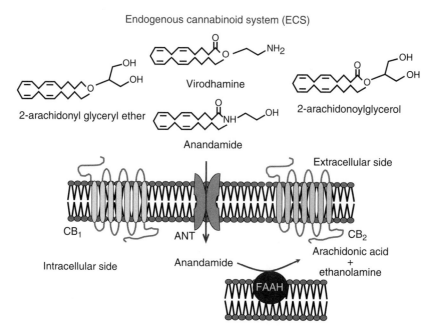

Figure 9.2 Overview of the endocannabinoid system (ECS). AEA, anandamide; ANT, anandamide transporter; FAAH, fatty acid amide hydrolase; CB_1, cannabinoid receptor type 1; CB_2, cannabinoid receptor type 2.

biosynthesized AEA. Importantly, endocannabinoids act on their receptors locally because of their high lipophilicity, and are immediately inactivated under physiological conditions. Two cannabinoid receptors have been identified: the cannabinoid CB_1 receptor (Devane et al., 1988; Matsuda et al., 1990) mainly expressed in the CNS and the CB_2 receptor (Munro et al., 1993) mainly expressed in cells of the immune system (Galiègue et al., 1995). CB_1 receptors are expressed by many different subtypes of neurons and at lower levels by other type of cells, including immunocompetent cells (Howlett et al., 2002; Matsuda et al., 1990; Tsou et al., 1998). In the CNS, the high levels of CB_1 receptor expression correspond to GABAergic interneurons. Glutamatergic principal neurons also express CB_1 receptor which seems to play a critical role as its selective deletion in this neuronal subpopulation impairs the classic behavioral tetrad induced by its activation (Monory et al., 2007). Both receptors CB_1 and CB_2 are members of the large seven-transmembrane G protein-coupled receptors family. CB_1 receptors are coupled to $G_{i/o}$ proteins and can affect the activity of several ion channels and intracellular signaling pathways (for revision, see Howlett, 2005). CB_2 receptors are mainly coupled to G_i proteins, and show a different expression pattern than CB_1 receptors as commented before. Within the brain it has been reported the lack of CB_2 receptor expression in healthy

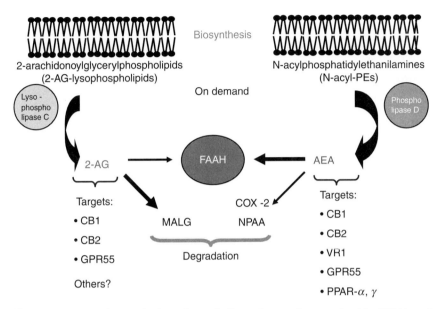

Figure 9.3 Main biosynthetic and metabolic pathways for anandamide (AEA) and 2-arachidonoylglycerol (2-AG). Receptor targets: CB_1 receptor, CB_2 receptor; TPRV1, vanilloid receptor type 1; GPR55, G-coupled orphan receptor; PPAR, peroxisome proliferator-activating receptors.

conditions, with the exception of a small population of brainstem neurons (Van Sickle et al., 2005). In addition, there is evidence indicating the possible presence of yet uncloned cannabinoid receptors on the basis of pharmacological studies using CB_1 and CB_2 receptors-deficient mice (Breivogel et al., 2001; Di Marzo et al., 2000). In the case of double-knockout mice for CB_1 and CB_2 receptors, Kunos and coworkers have identified the abnormal-cannabidiol (abn-CBD) receptors (Járai et al., 1999). These receptors also known as AEA receptors are expressed by endothelial cell of blood vessels, and regulate blood pressure (Offertaler et al., 2003). However, so far there is no knowledge about the molecular characterization of abn-CBD receptor and thus, its physiological relevance is not clear. The endocannabinoid AEA is capable to activate both receptors CB_1 and CB_2 with relatively high affinity and modulates their effector proteins (Felder et al., 1993; Vogel et al., 1993), but also interacts with several non-CB receptors, the best known of which is the transient receptor potential, vanilloid subtype 1 (TRPV1) channel (Zygmunt et al., 1999). Recent data reported that AEA and 2-AG activate the GPR55, an orphan G protein-coupled receptor (Waldeck-Weiermair et al., 2008), but other results do not support the occurrence of that interaction. Evidence for interactions of AEA and 2-AG with peroxisome proliferator-activating receptors (PPARs) α and γ has also been suggested (O'Sullivan, 2007). Additionally, AEA under several conditions, for instance, in situations in which the activity of FAAH is suppressed or the activity of cyclooxygenase-2 (COX-2) is overexpressed, might become substrate for COX-2 producing hydroperoxy derivatives that then can be converted to prostaglandins ethanolamides (prostamides) by prostaglandins synthases (Ross et al., 2002). The physiological relevance of these pathways is still not fully understood, but is subject of intense investigation (Matias et al., 2004; for revision see Fowler, 2007). The ECS is best known for affecting CNS functions by producing euphoria, alterations in cognition and analgesia, having anticonvulsant properties, and affecting temperature regulation, nociception, sleep, and appetite and in general in modulating synaptic plasticity. Additionally, the ECS possess also immunomodulatory activity, and has been shown to reduce inflammatory responses. Many diseases of the CNS including Alzheimer's disease, Parkinson's disease, AIDS dementia, and mainly, multiple sclerosis involve inflammation and cause upregulation of cytokines and other inflammatory mediators. Glial cells, in particular microglia and astroglia, participate in brain immune responses. Under pathological conditions, such as the case of MS, the blood–brain barrier is less restrictive to the migration of activated monocytes, T and B lymphocytes, and other immune cells and a cross-dialogue can be established between immune-derived cells and glial cells, through soluble factors (cytokines, chemokines, etc.) and by direct cell-to-cell contact.

II. AEA as a Neuroimmune Signal

A. AEA as modulator of immune function

The role of ECS in the regulation of immune responses has acquired a great interest because of their possible usefulness in chronic inflammatory diseases. Activation of CB_2 receptor has been associated with most of the immunomodulatory activity of CBs, but several reports indicate that activation of CB_1 also may be linked to CB-mediated alterations of immune cell reactivity. The cannabinoid CB_2 receptor is expressed abundantly in various types of inflammatory cells and immune competent cells at levels 10–100 times higher than CB_1 receptor mRNA (Carlisle et al., 2002; Galiègue et al., 1995). One of the most significant observations is that the expression level of CB_2 receptor depends on the state of activation of the cells. As an example, differentiation of B cells is followed by decreased expression of CB_2 receptor, and their activation increases its levels (Carayon et al., 1998). Most of the actions of CBs have been related to a downregulation of immune system (for revision, see Klein, 2005, 2007), but there are several exceptions in particular at low doses. Cannabinoids exhibit immunosuppressive properties by interfering with humoral immunity, cell-mediated immunity, and cellular defenses against infectious agents. A number of *in vitro* studies reported that cannabinoids inhibit T-cell mitogenesis and IL-2 production from lymphocyte cell lines (Kaplan *et al.*, 2003). In general, IL-2 regulates both antigen-specific and nonantigen-specific proliferation of T cells, natural killer cells, and B cells. It is well known that Th1 and Th2 and more recently Th17 responses are critical in regulating immune responses, inflammation, and ultimately repair on a variety of CNS diseases. Many laboratories have been shown that THC modulates Th1 and Th2 cytokine responses (Kaplan *et al.*, 2003; Klein *et al.*, 2004; Smith *et al.*, 2000; Yuan *et al.*, 2002). Recent data support that CBs bias toward Th2-type immunity by inducing B-cell class switching from IgM to IgE through a mechanism involving CB_2 receptors (Agudelo *et al.*, 2008). Cannabinoid receptor CB_2 also modulates the CXCL12/CXCR4-mediated chemotaxis of T lymphocytes (Ghosh *et al.*, 2006). Peripheral and bone marrow-derived dendritic cells as well as macrophages, which play a key role in the initiation and development of the immune response, express CB_2 receptors which may be related to the reduction of antigen presentation by cannabinoids described several years ago (McCoy *et al.*, 1999). Levels of CB_2 receptors in cells of myeloid lineage undergo also changes depending on cell activation by increasing its expression under inflammatory conditions (Carlisle, 2002; Derocq *et al.*, 2000). Moreover, dendritic cells as well as macrophages generate AEA and 2-AG during inflammatory conditions, express both type of CB receptors and the enzymes responsible for their hydrolysis, FAAH for AEA and monoacylglycerol lipase (MGL) for 2-AG, pointing out the existence of a complete ECS (Di Marzo *et al.*, 1996; Matias *et al.*, 2002). Nevertheless, specific information about the role played by the endogenous ligands, in particular by AEA in the regulation of immune function is still scarce. Here, we described the main findings about the actions of AEA on immune reactivity. One of the most important is the inhibition of IL-2 secretion from lymphocytes by AEA that appears to be independent of the activation of CB_1 or CB_2 receptors. AEA-elicited inhibition of IL-2 is mediated by PPAR-γ receptor which is activated by a COX-2 metabolite of AEA (Rockwell and Kaminsky, 2004). The endocannabinoid AEA has also been described to inhibit human neutrophil migration at nanomolar concentrations in a biphasic manner (McHugh *et al.*, 2008), which seems to be important in several diseases such as human fibromyalgia. In addition, AEA induced a dose-related immunosuppression in both the primary and secondary *in vitro* plaque-forming cell assays of antibody formation that was blocked by SR144528, an antagonist specific for the CB_2 receptor (Eisenstein *et al.*, 2007). The above data suggest that ECS can play a physiological role in the regulation of immune responses, which may have important implications in pathological inflammatory conditions including brain immune-related disorders.

B. AEA as modulator of cytokine network in glial cells

Cannabinoids exert a variety of modulatory effects on the immune reactivity and inflammation in the CNS (Arévalo-Martín et al., 2008; Mestre et al., 2009). AEA displays some of its immunomodulatory and anti-inflammatory activities through actions in both central and peripheral tissues with the participation of CB_1 and CB_2 receptors. In the brain, glial cells and endothelial cells are the main targets for immunomodulatory activities of cannabinoids, an hypothesis supported by the expression of CB_1 and CB_2 receptors in astrocytes and astrocytoma cells (Molina-Holgado et al., 2002a; Sánchez et al., 1998), microglia (Carlisle et al., 2002; Fachinetti et al., 2003; Walter et al., 2003), oligodendrocytes (Molina-Holgado et al., 2002b), and endothelial cells (Mestre et al., 2006). Astrocytes are the major glial cells within the CNS and play a critical role in the maintenance of CNS homeostasis. Astrocytes, microglia, and endothelial cells are capable to generate a wide array of chemokines and cytokines that may impact the development of a variety of neurological diseases. On this context, endocannabinoids have been shown to regulate cytokine production in astrocytes and microglial cells. For example, in a cell culture preparation of Theiler's virus-infected astrocytes AEA enhanced the synthesis of IL-6, a potentially anti-inflammatory cytokine (Molina-Holgado et al., 1998), through the activation of CB_1 receptor and suppressed the production of the proinflammatory cytokine TNF-α (Molina-Holgado et al., 1997). LPS-activated astrocytes also decreased the production of nitric oxide and the cytokines IL-1β and TNF-α by exposure to an inhibitor to AEA uptake (Ortega-Gutiérrez et al., 2005a). Microglial cells constitute a population of "facultative" macrophages adapted to neural environment which play a critical role in CNS immune surveillance (Hanisch and Kettenmann, 2007; Raivich, 2005). Most of neurological disorders involve activation and even, dysregulation of microglial cells that in addition to their phagocytic function participate in the regulation of nonspecific inflammation as well as in adaptive immune responses. Under CNS cell-damaged conditions, local microglia changes from resting to an activated phenotype aimed at repairing damaged cells and eliminating debris and pathogens. Multiple laboratories agree that in vitro-activated microglia express CB_2 receptors (Carlisle et al., 2002; Carrier et al., 2004; Ramirez et al., 2005). However, it is probably that in the healthy brain microglial cells do not express CB_2 receptors, and that the cells upregulate their expression in diseased brain tissue. As observed in immunocompetent cells, the microglial expression level of CB_2 receptor is regulated by the activation state of the cells (Cabral and Marciano Cabral, 2005). The exposure to LPS reduces CB_2 receptor expression in cultured microglial cells, but IFN-γ plus GM-CSF increase it (Carlisle et al., 2002; Maresz et al., 2005). Therefore, the microglial expression level of CB_2 receptors can be regulated by pathogens and cytokines. In neuropathological situations, activated microglia also express CB_2 receptors, as evaluated in postmortem brain tissue from patients with Alzheimer's disease, MS, and amyotrophic lateral sclerosis, as well as in animal models of the above pathologies (Benito et al., 2003, 2007; Ramirez et al., 2005; Yiangou et al., 2006). Moreover, astrocytes and microglia produce AEA and 2-AG under several stimulatory conditions and in response to tissue damage (Walter et al., 2002, 2003; for revision see Stella, 2004, 2009). The enzymes that metabolize endocannabinoids, FAAH, and MGL are also expressed in glial cells (Benito et al., 2007; Witting et al., 2004), and recent studies suggest the existence of novel enzymes capable of hydrolyzing 2-AG in microglial cell lines (Muccioli et al., 2007). However, the role of glial endocannabinoids in the modulation of neuroinflammation awaits a more complete analysis of glial endocannabinoid expression patterns mainly *in vivo* glial cells. Most of information about the regulation of glial cytokine network comes from *in vitro* studies in which synthetic or plant-derived cannabinoids are administered (Fachinetti et al., 2003; Molina-Holgado et al., 2003; Puffenbarger et al., 2000). Nevertheless, if we consider the ability of glial cells to produce local increases in endocannabinoids under neuroinflammation or brain injury, together with the effects of 2-AG by promoting microglial cell migration (Walter et al., 2003) and proliferation (Carrier et al., 2004), as well as the actions of AEA by reducing the production of nitric oxide and proinflammatory cytokines (Mestre et al., 2005; Ortega-Gutiérrez et al., 2005a), the scenario we propose is that due to endocannabinoid actions less harmful microglia will be accumulated at the lesion site. In support of this, AEA has been shown to protect neurons from inflammatory damage by CB_1 and CB_2 receptor-mediated rapid induction of mitogen-activated protein kinase phosphatase-1 (MKP-1) by reducing the release of TNF-α and free radicals from microglial cells (Eljaschewitsch et al., 2006). Additionally, AEA has been shown to regulate inflammation through NF-κB pathway inhibition independently of CB_1 or CB_2 receptor activation (Nakajima et al., 2006; Sancho et al., 2003).

C. Regulatory role of anandamide on the production of cytokines of the IL-12 family

Dysregulation of IL-12 gene expression may contribute to the initiation and perpetuation of various autoimmune and chronic inflammatory diseases such as MS. IL-12 is a heterodimeric cytokine composed of p35 and p40 subunits, especially important because its expression regulates innate immunity and determines the type and duration of adaptive immune response (Trinchieri, 2003). IL-12 is produced mainly by monocytes, macrophages, and dendritic cells, but also by brain microglia (Li *et al.*, 2003). It plays essential functions in cell-mediated immunity and in Th1 T-cell differentiation by coordinating innate and adaptive immunity (Fig. 9.4). Besides forming IL-12p70 heterodimer, p35 and p40 may dimerize with alternate partners to form distinct bimolecular complexes such as IL-23 and IL-27. Microglial cells express the p19, p35, and p40 subunits, therefore making possible the formation of the heterodimeric cytokines, IL-12 and IL-23, and then, raising the possibility that microglia may contribute to Th1-skewing within the CNS and hence to the pathogenesis of Th1-mediated disorders (Correa *et al.*, 2008). Our lab described for the first time that AEA is a potent inhibitor of the expression of mRNA for p19, p35, and p40 subunits and of IL-12 and IL-23 protein release by activated microglia with the involvement of CB_2 receptors as shown in Fig. 9.5 (Correa *et al.*, 2009). Note that stimulation of CB_2 receptors by AEA in microglia reduces the release of key cytokines for the initiation and maintenance of chronic inflammation and autoimmune processes. Therefore, the activation of CB_2 receptors is being considered as an emerging therapeutic strategy in MS (Arevalo-Martín *et al.*, 2008; Docagne *et al.*, 2008). Additionally, we provided evidence that AEA suppressed the transcriptional activity of *IL-12p40* gene in activated

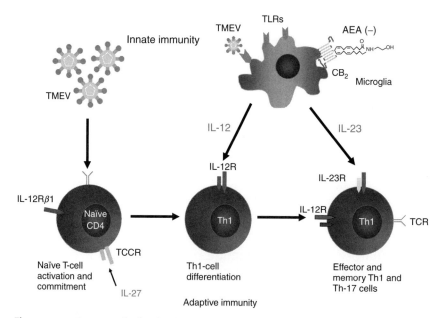

Figure 9.4 Innate and adaptive immune responses to Theiler's virus (TMEV) infection: IL-12 family members on the orchestration of Th1 responses. Regulation of Th-cell differentiation by members of the IL-12 family by macrophage/microglial cells infected with the TMEV. Anandamide (AEA) interacts with CB_2 receptors in macrophages/microglia to downregulate IL-12 production.

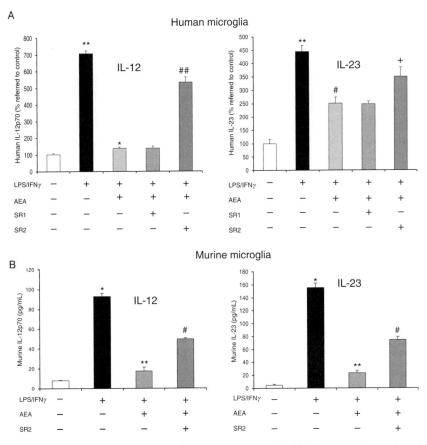

Figure 9.5 Production of biologically active IL-12 and IL-23 by LPS/IFN-γ-activated microglia can be regulated by AEA by a partially CB_2 receptor-dependent mechanism. Human (A) and murine (B) microglial cells stimulated overnight with LPS/INF-γ (50 ng/ml and 100 U/ml, respectively) showed an induction on the expression of IL-12 and IL-23. This induction was downregulated by cotreatment with AEA 10 μM. The previous treatment with the CB_2 antagonist, SR2, partially reversed the effects of AEA, suggesting an involvement of this cannabinoid receptor in AEA's regulation of IL-12 and IL-23 production. Statistics: $*p < 0.005$ versus control; $**p < 0.001$ versus LPS/IFN-γ; $^{\#}p < 0.01$ versus LPS/IFN-γ + AEA.

transfected cells independently of the activation of CB or vanilloid receptors. Experiments performed with site-directed mutagenesis of the different element sites of the p40 promoter showed that AEA regulates IL-12p40 expression by enhancing the activity of the repressor site GA-12. Prostamide E_2, a product recognized as a putative metabolite of AEA by COX-2 oxygenation, was also able to inhibit the activity of IL-12p40 promoter by acting at the repressor site. The effects of AEA and prostamide E_2 on p40 transcription were partially reversed by an antagonist of the prostanoid receptor, EP_2 raising the possibility that prostamide E_2 may contribute to AEA effects on *IL-12p40* gene regulation, provide new mechanistic insights into the activities of AEA to manage immune-related disorders. However, since AEA-induced inhibition of IL-12p40 by primary microglia was partially dependent on stimulation of CB_2 receptors, we hypothesized that under early inflammatory stimuli the activation of CB_2 receptors and its corresponding signaling pathways by AEA play a major role in blocking the production of IL-12p40, while in long-lasting inflammation, concurrent mechanisms may be engaged, such as the induction of COX-2 and then, the generation of prostamide E_2 that may in turn contribute to the inhibition of *IL-12p40* gene transcription (Fig. 9.6). The inhibition of IL-12p40 may help to control and limit local immune response in order to prevent overactivation of cellular immunity. It is also suggested that excessive COX-2 resulting from inflammation will have significant impact on endocannabinoids-derived prostanoids signaling in immune reactivity. Although the physiological relevance of these findings needs to be established, the identification of mechanisms underlying the limiting effects of

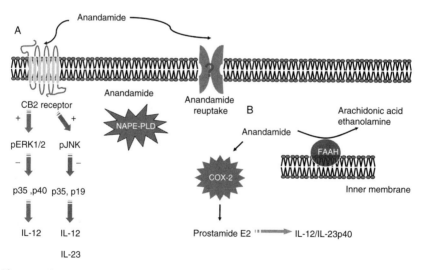

Figure 9.6 Scheme proposed for the different mechanisms of action of AEA in the regulation of IL-12 and IL-23 by activated microglial cells. (A) At the initial phase of an inflammatory response, AEA would act basically by activating the CB_2 receptor and the coupled intracellular signaling such as the MAPKs cascade (ERK1/2 and JNK) to reduce IL-12 and IL-23 production. (B) In a lasting inflammatory response, when different "mediators of inflammation" (for instance, COX-2) appear, AEA would use alternative mechanisms of actions. In this case, we propose that AEA can be metabolized either by FAAH to render ethanolamide and arachidonic acid (which can give prostaglandin E_2) or by COX-2 to render prostamide E_2.

endocannabinoids on the production of IL-12 cytokine family during neuroinflammation is critical to open new avenues of therapeutic intervention in pathologies such as MS.

III. Anandamide and Multiple Sclerosis

A. Inhibitors of AEA uptake and metabolism

MS is a chronic inflammatory disease of the CNS characterized by autoimmune responses against myelin proteins that eventually impair the normal neurotransmission leading to sensory deficits and deteriorated motor coordination. Furthermore, neuroimaging studies highlighted the importance of axonal damage and neurodegeneration in this disease. Exogenous administration of synthetic CB_1 and CB_2 agonists exerts a direct effect on the motor neurotransmission pathways, diminishes the neuroinflammation (Arévalo-Martín et al., 2003; Croxford and Miller, 2003), and inhibits neurodegeneration and axonal damage (Docagne et al., 2007; Pryce et al., 2003) leading to benefits in animal models of MS. Studies from the lab of David Baker showed that endocannabinoids tonically control spasticity in animal models of MS (Baker et al., 2001). The use of selective inhibitors of the cellular uptake of AEA, such as the compounds OMDM1 and OMDM2 synthetized in the lab of Dr. Di Marzo (Ortar et al., 2003) that increases the levels of AEA, without affecting 2-AG levels, in the spinal cord of diseased mice improves motor function in Theiler's murine encephalomyelitis virus-induced demyelinating disease (TMEV-IDD), a well-characterized viral model of MS by diminishing neuroinflammation (Mestre et al., 2005). In a similar way, the potent and selective inhibitor of AEA uptake, UCM-707 (López-Rodriguez et al., 2003) also decreases the severity of symptoms in the TMEV-IDD model and decreases microglial activation, immune cellular infiltrates, and microglial MHC class II antigen expression (Ortega-Gutiérrez et al., 2005b). In an acute rat model of EAE, the elevation of endocannabinoid activity following inhibition of endocannabinoids uptake (OMDM2) reduced the magnitude of neurological impairments (Cabranes et al., 2005). The enzyme FAAH is a key regulator of the endogenous levels of the fatty acid amides that include AEA, oleoylethanolamide (OEA), and palmitoylethanolamide (PEA) and mice lacking FAAH exhibited a more substantial clinical remission than wild-type animals (Webb et al., 2008). Other data confirm the potential utility of selective AEA uptake inhibitors as antispasticity drugs in the CREAE model of MS and, given the very subtle chemical differences between potent and weak inhibitors of uptake, support further the existence of a specific mechanism for this process (Ligresti et al., 2006). The strategy of manipulating the ECS might open new avenues in the development of therapeutic approaches for a number of disorders, both central and peripheral, that lack as yet effective treatments. The results above described confirm the therapeutic potential of inhibitors of AEA uptake as well as of inhibitors of its degradation by FAAH, for the treatment of MS by the regulation of ECS activity and suggest that these compounds can be used to efficaciously treat MS. This type of treatment has the advantage of avoiding at least part of the undesirable side effects induced by direct activation of cannabinoid central receptors, since it results in the enhancement of the extracellular levels of endocannabinoids, available for cannabinoid receptor activation, only at those sites where there is an ongoing biosynthesis of endocannabinoids. Of particular interest is that cannabinoid receptors and FAAH expression have been considered as markers of plaque cell subtypes in human MS (Benito et al., 2007). Particularly, CB_2-positive microglial cells were evenly distributed within active plaques but were located in the periphery of chronic active plaques while FAAH expression was restricted to neurons and hypertrophic astrocytes. As seen for other neuroinflammatory conditions (Benito, 2003), selective glial expression of cannabinoid CB_1 and CB_2 receptors and FAAH enzyme is induced in MS, thus, supporting a role for the ECS in the pathogenesis and/or evolution of this disease. At this point, it is noteworthy that there are some controversial about the cellular players mainly involved in the regulation of pathogenic mechanisms by the ECS in MS and in their experimental models. While several authors (Maresz et al., 2007) claim that CB_1 receptor expression by neurons, but not T cells, is required for cannabinoid-mediated EAE suppression and that CB_2 receptor expression by encephalitogenic T cells is critical for controlling inflammation, other studies highlight the importance of CB_2 receptors activation in macrophages/microglia to suppress EAE (Palazuelos et al., 2008). Taken together the findings from the above studies demonstrate that the ECS within the CNS plays a critical role in regulating autoimmune inflammation, but that it is necessary more work to gain better insight especially from a mechanistic point of view.

B. Anandamide upregulation under neuroinflammatory condition

Endocannabinoid levels, mainly AEA levels, are usually elevated following inflammatory stimuli. Macrophages/microglia infected with Theiler's virus (TMEV) increased their production of AEA (Mestre et al., 2005). Cells of the myeloid lineage such as macrophages, dendritic cells, and microglial cells produce important amounts of AEA (Liu et al., 2003, 2006) that can act in the same cell in an autocrine way probably as an adaptive reaction aimed at reducing inflammation. For instance, CNS tissue from inflammatory active lesions of MS patients shows higher levels of AEA, without changes in

2-AG, in comparison with chronic silent lesions or with tissue from healthy controls (Eljaschewitsch et al., 2006). A recent study showed that MS was associated with significant alterations of AEA, but not of 2-AG, since AEA was increased in the CSF of relapsing MS patients (Centonze et al., 2007). The levels of AEA were also higher in peripheral lymphocytes of MS patients due to its increased synthesis and reduced degradation. In the same study, increased synthesis, reduced degradation, and higher levels of AEA were also detected in the brains of EAE mice in the acute phase of the disease (Centonze et al., 2007). By contrast, the results of other study showed that brain levels of AEA and 2-AG were not significantly increased in the EAE model of MS, but in this case the severity of disease in terms of clinical score symptoms was mild (Witting et al., 2006). In the TMEV-IDD model, AEA was only upregulated in the spinal cord following the inactivation of its uptake (Mestre et al., 2005), while spinal cord levels of PEA and 2-AG were increased at 60 days postviral infection (Loría et al., 2008). In spinal cord injury, during early stages lesion induced increases of AEA and PEA levels, an upregulation of the synthesizing enzyme NAPE-phospholipase D and a downregulation of the degradative enzyme FAAH, but in delayed stages, lesion induced increases in 2-AG and a strong upregulation of the synthesizing enzyme DAGL-α (García-Ovejero et al., 2009). It is important to note that the balance between endocannabinoid production and inactivation determines the extent of their accumulation in the tissue in a specific period of time. In the case of chronic inflammatory diseases such as MS and their experimental models, it is difficult to exactly know the timing changes on the levels of AEA and 2-AG as well as other cannabimimetics substances and their significance on the evolution/progression of the disease. In addition, the different analytical methods for measurements of endocannabinoids used in the distinct laboratories may contribute to the heterogeneity of results mainly observed in experimental models of MS. Without a doubt, the improvement of analytical methods for detection of endocannabinoids will contribute to a better understanding of the functional role of endocannabinoid changes associated to neurological disorders. In any case, there is agreement that microglia cells produce large amounts of endocannabinoids. In particular, microglial cells produce AEA in response to immune and inflammatory stimuli (LPS, TMEV infection, HIV), but also AEA and 2-AG in response to ATP through the activation of microglial P2X7 ionotropic receptors (Witting et al., 2004). Therefore, microglia is likely to contribute to the increased levels of endocannabinoids under inflammatory/neuropathological conditions. In the context of neuroimmune interactions, recent reports suggest that elements of the peripheral ECS mirror central dysfunctions of endocannabinoid signaling as revealed by studying alterations of the ECS within the CNS and in peripheral blood mononuclear cells (Centonze et al., 2008).

C. AEA as a neuroprotective agent

It is well known that during neuronal activity, AEA is formed by glutamate receptor agonists in a Ca^{2+}-dependent manner, and that mediates a localized signaling mechanism through the neurons may modify the strength of synaptic inputs (Giufrida et al., 1999). Activation of CB_1 receptors located in presynaptic sites mediates retrograde inhibition of glutamate release, controls neuronal excitability and regulates synaptic plasticity. Therefore, endocannabinoids can exert direct neuroprotective effects by limiting glutamate release and excitotoxic damage in several neurodegenerative diseases and cell models (Marsicano et al., 2003; Nagayama et al., 1999) Activation of CB_1 receptors has also been shown to increase the expression of brain-derived neurotrophic factor (BDNF) in neurons, enhancing cell survival (Khaspekov et al., 2004). Coupling of neuronal CB_1 receptors to cell survival routes such as the phosphatidylinositol 3-kinase/Akt and extracellular signal-regulated kinase pathways may contribute to cannabinoid neuroprotective action. These prosurvival signals also occur in oligodendroglial cells (Molina-Holgado et al., 2002b) and, at least in part, involve the crosstalk between CB_1 receptors and growth factor tyrosine kinase receptors. In addition, signaling coupled to activation of CB_2 receptors mainly on microglial cells may also participate in neuroprotection by limiting the extent of neuroinflammation, as we and other groups have described. As commented before AEA is released after several forms of injury, particularly after those involved inflammation and protects neurons from inflammatory damage (Eljaschewitsch et al., 2006). It is tempting to hypothesize that during immune-mediated attack of the brain, activation of endocannabinoids represents a protective mechanism, aimed at reducing both neurodegenerative and inflammatory damage through various and partially converging mechanisms that involve neuronal and immune cells. In fact, the activation of CB_2 receptors has been related to a delayed progression of neurodegenerative events, in particular, those related to the toxic influence of microglial cells on neuronal homeostasis (Fernández-Ruiz et al., 2008). Multiple lines of evidence show that AEA, PEA, and 2-AG act as endogenous protective factors of the brain, using different pathways of neuroprotection against neuronal damage. The consideration of ECS as a natural system of neuroprotection is based on some of the properties of endocannabinoids and in particular of AEA, as their vasodilatory effect, the inhibition of the release of excitotoxic amino acids and detrimental cytokines, and the modulation of oxidative stress and toxic production of nitric oxide. Increasing evidence supports the involvement of AEA in promoting neuroprotection on the basis of its antiexcitotoxic properties. Using a well-defined rodent model of neonatal excitotoxic brain lesions, AEA provided dose-dependent and long-lasting protection of developing white matter and cortical plate reducing the size of lesions (Shouman et al., 2006). AEA-induced neuroprotection against AMPA–kainate receptor-mediated brain lesions were blocked by a CB_1 antagonist but not by a CB_2 antagonist. Furthermore, AEA effects were mimicked by a CB_1 agonist but not by a CB_2 agonist. Interestingly, the neuroprotective effects of AEA in white matter involved increased survival of preoligodendrocytes and better preservation of myelination. Excitotoxicity plays a crucial role also in MS associated neuronal damage and glutamate levels are increased in the CSF (Sharchielli et al., 2003) and in the brains of MS patients (Srinivasan et al., 2005). In the case of experimental models of MS, our group reported for the first time that excitotoxicity contributes to TMEV-IDD pathology (Docagne et al., 2007) in agreement with previous studies showing that glutamate antagonists exert beneficial effects in the model of EAE (Pitt et al., 2000). Experiments from our laboratory showed that nonselective CB_1 or CB_2 synthetic agonist reduced AMPA-induced excitotoxicity both in vivo and in vitro through the obligatory activation of both CB_1 and CB_2 receptors as we demonstrated in neurons–astrocyte cocultures. This is an important point since the involvement of both receptors in the antiexcitotoxicity displayed by cannabinoids was also reported in mixed glial cell cultures (Molina-Holgado et al., 2003). On the MS context, we have evidence suggesting that AEA also acts as a neuroprotective agent against excitotoxicity (unpublished data) and the molecular mechanisms involved in the ability of AEA to diminish excitotoxicity are subject of our current research. In summary, the above studies suggest that AEA protects neurons from inflammatory and excitotoxic damage and support the interest of the pharmacological modulation of the AEA synthesis, transport, and degradation as a therapeutic approach to treat MS (Fig. 9.6).

IV. Concluding Remarks

In the past years, multiple lines of evidence suggest that the ECS participates crucially in the immune control and protection of the CNS. We discuss here the AEA is a key player of the crosstalk between brain and the immune system as well as its potential as a therapeutic target. The use of AEA uptake and metabolism inhibitors may prove efficient in the future as a novel therapeutic design in inflammatory CNS diseases as they may induce less undesired side effects than direct cannabinoid agonists. In effect, as observed in several animal models of CNS pathologies, endocannabinoids are produced by the inflamed brain and in damaged tissues in response to injury. In particular, rationale for the above AEA-based therapies on MS relies on the variety of properties that AEA displays which include immunomodulation and neuroprotection. The reported beneficial effects of pharmacological manipulation of ECS in experimental models of MS need further validation with clinical trials involving a large sample of MS patients.

ACKNOWLEDGMENT

This work was supported by the Ministerio de Ciencia e Innovación (SAF 2007/60038).

REFERENCES

Agudelo, M., Newton, C., Widen, R., Sherwood, T., Nong, L., Friedman, H., and Klein, T. W. (2008). Cannabinoid CB_2 receptor mediates immunoglobulin switching from IgM to IgE in cultures of murine purified B lymphocytes. *J. Neuroimmune Pharmacol.* **3**(1), 35–42.

Arévalo-Martín, A., Vela, J. M., Molina-Holgado, E., Borrell, J., and Guaza, C. (2003). Therapeutic action of cannabinoids in a murine model of multiple sclerosis. *J. Neurosci.* **23**, 2511–2516.

Arévalo-Martín, A., García-Ovejero, D., Gómez, O., Rubio-Araiz, A., Navarro-Galve, B., Guaza, C., Molina-Holgado, E., and Molina-Holgado, F. (2008). CB_2 cannabinoid receptors as an emerging target for demyelinating diseases: From neuroimmune interactions to cell replacement strategies. *Br. J. Pharmacol.* **153**, 216–225.

Baker, D., Pryce, G., Croxford, J. L., Brown, P., Pertwee, R. G., Makriyannis, A., Khanolkar, A., Layward, L., Fezza, F., Bisogno, T., and Di Marzo, V. (2001). Endocannabinoids control spasticity in a multiple sclerosis model. *FASEB J.* **15**, 300–302.

Benito, C., Nuñez, E., Tolón, R. M., Carrier, E. J., Rábano, A., Hillard, C. J., and Romero, J. (2003). Cannabinoid CB2 receptors and fatty cid amide hydrolase are selectively overexpressed in neuritis plaques-associated glia in Alzheimer disease. *J. Neurosci.* **23**, 11136–11141.

Benito, C., Romero, J. P., Tolon, R. M., Clemente, D., Docagne, F., Hillard, C. J., Guaza, C., and Romero, J. (2007). Cannabinoid CB_1 and CB_2 receptors and fatty acid amide hydrolase are specific markers of plaque cell subtypes in human multiple sclerosis. *J. Neurosci.* **27**, 2396–2402.

Breivogel, C. S., Griffin, G., Di Marzo, V., and Martin, B. R. (2001). Evidence for a new G protein-coupled cannabinoid receptor in mouse brain. *Mol. Pharmacol.* **60**, 155–163.

Cabral, G. A., and Marciano-Cabral, F. (2005). Cannabinoid receptors in microglia of the central nervous system: Immune functional relevance. *J. Leukoc. Biol.* **78**, 1192–1197.

Cabranes, A., Venderoka, K., de Lago, E., Fezza, F., Sanchez, A., Mestre, L., Valenti, M., García-Merino, A., Ramos, J. A., Di Marzo, V., and Fernandez-Ruiz, J. (2005). Decreased endocannabinoid levels in the brain and beneficial effects of agents activating cannabinoid and/or vanilloid receptors in a rat model of multiple sclerosis. *Neurobiol. Dis.* **20**, 207–217.

Carayon, P., Marcahnd, J., Dussossoy, D., Derocq, J. M., Jbilo, O., Bord, A., Bouaboula, M., Galiègue, S., Mondiere, P., Penarier, G., LeFur, G. L., Defrance, T., and Casellas, P. (1998). Modulation and functional involvement of peripheral CB_2 receptors during B-cell differentiation. *Blood* **92**, 3605–3615.

Carlisle, S. J., Marciano-Cabral, F., Staab, A., Ludwick, C., and Cabral, G. A. (2002). Differential expression of the CB_2 cannabinoid receptor by rodent macrophages and macrophage-like cells in relation to cell activation. *Int. Immunopharmacol.* **2**, 69–82.

Carrier, E. J., Kearn, C. S., and Barkmeier, A. J. (2004). Cultured rat microglial cells synthesize the endocannabinoid 2-arachidonylglycerol, which increases proliferation via a CB_2 receptor-dependent mechanism. *Mol. Pharmacol.* **65**, 999–1007.

Centonze, D., Bari, M., Rossi, S., Prosperetti, C., Furlan, R., Fezza, F., De Chiara, V., Battistini, L., Bernardi, G., Bernardini, S., Martino, G., and Maccarrone, M. (2007). The endocannabinoid system is dysregulated in multiple sclerosis and in experimental autoimmune encephalomyelitis. *Brain* **130**(10), 2543–2553.

Centonze, D., Battistinni, L., and Maccarrone, M. (2008). The endocannabinoid system in peripheral lymphocytes as a mirror of neuroinflammatory diseases. *Curr. Pharm. Des.* **14** (23), 2370–2420.

Correa, F., Docagne, F., Clemente, D., Mestre, L., Becker, C., and Guaza, C. (2008). Anandamide inhibits IL-12p40 production by acting on the promoter repressor element GA-12: Possible involvement of the COX-2 metabolite prostamide E(2). *Biochem. J.* **409**, 761–770.

Correa, F., Docagne, F., Mestre, L., Clemente, D., Hernangómez, M., Loría, F., and Guaza, C. (2009). A role for CB2 receptors in anandamide signalling pathways involved in the regulation of IL-12 and IL-23 in microglial cells. *Biochem. Pharmacol.* **77**, 86–100. doi:10.1016/j.bcp.2008.09.014.

Cravatt, B. J., Demarest, K., Patricelli, M., Bracey, M. H., Giang, D. K., Martin, B. R., and Lichtman, A. H. (2001). Supersensitivity to anandamide and enhanced endogenous cannabinoid signaling in mice lacking fatty acid amide hydrolase. *Proc. Natl. Acad. Sci. USA* **98**, 9371–9376.

Croxford, J. L., and Miller, S. D. (2003). Immunoregulation of a viral model of multiple sclerosis using the synthetic cannabinoid R+WIN55,212-2. *J. Clin. Invest.* **111**, 1231–1240.

Derocq, J. M., Jbilo, O., Bouaboula, M., Segui, M., Clère, C., and Casellas, P. (2000). Genomic and functional changes induced by the activation of the peripheral CB_2 in the promyelocytic cells HL-60. *J. Biol. Chem.* **275**, 15621–15628.

Devane, W. A., Dysarz, F. A., Johnson, M. R., Melvin, L. S., and Howlett, A. C. (1988). Determination and characterization of a cannabinoid receptor in a rat brain. *Mol. Pharmacol.* **34**, 605–613.

Devane, W. A., Hanus, L., Breuer, A., Pertwee, R. G., Stevenson, L. A., Griffin, G., et al. (1992). Isolation and structure of a brain constituent that binds to the cannabinoid receptor. *Science* **258**, 1946–1949.

Di Marzo, V., Breivogel, C. S., Tao, Q., Bridgen, D. T., Razdan, R. K., Zimmer, A. M., Zimmer, A., and Martin, B. R. (2000). Levels, Metabolism, and pharmacological activity of anandamide in CB(1) cannabinoid receptor knockout mice: Evidence for non-CB(1), non-CB(2) receptor-mediated actions of anandamide in mouse brain. *J. Neurochem.* **75**, 2434–2444.

Di Marzo, V., Fontana, A., Cadas, H., Schinelli, S., Cimino, G., Schwartz, J. C., and Piomelli, D. (1994). Formation and inactivation of endogenous cannabinoid anandamide in central neurons. *Nature* **372**, 686–691.

Di Marzo, V., De Petrocellis, L., Sepe, N., and Buono, A. (1996). Biosynthesis of anandamide and other acylethanolamides in mouse J744 macrophages and N18 neuroblastoma cells. *Biochem. J.* **316**(3), 977–984.

Docagne, F., Muñeton, V., Clemente, D., Ali, C., Loría, F., Correa, F., Hernangómez, M., Mestre, L., Vivian, D., and Guaza, C. (2007). Excitoxicity in a chronic model of multiple sclerosis: Neuroprotective effects of cannabinoids through CB_1 and CB_2 receptors activation. *Mol. Cell. Neurosci.* **34**(4), 551–561.

Docagne, F., Mestre, L., Loría, F., Hernangómez, M., Correa, F., and Guaza, C. (2008). Therapeutic potential of CB_2 targeting in multiple sclerosis. *Expert Opin. Ther. Targets* **12**, 185–195.

Eisenstein, T. K., Meissler, J. J., Wilson, Q., Gaughan, J. P., and Adler, M. W. (2007). Anandamide and delta-9-tetrahydrocannabinol directly inhibit cells of the immune system via CB_2 receptors. *J. Neuroimmunol.* **189**(1–2), 17–22.

Eljaschewitsch, E., Witting, A., Mawrin, C., Lee, T., Schmidt, P. M., Wolf, S., Hoertnalg, H., Raine, S. C., Scheneider-Stock, R., Nitsch, R., and Ulrich, O. (2006). The endocannabinoid anandamide protects neurons during CNS inflammation by induction of MKP-1 in microglial cells. *Neuron* **49**, 67–79.

Fachinetti, F., Del Giudice, G., Furegato, S., Passarotto, M., and Leon, M. (2003). Cannabinoids ablate release of TNF alpha in rat microglial cells stimulated with lypopolysaccharide. *Glia* **41,** 161–168.

Felder, C. C., Briley, E. M., Axelrod, J., Simpson, J. T., Mackie, K., and Devane, W. A. (1993). Anandamide, an endogenous cannabimimetic eicosanoid. Binds to the cloned human cannabinoid receptor and stimulates receptor-mediated signal transduction. *Proc. Natl. Acad. Sci. USA* **90,** 7656–7660.

Fernández-Ruiz, J., Pazos, M. R., García-Arencibia, M., Sagredo, O., and Ramos, J. A. (2008). Role of CB_2 receptors in neuroprotective effects of cannabinoids. *Mol. Cell. Endocrinol.* **286,** 91–96.

Fowler, C. J. (2007). The contribution of cyclooxygenase-2 to endocannabinoid metabolism and action. *Br. J. Pharmacol.* **152**(5), 744–750.

Galiègue, S., Mary, S., Marchand, J., Dussossoy, D., Carrière, D., Carayon, P., Bouaboula, M., Shire, D., Le Fur, G., and Casellas, P. (1995). Expression of central and peripheral cannabinoid receptors in human immune tissues and leukocyte subpopulations. *Eur. J. Biochem.* **232,** 54–61.

Gaoni, Y., and Mechoulam, R. (1964). Isolation, structure and partial synthesis of an active constituent of hashish. *J. Am. Chem. Soc.* **86,** 1646–1647.

García-Ovejero, D., Arévalo-Martin, A., Petrosino, S., Docagne, F., Hagen, C., Bisogno, T., Watanabe, M., Guaza, C., Di Marzo, V., and Molina-Holgado, E. (2009). The endocannabinoid system is modulated in response to spinal cord injury in rats. *Neurobiol. Dis.* **33,** 57–71.

Ghosh, S., Preet, A., Groopman, J. E., and Ganju, R. K. (2006). Cannabinoid receptor CB_2 modulates the CXCL12/CXCR4-mediated chemotaxis of T lymphocytes. *Mol. Immunol.* **43,** 2169–2179.

Giuffrida, A., Parsons, L. H., Kerr, T. M., Rodriguez de Fonseca, F., Navarro, M., and Piomelli, D. (1999). Dopamine activation of endogenous cannabinoid signaling in dorsal striatum. *Nat. Neurosci.* **2,** 358–363.

Hanisch, U. K., and Kettenmann, H. (2007). Microglia: Active sensor and versatile effector cells in the normal and pathologic brain. *Nat. Neurosci.* **10**(11), 1387–1394.

Howlett, A. C., Barth, F., Bonner, T. I., Cabral, G., Casellas, P., Devane, W. A., Felder, C. C., Herhkenman, M., Mackie, K., Martin, B. R., Mechoulam, R., and Pertwee, R. G. (2002). International Union of Pharmacology. XXVII. Classification of cannabinoid receptors. *Pharmacol. Rev.* **54,** 161–202.

Howlett, A. C. (2005). Cannabinoid receptor signalling. *Handb. Exp. Pharmacol.* **168,** 53–79.

Járai, Z., Wagner, J. A., Varga, K., Lake, K. D., Compton, D. R., Martin, B. R., Zimmer, A. M., Bonner, T. I., Buckley, N. E., Mezey, E., Razdan, R. K., Zimmer, A., and Kunos, G. (1999). Cannabinoid-induced mesenteric vasodilation through an endothelial site distinct from CB_1 or CB_2 receptors. *Proc. Natl. Acad. Sci. USA* **23**(24), 14136–14141.

Kaplan, B. L., Rockwell, C. E., and Kaminski, N. E. (2003). Evidence for cannabinoid dependent and independent mechanisms of actions on leukocytes. *J. Pharmacol. Exp. Ther.* **306**(3), 1077–1085.

Khaspekov, L. G., Brenz, M. S., Frumkina, L. E., Hermann, H., Marsicano, G., and Lutz, B. (2004). Involvement of brain-derived neurotrophic factor in cannabinoid-dependent protection against excitotoxicity. *Eur. J. Neurosci.* **19,** 1691–1698.

Klein, T. W. (2005). Cannabinoid-based drugs as anti-inflammatory therapeutics. *Nat. Rev. Immunol.* **5,** 400–411.

Klein, T. W., and Newton, C. A. (2007). Therapeutic potential of cannabinoid-based drugs. *Adv. Exp. Med. Biol.* **601,** 395–413.

Klein, T. W., Newton, C., Larsen, K., Chou, J., Perkins, I., Lu, L., Nong, L., and Friedman, H. (2004). Cannabinoid receptors and T helper cells. *J. Neuroimmunol.* **147,** 91–94.

Lambert, D. M., and Fowler, C. J. (2005). The endocannabinoid system: Drug targets, lead compounds and potential therapeutic applications. *J. Med. Chem.* **48,** 5059–5087.

Li, J., Gran, B., Zhang, G. X., Ventura, E. S., Siglienti, I., Rostami, A., and Kamoun, M. (2003). Differential expression and regulation of IL-23 and IL-12 subunits and receptors in adult mouse microglia. *J. Neurol. Sci.* **215,** 95–103.

Ligresti, A., Cascio, M. G., Pryce, G., Kulasegram, S., Beletskaya, I., De Petrocellis, L., Saha, B., Mahadevan, A., Visintin, C., Wiley, J. L., Baker, D., Martin, B. R., *et al.* (2006). New potent and selective inhibitors of anandamide reuptake with antispastic activity in a mouse model of multiple sclerosis. *Br. J. Pharmacol.* **147,** 83–91.

Liu, J., Batkai, S., Pacher, P., Harvey-White, J., Wagner, J. A., Cravatt, B. J., Gao, B., and Kunos, G. (2003). Lipopolysaccharide induces anandamide synthesis in macrophages via CD14/MAPK/phosphoinositide 3-kinase/NF-kappaB independently of platelet-activating factor. *J. Biol. Chem.* **278,** 45034–45039.

Liu, J., Wang, L., Harvey-White, J., Osei-Hyiaman, D., Razdan, R., Gong, Q., Chan, A. C., Zhou, Z., Huang, B. X., Kim, H. Y., and Kunos, G. (2006). A biosynthetic pathway for anandamide. *Proc. Natl. Acad. Sci. USA* **103,** 13345–13350.

López-Rodríguez, M. L., Viso, A., Ortega-Gutiérrez, S., Fowler, C. J., Tiger, G., de Lago, E., Fernández-Ruiz, J., and Ramos, J. A. (2003). Design, synthesis, and biological evaluation of new inhibitors of the endocannabinoid uptake: Comparison with effects on fatty acid amidohydrolase. *J. Med. Chem.* **46,** 1512–1522.

Loría, F., Petrosino, S., Mestre, L., Spagnolo, A., Correa, F., Hernangómez, M., Di Marzo, V., Guaza, C., and Docagne, F. (2008). Study of the regulation of the endocannabinoid system in a virus model of multiple sclerosis reveals a therapeutic effect of palmitoylethanolamide. *Eur. J. Neurosci.* **28,** 633–641.

Maresz, K., Carrier, E. J., Ponomarev, E. D., Hillard, C. J., and Dittel, B. N. (2005). Modulation of the cannabinoid CB2 receptor in microglial cells in response to inflammatory stimuli. *J. Neurochem.* **95,** 437–445.

Maresz, K., Pryce, G., Ponomarev, E. D., Marsicano, G., Croxford, J. L., Shriver, L. P., Ledent, C., Cheng, X., Carrier, E. J., Mann, M. K., Giovannoni, G., Pertwee, R. G., *et al.* (2007). Direct suppression of CNS autoimmune inflammation via the cannabinoid receptor CB1 on neurons and CB2 on autoreactive T cells. *Nat. Med.* **13,** 492–497.

Marsicano, G., Goodenough, S., Monory, K., Hermann, H., Eder, M., Cannich, A., Azad, S. C., Gracia-Cascio, M., Ortega-Gutiérrez, S., Van der Stelt, M., López-Rodriguez, M. L., Casanova, E., *et al.* (2003). CB_1 cannabinoid receptors on-demand defense against excitotoxicity. *Science* **302,** 84–88.

Matias, I., Pochard, P., Orlando, P., Salzet, M., Pestel, J., and Di Marzo, V. (2002). Presence and regulation of the endocannabinoid system in human dendritic cells. *Eur. J. Biochem.* **269,** 3771–3778.

Matias, I., Chen, J., De Petrocellis, L., Bisogno, T., Ligresti, A., Fezza, F., Krauss, A. H., Shi, L., Protzman, C. E., Li, C., Liang, Y., Nieves, A. L., *et al.* (2004). Prostaglandin ethanolamides (prostamides): *In vitro* pharmacology and metabolism. *J. Pharmacol. Exp. Ther.* **309,** 745–757.

Matsuda, L. A., Lolait, S. J., Brownstein, M. J., Young, A. C., and Bonner, T. I. (1990). Structure of a cannabinoid and functional expression of the cloned cDNA. *Nature* **346,** 561–564.

Mc Coy, K. L., Matmeyeva, M., Carlisle, S. J., and Cabral, G. A. (1999). Cannabinoid inhibition of the processing of intact lysozyme by macrophages: Evidence for CB2 receptor participation. *J. Pharmacol. Exp. Ther.* **289,** 1620–1625.

McHugh, D., Tanner, C., Mechoulam, R., Pertwee, R. G., and Ross, R. A. (2008). Inhibition of human neutrophil chemotaxis by endogenous cannabinoids and phytocannabinoids: Evidence for a site distinct of CB1 and CB2. *Mol. Pharmacol.* **73**(2), 441–450.

Mechoulam, R., Ben-Shabat, S., Hanus, L., Ligumsky, M., Kaminski, N. E., Schatz, A. R., Gopher, A., Almog, S., Martin, B. R., Compton, D. R., Pertwee, R. G., Griffin, G., et al. (1995). Identification of an endogenous 2-monoglyceride, present in canine gut, that binds to cannabinoid receptors. *Biochem. Pharmacol.* **50,** 83–90.

Mestre, L., Correa, F., Arévalo-Martín, A., Molina-Holgado, E., Valenti, M., Ortar, G., Di Marzo, V., and Guaza, C. (2005). Pharmacological modulation of the endocannabinoid system in a viral model of multiple sclerosis. *J. Neurochem.* **92,** 1327–1339.

Mestre, L., Correa, F., Docagne, F., Clemente, D., and Guaza, C. (2006). The synthetic cannabinoid Win 55-212, 2 increases COX-2 expression and PGE_2 release in murine brain derived endothelial cells following Theiler's virus infection. *Biochem. Pharmacol.* **72,** 869–880.

Mestre, L., Docagne, F., Correa, F., Loría, F., Hernangómez, M., Borrell, J., and Guaza, C. (2009). A cannabinoid agonist interferes with the progression of a chronic model of multiple sclerosis by downregulating adhesion molecules. *Mol. Cell Neurosci.* **40,** 258–266.

Molina-Holgado, F., Lledó, A., and Guaza, C. (1997). Anandamide suppresses nitric oxide and TNF-α responses to Theiler's virus or endotoxin in astrocytes. *Neuroreport* **8,** 1929–1933.

Molina-Holgado, F., Molina-Holgado, E., and Guaza, C. (1998). The endogenous cannabinoid anandamide potentiates IL-6 production by astrocytes infected with Theiler's murine encephalomyelitis virus by a receptor-mediated pathway. *FEBS Lett.* **433,** 139–142.

Molina-Holgado, F., Molina-Holgado, E., Guaza, C., and Rothwell, N. J. (2002a). Role of CB1 and CB2 receptors in the inhibitory effects of cannabinoids on lipopolysaccharide induced nitric oxide release in astrocyte cultures. *J. Neurosci. Res.* **67,** 829–836.

Molina-Holgado, E., Vela, J. M., Arévalo-Martín, A., Almazán, G., Molina-Holgado, F., Borrell, J., and Guaza, C. (2002b). Cannabinoids promote oligodendrocyte progenitor survival: Involvement of cannabinoid receptors and phosphatidylinositol-3 kinase/Akt signaling. *J. Neurosci.* **22**(22), 9742–9753.

Molina-Holgado, F., Pinteaux, E., Moore, J., Molina-Holgado, E., Guaza, C., Gibson, R. M., and Rothwell, N. (2003). Endogenous interleukin-1 receptor antagonist mediates anti-inflammatory and neuroprotective actions of cannabinoids in neurons and glial cells. *J. Neurosci.* **23,** 6470–6474.

Monory, K., Blaudzum, H., Massa, F., Kaiser, N., Lemberger, T., Schtz, G., Wotjak, C. T., Luzt, B., and Marsicano, G. (2007). Genetic dissection of behavioural and autonomic effects of delta-9-tetrahydrocannabinol in mice. *PLoS Biology* e269.

Muccioli, G. G., Xu, C., Odah, E., Cudaback, E., Cisneros, J. A., Lambert, D. M., López Rodríguez, M. L., Bajjalieh, S., and Stella, N. (2007). Identification of a novel endocannabinoid-hydrolyzing enzyme expressed by microglial cells. *J. Neurosci.* **27**(11), 2883–2890.

Munro, S., Thomas, K. L., and Abu-Shaar, M. (1993). Molecular characterization of a peripheral receptor for cannabinoids. *Nature* **365,** 61–65.

Nagayama, T., Sinor, A. D., Simon, R. P., Chen, J., Graham, S. H., Jin, K., and Greenberg, D. A. (1999). Cannabinoids and neuroprotection in focal and global cerebral ischemia and neuronal cultures. *J. Neurosci.* **19,** 2987–2995.

Nakajima, Y., Furuichim, Y., Biswas, K. K., Hashiguchi, T., Kawahara, K., Yamaji, K., Uchimura, T., Izumi, Y., and Maruyama, I. (2006). Endocannabinoid, anandamide in gingival tissue regulates the periodontal inflammation through NF-kappaB pathway inhibition. *FEBS Lett.* **580,** 613–619.

Offertaler, L., Mo, F. M., Batkai, S., Liu, J., Begg, M., Razdan, R. K., Martin, B. R., Bukoski, R. D., and Kunos, G. (2003). Selective ligands and cellular effectors of a G protein-coupled endothelial cannabinoid receptor. *Mol. Pharmacol.* **63,** 699–705.

Ortar, G., Ligresti, A., de Petrocellis, L., Morera, E., and Di Marzo, V. (2003). Novel selective and metabolically stable inhibitors of anandamide cellular uptake. *Biochem. Pharmacol.* **65,** 1475–1481.

Ortega-Gutiérrez, S., Molina-Holgado, E., and Guaza, C. (2005a). Effect of anandamide uptake inhibition in the production of nitric oxide and in the release of cytokines in astrocyte cultures. *Glia* **52,** 166–168.

Ortega-Gutiérrez, S., Molina-Holgado, E., Arévalo-Martín, A., Correa, F., Viso, A., López-Rodríguez, M. L., Di Marzo, V., and Guaza, C. (2005b). Activation of the endocannabinoid system as therapeutic approach in a murine model of multiple sclerosis. *FASEB J.* **19,** 1338–1340.

O'Sullivan, S. E. (2007). Cannabinoids go nuclear: Evidence for activation of peroxisome proliferator-activated receptors. *Br. J. Pharmacol.* **152**(5), 576–582.

Palazuelos, J., Davoust, N., Julien, B., Hatterer, E., Aguado, T., Mechoulam, R., Benito, C., Romero, J., Silva, A., Guzman, M., Nataf, S., and Glave-Roperh, I. (2008). The CB_2 cannabinoid receptor controls myeloid progenitor trafficking. Involvement in the pathogenesis of an animal model of multiple sclerosis. *J. Biol. Chem.* **283,** 13320–13329.

Pitt, D., Werner, P., and Raine, C. S. (2000). Glutamate excitoxicity in a model of multiple sclerosis. *Nat. Med.* **6,** 67–70.

Pryce, G., Ahmed, Z., Hankey, D. J., Jackson, S. J., Croxford, J. L., Pocock, J. M., Ledent, C., Petzold, A., Thompson, A. J., Giovannoni, G., Cuzner, M. L., and Baker, D. (2003). Cannabinoids inhibit neurodegeneration in models of multiple sclerosis. *Brain* **126,** 2191–2202.

Puffenbarger, R. A., Boothe, A. C., and Cabral, G. A. (2000). Cannabinoids inhibit LPS-inducible cytokine mRNA expression in rat microglial cells. *Glia* **29,** 58–69.

Raivich, G. (2005). Like cops of the beat: The active role of resting microglia. *Trends Neurosci.* **28,** 571–573.

Ramirez, B. G., Blazquez, C., Gómez del Pulgar, T., Guzman, M., and De Ceballos, M. (2005). Prevention of Alzheimer's disease pathology by cannabinoids: Neuroprotection mediated by blockade of microglial activation. *J. Neurosci.* **25,** 1904–1913.

Rockwell, C. E., and Kaminski, N. E. (2004). A cyclooxygenase metabolite of anandamide causes inhibition of interleukin-2 secretion in murine splenocytes. *J. Pharmacol. Exp. Ther.* **311,** 683–690.

Ross, R. A., Craib, S. J., Stevenson, L. A., Pertwee, R. G., Henderson, A., Toole, J., and Ellington, H. C. (2002). Pharmacological characterization of the anandamide cyclooxygenase metabolite: Prostaglandin E_2 ethanolamine. *J. Pharmacol. Exp. Ther.* **301,** 900–907.

Sanchez Galve-Roperth, I., Canova, C., Brachet, P., and Guzmán, M. (1998). Delta9-tetrahydrocannabinol induces apoptosis in C6 glioma cells. *FEBS Lett.* **436,** 6–10.

Sancho, R., Calzado, M. A., Di Marzo, V., Appendino, G., and Muñoz, E. (2003). Anandamide inhibits nuclear factor-kappa B activation through a cannabinoid receptor-independent pathway. *Mol. Pharmacol.* **63**(2), 429–438.

Sharchielli, P., Greco, L., Floridi, A., and Gallai, V. (2003). Excitatory amino acids and multiple sclerosis: Evidence from cerebrospinal fluid. *Arch. Neurol.* **60,** 1082–1088.

Shouman, B., Fontaine, R., Baud, O., Schwendimann, L., Keller, M., Sppedding, M., Lelievre, V., Gressens, P. (2006). Endocannabinoids potently protect the new born brain against AMPA-kaniate receptor- mediated excitotoxicity damage. *Br. J. Pharmacol.* **148,** 442–451.

Smith, S. R., Terminelli, C., and Denhardt, G. (2000). Effects of cannabinoid receptor agonist and antagonist ligands on production of inflammatory cytokines and anti-inflammatory interleukin-10 in endotoxemic mice. *J. Pharmacol. Exp. Ther.* **293,** 136–150.

Srinivasan, R., Sailasuta, N., Hurd, R., Nelson, S., and Pelleetier, D. (2005). Evidence of elevated glutamate in multiple sclerosis using magnetic resonante spectroscopy at 3 T. *Brain* **128,** 1016–1025.

Stella, N. (2004). Cannabinoid signalling in glial cells. *Glia* **48,** 267–277.

Stella, N. (2009). Endocannabinoid signalling in microglial cells. *Neuropharmacology* **1**, 244–253.

Sugiura, T., Kondo, S., Sukagawa, A., Nakane, S., Shinoda, A., Itoh, K., Yamashita, A., and Waku, K. (1995). 2-Arachidonoylglycerol: A possible endogenous cannabinoid receptor ligand in brain. *Biochem. Biophys. Res. Commun.* **215**, 89–97.

Trinchieri, G. (2003). Interleukin-12 and the regulation of innate resistance and adaptive immunity. *Nat. Rev. Immunol.* **3**(2), 133–146.

Tsou, K., Brown, S., Sanudo-Pena, M. C., Mackie, K., and Walker, J. M. (1998). Immunohistochemical distribution of cannabinoid CB_1 receptors in the rat central nervous system. *Neuroscience* **83**, 393–411.

Van Sickle, M. D., Duncan, M., Kingsley, P. J., Mouihate, A., Urbani, P., Mackie, K., Stella, N., Makryannis, A., Piomelli, D., Davison, J. S., Marnett, L. J., Di Marzo, V., et al. (2005). Identification and functional characterization of brainstem cannabinoid CB_2 receptors. *Science* **310**, 329–332.

Vogel, Z., Barg, J., Levy, R., Saya, D., Hedaman, E., and Mechoulam, R. (1993). Anandamide a brain endogenous compound interacts specifically with the cannabinoid receptors and inhibits adenylate cyclase. *J. Neurochem.* **61**, 352–355.

Waldeck-Weiermair, M., Zoratti, C., Osibow, K., Balenga, N., Goessnitzer, E., Waldhoer, M., Malli, R., and Graier, W. F. (2008). Integrin clustering enables anandamide-induced Ca2+ signalling in endothelial cells via GPR55 by protection against CB1-triggered repression. *J. Cell Sci.* **121**, 1704–1717.

Walter, L., Franklin, A., Witting, A., Moller, T., and Stella, N. (2002). Astrocytes in culture produce anandamide and other acylethanolamides. *J. Biol. Chem.* **277**, 20869–20876.

Walter, L., Franklin, A., Witting, A., Wade, C., Xie, Y., Kunos, G., and Stella, N. (2003). Non-psychotropic cannabinoid receptors regulate microglia cell migration. *J. Neurosci.* **23**, 1398–1405.

Webb, M., Luo, L., Ma, J. Y., and Tham, C. S. (2008). Genetic deletion of fatty acid amide hydrolase results in improved long-term outcome in chronic autoimmune encephalitis. *Neurosci. Lett.* **439**(1), 106–110.

Witting, P., Chen, L., Cudaback, E., Straiker, A., Walter, R., Rickman, B., Moller, T., Brosnan, C., and Stella, N. (2006). Experimental autoimmune encephalomyelitis disrupts endocannabinoid-mediated neuroprotection. *Proc. Natl. Acad. Sci. USA* **103**, 6362–6367.

Witting, A., Walter, R., Moller, T., and Stella, N. (2004). P2X7 receptors control 2-arachidonoyl production by microglial cells. *Proc. Natl. Acad. Sci. USA* **101**, 3214–3219.

Yiangou, Y., Facer, P., Durrenberger, P., Chessell, I. P., Naylor, A., Bountra, C., Banati, R. R., and Anand, P. (2006). COX-2, CB2, and P2X7-immunoreactivities are increased in activated microglia cells/macrophages of multiple sclerosis and amytrophic lateral sclerosis spinal cord. *BMC Neurol.* **6**, 12–16.

Yuan, M., Kiertscher, S. M., Cheng, Q., Zoumalan, R., Tashkin, D. P., and Roth, M. D. (2002). Delta 9 tetrahydrocannabinol regulates Th1/Th2 cytokine balance in activated human T cells. *J. Neuroimmunol.* **133**, 124–131.

Zygmunt, P. M., Petersson, J., Andersson, D. A., Chuang, H., Sorgard, M., Di Marzo, V., Julius, D., and Hogestatt, E. D. (1999). Vanilloid receptors on sensory nerves mediate the vasodilator action of anandamide. *Nature* **400**, 452–457.

CHAPTER TEN

Modulation of the Endocannabinoid-Degrading Enzyme Fatty Acid Amide Hydrolase by Follicle-Stimulating Hormone

Paola Grimaldi,[*,1] Gianna Rossi,[†,1] Giuseppina Catanzaro,[‡,§] and Mauro Maccarrone[‡,§]

Contents

I. Follicle-Stimulating Hormone: Signal Transduction and Molecular Targets	232
II. Sertoli Cells: Activities and Biological Relevance	237
III. Overview of the Endocannabinoid System	239
IV. The ECS in Sertoli Cells	241
V. Regulation of FAAH by FSH in Sertoli Cells	243
VI. FAAH Is an Integrator of Fertility Signals	244
VII. Conclusions	254
Acknowledgments	255
References	255

Abstract

Follicle-stimulating hormone (FSH) is a glycoprotein that transmits its signals via a G protein-coupled receptor. As yet, not many targets of FSH have been identified, able to justify the critical role of this hormone on reproductive events. On the other hand, among the biological activities of the endocannabinoid anandamide (AEA), growing interest has been attracted by the regulation of mammalian fertility. Recently, we have shown that treatment of mouse primary Sertoli cells with FSH enhances the activity of the AEA hydrolase (fatty acid amide hydrolase, FAAH), whereas it does not affect the enzymes that synthesize AEA, nor the level of the AEA-binding type-2 cannabinoid and type-1 vanilloid receptors. In addition,

[*] Department of Public Health and Cell Biology, University of Rome Tor Vergata, Rome, Italy
[†] Department of Health Sciences, University of L'Aquila, L'Aquila, Italy
[‡] Department of Biomedical Sciences, University of Teramo, Teramo, Italy
[§] European Center for Brain Research (CERC)/Santa Lucia Foundation, Rome, Italy
[1] These authors contributed equally to this work.

diacylglycerol lipase and monoacylglycerol lipase, which, respectively, synthesize and degrade the other major endocannabinoid 2-arachidonoylglycerol, were not regulated by FSH. Interestingly, FAAH stimulation by FSH occurred through protein kinase A and aromatase-dependent pathways that were able to modulate FAAH activity (via phosphorylation of accessory proteins) and *faah* gene expression (via an estrogen response element on the promoter region). Taken together, these data identify FAAH as the only target of FSH among the elements of the endocannabinoid system, with a critical impact on Sertoli cell proliferation, and thus spermatogenesis and male reproduction.

I. Follicle-Stimulating Hormone: Signal Transduction and Molecular Targets

Follicle-stimulating hormone (FSH) is a member of the glycoprotein hormone family that includes luteinizing hormone (LH), thyroid-stimulating hormone (TSH), and human chorionic gonadotropin (hCG) (Walker and Cheng, 2005). FSH is secreted from the pituitary gland under the control of the gonadotropin-releasing hormone (GnRH) and it is essential for normal reproduction in both male and female mammals. In female, FSH induces the maturation of ovarian follicles by targeting a receptor expressed specifically on granulosa cells, and by stimulating estradiol and inhibin secretion. High estradiol levels, produced by mature follicles, lead to a positive feedback on the hypothalamus, which causes the LH surge responsible for ovulation. Secretion of inhibin causes an inhibition of FSH release from the pituitary. In male, FSH is responsible for the initiation of spermatogenesis at puberty, whereas in adult it may play a role, together with testosterone, in maintaining sperm production. FSH binds to specific receptors that, in the testis, are present only on Sertoli cells. Mice lacking the FSH receptor (FORKO) show a reduction in sperm production, associated with a reduction in Sertoli cell number and germ cell number (Krishnamurthy *et al.*, 2001), thus indicating that FSH is required for normal Sertoli cell functions and for a quantitatively normal spermatogenesis.

FSH, as the other members of the glycoprotein hormones, is a disulfide-rich heterodimer consisting of noncovalently associated α- and β-subunits. In a given species, glycoprotein hormones share a common α-chain, but they have a unique β-chain that determines hormone specificity. FSH acts through a specific FSH receptor (FSHR) on target cells. FSHR belongs to the family of G protein-coupled receptors which are characterized by a domain of seven α-helices spanning the membrane (Heckert and Griswold, 2002). The interaction of FSH with its receptor on the target Sertoli cells determines the activation of different signal transduction pathways that are summarized below and depicted in Fig. 10.1.

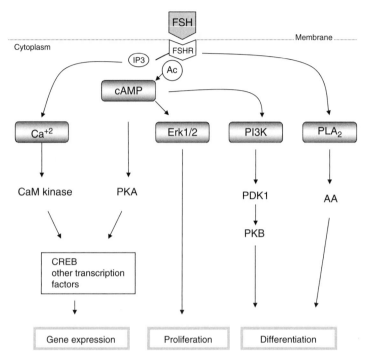

Figure 10.1 Signal transduction pathways activated by FSH in Sertoli cells. The scheme shows that FSH, upon binding its receptor, determines the activation of the adenylyl cyclase (Ac) and the accumulation of cAMP. The latter substance is able to activate different signals in the cells, through activation of PKA that phosphorylates proteins and transcription factors like CREB. See text for details.

(A) The classical cascade of biochemical events activated by FSH is the stimulation of Gs/adenylyl cyclase that leads to the accumulation of intracellular cAMP and to activation of the catalytic subunit of protein kinase A (PKA). PKA can phosphorylate target proteins thus regulating cell proliferation, survival, and differentiation. One of the best characterized targets of PKA is the transcription factor CREB, which binds to specific cAMP-responsive element (CRE) on the regulatory regions of several genes. CREB activation is determined by phosphorylation at residue Ser-133, allowing the binding to coactivator CREB-binding proteins (CBP/p300), and hence activation of transcription (Walker et al., 1995). At least six alternatively spliced transcripts of CREB have been identified and two of these isoforms are expressed in all tissues and cell lines tested (Ruppert et al., 1992). In addition, a protein highly homologous to CREB, called cAMP-responsive element modulator (CREM), has been cloned, of which three isoforms, encoded by alternative splicing, exist (Foulkes et al., 1991). In rat Sertoli cells,

CREB is expressed in a cyclic manner, depending on specific germ cell associations in the seminiferous tubules, being at high level from stage I to VIII and decreasing to low levels from stage IX to XIV (Waeber et al., 1991). It has been demonstrated that FSH increases CREB mRNA levels and that, in the CREB promoter, two CRE sequences are essential for its cAMP-induced transcription. Hence, CREB is able to positively autoregulate its expression by acting on its own promoter (Walker et al., 1995). FSH also stimulates, in Sertoli cells, expression of the ICER isoform of CREM, a suppressor of cAMP-induced transcription (Monaco et al., 1995). The increased levels of ICER correlate with downregulation of FSHR transcript, indicating that downregulation of CREB expression could be obtained indirectly by a modulation of the number of FSHR, thus desensitizing Sertoli cells to FSH (Monaco et al., 1995). Other cAMP-induced genes in Sertoli cells are regulated at the transcriptional level by CREB, such as the proto-oncogenes c-fos (Chaudhary et al., 1996), androgen-binding protein (Saxlund et al., 2004), transferrin (Chaudhary and Skinner, 2001), and α-inhibin (Ito et al., 2000). All these CRE-regulated genes have been shown to respond to cAMP with rapid and sustained kinetics. However, other FSH-responsive genes have been identified that show a slower kinetics and contain in their promoters cAMP-responsive regions distinct from the classical CRE. For example, FSH upregulates the expression of AMH through transcription factors AP2 and NFκB (Lukas-Croisier et al., 2003); the cAMP-stimulated expression of the RIIβ-subunit of PKA and of phosphodiesterase (PDE) 4D1/2 isoforms are both mediated by the CCAAT enhancer-binding protein C/EBP (Gronning et al., 1999); FSH-induced expression of the growth factor KL in Sertoli cells is mediated by a cell-specific factor that binds a GC-rich element in the proximal promoter, and cooperates together with nuclear factors binding TATA box and Sp1 sites (Grimaldi et al., 2003). The cAMP-responsive genes for which a regulation at transcriptional level has been demonstrated in Sertoli cells are summarized in Table 10.1.

(B) Binding of FSH to its receptor can also trigger phosphatidylinositol 3-kinase (PI3K) signaling pathway. This route has been first described in granulosa cells, where it has been reported a link between FSH, cAMP, and PI3K signaling (Gonzalez-Robayna et al., 2000). In these cells, FSH induces a rapid phosphorylation and activation of protein kinase B (PKB) through PI3K, in a way that is cAMP-dependent but PKA-independent. The phosphorylation of PKB by FSH in granulosa cells is associated with increased phosphorylation of serum and glucocorticoid-induced kinase 3 (SGK) (Gonzalez-Robayna et al., 2000).

Table 10.1 Transcription factors mediating FSH response of hormone-regulated genes in Sertoli cells

Gene	Transcription factor	References
c-fos	CREB	Chaudhary et al. (1996)
Transferrin	CREB, bHLH, p300/CBP	Chaudhary and Skinner (2001)
ABP	CREB, bHLH	Saxlund et al. (2004)
CREB	CREB	Walker et al. (1995)
Inhibin	CREB, SF1, p300/CBP	Ito et al. (2000)
RIIβ-subunit PKA	C/EBP	Gronning et al. (1999)
PDE4 D1/2 isoform	C/EBP	Gronning et al. (1999)
AMH	NFκB and AP2	Lukas-Croisier et al. (2003)
KL	Sp1 and GC element-binding protein	Grimaldi et al. (2003)
FSH receptor	ICER	Monaco et al. (1995)
Urokinase plasminogen activator	GC element-binding protein	Grimaldi et al. (1993)

See text for details.

Activation of PI3K by FSH has been also shown in mature Sertoli cells (Meroni et al., 2002). FSH increases PKB levels by a transduction pathway that seems very similar to that observed in granulosa cells, being PI3K-dependent and PKA-independent and probably involving increased intracellular levels of cAMP. However, the exact mechanism by which cAMP regulates PI3K remains largely unknown, although possible mediators could be cAMP-regulated, like guanine nucleotide exchange factors (cAMP-GEF) that can bind cAMP and activate PI3K (Richards, 2001). Yet, at present, a link between cAMP, GEF, and PI3K in rat Sertoli cells has not been demonstrated.

The PI3K/PKB signaling pathway activated by FSH seems to be required for important functions of Sertoli cells, such as transferrin and lactate production (Mita et al., 1982; Skinner and Griswold, 1982), as demonstrated by the fact that FSH stimulation of secretion of these two proteins is blocked by treatment with a PI3K inhibitor (Meroni et al., 2004). More recently, PI3K/PKB signaling has been shown to be involved in the FSH-induced aromatase (ARO) activity in rat Sertoli cells (McDonald et al., 2006). ARO is a terminal enzyme which transforms irreversibly androgens into estrogens and plays a key role in male reproduction (Carreau et al., 2006). Hence, 17β-estradiol (E_2) production in FSH-treated Sertoli cells seems to require PI3K/PKB signals for optimal induction of the *aro* gene.

(C) Several studies have also demonstrated that the interaction of FSH with its receptor induces a rapid Ca^{2+} influx. In rat Sertoli cells, this event is associated with a phosphatidylinositol/phospholipase C (PI/PLC)-dependent pathway, and not with the classical Gs/adenylyl cyclase-dependent route (Lin et al., 2006). The rapid Ca^{2+} influx is mediated by the Gαh/phospholipase C-δ1 (PLC-δ1) signaling through the production of inositol triphosphate (IP_3). It has been also clarified that Ca^{2+} influx involves T-type Ca^{2+} channels and is independent of intracellular calcium store (Lai et al., 2008). Increased levels of intracellular Ca^{2+} can activate the calmodulin and Ca^{2+}/calmodulin-dependent protein kinases (CaM kinases), and thus affecting the phosphorylation of target proteins. One of these proteins is the transcription factor CREB (Dash et al., 1991). Moreover, increased intracellular Ca^{2+} level can promote phosphorylation of cytoskeletal proteins, such as vimentin (Spruill et al., 1983), thus regulating the rearrangement of cell cytoskeleton that plays important functions in the dynamics of Sertoli–Sertoli adherent junction, as well as in Sertoli–germ cells junctions.

(D) Another signaling route activated by FSH upon its interaction with FSHR is the mitogen-activated protein kinase (MAPK) pathway. It has been reported that in rat Sertoli cells FSH activates extracellular signal-regulated protein kinases (ERK) and that this activation is dependent on PKA and Src kinase (Crepieux et al., 2001). The effect of FSH on ERK kinase activation is strictly correlated to the proliferative state of Sertoli cells during the prepuberal period. FSH stimulates Sertoli cell proliferation by activation of ERK kinase and induction of cyclin D1 and E2F. Yet, when cells mature and stop dividing, this pathway is no longer activated by FSH (Crepieux et al., 2001). Hence, the same hormone (FSH) can trigger both activation and inhibition of the MAPK pathway, depending on the maturation state of Sertoli cells.

(E) Finally, at least another signaling pathway involving the activation of phospholipase A_2 (PLA_2) has been shown to participate in the mechanism of action of FSH. The latter substance induces PLA_2 activation with the consequent release of arachidonic acid (AA) (Jannini et al., 1994). AA is involved in the regulation of Sertoli cell metabolic responses to FSH, such as lactate production and glucose transport, and it modulates mRNA levels and enzymatic activity of lactate dehydrogenase (Meroni et al., 2003).

II. Sertoli Cells: Activities and Biological Relevance

Spermatogenesis is a complex process by which mitotic germ cells undergo proliferation, differentiation, meiotic division, and cell morphological changes to produce mature sperm.

The process of spermatogenesis is dependent upon the somatic Sertoli cells, which provide support and factors necessary for the successful progression of germ cells into spermatozoa (Griswold, 1998). Sertoli cells comprise the main structural component of the seminiferous epithelium, forming a single-layered lamina and extending from the basal lamina toward the tubule lumen. With their cytoplasmic protrusions, they surround the single germ cells more or less completely, thus providing a unique protected environment within the seminiferous tubules for germ cell development.

In mouse, rapid growth of the testis occurs in the perinatal period, during which Sertoli cells proliferate quickly and continue to divide until about 12–16 days *postcoitum*, after which they become mature cells and cease to divide. Both the rate and the duration of the proliferative phase determine the final number of Sertoli cells in the adult testis. The number of Sertoli cells is strictly correlated to adult testis size and sperm production, because a limited number of germ cells can be supported by each Sertoli cell (Orth *et al.*, 1988). During the proliferative phase, adjacent Sertoli cells form tight junctions with each other, determining the formation of a blood–testis barrier. This barrier creates two separated compartments in the seminiferous epithelium: a basal one, in which the spermatogonia are lined up, and a luminal one, in which all the other stages of spermatogenesis are found. Germ cells are dependent on glycoproteins secreted by Sertoli cells (Mruk and Cheng, 2004), classified into different groups on the basis of their functions: (1) *hormones*, such as the mullerian-inhibiting substance (MIS) that regulates the development of the male reproductive tracts, and inhibin which blocks FSH production by the pituitary; (2) *growth factors*, such as glial cell line-derived neurotrophic factor (GDNF), kit ligand (KL), bone morphogenetic protein 4 (BMP4), transforming growth factor-β (TGFβ), insulin-like growth factor-I (IGF-I), fibroblast growth factor (FGF), and epidermal growth factor (EGF); and (3) *proteins with different functions*: transferrin for binding and transport of iron; androgen-binding protein (ABP), important for maintaining a high concentration of testosterone in the testis; proteases and their inhibitors, implicated in the events of germ cell movement; lactate which represents an energy substrate for germ cells and extracellular matrix components (Griswold, 1998). Synthesis and secretion of some glycoproteins by Sertoli cells change during testicular development: MIS is produced during fetal life and decrease after birth, while ABP and transferrin are present at higher level in adult testis. Moreover, the production of many proteins is regulated by hormonal signal, such as FSH and testosterone, whose receptors are distinctively expressed by Sertoli cells.

Somatic Sertoli cells provide also a biological compartment, known as the stem cell niche, which is a specialized microenvironment that promotes self-renewal and maintenance of germ stem cells (GSCs) in their undifferentiated state. Up to now only two Sertoli cell-specific proteins have been identified as essential for spermatogonial stem cell maintenance and self-renewal: one is the transcription factor Ets variant gene 5 (ETV5, or Ets-related molecule, ERM). Mice with targeted mutation of ETV5, which is exclusively expressed by Sertoli cells, show a failure of spermatogonial stem cell self-renewal and a complete depletion of germ cells in the adult testis (Chen et al., 2005). The other one is GDNF. This factor is secreted by Sertoli cells and interacts with membrane receptors designated Ret and GDNF family coreceptor α_1 (GFRα-1), which are expressed by spermatogonial stem cells, stimulating proliferation of this cell population. The essential role of GDNF in self-renewal and maintenance of GSCs has been demonstrated using mice heterozygous for a *Gdnf* null allele, which show partial loss of GSCs in adulthood, and mice overexpressing GDNF, which show accumulation of undifferentiated spermatogonia (Meng et al., 2000). The production of GDNF is under the influence of FSH (Tadokoro et al., 2002), but it is also dependent on Fgf2, tumor necrosis factor-α and interleukin-1β (Simon et al., 2007).

Proliferation and differentiation of undifferentiated mitotic germ cells are regulated by another growth factor expressed by Sertoli cells: BMP4. This is a member of the TGFβ–BMP superfamily. Its expression is high in prepuberal Sertoli cells and is dramatically lower in adult cells, suggesting that its action is important mainly during the first wave of spermatogenesis. BMP4 binds activin receptor-like kinase 3 (Alk3) and BMPIIR on undifferentiated spermatogonia, inducing both proliferation and differentiation (Pellegrini et al., 2003).

Another well-characterized paracrine growth factor produced by Sertoli cells under FSH regulation is kit ligand (KL). KL is expressed by Sertoli cells in both a soluble and a transmembrane form, which are generated by alternative splicing (Rossi et al., 1993). Interestingly, the two forms of KL are differentially expressed during testis development. Sertoli cells from prepuberal mice express the mRNA encoding for both transmembrane and soluble forms, but the latter is expressed at higher levels in coincidence with the beginning of the spermatogenic process, and the two transcripts are expressed at equivalent levels in the adult testis (Rossi et al., 1993).

The expression of KL is induced by FSH in prepuberal mouse Sertoli cells through increased cAMP levels and increased gene transcription (Grimaldi et al., 2003). KL plays an important role in regulating proliferation and survival of differentiating type A cells that express the c-kit receptor.

Addition of the soluble form of KL to *in vitro* cultured male mitotic germ cells stimulates their proliferation and significantly reduces their apoptosis (Dolci *et al.*, 2001). More recently, evidence has been reported for a role of KL in stimulating entry of spermatogonia into the meiotic program, by activating the expression of genes of early meiotic phase (Rossi *et al.*, 2008).

It is interesting to observe that most of paracrine growth factors secreted by Sertoli cells regulate mainly mitotic stage of spermatogenesis. A schematic representation of the best characterized paracrine growth factors controlling mouse male germ cell development is shown in Fig. 10.2. However, Sertoli cells also produce autocrine factors. Indeed, they express IGF-I receptors and secrete IGF-I and IGF-binding proteins (IGF-BPs) that may enhance or inhibit the effects of IGFs. IGF-I has been shown to exert multiple effects on immature Sertoli cells, such as stimulation of proliferation and differentiation, which is elicited by lactate production, glucose transport, transferrin production, and secretion of plasminogen activator (Khan *et al.*, 2002).

In summary, it appears evident the fundamental role played by Sertoli cells in supporting and regulating germ cell development.

III. Overview of the Endocannabinoid System

Endocannabinoids are endogenous ligands of cannabinoid receptors that mimic several actions of the natural *Cannabis sativa* component Δ^9-tetrahydrocannabinol (THC); this substance accounts for the majority of the reproductive hazards in marijuana users (Piomelli, 2004). The best characterized endocannabinoids are N-arachidonoylethanolamine (anandamide, AEA) and 2-arachidonoylglycerol (2-AG), which bind to and activate type-1 and type-2 cannabinoid receptors (CB1R and CB2R). The latter proteins belong to the rhodopsin family of G protein-coupled seven-transmembrane spanning receptors (Howlett *et al.*, 2002). CB1R has been found mainly in the central nervous system, but it is also present in ovary, uterine endometrium, testis, vas deferens, urinary bladder, and other peripheral endocrine and neurological tissues. CB2R has been identified mainly in immune cells, but is expressed also in brainstem (Van Sickle *et al.*, 2005). Signal transduction pathways regulated by CBR-coupled $G_{i/o}$ proteins include inhibition of adenylyl cyclase, regulation of ionic currents, activation of focal adhesion kinase, and of mitogen-activated protein kinase (Howlett *et al.*, 2004). Endocannabinoids can induce a biological activity also via other CB receptors, like a purported CB3 (GPR55) receptor (Lauckner *et al.*, 2008), via non-CB1/non-CB2 receptors, and via noncannabinoid receptors. Among the latter proteins, transient receptor potential vanilloid 1 (TRPV1) has emerged as a physiological target of AEA, which is also considered a true "endovanilloid" (Starowicz *et al.*, 2007). In contrast, 2-AG is unable to bind

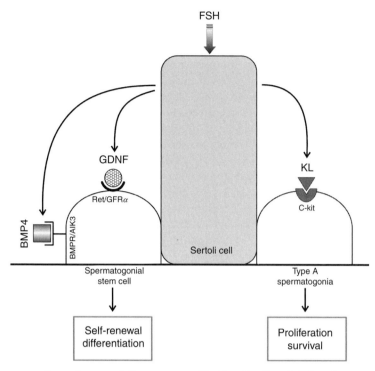

Figure 10.2 Paracrine growth factors secreted by Sertoli cells and their action on mitotic germ cells development. Sertoli cells produce growth factors, like GDNF that regulates spermatogonial stem cell self-renewal, and is important for the maintenance of the stem cell pool; BMP4 that stimulates proliferation and differentiation of the undifferentiated germ cells; and KL that regulates proliferation, survival, and meiotic entry of differentiating type-A spermatogonia that express c-kit receptor. See text for details.

to and activate TRPV1 (Szallasi and Blumberg, 1999). It should be stressed that, while the AEA–CBR interaction occurs at an extracellular binding site, the AEA–TRPV1 interaction occurs at a cytosolic-binding site (De Petrocellis et al., 2001; Jordt and Julius, 2002). In addition, recent evidence demonstrates that AEA, by binding to TRPV1, can regulate the endogenous tone and biological activity of 2-AG (Maccarrone et al., 2008).

Biochemical studies have revealed that AEA is produced by a transacylase-phosphodiesterase-mediated synthesis, starting from the precursor N-arachidonoyl-phosphatidylethanolamine (NArPE), through the action of N-acyltransferase (NAT) followed by the N-arachidonoyl-phosphatidylethanolamine (NAPE) hydrolysis catalyzed by a specific phospholipase D (NAPE-PLD) (Okamoto et al., 2004). More recently, additional pathways through which AEA can be generated have been described (Simon and Cravatt, 2008).

2-AG acts as a potent and full agonist for both CB1R and CB2R and, like AEA, it is not stored in intracellular compartments, but is produced on demand and is released from neuronal membranes through multiple biosynthetic pathways (Ligresti et al., 2005).

The major biosynthetic route of 2-AG provides for rapid hydrolysis of inositol phospholipids by a specific phospholipase C (PLC); this enzyme generates diacylglycerol (DAG), which is subsequentially converted to 2-AG by a sn-1-DAG lipase (Bisogno et al., 2003; Stella et al., 1997).

The biological effects of AEA and 2-AG depend on their lifespan in the extracellular space, which is limited by a rapid transport through the membrane. Both compounds have been proposed to be taken up by cells through specific carriers that might be represented by the same entity: the endocannabinoid membrane transporter (EMT) (Beltramo and Piomelli, 2000). However, there is as yet controversy regarding the identity of EMT and the actual mechanism by which AEA and 2-AG are transported (Mechoulam and Deutsch, 2005). At any rate, once inside the cells endocannabinoids can be metabolized by independent routes. AEA is a substrate for fatty acid amide hydrolase (FAAH) that breaks the amide bond and releases arachidonic acid and ethanolamine (Fezza et al., 2008; McKinney and Cravatt, 2005). 2-AG is degraded to AA and glycerol mainly by a specific monoacylglycerol lipase (MAGL). The latter enzyme has been characterized in rat (Dinh et al., 2002) and human brain (Ho et al., 2002) and, unlike FAAH that is bound to intracellular membranes (Oddi et al., 2005), it is localized in the cytosol (Dinh et al., 2002). Taken together, AEA and 2-AG, their congeners and metabolic enzymes, their purported transporters, and molecular targets form the "endocannabinoid system (ECS)."

IV. THE ECS IN SERTOLI CELLS

AEA has been identified in human reproductive fluids, as well as in invertebrate (Schuel et al., 2002a), amphibian (Cobellis et al., 2006), and mammalian sperm (Maccarrone et al., 2005; Ricci et al., 2007; Rossato et al., 2005; Schuel et al., 2002b). Comprehensive reviews have been recently published on the role of ECS in male reproduction (Battista et al., 2008a; Schuel and Burkman, 2005; Wang et al., 2006a).

It should be recalled that at all stages of differentiation the spermatogenic cells are in close contact with Sertoli cells, which are thought to provide structural and metabolic support to the developing sperm.

In recent years, it has been shown that Sertoli cells, obtained from 4- to 24-day-old mice, possess the biochemical machinery to synthesize, transport, degrade, and bind both AEA (Maccarrone et al., 2003a) and 2-AG (Rossi et al., 2007). The ECS elements that have been demonstrated in Sertoli cells are summarized in Table 10.2. In particular, Sertoli cells express on their surface functional CB2R, which is rather distinctive because the

Table 10.2 Elements of the ECS that have been demonstrated in mouse Sertoli cells

Member	Description	Function
AEA	Prototype member of fatty acid amides	Bioactive lipid that acts at cannabinoid and noncannabinoid receptors in the central nervous system and in the periphery
EMT	Endocannabinoid membrane transporter	So far putative entity responsible for the transport of AEA and 2-AG
NAT NAPE-PLD	Biosynthetic enzymes	Main responsible for the biosynthesis of AEA
DAGL	Biosynthetic enzyme	Main responsible for the biosynthesis of 2-AG
FAAH	Hydrolytic enzyme	Main responsible for AEA degradation
MAGL	Hydrolytic enzyme	Main responsible for 2-AG degradation
CB2R	Cannabinoid receptor	Main target of AEA and 2-AG
TRPV1	Vanilloid receptor	Target of AEA

See text for details.

most common type of CB receptors in reproductive cells is CB1R (for a recent review, see Maccarrone, 2007). The levels of CB2R in Sertoli cells remain constant for at least 16 days (Maccarrone et al., 2003a). On the other hand, FAAH activity and the uptake of AEA through EMT were found to decrease in an age-dependent manner, and in the case of FAAH this was due to downregulation of gene expression (Maccarrone et al., 2003a). Moreover, the transport of AEA across the plasma membrane was increased by nitric oxide (NO) donors, extending previous data on human peripheral cells (Maccarrone et al., 2003a). Since NO has a potential role in both normal spermatogenesis and inflammation-mediated male infertility (Rossi et al., 2007, and references therein), its effect on the upregulation of EMT in Sertoli cells might deserve further *in vivo* investigation.

FSH has been shown to increase FAAH activity in mouse Sertoli cells, thus reducing endogenous level and proapoptotic potential of AEA (Maccarrone et al., 2003a). Figure 10.3 recapitulates these findings. Later on, it was demonstrated that FAAH is the only target of FSH among the components of the ECS, and in fact the AEA-synthesizing enzymes NAT and NAPE-PLD, and the AEA-binding receptors CB2R and TRPV1 were not modulated by physiological doses of FSH (Rossi et al., 2007). The

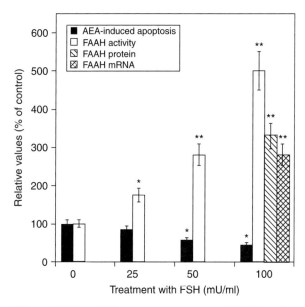

Figure 10.3 Effect of FSH on AEA-induced apoptosis and FAAH activity and expression in Sertoli cells. Apoptosis was evaluated as DNA fragmentation induced by 1 μM AEA in 16-day-old Sertoli cells treated for 24 h, alone or in the presence of increasing concentrations of FSH. FAAH activity and expression were determined under the same experimental conditions. $*p < 0.05$ and $**p < 0.01$ versus controls. Original data on apoptosis and FAAH activity were from Maccarrone et al. (2003a); those on FAAH expression were from Rossi et al. (2007).

analysis of the effect of FSH was further extended to the proteins that synthesize, degrade, and transport the other major endocannabinoid 2-AG. Neither DAGL, MAGL, or 2-AG transporter were found to be under control of FSH (Rossi et al., 2007). Next, it was also checked whether two key pathways triggered by FSH, like those dependent on PKA and ARO (Lecureuil et al., 2005; Simoni et al., 1997; Walker and Cheng, 2005), were implicated in the regulation of FAAH.

V. Regulation of FAAH by FSH in Sertoli Cells

In males, the expression of FSHR is limited to the testicular Sertoli cells (Rannikki et al., 1995), and in fact the primary role of FSH in spermatogenesis is stimulation of Sertoli cell proliferation during prepubertal development (Heckert and Griswold, 2002). Humans who lack functional FSHR develop smaller testis than do normal males, in agreement with the reduction in Sertoli cell number observed in the testis of mice lacking FSHR (Dierich et al., 1998). These men also exhibit disturbed spermatogenesis (oligozoospermia and teratozoospermia), although they remain fertile (Tapanainen et al., 1997). Previous studies have also demonstrated the acute effects of THC on the secretion of FSH at the hypothalamo-pituitary levels, with an inhibitory effect on gonadal function (Ayalon et al., 1977; Wenger et al., 1987). FSH also causes Ca^{2+} influx into Sertoli cells that is mediated by cAMP-dependent and perhaps PKA-dependent modifications of surface Ca^{2+} channels. In addition, the initiation of PI-3 kinase signaling by FSH stimulates *aro* expression, acting at the transcriptional level (McDonald et al., 2006). Testosterone and other androgens can undergo peripheral conversion to estrogen and its derivates through the action of ARO; for instance, the oral treatment with testolactone and anastrozole, that are ARO inhibitors, prevents the conversion of testosterone to estradiol and, thus, blocks the inhibitory effects of estrogen on spermatogenesis (Kumar et al., 2006).

Recently, both PKA- and ARO-mediated signaling pathways have been shown to be involved in the modulation of FAAH by FSH. In particular, using specific enzyme inhibitors it was demonstrated that the PKA-dependent pathway acts on the activity of FAAH, and only weakly on its expression (Rossi et al., 2007). Since FSH does not phosphorylate FAAH directly (Maccarrone et al., 2003a), it can be proposed that it triggers phosphorylation of transcription factors and, to a larger extent, of accessory proteins which, in turn, activate the enzyme. On the other hand, the ARO-dependent pathway increases FAAH expression at the transcriptional and translational levels (Rossi et al., 2007). Indeed, two potential estrogen-responsive elements (ERE) have been identified in the mouse *faah* promoter (Puffenbarger et al., 2001). In line with this, recent transient expression experiments performed in our laboratory (Fig. 10.4A) have confirmed that E_2 is indeed able to enhance *faah* expression at the transcriptional level (Fig. 10.4B). Taken together, available evidence on the overall regulation of FAAH by FSH in Sertoli cells can be depicted as shown in Fig. 10.5.

VI. FAAH Is an Integrator of Fertility Signals

The above data emphasize a main role for FAAH in the interaction between ECS and sex hormones during male reproduction, and underline its value as promising target to treat infertility problems. This concept can be further extended to other essential fertility signals, like progesterone. This hormone upregulates *faah* gene expression in human lymphocytes by binding to the *faah* promoter through the Ikaros transcription factor (Maccarrone et al., 2003b). Such an effect is synergistic with that of leptin, the 16-kDa nonglycosylated product of the *obese* gene that triggers STAT-3 (signal transducer and activator of transcription-3), which in turn binds to a

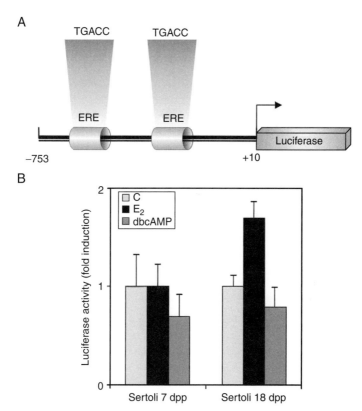

Figure 10.4 Activity of *faah* promoter in primary mouse Sertoli cells. (A) Schematic representation of the proximal promoter region of *faah* gene from nucleotide −753 to nucleotide +10 (+1 is the start of the open reading frame), cloned upstream the luciferase gene reporter in pGL3 basic vector. Two potential ERE sites are indicated. (B) Transient transfections of Sertoli cells from mice at different ages were performed with *faah* promoter construct. Data represent fold induction by E_2 or dibutyryl-cAMP (dbcAMP), compared to unstimulated cells (mean ± SEM of two independent transfections, each in duplicate). Treatment with E_2 increases reporter activity driven by *faah* promoter in Sertoli cells at 18 days *postpartum* (dpp), but not at 7 dpp.

CRE-like site in the promoter region of *faah* (Maccarrone et al., 2003b). Furthermore, FAAH activity is stimulated by profertility signals like T helper 2 (Th2) cytokines, whereas it is inhibited by antifertility Th1 cytokines, overall representing the key regulator of the interplay between endocannabinoids, steroids, and cytokines in the control of human reproduction (for a recent review, see Battista et al., 2008b). Overall, FAAH can be considered a molecular integrator of fertility signals, which acts as a "guardian angel" (Maccarrone and Finazzi-Agrò, 2004) or a "gate-keeper" (Wang et al., 2006b) of mammalian reproduction. In this context, it should

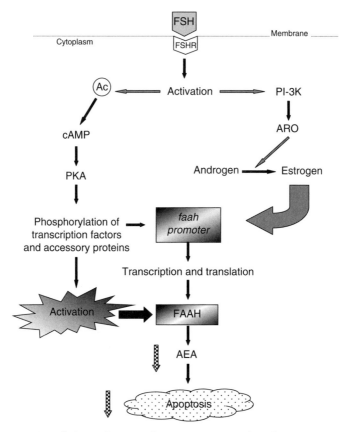

Figure 10.5 Regulation of FAAH by FSH. In Sertoli cells, FSH activates the AEA-degrading enzyme FAAH, through FSHR-dependent PKA and ARO pathways. The scheme suggests that FSHR triggers activation of PKA, which phosphorylates transcription factors and, more importantly, accessory proteins able to behave as enzyme activators. Independently of this pathway, FSH can trigger an ARO-dependent route, which enhances *faah* gene expression by acting directly at the promoter level. The final outcome is a reduction of AEA-induced apoptosis of Sertoli cells.

be recalled that analysis of the *faah* promoter has revealed differences among species (Maccarrone *et al.*, 2003b; Puffenbarger, 2005). Here, a further inspection into the promoter of human and mouse *faah* genes by the Clustal 2.0.3 multiple sequence alignment program has revealed that indeed FAAH can be regulated by a plethora of transcription factors, many of which are strictly related to fertility. Table 10.3 summarizes the results of this *in silico* analysis.

Table 10.3 Transcription factors that can modulate *faah* gene expression

	Transcription factor	General protein information	Process involved according to IEA
Mouse	HSF2 (heat shock factor 2)	1. HSF2 plays an important role in FGF-2-induced RANKL expression in stromal/preosteoblast cells 2. HSF2 deficiency has pleiotropic effects on gene expression during osteoblast differentiation 3. HSF2, most probably because its physiological roles are integrated into a redundant network of gene regulation and function, is indispensable for normal development, fertility, and postnatal psychomotor function 4. Brain abnormalities, defective meiotic chromosome synapsis and female subfertility have been reported in HSF2 null mice 5. Identify p35 as the first target gene for HSF2 in cortical development	1. Regulation of transcription 2. Response to stress 3. Spermatogenesis
	ER (estrogen receptor)	1. Estrogen, acting via ERα, regulates serum gonadotropin levels and pituitary gonadotropin subunit expression. However, the cellular pathways mediating this regulation are unknown. ERα signals through classical estrogen response	1. Regulation of transcription 2. Regulation of reproduction

(continued)

Table 10.3 (continued)

Transcription factor	General protein information	Process involved according to IEA
	element (ERE)-dependent genomic as well as nonclassical ERE-independent genomic and nongenomic pathways 2. Studies examined the relative roles of ERE-dependent and -independent estrogen signaling in estrogen regulation of LH, FSH, prolactin and activin/inhibin subunit gene expression, pituitary LH and FSH protein content, and serum FSH levels 3. ERE-independent signaling was not sufficient for estrogen to induce pituitary prolactin mRNA or suppress pituitary LHβ mRNA, LH content, or serum FSH in estrogen-treated ovariectomized mice. ERE-dependent and ERE-independent ERα pathways may distinctly regulate steps involved in the synthesis and secretion of FSH	
NFκB (nuclear factor of κ light polypeptide	1. Flavonoids protected against IL-1β/IFNγ-induced pancreatic β-cell damage by inhibiting activation of NFκB p50/p65 subunits	1. Protection against apoptosis 2. Negative regulation of interleukin-12 biosynthetic process

gene enhancer in B cells)

2. Transcription factors, NFκB, and iNOS have roles in ethylene glycol-induced crystal depositions in renal tubules
3. Data suggest that cell survival pathways other than those involving NFκB-inducible gene expression or other well-known pathways appear to be involved in protecting the dedifferentiated hepatoma variant cells from LPS-mediated apoptosis
4. Expression of p53, nuclear factor-κB, COX-1, and COX-2 was increased in the irradiated intestinal sections
5. Stimulation of cyclic adenosine monophosphate–protein kinase A signaling pathway by CCK-8 through CCK-1R and CCK-2R inhibits the LPS-induced activation of p38 kinase and NFκB to block the IL-1β production in rat pulmonary interstitial macrophages
6. Data suggest that γ-hexachlorocyclohexane-induced liver oxidative stress triggers DNA-binding activity of NFκB, with a consequent increase in expression of NFκB-dependent genes for tumor necrosis factor-α and interleukin-1α

3. Positive regulation of transcription from RNA polymerase II promoter
4. Regulation of transcription
5. Response to copper ion
6. Response to oxidative stress
7. Regulation of signal transduction

(*continued*)

Table 10.3 (continued)

Transcription factor		General protein information	Process involved according to IEA
		7. Data show that nuclear factor-κB is an activator of androgen receptor gene transcription in Sertoli cells	
	S8 (Prrx2 paired related homeobox 2)	1. Three-dimensional morphological reconstructions demonstrated that Prx2(−/−) embryos had malformations, including cuspal changes and ectopic epithelial projections 2. Results show that Prx I and II are constitutively expressed in the testis and their expression levels are decreased to some extent as the testis develops	1. Multicellular organismal development 2. Regulation of transcription 3. Formation of transcription factor complex
Human	Ikaros (IKAROS family zinc finger)	1. Showed Aiolos overexpression in primary lymphoma tissue 2. Role in unconventional potentiation of gene expression 3. Ikaros proteins function in myeloid as well as lymphoid differentiation, and specific Ikaros isoforms may play a role in regulating lineage commitment decisions 4. Ikaros has a role in progesterone activation of fatty acid amide hydrolase in human T lymphocytes	1. Mesoderm development 2. Regulation of transcription

	5. Ikaros is involved in migration and invasion of extravillous trophoblasts in early placentation	
SRY (sex-determining region Y)	1. Acetylation and deacetylation of SRY may be important mechanisms for regulating SRY activity during mammalian sex determination	1. Cell differentiation
2. Male sex determination
3. Regulation of transcription
	2. Data suggest the involvement of SRY gene in sex reversal, which further supports the relationship between SRY alterations, gonadal dysgenesis, and/or primary infertility	
	3. Familial mutation in the testis-determining gene SRY shared by an XY female and her normal father	
	4. Defective importin-β recognition and nuclear import of the sex-determining factor SRY are associated with XY sex-reversing mutations	
	5. The SRY gene encodes for a protein in the high mobility group that binds to DNA in the nucleus and it regulates the transcription of other genes necessary for testis determination by acting as a repressor or activator of this process	

(continued)

Table 10.3 (continued)

Transcription factor		General protein information	Process involved according to IEA
Human/ mouse	GATA-X,1,2 (GATA-binding protein)	1. Differential GATA-1 and GATA-2 factor stability is an important determinant of chromatin target site occupancy and therefore the establishment of genetic networks that control hematopoiesis 2. GATA-1 may represent a novel MAPK substrate that plays an essential role in a cytokine-mediated antiapoptotic response 3. GATA-1 testis activation region is essential for Sertoli cell-specific expression of GATA-1 gene 4. Investigation of IL-13 promoter region finds the GATA-1-binding site indispensable for IL-13 transcription in mast cells 5. Gonadotropins via cAMP negatively regulate the mRNA and protein levels of GATA-1; the inhibitory effect on GATA-1 gene expression was specific to testicular cells 6. FOG-2, in addition to GATA-4, has a role in early gonadal development and sexual differentiation, and FOG-1 at later fetal stages, while GATA-1 executes its action postnatally	1. Regulation of transcription

| AP1 (FOS v-fos FBJ murine osteosarcoma viral oncogene homolog) | 1. The Fos gene family consists of four members: FOS, FOSB, FOSL1, and FOSL2. These genes encode leucine zipper proteins that can dimerize with proteins of the JUN family, thereby forming the transcription factor complex AP1. As such, the FOS proteins have been implicated as regulators of cell proliferation, differentiation, and transformation. In some cases, expression of the FOS gene has also been associated with apoptotic cell death
2. Studies suggest that the cooperative interaction of the estrogen receptor with Fos and Jun proteins helps confer estrogen responsiveness to the endogenous progesterone receptor gene
3. AP1 and JNK have roles in reactive oxygen species activation in tobacco-induced mucin production in lung cells | 1. DNA methylation
2. Inflammatory response
3. Nervous system development;
4. Regulation of transcription from RNA polymerase II promoter |

Fertility-related factors and functions are shadowed. IEA (inferred from electronic annotation) identifies computationally assigned processes by searching through Gene Ontology databases.

VII. Conclusions

A role for the endogenous cannabinoid system in several aspects of human (patho)physiology has been proposed through the activation of cannabinoid and vanilloid receptors, and/or via nonreceptor-mediated actions. In the case of AEA, the role of FAAH in controlling its cellular activity seems more critical than that of NAT, NAPE-PLD, or EMT. Here, the activity of FSH, the biological relevance of Sertoli cells, and the main features of the ECS have been briefly described, to put in a better perspective the relevance of FAAH regulation by FSH for male reproduction. In particular, it is suggested that FAAH controls hormone level of AEA, by acting as a molecular integrator of fertility signals. In this context, since low FAAH levels in peripheral lymphocytes (Maccarrone *et al.*, 2000), and hence high levels of AEA in blood of pregnant women (Habayeb *et al.*, 2008), correlate with pregnancy failure, it can be suggested that FAAH is a critical "sensor" of reproductive abnormalities also on the female side of human gestation. Therefore, it can be proposed that substances able to enhance FAAH activity might become useful therapeutic tools for the treatment of human infertility. As yet, FSH is the only hormone known to upregulate FAAH. Two other natural compounds, like an as-yet elusive lipid substance released from mouse blastocysts (Maccarrone *et al.*, 2004), and the *Cannabis* component cannabidiol (Massi *et al.*, 2008), are known to behave as FAAH activators, whereas synthetic compounds with this activity are not available. This calls for attention to "the other side of the medal," because in the last few years most investigators have put efforts on the synthesis of inhibitors, rather than activators, of FAAH as novel therapeutic agents for the treatment of pain, neurodegenerative disorders, cancer, and anxiety (Di Marzo, 2008). If not as therapeutics *per se*, FAAH activators could be used together with other drugs to lower the doses or to shorten the treatment necessary *in vivo* to observe an effect, and hence minimize the possible side effects of substances used to treat infertility. On a final note, it should be recalled that any future development of selective FAAH modulators as therapeutics must take into account that:

(i) FAAH regulates the levels of several endogenous endocannabinoid (-like) compounds, whose functions are not yet understood. This leaves open the question of the biological consequences of their reduced/increased degradation upon treatment with FAAH activators/inhibitors.
(ii) Other enzymes have been recently identified that catalyze the synthesis (Simon and Cravatt, 2008) and hydrolysis of AEA (Wei *et al.*, 2006), and the synthesis and hydrolysis of 2-AG (Ligresti *et al.*, 2005). It remains to be elucidated how these pathways may contribute to the overall endocannabinoid tone.

(iii) Therapeutic benefits derived from curbing a hyperactive ECS could be obtained by developing not only FAAH activators, but also inhibitors of endocannabinoid biosynthesis, or selective CBR antagonists.

In conclusion, we have reviewed the evidence that FAAH is a distinct target of FSH within the ECS. Of particular interest seems the fact that FSH activates FAAH, both through phosphorylation of accessory proteins and through binding of estrogen to an ERE site in the *faah* promoter. This is noteworthy, because only a few compounds are known to activate FAAH; it can be anticipated that these substances, like the widely studied FAAH inhibitors, might become useful therapeutics for the treatment of human diseases caused by a high endocannabinoid tone.

ACKNOWLEDGMENTS

We wish to thank Dr. Nicoletta Pasquariello for the *in silico* analysis of *faah* promoter. This work was supported by Ministero dell'Università e della Ricerca (COFIN 2006), and by Fondazione TERCAS (Research Program 2005) to MM.

REFERENCES

Ayalon, D., Nir, I., Cordova, T., Bauminger, S., Puder, M., Naor, Z., Kashi, R., Zor, U., Harell, A., and Lindner, H. R. (1977). Acute effect of delta1-tetrahydrocannabinol on the hypothalamo-pituitary–ovarian axis in the rat. *Neuroendocrinology* **23,** 31–42.

Battista, N., Rapino, C., Di Tommaso, M., Bari, M., Pasquariello, N., and Maccarrone, M. (2008a). Regulation of male fertility by the endocannabinoid system. *Mol. Cell. Endocrinol.* **286,** S17–S23.

Battista, N., Pasquariello, N., Di Tommaso, M., and Maccarrone, M. (2008b). Interplay between endocannabinoids, steroids and cytokines in the control of human reproduction. *J. Neuroendocrinol.* **20,** 82–89.

Beltramo, M., and Piomelli, D. (2000). Carrier-mediated transport and enzymatic hydrolysis of the endogenous cannabinoid 2-arachidonylglycerol. *Neuroreport* **11,** 1231–1235.

Bisogno, T., Howell, F., Williams, G., Minassi, A., Cascio, M. G., Ligresti, A., Matias, I., Schiano-Moriello, A., Paul, P., Williams, E. J., Gangadharan, U., Hobbs, C., *et al.* (2003). Cloning of the first *sn*1-DAG lipases points to the spatial and temporal regulation of endocannabinoid signaling in the brain. *J. Cell Biol.* **163,** 463–468.

Carreau, S., Delalande, C., Silandre, D., Bourguiba, S., and Lambard, S. (2006). Aromatase and estrogen receptors in male reproduction. *Mol. Cell. Endocrinol.* **246,** 65–68.

Chaudhary, J., and Skinner, M. K. (2001). Role of the transcriptional coactivator CBP/p300 in linking basic helix-loop-helix and CREB responses for follicle-stimulating hormone-mediated activation of the transferrin promoter in Sertoli cells. *Biol. Reprod.* **65,** 568–574.

Chaudhary, J., Whaley, P. D., Cupp, A., and Skinner, M. K. (1996). Transcriptional regulation of Sertoli cell differentiation by follicle-stimulating hormone at the level of the c-fos and transferrin promoters. *Biol. Reprod.* **54,** 692–699.

Chen, C., Ouyang, W., Grigura, V., Zhou, Q., Carnes, K., Lim, H., Zhao, G. Q., Arber, S., Kurpios, N., Murphy, T. L., Cheng, A. M., Hassell, J. A., *et al.* (2005). ERM is required for transcriptional control of the spermatogonial stem cell niche. *Nature* **436,** 1030–1034.

Cobellis, G., Cacciola, G., Scarpa, D., Meccariello, R., Chianese, R., Franzoni, M. F., Mackie, K., Pierantoni, R., and Fasano, S. (2006). Endocannabinoid system in frog and rodent testis: Type-1 cannabinoid receptor and fatty acid amide hydrolase activity in male germ cells. *Biol. Reprod.* **75,** 82–89.

Crepieux, P., Marion, S., Martinat, N., Fafeur, V., Vern, Y. L., Kerboeuf, D., Guillou, F., and Reiter, E. (2001). The ERK-dependent signalling is stage-specifically modulated by FSH, during primary Sertoli cell maturation. *Oncogene* **20,** 4696–4709.

Dash, P. K., Karl, K. A., Colicos, M. A., Prywes, R., and Kandel, E. R. (1991). cAMP response element-binding protein is activated by Ca^{2+}/calmodulin—As well as cAMP-dependent protein kinase. *Proc. Natl. Acad. Sci. USA* **88,** 5061–5065.

De Petrocellis, L., Bisogno, T., Maccarrone, M., Davis, J. B., Finazzi-Agrò, A., and Di Marzo, V. (2001). The activity of anandamide at vanilloid VR1 receptors requires facilitated transport across the cell membrane and is limited by intracellular metabolism. *J. Biol. Chem.* **276,** 12856–12863.

Dierich, A., Sairam, M. R., Monaco, L., Fimia, G. M., Gansmuller, A., LeMeur, M., and Sassone-Corsi, P. (1998). Impairing follicle-stimulating hormone (FSH) signaling *in vivo*: Targeted disruption of the FSH receptor leads to aberrant gametogenesis and hormonal imbalance. *Proc. Natl. Acad. Sci. USA* **95,** 13612–13617.

Di Marzo, V. (2008). Targeting the endocannabinoid system: To enhance or reduce? *Nat. Rev. Drug Discov.* **7,** 438–455.

Dinh, T. P., Carpenter, D., Leslie, F. M., Freund, T. F., Katona, I., Sensi, S. L., Kathuria, S., and Piomelli, D. (2002). Brain monoglyceride lipase participating in endocannabinoid inactivation. *Proc. Natl. Acad. Sci. USA* **99,** 10819–10824.

Dolci, S., Pellegrini, M., Di Agostino, S., Geremia, R., and Rossi, P. (2001). Signaling through extracellular signal-regulated kinase is required for spermatogonial proliferative response to stem cell factor. *J. Biol. Chem.* **276,** 40225–40233.

Fezza, F., De Simone, C., Amadio, D., and Maccarrone, M. (2008). Fatty acid amide hydrolase: A gate-keeper of the endocannabinoid system. *Subcell. Biochem.* **49,** 101–132.

Foulkes, N. S., Borrelli, E., and Sassone-Corsi, P. (1991). CREM gene: Use of alternative DNA-binding domains generates multiple antagonists of cAMP-induced transcription. *Cell* **64,** 739–749.

Gonzalez-Robayna, I. J., Falender, A. E., Ochsner, S., Firestone, G. L., and Richards, J. S. (2000). Follicle-stimulating hormone (FSH) stimulates phosphorylation and activation of protein kinase B (PKB/Akt) and serum and glucocorticoid-induced kinase (Sgk): Evidence for A kinase-independent signaling by FSH in granulosa cells. *Mol. Endocrinol.* **14,** 1283–1300.

Grimaldi, P., Piscitelli, D., Albanesi, C., Blasi, F., Geremia, R., and Rossi, P. (1993). Identification of 3′,5′-cyclic adenosine monophosphate-inducible nuclear factors binding to the human urokinase promoter in mouse Sertoli cells. *Mol. Endocrinol.* **7,** 1217–1225.

Grimaldi, P., Capolunghi, F., Geremia, R., and Rossi, P. (2003). Cyclic adenosine monophosphate (cAMP) stimulation of the kit ligand promoter in Sertoli cells requires an Sp1-binding region, a canonical TATA box, and a cAMP-induced factor binding to an immediately downstream GC-rich element. *Biol. Reprod.* **69,** 1979–1988.

Griswold, M. D. (1998). The central role of Sertoli cells in spermatogenesis. *Semin. Cell Dev. Biol.* **9,** 411–416.

Gronning, L. M., Dahle, M. K., Tasken, K. A., Enerback, S., Hedin, L., Tasken, K., and Knutsen, H. K. (1999). Isoform-specific regulation of the CCAAT/enhancer-binding protein family of transcription factors by 3′,5′-cyclic adenosine monophosphate in Sertoli cells. *Endocrinology* **140,** 835–843.

Habayeb, M. H., Taylor, A. H., Finney, M., Evans, M. D., and Konje, J. C. (2008). Plasma anandamide concentration and pregnancy outcome in women with threatened miscarriage. *JAMA* **299,** 1135–1136.

Heckert, L. L., and Griswold, M. D. (2002). The expression of the follicle-stimulating hormone receptor in spermatogenesis. *Recent Prog. Horm. Res.* **57,** 129–148.

Ho, S. Y., Delgado, L., and Storch, J. (2002). Monoacylglycerol metabolism in human intestinal Caco-2 cells: Evidence for metabolic compartmentation and hydrolysis. *J. Biol. Chem.* **277,** 1816–1823.

Howlett, A. C., Barth, F., Bonner, T. I., Cabral, G., Casellas, P., Devane, W. A., Felder, C. C., Herkenham, M., Mackie, K., Martin, B. R., Mechoulam, R., and Pertwee, R. G. (2002). International Union of Pharmacology. XXVII. Classification of cannabinoid receptors. *Pharmacol. Rev.* **54,** 161–202.

Howlett, A. C., Breivogel, C. S., Childers, S. R., Deadwyler, S. A., Hampson, R. E., and Porrino, L. J. (2004). Cannabinoid physiology and pharmacology: 30 years of progress. *Neuropharmacology* **47,** 345–358.

Ito, M., Park, Y., Weck, J., Mayo, K. E., and Jameson, J. L. (2000). Synergistic activation of the inhibin alpha-promoter by steroidogenic factor-1 and cyclic adenosine $3',5'$-monophosphate. *Mol. Endocrinol.* **14,** 66–81.

Jannini, E. A., Ulisse, S., Cecconi, S., Cironi, L., Colonna, R., D'Armiento, M., Santoni, A., and Cifone, M. G. (1994). Follicle-stimulating hormone-induced phospholipase A2 activity and eicosanoid generation in rat Sertoli cells. *Biol. Reprod.* **51,** 140–145.

Jordt, S. E., and Julius, D. (2002). Molecular basis for species-specific sensitivity to "hot" chili peppers. *Cell* **108,** 421–430.

Khan, S. A., Ndjountche, L., Pratchard, L., Spicer, L. J., and Davis, J. S. (2002). Follicle stimulating hormone amplifies insulin-like growth factor I-mediated activation of AKT/protein kinase B signaling in immature rat Sertoli cells. *Endocrinology* **143,** 2259–2267.

Krishnamurthy, H., Babu, P. S., Morales, C. R., and Sairam, M. R. (2001). Delay in sexual maturity of the follicle-stimulating hormone receptor knockout male mouse. *Biol. Reprod.* **65,** 522–531.

Kumar, R., Gautam, G., and Gupta, N. P. (2006). Drug therapy for idiopathic male infertility: Rationale versus evidence. *J. Urol.* **176,** 1307–1312.

Lai, T. H., Lin, Y. F., Wu, F. C., and Tsai, Y. H. (2008). Follicle-stimulating hormone-induced Galphah/phospholipase C-delta1 signaling mediating a noncapacitative Ca^{2+} influx through T-type Ca^{2+} channels in rat Sertoli cells. *Endocrinology* **149,** 1031–1037.

Lauckner, J. E., Jensen, J. B., Chen, H. Y., Lu, H. C., Hille, B., and Mackie, K. (2008). GPR55 is a cannabinoid receptor that increases intracellular calcium and inhibits M current. *Proc. Natl. Acad. Sci. USA* **105,** 2699–2704.

Lecureuil, C., Tesseraud, S., Kara, E., Martinat, N., Sow, A., Fontaine, I., Gauthier, C., Reiter, E., Guillou, F., and Crepieux, P. (2005). Follicle-stimulating hormone activates p70 ribosomal protein S6 kinase by protein kinase A-mediated dephosphorylation of Thr 421/Ser 424 in primary Sertoli cells. *Mol. Endocrinol.* **19,** 1812–1820.

Ligresti, A., Cascio, M. G., and Di Marzo, V. (2005). Endocannabinoid metabolic pathways and enzymes. *Curr. Drug Targets CNS Neurol. Disord.* **4,** 615–623.

Lin, Y. F., Tseng, M. J., Hsu, H. L., Wu, Y. W., Lee, Y. H., and Tsai, Y. H. (2006). A novel follicle-stimulating hormone-induced G alpha h/phospholipase C-delta1 signaling pathway mediating rat Sertoli cell Ca^{2+}-influx. *Mol. Endocrinol.* **20,** 2514–2527.

Lukas-Croisier, C., Lasala, C., Nicaud, J., Bedecarras, P., Kumar, T. R., Dutertre, M., Matzuk, M. M., Picard, J. Y., Josso, N., and Rey, R. (2003). Follicle-stimulating hormone increases testicular Anti-Mullerian hormone (AMH) production through Sertoli cell proliferation and a nonclassical cyclic adenosine $5'$-monophosphate-mediated activation of the AMH Gene. *Mol. Endocrinol.* **17,** 550–561.

Maccarrone, M. (2007). CB_2 receptors in reproduction. *Br. J. Pharmacol.* **153,** 189–198.

Maccarrone, M., and Finazzi-Agrò, A. (2004). Anandamide hydrolase: A guardian angel of human reproduction? *Trends Pharmacol. Sci.* **25,** 353–357.

Maccarrone, M., Valensise, H., Bari, M., Lazzarin, N., Romanini, C., and Finazzi-Agrò, A. (2000). Relation between decreased anandamide hydrolase concentrations in human lymphocytes and miscarriage. *Lancet* **355**, 1326–1329.
Maccarrone, M., Cecconi, S., Rossi, G., Battista, N., Pauselli, R., and Finazzi-Agrò, A. (2003a). Anandamide activity and degradation are regulated by early postnatal aging and follicle-stimulating hormone in mouse Sertoli cells. *Endocrinology* **144**, 20–28.
Maccarrone, M., Bari, M., Di Rienzo, M., Finazzi-Agrò, A., and Rossi, A. (2003b). Progesterone activates fatty acid amide hydrolase (FAAH) promoter in human T lymphocytes through the transcription factor Ikaros. Evidence for a synergistic effect of leptin. *J. Biol. Chem.* **278**, 32726–32732.
Maccarrone, M., De Felici, M., Klinger, F. G., Battista, N., Fezza, F., Dainese, E., Siracusa, G., and Finazzi-Agrò, A. (2004). Mouse blastocysts release a lipid which activates anandamide hydrolase in intact uterus. *Mol. Hum. Reprod.* **10**, 215–221.
Maccarrone, M., Barboni, B., Paradisi, A., Bernabò, N., Gasperi, V., Pistilli, M. G., Fezza, F., Lucidi, P., and Mattioli, M. (2005). Characterization of the endocannabinoid system in boar spermatozoa and implications for sperm capacitation and acrosome reaction. *J. Cell Sci.* **118**, 4393–4404.
Maccarrone, M., Rossi, S., Bari, M., De Chiara, V., Fezza, F., Musella, A., Gasperi, V., Prosperetti, C., Bernardi, G., Finazzi-Agrò, A., Cravatt, B. F., and Centonze, D. (2008). Anandamide inhibits metabolism and physiological actions of 2-arachidonoylglycerol in the striatum. *Nat. Neurosci.* **11**, 152–159.
Massi, P., Valenti, M., Vaccani, A., Gasperi, V., Perletti, G., Marras, E., Fezza, F., Parolaro, D., and Maccarrone, M. (2008). 5-Lipoxygenase and anandamide hydrolase (FAAH) mediate the antitumor activity of cannabidiol, a non-psychoactive cannabinoid. *J. Neurochem.* **104**, 1091–1100.
McDonald, C., Millena, A. C., Reddy, S., Finlay, S., Vizcarra, J., Khan, S. A., and Davis, J. S. (2006). Follicle-stimulating hormone-induced aromatase in immature rat Sertoli cells requires an active phosphatidylinositol 3-kinase pathway and is inhibited via the mitogen-activated protein kinase signalling pathway. *Mol. Endocrinol.* **20**, 608–618.
McKinney, M. K., and Cravatt, B. F. (2005). Structure and function of fatty acid amide hydrolase. *Annu. Rev. Biochem.* **74**, 411–432.
Mechoulam, R., and Deutsch, D. G. (2005). Toward an anandamide transporter. *Proc. Natl. Acad. Sci. USA* **102**, 17541–17542.
Meng, X., Lindahl, M., Hyvonen, M. E., Parvinen, M., de Rooij, D. G., Hess, M. W., Raatikainen-Ahokas, A., Sainio, K., Rauvala, H., Lakso, M., Pichel, J. G., and Westphal, H. (2000). Regulation of cell fate decision of undifferentiated spermatogonia by GDNF. *Science* **287**, 1489–1493.
Meroni, S. B., Riera, M. F., Pellizzari, E. H., and Cigorraga, S. B. (2002). Regulation of rat Sertoli cell function by FSH: Possible role of phosphatidylinositol 3-kinase/protein kinase B pathway. *J. Endocrinol.* **174**, 195–204.
Meroni, S. B., Riera, M. F., Pellizzari, E. H., Schteingart, H. F., and Cigorraga, S. B. (2003). Possible role of arachidonic acid in the regulation of lactate production in rat Sertoli cells. *Int. J. Androl.* **26**, 310–317.
Meroni, S. B., Riera, M. F., Pellizzari, E. H., Galardo, M. N., and Cigorraga, S. B. (2004). FSH activates phosphatidylinositol 3-kinase/protein kinase B signaling pathway in 20-day-old Sertoli cells independently of IGF-I. *J. Endocrinol.* **180**, 257–265.
Mita, M., Price, J. M., and Hall, P. F. (1982). Stimulation by follicle-stimulating hormone of synthesis of lactate by Sertoli cells from rat testis. *Endocrinology* **110**, 1535–1541.
Monaco, L., Foulkes, N. S., and Sassone-Corsi, P. (1995). Pituitary follicle-stimulating hormone (FSH) induces CREM gene expression in Sertoli cells: Involvement in long-term desensitization of the FSH receptor. *Proc. Natl. Acad. Sci. USA* **92**, 10673–10677.

Mruk, D. D., and Cheng, C. Y. (2004). Sertoli–Sertoli and Sertoli–germ cell interactions and their significance in germ cell movement in the seminiferous epithelium during spermatogenesis. *Endocr. Rev.* **25,** 747–806.

Oddi, S., Bari, M., Battista, N., Barsacchi, D., Cozzani, I., and Maccarrone, M. (2005). Confocal microscopy and biochemical analysis reveal spatial and functional separation between anandamide uptake and hydrolysis in human keratinocytes. *Cell. Mol. Life Sci.* **62,** 386–395.

Okamoto, Y., Morishita, J., Tsuboi, K., Tonai, T., and Ueda, N. (2004). Molecular characterization of a phospholipase D generating anandamide and its congeners. *J. Biol. Chem.* **279,** 5298–5305.

Orth, J. M., Gunsalus, G. L., and Lamperti, A. A. (1988). Evidence from Sertoli cell-depleted rats indicates that spermatid number in adults depends on numbers of Sertoli cells produced during perinatal development. *Endocrinology* **122,** 787–794.

Pellegrini, M., Grimaldi, P., Rossi, P., Geremia, R., and Dolci, S. (2003). Developmental expression of BMP4/ALK3/SMAD5 signaling pathway in the mouse testis: A potential role of BMP4 in spermatogonia differentiation. *J. Cell Sci.* **116,** 3363–3372.

Piomelli, D. (2004). The endogenous cannabinoid system and the treatment of marijuana dependence. *Neuropharmacology* **47,** 359–367.

Puffenbarger, R. A. (2005). Molecular biology of the enzymes that degrade endocannabinoids. *Curr. Drug Targets CNS Neurol. Disord.* **4,** 625–631.

Puffenbarger, R. A., Kapulina, O., Howell, J. M., and Deutsch, D. G. (2001). Characterization of the 5′-sequence of the mouse fatty acid amide hydrolase. *Neurosci. Lett.* **314,** 21–24.

Rannikki, A. S., Zhang, F. P., and Huhtaniemi, I. T. (1995). Ontogeny of follicle-stimulating hormone receptor gene expression in the rat testis and ovary. *Mol. Cell. Endocrinol.* **107,** 199–208.

Ricci, G., Cacciola, G., Altucci, L., Meccariello, R., Pierantoni, R., Fasano, S., and Cobellis, G. (2007). Endocannabinoid control of sperm motility: The role of epididymus. *Gen. Comp. Endocrinol.* **153,** 320–322.

Richards, J. S. (2001). New signaling pathways for hormones and cyclic adenosine 3′,5′-monophosphate action in endocrine cells. *Mol. Endocrinol.* **15,** 209–218.

Rossato, M., Ion Popa, F., Ferigo, M., Clari, G., and Foresta, C. (2005). Human sperm express cannabinoid receptor Cb1, the activation of which inhibits motility, acrosome reaction, and mitochondrial function. *J. Clin. Endocrinol. Metab.* **90,** 984–991.

Rossi, P., Dolci, S., Albanesi, C., Grimaldi, P., Ricca, R., and Geremia, R. (1993). Follicle-stimulating hormone induction of steel factor (SLF) mRNA in mouse Sertoli cells and stimulation of DNA synthesis in spermatogonia by soluble SLF. *Dev. Biol.* **155,** 68–74.

Rossi, G., Gasperi, V., Paro, R., Barsacchi, D., Cecconi, S., and Maccarrone, M. (2007). Follicle-stimulating hormone activates fatty acid amide hydrolase by protein kinase A and aromatase-dependent pathways in mouse primary Sertoli cells. *Endocrinology* **148,** 1431–1439.

Rossi, P., Lolicato, F., Grimaldi, P., Dolci, S., Di Sauro, A., Filipponi, D., and Geremia, R. (2008). Transcriptome analysis of differentiating spermatogonia stimulated with kit ligand. *Gene Expr. Patterns* **8,** 58–70.

Ruppert, S., Cole, T. J., Boshart, M., Schmid, E., and Schutz, G. (1992). Multiple mRNA isoforms of the transcription activator protein CREB: Generation by alternative splicing and specific expression in primary spermatocytes. *EMBO J.* **11,** 1503–1512.

Saxlund, M. A., Sadler-Riggleman, I., and Skinner, M. K. (2004). Role of basic helix-loop-helix (bHLH) and CREB transcription factors in the regulation of Sertoli cell androgen-binding protein expression. *Mol. Reprod. Dev.* **68,** 269–278.

Schuel, H., and Burkman, L. J. (2005). A tale of two cells: Endocannabinoid-signaling regulates functions of neurons and sperm. *Biol. Reprod.* **73,** 1078–1086.

Schuel, H., Burkman, L. J., Lippes, J., Crickard, K., Mahony, M. C., Giuffrida, A., Picone, R. P., and Makriyannis, A. (2002a). Evidence that anandamide-signaling regulates human sperm functions required for fertilization. *Mol. Reprod. Dev.* **63**, 376–387.

Schuel, H., Burkman, L. J., Lippes, J., Crickard, K., Forester, E., Piomelli, D., and Giuffrida, A. (2002b). N-Acylethanolamines in human reproductive fluids. *Chem. Phys. Lipids* **121**, 211–227.

Simon, G. M., and Cravatt, B. F. (2008). Anandamide biosynthesis catalyzed by the phosphodiesterase GDE1 and detection of glycerophospho-N-acyl ethanolamine precursors in mouse brain. *J. Biol. Chem.* **283**, 9341–9349.

Simon, L., Ekman, G. C., Tyagi, G., Hess, R. A., Murphy, K. M., and Cooke, P. S. (2007). Common and distinct factors regulate expression of mRNA for ETV5 and GDNF, Sertoli cell proteins essential for spermatogonial stem cell maintenance. *Exp. Cell Res.* **313**, 3090–3099.

Simoni, M., Gromoll, J., and Nieschlag, E. (1997). The follicle-stimulating hormone receptor: Biochemistry, molecular biology, physiology, and pathophysiology. *Endocr. Rev.* **18**, 739–773.

Skinner, M. K., and Griswold, M. D. (1982). Secretion of testicular transferrin by cultured Sertoli cells is regulated by hormones and retinoids. *Biol. Reprod.* **27**, 211–221.

Spruill, W. A., Zysk, J. R., Tres, L. L., and Kierszenbaum, A. L. (1983). Calcium/calmodulin-dependent phosphorylation of vimentin in rat Sertoli cells. *Proc. Natl. Acad. Sci. USA* **80**, 760–764.

Starowicz, K., Nigam, S., and Di Marzo, V. (2007). Biochemistry and pharmacology of endovanilloids. *Pharmacol. Ther.* **114**, 13–33.

Stella, N., Schweitzer, P., and Piomelli, D. (1997). A second endogenous cannabinoid that modulates long-term potentiation. *Nature* **388**, 773–778.

Szallasi, A., and Blumberg, P. M. (1999). Vanilloid (capsaicin) receptors and mechanisms. *Pharmacol. Rev.* **51**, 159–212.

Tadokoro, Y., Yomogida, K., Ohta, H., Tohda, A., and Nishimune, Y. (2002). Homeostatic regulation of germinal stem cell proliferation by the GDNF/FSH pathway. *Mech. Dev.* **113**, 29–39.

Tapanainen, J. S., Aittomäki, K., Min, J., Vaskivuo, T., and Huhtaniemi, I. T. (1997). Men homozygous for an inactivating mutation of the follicle-stimulating hormone (FSH) receptor gene present variable suppression of spermatogenesis and fertility. *Nat. Genet.* **15**, 205–206.

Van Sickle, M. D., Duncan, M., Kingsley, P. J., Mouihate, A., Urbani, P., Mackie, K., Stella, N., Makriyannis, A., Piomelli, D., Davison, J. S., Marnett, L. J., Di Marzo, V., et al. (2005). Identification and functional characterization of brainstem cannabinoid CB2 receptors. *Science* **310**, 329–332.

Waeber, G., Meyer, T. E., LeSieur, M., Hermann, H. L., Gerard, N., and Habener, J. F. (1991). Developmental stage-specific expression of cyclic adenosine 3′,5′-monophosphate response element-binding protein CREB during spermatogenesis involves alternative exon splicing. *Mol. Endocrinol.* **5**, 1418–1430.

Walker, W. H., and Cheng, J. (2005). FSH and testosterone signaling in Sertoli cells. *Reproduction* **130**, 15–28.

Walker, W. H., Fucci, L., and Habener, J. F. (1995). Expression of the gene encoding transcription factor cyclic adenosine 3′,5′-monophosphate (cAMP) response element-binding protein (CREB): Regulation by follicle-stimulating hormone-induced cAMP signaling in primary rat Sertoli cells. *Endocrinology* **136**, 3534–3545.

Wang, H., Dey, S. K., and Maccarrone, M. (2006a). Jekyll and Hyde: Two faces of cannabinoid signaling in male and female fertility. *Endocr. Rev.* **27**, 427–448.

Wang, H., Xie, H., Guo, Y., Zhang, H., Takahashi, T., Kingsley, P. J., Marnett, L. J., Das, S. K., Cravatt, B. F., and Dey, S. K. (2006b). Fatty acid amide hydrolase deficiency limits early pregnancy events. *J. Clin. Invest.* **116,** 2122–2131.

Wei, B. Q., Mikkelsen, T. S., McKinney, M. K., Lander, E. S., and Cravatt, B. F. (2006). A second fatty acid amide hydrolase with variable distribution among placental mammals. *J. Biol. Chem.* **281,** 36569–36578.

Wenger, T., Rettori, V., Snyder, G. D., Dalterio, S., and McCann, S. M. (1987). Effects of delta-9-tetrahydrocannabinol on the hypothalamic–pituitary control of luteinizing hormone and follicle-stimulating hormone secretion in adult male rats. *Neuroendocrinology* **46,** 488–493.

CHAPTER ELEVEN

Glucocorticoid-Regulated Crosstalk Between Arachidonic Acid and Endocannabinoid Biochemical Pathways Coordinates Cognitive-, Neuroimmune-, and Energy Homeostasis-Related Adaptations to Stress

Renato Malcher-Lopes[*,†] and Marcelo Buzzi[‡]

Contents

I. Introduction	264
II. The Arachidonic Acid Cascade	266
III. Glucocorticoid-Mediated Inhibition of cPLA$_2$-Dependent AA Release from Membrane Phospholipids	269
IV. Biosynthesis of the AA-Containing Endocannabinoids AEA and 2-AG	271
V. Nongenomic Glucocorticoid-Induced Activation of Endocannabinoid Biosynthesis	274
VI. Endocannabinoids Metabolization	278
VII. Crosstalk Between GCs and COX$_2$ in the Control of Neuroinflammation and Neuroprotection	282
VIII. Crosstalk Between GCs and COX$_2$ in the Control of Synaptic Plasticity and Learning Processes	285
IX. Coordination of GC-Mediated Control of the Neuroimmune Response and Energy Homeostasis Control	288
Acknowledgments	295
References	295

[*] Laboratory of Mass Spectrometry, EMBRAPA-Center for Genetic Resources and Biotechnology, Brasília-DF, Brazil
[†] Genomic Sciences and Biotechnology Graduation Program of the Universidade Católica de Brasília
[‡] Laboratory of Analytical Biochemistry, Molecular Pathology Division, SARAH Network of Rehabilitation Hospitals, Brasília-DF, Brazil

Abstract

Arachidonic acid and its derivatives constitute the major group of signaling molecules involved in the innate immune response and its communication with all cellular and systemic aspects involved on homeostasis maintenance. Glucocorticoids spread throughout the organism their influences over key enzymatic steps of the arachidonic acid biochemical pathways, leading, in the central nervous system, to a shift favoring the synthesis of anti-inflammatory endocannabinoids over proinflammatory metabolites, such as prostaglandins. This shift modifies local immune–inflammatory response and neuronal activity to ultimately coordinate cognitive, behavioral, neuroendocrine, neuroimmune, physiological, and metabolic adjustments to basal and stress conditions. In the hypothalamus, a reciprocal feedback between glucocorticoids and arachidonate-containing molecules provides a mechanism for homeostatic control. This neurochemical switch is susceptible to fine-tuning by neuropeptides, cytokines, and hormones, such as leptin and interleukin-1β, assuring functional integration between energy homeostasis control and the immune/stress response.

I. Introduction

Ultimately, the central nervous system (CNS) has evolved to oversee and control all organic functions of animal systems in order to guarantee that each cell will receive proper supply of matter and energy necessary to keep the stability of their internal milieu (homeostasis) in face to an ever-changing external environment (Cannon, 1939; Recordati, 2003). While interacting with their surroundings, animals must avoid any threat to their physical and biochemical integrity in order to sustain the delicate balance between acquiring nutrients and escaping themselves from being used as a source of matter and energy for predators, parasites, and pathogens. To do so, hypothalamic circuits continually integrate humoral, visceral, sensory, emotional, and cognitive information in order to coordinate any metabolic, physiological, or behavioral adjustments necessary to maintain the homeostasis (Swanson and Sawchenko, 1980), which is a condition to stay alive. Such adjustments require a judicious administration of endogenous and external sources of energy. This feature is especially critical whenever there is an ongoing or perceived potential threat, situation in which the organism as a whole must put on motion a complex set of adaptations to prevent, reduce or remediate damage and subsequent loss of homeostasis.

The responses carried on by the innate and acquired immune systems are among the most important physiological functions that integrate the ample scope of the stress response orchestrated by the so-called stress hormones, catecholamines, and glucocorticoids (GCs) (Elenkov *et al.*, 2000;

Sapolsky et al., 2000; Selye, 1950). The circulating concentration of GCs is the main hormonal signal coordinating permissive, suppressive, stimulatory, and preparative aspects of the central and peripheral adaptations to any kind of stress (Sapolsky et al., 2000), as well as the homeostatic adaptations that take place in the transitions between the fasted and the satiated states (Dallman et al., 1993; Malcher-Lopes et al., 2006). The secretion of GCs is controlled by the hypothalamic–pituitary–adrenal (HPA) axis, which is driven by neurosecretory cells located in the paraventricular nucleus of the hypothalamus (PVN). Once activated, these neurons liberate corticotropin-releasing hormone (CRH) and/or vasopressin into the portal blood of the anterior pituitary gland from where they gain access to corticotroph endocrine cells to activate the secretion of the adrenocorticotropic hormone (ACTH) (Chrousos, 1998). ACTH, in turn, gains access to the general circulation and activates the secretion of GCs (corticosterone in rodents and cortisol in humans) from the cortical adrenal gland. As their circulating levels rise, GCs cross the blood–brain barrier to regulate several neuronal functions, including the HPA axis activity. The HPA axis activity is normally under an ultradian control, which in mammals tends to synchronize GCs peak, and their metabolic effects, with feeding behavior, while preparing the organism for potentially stressful events when the animal is foraging. Additionally, in response to different stressors, GC release can be significantly increased (Chrousos, 1998; Dallman et al., 1993, 1995; Sapolsky et al., 2000).

Arachidonic acid (AA, $20:4n - 6$), a polyunsaturated fatty acid, is perhaps the single most important signaling molecule connecting the innate immune system with the cellular and systemic adaptations to stress. AA is stored as arachidonate in membrane phospholipids of animal cells, from where it can be mobilized or transferred among signaling molecules, to assume several "biochemical identities," each one with a distinct biological function. For instance, besides its functions as a free fatty acid, AA serves as a precursor for prostaglandins, prostacyclins, leukotrienes, thromboxanes, lipoxins, epoxins, endocannabinoids, endocannabinoid-derived prostamides, and endocannabinoid-derived glycerylprostaglandins, among other molecules (Bazan, 2005; Bosetti, 2007; Rouzer and Marnett, 2008; Tassoni et al., 2008). Once hydrolyzed, these molecules liberate free AA, allowing its recycling via reincorporation into membrane phospholipids. In the CNS, AA and AA-containing molecules regulate, for instance, neurotransmission, cognitive processes, local neuroinflammatory response, neuroprotection, brain microcirculation, and blood–brain barrier permeability. They also modulate the central regulation of neuroimmune response, neuroendocrine and sympathoadrenomedullary outflow, fluid balance, energy homeostasis, and virtually all aspects of the CNS-mediated control of general homeostasis under basal and stress conditions (Bazan, 2005; Bosetti, 2007; Cota, 2008; Malcher-Lopes et al., 2008; Tassoni et al., 2008; Yang and Chen, 2008). Each specific situation requires an appropriate combination of adjustments, and the decision about the nature and intensity of each aspect of such adaptation is ultimately determined by which "biochemical identities" AA will assume on the signaling flow underlying the integrated safeguard of homeostasis.

Recent findings provide support for the concept that acute (within 30 min or less) GC-mediated control of key enzymatic steps favors the synthesis of ant-inflammatory over proinflammatory AA-containing signaling molecules (Malcher-Lopes et al., 2008), thereby coordinating several appropriated adjustments in neuroimmune, physiological, metabolic, cognitive, and behavioral aspects of the general homeostasis. A reciprocal feedback regulation between GCs and AA-containing compounds provides a mechanism for self-regulation of this system, and establishes a biochemical switch subjected to fine-tuning by both neuronal and hormonal modulation, assuring integrated temporal and functional coordination between these major aspects. For instance, the central crosstalk between this biochemical switch with the signaling triggered by the cytokines leptin and interleukin-1β coordinates energy homeostasis control with the immune–inflammatory response.

II. The Arachidonic Acid Cascade

Neurons are rich in arachidonate-containing lipids but are not capable of synthesizing AA *de novo* from its precursor linoleic acid (18:$2n-6$) (DeMar et al., 2006), a function that is fulfilled in the brain by astrocytes and endothelial cells (Calder, 2005; Iversen and Kragballe, 2000). Neurons may also be supplied with plasma unesterified AA (Robinson et al., 1992). Proinflammatory stimuli such as histamine, cytokines, and bacterial infection stimulate the release of AA as part of the inflammatory response. In an experimental model of brain inflammation, intracerebroventricular (i.c.v.) infusion of bacterial lipopolysaccharides (LPS) for 6 days in rats increases whole brain activities of cytosolic and secretory PLA$_2$ (cPLA$_2$ and sPLA$_2$, respectively), leading to higher concentrations of unesterified AA and its metabolic products prostaglandin E2 (PGE2) and prostaglandin D2 (PGD2). This effect was accompanied by increased AA turnover rates in brain phospholipids and by an elevation on the coefficient of incorporation of plasma AA into brain phospholipids (Lee et al., 2004; Rosenberger et al., 2004). In the CNS, AA can also be released in response to neurotransmitters, neuromodulators, and neuropeptides. For instance, AA is liberated from membrane phospholipids following ligand binding to a number of neuroreceptors, including cholinergic and dopamine receptors that are coupled to AA-selective Ca^{2+}-dependent cPLA$_2$ (Basselin et al., 2007; Bhattacharjee et al., 2005). Nitric oxide (NO) stimulates the release of AA from phosphatidylinositol (PI) and phosphatidylcoline (PC) through alteration of cPLA$_2$ and AA-CoA acyltransferase activities (Chalimoniuk et al., 2006). Glutamate acting at a metabotropic receptor also evokes the release of AA from mouse brain astrocytes (Stella et al., 1994).

Upon stimulation, free arachidonate is released from the cell membrane by three main distinct pathways: (1) phospholipase-D (PLD)-catalyzed production of phosphatidic acid (PA) from phosphatidylethanolamine (PE) or PC, followed by formation of diglyceride, monoglyceride, and AA; (2) phospholipase-C (PLC)-mediated conversion of PI into diacylglycerol (DAG), followed by the action of DAG lipase and monoglyceride lipase to produce AA and glycerol; and (3) the phospholipase-A$_2$ (PLA$_2$)-catalyzed release of AA from the sn-2 position of membrane phospholipids (Calder, 2005; Iversen and Kragballe, 2000; Seeds and Bass, 1999) (Fig. 11.1). Once released, free AA has a short lifespan, during which it can be rapidly esterified and recycled into membrane phospholipids, be metabolized into eicosanoids, or diffuse outside the cells, where it can modulate the activity of ion channels and protein kinases (DeGeorge et al., 1989; Horrocks, 1989). Free arachidonate can be metabolized in the brain by the action of three distinct classes of enzymes, cyclooxygenases (COXs), lipoxygenases (LOXs), and cytochrome P450 (Fig. 11.1), all of which are expressed by neurons (Adesuyi et al., 1985; Alkayed et al., 1996; Dembinska-Kiec et al., 1984a,b). COXs are the rate-limiting enzymes in AA metabolism to prostaglandins, prostacyclins, or thromboxanes (Kozak et al., 2002a). Different genes under specific regulation encode two COXs isoenzymes, COX$_1$ and COX$_2$. COX$_1$ is constitutively expressed while COX$_2$ is an immediate early gene, which is expressed at low basal levels but is rapidly induced in response to a variety of inflammatory and proliferative stimuli, including cytokines, growth factors, and tumor promoters. Prostaglandin biosynthesis from AA in inflammatory cells and in the CNS depends on three distinct enzymatic steps: (1) phospholipase-dependent release of AA from phospholipid pools; (2) COX-mediated oxygenation of AA to generate the prostaglandin hydroxy-endoperoxide (PGH$_2$); and (3) the conversion of PGH$_2$ to prostaglandins, thromboxanes or prostacyclins by distinct synthases. Thromboxanes derive from AA preferentially via COX$_1$, whereas prostaglandins and prostacyclins derive from AA preferentially via COX$_2$-mediated pathways (Bosetti et al., 2004; Brock et al., 1999; Dubois et al., 1998; Pepicelli et al., 2005; Smith et al., 2000; Vane et al., 1998). The LOX isozymes LOX$_5$, LOX$_{12}$, and LOX$_{15}$ convert AA into the corresponding hydroperoxyeicosatetraenoic acids (5-, 12-, and 15-HPETEs) and dihydroxyeicosatetraenoic acid, which are subsequently converted to hydroxyeicosatetraenoic acids (HETEs) by peroxidases, to leukotrienes by hydrase and glutathione S-transferase, or to lipoxins by LOXs (Yamamoto, 1992). Epoxyeicosatrienoic acid (EET) and dihydroxy acids are formed by cytochrome p450 epoxygenase (McGiff, 1991).

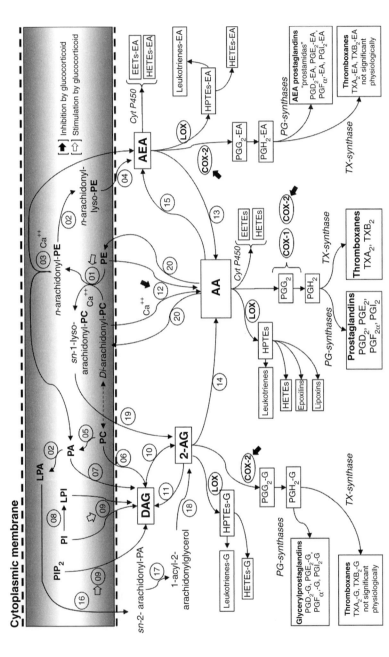

Figure 11.1 Glucocorticoid-mediated control of the crosstalk between arachidonic acid and endocannabinoid biochemical pathways. Glucocorticoids stimulate key enzymatic steps that generate different endocannabinoid precursors. On the other hand, glucocorticoids reduce COX_2 activity, thereby preventing endocannabinoid metabolism into endocannabinoid-derived prostanoids. Additionally, by

III. GLUCOCORTICOID-MEDIATED INHIBITION OF cPLA₂-DEPENDENT AA RELEASE FROM MEMBRANE PHOSPHOLIPIDS

One way by which GCs suppress inflammation is by preventing release of free AA from membrane phospholipids such as PC, PI, and PE, making it unavailable for subsequent prostaglandin synthesis. In a variety of cells, GC-mediated suppression of both basal and stimulated AA formation, albeit incomplete, significantly impacts downstream synthesis of the AA-containing proinflammatory prostaglandins (Bailey, 1991; Becker *et al.*, 1988; De Caterina and Weksler, 1986; Doherty, 1981; Floman and Zor, 1976; Lewis *et al.*, 1986; Potestio and Olson, 1990; Schleimer *et al.*, 1986; Tanaka *et al.*, 1995; Yousufzai and Abdel-latif, 1997). In many instances, GC inhibitory effect on prostaglandin biosynthesis took a few hours to be observed, and was prevented by protein synthesis inhibition or by the addition of exogenous AA (De Caterina and Weksler, 1986; Lewis *et al.*, 1986; Schleimer *et al.*, 1986; Vishwanath *et al.*, 1993). Accordingly, evidences indicate that GCs cause downregulation of group II PLA₂ expression with concomitant upregulation of lipocortin-1, a PLA₂ inhibitor. For instance, systemic GC depletion in adrenalectomized rats strongly upregulates the expression of group II PLA₂ in different tissues (including lung, spleen, liver, and kidney), while the expression of lipocortin-1 is downregulated in the same tissues (Vishwanath *et al.*, 1993). Additionally, it was shown that a neutralizing antibody to lipocortin-1 prevents GCs from suppressing the rapid endothelial growth factor (EGF)-induced stimulation of cPLA₂, and the subsequent AA release from A549 cells (Croxtall *et al.*, 1995, 1996, 2000, 2002). Interestingly, the synthetic GC dexamethasone (DEX) inhibitory effect increased from 40% with short treatments (<5 min) to 70% with longer treatments. The fast DEX effect was inhibited by the cytosolic GC receptor antagonist RU486, but not by blocking gene

---

inhibiting both cPLA₂ and COX₂, glucocorticoids also prevent the use of endocannabinoid precursors to form arachidonic acid and arachidonate-containing prostanoids. The overall effect is a shift in the arachidonic acid metabolism away from the synthesis of proinflammatory signaling molecules, while the formation of the anti-inflammatory endocannabinoids is favored. 1, N-acyltransferase; 2, secretory phospholipase-A₂ (sPLA₂); 3, NAPE-phospholipase-D; 4, lysophospholipase-D; 5, phospholipase-D; 6, phosphatidylcholine phospholipase-C (PC-PLC); 7, phosphatidic acid phosphatase; 8, phospholipase-A₁; 9, phosphatidylinositol phospholipase-C (PI-PLC); 10, diacylglycerol lipase (*sn*-1-DAG lipase); 11, monoglycerol acyltransferase; 12, cytosolic phospholipase-A₂; 13, fatty acid amidohydrolase; 14, monoacylglycerol lipase; 15, anandamide synthase; 16, acyl-CoA acyltransferase-mediated LPA remodeling; 17, phosphatidic acid phosphohydrolase; 18, diacylglycerol; 19, phospholipase-C EGTA insensitive; 20, phosphatidylcholine/phosphatidylethanolamine transacylase.

transcription, whereas the delayed effect was inhibited by blocking gene transcription, demonstrating a combination of nongenomic and genomic inhibition of EGF-induced AA release by GCs. Further investigation demonstrated that the nongenomic component involved inhibition of the mitogen-activated protein kinase (MAPK) and extracellular signal-regulated kinase 1 (ERK_1)-mediated phosphorylation of $cPLA_2$, thus preventing $cPLA_2$ translocation to the membrane (Croxtall et al., 2000; Lin et al., 1993).

The suppression of IL-1β-induced AA and PGE2 release from A549 cell line by DEX was mimicked by the GC receptor antagonist RU486 (Chivers et al., 2004), which is reminiscent of the nongenomic immunosuppressive effect mediated by both RU486 and DEX described for T cells (Lowenberg et al., 2005). In T cells, both RU486 and DEX binding cause disassembly of a GC receptor-containing multiprotein complex. Therefore, a similar mechanism is likely to underlay the nongenomic DEX-induced suppression of $cPLA_2$-mediated AA release. Demonstration of the nongenomic GC-mediated control of AA release in physiological conditions was provided by a study, showing that the endotoxin-induced, PGE2-mediated diarrhea is prevented in mice by DEX-induced inhibition AA release, which was unaffected by protein synthesis blockade (Doherty, 1981).

Taken together, these studies indicate that, in addition to its lipocortin-1-mediated effect, GCs prevent AA release through a cytosolic receptor-dependent mechanism that nongenomically abrogates $cPLA_2$ activation. However, in some cells GCs seem to act through alternative pathways to suppress stimulated release of AA. In cultured pituitary corticotroph cell (AtT-20 cells), 2-h DEX treatment caused robust inhibition (up to 80%) of the early (within 1 min) AA release induced by the bee venom peptide melittin (Pompeo et al., 1997). Although DEX inhibitory effect was not detected before 30 min, preincubation with the inhibitor of protein synthesis, cycloheximide, further increased the suppressive effect of DEX by 15%, ruling out the involvement of de novo protein synthesis. Melittin-induced AA is traditionally thought to require activation of $sPLA_2$ types I, II, and III, with no involvement of $cPLA_2$ (Cajal and Jain, 1997; Clark et al., 1987; Emmerling et al., 1993; Mollay et al., 1976; Rao, 1992; Steiner et al., 1993; Suzuki et al., 1991; Zeitler et al., 1991). However, additional evidences indicate that melittin effect is mediated by a broader combination of lipase activities (Lee et al., 2001). In skeletal muscle cells, for instance, the target for melittin effects on phospholipid metabolism proved to be PLC and, when used at higher concentrations, melittin's effect also involved triglyceride lipases (Fletcher et al., 1991). Accordingly, melittin was shown to increase not only the release of AA from cultured mouse lymphocytic leukemia cells, but also several other free fatty acids, including palmitoleic, myristic, oleic, palmitic, linoleic, and stearic acids (Lee et al., 2001). In this case, different PLA_2-specific inhibitors (other than DEX) had only minimal effects, whereas PI-PLC and DAG lipase inhibitors exerted robust inhibitory effects on melittin-induced release of AA. Hence, GC inhibition of melittin-induced release of AA in AtT-20 cells is unlikely to depend only on the blockade of any positive effect of melittin over PLA_2 isozymes. However, it has been shown that GCs have a rather stimulatory effect on PI-PLC and DAG lipase in different cells (Cifone *et al.*, 1999; Marchetti *et al.*, 2003; Tong *et al.*, 2004). One explanation for this apparent paradox is that, even without blocking melittin-induced activation of PI-PLC, Ca^{2+} influx, and DAG lipase activity (Lee *et al.*, 2001), GCs may facilitate the transfer of free AA or arachidonate from phospholipid stores to other AA-containing molecules, thereby reducing AA release as a free fatty acid (Malcher-Lopes *et al.*, 2008). In principle, melittin-induced activation of the PI-PLC-Ca^{2+}-DAG lipase pathway is favorable for the synthesis, for instance, of the AA-containing endocannabinoids N-arachidonoylethanolamide (AEA) and arachidonoyl *sn* glycerol (2-AG) (see Section IV). There is no direct evidence so far that corticotroph cells produce endocannabinoids upon GC stimulation; however, the presence of endocannabinoids and cannabinoid CB1 receptor has been demonstrated in the anterior pituitary, where these cells are found (Gonzalez *et al.*, 1999; Wenger *et al.*, 1999). Furthermore, a high-affinity, G-coupled membrane GC receptor, which does not bind RU486, has been found in corticotroph AtT-20 cells (Harrison *et al.*, 1979; Maier *et al.*, 2005). This is striking because, as it will be discussed in detail in the Section V, a receptor with such characteristics was shown to mediate nongenomic, GC-induced endocannabinoid biosynthesis and release in the hypothalamus (Di *et al.*, 2003, 2005; Malcher-Lopes *et al.*, 2006).

IV. BIOSYNTHESIS OF THE AA-CONTAINING ENDOCANNABINOIDS AEA AND 2-AG

From the nonoxidative metabolism of AA derive N-acylethanolamides (NAEs) and monoglycerides (Schmid *et al.*, 1990). NAEs are ethanolamides with long-chain fatty acids, which have been involved in anti-inflammatory, membrane-stabilizing actions and apoptotic processes in different mammalian tissues (Epps *et al.*, 1979, 1980; Hansen *et al.*, 1995, 2000; Kondo *et al.*, 1998; Schmid, 2000; Schmid and Berdyshev, 2002; Schmid *et al.*, 1990, 1995; Sugiura *et al.*, 2002). The AA-containing NAE, AEA (also known as anandamide) was the first compound identified as an endogenous ligand for the G-coupled membrane cannabinoid receptors CB1 and CB2 (Devane *et al.*, 1992), and for the transient receptor potential vanilloid subtype 1 (TRPV1, also known as vanilloid receptor VR1), after being internalized (Di Marzo *et al.*, 2002). Therefore, AEA was the first endogenous compound shown to mimic various behavioral, physiological, and psychological effects of the *Cannabis sativa*-derived cannabinoid Δ^9-tetrahydrocannabinol (THC) (Di Marzo, 1998; Mechoulam *et al.*, 1998). The second endocannabinoid discovered was 2-AG and exhibits higher selectivity and efficacy for both cannabinoid CB1 and CB2 receptors than AEA, and is considered to be a full agonist for these receptors, whereas AEA is a partial agonist (Di Marzo and Petrosino, 2007).

AEA can be synthesized by the direct condensation of free AA and ethanolamine catalyzed by a brain AEA synthase, which may actually be an amidohydrolase working "in reverse" when high concentrations of substrates are available (Devane and Axelrod, 1994; Kruszka and Gross, 1994). Brain neurons, however, can also produce AEA by a pathway that does not require PLA_2-mediated AA release, in which AA is transferred to an *N*-acylphosphatidylethanolamine intermediate, which in turn serves as a direct AEA precursor (Cadas *et al.*, 1996; Di Marzo *et al.*, 1994; Sugiura *et al.*, 1996). Thus, when cultured neurons were stimulated with membrane-depolarizing agents, such as ionomycin, kainate, or high extracellular concentration of K^+, *N*-arachidonoyl-PE (NArPE) was generated by transferring the AA esterified on the *sn*-1 position of membrane phospholipids onto the primary amine of PE, in a reaction catalyzed by an *N*-acyltransferase activity (Cadas *et al.*, 1996; Di Marzo *et al.*, 1994). Both intracellular Ca^{2+} and PKA-mediated phosphorylation may stimulate this enzyme (Cadas *et al.*, 1996; Di Marzo *et al.*, 1994); consequently, PKA activity may be necessary for *in vivo* endocannabinoid biosynthesis through this pathway whenever intracellular Ca^{2+} levels are below a certain threshold level. Additionally, a Ca^{2+}-independent *N*-acyltransferase that catalyzes NArPE formation has been recently cloned (Jin *et al.*, 2007). A neuronal PLD-like enzyme, named *N*-acylphosphatidylethanolamine-selective PLD (NAPE-PLD), promotes the subsequent conversion of NArPE into AEA and like the upstream *N*-acyltransferase, is stimulated by Ca^{2+} (Okamoto *et al.*, 2004). However, targeted disruption of the NAPE-PLD gene generated a mutant mouse strain with normal brain levels of AEA and other polyunsaturated NAEs, but with lower levels of long-chain ($C \geq 16$) saturated NAEs (Leung *et al.*, 2006). In fact, a few alternative pathways for the generation of AEA from NArPE have been demonstrated. For instance, several $sPLA_2$ isozymes, but not $cPLA_{2\alpha}$, can generate *N*-acyl-lyso-PE from NArPE, which is then converted to AEA by a lyso-PLD enzyme (Natarajan *et al.*, 1984; Sun *et al.*, 2004). A PLA_2/PLA_1-like enzyme targeting *N*-palmitoyl-PE is broadly distributed in different rat tissues, but it is present in relatively low levels in the brain (Sun *et al.*, 2004). In mouse macrophages, LPS-induced AEA synthesis initiates with the PLC-catalyzed cleavage of NArPE to generate phosphoarachidonylethanolamide (phospho-AEA), followed by its subsequent dephosphorylation by phosphatases, such as the protein tyrosine phosphatase $PTPN_{22}$ (Liu *et al.*, 2006).

Phospho-AEA was also isolated from the mouse brain, suggesting that this pathway may also operate in the CNS (Liu et al., 2006). A third alternative route for AEA formation from NArPE, which was demonstrated in mouse brain homogenates and peripheral tissues, requires serine hydrolase-mediated cleavage of both O-acyl chains of NArPE (and other N-acyl-PEs) to generate glycerophospho-AEA (and other glycerophospho-NAEs), which is then converted into AEA and glycerol-3-phosphate by a metal-dependent phosphodiesterase (Simon and Cravatt, 2006) (Fig. 11.1).

The endocannabinoid 2-AG is found in the brain at higher concentration than AEA, reflecting the fact that 2-AG and AEA can be synthesized by separated pathways which, therefore, can be differentially regulated (Di Marzo and Petrosino, 2007). Nevertheless, among the known possible biosynthetic pathways for 2-AG some are coupled to AEA biosynthesis. For instance, if a diarachidonoyl-PC and -PE are used as substrate for the Ca^{2+}-dependent action of N-acyltransferase, NArPE, and 1-lysophosphoglycerides are formed. Then, a EGTA-insensitive PLC isoenzyme can use 1-lysophosphoglycerides as direct precursors for the formation of 2-AG (Di Marzo et al., 1996), indicating that N-acyltransferase-catalyzed reaction can be a common step in the biosynthesis of AEA and 2-AG from AA-containing membrane phospholipids. Furthermore, the same reaction catalyzed by NAPE-PLD to form AEA from NArPE, also generates PA, which is a precursor for DAG that, in turn, can be a major direct 2-AG precursor if it contains an arachidonate chain. In fact, a PLC-independent pathway involving arachidonate-containing DAG was observed in ionomycin-stimulated neuroblastoma cells, where DAG serving as 2-AG precursors can be produced through the hydrolysis of *sn*-2-arachidonate-containing PA (Bisogno et al., 1999b). In this case, type II $sPLA_2$-mediated production of lyso-PA (LPA) from AA is followed by the conversion of LPA into *sn*-2-arachidonate-containing PA via the LPA acyl-CoA: acyltransferase-mediated PA remodeling. The *sn*-2-arachidonate-containing PA is first converted to a DAG (1-acyl-2-arachidonoylglycerol) by the PA phosphohydrolase and then to 2-AG by a *sn*-1-selective DAG lipase (*sn*-1-DAG lipase) (Bisogno et al., 1999b; Nakane et al., 2002). Several distinct PLC-dependent pathways converging to DAG formation can also participate in 2-AG biosynthesis. Thus, DAG can be generated by the hydrolytic reactions catalyzed by PLC activities on either AA-containing PC, PI or phosphatidylinositol bisphosphate (PIP2). The later one is catalyzed by the PIP2-selective PLCβ (PI-PLCβ). DAG is then hydrolyzed to 2-AG by either one of the two plasma membrane *sn*-1-selective, Ca^{2+}-stimulated DAG lipases, DAGLα and DAGLβ (Bisogno et al., 2003; Carrier et al., 2004; Di Marzo et al., 1996; Oka et al., 2005; Prescott and Majerus, 1983; Stella et al., 1997; Sugiura et al., 2002, 2004). The formation of 2-AG from its stereoisomer 1(3)-arachidonoylglycerol was also demonstrated in several nervous cells stimulated with agonists coupled to PI-PLC (Di Marzo and Deutsch, 1998). The stereoisomer precursor is formed by the sequential hydrolysis of phosphatidylinositols and DAGs catalyzed, respectively, by the G protein-coupled PI-PLC and a sn-1-selective DAG lipase. It has also been proposed that 2-AG can be generated by the hydrolysis of PI by a PI-specific PLA_1 ($PI-PLA_1$) to produce lysophosphatidylinositol (LPI), which is then hydrolyzed to 2-AG by the LPI-specific PLC, which has been isolated from synaptosomes (Tsutsumi et al., 1994, 1995; Ueda et al., 1993) (Fig. 11.1).

V. Nongenomic Glucocorticoid-Induced Activation of Endocannabinoid Biosynthesis

The HPA axis activity is homeostatically controlled by means of a negative feedback mechanism comprised of both rapid (nongenomic) and delayed (genomic) components, resulting in the reduction of CRH-induced secretion of ACTH from the pituitary gland (Dallman and Jones, 1973; Dayanithi and Antoni, 1989; Jingami et al., 1985; Keller-Wood and Dallman, 1984; Keller-Wood et al., 1988). There are conflicting reports on the genomic versus nongenomic mechanisms controlling fast feedback at the pituitary gland (Dayanithi and Antoni, 1989; Hinz and Hirschelmann, 2000). On the other hand, recent works done in acute hypothalamic slices demonstrated that GCs dose-dependently (Di et al., 2003) and within 1 min (Malcher-Lopes et al., 2006) suppressed excitatory neuronal inputs to all kinds of PVN neuroendocrine cells, including those that regulate the HPA axis, the hypothalamic–pituitary–thyroid (HPT) axis, and the circulating levels of oxytocin and vasopressin. The threshold concentration for this *GC-induced suppression of excitation* (GSE) was between 10 and 100 nM and the EC_{50} was between 300 and 500 nM (Di et al., 2003), which are physiological concentrations within the range achieved during stress-induced activation of the HPA axis. The suppression of presynaptic glutamate release was prevented by postsynaptic blockade of G protein-coupled signaling and by cannabinoid CB1 receptor antagonists, and was mimicked and occluded by exogenous cannabinoid application (Di et al., 2003, 2005; Malcher-Lopes et al., 2006), indicating that the effect was mediated by retrograde release of endocannabinoids. GSE was not mimicked by intracellular application of the GCs, was insensitive to "classical" cytosolic GC and mineralocorticoid receptor-specific antagonists, and was maintained when DEX was conjugated to the membrane-impermeant protein bovine serum albumin, indicating that it was mediated by a G protein-coupled membrane receptor (Di et al, 2003).

Supporting the role of this mechanism on the fast GC-negative feedback on the HPA axis, pretreatment of mice with the CB1 antagonist SR141716 strongly potentiated restraint-induced PVN cells activation (Fos expression) and corticosterone release (Patel et al., 2004). On the other hand, pretreatment with CB1 agonists, or endocannabinoid transport inhibitor AM404, or an antagonist of fatty acid amidohydrolase (FAAH) (an enzyme that hydrolyzes endocannabinoids) significantly reduced or eliminated restraint-induced corticosterone release (Patel et al., 2004). These results indicate that a major part of the nongenomic component of GC-negative feedback operated through the CB1-mediated suppression of glutamate release from presynaptic terminals controlling PVN neuroendocrine cell excitation (Cota, 2008; Di et al., 2003, 2005).

Mass spectrometry analyses provided direct biochemical evidence that GCs robustly activate the biosynthetic pathways leading to the release of both AEA and 2-AG in the PVN even when synaptic activity is suppressed (Malcher-Lopes et al., 2006). Both the increase on tissue levels of endocannabinoids triggered by GCs and GSE, the synaptic effect mediated by the endocannabinoids, were completely blocked by the cytokine-like adipocyte peptide leptin (Malcher-Lopes et al., 2006). Leptin has been shown to prevent fasting-induced hyperphagia in rats by blocking the rise on hypothalamic endocannabinoid levels that normally takes place during caloric deficit (Di Marzo et al., 2001; Kirkham et al., 2002). Leptin effect on GSE was mediated by reducing cAMP levels via phosphodiesterase 3B (PDE$_{3B}$) activation, fact that conduced to the subsequent demonstration that GCs trigger the synthesis and release of AEA and 2-AG through a Gα_s-cAMP-PKA-dependent mechanism (Malcher-Lopes et al., 2006) (Fig. 11.2). Even though the GC-induced activation of PKA may, in principle, favor the N-acyltransferase-catalyzed production of precursors for both AEA and 2-AG biosynthesis (Cadas et al., 1996; Di Marzo and Deutsch, 1998), GSE was also prevented by GF109203X, a PKC blocker (Di et al., 2003). PKC itself has been shown to suppress CB1-mediated signaling in pituitary cell line AtT-20 (Garcia et al., 1998). On the other hand, AEA is released from dorsal root ganglia sensory neurons in response to both forskolin, an adenylate cyclase stimulant, and the PKC activator phorbol-myristyl-acetate, even in the presence of the Ca^{2+} chelator BAPTA (Vellani et al., 2008). Therefore, it seems that, like PKA, PKC mediates GC-induced biosynthesis and release of endocannabinoids from the postsynaptic cells in the PVN.

The orexigenic hormone ghrelin was shown to cause retrograde release of 2-AG, but not AEA, leading to suppression of excitation in the PVN, an effect that was prevented by blockade of DAG lipase (Kola et al., 2008). Even though PKC involvement was not tested by this study, ghrelin is known to be an agonist of the growth hormone secretagogue receptor type 1a, which is a G$\alpha_{q/11}$-PKC pathway-coupled receptor. In this case, however, intracellular application of BAPTA did block the synaptic effects of ghrelin (Kola et al., 2008). Hence, it was proposed that ghrelin causes an increase on the intracellular levels of Ca^{2+}, leading downstream to Ca^{2+}- and/or PKC-mediated activation of membrane DAG lipase and subsequent

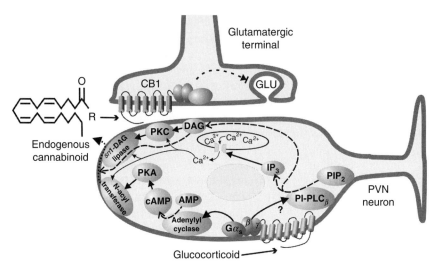

Figure 11.2 A model for the nongenomic, glucocorticoid-induced biosynthesis and retrograde release of endocannabinoids from PVN neurons. Glucocorticoids act at a putative G protein-coupled membrane glucocorticoid receptor to activate both $G\alpha_s$-cAMP-PKA- and $G\beta\gamma$-PLCβ-PKC-DAG lipase-mediated intracellular signaling pathways. Mobilization of Ca^{2+} from endogenous stores, and the stimulatory crosstalk between the two pathways, makes it possible for the activity-independent synthesis and release of both AEA and 2-AG. Endocannabinoids retrogradely released from the postsynaptic neuron act at CB1 receptors to inhibit its presynaptic glutamatergic inputs, thereby suppressing neuronal activation.

2-AG release (Kola et al., 2008). The combination of results obtained for GCs and ghrelin suggests, therefore, that hormone-stimulated release of endocannabinoids (at least 2-AG) from PVN neurons may depend on both PKC and DAG lipase. In agreement with this postulation, GC-induced activation of the PI-PLC-initiated pathway, leading downstream to DAG lipase activation, has been shown in different cells (Cifone et al., 1999; Marchetti et al., 2003; Tong et al., 2004). In thymocytes (T-cell precursors), for instance, DEX rapid stimulation of DAG generation is mediated by G protein-dependent PI-PLC and PKC activation (Cifone et al., 1999). Moreover, GC-induced activation of this pathway is likely to favor the N-acyltransferase-mediated endocannabinoid biosynthesis, by stimulating the release of intracellular Ca^{2+} via IP3. This is especially relevant if we consider that both 2-AG and AEA were released in response to GCs by PVN neurons even in the presence of tetrodotoxin (Malcher-Lopes et al., 2006), which completely suppresses action potentials, thereby virtually abolishing activity-dependent Ca^{2+} influx. On the other hand, simultaneous PKA activation triggered by GCs may further enhance sn-1-DAG lipase activity in neurons in the same way it does for DAG lipase isozymes found in rod membranes (Bisogno et al., 2003; Perez Roque et al., 1998).

Although PKC blockade prevented GSE, a $G\alpha_{q/11}$-mediated pathway does not seem to be required by GC-stimulated endocannabinoid biosynthesis, since intracellular application of antiserum against this subunit did not block GSE (Malcher-Lopes et al., 2006). Transgenic mice lacking $G\alpha_{q/11}$ proteins do exhibit normal basal brain levels of 2-AG (Wettschureck et al., 2006), but they show impaired activity-dependent 2-AG synthesis (Wettschureck et al., 2006). Somewhat surprisingly, these results indicate that $G\alpha_{q/11}$ is critical for activity-dependent 2-AG biosynthesis. On the other hand, they demonstrate that there are $G\alpha_{q/11}$-independent pathways for 2-AG biosynthesis under physiological conditions, which seems to be the case when both AEA and 2-AG are released by PVN neurons in response to GCs by an activity-independent mechanism. Many different hormone receptors activate PKC through its "classical" pathway in which $G\alpha_{q/11}$ stimulates the hydrolysis of PIP2 by PI-PLC, yielding the second messengers IP3 and DAG. IP3 acts on specific endoplasmic reticular receptors, thus evoking Ca^{2+} release into the cytoplasm, which along with DAG leads to the activation and translocation of cytosolic PKC to the cell membrane (Smrcka et al., 1991). Nevertheless, PI-PLCβ isozymes can also be activated by G$\beta\gamma$-subunit, albeit somewhat less effectively than by $G\alpha_{q/11}$-subunit in some cases (Camps et al., 1992; Gudermann et al., 1996; Hepler et al., 1993; Jin et al., 1998; Katz et al., 1992; Lee et al., 1993; Meldrum et al., 1991; Schnabel et al., 1993). The PI-PLCβ isozymes have distinct sites of interaction with Gα_q and G$\beta\gamma$, hence Gα_q- and G$\beta\gamma$-subunits may concomitantly modulate a single PI-PLCβ molecule (Rhee and Bae, 1997), which probably enhances the efficacy of Gα_q-coupled receptors in activating PI-PLCβ. However, there are also some Gα_s-coupled receptors connected to signaling cascades via G$\beta\gamma$-subunits. This is the case, for instance, of the β_2-adrenergic receptors, which potentiates cytosolic GC receptor transactivation in a hippocampal cell line (HT22) through a G$\beta\gamma$-dependent pathway (Schmidt et al., 2001). Therefore, in principle, the same putative Gα_s-coupled membrane GC receptor may activate both Gα_s-cAMP-PKA-mediated and G$\beta\gamma$-PLCβ-PKC-DAG lipase-mediated pathways, leading to concomitant, activity-independent release of both AEA and 2-AG.

In fact, there are some indications that GCs activate PLCβ. The nongenomic DEX-mediated stimulation of PI-PLC both on lymphoblastoid and in T cells is biphasic, starts within 15 s, decreases to the control levels after about 2 min and peaks again at 5 min (Bamberger et al., 1999; Graber and Losa, 1995; Tong et al., 2004). This biphasic kinetics is strikingly similar to that of different G-coupled, PLCβ-linked receptors, such as α-thrombin and angiotensin II receptors (Hwang et al., 2005; Schelling et al., 1997; Wright et al., 1988). In thymocytes, the nongenomic DEX-induced activation of PI-PLC was strongly reduced by RU486, but DEX could still increase PI-PLC activity by about two fold (Cifone et al., 1999). Therefore, it seems that GCs can nongenomically stimulate PI-PLCβ activity through at least two distinct signaling pathways, and one of them does not depend on the cytosolic GC receptor. Evidences also indicate that PI-PLCβ is the PI-PLC isozyme that mediates 2-AG production (Hashimotodani *et al.*, 2005; Jung *et al.*, 2005). This is the case, for instance, of 2-AG release triggered by the activation of $G_{q/11}$ protein-coupled receptors belonging to group I metabotropic glutamate receptors (mGluR) in primary cultures of corticostriatal and hippocampal slices (Jung *et al.*, 2005).

At this time, there is no direct evidence showing the involvement of G$\beta\gamma$ or PI-PLCβ isoform in GSE. Nevertheless, the current direct and indirect evidences discussed above support a model in which GCs act at a putative membrane G protein-coupled receptor to activate both Gα_s-cAMP-PKA- and G$\beta\gamma$-PLCβ-PKC-DAG lipase-mediated pathways, leading downstream to a mutual activation of AEA and 2-AG biosynthesis (Fig. 11.2). The activation of PI-PLCβ by G$\beta\gamma$ would allow for G$\alpha_{q/11}$-independent intracellular raise in Ca^{2+}, which is probably required for the N-acyltransferase-mediated formation of NArPE and sn-1-lyso-2-arachidonoyl-PC, by the subsequent NAPE-PLD-mediated formation of AEA and PA from NArPE, and by DAG lipase-mediated 2-AG formation from DAG (Fig. 11.1). Due to the relatively low efficacy of the G$\beta\gamma$-mediated pathway, it is possible that PKA-mediated phosphorylation will be necessary to fully activate N-acyltransferase and put on motion downstream biosynthetic steps for both AEA and 2-AG. Likewise, PKC activity may be required for a complete DAG lipase activation. The PA formed by NAPE-PLD may be then converted to 2-AG by the sequential actions of PA phosphatase and DAG lipase (Sugiura *et al.*, 2004). The removal of PA by this pathway is likely to increase the rate of NArPE and sn-1-lyso-2-arachidonoyl-PC formation, which in turn may facilitate 2-AG formation by the EGTA-insensitive PLC-mediated pathway (Di Marzo *et al.*, 1996). Since GCs stimulate the release of both AEA and 2-AG, this helps to explain why PKC blockade prevents GSE. In partial agreement with this model, recent data from Jeffrey Tasker's laboratory demonstrated that intracellular application of an antibody against Gbg subunit reduced but did not block GSE (Di *et al.*, 2009). Indicating that an additional mechanism is likely to be involved in PKC activation by GCs. Additionally, PKC activity has been shown to potentiate neurotransmitter-stimulated cAMP accumulation in the brain, as well as the Gα_s-coupled, β_2-adrenoceptor-stimulated cAMP formation in rat pinealocytes (Ho *et al.*, 1988; Karbon *et al.*, 1986).

VI. ENDOCANNABINOIDS METABOLIZATION

Several studies suggest that both AEA and 2-AG are internalized via a common carrier-mediated, facilitated diffusion process (Beltramo and Piomelli, 2000; Hillard and Jarrahian, 2000; Piomelli *et al.*, 1999), which in many instances is the rate-limiting step on the inactivation of endocannabinoid signaling. This carrier is expressed in several cell types, including glial cells, neurons, endothelial cells, and macrophages (Hillard and Jarrahian, 2000). It has been proposed that such a transporter may also be involved on the release of endocannabinoids to the extracellular space (Hillard and Jarrahian, 2000). The transporter activity is positively modulated by nitric oxide donors (Bisogno et al., 2001; Maccarrone et al., 2000). The expression of neuronal-type nitric oxide synthase (NOS) is upregulated in neuroblastoma cells by DEX (Schwarz et al., 1998), indicating that GCs may indirectly influence endocannabinoid levels by increasing NO-mediated activation of endocannabinoid reuptake. Once internalized, both AEA and 2-AG can be hydrolyzed by the endoplasmic reticular integral membrane-bound FAAH to form AA and ethanolamine from AEA, or AA and glycerol from 2-AG (Bisogno et al., 1999a, 2002; Deutsch et al., 2001; Di Marzo et al., 1999; Goparaju et al., 1998). For AEA, this seems to be the main degrading pathway in most of the cells, including brain neurons (Cravatt et al., 2001), and it seems to be an important metabolic pathway for 2-AG in macrophages (Goparaju et al., 1998). However, in the brain, it is the cytosolic monoacylglycerol (MAG) lipase that responds for most (~85%) of 2-AG hydrolysis (Blankman et al., 2007; Dinh et al., 2002; Nomura et al., 2008; Saario et al., 2005).

The cytochrome $p450_{2D6}$ is one of the major brain cytochrome p450 isoforms and has been implicated in neurodegeneration and psychosis (Snider et al., 2008), and has been shown to convert AEA to 20-HETE-EA and to 5, 6-, 8, 9-, 11, 12-, and 14, 15-EET-EAs in vitro with low K_m values (Snider et al., 2008). The cytochrome $P450_{4\times1}$, which is expressed in several human tissues, including heart, brain, and cerebellum, but has an yet unknown biological function, was shown to convert AEA into 14, 15-EET-EA in vitro, whereas AA and 2-AG were not efficiently metabolized by this cytochrome, indicating its specificity (Stark et al., 2008). In the liver, $P450_{4F2}$ is thought to be the isoform responsible for 20-HETE-EA formation, whereas $P450_{3A4}$ was identified as the primary enzyme responsible for AEA conversion into EET-EAs in the liver (Snider et al., 2007) and, perhaps in the brain (Bornheim et al., 1995). GCs induce the expression of hepatic cytochromes from the $P450_{3A}$ group (Hoen et al., 2000), suggesting that GCs may also regulate endocannabinoid metabolism via cytochrome p450 in the brain.

Leukocyte 12-lipoxygenase (12-LOX) and 15-lipoxygenase (15-LOX) oxygenate 2-AG, producing the hydroperoxyeicosatetraenoic acid glyceryl esters 12-HETE-G and 15-HETE-G, respectively (Kozak et al., 2002b; Moody et al., 2001). 2-AG is metabolized by 12-LOX 40% as efficiently as AA, whereas 15-LOX oxygenates 2-AG comparably or preferably to AA. 15-HETE-G is an agonist of the peroxisome proliferator-activated receptor α, a ligand-activated transcription factor belonging to the steroid hormone receptor superfamily (Kozak et al., 2002b), which has been implicated in anti-inflammatory and antiatherogenic actions in vascular cells, and in the control of genes involved in systemic lipid metabolism (Marx et al., 2004). The brain metabolite, 12-hydroxy-AEA, produced by direct oxygenation of AEA by 12-LOX has an affinity to CB1 receptor corresponding to twice that of AEA, although less active than the parent compound in inhibiting the murine vas deferens twitch response (Hampson et al., 1995). LOX inhibitors reduce the AEA-induced activation of vanilloid receptors in the guinea-pig bronchus (Craib et al., 2001). Likewise, AEA evoked depolarization of guinea-pig vagus nerve via activation of vanilloid receptors is partially suppressed by blockade of LOX activity (Kagaya et al., 2002). However, it is not clear whether the LOX products involved in vanilloid receptor activation in these cases are produced directly from AEA or from AEA-derived AA. GCs suppress EGF-induced expression of 12-LOX in human epidermoid carcinoma A431 cells through classical GC receptor activation (Chang et al., 1995), indicating that in some tissues GCs may regulate LOX-mediated metabolism of endocannabinoids via 12-LOX.

For more than a century, paracetamol has been among the most popular medicines, being widely used as analgesic, antipyretic, and nonsteroidal anti-inflammatory drug. Its mode of action, however, remained a mystery until 2005, when it was demonstrated that it indirectly causes the activation of cannabinoid receptors (Hogestatt et al., 2005). In the brain and spinal cord, paracetamol, following deacetylation to its primary amine (p-aminophenol), is conjugated with AA to form N-arachidonoylphenolamine, also known as AM404, which is a blocker for the endocannabinoid transporter. Therefore, paracetamol causes CB1 and CB2 receptors activation by preventing endocannabinoids reuptake and subsequent metabolization. Interestingly, AM404 is also an inhibitor of COX enzymes in the brain (Bertolini et al., 2006), which is likely to further contribute to endocannabinoid built-up, since COX is implicated in the nonhydrolytic, metabolization of endocannabinoids in the CNS and peripheral tissues. For instance, the vasodilator action of AEA and 2-AG in rat small mesenteric arteries is limited by local endothelial activity of FAAH, MAG lipase, or COX_2 (Ho and Randall, 2007), so that inhibition of COX_2 by nimesulide potentiated AEA-induced vascular relaxation (Ho and Randall, 2007). In some cases, it seems that FAAH removes AA from endocannabinoids, which are then converted to prostaglandins by COX_2 (Ho and Randall, 2007). Several cannabinoids are known to elicit systemic vasodilatation, mainly via cannabinoid CB1 and vanilloid TRPV1 receptors. However, both AEA and 2-AG are also able to increase pulmonary arterial pressure by a mechanism that is not mediated by CB1, CB2, or TRPV1 receptors (Wahn et al., 2005). The unspecific COX inhibitor acetylsalicylic acid (100 μM), the specific COX_2 inhibitor nimesulide (10 μM), the specific EP1 prostanoid receptor antagonist SC-19220 (100 μM), and FAAH inhibitor methyl arachidonyl fluorophosphonate (0.1 μM) all prevented this effect, indicating that endocannabinoids increase pulmonary arterial pressure via COX_2 metabolites of AA (Wahn et al., 2005).

COX_2 may also promote bioactive molecules synthesis using endocannabinoids as substrate. For instance, AEA-induced inhibition of interleukin-2 (IL-2) in primary splenocytes is not mediated by the cannabinoid CB1 or CB2 receptors, nor by the endocannabinoid membrane transporter or FAAH, but it was reduced by COX inhibitors (Rockwell and Kaminski, 2004), suggesting the involvement of oxygenated endocannabinoid products. In another example, AEA inhibited the growth of colorectal carcinoma cell lines HT29 and HCA7/C29, which express COX_2, but had little effect on the SW480 carcinoma cell line, which express very low COX_2 (Patsos et al., 2005). The induction of cell death by AEA in the COX_2-expressing cell lines was abrogated by the COX_2-selective inhibitor NS398, whereas inhibition of FAAH potentiated this effect, indicating that AEA-induced cell death was mediated via direct oxygenation of AEA by COX_2 (Patsos et al., 2005). In fact, both endocannabinoids are efficiently oxygenated by COX_2, but not COX_1, to produce hydroxy endoperoxides analogous to AA-derived PGH_2: $PG-H_2$-ethanolamide ($PG-H_2$-EA) and PGH_2-glycerol (PGH_2-G) from AEA and 2-AG, respectively. 2-AG is oxygenated by COX_2 as effectively as AA, whereas AEA oxygenation by COX_2 occurs with a relatively high (micromolar) K_m, suggesting that this reaction may only occur in tissues in which high amounts of AEA are found (Kozak and Marnett, 2002). After COX_2-mediated oxygenation, $PG-H_2$-EA and PGH_2-G can be converted into a range of prostaglandin ethanolamides (prostamides) and prostaglandin glyceryl esters (glycerylprostaglandins), respectively (Kozak et al., 2000; Rouzer and Marnett, 2008; Woodward et al., 2008; Yu et al., 1997). In murine RAW cells, the endocannabinoid-derived prostanoids $PG-H_2$-EA and PGH_2-G serve as precursors for glycerol esters and ethanolamides of the prostaglandin E2, prostaglandin D2, and prostaglandin F2α in reactions catalyzed by the respective prostaglandin synthases (Kozak et al., 2002a). Similarly, PGH_2-G and $PG-H_2$-EA served as substrates to produce the corresponding endocannabinoid-derived prostacyclin derivatives by prostacyclin synthases. It was also demonstrated that the sequential action of COX_2 and thromboxane synthase on AEA and 2-AG could generate thromboxane A2 ethanolamide and thromboxane A2 glycerol ester, respectively. However, endocannabinoid-derived thromboxanes are produced at much reduced levels when compared with AA-derived thromboxanes, indicating that this pathway may not be as physiologically significant (Kozak et al., 2002a).

The endocannabinoid-derived prostanoids PGH_2-G and $PG-H_2$-EA and their metabolites represent a unique class of lipids, and the scope of their biological functions is only beginning to be elucidated. Zymosan, a preparation of protein–carbohydrate complexes from yeast cell wall, which is used to induce experimental sterile inflammation, has been shown to stimulate PGH_2-G synthesis by resident peritoneal macrophages (Rouzer and Marnett, 2005), indicating its involvement in innate immune and inflammatory responses. However, prostamides and glycerylprostaglandins do not interact with any known natural or recombinant prostanoid receptors involved in such responses. In most cases, their biological effect can be achieved with very small concentrations, implicating high-affinity interaction between these molecules and unidentified specific receptors. In RAW macrophage cells, for example, picomolar concentrations of PGE2-G dose-dependently induces Ca^{2+} mobilization from endoplasmic reticulum stores and activation of PLCβ-PI3-PKC pathway, leading to the phosphorylation of ERK_1 and ERK_2 (Nirodi et al., 2004). PGE2-G and its chemically stable synthetic analog PGE2-G-serinolamide are potent reducers of intraocular pressure in dogs (Woodward et al., 2008), whereas the prostamides PGF2α-EA, PGD2-EA, and PGE2-EA are potent inducers of cat iris sphincter contraction (Matias et al., 2004). Subsequent work has demonstrated that the compound AGN204396 blocked prostamides-induced, but not PGE2-G-induced iris contraction, indicating separate receptors for prostamides and glycerylprostaglandins (Woodward et al., 2007). PGE2-G was detected in vivo and was shown to induce mechanical allodynia and thermal hyperalgesia in rats (Hu et al., 2008). Even though PGE2-G can be quickly metabolized into PGE2, these effects were not blocked by prostanoid receptors antagonists (Hu et al., 2008). Concerning pain sensitivity, these are the opposite effects of the parent endocannabinoid 2-AG, indicating that COX_2 can modulate pain sensation by oxygenating 2-AG (Hu et al., 2008). Given the regulation of COX_2 expression by GCs, this implies that COX_2 activity may represent a context-sensitive enzymatic switch that converts the anti-inflammatory and antinociceptive mediator 2-AG into a pronociceptive and proinflammatory prostanoid.

VII. Crosstalk Between GCs and COX_2 in the Control of Neuroinflammation and Neuroprotection

Probably, the most striking evidences currently available for the biological relevance of endocannabinoid-derived prostanoids are their role in the modulation of synaptic activity in the CNS. In contrast to their endocannabinoid precursors, AEA and 2-AG, which cause suppression of inhibitory transmission in hippocampal neurons, the glycerylprostaglandins PGE2-G, PGD2-G, and PGF2α-G, and the prostamide PGD2-EA (but not PGE2-EA or PGE2α-EA), stimulate inhibitory (GABAergic) synaptic transmission in cultured hippocampal neurons (Sang et al., 2006). This concentration-dependent effect was CB1 receptor-independent, but was attenuated by IP3 and MAPK inhibitors, which is reminiscent of the above-mentioned PGE2-G-mediated Ca^{2+} mobilization and IP_3 production in RAW macrophage cells (Nirodi et al., 2004). Likewise, these synaptic effects were not mediated by the same prostanoid receptors known to bind the corresponding AA-derived prostanoids. Furthermore, inhibition of COX_2 activity reduced spontaneous inhibitory synaptic activity and augmented the depolarization-induced suppression of inhibition (DSI, the CB1-mediated inhibition of presynaptic GABAergic neurons by depolarization-induced release of endocannabinoids from the postsynaptic neurons), indicating that tonic COX_2 activity may control endocannabinoid levels both during basal conditions and after depolarization-induced endocannabinoid biosynthesis. In agreement with this assumption, enhancement of COX_2 activity not only stimulated GABAergic synaptic transmission but also abolished DSI (Sang et al., 2006). PGE2-G was also shown to directly stimulate excitatory (glutamatergic) synaptic activity in the hippocampus, as revealed by CB1-independent increase in the frequency of miniature excitatory postsynaptic currents (Sang et al., 2006). Hence, PGE2-G is consistently involved in neuronal stimulation, increasing activity in both GABAergic and glutamatergic neurons, which is the opposite of the direct inhibitory synaptic effects triggered by the parent endocannabinoids, AEA and 2-AG, via CB1 receptors in the CNS (Lovinger, 2008). This highlights the functional relevance of COX_2-mediated oxygenation of endocannabinoids and its regulation by GCs.

The COXs isozymes are among the most preferred pharmacological targets for both nonsteroidal (e.g., acetylsalicylic acid, ibuprofen, and paracetamol) and steroidal anti-inflammatory drugs, such as DEX (Barnes, 1995). Overnight treatment of macrophages with DEX causes a relatively weak inhibition of PLA_2 (~23–27%) along with a much stronger reduction of COX activity (~40%), resulting in the suppression of AA-stimulated PGE2 synthesis in these cells (Goppelt-Struebe et al., 1989). This is likely a consequence of GC-induced downregulation of COX_2 mRNA translation, which proved to be the case in human pulmonary cells, where IL-β-induced PGE2 release is completely suppressed by DEX (Mitchell et al., 1994). In these cells, DEX reduced COX_2 gene transcription rate by 25–40%, and almost completely blocked COX_2 mRNA translation by reducing COX_2 mRNA half-life (Newton et al., 1998). This effect was mediated by the cytosolic GC receptor and was transcription-dependent, suggesting the involvement of GC-induced cellular factors in the rapid degradation of COX_2 mRNA (Newton et al., 1998; Ristimaki et al., 1996). Although COX_2 may be constitutively expressed in brain neurons at relatively low levels, its expression is upregulated several fold following the exposure to stressors or pathological conditions, such as immobilization, seizures, ischemia, and neurodegenerative diseases (Madrigal et al., 2003). In principle, therefore, GCs can override any ongoing stimulatory influence over COX_2 gene transcription by stimulating degradation of COX_2 mRNA.

The upregulation of COX_2 expression in cortical and hippocampal neurons following psychological stress (immobilization for 6 h) is prevented by blockade of a postsynaptic glutamate-activated receptor of the N-methyl-D-aspartate-sensitive (NMDA) type. NMDA receptors are Ca^{2+}-permeant channels involved in neuronal plasticity and learning processes (MacDonald et al., 2006), indicating the importance of neuronal activation for the induction of COX_2 in the brain, and the involvement of this enzyme in cognitive processes. Moreover, pretreatment with a specific COX_2 inhibitor decreased stress-induced glutathione oxidation in the cortex, indicating that COX_2 contributed substantially to the oxidative state of the brain and active-dependent neurotoxicity (Madrigal et al., 2003). Indeed, the increased excitatory synaptic transmission induced by PGE2-G in the hippocampus was shown to cause NMDA receptor-dependent, Ca^{2+}-mediated neurotoxic effects (Sang et al., 2006). Taking into account the antioxidant, anti-inflammatory, and neuroprotective effects of endocannabinoids during CNS inflammation (Carrier et al., 2005; Correa et al., 2005; Eljaschewitsch et al., 2006; Ullrich et al., 2006), the neurotoxic effect of PGE2-G indicates that COX_2 metabolism of endocannabinoids may contribute to inflammation-induced neurodegeneration, and that GC activity is likely to counteract these effects. In fact, the levels of COX_2 detected in the brain under different stress conditions, namely infectious (LPS treatment), metabolic (hypoglycemia), and psychological (restraint for 1 h), were much enhanced by depletion of endogenous GCs after adrenalectomy, and this increase was counteracted by exogenous corticosterone application (Matsuwaki et al., 2006). Both neuronal and glial cells express COX_2 under the opposing control of proinflammatory factors and GCs. For instance, basal COX_2 mRNA and protein, as well as PGE2 formation, are almost undetectable in unstimulated microglial cultures, but are strongly upregulated in response to LPS, an effect that is substantially suppressed by DEX (Bauer et al., 1997). DEX also suppresses IL-1β-induced COX_2 immunoreactivity and PGE2 release by human neuroblastoma cells (Hoozemans et al., 2001), and by murine astrocytes in culture (O'Banion et al., 1996). In astrocytes, DEX also blocks TNF-α-induced COX_2 expression (O'Banion et al., 1996). Brain injury and neurodegenerative diseases lead to activation of microglia and astrocytes, which in turn release cytokines and other intercellular signaling molecules involved in innate neuroinflammatory responses (Hauwel et al., 2005). Brain neuritic plaques characteristic of Alzheimer's disease are closely associated with IL-1α-expressing microglia (Griffin et al., 1995), and increased levels of IL-1β and IL-6 are found in the cerebrospinal fluid of both Alzheimer's and Parkinson's disease patients (Blum-Degen et al., 1995). For this reason, it has been suggested an interplay between glial-derived cytokines, such as IL-1α and IL-1β, and neuronal upregulation of COX_2 expression in these chronic neurodegenerative diseases (Hoozemans et al., 2001).

The crucial importance of a fine regulatory mechanism for COX_2 activity during the CNS response to stressors, injuries, infections, and other insults is corroborated by studies with COX_2 knockout ($COX_2^{-/-}$) mice (Bosetti, 2007). These animals are less susceptible to brain injury produced by middle cerebral artery occlusion, by bilateral common carotid artery occlusion, or by intracerebral NMDA-mediated neurotoxicity (Iadecola et al., 2001; Kunz et al., 2007; Sasaki et al., 2004), despite a compensatory increase in the expression of COX_1 (Bosetti et al., 2004). Similar reduction on damage caused by brain physical injury or neurotoxic cell death (induced by the glutamate receptor activator kainic acid) is obtained with specific COX_2 inhibitors (Kunz and Oliw, 2001; Kunz et al., 2006). The relative contribution of each COX isozyme is revealed by the fact that despite the increase in COX_1 expression, basal brain PGE2 level is decreased by about 50% in the $COX_2^{-/-}$ as compared with wild-type mice (Bosetti et al., 2004). On the other hand, transgenic mice overexpressing COX_2 show increased susceptibility to death caused by the neurotoxic effect of excessive neuronal activation after central injection of kainic acid (Kelley et al., 1999), and significant increase in brain infarct size after middle cerebral artery occlusion (Dore et al., 2003).

The GC-induced suppression of COX_2-mediated oxygenation of AA and AA-containing endocannabinoids is likely to reduce the release of proinflammatory lipids and allow for concomitant accumulation of endocannabinoids. Additionally, it is likely that the direct GC-induced synthesis of endocannabinoids observed in the hypothalamus may also occur in other brain areas rich in CB1 receptors and known to be sensitive to GC influence, such as the neocortex, hippocampus, amygdale, and the nucleus accumbens. On the other hand, in principle, GCs may homeostaticaly limit the duration of CB1 activation by stimulating NO-dependent endocannabinoid re-uptake and may even lead to a NO-dependent shift in phospholipid metabolism toward AA formation (Bisogno et al., 2001; Chalimoniuk et al., 2006; Di et al., 2009; Maccarrone et al., 2000; Schwarz et al., 1998). This system may represent a key neuroprotective mechanism against neurotoxicity during inflammation or after the excessive activation of excitatory neuronal circuits in the acute phase of a stress response, during seizures, or following ischemia. On the other hand, deregulation of this system may contribute to the etiology of neurodegenerative diseases.

VIII. Crosstalk Between GCs and COX_2 in the Control of Synaptic Plasticity and Learning Processes

The connection between cognitive deficits and impaired neuroprotection is clearly illustrated by the finding that site directed overexpression of functional COX_2 in neurons of the hippocampus, cortex, and amygdale resulted in cognitive deficits that are associated with a parallel age-dependent increase in neuronal apoptosis (Andreasson et al., 2001). Nevertheless, the relevance of COX_2 in cognitive processes is not limited to pathological deficits involving neuronal death. For instance, COX_2-selective inhibitors impair hippocampal-dependent, but not dorsal striate-dependent, memory consolidation (Rall et al., 2003; Teather et al., 2002). This effect was only observed when COX blockers were injected immediately after task training but not when they were administered 2 h later (Teather et al., 2002), suggesting a role of COX in the retention of short-term memory for subsequent consolidation. Other studies showed that the retention of hippocampus-dependent spatial learning was impaired by a COX_2-selective inhibitor, but not by a COX_1-selective inhibitor, suggesting the involvement of endocannabinoids. A physiological function for COX_2 in the learning processes is further supported by the involvement of this enzyme and its products on hippocampal synaptic plasticity. For instance, COX_2 inhibition enhances DSI in the hippocampus, which is an endocannabinoid-mediated processes (Sang et al., 2006; Wilson and Nicoll, 2001). In fact, inhibition of COX_2, but not of the AEA-degrading enzyme FAAH, prolongs DSI, indicating the importance of COX_2 on limiting the duration of endocannabinoid actions in the hippocampus (Kim and Alger, 2004). Moreover, by oxygenating AEA and 2-AG, besides turning off CB1-mediated signaling, COX_2 is likely to promote, via endocannabinoid-derived prostamides and glycerylprostaglandins, synaptic effects that are the opposite of those triggered by cannabinoid CB1 receptors in the hippocampus (Sang et al., 2006). Accordingly, exogenous PGE2-G application, or the promotion of its endogenous synthesis when COX_2 expression is increased by LPS application, leads to enhanced basal synaptic activity and augmented long-term potentiation (LTP) in the mouse hippocampus (Yang et al., 2008). In fact, most of the COX_2 oxidative metabolites of AEA and 2-AG, namely PGE2-EA, PGF2α-EA, PGE2-G, but not PGD2-EA, PGD2-G, and PGF2α-G, increased LTP when individually applied on hippocampal slices (Yang et al., 2008). Therefore, increased COX_2 expression enhances LTP, and abolishes DSI, which are the symmetrical effects of endocannabinoids in the hippocampus in some circumstances (Auclair et al., 2000; Chevaleyre and Castillo, 2004; Diana and Marty, 2004; Paton et al., 1998; Slanina et al., 2005; Terranova et al., 1995). On the other hand, inhibition of COX_2 reduces LTP and potentiates endocannabinoid-dependent DSI in the hippocampus (Yang et al., 2008). Since LTP is a persistent increase in synaptic strength associated with learning (MacDonald et al., 2006), these changes are likely to connect stress-related signaling with learning processes.

The hippocampus is one of the brain areas with the highest concentration of CB1 receptors. It is a major center for spatial learning and for temporary storage of information, which may be eventually consolidated in the neocortex (Ambrogi Lorenzini et al., 1999; Dash et al., 2004; Gatley et al., 1996; Knowlton and Fanselow, 1998; Roozendaal, 2003). Endocannabinoids are key players in the processes of memory extinction and reverse learning (Bitencourt et al., 2008; Riedel and Davies, 2005; Rueda-Orozco et al., 2008; Shiflett et al., 2004; Takahashi et al., 2005; Varvel et al., 2005, 2007; Wise et al., 2007; Wolff and Leander, 2003). When endocannabinoid signaling is systemically blocked, memories that refer to a previously learned answer, which is no longer valid, cannot be "erased" to accommodate the new ecologically relevant reality (Shiflett et al., 2004). Additionally, endocannabinoid system regulation is also essential for the extinction of aversive memories (Marsicano et al., 2002; Rueda-Orozco et al., 2008). Hence, the crosstalk between GCs, COX_2, and endocannabinoids in the hippocampus, in the amygdale (Albrecht, 2007; Quan et al., 1998; Yang et al., 2006) and, perhaps, in the dorsolateral striatum (Rossi et al., 2008), is probably critical for the coordination between learning processes associated with behavioral changes, including the acquisition of aversive memories related to ecologically relevant events. To better illustrate this hypothesis, let us take two hypothetic situations, one in which the animal drinks water from a pond contaminated with a pathogenic bacteria, and the other in which the animal escapes from a predator attack while foraging in a risky area. Simply put, on the first example, the animal should learn that drinking water from that specific pond or eating in that risky spot are not good responses anymore for thirst and hungry. Therefore, the animal should "unlearn" the old responses while searching for new ones. On the other hand, if predators eventually go away and never came back, before feeling safe to search for food in that same area again, the animal must now "unlearn" the newly acquired cognitive association between that spot and the menacing presence of predators. Here, "unlearn" does not necessarily means that the memory should be completely erased, but rather that the synapses, supposedly underlying the cognitive association between the problem and the older behavioral response, must be weakened relatively to those synapses connecting the same problem with a new response. From the surviving perspective, such a mechanism might be especially critical when a rewarding memory must be overridden by an aversive one, and *vice versa*. On the one hand, without endocannabinoids, these behavioral changes would be impaired. On the other hand, however, abnormally high levels of endocannabinoids would facilitate memory extinction but would also impair the subsequent reacquisition phase. This apparently paradoxical role of endocannabinoids is reflected in their ability to regulate selectively synaptic plasticity in brain areas such as the hippocampus. By inducing long-term depression of inhibitory synapses in restricted areas of the dendritic tree of hippocampal neurons, endocannabinoids selectively "prime" nearby excitatory synapses, thereby facilitating subsequent induction of LTP (Carlson et al., 2002; Chevaleyre and Castillo, 2004). This effect was observed in experimental conditions in which the cannabinoid CB1 receptor-mediated effect is induced by the release of endogenous cannabinoids. On the other hand, other reports show that exogenous application of endocannabinoids may actually suppress the formation of LTP (Paton *et al.*, 1998; Terranova *et al.*, 1995), leading to the assumption that endocannabinoid-mediated suppression of LTP in the hippocampus is mediated by a mechanism other than the suppression of inhibition (Paton *et al.*, 1998). The exact mechanisms underlying endocannabinoid-mediated synaptic plasticity in the hippocampus, and its implications for the cognitive process are still far from being completely understood; however, there are enough evidences to support their protagonist role in selective memory extinction and/or consolidation. Therefore, GC inhibitory effect on COX_2 activity, and perhaps its stimulatory effect on endocannabinoids synthesis, may contribute to deficits on reversal learning caused by chronic stress (Bondi *et al.*, 2008; Lapiz-Bluhm *et al.*, 2009), which is associated with a sustained elevation of brain GCs (Song *et al.*, 2006). Likewise, an increase on endocannabinoid signaling is likely to contribute to the deficit in spatial learning caused in rats by chronic stress (Song *et al.*, 2006) or by treatments with drugs that, like GCs, inhibit COX_2 activity, such as ibuprofen (Shaw *et al.*, 2003).

IX. Coordination of GC-Mediated Control of the Neuroimmune Response and Energy Homeostasis Control

Both basal and stress-related *homeostasis* is centrally coordinated by the sympathoadrenomedullary system (SAS), and by the HPA axis. When *homeostasis* is disturbed or threatened by immune challenge or by some stressors both the SAS and HPA axis become centrally activated, resulting in increased release of catecholamines from the adrenal medulla and peripheral sympathetic terminals, and increased GCs release from the adrenal cortex (Elenkov *et al.*, 2000). In the periphery, the HPA axis is functionally connected with the adrenal medulla by GCs produced in the adrenal cortex, which, via an intra-adrenal portal vascular system, gain access to the adrenal medulla, where they induce the enzyme phenylethanolamine-N-methyltransferase, which controls the synthesis of adrenaline and noradrenalin (Wurtman, 2002). Additionally, both SAS and the HPA axis are centrally controlled by the PVN, which integrates humoral, visceral, sensorial, emotional, and cognitive information to operate, via neuroendocrine and preautonomic pathways, the coordinated control of basal energy and fluid homeostasis, as well as the immune, metabolic, physiological, and behavioral adaptations to different stressors (Swanson and Sawchenko, 1980; Yang *et al.*, 1997). Preautonomic sympathetic neurons located in the

PVN send axons to relay centers in the brainstem, which give rise to preganglionic efferent fibers terminating in the spinal column ganglia. From these ganglia, postganglionic noradrenergic sympathetic fibers innervate virtually all systems, except the skeletal muscle, to ultimately stimulate the endocrine release of catecholamines from the adrenal medulla, and regulate the visceral function. The adrenal medulla has two separate populations of adrenaline-containing cells (A cells) and noradrenalin-containing cells (NA cells), which are innervated by separate groups of the preganglionic sympathetic neurons (Edwards et al., 1996; Yamaguchi-Shima et al., 2007). Most of the circulating adrenaline is secreted by adrenal A cells, while circulating noradrenalin seems to reflect the secretion from adrenal NA cells in addition to the release from sympathetic nerves (Edwards et al., 1996; Folkow and Von Euler, 1954; Vollmer et al., 1997; Wurtman, 2002). Circulating catecholamines and the noradrenalin locally released from sympathetic nonsynaptic varicosities into the lymphoid organs modulate the immune response by affecting lymphocyte traffic, circulation, and proliferation, and by modulating cytokine production (Elenkov et al., 2000). Circulating GCs and catecholamines are also critical for metabolic adjustments required during periods of fasting related to lack of food availability or due to the impossibility to search for food, if the animal is ill, or due to the impossibility of food digestion, in case of food poisoning or other causes for sickness. In such situations, metabolic energy stores, such as glycogen and fatty, must be mobilized as glucose and fatty acids in order to provide metabolic energy until the animal recovers conditions to acquire nutrients from the environment (Boyle et al., 1989; John et al., 1990; Peinado-Onsurbe et al., 1991).

When either the bee venom melittin, which stimulates AA release, or AA itself, is injected intracerebroventricularly (i.c.v.), they dose-dependently elevate plasma levels of both adrenaline and noradrenalin. The COX_2 inhibitor indomethacin abolishes the responses to both melittin and AA, indicating that COX_2 products mediate the activation of the SAS outflow in response to this central insult. More specifically, i.c.v. administration of PGE2, but not PGD2, PGF2α, or PGI2, significantly elevates plasma levels of noradrenalin, but not adrenaline. On the other hand, a thromboxane A2 mimetic microinjected into the PVN elevates plasma adrenaline, but has little effect on noradrenalin. Accordingly, inhibition of thromboxane A(2) synthase abolishes only the elevation of adrenaline, but not of noradrenalin, induced by either i.c.v. melittin or AA (Yokotani et al., 2000). Several signaling factors that stimulate SAS outflow via prostanoids are also involved in the central suppression of appetite in response to stress and/or sickness. For instance, i.c.v. injection of the neuropeptides urocortin, CRH, or glucagon-like peptide-1 (GLP-1) causes COX_2-dependent peripheral release of both catecholamines (Arai et al., 2008; Murakami et al., 2002). In all these cases, the release of adrenalin also proved to be mediated by thromboxane A2 (Arai et al., 2008; Yokotani et al., 2001). Both urocortin and CRH suppress appetite and stimulate SAS by activating the same CRH1 receptor in the PVN (Yokotani et al., 2001). Likewise, when synaptically released onto the PVN, GLP-1 activates CRH-expressing neurons, leading downstream to CRH-mediated suppression of food intake (Rowland et al., 1997; Turton et al., 1996). Centrally administered IL-1β also causes COX-dependent stimulation of the SAS- and CRH-dependent appetite suppression (Callahan and Piekut, 1997; Luheshi et al., 1999; Mrosovsky et al., 1989; Uehara et al., 1989; Yokotani et al., 1995). Receptors located in the PVN are also involved in the central anorexigenic effect of histamine and bombesin-like peptides, which also induces COX-dependent elevation of circulating catecholamines (Arora and Anubhuti, 2006; Ookuma et al., 1989; Sakata et al., 1997; Shimizu et al., 2005, 2007).

The central stimulation of SAS outflow by vasopressin, histamine, bombesin, or CRH were not affected by PLA_2 inhibitors, but were dose-dependently reduced by inhibitors of PI-PLC and DAG lipase blockers (Okada et al., 2002, 2003; Shimizu et al., 2004). Furthermore, vasopressin effect was attenuated by a MAG lipase inhibitor. Together these results indicate that 2-AG could provide some of the neuropeptide-induced AA necessary for COX_2-dependent SAS activation (Shimizu and Yokotani, 2008). In fact, central administration of 2-AG itself caused MAG lipase-dependent increase in circulating catecholamines (Shimizu and Yokotani, 2008). On the other hand, central cannabinoid CB1 receptor activation by a synthetic cannabinoid (WIN 55212-2) dose-dependently reduced the vasopressin-induced response, whereas the CB1 antagonist AM251 led to the opposite effect (Shimizu and Yokotani, 2008). These results indicate that the endogenous hypothalamic 2-AG (and perhaps AEA) is likely to suppress SAS outflow by acting at CB1 receptor. However, once metabolized by MAG lipase, 2-AG provides AA for COX_2-mediated synthesis of prostanoids that, in turn, activate SAS outflow. It is also possible that COX_2-mediated formation of PGE2-G will activate preautonomic neurons, leading to SAS activation. On the other hand, since GCs cause the release of 2-AG in the PVN (Malcher-Lopes et al., 2006), it seems very likely that GC-induced endocannabinoids may reduce synaptic activation of the SAS outflow in the same way as they do for the PVN neuroendocrine cells.

Both GCs (Tempel et al., 1993) and the cannabis-derived cannabinoid THC (Verty et al., 2005) cause increased appetite and weight gain when injected directly into the PVN. In fact, the PVN is the only hypothalamic center in which local GC application leads to hyperphagia (Tempel et al., 1993). The fact that endocannabinoids mediate the inhibitory effects of GCs in the PVN, along with several indirect evidences, indicates that the orexigenic effect of GCs are mediated, at least in part, by endocannabinoids via cannabinoid CB1 receptor activation in the PVN, which also contributes to explain the orexigenic effect of marijuana (Di et al., 2003;

Malcher-Lopes et al., 2006). On the other hand, GC-induced biosynthesis of both 2-AG and AEA in the PVN is completely blocked by leptin (Malcher-Lopes et al., 2006), which is an anorexigenic cytokine-like peptide produced by the adipose tissue, functioning as a major peripheral signal informing the CNS about the current nutritional state (Ahima, 2000; Ahima and Osei, 2004; Ahima et al., 1996). Leptin was also shown to prevent fasting-induced hyperphagia by blocking fasting-induced biosynthesis of endocannabinoids in the hypothalamus (Di Marzo et al., 2001; Kirkham et al., 2002). There are evidences that leptin-promoted short-term suppression of appetite and meal termination involves the activation of preautonomic oxytocinergic PVN neurons (Blevins et al., 2004). Leptin can also stimulate SAS outflow when injected i.c.v.; however, the primary hypothalamic site for this action seems to be the ventromedial hypothalamus, not the PVN (Satoh et al., 1999).

Leptin is released by adipose tissue in response to AA or PGE2, and circulating leptin is involved as a mediator for the LPS-induced anorexia and fever in rats (Fain et al., 2001; Sachot et al., 2004), which supports a key role of this peptide in the peripheral coordination between energy homeostasis and the adaptation to stress and inflammation. Additionally, in the CNS, this peptide acts as an indirect inflammatory signal by causing IL-1β release from microglial cells, by inducing IL-1β expression in macrophages located in the meninges and perivascular space, and by inducing COX_2 expression in endothelial cells (Inoue et al., 2006; Pinteaux et al., 2007). Moreover, leptin has been shown to stimulate $cPLA_2$ activity in bone marrow stromal cells, in alveolar macrophages and in muscle cells, leading, in the later two cases, to increased AA release (Bendinelli et al., 2005; Kim et al., 2003; Mancuso et al., 2004). Accordingly, leptin was shown to cause the release of PGE2 and PGF2α in the neonatal hypothalamus (Brunetti et al., 1999). Hence, acting as a proinflammatory factor, leptin may promote a shift toward AA mobilization and, perhaps, AA-containing endocannabinoid metabolism by COX_2, favoring prostanoids production in the endothelial cells and in the hypothalamic cells under the influence of IL-1β secreted by glial cells and macrophages. As a nutritional state-dependent modulator of energy homeostasis, leptin acts in the PVN to suppress GC-induced release of endocannabinoids and in the ventromedial hypothalamus to prevent activity-dependent release of endocannabinoids (Jo et al., 2005; Malcher-Lopes et al., 2006). In all these cases, leptin actions tend toward the prevention of endocannabinoid accumulation. Additionally, leptin has been shown to stimulate the expression and the activity of FAAH in lymphocytes (Maccarrone et al., 2003), and, in the uterus of leptin knockout (ob/ob) mice, levels of both AEA and 2-AG are significantly elevated as compared to wild-type littermates, due to reduced activity of both FAAH and MAG lipase (Maccarrone et al., 2005). There are indications, however, that leptin stimulatory effect on FAAH may not occur in hypothalamic neurons, since it was absent in neuroblastoma cells (Gasperi et al., 2005), and FAAH activity was not increased in leptin treated rat hypothalamus as accessed by postmortem assays with tissue extracts (Di Marzo et al., 2001). Nevertheless, it is still an open question whether or not leptin modulates MAG activity in the CNS.

The cytokine IL-1β is released in the brain by glial cells (Griffin et al., 1995) and by activated macrophages in the periphery, functioning as a major mediator of the humoral communication between the peripheral immune system and the CNS. IL-1β is transported from blood to brain across the blood–brain barrier by a saturable system (Banks et al., 1989; Coceani et al., 1988; McLay et al., 2000). IL-1b has been shown to activate the HPA axis, however, the receptors for IL-1β (IL-1R) do not seem to be localized on PVN neurons, leading to the assumption that it may act through IL-1R receptors located on endothelial cells, which are abundant on PVN blood vessels, to stimulate local release of prostaglandins (Parsadaniantz et al., 2000; Quan et al., 2003). Prostaglandin in turn stimulates the HPA axis and other neuroendocrine systems (Berkenbosch et al., 1987; Besedovsky et al., 1986; Del Rey et al., 1987; Sapolsky et al., 1987). In respect to IL-1b role in the coordination between energy homeostasis and immune response, it is striking that, besides being stimulated by the anorexigenic hormone leptin, IL-1β itself stimulates neuronal activity to induce anorexia and loss of body weight when injected into the PVN (Avitsur et al., 1997). Electrophysiological studies showed that IL-1β inhibits local GABA synaptic inputs into both parvocellular and magnocellular PVN neurons, thereby disinhibiting them and facilitating their depolarization (Ferri and Ferguson, 2003; Ferri et al., 2005), an effect that was abolished by COX inhibitors, indicating the participation of prostaglandins as downstream mediators of IL-1β synaptic effects. Furthermore, both IL-1β and PGE2 also caused hyperpolarization (inhibition) in putative GABAergic hypothalamic neurons surrounding the PVN (Ferri and Ferguson, 2005), which are probably inhibitory interneurons upstream from the PVN neuroendocrine cells (Boudaba et al., 1996). In fact, COX inhibition abolished the activation of the HPA axis by both central and peripheral administration of IL-1β (Katsuura et al., 1988; Parsadaniantz et al., 2000). COX inhibitors also blocked IL-1-, IL-2-, and IL-6-induced CRH release from medial basal hypothalamic explants (Karanth et al., 1995; Navarra et al., 1991).

Therefore, it seems that the integrative role of the hypothalamus, especially of the PVN, in coordinating metabolic, physiological, and behavioral adaptations to different stressors strongly relies on a regulatory switch centered at the GC-dependent determination of the AA metabolic fate in a nutritional and inflammatory state-dependent fashion, as signaled by hypothalamic levels of leptin and IL-1β. Thus, low levels of leptin and/or IL-1β are permissive for the GC-induced production of AA-containing endocannabinoids in the hypothalamus. In the PVN, the subsequent CB1

Glucocorticoids, Endocannabinoids, and Homeostasis 293

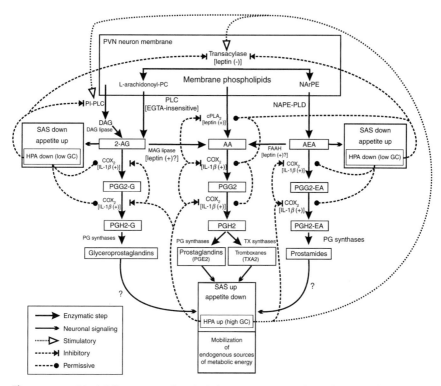

Figure 11.3 Model for a neurochemical, homeostatic switch in the central coordination between energy homeostasis and the neuroimmune/stress response. Different stressors require the mobilization of internal sources of metabolic energy. This is the case, for instance, during illness, which imposes rest (energy conservation) and impairs the search for food. This is also the case during dehydration, when eating must be avoided to prevent further increase of blood osmolarity, or after food poisoning, when food already ingested must not be absorbed. Depending on the case, distinct humoral and/or neuronal signaling (neuropeptides) act through the hypothalamic circuits to switch the metabolism of neuronal membrane phospholipids toward the production of arachidonic acid and arachidonate-containing proinflammatory signaling molecules, which in turn activate neuroendocrine and preautonomic PVN neurons, leading to the stimulation of both the sympathoadrenomedullary system (SAS) and the hypothalamic–pituitary–adrenal (HPA) axis. The subsequent increase on the circulating levels of catecholamines and glucocorticoids promotes the mobilization of glucose and fatty acid from endogenous stores and switches the metabolism to favor peripheral use of fatty acids as the main source of energy, while saving glucose for the brain. Activation of the sympathetic system also stimulates the immune response and contributes to appetite suppression. At the PVN, the cytokines IL-1β and leptin modulate this system, contributing to the maintenance of both the anorexia and proinflammatory response. The hypoglycemia and the caloric deficit associated with prolonged fasting are also a signal for the mobilization of endogenous sources of energy. However, in the absence of concurrent infection, the lower levels of leptin and the higher levels of glucocorticoids, which occur during caloric deficit, favor glucocorticoid-induced production of endocannabinoids in the PVN. Likewise, a drop on the PVN levels of IL-1β after

activation by AEA and 2-AG is likely to turn down HPA axis-mediated response to stress, to increase appetite, to reduce metabolic rate and energy expenditure by suppressing the HPT axis, and to reduce SAS-stimulated proinflammatory responses and mobilization of endogenous sources of metabolic energy. On the other hand, if leptin levels are elevated, GC-induced endocannabinoid synthesis is suppressed and MAG lipase activity may be increased, whereas, if IL-1β is elevated, COX_2 activation is likely to metabolize GC-induced endocannabinoids, and/or AA derived from endocannabinoids, to produce proinflammatory, anorexigenic prostanoids. Neuropeptides released by neuronal inputs from distinct CNS circuits and inflammatory factors constitute a diversified repertoire of neuromodulators, which can fine-tune this neurochemical switch in a context-dependent manner. This may be necessary to acutely produce a specific integrated response, which takes into account internal and external information in order to sustain general homeostasis and optimize survival in any given situation. Additionally, genomic effects of each endocannabinoid may differentially control COX_2 activity, even though at this time there is no direct evidence of such interactions in the PVN. Nevertheless, 2-AG has been shown to reduce COX_2 expression in the hippocampus after proinflammatory or neurotoxic (kainic acid) stimuli (Zhang and Chen, 2008), and to reduce the expression of proinflammatory cytokines TNF-a, IL-1b, and IL-6 after closed head injury (Panikashvili et al., 2006). By contrast, AEA was shown to induce COX_2 expression in cerebral microvascular endothelium in a concentration-dependent fashion (Chen et al., 2005).

In summary, on the one hand, the induction of COX_2 by cytokines favors the production of prostaglandins and endocannabinoid-derived prostanoids, reducing CB1-mediated signaling. On the other hand, elevated GCs would shift membrane lipid metabolism, leading to reduction of prostaglandins, thromboxanes, and endocannabinoid-derived prostanoids, with the concomitant accumulation of endocannabinoids. Subsequently, this increase on PVN endocannabinoids leads to suppression of the HPA axis, thereby reducing GC levels, which, in turn, reduces endocannabinoid

---

recovery from an infection, for instance, would also be permissive for such a switch. In both cases, the anorexigenic, proinflammatory hypothalamic mode of operation is shifted toward the anti-inflammatory and orexigenic state promoted by the increase on endocannabinoids release and on the CB1-mediated regulation of both the neuroendocrine and preautonomic function. From this regulatory loop, controlled by the opposing influences of glucocorticoids and cytokines on the crosstalk between arachidonic acid and endocannabinoid biochemical pathways, emerges a reciprocal feedback mechanism connecting endocannabinoid-mediated synaptic modulation and the HPA axis activity. [leptin (+)] and [leptin (−)], leptin stimulatory and inhibitory effects, respectively. [IL 1b (+)] and [IL 1b (−)], IL 1b stimulatory and inhibitory effects, respectively.

release and is permissive for COX_2 expression, favoring endocannabinoid-mediated signaling termination and, in some circumstances, the reversal of their effects (Fig. 11.3).

ACKNOWLEDGMENTS

We would like to thank Dr. Carlos Bloch Jr. for our insightful discussions and for providing us with a critical revision of this manuscript. We also would like to thank Dr. Beatriz Guimarães for her critical revision and valuable suggestions. This work was supported by the *Conselho Nacional de Desenvolvimento Científico e Tecnológico* (National Counsel of Technological and Scientific Development, CNPq) fellowship No. 151789/2007-0.

REFERENCES

Adesuyi, S. A., Cockrell, C. S., Gamache, D. A., and Ellis, E. F. (1985). Lipoxygenase metabolism of arachidonic acid in brain. *J. Neurochem.* **45,** 770–776.
Ahima, R. S. (2000). Leptin and the neuroendocrinology of fasting. *Front. Horm. Res.* **26,** 42–56.
Ahima, R. S., and Osei, S. Y. (2004). Leptin signaling. *Physiol. Behav.* **81,** 223–241.
Ahima, R. S., Prabakaran, D., Mantzoros, C., Qu, D., Lowell, B., Maratos-Flier, E., and Flier, J. S. (1996). Role of leptin in the neuroendocrine response to fasting. *Nature* **382,** 250–252.
Albrecht, D. (2007). Angiotensin-(1–7)-induced plasticity changes in the lateral amygdala are mediated by COX-2 and NO. *Learn. Mem.* **14,** 177–184.
Alkayed, N. J., Narayanan, J., Gebremedhin, D., Medhora, M., Roman, R. J., and Harder, D. R. (1996). Molecular characterization of an arachidonic acid epoxygenase in rat brain astrocytes. *Stroke* **27,** 971–979.
Ambrogi Lorenzini, C. G., Baldi, E., Bucherelli, C., Sacchetti, B., and Tassoni, G. (1999). Neural topography and chronology of memory consolidation: A review of functional inactivation findings. *Neurobiol. Learn. Mem.* **71,** 1–18.
Andreasson, K. I., Savonenko, A., Vidensky, S., Goellner, J. J., Zhang, Y., Shaffer, A., Kaufmann, W. E., Worley, P. F., Isakson, P., and Markowska, A. L. (2001). Age-dependent cognitive deficits and neuronal apoptosis in cyclooxygenase-2 transgenic mice. *J. Neurosci.* **21,** 8198–8209.
Arai, J., Okada, S., Yamaguchi-Shima, N., Shimizu, T., Sasaki, T., Yorimitsu, M., Wakiguchi, H., and Yokotani, K. (2008). Role of brain prostanoids in glucagon-like peptide-1-induced central activation of sympatho-adrenomedullary outflow in rats. *Clin. Exp. Pharmacol. Physiol.* **35,** 965–970.
Arora, S., and Anubhuti, S. (2006). Role of neuropeptides in appetite regulation and obesity—A review. *Neuropeptides* **40,** 375–401.
Auclair, N., Otani, S., Soubrie, P., and Crepel, F. (2000). Cannabinoids modulate synaptic strength and plasticity at glutamatergic synapses of rat prefrontal cortex pyramidal neurons. *J. Neurophysiol.* **83,** 3287–3293.
Avitsur, R., Pollak, Y., and Yirmiya, R. (1997). Administration of interleukin-1 into the hypothalamic paraventricular nucleus induces febrile and behavioral effects. *Neuroimmunomodulation* **4,** 258–265.
Bailey, J. M. (1991). New mechanisms for effects of anti-inflammatory glucocorticoids. *Biofactors* **3,** 97–102.

Bamberger, C. M., Else, T., Bamberger, A. M., Beil, F. U., and Schulte, H. M. (1999). Dissociative glucocorticoid activity of medroxyprogesterone acetate in normal human lymphocytes. *J. Clin. Endocrinol. Metab.* **84,** 4055–4061.

Banks, W. A., Kastin, A. J., and Durham, D. A. (1989). Bidirectional transport of interleukin-1 alpha across the blood-brain barrier. *Brain Res. Bull.* **23,** 433–437.

Barnes, P. J. (1995). Inhaled glucocorticoids for asthma. *N. Engl. J. Med.* **332,** 868–875.

Basselin, M., Villacreses, N. E., Lee, H. J., Bell, J. M., and Rapoport, S. I. (2007). Flurbiprofen, a cyclooxygenase inhibitor, reduces the brain arachidonic acid signal in response to the cholinergic muscarinic agonist, arecoline, in awake rats. *Neurochem. Res.* **32,** 1857–1867.

Bauer, M. K., Lieb, K., Schulze-Osthoff, K., Berger, M., Gebicke-Haerter, P. J., Bauer, J., and Fiebich, B. L. (1997). Expression and regulation of cyclooxygenase-2 in rat microglia. *Eur. J. Biochem.* **243,** 726–731.

Bazan, N. G. (2005). Lipid signaling in neural plasticity, brain repair, and neuroprotection. *Mol. Neurobiol.* **32,** 89–103.

Becker, J. L., Grasso, R. J., and Davis, J. S. (1988). Dexamethasone action inhibits the release of arachidonic acid from phosphatidylcholine during the suppression of yeast phagocytosis in macrophage cultures. *Biochem. Biophys. Res. Commun.* **153,** 583–590.

Beltramo, M., and Piomelli, D. (2000). Carrier-mediated transport and enzymatic hydrolysis of the endogenous cannabinoid 2-arachidonylglycerol. *Neuroreport* **11,** 1231–1235.

Bendinelli, P., Piccoletti, R., and Maroni, P. (2005). Leptin rapidly activates PPARs in C2C12 muscle cells. *Biochem. Biophys. Res. Commun.* **332,** 719–725.

Berkenbosch, F., van Oers, J., del Rey, A., Tilders, F., and Besedovsky, H. (1987). Corticotropin-releasing factor-producing neurons in the rat activated by interleukin-1. *Science* **238,** 524–526.

Bertolini, A., Ferrari, A., Ottani, A., Guerzoni, S., Tacchi, R., and Leone, S. (2006). Paracetamol: New vistas of an old drug. *CNS Drug Rev.* **12,** 250–275.

Besedovsky, H., del Rey, A., Sorkin, E., and Dinarello, C. A. (1986). Immunoregulatory feedback between interleukin-1 and glucocorticoid hormones. *Science* **233,** 652–654.

Bhattacharjee, A. K., Chang, L., Lee, H. J., Bazinet, R. P., Seemann, R., and Rapoport, S. I. (2005). D2 but not D1 dopamine receptor stimulation augments brain signaling involving arachidonic acid in unanesthetized rats. *Psychopharmacology (Berl.)* **180,** 735–742.

Bisogno, T., Delton-Vandenbroucke, I., Milone, A., Lagarde, M., and Di Marzo, V. (1999a). Biosynthesis and inactivation of N-arachidonoylethanolamine (anandamide) and N-docosahexaenoylethanolamine in bovine retina. *Arch. Biochem. Biophys.* **370,** 300–307.

Bisogno, T., Melck, D., De Petrocellis, L., and Di Marzo, V. (1999b). Phosphatidic acid as the biosynthetic precursor of the endocannabinoid 2-arachidonoylglycerol in intact mouse neuroblastoma cells stimulated with ionomycin. *J. Neurochem.* **72,** 2113–2119.

Bisogno, T., MacCarrone, M., De Petrocellis, L., Jarrahian, A., Finazzi-Agro, A., Hillard, C., and Di Marzo, V. (2001). The uptake by cells of 2-arachidonoylglycerol, an endogenous agonist of cannabinoid receptors. *Eur. J. Biochem.* **268,** 1982–1989.

Bisogno, T., De Petrocellis, L., and Di Marzo, V. (2002). Fatty acid amide hydrolase, an enzyme with many bioactive substrates. Possible therapeutic implications. *Curr. Pharm. Des.* **8,** 533–547.

Bisogno, T., et al. (2003). Cloning of the first sn1-DAG lipases points to the spatial and temporal regulation of endocannabinoid signaling in the brain. *J. Cell Biol.* **163,** 463–468.

Bitencourt, R. M., Pamplona, F. A., and Takahashi, R. N. (2008). Facilitation of contextual fear memory extinction and anti-anxiogenic effects of AM404 and cannabidiol in conditioned rats. *Eur. Neuropsychopharmacol.* **18,** 849–859.

Blankman, J. L., Simon, G. M., and Cravatt, B. F. (2007). A comprehensive profile of brain enzymes that hydrolyze the endocannabinoid 2-arachidonoylglycerol. *Chem. Biol.* **14,** 1347–1356.

Blevins, J. E., Schwartz, M. W., and Baskin, D. G. (2004). Evidence that paraventricular nucleus oxytocin neurons link hypothalamic leptin action to caudal brain stem nuclei controlling meal size. *Am. J. Physiol. Regul. Integr. Comp. Physiol.* **287,** R87–96.

Blum-Degen, D., Muller, T., Kuhn, W., Gerlach, M., Przuntek, H., and Riederer, P. (1995). Interleukin-1 beta and interleukin-6 are elevated in the cerebrospinal fluid of Alzheimer's and de novo Parkinson's disease patients. *Neurosci. Lett.* **202,** 17–20.

Bondi, C. O., Rodriguez, G., Gould, G. G., Frazer, A., and Morilak, D. A. (2008). Chronic unpredictable stress induces a cognitive deficit and anxiety-like behavior in rats that is prevented by chronic antidepressant drug treatment. *Neuropsychopharmacology* **33,** 320–331.

Bornheim, L. M., Kim, K. Y., Chen, B., and Correia, M. A. (1995). Microsomal cytochrome P450-mediated liver and brain anandamide metabolism. *Biochem. Pharmacol.* **50,** 677–686.

Bosetti, F. (2007). Arachidonic acid metabolism in brain physiology and pathology: Lessons from genetically altered mouse models. *J. Neurochem.* **102,** 577–586.

Bosetti, F., Langenbach, R., and Weerasinghe, G. R. (2004). Prostaglandin E2 and microsomal prostaglandin E synthase-2 expression are decreased in the cyclooxygenase-2-deficient mouse brain despite compensatory induction of cyclooxygenase-1 and Ca^{2+}-dependent phospholipase A2. *J. Neurochem.* **91,** 1389–1397.

Boudaba, C., Szabo, K., and Tasker, J. G. (1996). Physiological mapping of local inhibitory inputs to the hypothalamic paraventricular nucleus. *J. Neurosci.* **16,** 7151–7160.

Boyle, P. J., Shah, S. D., and Cryer, P. E. (1989). Insulin, glucagon, and catecholamines in prevention of hypoglycemia during fasting. *Am. J. Physiol.* **256,** E651–E661.

Brock, T. G., McNish, R. W., and Peters-Golden, M. (1999). Arachidonic acid is preferentially metabolized by cyclooxygenase-2 to prostacyclin and prostaglandin E2. *J. Biol. Chem.* **274,** 11660–11666.

Brunetti, L., Orlando, G., Michelotto, B., Ragazzoni, E., and Vacca, M. (1999). Leptin stimulates prostaglandin E2 and F2alpha, but not nitric oxide production in neonatal rat hypothalamus. *Eur. J. Pharmacol.* **369,** 299–304.

Cadas, H., Gaillet, S., Beltramo, M., Venance, L., and Piomelli, D. (1996). Biosynthesis of an endogenous cannabinoid precursor in neurons and its control by calcium and cAMP. *J. Neurosci.* **16,** 3934–3942.

Cajal, Y., and Jain, M. K. (1997). Synergism between mellitin and phospholipase A2 from bee venom: Apparent activation by intervesicle exchange of phospholipids. *Biochemistry* **36,** 3882–3893.

Calder, P. C. (2005). Polyunsaturated fatty acids and inflammation. *Biochem. Soc. Trans.* **33,** 423–427.

Callahan, T. A., and Piekut, D. T. (1997). Differential Fos expression induced by IL-1beta and IL-6 in rat hypothalamus and pituitary gland. *J. Neuroimmunol.* **73,** 207–211.

Camps, M., Hou, C., Sidiropoulos, D., Stock, J. B., Jakobs, K. H., and Gierschik, P. (1992). Stimulation of phospholipase C by guanine-nucleotide-binding protein beta gamma subunits. *Eur. J. Biochem.* **206,** 821–831.

Cannon, W. B. (1939). "The Wisdom of the Body." WW Norton & Company, New York.

Carlson, G., Wang, Y., and Alger, B. E. (2002). Endocannabinoids facilitate the induction of LTP in the hippocampus. *Nat. Neurosci.* **5,** 723–724.

Carrier, E. J., Kearn, C. S., Barkmeier, A. J., Breese, N. M., Yang, W., Nithipatikom, K., Pfister, S. L., Campbell, W. B., and Hillard, C. J. (2004). Cultured rat microglial cells synthesize the endocannabinoid 2-arachidonylglycerol, which increases proliferation via a CB2 receptor-dependent mechanism. *Mol. Pharmacol.* **65,** 999–1007.

Carrier, E. J., Patel, S., and Hillard, C. J. (2005). Endocannabinoids in neuroimmunology and stress. *Curr. Drug Targets CNS Neurol. Disord.* **4,** 657–665.

Chalimoniuk, M., Glowacka, J., Zabielna, A., Eckert, A., and Strosznajder, J. B. (2006). Nitric oxide alters arachidonic acid turnover in brain cortex synaptoneurosomes. *Neurochem. Int.* **48**, 1–8.

Chang, W. C., Kao, H. C., and Liu, Y. W. (1995). Down-regulation of epidermal growth factor-induced 12-lipoxygenase expression by glucocorticoids in human epidermoid carcinoma A431 cells. *Biochem. Pharmacol.* **50**, 947–952.

Chen, P., Hu, S., Yao, J., Moore, S. A., Spector, A. A., and Fang, X. (2005). Induction of cyclooxygenase-2 by anandamide in cerebral microvascular endothelium. *Microvasc. Res.* **69**, 28–35.

Chevaleyre, V., and Castillo, P. E. (2004). Endocannabinoid-mediated metaplasticity in the hippocampus. *Neuron* **43**, 871–881.

Chivers, J. E., Cambridge, L. M., Catley, M. C., Mak, J. C., Donnelly, L. E., Barnes, P. J., and Newton, R. (2004). Differential effects of RU486 reveal distinct mechanisms for glucocorticoid repression of prostaglandin E release. *Eur. J. Biochem.* **271**, 4042–4052.

Chrousos, G. P. (1998). Ultradian, circadian, and stress-related hypothalamic-pituitary-adrenal axis activity—A dynamic digital-to-analog modulation. *Endocrinology* **139**, 437–440.

Cifone, M. G., Migliorati, G., Parroni, R., Marchetti, C., Millimaggi, D., Santoni, A., and Riccardi, C. (1999). Dexamethasone-induced thymocyte apoptosis: Apoptotic signal involves the sequential activation of phosphoinositide-specific phospholipase C, acidic sphingomyelinase, and caspases. *Blood* **93**, 2282–2296.

Clark, M. A., Conway, T. M., Shorr, R. G., and Crooke, S. T. (1987). Identification and isolation of a mammalian protein which is antigenically and functionally related to the phospholipase A2 stimulatory peptide melittin. *J. Biol. Chem.* **262**, 4402–4406.

Coceani, F., Lees, J., and Dinarello, C. A. (1988). Occurrence of interleukin-1 in cerebrospinal fluid of the conscious cat. *Brain Res.* **446**, 245–250.

Correa, F., Mestre, L., Molina-Holgado, E., Arevalo-Martin, A., Docagne, F., Romero, E., Molina-Holgado, F., Borrell, J., and Guaza, C. (2005). The role of cannabinoid system on immune modulation: Therapeutic implications on CNS inflammation. *Mini. Rev. Med. Chem.* **5**, 671–675.

Cota, D. (2008). The role of the endocannabinoid system in the regulation of hypothalamic-pituitary-adrenal axis activity. *J. Neuroendocrinol.* **20**(Suppl. 1), 35–38.

Craib, S. J., Ellington, H. C., Pertwee, R. G., and Ross, R. A. (2001). A possible role of lipoxygenase in the activation of vanilloid receptors by anandamide in the guinea-pig bronchus. *Br. J. Pharmacol.* **134**, 30–37.

Cravatt, B. F., Demarest, K., Patricelli, M. P., Bracey, M. H., Giang, D. K., Martin, B. R., and Lichtman, A. H. (2001). Supersensitivity to anandamide and enhanced endogenous cannabinoid signaling in mice lacking fatty acid amide hydrolase. *Proc. Natl. Acad. Sci. USA* **98**, 9371–9376.

Croxtall, J. D., Choudhury, Q., Tokumoto, H., and Flower, R. J. (1995). Lipocortin-1 and the control of arachidonic acid release in cell signalling. Glucocorticoids (changed from glucorticoids) inhibit G protein-dependent activation of cPLA2 activity. *Biochem. Pharmacol.* **50**, 465–474.

Croxtall, J. D., Choudhury, Q., Newman, S., and Flower, R. J. (1996). Lipocortin 1 and the control of cPLA2 activity in A549 cells. Glucocorticoids block EGF stimulation of cPLA2 phosphorylation. *Biochem. Pharmacol.* **52**, 351–356.

Croxtall, J. D., Choudhury, Q., and Flower, R. J. (2000). Glucocorticoids act within minutes to inhibit recruitment of signalling factors to activated EGF receptors through a receptor-dependent, transcription-independent mechanism. *Br. J. Pharmacol.* **130**, 289–298.

Croxtall, J. D., van Hal, P. T., Choudhury, Q., Gilroy, D. W., and Flower, R. J. (2002). Different glucocorticoids vary in their genomic and non-genomic mechanism of action in A549 cells. *Br. J. Pharmacol.* **135**, 511–519.

Dallman, M. F., and Jones, M. T. (1973). Corticosteroid feedback control of ACTH secretion: Effect of stress-induced corticosterone ssecretion on subsequent stress responses in the rat. *Endocrinology* **92**, 1367–1375.

Dallman, M. F., Strack, A. M., Akana, S. F., Bradbury, M. J., Hanson, E. S., Scribner, K. A., and Smith, M. (1993). Feast and famine: Critical role of glucocorticoids with insulin in daily energy flow. *Front. Neuroendocrinol.* **14**, 303–347.

Dallman, M. F., Akana, S. F., Strack, A. M., Hanson, E. S., and Sebastian, R. J. (1995). The neural network that regulates energy balance is responsive to glucocorticoids and insulin and also regulates HPA axis responsivity at a site proximal to CRF neurons. *Ann. NY Acad. Sci.* **771**, 730–742.

Dash, P. K., Hebert, A. E., and Runyan, J. D. (2004). A unified theory for systems and cellular memory consolidation. *Brain Res. Rev.* **45**, 30–37.

Dayanithi, G., and Antoni, F. A. (1989). Rapid as well as delayed inhibitory effects of glucocorticoid hormones on pituitary adrenocorticotropic hormone release are mediated by type II glucocorticoid receptors and require newly synthesized messenger ribonucleic acid as well as protein. *Endocrinology* **125**, 308–313.

De Caterina, R., and Weksler, B. B. (1986). Modulation of arachidonic acid metabolism in human endothelial cells by glucocorticoids. *Thromb. Haemost.* **55**, 369–374.

DeGeorge, J. J., Noronha, J. G., Bell, J., Robinson, P., and Rapoport, S. I. (1989). Intravenous injection of [1-14C]arachidonate to examine regional brain lipid metabolism in unanesthetized rats. *J. Neurosci. Res.* **24**, 413–423.

Del Rey, A., Besedovsky, H., Sorkin, E., and Dinarello, C. A. (1987). Interleukin-1 and glucocorticoid hormones integrate an immunoregulatory feedback circuit. *Ann. NY Acad. Sci.* **496**, 85–90.

DeMar, J. C. Jr., Lee, H. J., Ma, K., Chang, L., Bell, J. M., Rapoport, S. I., and Bazinet, R. P. (2006). Brain elongation of linoleic acid is a negligible source of the arachidonate in brain phospholipids of adult rats. *Biochim. Biophys. Acta* **1761**, 1050–1059.

Dembinska-Kiec, A., Korbut, R., Zmuda, A., Kostka-Trabka, E., Simmet, T., and Peskar, B. A. (1984a). Formation of lipoxygenase and cyclooxygenase metabolites of arachidonic acid by brain tissue. *Biomed. Biochim. Acta* **43**, S222–S226.

Dembinska-Kiec, A., Simmet, T., and Peskar, B. A. (1984b). Formation of leukotriene C4-like material by rat brain tissue. *Eur. J. Pharmacol.* **99**, 57–62.

Deutsch, D. G., Glaser, S. T., Howell, J. M., Kunz, J. S., Puffenbarger, R. A., Hillard, C. J., and Abumrad, N. (2001). The cellular uptake of anandamide is coupled to its breakdown by fatty-acid amide hydrolase. *J. Biol. Chem.* **276**, 6967–6973.

Devane, W. A., and Axelrod, J. (1994). Enzymatic synthesis of anandamide, an endogenous ligand for the cannabinoid receptor, by brain membranes. *Proc. Natl. Acad. Sci. USA* **91**, 6698–6701.

Devane, W. A., Hanus, L., Breuer, A., Pertwee, R. G., Stevenson, L. A., Griffin, G., Gibson, D., Mandelbaum, A., Etinger, A., and Mechoulam, R. (1992). Isolation and structure of a brain constituent that binds to the cannabinoid receptor. *Science* **258**, 1946–1949.

Di, S., Malcher-Lopes, R., Halmos, K. C., and Tasker, J. G. (2003). Nongenomic glucocorticoid inhibition via endocannabinoid release in the hypothalamus: A fast feedback mechanism. *J. Neurosci.* **23**, 4850–4857.

Di, S., Malcher-Lopes, R., Marcheselli, V. L., Bazan, N. G., and Tasker, J. G. (2005). Rapid glucocorticoid-mediated endocannabinoid release and opposing regulation of glutamate and gamma-aminobutyric acid inputs to hypothalamic magnocellular neurons. *Endocrinology* **146**, 4292–4301.

Di, S., Maxson, M. M., Franco, A., and Tasker, J. G. (2009). Glucocorticoids regulate glutamate and GABA synapse-specific retrograde transmission via divergent nongenomic signaling pathways. *J. Neurosci.* **29**(2), 393–401.

Diana, M. A., and Marty, A. (2004). Endocannabinoid-mediated short-term synaptic plasticity: Depolarization-induced suppression of inhibition (DSI) and depolarization-induced suppression of excitation (DSE). *Br. J. Pharmacol.* **142**, 9–19.

Di Marzo, V. (1998). "Endocannabinoids" and other fatty acid derivatives with cannabimimetic properties: Biochemistry and possible physiopathological relevance. *Biochim. Biophys. Acta* **1392**, 153–175.

Di Marzo, V., and Deutsch, D. G. (1998). Biochemistry of the endogenous ligands of cannabinoid receptors. *Neurobiol. Dis.* **5**, 386–404.

Di Marzo, V., and Petrosino, S. (2007). Endocannabinoids and the regulation of their levels in health and disease. *Curr. Opin. Lipidol.* **18**, 129–140.

Di Marzo, V., Fontana, A., Cadas, H., Schinelli, S., Cimino, G., Schwartz, J. C., and Piomelli, D. (1994). Formation and inactivation of endogenous cannabinoid anandamide in central neurons. *Nature* **372**, 686–691.

Di Marzo, V., De Petrocellis, L., Sugiura, T., and Waku, K. (1996). Potential biosynthetic connections between the two cannabimimetic eicosanoids, anandamide and 2-arachidonoyl-glycerol, in mouse neuroblastoma cells. *Biochem. Biophys. Res. Commun.* **227**, 281–288.

Di Marzo, V., Bisogno, T., De Petrocellis, L., Melck, D., Orlando, P., Wagner, J. A., and Kunos, G. (1999). Biosynthesis and inactivation of the endocannabinoid 2-arachidonoylglycerol in circulating and tumoral macrophages. *Eur. J. Biochem.* **264**, 258–267.

Di Marzo, V., Goparaju, S. K., Wang, L., Liu, J., Batkai, S., Jarai, Z., Fezza, F., Miura, G. I., Palmiter, R. D., Sugiura, T., and Kunos, G. (2001). Leptin-regulated endocannabinoids are involved in maintaining food intake. *Nature* **410**, 822–825.

Di Marzo, V., De Petrocellis, L., Fezza, F., Ligresti, A., and Bisogno, T. (2002). Anandamide receptors. *Prostaglandins Leukot. Essent. Fatty Acids* **66**, 377–391.

Dinh, T. P., Freund, T. F., and Piomelli, D. (2002). A role for monoglyceride lipase in 2-arachidonoylglycerol inactivation. *Chem. Phys. Lipids* **121**, 149–158.

Doherty, N. S. (1981). Inhibition of arachidonic acid release as the mechanism by which glucocorticoids inhibit endotoxin-induced diarrhoea. *Br. J. Pharmacol.* **73**, 549–554.

Dore, S., Otsuka, T., Mito, T., Sugo, N., Hand, T., Wu, L., Hurn, P. D., Traystman, R. J., and Andreasson, K. (2003). Neuronal overexpression of cyclooxygenase-2 increases cerebral infarction. *Ann. Neurol.* **54**, 155–162.

Dubois, R. N., Abramson, S. B., Crofford, L., Gupta, R. A., Simon, L. S., Van De Putte, L. B., and Lipsky, P. E. (1998). Cyclooxygenase in biology and disease. *FASEB J.* **12**, 1063–1073.

Edwards, S. L., Anderson, C. R., Southwell, B. R., and McAllen, R. M. (1996). Distinct preganglionic neurons innervate noradrenaline and adrenaline cells in the cat adrenal medulla. *Neuroscience* **70**, 825–832.

Elenkov, I. J., Wilder, R. L., Chrousos, G. P., and Vizi, E. S. (2000). The sympathetic nerve—An integrative interface between two supersystems: The brain and the immune system. *Pharmacol. Rev.* **52**, 595–638.

Eljaschewitsch, E., Witting, A., Mawrin, C., Lee, T., Schmidt, P. M., Wolf, S., Hoertnagl, H., Raine, C. S., Schneider-Stock, R., Nitsch, R., and Ullrich, O. (2006). The endocannabinoid anandamide protects neurons during CNS inflammation by induction of MKP-1 in microglial cells. *Neuron* **49**, 67–79.

Emmerling, M. R., Moore, C. J., Doyle, P. D., Carroll, R. T., and Davis, R. E. (1993). Phospholipase A2 activation influences the processing and secretion of the amyloid precursor protein. *Biochem. Biophys. Res. Commun.* **197**, 292–297.

Epps, D. E., Schmid, P. C., Natarajan, V., and Schmid, H. H. (1979). N-Acylethanolamine accumulation in infarcted myocardium. *Biochem. Biophys. Res. Commun.* **90**, 628–633.

Epps, D. E., Natarajan, V., Schmid, P. C., and Schmid, H. O. (1980). Accumulation of N-acylethanolamine glycerophospholipids in infarcted myocardium. *Biochim. Biophys. Acta* **618**, 420–430.

Fain, J. N., Leffler, C. W., Cowan, G. S. Jr., Buffington, C., Pouncey, L., and Bahouth, S. W. (2001). Stimulation of leptin release by arachidonic acid and prostaglandin E(2) in adipose tissue from obese humans. *Metabolism* **50**, 921–928.

Ferri, C. C., and Ferguson, A. V. (2003). Interleukin-1 beta depolarizes paraventricular nucleus parvocellular neurones. *J. Neuroendocrinol.* **15**, 126–133.

Ferri, C. C., and Ferguson, A. V. (2005). Prostaglandin E2 mediates cellular effects of interleukin-1beta on parvocellular neurones in the paraventricular nucleus of the hypothalamus. *J. Neuroendocrinol.* **17**, 498–508.

Ferri, C. C., Yuill, E. A., and Ferguson, A. V. (2005). Interleukin-1beta depolarizes magnocellular neurons in the paraventricular nucleus of the hypothalamus through prostaglandin-mediated activation of a non selective cationic conductance. *Regul. Pept.* **129**, 63–71.

Fletcher, J. E., Jiang, M. S., Gong, Q. H., and Smith, L. A. (1991). Snake venom cardiotoxins and bee venom melittin activate phospholipase C activity in primary cultures of skeletal muscle. *Biochem. Cell Biol.* **69**, 274–281.

Floman, Y., and Zor, U. (1976). Mechanism of steroid action in inflammation: Inhibition of prostaglandin synthesis and release. *Prostaglandins* **12**, 403–413.

Folkow, B., and Von Euler, U. S. (1954). Selective activation of noradrenaline and adrenaline producing cells in the cat's adrenal gland by hypothalamic stimulation. *Circ. Res.* **2**, 191–195.

Garcia, D. E., Brown, S., Hille, B., and Mackie, K. (1998). Protein kinase C disrupts cannabinoid actions by phosphorylation of the CB1 cannabinoid receptor. *J. Neurosci.* **18**, 2834–2841.

Gasperi, V., Fezza, F., Spagnuolo, P., Pasquariello, N., and Maccarrone, M. (2005). Further insights into the regulation of human FAAH by progesterone and leptin implications for endogenous levels of anandamide and apoptosis of immune and neuronal cells. *Neurotoxicology* **26**, 811–817.

Gatley, S. J., Gifford, A. N., Volkow, N. D., Lan, R., and Makriyannis, A. (1996). 123I-labeled AM251: A radioiodinated ligand which binds *in vivo* to mouse brain cannabinoid CB1 receptors. *Eur. J. Pharmacol.* **307**, 331–338.

Gonzalez, S., Manzanares, J., Berrendero, F., Wenger, T., Corchero, J., Bisogno, T., Romero, J., Fuentes, J. A., Di Marzo, V., Ramos, J. A., and Fernandez-Ruiz, J. (1999). Identification of endocannabinoids and cannabinoid CB(1) receptor mRNA in the pituitary gland. *Neuroendocrinology* **70**, 137–145.

Goparaju, S. K., Ueda, N., Yamaguchi, H., and Yamamoto, S. (1998). Anandamide amidohydrolase reacting with 2-arachidonoylglycerol, another cannabinoid receptor ligand. *FEBS Lett.* **422**, 69–73.

Goppelt-Struebe, M., Wolter, D., and Resch, K. (1989). Glucocorticoids inhibit prostaglandin synthesis not only at the level of phospholipase A2 but also at the level of cyclooxygenase/PGE isomerase. *Br. J. Pharmacol.* **98**, 1287–1295.

Graber, R., and Losa, G. A. (1995). Changes in the activities of signal transduction and transport membrane enzymes in CEM lymphoblastoid cells by glucocorticoid-induced apoptosis. *Anal. Cell Pathol.* **8**, 159–175.

Griffin, W. S., Sheng, J. G., Roberts, G. W., and Mrak, R. E. (1995). Interleukin-1 expression in different plaque types in Alzheimer's disease: Significance in plaque evolution. *J. Neuropathol. Exp. Neurol.* **54**, 276–281.

Gudermann, T., Kalkbrenner, F., and Schultz, G. (1996). Diversity and selectivity of receptor-G protein interaction. *Annu. Rev. Pharmacol. Toxicol.* **36**, 429–459.

Hampson, A. J., Hill, W. A., Zan-Phillips, M., Makriyannis, A., Leung, E., Eglen, R. M., and Bornheim, L. M. (1995). Anandamide hydroxylation by brain lipoxygenase: Metabolite structures and potencies at the cannabinoid receptor. *Biochim. Biophys. Acta* **1259**, 173–179.

Hansen, H. S., Lauritzen, L., Strand, A. M., Moesgaard, B., and Frandsen, A. (1995). Glutamate stimulates the formation of N-acylphosphatidylethanolamine and N-acylethanolamine in cortical neurons in culture. *Biochim. Biophys. Acta* **1258**, 303–308.

Hansen, H. S., Moesgaard, B., Hansen, H. H., and Petersen, G. (2000). N-Acylethanolamines and precursor phospholipids—Relation to cell injury. *Chem. Phys. Lipids* **108**, 135–150.

Harrison, R. W., Balasubramanian, K., Yeakley, J., Fant, M., Svec, F., and Fairfield, S. (1979). Heterogeneity of AtT-20 cell glucocorticoid binding sites: Evidence for a membrane receptor. *Adv. Exp. Med. Biol.* **117**, 423–440.

Hashimotodani, Y., Ohno-Shosaku, T., Tsubokawa, H., Ogata, H., Emoto, K., Maejima, T., Araishi, K., Shin, H. S., and Kano, M. (2005). Phospholipase Cbeta serves as a coincidence detector through its Ca^{2+} dependency for triggering retrograde endocannabinoid signal. *Neuron* **45**, 257–268.

Hauwel, M., Furon, E., Canova, C., Griffiths, M., Neal, J., and Gasque, P. (2005). Innate (inherent) control of brain infection, brain inflammation and brain repair: The role of microglia, astrocytes, "protective" glial stem cells and stromal ependymal cells. *Brain Res. Brain Res. Rev.* **48**, 220–233.

Hepler, J. R., Kozasa, T., Smrcka, A. V., Simon, M. I., Rhee, S. G., Sternweis, P. C., and Gilman, A. G. (1993). Purification from Sf9 cells and characterization of recombinant Gq alpha and G11 alpha. Activation of purified phospholipase C isozymes by G alpha subunits. *J. Biol. Chem.* **268**, 14367–14375.

Hillard, C. J., and Jarrahian, A. (2000). The movement of N-arachidonoylethanolamine (anandamide) across cellular membranes. *Chem. Phys. Lipids* **108**, 123–134.

Hinz, B., and Hirschelmann, R. (2000). Rapid non-genomic feedback effects of glucocorticoids on CRF-induced ACTH secretion in rats. *Pharm. Res.* **17**, 1273–1277.

Ho, W. S., and Randall, M. D. (2007). Endothelium-dependent metabolism by endocannabinoid hydrolases and cyclooxygenases limits vasorelaxation to anandamide and 2-arachidonoylglycerol. *Br. J. Pharmacol.* **150**, 641–651.

Ho, A. K., Chik, C. L., and Klein, D. C. (1988). Effects of protein kinase inhibitor (1-(5-isoquinolinesulfonyl)-2-methylpiperazine (H7) on protein kinase C activity and adrenergic stimulation of cAMP and cGMP in rat pinealocytes. *Biochem. Pharmacol.* **37**, 1015–1020.

Hoen, P. A., Commandeur, J. N., Vermeulen, N. P., Van Berkel, T. J., and Bijsterbosch, M. K. (2000). Selective induction of cytochrome P450 3A1 by dexamethasone in cultured rat hepatocytes: Analysis with a novel reverse transcriptase-polymerase chain reaction assay section sign. *Biochem. Pharmacol.* **60**, 1509–1518.

Hogestatt, E. D., Jonsson, B. A., Ermund, A., Andersson, D. A., Bjork, H., Alexander, J. P., Cravatt, B. F., Basbaum, A. I., and Zygmunt, P. M. (2005). Conversion of acetaminophen to the bioactive N-acylphenolamine AM404 via fatty acid amide hydrolase-dependent arachidonic acid conjugation in the nervous system. *J. Biol. Chem.* **280**, 31405–31412.

Hoozemans, J. J., Veerhuis, R., Janssen, I., Rozemuller, A. J., and Eikelenboom, P. (2001). Interleukin-1beta induced cyclooxygenase 2 expression and prostaglandin E2 secretion by human neuroblastoma cells: Implications for Alzheimer's disease. *Exp. Gerontol.* **36**, 559–570.

Horrocks, L. A. (1989). Sources for brain arachidonic acid uptake and turnover in glycerophospholipids. *Ann. NY Acad. Sci.* **559**, 17–24.

Hu, S. S., Bradshaw, H. B., Chen, J. S., Tan, B., and Walker, J. M. (2008). Prostaglandin E2 glycerol ester, an endogenous COX-2 metabolite of 2-arachidonoylglycerol, induces hyperalgesia and modulates NFkappaB activity. *Br. J. Pharmacol.* **153**, 1538–1549.

Hwang, J. I., Shin, K. J., Oh, Y. S., Choi, J. W., Lee, Z. W., Kim, D., Ha, K. S., Shin, H. S., Ryu, S. H., and Suh, P. G. (2005). Phospholipase C-beta3 mediates the thrombin-induced Ca^{2+} response in glial cells. *Mol. Cells* **19**, 375–381.

Iadecola, C., Niwa, K., Nogawa, S., Zhao, X., Nagayama, M., Araki, E., Morham, S., and Ross, M. E. (2001). Reduced susceptibility to ischemic brain injury and N-methyl-D-aspartate-mediated neurotoxicity in cyclooxygenase-2-deficient mice. *Proc. Natl. Acad. Sci. USA* **98,** 1294–1299.

Inoue, W., Poole, S., Bristow, A. F., and Luheshi, G. N. (2006). Leptin induces cyclooxygenase-2 via an interaction with interleukin-1beta in the rat brain. *Eur. J. Neurosci.* **24,** 2233–2245.

Iversen, L., and Kragballe, K. (2000). Arachidonic acid metabolism in skin health and disease. *Prostaglandins Other Lipid Mediat.* **63,** 25–42.

Jin, T., Amzel, M., Devreotes, P. N., and Wu, L. (1998). Selection of gbeta subunits with point mutations that fail to activate specific signaling pathways *in vivo*: Dissecting cellular responses mediated by a heterotrimeric G protein in Dictyostelium discoideum. *Mol. Biol. Cell* **9,** 2949–2961.

Jin, X. H., Okamoto, Y., Morishita, J., Tsuboi, K., Tonai, T., and Ueda, N. (2007). Discovery and characterization of a Ca^{2+}-independent phosphatidylethanolamine N-acyltransferase generating the anandamide precursor and its congeners. *J. Biol. Chem.* **282,** 3614–3623.

Jingami, H., Matsukura, S., Numa, S., and Imura, H. (1985). Effects of adrenalectomy and dexamethasone administration on the level of prepro-corticotropin-releasing factor messenger ribonucleic acid (mRNA) in the hypothalamus and adrenocorticotropin/beta-lipotropin precursor mRNA in the pituitary in rats. *Endocrinology* **117,** 1314–1320.

Jo, Y. H., Chen, Y. J., Chua, S. C. Jr., Talmage, D. A., and Role, L. W. (2005). Integration of endocannabinoid and leptin signaling in an appetite-related neural circuit. *Neuron* **48,** 1055–1066.

John, G. W., Doxey, J. C., Walter, D. S., and Reid, J. L. (1990). The role of alpha- and beta-adrenoceptor subtypes in mediating the effects of catecholamines on fasting glucose and insulin concentrations in the rat. *Br. J. Pharmacol.* **100,** 699–704.

Jung, K. M., Mangieri, R., Stapleton, C., Kim, J., Fegley, D., Wallace, M., Mackie, K., and Piomelli, D. (2005). Stimulation of endocannabinoid formation in brain slice cultures through activation of group I metabotropic glutamate receptors. *Mol. Pharmacol.* **68,** 1196–1202.

Kagaya, M., Lamb, J., Robbins, J., Page, C. P., and Spina, D. (2002). Characterization of the anandamide induced depolarization of guinea-pig isolated vagus nerve. *Br. J. Pharmacol.* **137,** 39–48.

Karanth, S., Lyson, K., Aguila, M. C., and McCann, S. M. (1995). Effects of luteinizing-hormone-releasing hormone, alpha-melanocyte-stimulating hormone, naloxone, dexamethasone and indomethacin on interleukin-2-induced corticotropin-releasing factor release. *Neuroimmunomodulation* **2,** 166–173.

Karbon, E. W., Shenolikar, S., and Enna, S. J. (1986). Phorbol esters enhance neurotransmitter-stimulated cyclic AMP production in rat brain slices. *J. Neurochem.* **47,** 1566–1575.

Katsuura, G., Gottschall, P. E., Dahl, R. R., and Arimura, A. (1988). Adrenocorticotropin release induced by intracerebroventricular injection of recombinant human interleukin-1 in rats: Possible involvement of prostaglandin. *Endocrinology* **122,** 1773–1779.

Katz, A., Wu, D., and Simon, M. I. (1992). Subunits beta gamma of heterotrimeric G protein activate beta 2 isoform of phospholipase C. *Nature* **360,** 686–689.

Keller-Wood, M. E., and Dallman, M. F. (1984). Corticosteroid inhibition of ACTH secretion. *Endocr. Rev.* **5,** 1–24.

Keller-Wood, M., Leeman, E., Shinsako, J., and Dallman, M. F. (1988). Steroid inhibition of canine ACTH: *In vivo* evidence for feedback at the corticotrope. *Am. J. Physiol.* **255,** E241–E246.

Kelley, K. A., Ho, L., Winger, D., Freire-Moar, J., Borelli, C. B., Aisen, P. S., and Pasinetti, G. M. (1999). Potentiation of excitotoxicity in transgenic mice overexpressing neuronal cyclooxygenase-2. *Am. J. Pathol.* **155,** 995–1004.

Kim, J., and Alger, B. E. (2004). Inhibition of cyclooxygenase-2 potentiates retrograde endocannabinoid effects in hippocampus. *Nat. Neurosci.* **7,** 697–698.

Kim, G. S., Hong, J. S., Kim, S. W., Koh, J. M., An, C. S., Choi, J. Y., and Cheng, S. L. (2003). Leptin induces apoptosis via ERK/cPLA2/cytochrome *c* pathway in human bone marrow stromal cells. *J. Biol. Chem.* **278,** 21920–21929.

Kirkham, T. C., Williams, C. M., Fezza, F., and Di Marzo, V. (2002). Endocannabinoid levels in rat limbic forebrain and hypothalamus in relation to fasting, feeding and satiation: Stimulation of eating by 2-arachidonoyl glycerol. *Br. J. Pharmacol.* **136,** 550–557.

Knowlton, B. J., and Fanselow, M. S. (1998). The hippocampus, consolidation and on-line memory. *Curr. Opin. Neurobiol.* **8,** 293–296.

Kola, B., *et al.* (2008). The orexigenic effect of ghrelin is mediated through central activation of the endogenous cannabinoid system. *PLoS ONE* **3,** e1797.

Kondo, S., Sugiura, T., Kodaka, T., Kudo, N., Waku, K., and Tokumura, A. (1998). Accumulation of various *N*-acylethanolamines including *N*-arachidonoylethanolamine (anandamide) in cadmium chloride-administered rat testis. *Arch. Biochem. Biophys.* **354,** 303–310.

Kozak, K. R., and Marnett, L. J. (2002). Oxidative metabolism of endocannabinoids. *Prostaglandins Leukot. Essent. Fatty Acids* **66,** 211–220.

Kozak, K. R., Rowlinson, S. W., and Marnett, L. J. (2000). Oxygenation of the endocannabinoid, 2-arachidonylglycerol, to glyceryl prostaglandins by cyclooxygenase-2. *J. Biol. Chem.* **275,** 33744–33749.

Kozak, K. R., Crews, B. C., Morrow, J. D., Wang, L. H., Ma, Y. H., Weinander, R., Jakobsson, P. J., and Marnett, L. J. (2002a). Metabolism of the endocannabinoids, 2-arachidonylglycerol and anandamide, into prostaglandin, thromboxane, and prostacyclin glycerol esters and ethanolamides. *J. Biol. Chem.* **277,** 44877–44885.

Kozak, K. R., Gupta, R. A., Moody, J. S., Ji, C., Boeglin, W. E., DuBois, R. N., Brash, A. R., and Marnett, L. J. (2002b). 15-Lipoxygenase metabolism of 2-arachidonylglycerol. Generation of a peroxisome proliferator-activated receptor alpha agonist. *J. Biol. Chem.* **277,** 23278–23286.

Kruszka, K. K., and Gross, R. W. (1994). The ATP- and CoA-independent synthesis of arachidonoylethanolamide. A novel mechanism underlying the synthesis of the endogenous ligand of the cannabinoid receptor. *J. Biol. Chem.* **269,** 14345–14348.

Kunz, T., and Oliw, E. H. (2001). The selective cyclooxygenase-2 inhibitor rofecoxib reduces kainate-induced cell death in the rat hippocampus. *Eur. J. Neurosci.* **13,** 569–575.

Kunz, T., Marklund, N., Hillered, L., and Oliw, E. H. (2006). Effects of the selective cyclooxygenase-2 inhibitor rofecoxib on cell death following traumatic brain injury in the rat. *Restor. Neurol. Neurosci.* **24,** 55–63.

Kunz, A., Anrather, J., Zhou, P., Orio, M., and Iadecola, C. (2007). Cyclooxygenase-2 does not contribute to postischemic production of reactive oxygen species. *J. Cereb. Blood Flow Metab.* **27,** 545–551.

Lapiz-Bluhm, M. D., Soto-Pina, A. E., Hensler, J. G., and Morilak, D. A. (2009). Chronic intermittent cold stress and serotonin depletion induce deficits of reversal learning in an attentional set-shifting test in rats. *Psychopharmacology (Berl.)* **202**(1-3), 329–341.

Lee, S. B., Shin, S. H., Hepler, J. R., Gilman, A. G., and Rhee, S. G. (1993). Activation of phospholipase C-beta 2 mutants by G protein alpha q and beta gamma subunits. *J. Biol. Chem.* **268,** 25952–25957.

Lee, S. Y., Park, H. S., Lee, S. J., and Choi, M. U. (2001). Melittin exerts multiple effects on the release of free fatty acids from L1210 cells: Lack of selective activation of phospholipase A2 by melittin. *Arch. Biochem. Biophys.* **389,** 57–67.

Lee, H., Villacreses, N. E., Rapoport, S. I., and Rosenberger, T. A. (2004). *In vivo* imaging detects a transient increase in brain arachidonic acid metabolism: A potential marker of neuroinflammation. *J. Neurochem.* **91**(4), 936–945.

Leung, D., Saghatelian, A., Simon, G. M., and Cravatt, B. F. (2006). Inactivation of N-acyl phosphatidylethanolamine phospholipase D reveals multiple mechanisms for the biosynthesis of endocannabinoids. *Biochemistry* **45**, 4720–4726.

Lewis, G. D., Campbell, W. B., and Johnson, A. R. (1986). Inhibition of prostaglandin synthesis by glucocorticoids in human endothelial cells. *Endocrinology* **119**, 62–69.

Lin, L. L., Wartmann, M., Lin, A. Y., Knopf, J. L., Seth, A., and Davis, R. J. (1993). cPLA2 is phosphorylated and activated by MAP kinase. *Cell* **72**(2), 269–278.

Liu, J., Wang, L., Harvey-White, J., Osei-Hyiaman, D., Razdan, R., Gong, Q., Chan, A. C., Zhou, Z., Huang, B. X., Kim, H. Y., and Kunos, G. (2006). A biosynthetic pathway for anandamide. *Proc. Natl. Acad. Sci. USA* **103**, 13345–13350.

Lovinger, D. M. (2008). Presynaptic modulation by endocannabinoids. *Handb. Exp. Pharmacol.* 435–477.

Lowenberg, M., Tuynman, J., Bilderbeek, J., Gaber, T., Buttgereit, F., van Deventer, S., Peppelenbosch, M., and Hommes, D. (2005). Rapid immunosuppressive effects of glucocorticoids mediated through Lck and Fyn. *Blood* **106**, 1703–1710.

Luheshi, G. N., Gardner, J. D., Rushforth, D. A., Loudon, A. S., and Rothwell, N. J. (1999). Leptin actions on food intake and body temperature are mediated by IL-1. *Proc. Natl. Acad. Sci. USA* **96**, 7047–7052.

Maccarrone, M., Bari, M., Lorenzon, T., Bisogno, T., Di Marzo, V., and Finazzi-Agro, A. (2000). Anandamide uptake by human endothelial cells and its regulation by nitric oxide. *J. Biol. Chem.* **275**, 13484–13492.

Maccarrone, M., Di Rienzo, M., Finazzi-Agro, A., and Rossi, A. (2003). Leptin activates the anandamide hydrolase promoter in human T lymphocytes through STAT3. *J. Biol. Chem.* **278**, 13318–13324.

Maccarrone, M., Fride, E., Bisogno, T., Bari, M., Cascio, M. G., Battista, N., Finazzi Agro, A., Suris, R., Mechoulam, R., and Di Marzo, V. (2005). Up-regulation of the endocannabinoid system in the uterus of leptin knockout (ob/ob) mice and implications for fertility. *Mol. Hum. Reprod.* **11**, 21–28.

MacDonald, J. F., Jackson, M. F., and Beazely, M. A. (2006). Hippocampal long-term synaptic plasticity and signal amplification of NMDA receptors. *Crit. Rev. Neurobiol.* **18**, 71–84.

Madrigal, J. L., Moro, M. A., Lizasoain, I., Lorenzo, P., Fernandez, A. P., Rodrigo, J., Bosca, L., and Leza, J. C. (2003). Induction of cyclooxygenase-2 accounts for restraint stress-induced oxidative status in rat brain. *Neuropsychopharmacology* **28**, 1579–1588.

Maier, C., Runzler, D., Schindelar, J., Grabner, G., Waldhausl, W., Kohler, G., and Luger, A. (2005). G-protein-coupled glucocorticoid receptors on the pituitary cell membrane. *J. Cell Sci.* **118**, 3353–3361.

Malcher-Lopes, R., Di, S., Marcheselli, V. S., Weng, F. J., Stuart, C. T., Bazan, N. G., and Tasker, J. G. (2006). Opposing crosstalk between leptin and glucocorticoids rapidly modulates synaptic excitation via endocannabinoid release. *J. Neurosci.* **26**, 6643–6650.

Malcher-Lopes, R., Franco, A., and Tasker, J. G. (2008). Glucocorticoids shift arachidonic acid metabolism toward endocannabinoid synthesis: A non-genomic anti-inflammatory switch. *Eur. J. Pharmacol.* **583**, 322–339.

Mancuso, P., Canetti, C., Gottschalk, A., Tithof, P. K., and Peters-Golden, M. (2004). Leptin augments alveolar macrophage leukotriene synthesis by increasing phospholipase activity and enhancing group IVC iPLA2 (cPLA2gamma) protein expression. *Am. J. Physiol. Lung Cell Mol. Physiol.* **287**, L497–L502.

Marchetti, M. C., Di Marco, B., Cifone, G., Migliorati, G., and Riccardi, C. (2003). Dexamethasone-induced apoptosis of thymocytes: Role of glucocorticoid receptor-associated Src kinase and caspase-8 activation. *Blood* **101**, 585–593.

Marsicano, G., *et al.* (2002). The endogenous cannabinoid system controls extinction of aversive memories. *Nature* **418**, 530–534.

Marx, N., Duez, H., Fruchart, J. C., and Staels, B. (2004). Peroxisome proliferator-activated receptors and atherogenesis: Regulators of gene expression in vascular cells. *Circ. Res.* **94**, 1168–1178.

Matias, I., *et al.* (2004). Prostaglandin ethanolamides (prostamides): *In vitro* pharmacology and metabolism. *J. Pharmacol. Exp. Ther.* **309**, 745–757.

Matsuwaki, T., Kayasuga, Y., Yamanouchi, K., and Nishihara, M. (2006). Maintenance of gonadotropin secretion by glucocorticoids under stress conditions through the inhibition of prostaglandin synthesis in the brain. *Endocrinology* **147**, 1087–1093.

McGiff, J. C. (1991). Cytochrome P-450 metabolism of arachidonic acid. *Annu. Rev. Pharmacol. Toxicol.* **31**, 339–369.

McLay, R. N., Kastin, A. J., and Zadina, J. E. (2000). Passage of interleukin-1-beta across the blood-brain barrier is reduced in aged mice: A possible mechanism for diminished fever in aging. *Neuroimmunomodulation* **8**, 148–153.

Mechoulam, R., Fride, E., and Di Marzo, V. (1998). Endocannabinoids. *Eur. J. Pharmacol.* **359**, 1–18.

Meldrum, E., Parker, P. J., and Carozzi, A. (1991). The PtdIns-PLC superfamily and signal transduction. *Biochim. Biophys. Acta* **1092**, 49–71.

Mitchell, J. A., Belvisi, M. G., Akarasereenont, P., Robbins, R. A., Kwon, O. J., Croxtall, J., Barnes, P. J., and Vane, J. R. (1994). Induction of cyclo-oxygenase-2 by cytokines in human pulmonary epithelial cells: Regulation by dexamethasone. *Br. J. Pharmacol.* **113**, 1008–1014.

Mollay, C., Kreil, G., and Berger, H. (1976). Action of phospholipases on the cytoplasmic membrane of Escherichia coli. Stimulation by melittin. *Biochim. Biophys. Acta* **426**, 317–324.

Moody, J. S., Kozak, K. R., Ji, C., and Marnett, L. J. (2001). Selective oxygenation of the endocannabinoid 2-arachidonylglycerol by leukocyte-type 12-lipoxygenase. *Biochemistry* **40**, 861–866.

Mrosovsky, N., Molony, L. A., Conn, C. A., and Kluger, M. J. (1989). Anorexic effects of interleukin 1 in the rat. *Am. J. Physiol.* **257**, R1315–R1321.

Murakami, Y., Okada, S., Nishihara, M., and Yokotani, K. (2002). Roles of brain prostaglandin E2 and thromboxane A2 in the activation of the central sympatho-adrenomedullary outflow in rats. *Eur. J. Pharmacol.* **452**, 289–294.

Nakane, S., Oka, S., Arai, S., Waku, K., Ishima, Y., Tokumura, A., and Sugiura, T. (2002). 2-Arachidonoyl-*sn*-glycero-3-phosphate, an arachidonic acid-containing lysophosphatidic acid: Occurrence and rapid enzymatic conversion to 2-arachidonoyl-*sn*-glycerol, a cannabinoid receptor ligand, in rat brain. *Arch. Biochem. Biophys.* **402**, 51–58.

Natarajan, V., Schmid, P. C., Reddy, P. V., and Schmid, H. H. (1984). Catabolism of N-acylethanolamine phospholipids by dog brain preparations. *J. Neurochem.* **42**, 1613–1619.

Navarra, P., Tsagarakis, S., Faria, M. S., Rees, L. H., Besser, G. M., and Grossman, A. B. (1991). Interleukins-1 and -6 stimulate the release of corticotropin-releasing hormone-41 from rat hypothalamus *in vitro* via the eicosanoid cyclooxygenase pathway. *Endocrinology* **128**, 37–44.

Newton, R., Seybold, J., Kuitert, L. M., Bergmann, M., and Barnes, P. J. (1998). Repression of cyclooxygenase-2 and prostaglandin E2 release by dexamethasone occurs by transcriptional and post-transcriptional mechanisms involving loss of polyadenylated mRNA. *J. Biol. Chem.* **273**, 32312–32321.

Nirodi, C. S., Crews, B. C., Kozak, K. R., Morrow, J. D., and Marnett, L. J. (2004). The glyceryl ester of prostaglandin E2 mobilizes calcium and activates signal transduction in RAW264.7 cells. *Proc. Natl. Acad. Sci. USA* **101**, 1840–1845.

Nomura, D. K., Hudak, C. S., Ward, A. M., Burston, J. J., Issa, R. S., Fisher, K. J., Abood, M. E., Wiley, J. L., Lichtman, A. H., and Casida, J. E. (2008). Monoacylglycerol

lipase regulates 2-arachidonoylglycerol action and arachidonic acid levels. *Bioorg. Med. Chem. Lett.* **18,** 5875–5878.
O'Banion, M. K., Miller, J. C., Chang, J. W., Kaplan, M. D., and Coleman, P. D. (1996). Interleukin-1 beta induces prostaglandin G/H synthase-2 (cyclooxygenase-2) in primary murine astrocyte cultures. *J. Neurochem.* **66,** 2532–2540.
Oka, S., Yanagimoto, S., Ikeda, S., Gokoh, M., Kishimoto, S., Waku, K., Ishima, Y., and Sugiura, T. (2005). Evidence for the involvement of the cannabinoid CB2 receptor and its endogenous ligand 2-arachidonoylglycerol in 12-O-tetradecanoylphorbol-13-acetate-induced acute inflammation in mouse ear. *J. Biol. Chem.* **280,** 18488–18497.
Okada, S., Murakami, Y., Nakamura, K., and Yokotani, K. (2002). Vasopressin V(1) receptor-mediated activation of central sympatho-adrenomedullary outflow in rats. *Eur. J. Pharmacol.* **457,** 29–35.
Okada, S., Shimizu, T., and Yokotani, K. (2003). Brain phospholipase C and diacylglycerol lipase are involved in corticotropin-releasing hormone-induced sympatho-adrenomedullary outflow in rats. *Eur. J. Pharmacol.* **475,** 49–54.
Okamoto, Y., Morishita, J., Tsuboi, K., Tonai, T., and Ueda, N. (2004). Molecular characterization of a phospholipase D generating anandamide and its congeners. *J. Biol. Chem.* **279,** 5298–5305.
Ookuma, K., Yoshimatsu, H., Sakata, T., Fujimoto, K., and Fukagawa, F. (1989). Hypothalamic sites of neuronal histamine action on food intake by rats. *Brain Res.* **490,** 268–275.
Panikashvili, D., Shein, N. A., Mechoulam, R., Trembovler, V., Kohen, R., Alexandrovich, A., and Shohami, E. (2006). The endocannabinoid 2-AG protects the blood-brain barrier after closed head injury and inhibits mRNA expression of proinflammatory cytokines. *Neurobiol. Dis.* **22,** 257–264.
Parsadaniantz, S. M., Lebeau, A., Duval, P., Grimaldi, B., Terlain, B., and Kerdelhue, B. (2000). Effects of the inhibition of cyclo-oxygenase 1 or 2 or 5-lipoxygenase on the activation of the hypothalamic-pituitary-adrenal axis induced by interleukin-1beta in the male Rat. *J. Neuroendocrinol.* **12,** 766–773.
Patel, S., Roelke, C. T., Rademacher, D. J., Cullinan, W. E., and Hillard, C. J. (2004). Endocannabinoid signaling negatively modulates stress-induced activation of the hypothalamic-pituitary-adrenal axis. *Endocrinology* **145,** 5431–5438.
Paton, G. S., Pertwee, R. G., and Davies, S. N. (1998). Correlation between cannabinoid mediated effects on paired pulse depression and induction of long term potentiation in the rat hippocampal slice. *Neuropharmacology* **37,** 1123–1130.
Patsos, H. A., Hicks, D. J., Dobson, R. R., Greenhough, A., Woodman, N., Lane, J. D., Williams, A. C., and Paraskeva, C. (2005). The endogenous cannabinoid, anandamide, induces cell death in colorectal carcinoma cells: A possible role for cyclooxygenase 2. *Gut* **54,** 1741–1750.
Peinado-Onsurbe, J., Soler, C., Galan, X., Poveda, B., Soley, M., Llobera, M., and Ramirez, I. (1991). Involvement of catecholamines in the effect of fasting on hepatic endothelial lipase activity in the rat. *Endocrinology* **129,** 2599–2606.
Pepicelli, O., Fedele, E., Berardi, M., Raiteri, M., Levi, G., Greco, A., Ajmone-Cat, M. A., and Minghetti, L. (2005). Cyclo-oxygenase-1 and -2 differently contribute to prostaglandin E2 synthesis and lipid peroxidation after *in vivo* activation of N-methyl-D-aspartate receptors in rat hippocampus. *J. Neurochem.* **93,** 1561–1567.
Perez Roque, M. E., Pasquare, S. J., Castagnet, P. I., and Giusto, N. M. (1998). Can phosphorylation and dephosphorylation of rod outer segment membranes affect phosphatidate phosphohydrolase and diacylglycerol lipase activities? *Comp. Biochem. Physiol. B Biochem. Mol. Biol.* **119,** 85–93.
Pinteaux, E., Inoue, W., Schmidt, L., Molina-Holgado, F., Rothwell, N. J., and Luheshi, G. N. (2007). Leptin induces interleukin-1beta release from rat microglial cells through a caspase 1 independent mechanism. *J. Neurochem.* **102,** 826–833.

Piomelli, D., Beltramo, M., Glasnapp, S., Lin, S. Y., Goutopoulos, A., Xie, X. Q., and Makriyannis, A. (1999). Structural determinants for recognition and translocation by the anandamide transporter. *Proc. Natl. Acad. Sci. USA* **96,** 5802–5807.

Pompeo, A., Luini, A., and Buccione, R. (1997). Functional dissociation between glucocorticoid-induced decrease in arachidonic acid release and inhibition of adrenocorticotropic hormone secretion in AtT-20 corticotrophs. *J. Steroid Biochem. Mol. Biol.* **60,** 51–57.

Potestio, F. A., and Olson, D. M. (1990). Arachidonic acid release from cultured human amnion cells: The effect of dexamethasone. *J. Clin. Endocrinol. Metab.* **70,** 647–654.

Prescott, S. M., and Majerus, P. W. (1983). Characterization of 1,2-diacylglycerol hydrolysis in human platelets. Demonstration of an arachidonoyl-monoacylglycerol intermediate. *J. Biol. Chem.* **258,** 764–769.

Quan, N., Whiteside, M., and Herkenham, M. (1998). Cyclooxygenase 2 mRNA expression in rat brain after peripheral injection of lipopolysaccharide. *Brain Res.* **802,** 189–197.

Quan, N., He, L., and Lai, W. (2003). Endothelial activation is an intermediate step for peripheral lipopolysaccharide induced activation of paraventricular nucleus. *Brain Res. Bull.* **59,** 447–452.

Rall, J. M., Mach, S. A., and Dash, P. K. (2003). Intrahippocampal infusion of a cyclooxygenase-2 inhibitor attenuates memory acquisition in rats. *Brain Res.* **968,** 273–276.

Rao, N. M. (1992). Differential susceptibility of phosphatidylcholine small unilamellar vesicles to phospholipases A2, C and D in the presence of membrane active peptides. *Biochem. Biophys. Res. Commun.* **182,** 682–688.

Recordati, G. (2003). A thermodynamic model of the sympathetic and parasympathetic nervous systems. *Auton. Neurosci.* **103,** 1–12.

Rhee, S. G., and Bae, Y. S. (1997). Regulation of phosphoinositide-specific phospholipase C isozymes. *J. Biol. Chem.* **272,** 15045–15048.

Riedel, G., and Davies, S. N. (2005). Cannabinoid function in learning, memory and plasticity. *Handb. Exp. Pharmacol.* 445–477.

Ristimaki, A., Narko, K., and Hla, T. (1996). Down-regulation of cytokine-induced cyclooxygenase-2 transcript isoforms by dexamethasone: Evidence for post-transcriptional regulation. *Biochem. J.* **318**(Pt. 1), 325–331.

Robinson, P. J., Noronha, J., DeGeorge, J. J., Freed, L. M., Nariai, T., and Rapoport, S. I. (1992). A quantitative method for measuring regional *in vivo* fatty-acid incorporation into and turnover within brain phospholipids: Review and critical analysis. *Brain Res. Brain Res. Rev.* **17,** 187–214.

Rockwell, C. E., and Kaminski, N. E. (2004). A cyclooxygenase metabolite of anandamide causes inhibition of interleukin-2 secretion in murine splenocytes. *J. Pharmacol. Exp. Ther.* **311,** 683–690.

Roozendaal, B. (2003). Systems mediating acute glucocorticoid effects on memory consolidation and retrieval. *Prog. Neuropsychopharmacol. Biol. Psychiatry* **27,** 1213–1223.

Rosenberger, T. A., Villacreses, N. E., Hovda, J. T., Bosetti, F., Weerasinghe, G., Wine, R. N., Harry, G. J., and Rapoport, S. I. (2004). Rat brain arachidonic acid metabolism is increased by a 6-day intracerebral ventricular infusion of bacterial lipopolysaccharide. *J. Neurochem.* **88**(5), 1168–1178.

Rossi, S., De Chiara, V., Musella, A., Kusayanagi, H., Mataluni, G., Bernardi, G., Usiello, A., and Centonze, D. (2008). Chronic psychoemotional stress impairs cannabinoid-receptor-mediated control of GABA transmission in the striatum. *J. Neurosci.* **28,** 7284–7292.

Rouzer, C. A., and Marnett, L. J. (2005). Glycerylprostaglandin synthesis by resident peritoneal macrophages in response to a zymosan stimulus. *J. Biol. Chem.* **280,** 26690–26700.

Rouzer, C. A., and Marnett, L. J. (2008). Non-redundant functions of cyclooxygenases: Oxygenation of endocannabinoids. *J. Biol. Chem.* **283,** 8065–8069.

Rowland, N. E., Crews, E. C., and Gentry, R. M. (1997). Comparison of Fos induced in rat brain by GLP-1 and amylin. *Regul. Pept.* **71,** 171–174.

Rueda-Orozco, P. E., Montes-Rodriguez, C. J., Soria-Gomez, E., Mendez-Diaz, M., and Prospero-Garcia, O. (2008). Impairment of endocannabinoids activity in the dorsolateral striatum delays extinction of behavior in a procedural memory task in rats. *Neuropharmacology* **55,** 55–62.

Saario, S. M., Salo, O. M., Nevalainen, T., Poso, A., Laitinen, J. T., Jarvinen, T., and Niemi, R. (2005). Characterization of the sulfhydryl-sensitive site in the enzyme responsible for hydrolysis of 2-arachidonoyl-glycerol in rat cerebellar membranes. *Chem. Biol.* **12,** 649–656.

Sachot, C., Poole, S., and Luheshi, G. N. (2004). Circulating leptin mediates lipopolysaccharide-induced anorexia and fever in rats. *J. Physiol.* **561,** 263–272.

Sakata, T., Yoshimatsu, H., and Kurokawa, M. (1997). Hypothalamic neuronal histamine: Implications of its homeostatic control of energy metabolism. *Nutrition* **13,** 403–411.

Sang, N., Zhang, J., and Chen, C. (2006). PGE2 glycerol ester, a COX-2 oxidative metabolite of 2-arachidonoyl glycerol, modulates inhibitory synaptic transmission in mouse hippocampal neurons. *J. Physiol.* **572,** 735–745.

Sapolsky, R., Rivier, C., Yamamoto, G., Plotsky, P., and Vale, W. (1987). Interleukin-1 stimulates the secretion of hypothalamic corticotropin-releasing factor. *Science* **238,** 522–524.

Sapolsky, R. M., Romero, L. M., and Munck, A. U. (2000). How do glucocorticoids influence stress responses? Integrating permissive, suppressive, stimulatory, and preparative actions. *Endocr. Rev.* **21,** 55–89.

Sasaki, T., Kitagawa, K., Yamagata, K., Takemiya, T., Tanaka, S., Omura-Matsuoka, E., Sugiura, S., Matsumoto, M., and Hori, M. (2004). Amelioration of hippocampal neuronal damage after transient forebrain ischemia in cyclooxygenase-2-deficient mice. *J. Cereb. Blood Flow Metab.* **24,** 107–113.

Satoh, N., Ogawa, Y., Katsuura, G., Numata, Y., Tsuji, T., Hayase, M., Ebihara, K., Masuzaki, H., Hosoda, K., Yoshimasa, Y., and Nakao, K. (1999). Sympathetic activation of leptin via the ventromedial hypothalamus: Leptin-induced increase in catecholamine secretion. *Diabetes* **48,** 1787–1793.

Schelling, J. R., Nkemere, N., Konieczkowski, M., Martin, K. A., and Dubyak, G. R. (1997). Angiotensin II activates the beta 1 isoform of phospholipase C in vascular smooth muscle cells. *Am. J. Physiol.* **272,** C1558–C1566.

Schleimer, R. P., Davidson, D. A., Lichtenstein, L. M., and Adkinson, N. F., Jr. (1986). Selective inhibition of arachidonic acid metabolite release from human lung tissue by antiinflammatory steroids. *J. Immunol.* **136,** 3006–3011.

Schmid, H. H. (2000). Pathways and mechanisms of N-acylethanolamine biosynthesis: Can anandamide be generated selectively?. *Chem. Phys. Lipids* **108,** 71–87.

Schmid, H. H., and Berdyshev, E. V. (2002). Cannabinoid receptor-inactive N-acylethanolamines and other fatty acid amides: Metabolism and function. *Prostaglandins Leukot. Essent. Fatty Acids* **66,** 363–376.

Schmid, H. H., Schmid, P. C., and Natarajan, V. (1990). N-acylated glycerophospholipids and their derivatives. *Prog. Lipid Res.* **29,** 1–43.

Schmid, P. C., Krebsbach, R. J., Perry, S. R., Dettmer, T. M., Maasson, J. L., and Schmid, H. H. (1995). Occurrence and postmortem generation of anandamide and other long-chain N-acylethanolamines in mammalian brain. *FEBS Lett.* **375,** 117–120.

Schmidt, P., Holsboer, F., and Spengler, D. (2001). Beta(2)-adrenergic receptors potentiate glucocorticoid receptor transactivation via G protein beta gamma-subunits and the phosphoinositide 3-kinase pathway. *Mol. Endocrinol.* **15,** 553–564.

Schnabel, P., Camps, M., Carozzi, A., Parker, P. J., and Gierschik, P. (1993). Mutational analysis of phospholipase C-beta 2. Identification of regions required for membrane association and stimulation by guanine-nucleotide-binding protein beta gamma subunits. *Eur. J. Biochem.* **217,** 1109–1115.

Schwarz, P. M., Gierten, B., Boissel, J. P., and Forstermann, U. (1998). Expressional down-regulation of neuronal-type nitric oxide synthase I by glucocorticoids in N1E-115 neuroblastoma cells. *Mol. Pharmacol.* **54,** 258–263.

Seeds, M. C., and Bass, D. A. (1999). Regulation and metabolism of arachidonic acid. *Clin. Rev. Allergy Immunol.* **17,** 5–26.

Selye, H. (1950). Stress and the general adaptation syndrome. *Br. Med. J.* **1,** 1383–1392.

Shaw, K. N., Commins, S., and O'Mara, S. M. (2003). Deficits in spatial learning and synaptic plasticity induced by the rapid and competitive broad-spectrum cyclooxygenase inhibitor ibuprofen are reversed by increasing endogenous brain-derived neurotrophic factor. *Eur. J. Neurosci.* **17,** 2438–2446.

Shiflett, M. W., Rankin, A. Z., Tomaszycki, M. L., and DeVoogd, T. J. (2004). Cannabinoid inhibition improves memory in food-storing birds, but with a cost. *Proc. Biol. Sci.* **271,** 2043–2048.

Shimizu, T., and Yokotani, K. (2008). Bidirectional roles of the brain 2-arachidonoyl-*sn*-glycerol in the centrally administered vasopressin-induced adrenomedullary outflow in rats. *Eur. J. Pharmacol.* **582,** 62–69.

Shimizu, T., Okada, S., Yamaguchi-Shima, N., and Yokotani, K. (2004). Brain phospholipase C-diacylglycerol lipase pathway is involved in vasopressin-induced release of noradrenaline and adrenaline from adrenal medulla in rats. *Eur. J. Pharmacol.* **499,** 99–105.

Shimizu, T., Okada, S., Yamaguchi, N., Arai, J., Wakiguchi, H., and Yokotani, K. (2005). Brain phospholipase C/diacylglycerol lipase are involved in bombesin BB2 receptor-mediated activation of sympatho-adrenomedullary outflow in rats. *Eur. J. Pharmacol.* **514,** 151–158.

Shimizu, T., Yamaguchi, N., Okada, S., Lu, L., Sasaki, T., and Yokotani, K. (2007). Roles of brain phosphatidylinositol-specific phospholipase C and diacylglycerol lipase in centrally administered histamine-induced adrenomedullary outflow in rats. *Eur. J. Pharmacol.* **571,** 138–144.

Simon, G. M., and Cravatt, B. F. (2006). Endocannabinoid biosynthesis proceeding through glycerophospho-*N*-acyl ethanolamine and a role for alpha/beta-hydrolase 4 in this pathway. *J. Biol. Chem.* **281,** 26465–26472.

Slanina, K. A., Roberto, M., and Schweitzer, P. (2005). Endocannabinoids restrict hippocampal long-term potentiation via CB1. *Neuropharmacology* **49,** 660–668.

Smith, W. L., DeWitt, D. L., and Garavito, R. M. (2000). Cyclooxygenases: Structural, cellular, and molecular biology. *Annu. Rev. Biochem.* **69,** 145–182.

Smrcka, A. V., Hepler, J. R., Brown, K. O., and Sternweis, P. C. (1991). Regulation of polyphosphoinositide-specific phospholipase C activity by purified Gq. *Science* **251,** 804–807.

Snider, N. T., Kornilov, A. M., Kent, U. M., and Hollenberg, P. F. (2007). Anandamide metabolism by human liver and kidney microsomal cytochrome p450 enzymes to form hydroxyeicosatetraenoic and epoxyeicosatrienoic acid ethanolamides. *J. Pharmacol. Exp. Ther.* **321,** 590–597.

Snider, N. T., Sikora, M. J., Sridar, C., Feuerstein, T. J., Rae, J. M., and Hollenberg, P. F. (2008). The endocannabinoid anandamide is a substrate for the human polymorphic cytochrome P450 2D6. *J. Pharmacol. Exp. Ther.* **327,** 538–545.

Song, L., Che, W., Min-Wei, W., Murakami, Y., and Matsumoto, K. (2006). Impairment of the spatial learning and memory induced by learned helplessness and chronic mild stress. *Pharmacol. Biochem. Behav.* **83,** 186–193.

Stark, K., Dostalek, M., and Guengerich, F. P. (2008). Expression and purification of orphan cytochrome P450 4X1 and oxidation of anandamide. *FEBS J.* **275,** 3706–3717.

Steiner, M. R., Bomalaski, J. S., and Clark, M. A. (1993). Responses of purified phospholipases A2 to phospholipase A2 activating protein (PLAP) and melittin. *Biochim. Biophys. Acta* **1166,** 124–130.
Stella, N., Tence, M., Glowinski, J., and Premont, J. (1994). Glutamate-evoked release of arachidonic acid from mouse brain astrocytes. *J. Neurosci.* **14,** 568–575.
Stella, N., Schweitzer, P., and Piomelli, D. (1997). A second endogenous cannabinoid that modulates long-term potentiation. *Nature* **388,** 773–778.
Sugiura, T., Kondo, S., Sukagawa, A., Tonegawa, T., Nakane, S., Yamashita, A., Ishima, Y., and Waku, K. (1996). Transacylase-mediated and phosphodiesterase-mediated synthesis of N-arachidonoylethanolamine, an endogenous cannabinoid-receptor ligand, in rat brain microsomes. Comparison with synthesis from free arachidonic acid and ethanolamine. *Eur. J. Biochem.* **240,** 53–62.
Sugiura, T., Kobayashi, Y., Oka, S., and Waku, K. (2002). Biosynthesis and degradation of anandamide and 2-arachidonoylglycerol and their possible physiological significance. *Prostaglandins Leukot. Essent. Fatty Acids* **66,** 173–192.
Sugiura, T., Oka, S., Gokoh, M., Kishimoto, S., and Waku, K. (2004). New perspectives in the studies on endocannabinoid and cannabis: 2-arachidonoylglycerol as a possible novel mediator of inflammation. *J. Pharmacol. Sci.* **96,** 367–375.
Sun, Y. X., Tsuboi, K., Okamoto, Y., Tonai, T., Murakami, M., Kudo, I., and Ueda, N. (2004). Biosynthesis of anandamide and N-palmitoylethanolamine by sequential actions of phospholipase A2 and lysophospholipase D. *Biochem. J.* **380,** 749–756.
Suzuki, A., Mitsuda, S., Higashio, K., and Kumasaka, T. (1991). Participation of phospholipase A2 in induction of tissue plasminogen activator (t-PA) production by human fibroblast, IMR-90 cells, stimulated by proteose peptone. *Thromb. Res.* **64,** 191–202.
Swanson, L. W., and Sawchenko, P. E. (1980). Paraventricular nucleus: A site for the integration of neuroendocrine and autonomic mechanisms. *Neuroendocrinology* **31,** 410–417.
Takahashi, R. N., Pamplona, F. A., and Fernandes, M. S. (2005). The cannabinoid antagonist SR141716A facilitates memory acquisition and consolidation in the mouse elevated T-maze. *Neurosci. Lett.* **380,** 270–275.
Tanaka, H., Watanabe, K., Tamaru, N., and Yoshida, M. (1995). Arachidonic acid metabolites and glucocorticoid regulatory mechanism in cultured porcine tracheal smooth muscle cells. *Lung* **173,** 347–361.
Tassoni, D., Kaur, G., Weisinger, R. S., and Sinclair, A. J. (2008). The role of eicosanoids in the brain. *Asia Pac. J. Clin. Nutr.* **17**(Suppl. 1), 220–228.
Teather, L. A., Packard, M. G., and Bazan, N. G. (2002). Post-training cyclooxygenase-2 (COX-2) inhibition impairs memory consolidation. *Learn. Mem.* **9,** 41–47.
Tempel, D. L., Kim, T., and Leibowitz, S. F. (1993). The paraventricular nucleus is uniquely responsive to the feeding stimulatory effects of steroid hormones. *Brain Res.* **614,** 197–204.
Terranova, J. P., Michaud, J. C., Le Fur, G., and Soubrie, P. (1995). Inhibition of long-term potentiation in rat hippocampal slices by anandamide and WIN55212-2: Reversal by SR141716 A, a selective antagonist of CB1 cannabinoid receptors. *Naunyn Schmiedebergs Arch. Pharmacol.* **352,** 576–579.
Tong, D. C., Buck, S. M., Roberts, B. R., Klein, J. D., and Tumlin, J. A. (2004). Calcineurin phosphatase activity: Activation by glucocorticoids and role of intracellular calcium. *Transplantation* **77,** 259–267.
Tsutsumi, T., Kobayashi, T., Ueda, H., Yamauchi, E., Watanabe, S., and Okuyama, H. (1994). Lysophosphoinositide-specific phospholipase C in rat brain synaptic plasma membranes. *Neurochem. Res.* **19,** 399–406.
Tsutsumi, T., Kobayashi, T., Miyashita, M., Watanabe, S., Homma, Y., and Okuyama, H. (1995). A lysophosphoinositide-specific phospholipase C distinct from other phospholipase C families in rat brain. *Arch. Biochem. Biophys.* **317,** 331–336.

Turton, M. D., et al. (1996). A role for glucagon-like peptide-1 in the central regulation of feeding. *Nature* **379,** 69–72.

Ueda, H., Kobayashi, T., Kishimoto, M., Tsutsumi, T., and Okuyama, H. (1993). A possible pathway of phosphoinositide metabolism through EDTA-insensitive phospholipase A1 followed by lysophosphoinositide-specific phospholipase C in rat brain. *J. Neurochem.* **61,** 1874–1881.

Uehara, A., Sekiya, C., Takasugi, Y., Namiki, M., and Arimura, A. (1989). Anorexia induced by interleukin 1: Involvement of corticotropin-releasing factor. *Am. J. Physiol.* **257,** R613–R617.

Ullrich, O., Schneider-Stock, R., and Zipp, F. (2006). Cell-cell communication by endocannabinoids during immune surveillance of the central nervous system. *Results Probl. Cell Differ.* **43,** 281–305.

Vane, J. R., Bakhle, Y. S., and Botting, R. M. (1998). Cyclooxygenases 1 and 2. *Annu. Rev. Pharmacol. Toxicol.* **38,** 97–120.

Varvel, S. A., Anum, E. A., and Lichtman, A. H. (2005). Disruption of CB(1) receptor signaling impairs extinction of spatial memory in mice. *Psychopharmacology (Berl.)* **179,** 863–872.

Varvel, S. A., Wise, L. E., Niyuhire, F., Cravatt, B. F., and Lichtman, A. H. (2007). Inhibition of fatty-acid amide hydrolase accelerates acquisition and extinction rates in a spatial memory task. *Neuropsychopharmacology* **32,** 1032–1041.

Vellani, V., Petrosino, S., De Petrocellis, L., Valenti, M., Prandini, M., Magherini, P. C., McNaughton, P. A., and Di Marzo, V. (2008). Functional lipidomics. Calcium-independent activation of endocannabinoid/endovanilloid lipid signalling in sensory neurons by protein kinases C and A and thrombin. *Neuropharmacology* **55,** 1274–1279.

Verty, A. N., McGregor, I. S., and Mallet, P. E. (2005). Paraventricular hypothalamic CB(1) cannabinoid receptors are involved in the feeding stimulatory effects of Delta(9)-tetrahydrocannabinol. *Neuropharmacology* **49,** 1101–1109.

Vishwanath, B. S., Frey, F. J., Bradbury, M. J., Dallman, M. F., and Frey, B. M. (1993). Glucocorticoid deficiency increases phospholipase A2 activity in rats. *J. Clin. Invest.* **92,** 1974–1980.

Vollmer, R. R., Balcita, J. J., Sved, A. F., and Edwards, D. J. (1997). Adrenal epinephrine and norepinephrine release to hypoglycemia measured by microdialysis in conscious rats. *Am. J. Physiol.* **273,** R1758–R1763.

Wahn, H., Wolf, J., Kram, F., Frantz, S., and Wagner, J. A. (2005). The endocannabinoid arachidonyl ethanolamide (anandamide) increases pulmonary arterial pressure via cyclooxygenase-2 products in isolated rabbit lungs. *Am. J. Physiol. Heart Circ. Physiol.* **289,** H2491–H2496.

Wenger, T., Fernandez-Ruiz, J. J., and Ramos, J. A. (1999). Immunocytochemical demonstration of CB1 cannabinoid receptors in the anterior lobe of the pituitary gland. *J. Neuroendocrinol.* **11,** 873–878.

Wettschureck, N., van der Stelt, M., Tsubokawa, H., Krestel, H., Moers, A., Petrosino, S., Schutz, G., Di Marzo, V., and Offermanns, S. (2006). Forebrain-specific inactivation of Gq/G11 family G proteins results in age-dependent epilepsy and impaired endocannabinoid formation. *Mol. Cell Biol.* **26,** 5888–5894.

Wilson, R. I., and Nicoll, R. A. (2001). Endogenous cannabinoids mediate retrograde signalling at hippocampal synapses. *Nature* **410,** 588–592.

Wise, L. E., Iredale, P. A., Stokes, R. J., and Lichtman, A. H. (2007). Combination of rimonabant and donepezil prolongs spatial memory duration. *Neuropsychopharmacology* **32,** 1805–1812.

Wolff, M. C., and Leander, J. D. (2003). SR141716A, a cannabinoid CB1 receptor antagonist, improves memory in a delayed radial maze task. *Eur. J. Pharmacol.* **477,** 213–217.

Woodward, D. F., Krauss, A. H., Wang, J. W., Protzman, C. E., Nieves, A. L., Liang, Y., Donde, Y., Burk, R. M., Landsverk, K., and Struble, C. (2007). Identification of an antagonist that selectively blocks the activity of prostamides (prostaglandin-ethanolamides) in the feline iris. *Br. J. Pharmacol.* **150,** 342–352.

Woodward, D. F., Carling, R. W., Cornell, C. L., Fliri, H. G., Martos, J. L., Pettit, S. N., Liang, Y., and Wang, J. W. (2008). The pharmacology and therapeutic relevance of endocannabinoid derived cyclo-oxygenase (COX)-2 products. *Pharmacol. Ther.* **120,** 71–80.

Wright, T. M., Rangan, L. A., Shin, H. S., and Raben, D. M. (1988). Kinetic analysis of 1,2-diacylglycerol mass levels in cultured fibroblasts. Comparison of stimulation by alpha-thrombin and epidermal growth factor. *J. Biol. Chem.* **263,** 9374–9380.

Wurtman, R. J. (2002). Stress and the adrenocortical control of epinephrine synthesis. *Metabolism* **51,** 11–14.

Yamaguchi-Shima, N., Okada, S., Shimizu, T., Usui, D., Nakamura, K., Lu, L., and Yokotani, K. (2007). Adrenal adrenaline- and noradrenaline-containing cells and celiac sympathetic ganglia are differentially controlled by centrally administered corticotropin-releasing factor and arginine-vasopressin in rats. *Eur. J. Pharmacol.* **564,** 94–102.

Yamamoto, S. (1992). Mammalian lipoxygenases: Molecular structures and functions. *Biochim. Biophys. Acta* **1128,** 117–131.

Yang, H., and Chen, C. (2008). Cyclooxygenase-2 in synaptic signaling. *Curr. Pharm. Des.* **14,** 1443–1451.

Yang, H., Wang, L., and Ju, G. (1997). Evidence for hypothalamic paraventricular nucleus as an integrative center of neuroimmunomodulation. *Neuroimmunomodulation* **4,** 120–127.

Yang, Y. L., Chao, P. K., and Lu, K. T. (2006). Systemic and intra-amygdala administration of glucocorticoid agonist and antagonist modulate extinction of conditioned fear. *Neuropsychopharmacology* **31,** 912–924.

Yang, H., Zhang, J., Andreasson, K., and Chen, C. (2008). COX-2 oxidative metabolism of endocannabinoids augments hippocampal synaptic plasticity. *Mol. Cell Neurosci.* **37,** 682–695.

Yokotani, K., Okuma, Y., and Osumi, Y. (1995). Recombinant interleukin-1 beta inhibits gastric acid secretion by activation of central sympatho-adrenomedullary outflow in rats. *Eur. J. Pharmacol.* **279,** 233–239.

Yokotani, K., Wang, M., Murakami, Y., Okada, S., and Hirata, M. (2000). Brain phospholipase A(2)-arachidonic acid cascade is involved in the activation of central sympatho-adrenomedullary outflow in rats. *Eur. J. Pharmacol.* **398,** 341–347.

Yokotani, K., Murakami, Y., Okada, S., and Hirata, M. (2001). Role of brain arachidonic acid cascade on central CRF1 receptor-mediated activation of sympatho-adrenomedullary outflow in rats. *Eur. J. Pharmacol.* **419,** 183–189.

Yousufzai, S. Y., and Abdel-latif, A. A. (1997). Endothelin-1 stimulates the release of arachidonic acid and prostaglandins in cultured human ciliary muscle cells: Activation of phospholipase A2. *Exp. Eye Res.* **65,** 73–81.

Yu, M., Ives, D., and Ramesha, C. S. (1997). Synthesis of prostaglandin E2 ethanolamide from anandamide by cyclooxygenase-2. *J. Biol. Chem.* **272,** 21181–21186.

Zeitler, P., Wu, Y. Q., and Handwerger, S. (1991). Melittin stimulates phosphoinositide hydrolysis and placental lactogen release: Arachidonic acid as a link between phospholipase A2 and phospholipase C signal-transduction pathways. *Life Sci.* **48,** 2089–2095.

Zhang, J., and Chen, C. (2008). Endocannabinoid 2-arachidonoylglycerol protects neurons by limiting COX-2 elevation. *J. Biol. Chem.* **283,** 22601–22611.

CHAPTER TWELVE

Modulation of the Cys-Loop Ligand-Gated Ion Channels by Fatty Acid and Cannabinoids

Li Zhang *and* Wei Xiong

Contents

I. CB Receptor-Dependent and -Independent Effects of Endocannabinoids	316
II. Structure and Function of the Cys-Loop LGICs	317
III. Inhibition of 5-HT_3 Receptors by Cannabinoids	318
IV. Modulation of Gly Receptor Function by Cannabinoids	324
V. Inhibition of nACh Receptors by Endocannabinoids	328
VI. Modulation of $GABA_A$ Receptor Function by Fatty Acids	329
VII. Concluding Discussion	330
References	331

Abstract

The Cys-loop ligand-gated ion channel (LGIC) family comprises a group of membrane ion channel receptors that play a crucial role in fast synaptic neurotransmission in the central and peripheral nervous system. The members of this superfamily include γ-aminobutyric acid type A ($GABA_A$), neuronal nicotinic acetylcholine (nACh), 5-HT_3, and glycine receptors. These receptors serve as therapeutic sites for general anesthetic, antipsychoactive, antinociceptive, and anxiolytic drugs in the brain. These receptors are also thought to be primary targets of alcohol and other drugs of abuses. A number of studies reported that fatty acids affected the function of $GABA_A$ receptors in the early nineties. Accumulating evidence has suggested that the derivatives of arachidonic acid (AA), such as anandamide (*N*-arachidonoylethanolamine, AEA) and arachidonoylglycerol (2-AG), can critically regulate the other members of the Cys-loop LGIC superfamily through a cannabinoid receptor-independent mechanism. This chapter focuses on the results of recent studies showing that the Cys-loop LGICs

Laboratory for Integrative Neuroscience, National Institute on Alcohol Abuse and Alcoholism, National Institutes of Health, Rockville, Maryland, USA

could be additional molecular targets for fatty acid and endocannabinoid action in the central and peripheral nervous system. Some of these targets may mediate behavioral effects for cannabinoids to alter neuronal function.

Abbreviations

AEA	anandamide
ANOVA	analysis of variance
DA	dopamine
5-HT	5-hydroxytryptophan
GABA	γ-aminobutyric acid
GlyR	glycine receptor
I_{Gly}	glycine-activated current
THC	Δ^9-tetrahydrocannabinol
TM	transmembrane domain
VTA	ventral tegmental area
WT	wild type

I. CB Receptor-Dependent and -Independent Effects of Endocannabinoids

Recent research interest has turned to the role of endocannabinoids as retrograde neurotransmitters in synaptic modulations. Among five identified endocannabinoids, AEA, the amide between arachidonic acid (AA) and ethanolamine, was the first endocannabinoid named as an endogenous ligand for cannabinoid (CB) receptors. Unlike classical neurotransmitters, the endocannabinoids can travel backward to presynaptic neurons after released from the postsynaptic neurons upon depolarization. The endocannabinoids bind to and activate the presynaptic CB1 receptors (Pacher *et al.*, 2006; Wilson and Nicoll, 2001). Activation of the CB1 receptors in turn inhibits neurotransmitter release by inducing hyperpolarization and/or shutting voltage sensitive calcium channels. This retrograde endocannabinoid signaling plays an essential role in synaptic plasticity. On the other hand, accumulating evidence has emerged to show that some pharmacological effects produced by endogenous and exogenous CB receptor ligands are not mediated by CB1 or CB2 receptors (Oz, 2006; van der Stelt and Di Marzo, 2005). For instance, AEA and fatty acids have been found to modulate the functional properties of K^+, Na^+, and Ca^{2+} channels in a CB receptor-independent manner. AEA has also been shown to activate the vanilloid type 1 transient receptor potential channel

(TRPV1) in native neurons and in recombinant cell lines expressing TRPV1 channels (van der Stelt and Di Marzo, 2005). Because the cell lines such as HEK-293 cells and *Xenopus* oocytes that are used to express TRPV1 channels do not contain any of known CB receptors, the action of AEA on TRPV1 is thought to be a direct effect or (CB) receptor-independent effect. The receptor-independent effects of endocannabinoids on voltage-gated ion channels and TRPV1 channels were reviewed extensively (Oz, 2006). This chapter is intent to focus on the receptor-independent effects of fatty acids, endocannabinoids, and cannabimimetic lipids on the Cys-loop ligand-gated ion channels (LGICs).

II. Structure and Function of the Cys-Loop LGICs

The Cys-loop LGICs are membrane ion channel receptors activated by extracellular ligands. They are divided into cationic ion channels activated by acetylcholine, nicotine, and 5-HT and anionic ion channels activated by γ-aminobutyric acid (GABA) and glycine (Gly). This superfamily therefore includes $5-HT_3$, nicotinic acetylcholine (nACh), $GABA_A$, and Gly receptors. It has been well accepted that the functional form of the Cys-loop LGICs comprises homomeric or heteromeric pentameric oligomers (Unwin, 1993). Each of the five subunits symmetrically lines surround a central ion-conducting pore (Fig. 12.1). All members of the Cys-loop

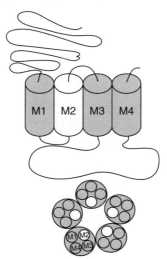

Figure 12.1 A schematic illustration of a typical Cys-loop subunit. The diagram below is a cross section of the channel shown from above and reveals the association of the five subunits within the membrane. The M2 domains (*while circles*) line the pore.

superfamily share a similar architecture; a large extracellular N-terminal domain, four-transmembrane (TM) domains and an extracellular C-terminal domain, and a large cytoplasmic domain (LCD) between TM3 and TM4 (Unwin, 1993). The extracellular N-terminal domain bears the Cys-loop signature and the specific binding sites for agonists and antagonist. The TM2 domain is thought to be a key channel-lining component, determining channel properties such as conductance, rectification, and desensitization (Corringer et al., 2000; Miyazawa et al., 2003). However, emerging evidence has suggested that the LCD between TM3 and TM4 may also contribute to the biophysical properties of these channels (Kelley et al., 2003; Peters et al., 2004, 2005). The members of this superfamily provide a wide range of important physiological functions from memory to sensory formation (Changeux et al., 1998; Lynch, 2009; Macdonald and Olsen, 1994; Zhang and Lummis, 2006). These receptors also serve as bona fide targets for anxiolytic, antinociceptive, antiemetic, and anesthetic agents. Many amino acid polymorphisms that occur on the structure of these receptors result in devastating pathological states (Frank et al., 2004; Lena and Changeux, 1997; Lynch, 2009) (Table 12.1).

III. Inhibition of 5-HT$_3$ Receptors by Cannabinoids

The 5-HT$_3$ receptor was one of the earliest serotonin-activated receptors defined by Gaddum and Picarelli a half century ago (Gaddum and Hameed, 1954). Seven distinct subtypes of 5-HT receptors have now been classified (Zhang and Lummis, 2006). However, unlike all other known serotonin receptors whose actions are mediated via G proteins, the 5-HT$_3$ receptor is the only 5-HT-gated ion channel (Barrera et al., 2005; Maricq et al., 1991; Yakel and Jackson, 1988). Its structure is more closely related to the nACh receptor than to any of the other classes of 5-HT receptor (Maricq et al., 1991). The 5-HT$_3$ receptors are expressed in post- and presynaptic sites in the peripheral and central nervous systems (Morales et al., 1998; Thompson et al., 2006; Zhang and Lummis, 2006). These receptors are located predominantly at interneurons in several brain areas including cortex, hippocampus, nucleus accumbens, substantiate nigra, and ventral tegmental area (VTA), although the highest levels are in the brain stem, especially the nucleus tractus solitarius and area postrema (Huang et al., 2004; Tecott et al., 1993; Waeber et al., 1988, 1989, 1990). In some of these regions, activation of 5-HT$_3$ receptors has been shown to modulate neurotransmitter release (van Hooft and Vijverberg, 2000). 5-HT$_3$ receptors have also been found to colocalize with CB1 receptors in rat brain neurons and a high proportion of 5-HT$_{3A}$/CB1-expressing neurons

Table 12.1 The CB receptor-independent effects of cannabinoids and AA on LGICs

Receptor		Assay	Preparation	Effect	Concentration and drug	References
GABA$_A$ receptor	Native	Radioligand binding	Synaptic plasma membrane, synaptoneurosomes from brain cortex	Increase	Arachidonic acid, oleic acid, and docosahexaenoic acid	Koenig and Martin (1992), Nielsen et al. (1988), Samochocki and Strosznajder (1993), Witt and Nielsen (1994), and Witt et al. (1996)
		Muscimol-induced ^{36}Cl$^-$ uptake	Cerebral cortical synaptoneurosomes	Inhibition	IC$_{50}$ = 40 μM arachidonic acid	Schwartz and Yu (1992)
		Ion current	Rat substantia nigra neuron	Inhibition	Docosahexaenoic acid, arachidonic acid, and docosapentaenoic acid	Hamano et al. (1996)
				Little or no effect	Docosatrienoic acid, docosatetraenoic acid, palmitic acid, and oleic acid	
	$\alpha_1\beta_2\gamma_2$	Ion current	Oocytes	Inhibition	IC$_{50}$ = 50 μM C$_8$-CTFE	DelRaso et al. (1996)
		Ion current	Sf-9 insect cell	Potentiation	Docosahexaenoic acid 1.0 μM	Nabekura et al. (1998)
		Ion current	HEK-293 cells	Inhibition	Docosahexaenoic acid >3.0 μM	
					IC$_{50}$ = 34 μM arachidonic acid	Saxena (2000)

(*continued*)

Table 12.1 (continued)

Receptor		Assay	Preparation	Effect	Concentration and drug	References
GABA-$\rho1$		Ion current	Oocytes	Inhibition	$IC_{50} = 300\ \mu M$ C_8-CTFE	DelRaso et al. (1996)
5-HT receptors	3A	Ion current	Oocytes and HEK-293 cells	Inhibition	$IC_{50} = 0.3\ \mu M$ AEA	Xiong et al. (2008)
	3A	Ion current	HEK-293 cells		$IC_{50} = 38.4$ nM THC $IC_{50} = 129.6$ nM AEA	Barann et al. (2002)
	Native	Radioligand binding	Brain membranes	Inhibition	$IC_{50} = 1$–$10\ \mu M$ AEA	
Glycine receptor	α_1	Ion current	Oocytes	Potentiation	$IC_{50} = 86$ nM THC $IC_{50} = 319$ nM AEA	Hejazi et al. (2006)
	$\alpha_1\beta_1$	Ion current	Oocytes	Potentiation	$IC_{50} = 73$ nM THC $IC_{50} = 318$ nM AEA	
	Native	Ion current	VTA neurons	Potentiation	$IC_{50} = 115$ nM THC $IC_{50} = 320$ nM AEA	
nACh receptor	α_7	Ion current	Oocytes	Inhibition	$IC_{50} = 168$ nM AEA $IC_{50} = 183$ nM R-methanandamide	Oz et al. (2004)
	$\alpha_4\beta_2$	Ion current	SH-EP1 cells	Inhibition	AEA	Spivak et al. (2007)
	α_7 and $\alpha_4\beta_2$	Ion current	Oocytes	Potentiation	$IC_{50} = 3\ \mu M$ arachidonic acid	Nishizaki et al. (1999)
	$\alpha\beta\gamma\delta$	Ion current	Oocytes	Short-term depression and long-term enhancement	linoleic and linolenic acid	Nishizaki et al. (1997b)
	$\alpha\beta\gamma\delta$	Ion current	Oocytes	Potentiation	Oleic acid	Nishizaki et al. (1997a)
	α_7	Ion current	Oocytes	Inhibition	$IC_{50} = 1\ \mu M$ arachidonic acid	Vijayaraghavan et al. (1995)
	Native	$^{86}Rb^+$ efflux	Thalamic synaptosomes	Inhibition	$IC_{50} = 0.9\ \mu M$ AEA	Butt et al. (2008)

contained the inhibitory neurotransmitter GABA, indicating a possible interaction between the CB1 and 5-HT$_{3A}$ receptors and their involvement in the regulation of GABA neurotransmission in brain (Morales et al., 2004). Growing evidence has suggested that 5-HT$_3$ receptors play roles in brain reward mechanisms and in neurological phenomena such as anxiety, psychosis, nociception, and cognitive function (Grant, 1995; Zhang and Lummis, 2006). In addition, there is some evidence for the therapeutic potential of 5-HT$_3$ receptor antagonists for antipsychotic, antinociceptive, and other psychiatric disorders (Zhang and Lummis, 2006). 5-HT$_3$ receptor antagonists have been widely used as first line drugs against emesis in cancer chemotherapy and irritable bowel syndrome (Zhang and Lummis, 2006).

The inhibitory effect of AEA on 5-HT$_3$ receptors was revealed by a previous study in nodose ganglion neurons (NGN) (Fan, 1995). AEA was shown to inhibit 5-HT-activated current in a concentration-dependent manner with IC$_{50}$ of 94 nM. In addition, several synthetic CB receptor agonists such as WIN55,212-2 and CP55,940 were found to inhibit 5-HT$_3$ receptor-mediated responses in NGN, although the potencies of these synthetic cannabinoids to inhibit 5-HT$_3$ receptors were less than that of AEA. The inhibition by all three chemicals was independent of agonist concentrations. Moreover, an enantiomer of CP55,940, CP56,667, also produced an inhibitory effect on 5-HT-activated current in a manner similar to that of CP55,940. Intracellular application of Sp-cAMP, an inhibitor of cAMP, and GDP-βS, an inhibitor of G proteins, did not alter the magnitude of AEA inhibition of 5-HT$_3$ receptors. Collectively, the author suggested that the inhibition of 5-HT$_3$ receptors by cannabinoids is unlikely due to activation of endogenous CB receptors expressed in NGN. However, this study did not use selective antagonists to block endogenous CB receptors in native neurons. Studies of 5-HT$_{3A}$ receptors expressed Xenopus oocytes and HEK-293 cells have provided the following evidence to indicate that AEA inhibition of 5-HT$_3$ receptors is independent of G protein-coupled CB receptors (Barann et al., 2002; Oz et al., 2002). First, there are no CB1 and CB2 receptors endogenously expressed in both Xenopus oocytes and HEK-293 cells (Pertwee, 2005). Second, SR141716A, a selective CB1 receptor antagonist, did not affect the extent of inhibition of 5-HT$_3$ receptors by AEA and other cannabinoids. Moreover, there was no specific binding of [^{3}H]-SR141716A and [^{3}H]-CP55,940 detected in HEK-293 cells (Barann et al., 2002). This is consistent with the idea that HEK-293 cells do not have endogenous CB1 receptors. Figure 12.2 illustrates AEA inhibition of the function of 5-HT$_{3A}$ receptors expressed in Xenopus oocytes.

The precise mechanism of AEA inhibition of 5-HT$_3$ receptors is unclear. Consistent with the observations in NGN, all cannabinoids tested in Xenopus oocytes and HEK-293 cells produced a noncompetitive inhibition of 5-HT$_3$ receptors (Barann et al., 2002; Oz et al., 2002). Radioligand binding experiments showed that three CB receptor agonists did not alter the concentration-dependent displacement of [^{3}H]-GR65630, a selective

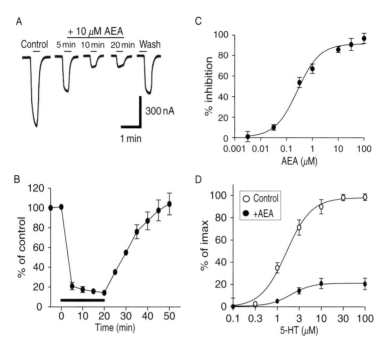

Figure 12.2 AEA inhibition of $I_{5\text{-HT}}$ in *Xenopus* oocytes expressing 5-HT$_{3A}$ receptors. (A) AEA inhibition of $I_{5\text{-HT}}$. Tracing records showing currents activated by 1 μM 5-HT in the absence and presence of 10 μM AEA. The solid bar on the top of each trace record represents the time of 5-HT application. A long solid bar indicates the time of continuous application of AEA. (B) Time course of AEA inhibition of $I_{5\text{-HT}}$ in cells previously injected with cRNA of mouse 5-HT$_{3A}$ subunit. The solid bar indicates the application time of 10 μM AEA. (C) The concentration–response curve of AEA inhibition of $I_{5\text{-HT}}$ in oocytes previously injected with cRNA of the mouse 5-HT$_{3A}$ subunit. (D) Non-competitive inhibition of 5-HT$_{3A}$ receptor-mediated responses by AEA. 5-HT concentration–response curves in the absence (*open circle*) and in the presence of 10 μM AEA (*solid circle*).

5-HT$_3$ receptor antagonist, binding by 5-HT in membranes isolated from HEK-293 cells transfected with 5-HT$_3$ receptor cDNA (Barann et al., 2002). One hypothesis suggests that 5-HT$_3$ receptors may contain a motif structurally similar to the binding pockets of CB receptor agonists since all three CB receptor agonists, THC, WIN55,212-2, and AEA, inhibited 5-HT$_3$ receptors with a remarkable potency (38–129 nM). These values are close to the binding affinity of the ligands for CB1 and CB2 receptors (Howlett et al., 2002). However, the IC$_{50}$ values of AEA inhibition of 5-HT$_3$ receptors varied significantly over a range from 94 nM in NGN to 3.7 μM in *Xenopus* oocytes (Barann et al., 2002; Fan, 1995; Oz et al., 2002; Xiong et al., 2008). A recent study from our laboratory has provided evidence to suggest that AEA inhibition of 5-HT$_3$ receptors varies with expression levels of receptor proteins at cell membrane surfaces (Xiong et al., 2008). The extent of AEA inhibition of 5-HT$_3$ receptors was inversely correlated with surface expression of 5-HT$_3$ receptors expressed in both *Xenopus* oocytes and HEK-293 cells. The magnitude of AEA inhibition was strongest at lower receptor expression levels.

The IC$_{50}$ values of AEA inhibition differed by nearly 120-fold between oocytes previously injected with 1 and 50 ng of cRNAs (Fig. 12.3). For instance, the EC$_{50}$ value for AEA inhibition was 167 ± 12 nM in cells injected with 1 ng of cRNA, whereas the EC$_{50}$ value for AEA inhibition was 20 ± 2 μM in cells injected with 50 ng of cRNA. Consistent with this idea, pretreatment with actinomycin D, which inhibits transcription, decreased the amplitude of current activated by maximal concentrations of 5-HT and increased the magnitude of AEA inhibition. Further

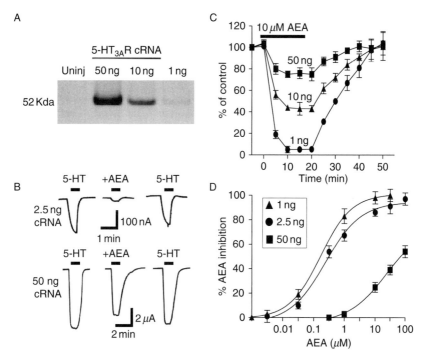

Figure 12.3 AEA inhibition depends on expression level of receptor proteins at the cell surface. (A) Western blot of 5-HT$_{3A}$ receptor protein at the surface membrane of *Xenopus* oocytes. A representative gel of Western blot showing the abundance of surface proteins from cells previously injected with various concentrations of 5-HT$_{3A}$ receptor cRNAs. (B) Tracing records showing AEA inhibition of 5-HT-activated currents in cells previously injected with 2.5 ng (*upper*) and 50 ng (*lower*) 5-HT$_{3A}$ receptor cRNAs. (C) Time course of AEA inhibition of I_{5-HT} in *Xenopus* oocytes previously injected with 1, 20, and 50 ng of the 5-HT$_{3A}$ receptor cRNAs. (D) Concentration–response curves of AEA inhibition of 5-HT-activated current in cells previously injected with 1, 2.5, and 50 ng of 5-T$_{3A}$ receptor cRNAs.

investigation into the mechanisms underlying AEA inhibition of 5-HT$_3$ receptors has shown that the low density 5-HT$_{3A}$ receptors at the cell surface were more readily than the high density receptors to enter the desensitized state, which appeared to be the key factor determining the sensitivity of 5-HT$_{3A}$ receptors to AEA-induced inhibition. Conversely, slowing 5-HT$_3$ receptor desensitization by 5-hydroxyindole and microtubule disruptors reduced AEA inhibition. This idea is consistent with a recent study showing that 5-HT$_{3A}$ receptor desensitization was regulated by the light chain of the microtubule-associated protein 1B. MAP1B-LC1 reduced steady-state receptor density at the cell surface and accelerated receptor desensitization kinetics at the steady state (Sun et al., 2008).

5-HT$_3$ receptor antagonists produced some pharmacological actions similar to that of cannabinoids (Pacher et al., 2006; Zhang and Lummis, 2006). For instance, both of them induced antinociceptive and antiemetic effects. There is also evidence that CB1 receptor agonists can directly inhibit the function of the peripheral 5-HT$_3$ receptors in vivo (Godlewski et al., 2003). In this study, WIN55,212-2 and CP55,940 inhibited peripheral 5-HT$_3$ receptor-mediated Bezold–Jarisch reflex in rats, whereas the TRPV1 receptor-mediated Bezold–Jarisch reflex, which is a decrease in heart rate, was unaffected. This effect was not mediated through endogenous CB receptors since the animals were pretreated with SR141716A. Similarly, a recent study has shown that WIN55,212-2 inhibited cocaine-induced hyperlocomotion in rats. The inhibiting effect induced by WIN55,212-2 was reduced by ondansetron, a selective 5-HT$_3$ receptor antagonist, but not by a selective CB1 receptor antagonist, suggesting an involvement of 5-HT$_3$ receptors (Przegalinski et al., 2005).

IV. Modulation of Gly Receptor Function by Cannabinoids

The Gly receptors are highly expressed in the spinal cord and many brain areas such as ventral tegmental area, nucleus accumbens, cerebellum, and brain stem (Lynch, 2009). In these regions, activation of Gly receptors triggers a rapid increase in Cl$^-$ ion conductance, hyperpolarization of the cell membrane, and shunting of excitatory current. This inhibitory action of Gly receptors regulates several important physiological processes such as pain transmission, respiratory rhythms, motor coordination, and development (Lynch, 2004; Lynch and Callister, 2006). There is evidence to show that Gly receptors are also involved in the regulation of dopamine release in nucleus accumbens and the VTA (Molander and Soderpalm, 2005a,b). These observations have contributed to the idea that Gly receptors may play a role in drug addiction and reward mechanisms.

Both THC and AEA potentiated, in a concentration-dependent manner, the amplitude of currents activated by low concentrations of Gly (EC_5) in *Xenopus* oocytes expressing α_1 homomeric and $\alpha_1\beta_1$ heteromeric Gly receptors and in acutely isolated VTA neurons (Hejazi et al., 2006). THC was more potent than AEA on both recombinant and native Gly receptors (Figs. 12.4 and 12.5). The EC_{50} values for the effect of THC on heteromeric Gly receptors (73 nM) and native Gly receptors (115 nM) are in the pharmacological ranges that induce psychotropic and antinociceptive effects in humans (Huestis and Cone, 2004). Similarly, the EC_{50} values for the AEA-induced potentiating effect on recombinant and native Gly receptors are in the range of 230–318 nM, comparable to AEA's affinity (89–300 nM) for CB1 receptor (Howlett et al., 2002). SR141716A, CB1 receptor antagonist, did not prevent AEA or THC-induced potentiation of I_{Gly} in both *Xenopus* oocytes and VTA neurons.

This suggests that the potentiating effects induced by both THC and AEA are not mediated by CB1 receptors. The findings in VTA neurons are

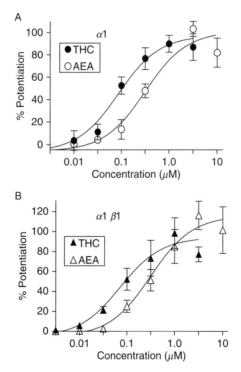

Figure 12.4 The concentration–response curves of THC and AEA potentiation of the α_1- and $\alpha_1\beta_1$-subunits expressed in Xenopus oocytes. (A, B) Graph plotting average percentage potentiation by THC and AEA as a function of THC and AEA concentration. Each data point represents the average of 5–10 oocytes. The curves shown are the best fits of the data using the Hill equation. Error bars not visible are smaller than the size of symbols.

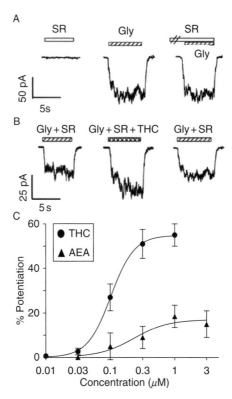

Figure 12.5 THC and AEA potentiation of Gly-activated current in isolated VTA neurons. (A) Membrane currents recorded after 1 μM SR141716A (SR) application (*left*), after 5 μM Gly was applied alone and after coapplication of SR and Gly. (B) I_{Gly} (induced by 5 μM Gly) recorded before (*left*), during (*middle*), and after (*right*) the application of 300 nM THC in the absence and presence of 1 μM SR. The SR was applied 15 s before the application of the mixture of Gly + THC + SR. (C) Concentration–response curves of the THC and AEA potentiation of I_{Gly} (activated by 5 μM Gly) in the presence of 1 μM SR. The error bars not visible are smaller than the size of symbols.

consistent with a previous study that Gly receptor-mediated synaptic currents in hypoglossal motoneurons were potentiated directly by conditions that promote Ca^{2+}-dependent endocannabinoid production (Diana and Bregestovski, 2005). This result indicates that in native neurons production of endocannabinoids potentiates the function of Gly receptors in a CB1 receptor-independent mechanism. However, these observations appeared to contradict to a previous study in which AEA and 2-arachidonoylglycerol (2-AG) at 1 μM were found to inhibit the amplitude of current activated by 100 μM Gly (I_{Gly}) in isolated rat hippocampal pyramidal and Purkinje cerebellar neurons (Lozovaya et al., 2005). In these neurons, AEA and 2-AG were also found to accelerate desensitization of I_{Gly}. The inhibitory action induced by endocannabinoids on Gly receptors was not affected by either CB1 and TRPV1 receptor antagonists or cytoplasmic application of G protein inhibitor GDP-βS, suggesting a direct action of cannabinoids on Gly receptors. A number of reasons may cause the discrepancies observed between two studies. One possibility is that the inhibition of I_{Gly} by AEA and 2-AG was produced at relative high concentrations of Gly (>100 μM), whereas AEA-induced potentiating effect on I_{Gly} was observed at low concentrations of Gly (less than EC_{10}). At high concentrations of Gly, I_{Gly} tends to desensitize, which could further complicate the mechanisms of endocannabinoid modulation of Gly receptor function. Consistent with this, a recent study from the very same laboratory has reported that a synthetic cannabinoid analogue, WIN55,212-2, increased I_{Gly} in acutely isolated hippocampal CA1/3 pyramidal neurons in the presence of AM251, a selective CB1 receptor antagonist (Iatsenko et al., 2007). The authors of this study also observed that the magnitude of cannabinoid potentiation of I_{Gly} was inversely correlated with the concentrations of Gly. Alternatively, the discrepancy could be due to different compositions of Gly receptor α-subunits expressed in different types of neurons. In favor of this idea, a very recent study has shown that AEA and other synthetic cannabinoids potentiated the α_1- and α_1/β-containing Gly receptors and inhibited the α_2- and α_3-subunits expressed in HEK-293 cells (Yang et al., 2008). Consistent with a previous observation (Hejazi et al., 2006), the potentiation by THC and WIN55,212-2 of I_{Gly} was dependent on concentrations of Gly. The maximal potentiation of I_{Gly} by cannabinoids was observed at the lowest concentrations of Gly (3–10 μM). There is strong evidence to suggest that synaptic and extrasynaptic Gly is likely at the concentrations that produce low occupancy of Gly receptors. For example, the affinity of glycine binding at NMDA receptors is 100-fold higher than that of its binding at Gly receptors. However, these sites at NMDA receptors are unlikely to be saturated at the synapse (Bradaia et al., 2004; Eulenburg et al., 2005; Gomeza et al., 2003). Extracellular Gly concentrations in rat spinal cord were determined to be in the range of 2–3 μM, and the Gly content of cerebrospinal fluid was around 6 μM (Whitehead et al., 2001). In this regard, cannabinoid-induced potentiation of Gly receptors at low Gly occupancy could have physiological roles in neuronal excitability. The modulation of Gly receptor function by cannabinoids may implicate in some physiological and pathological processes. The involvement of Gly receptors in pain transmission suggests a potential role for cannabinoid and endocannabinoid effects on the receptor. This idea is supported by the observations that THC-induced analgesia in the tail-flick test and AEA-induced cannabimimetic activity in three analgesic tests remained intact in CB1 knockout mice (Di Marzo et al., 2000; Zimmer et al., 1999). Moreover, we have recently observed that THC and AEA significantly potentiated I_{Gly} in cultured rat spinal cord neurons in the presence of both CB1 and CB2 receptor antagonists (unpublished data). The VTA is the origin of mesolimbic dopamine system, which mediates the reinforcing properties of cannabinoids (Gerdeman et al., 2002). Potentiation of the Gly receptor function by THC and AEA may have a modulatory role in CB1 receptor-independent release of dopamine in the mesolimbic system.

V. Inhibition of nACh Receptors by Endocannabinoids

The nACh receptors are the primary targets for nicotine in the brain. These receptors are highly conserved across species and expressed in most brain areas (Dani and Bertrand, 2007). The nACh receptors can be detected on presynaptic terminals, cell bodies, and dendrites of many neuronal subtypes (Dani and Bertrand, 2007). Activation of nACh receptors can potentiate neurotransmitter release and neuronal excitability throughout the brain, thereby regulating synaptic transmission and plasticity (Dani and Bertrand, 2007). The nACh receptors play crucial roles in many brain functions such as pain sensation, sleep pattern, feeding, learning, and reward mechanisms (Hogg et al., 2003). The most abundant and widespread subtypes in the brain are α_7- and α_4/β_2-containing nACh receptors (Lindstrom et al., 1996).

An initial study reported that activation of nACh receptors triggered release of ^{3}H-AA in chick ciliary ganglion neurons (Vijayaraghavan et al., 1995). On the other hand, low concentrations of AA inhibited the amplitude of current activated by nicotine in Xenopus oocytes expressing the nACh α_7 receptors (Vijayaraghavan et al., 1995). These observations suggest that AA may be a negative feedback inhibitor of nACh receptors in neurons. Following studies showed that certain unsaturated fatty acids such as linoleic and linolenic acids potentiated ACh-activated currents in Xenopus oocytes expressing different combinations of recombinant nACh receptor subunits (Nishizaki et al., 1997a,b, 1999). AA enhanced the amplitudes of currents mediated by nACh $\alpha_4\beta_2$- and α_7-subunits. The potentiation by AA was a long-lasting effect and was prevented by a protein kinase C inhibitor (Nishizaki et al., 1999). Similar effect was observed in cultured hippocampal neurons in which AA produced a long-lasting facilitation of hippocampal synaptic transmission through a nACh receptor-dependent mechanism (Nishizaki et al., 1999). Recent studies have shown that endocannabinoid AEA potently inhibited the peak amplitudes of currents mediated by α_7 nACh receptors expressed in Xenopus oocytes and by the nACh $\alpha_4\beta_2$-receptors expressed in SH-EP1 cell (Oz et al., 2003, 2004; Spivak et al., 2007). The IC$_{50}$ values for AEA inhibition the nACh α_7 and $\alpha_4\beta_2$ receptors were 229 and 300 nM. The inhibition of nACh receptors by AEA was reversible and independent of agonist concentrations. Consistent with the noncompetitive nature of AEA, AEA did not inhibit specific binding of [^{3}H]-nicotine in human frontal cortex membranes (Lagalwar et al., 1999). The CB receptor antagonist SR141716A and SR144528 did not affect the AEA inhibition of nACh receptor function, suggesting that AEA may act directly at the nACh receptor (Oz et al., 2003). However, THC was ineffective in modulating both nACh α_7 and $\alpha_4\beta_2$ receptors-mediated ion currents, indicating that the inhibitory effect induced by AEA and 2-AG was relatively selective for endocannabinoids. AEA reversibly inhibited ACh-induced ^{86}Rb$^+$ efflux with an IC$_{50}$ value of 0.9 μM in thalamic synaptosomes (Butt et al., 2008). Such inhibition was not affected by pretreatment with the CB$_1$/CB$_2$ receptor antagonists and pertussis toxin (PTX), consistent with an idea that AEA inhibits the function of native nACh receptors in thalamic synaptosomes via a CB-independent mechanism (Butt et al., 2008). There is also evidence to suggest a possible role of endogenous fatty acid-like lipids in regulation of nACh receptor function. Pretreatment of synaptosomes with 1% BSA, which removes endogenous fatty acids from cell membrane preparations, was found to be critical for AEA-induced inhibiting effect on ACh-induced ^{86}Rb$^+$ efflux (Butt et al., 2008). The inhibition of nACh receptor function by AA is unlikely due to its metabolic products because the IC$_{50}$ value for AA was significantly lower than that of 2-AG and AEA, and ethanolamine and glycerol did not significantly affect nACh α_7-receptor-mediated response (Oz et al., 2003; Vijayaraghavan et al., 1995).

VI. Modulation of GABA$_A$ Receptor Function by Fatty Acids

The GABA$_A$ receptor is the major inhibitory neurotransmitter receptor and plays an essential role in maintaining normal neurotransmission in the brain (Olsen and Tobin, 1990). GABA$_A$ receptors are also the primary targets for a variety of therapeutic agents such as barbiturates, steroids, anesthetics, and benzodiazepines (Olsen and Tobin, 1990). Molecular cloning has identified a number of receptor subunits including six α-, four β-, four γ-, one δ-, one ζ-, and one π-subunits (Sieghart, 2006). Various actions of polyunsaturated fatty acids on the functional properties of GABA$_A$ have been reported. Docosahexaenoic acid (DHA) was found to significantly suppress the amplitude of GABA-activated currents in the rat substantia nigra neurons (Hamano et al., 1996; Samochocki and Strosznajder, 1993; Witt and Nielsen, 1994; Witt et al., 1996). Consistent with this observation, AA and its metabolites were also found to inhibit muscimol-induced ^{36}Cl uptake in rat cerebral cortical synaptoneurosomes (Schwartz and Yu, 1992). The effect of arachidonic acid on muscimol responses was inhibited by BSA, and BSA enhanced muscimol responses directly, indicating the generation of endogenous AA in the synaptoneurosome preparation.

Polysaturated fatty acids, AA and linoleic acid, inhibited GABA current in a concentration-dependent manner over a concentration range of 0.1–100 μM in HEK-293 cells expressing a brain-abundant form of the $\alpha_1\beta_2\gamma_2$-subunits (Saxena, 2000). The inhibition of $GABA_A$ currents appeared to be selective for unsaturated fatty acids since saturated (arachidic, palmitic) fatty acids had no effect on GABA current under the same concentration range tested for unsaturated fatty acids. These results are in line with a previous study in which DHA was found to inhibit GABA current in Sf-9 insect cells expressing the $\alpha_1\beta_2\gamma_2$-subunits (Nabekura et al., 1998). The effect of unsaturated fatty acids was dependent on the subunit composition of $GABA_A$ receptor complexes (Nabekura et al., 1998).

The mechanism by which fatty acids modulate $GABA_A$ receptor function remains undetermined. There is evidence to suggest that active oxygen radicals might mediate inhibition of $GAGA_A$ receptor currents by unsaturated fatty acids (Saxena, 2000). Unsaturated fatty acids were also found to enhance both [^{3}H]-muscimol and [^{3}H]-diazepam binding by 150–250% of control binding in brain tissues isolated from different animal species (Koenig and Martin, 1992; Nielsen et al., 1988; Samochocki and Strosznajder, 1993; Witt and Nielsen, 1994; Witt et al., 1996). Surprisingly, AEA had no effect on GABA-activated current in VTA neurons and cells expressing recombinant $GABA_A$ receptors (Hejazi et al., 2006; Lozovaya et al., 2005). In addition to $GABA_A$ receptors, $GABA_C$ receptors were also found to be the target for the action of certain oligomer fatty acids when these receptors expressed in *Xenopus* oocytes (DelRaso et al., 1996). Halogenated fatty acids, such as chlorotrifluoroethylene and perfluorinated oligomer acids, inhibited $GABA_C$ receptor-mediated response in oocytes expressing human $GABA_C$ subunits or $GABA_A$ $\alpha_1\beta_2\gamma_2$ receptors (DelRaso et al., 1996). The inhibition of $GABA_A$ neurotransmission by AA and its metabolites can lead to increased neuronal excitability. This mechanism may play an important role in the development of neuronal damage following seizures or cerebral ischemia.

VII. Concluding Discussion

It is clear from the data presented in this chapter that fatty acids and cannabinoids critically regulate the functions of the Cys-loop LGICs through CB1/2 receptor-independent mechanisms. The receptor-independent effects of cannabinoids should be given enough attention since most of the EC_{50} values of cannabinoid action on the Cys-loop LGICs are either clinically relevant or close to the EC_{50} values of these ligands to bind to CB receptors. For instance, the EC_{50} values of AEA to modulate the function of nACh α_7, 5-HT$_3$, and Glyα_1 receptors are within a range of 38–319 nM (Barann et al., 2002; Hejazi et al., 2006; Oz et al., 2003; Yang *et al.*, 2008). This is close to the range of the K_d values of AEA to binding to CB1 receptors (Howlett *et al.*, 2002). Similarly, THC has been found to inhibit and potentiate 5-HT$_{3A}$ and Glyα_1 receptor-mediated responses by 110 and 73 nM (Barann *et al.*, 2002; Hejazi *et al.*, 2006). These values are even lower than plasma THC concentration (162 nM) detected 1 h after low dose cannabis smoking (Huestis and Cone, 2004). Many cannabinoid-induced modulatory effects on LGICs appear to be structurally selective. A notable example is that unsaturated fatty acids suppressed GABA current, whereas saturated fatty acids and endocannabinoid AEA had no effect on GABA current in native neurons and in cell lines expressing recombinant GABA$_A$ receptors (Hejazi *et al.*, 2006; Lozovaya *et al.*, 2005; Nabekura *et al.*, 1998). The mechanisms and physiological and pharmacological implications of cannabinoid actions on the Cys-loop LGICs are poorly understood. Future studies are needed to address these important issues.

REFERENCES

Barann, M., Molderings, G., Bruss, M., Bonisch, H., Urban, B. W., and Gothert, M. (2002). Direct inhibition by cannabinoids of human 5-HT3A receptors: Probable involvement of an allosteric modulatory site. *Br. J. Pharmacol.* **137**(5), 589–596.

Barrera, N. P., Herbert, P., Henderson, R. M., Martin, I. L., and Edwardson, J. M. (2005). Atomic force microscopy reveals the stoichiometry and subunit arrangement of 5-HT3 receptors. *Proc. Natl. Acad. Sci. USA* **102**(35), 12595–12600.

Bradaia, A., Schlichter, R., and Trouslard, J. (2004). Role of glial and neuronal glycine transporters in the control of glycinergic and glutamatergic synaptic transmission in lamina X of the rat spinal cord. *J. Physiol.* **559**(Pt. 1), 169–186.

Butt, C., Alptekin, A., Shippenberg, T., and Oz, M. (2008). Endogenous cannabinoid anandamide inhibits nicotinic acetylcholine receptor function in mouse thalamic synaptosomes. *J. Neurochem.* **105**(4), 1235–1243.

Changeux, J. P., Bertrand, D., Corringer, P. J., Dehaene, S., Edelstein, S., Lena, C., Le Novere, N., Marubio, L., Picciotto, M., and Zoli, M. (1998). Brain nicotinic receptors: Structure and regulation, role in learning and reinforcement. *Brain Res. Brain Res. Rev.* **26**(2–3), 198–216.

Corringer, P. J., Le Novere, N., and Changeux, J. P. (2000). Nicotinic receptors at the amino acid level. *Annu. Rev. Pharmacol. Toxicol.* **40**, 431–458.

Dani, J. A., and Bertrand, D. (2007). Nicotinic acetylcholine receptors and nicotinic cholinergic mechanisms of the central nervous system. *Annu. Rev. Pharmacol. Toxicol.* **47**, 699–729.

DelRaso, N. J., Huang, Y., and Lu, L. (1996). Effects of chlorotrifluoroethylene oligomer fatty acids on recombinant GABA receptors expressed in *Xenopus* oocytes. *J. Membr. Biol.* **149**(1), 33–40.

Diana, M. A., and Bregestovski, P. (2005). Calcium and endocannabinoids in the modulation of inhibitory synaptic transmission. *Cell Calcium* **37**(5), 497–505.

Di Marzo, V., Breivogel, C. S., Tao, Q., Bridgen, D. T., Razdan, R. K., Zimmer, A. M., Zimmer, A., and Martin, B. R. (2000). Levels, metabolism, and pharmacological activity of anandamide in CB(1) cannabinoid receptor knockout mice: Evidence for non-CB(1), non-CB(2) receptor-mediated actions of anandamide in mouse brain. *J. Neurochem.* **75**(6), 2434–2444.

Eulenburg, V., Armsen, W., Betz, H., and Gomeza, J. (2005). Glycine transporters: Essential regulators of neurotransmission. *Trends Biochem. Sci.* **30**(6), 325–333.

Fan, P. (1995). Cannabinoid agonists inhibit the activation of 5-HT3 receptors in rat nodose ganglion neurons. *J. Neurophysiol.* **73**(2), 907–910.

Frank, B., Niesler, B., Nothen, M. M., Neidt, H., Propping, P., Bondy, B., Rietschel, M., Maier, W., Albus, M., and Rappold, G. (2004). Investigation of the human serotonin receptor gene HTR3B in bipolar affective and schizophrenic patients. *Am. J. Med. Genet. B Neuropsychiatr. Genet.* **131B**(1), 1–5.

Gaddum, J. H., and Hameed, K. A. (1954). Drugs which antagonize 5-hydroxytryptamine. *Br. J. Pharmacol. Chemother.* **9**(2), 240–248.

Gerdeman, G. L., Ronesi, J., and Lovinger, D. M. (2002). Postsynaptic endocannabinoid release is critical to long-term depression in the striatum. *Nat. Neurosci.* **5**(5), 446–451.

Godlewski, G., Gothert, M., and Malinowska, B. (2003). Cannabinoid receptor-independent inhibition by cannabinoid agonists of the peripheral 5-HT3 receptor-mediated von Bezold–Jarisch reflex. *Br. J. Pharmacol.* **138**(5), 767–774.

Gomeza, J., Hulsmann, S., Ohno, K., Eulenburg, V., Szoke, K., Richter, D., and Betz, H. (2003). Inactivation of the glycine transporter 1 gene discloses vital role of glial glycine uptake in glycinergic inhibition. *Neuron* **40**(4), 785–796.

Grant, K. A. (1995). The role of 5-HT3 receptors in drug dependence. *Drug Alcohol Depend.* **38**(2), 155–171.

Hamano, H., Nabekura, J., Nishikawa, M., and Ogawa, T. (1996). Docosahexaenoic acid reduces GABA response in substantia nigra neuron of rat. *J. Neurophysiol.* **75**(3), 1264–1270.

Hejazi, N., Zhou, C., Oz, M., Sun, H., Ye, J. H., and Zhang, L. (2006). {Delta}9-tetrahydrocannabinol and endogenous cannabinoid anandamide directly potentiate the function of glycine receptors. *Mol. Pharmacol.* **69**(3), 991–997.

Hogg, R. C., Raggenbass, M., and Bertrand, D. (2003). Nicotinic acetylcholine receptors: From structure to brain function. *Rev. Physiol. Biochem. Pharmacol.* **147**, 1–46.

Howlett, A. C., Barth, F., Bonner, T. I., Cabral, G., Casellas, P., Devane, W. A., Felder, C. C., Herkenham, M., Mackie, K., Martin, B. R., Mechoulam, R., and Pertwee, R. G. (2002). International Union of Pharmacology. XXVII. Classification of cannabinoid receptors. *Pharmacol. Rev.* **54**(2), 161–202.

Huang, J., Spier, A. D., and Pickel, V. M. (2004). 5-HT3A receptor subunits in the rat medial nucleus of the solitary tract: Subcellular distribution and relation to the serotonin transporter. *Brain Res.* **1028**(2), 156–169.

Huestis, M. A., and Cone, E. J. (2004). Relationship of Delta 9-tetrahydrocannabinol concentrations in oral fluid and plasma after controlled administration of smoked cannabis. *J. Anal. Toxicol.* **28**(6), 394–399.

Iatsenko, N. M., Tsintsadze, T., and Lozova, N. O. (2007). The synthetic cannabinoid analog WIN 55,212-2 potentiates the amplitudes of glycine-activated currents. *Fiziol. Zh.* **53**(3), 31–37.

Kelley, S. P., Dunlop, J. I., Kirkness, E. F., Lambert, J. J., and Peters, J. A. (2003). A cytoplasmic region determines single-channel conductance in 5-HT3 receptors. *Nature* **424**(6946), 321–324.

Koenig, J. A., and Martin, I. L. (1992). Effect of free fatty acids on GABAA receptor ligand binding. *Biochem. Pharmacol.* **44**(1), 11–15.

Lagalwar, S., Bordayo, E. Z., Hoffmann, K. L., Fawcett, J. R., and Frey, W. H., II. (1999). Anandamides inhibit binding to the muscarinic acetylcholine receptor. *J. Mol. Neurosci.* **13**(1–2), 55–61.

Lena, C., and Changeux, J. P. (1997). Pathological mutations of nicotinic receptors and nicotine-based therapies for brain disorders. *Curr. Opin. Neurobiol.* **7**(5), 674–682.

Lindstrom, J., Anand, R., Gerzanich, V., Peng, X., Wang, F., and Wells, G. (1996). Structure and function of neuronal nicotinic acetylcholine receptors. *Prog. Brain Res.* **109**, 125–137.

Lozovaya, N., Yatsenko, N., Beketov, A., Tsintsadze, T., and Burnashev, N. (2005). Glycine receptors in CNS neurons as a target for nonretrograde action of cannabinoids. *J. Neurosci.* **25**(33), 7499–7506.

Lynch, J. W. (2004). Molecular structure and function of the glycine receptor chloride channel. *Physiol. Rev.* **84**(4), 1051–1095.

Lynch, J. W. (2009). Native glycine receptor subtypes and their physiological roles. *Neuropharmacology* **56**(1), 303–309.

Lynch, J. W., and Callister, R. J. (2006). Glycine receptors: A new therapeutic target in pain pathways. *Curr. Opin. Investig. Drugs* **7**(1), 48–53.

Macdonald, R. L., and Olsen, R. W. (1994). GABAA receptor channels. *Annu. Rev. Neurosci.* **17**, 569–602.

Maricq, A. V., Peterson, A. S., Brake, A. J., Myers, R. M., and Julius, D. (1991). Primary structure and functional expression of the 5HT3 receptor, a serotonin-gated ion channel. *Science* **254**(5030), 432–437.

Miyazawa, A., Fujiyoshi, Y., and Unwin, N. (2003). Structure and gating mechanism of the acetylcholine receptor pore. *Nature* **423**(6943), 949–955.

Molander, A., and Soderpalm, B. (2005a). Accumbal strychnine-sensitive glycine receptors: An access point for ethanol to the brain reward system. *Alcohol. Clin. Exp. Res.* **29**(1), 27–37.

Molander, A., and Soderpalm, B. (2005b). Glycine receptors regulate dopamine release in the rat nucleus accumbens. *Alcohol. Clin. Exp. Res.* **29**(1), 17–26.

Morales, M., Battenberg, E., and Bloom, F. E. (1998). Distribution of neurons expressing immunoreactivity for the 5HT3 receptor subtype in the rat brain and spinal cord. *J. Comp. Neurol.* **402**(3), 385–401.

Morales, M., Wang, S. D., Diaz-Ruiz, O., and Jho, D. H. (2004). Cannabinoid CB1 receptor and serotonin 3 receptor subunit A (5-HT3A) are co-expressed in GABA neurons in the rat telencephalon. *J. Comp. Neurol.* **468**(2), 205–216.

Nabekura, J., Noguchi, K., Witt, M. R., Nielsen, M., and Akaike, N. (1998). Functional modulation of human recombinant gamma-aminobutyric acid type A receptor by docosahexaenoic acid. *J. Biol. Chem.* **273**(18), 11056–11061.

Nielsen, M., Witt, M. R., and Thogersen, H. (1988). [3H]diazepam specific binding to rat cortex in vitro is enhanced by oleic, arachidonic and docosahexenoic acid isolated from pig brain. *Eur. J. Pharmacol.* **146**(2–3), 349–353.

Nishizaki, T., Ikeuchi, Y., Matsuoka, T., and Sumikawa, K. (1997a). Oleic acid enhances ACh receptor currents by activation of Ca^{2+}/calmodulin-dependent protein kinase II. *Neuroreport* **8**(3), 597–601.

Nishizaki, T., Ikeuchi, Y., Matsuoka, T., and Sumikawa, K. (1997b). Short-term depression and long-term enhancement of ACh-gated channel currents induced by linoleic and linolenic acid. *Brain Res.* **751**(2), 253–258.

Nishizaki, T., Nomura, T., Matsuoka, T., Enikolopov, G., and Sumikawa, K. (1999). Arachidonic acid induces a long-lasting facilitation of hippocampal synaptic transmission by modulating PKC activity and nicotinic ACh receptors. *Brain Res. Mol. Brain Res.* **69**(2), 263–272.

Olsen, R. W., and Tobin, A. J. (1990). Molecular biology of GABAA receptors. *FASEB J.* **4**(5), 1469–1480.

Oz, M. (2006). Receptor-independent actions of cannabinoids on cell membranes: Focus on endocannabinoids. *Pharmacol. Ther.* **111**(1), 114–144.

Oz, M., Zhang, L., and Morales, M. (2002). Endogenous cannabinoid, anandamide, acts as a noncompetitive inhibitor on 5-HT3 receptor-mediated responses in *Xenopus* oocytes. *Synapse* **46**(3), 150–156.

Oz, M., Ravindran, A., Diaz-Ruiz, O., Zhang, L., and Morales, M. (2003). The endogenous cannabinoid anandamide inhibits alpha7 nicotinic acetylcholine receptor-mediated responses in *Xenopus* oocytes. *J. Pharmacol. Exp. Ther.* **306**(3), 1003–1010.

Oz, M., Zhang, L., Ravindran, A., Morales, M., and Lupica, C. R. (2004). Differential effects of endogenous and synthetic cannabinoids on alpha7-nicotinic acetylcholine receptor-mediated responses in *Xenopus* oocytes. *J. Pharmacol. Exp. Ther.* **310**(3), 1152–1160.

Pacher, P., Batkai, S., and Kunos, G. (2006). The endocannabinoid system as an emerging target of pharmacotherapy. *Pharmacol. Rev.* **58**(3), 389–462.

Pertwee, R. G. (2005). Pharmacological actions of cannabinoids. *Handb. Exp. Pharmacol.* **168**, 1–51.

Peters, J. A., Kelley, S. P., Dunlop, J. I., Kirkness, E. F., Hales, T. G., and Lambert, J. J. (2004). The 5-hydroxytryptamine type 3 (5-HT3) receptor reveals a novel determinant of single-channel conductance. *Biochem. Soc. Trans.* **32**(Pt. 3), 547–552.

Peters, J. A., Hales, T. G., and Lambert, J. J. (2005). Molecular determinants of single-channel conductance and ion selectivity in the Cys-loop family: Insights from the 5-HT3 receptor. *Trends Pharmacol. Sci.* **26**(11), 587–594.

Przegalinski, E., Gothert, M., Frankowska, M., and Filip, M. (2005). WIN 55,212-2-induced reduction of cocaine hyperlocomotion: Possible inhibition of 5-HT(3) receptor function. *Eur. J. Pharmacol.* **517**(1–2), 68–73.

Samochocki, M., and Strosznajder, J. (1993). Modulatory action of arachidonic acid on GABAA/chloride channel receptor function in adult and aged brain cortex membranes. *Neurochem. Int.* **23**(3), 261–267.

Saxena, N. C. (2000). Inhibition of GABA(A) receptor (GABAR) currents by arachidonic acid in HEK 293 cells stably transfected with alpha1beta2gamma2 GABAR subunits. *Pflugers Arch.* **440**(3), 380–392.

Schwartz, R. D., and Yu, X. (1992). Inhibition of GABA-gated chloride channel function by arachidonic acid. *Brain Res.* **585**(1–2), 405–410.

Sieghart, W. (2006). Structure, pharmacology, and function of GABAA receptor subtypes. *Adv. Pharmacol.* **54**, 231–263.

Spivak, C. E., Lupica, C. R., and Oz, M. (2007). The endocannabinoid anandamide inhibits the function of alpha4beta2 nicotinic acetylcholine receptors. *Mol. Pharmacol.* **72**(4), 1024–1032.

Sun, H., Hu, X. Q., Emerit, M. B., Schoenebeck, J. C., Kimmel, C. E., Peoples, R. W., Miko, A., and Zhang, L. (2008). Modulation of 5-HT3 receptor desensitization by the light chain of microtubule-associated protein 1B expressed in HEK 293 cells. *J. Physiol.* **586**(3), 751–762.

Tecott, L. H., Maricq, A. V., and Julius, D. (1993). Nervous system distribution of the serotonin 5-HT3 receptor mRNA. *Proc. Natl. Acad. Sci. USA* **90**(4), 1430–1434.

Thompson, A. J., Zhang, L., and Lummis, S. R. (2006). "5-HT3 Receptors." Humana Press, Totowa, NJ.

Unwin, N. (1993). Nicotinic acetylcholine receptor at 9 A resolution. *J. Mol. Biol.* **229**(4), 1101–1124.

van der Stelt, M., and Di Marzo, V. (2005). Anandamide as an intracellular messenger regulating ion channel activity. *Prostaglandins Other Lipid Mediat.* **77**(1–4), 111–122.

van Hooft, J. A., and Vijverberg, H. P. (2000). 5-HT(3) receptors and neurotransmitter release in the CNS: A nerve ending story? *Trends Neurosci.* **23**(12), 605–610.

Vijayaraghavan, S., Huang, B., Blumenthal, E. M., and Berg, D. K. (1995). Arachidonic acid as a possible negative feedback inhibitor of nicotinic acetylcholine receptors on neurons. *J. Neurosci.* **15**(5 Pt. 1), 3679–3687.

Waeber, C., Dixon, K., Hoyer, D., and Palacios, J. M. (1988). Localisation by autoradiography of neuronal 5-HT3 receptors in the mouse CNS. *Eur. J. Pharmacol.* **151**(2), 351–352.

Waeber, C., Hoyer, D., and Palacios, J. M. (1989). 5-hydroxytryptamine3 receptors in the human brain: Autoradiographic visualization using [3H]ICS 205-930. *Neuroscience* **31**(2), 393–400.

Waeber, C., Pinkus, L. M., and Palacios, J. M. (1990). The (S)-isomer of [3H]zacopride labels 5-HT3 receptors with high affinity in rat brain. *Eur. J. Pharmacol.* **181**(3), 283–287.

Whitehead, K. J., Manning, J. P., Smith, C. G., and Bowery, N. G. (2001). Determination of the extracellular concentration of glycine in the rat spinal cord dorsal horn by quantitative microdialysis. *Brain Res.* **910**(1–2), 192–194.

Wilson, R. I., and Nicoll, R. A. (2001). Endogenous cannabinoids mediate retrograde signalling at hippocampal synapses. *Nature* **410**(6828), 588–592.

Witt, M. R., and Nielsen, M. (1994). Characterization of the influence of unsaturated free fatty acids on brain GABA/benzodiazepine receptor binding *in vitro*. *J. Neurochem.* **62**(4), 1432–1439.

Witt, M. R., Westh-Hansen, S. E., Rasmussen, P. B., Hastrup, S., and Nielsen, M. (1996). Unsaturated free fatty acids increase benzodiazepine receptor agonist binding depending on the subunit composition of the GABAA receptor complex. *J. Neurochem.* **67**(5), 2141–2145.

Xiong, W., Hosoi, M., Koo, B. N., and Zhang, L. (2008). Anandamide inhibition of 5-HT3A receptors varies with receptor density and desensitization. *Mol. Pharmacol.* **73**(2), 314–322.

Yakel, J. L., and Jackson, M. B. (1988). 5-HT3 receptors mediate rapid responses in cultured hippocampus and a clonal cell line. *Neuron* **1**(7), 615–621.

Yang, Z., Aubrey, K. R., Alroy, I., Harvey, R. J., Vandenberg, R. J., and Lynch, J. W. (2008). Subunit-specific modulation of glycine receptors by cannabinoids and N-arachidonyl-glycine. *Biochem. Pharmacol.* **76**(8), 1014–1023.

Zhang, L., and Lummis, S. C. (2006). "5-HT3 Receptors", 6 pp. Marcel Dekker, New York.

Zimmer, A., Zimmer, A. M., Hohmann, A. G., Herkenham, M., and Bonner, T. I. (1999). Increased mortality, hypoactivity, and hypoalgesia in cannabinoid CB1 receptor knockout mice. *Proc. Natl. Acad. Sci. USA* **96**(10), 5780–5785.

CHAPTER THIRTEEN

Endogenous Cannabinoids and Neutrophil Chemotaxis

Douglas McHugh* and Ruth A. Ross[†]

Contents

I. Cellular Motility and Neutrophils	338
II. The Endogenous Cannabinoid System	339
III. Cannabinoids Modulate Cell Migration	342
IV. Endocannabinoid Effects on Basal Locomotion of Neutrophils	343
V. Endocannabinoid Effects on Induced Migration of Neutrophils	343
VI. Cannabinoid Receptor Expression in Neutrophils	344
VII. Inhibition of Induced Migration: Which Receptors are Involved?	345
A. CB_1 cannabinoid receptors	346
B. CB_2 cannabinoid receptors	347
C. Novel cannabinoid receptors	347
VIII. Inhibitory Signal Transduction Mechanisms: Receptor Crosstalk	350
IX. Inhibitory Signal Transduction Mechanisms: Disruption of the Actin Cytoskeleton	351
X. Conclusion	356
References	357

Abstract

Neutrophils are the earliest inflammatory cell to infiltrate tissue, playing an important role in early phagocytosis. Under pathological conditions, proinflammatory actions of neutrophils contribute to the development of various inflammatory diseases. G_i protein-coupled cell-surface receptors are an essential component of pro-migratory responses in leukocytes; however, few investigations regarding inhibitors of cell migration have been reported. Kurihara *et al.* (2006) and McHugh *et al.* (2008) have revealed that certain endogenous cannabinoids and lipids are potent inhibitors of induced human neutrophil migration. McHugh *et al.* implicate a novel SR141716A-sensitive pharmacological target distinct from cannabinoid CB_1 and CB_2 receptors, which is antagonized by *N*-arachidonoyl-L-serine; and that the CB_2 receptor exerts negative

---

* Department of Psychological and Brain Sciences, Indiana University, Bloomington, Indiana, USA
[†] Institute of Medical Sciences, University of Aberdeen, Aberdeen, Scotland, United Kingdom

Vitamins and Hormones, Volume 81   © 2009 Elsevier Inc.
ISSN 0083-6729, DOI: 10.1016/S0083-6729(09)81013-3   All rights reserved.

co-operativity upon this receptor. Kurihara et al. demonstrate that fMLP-induced RhoA activity is decreased following endocannabinoid pretreatment, disrupting the front/rear polarization necessary for neutrophils to engage in chemotaxis. The therapeutic potential of exploiting endocannabinoids as neutrophilic chemorepellants is plain to see.

I. Cellular Motility and Neutrophils

Cellular motility is a critical feature of many different processes at various stages of human development. For instance, the correct positioning of differentiated cells during embryogenesis is entirely dependent upon effective cell migration. As a human matures toward adulthood, most cells switch off their motile capability, leaving only a few specialized cells able to migrate autonomously in order to fulfill their ongoing functions, such as stem cells in tissue regeneration (Kayali et al., 2003; Thiele et al., 2004), fibroblasts in wound healing (Gillitzer and Goebeler, 2001; Senior et al., 1983), and leukocytes in the immune response. In fact, the coordinated migration of leukocyte subsets is a prerequisite for effective host immunity, that is, lymphocytes continuously patrol the body for pathogenic invaders (Moser and Ebert, 2003), dendritic cells migrate into the secondary lymphatic organs after antigen uptake (Sallusto et al., 1998) and neutrophils are quickly recruited to sites of inflammation (Imhof and Dunon, 1997; Kucharzik and Williams, 2003).

Neutrophils are the most abundant type of white blood cell present in humans, and because they are much more numerous than the longer-lived monocytes or macrophages, the first phagocyte a pathogen is likely to encounter is a neutrophil. In healthy human blood, neutrophils normally exist in a quiescent state with an average half-life of 12 h; however, upon interaction with extracellular chemical cues, they undergo activation and become highly motile phagocytic cells able to pass through the walls of capillaries and enter tissue spaces to engulf and destroy disease-producing bacteria. Antibacterial and digestive enzymes within their cytoplasmic granules subsequently digest the phagocytosed particles. As a result of this action, neutrophils form the first line of defense against infection and are transported rapidly to specific areas of inflammation, where they have the ability to move through the tissues by an amoeboid action at speeds of up to 40 μm min^{-1} (Stevens and Lowe, 1996). To actually reach an area of infection or tissue damage, neutrophils marginate (position themselves adjacent to the blood vessel endothelium) and undergo selectin-dependent capture followed by, in most cases, integrin-dependent adhesion after which they move through the endothelium and basement membrane to the tissue of interest where they survive for 1–2 days. Once in the support tissue, neutrophils respond with chemotaxis, the process whereby cells sense soluble molecules and follow them along a concentration gradient to their source. Despite the variety of different chemotactic molecules, for example, sugars, peptides, cell metabolites, or membrane lipids, the migratory signals they deliver are transduced into the generation of locomotory force via G protein-coupled receptors (GPCRs). GPCRs are coupled on the cytosolic side of the neutrophil plasma membrane to heterotrimeric G proteins (Neer, 1995), which in turn activate a complex and as yet not fully defined array of signaling cascades, culminating in the assembly and disassembly of cellular actin filaments; generating cell motility (Hwang *et al.*, 2004; Neptune and Bourne, 1997; Wenzel-Seifert *et al.*, 1998; Yokomizo *et al.*, 2000). Studies involving chemotaxis of HEK293 cells, as well as those from a lymphocyte cell line (300-19), stably transfected with GPCRs that coupled exclusively to G_q, G_i, and G_s G proteins, found that cells expressing receptors coupled to G_i, but not to G_q or G_s proteins migrated in response to a concentration gradient of the appropriate agonist. This prompted the hypothesis that many G_i-coupled receptors share the ability to mediate cell migration (Arai *et al.*, 1997; Neptune and Bourne, 1997). The two most prominent groups of ligands for the G_i-coupled GPCRs involved in migratory processes are the chemokines and the neurotransmitters; the latter includes the endogenous cannabinoids.

II. The Endogenous Cannabinoid System

In the early 1990s, molecular cloning identified two distinct G protein-coupled cannabinoid receptors, which were given the nomenclature CB_1 and CB_2 receptors (Matsuda *et al.*, 1990; Munro *et al.*, 1993). The existence in mammalian cells of specific membrane receptors for plant-derived cannabinoids triggered a search for an endogenous ligand, which culminated in 1992 in the identification of *N*-arachidonylethanolamide or anandamide (AEA) (Devane, 1992). Three years later, a second endogenous cannabinoid, 2-arachidonylglycerol (2-AG), was isolated from gut (Mechoulam *et al.*, 1995) and brain tissue (Sugiura *et al.*, 1995). By definition cannabinoids are molecules that are able to bind and activate cannabinoid receptors. They encompass a wide range of compounds, which can be categorized, according to chemical structure, into the following groups: classical, nonclassical, aminoalkylindole, and eicosanoid cannabinoids. Alternative categorization groups the cannabinoids into three main classes according to their source. The first, the endocannabinoids are endogenous long-chain fatty acid derivatives, for example, AEA and 2-AG. Other putative members of this class include noladin ether, virodhamine, oleamide, *N*-arachidonoyldopamine (NADA), *N*-oleoyldopamine, *N*-homo-γ-linoenylethanolamide, and *N*-docosatetraenylethanolamide;

however, their activity has not yet been well characterized (Fezza et al., 2002; Hanus et al., 1993; Walker et al., 2002). The second, the phytocannabinoids comprise more than 60 compounds derived from the *Cannabis* plant family, for example, the psychoactive principle of *Cannabis sativa*, Δ^9-tetrahydrocannabinol (Δ^9-THC), and the primary nonpsychoactive constituent, cannabidiol (CBD). The third, the synthetic cannabinoids were developed from various SAR-based studies, for example, CP55940, WIN55212-2, JWH-133, AM630, and so on.

CB_1 receptors are predominantly expressed by neurons in the brain, spinal cord, and peripheral nervous system; however, distribution is not homogenous. CB_1 expression is greater in the cerebral cortex, hippocampus, basal ganglia, and cerebellum than the hypothalamus and spinal cord, where receptor numbers are significantly lower (Howlett et al., 2002). With regard to the peripheral nervous system, CB_1 mRNA transcripts have been detected in the following tissues: pituitary gland, immune cells, reproductive tissues, gastrointestinal tract, superior cervical ganglion, heart, lung, urinary bladder, and adrenal gland (Pertwee, 1997). On balance, CB_1 receptor levels are considerably lower in the peripheral tissues compared with the central nervous system, although some tissues contain high concentrations of CB_1 receptors localized in discreet regions, for example, nerve terminals, which form a small percentage of the total mass (Pertwee, 2001).

Current opinion holds that the CB_2 receptor is less widely expressed than CB_1, having been only confidently detected outwith the central nervous system (Galiègue et al., 1995; Munro et al., 1993). The CB_2 receptor is present principally in the immune system where mRNA transcripts are 10–100-fold more abundant than those of CB_1 (Galiègue et al., 1995; Kaminski et al., 1992; Massi et al., 1997). Moreover, CB_2 receptors have been located in tonsils, thymus, bone marrow, adrenal glands, heart, lung, prostate gland, uterus, pancreas, ovary, and testis (Galiègue et al., 1995; Howlett et al., 2002). CB_2 receptor protein has also been identified in the retina of adult rats (Lu et al., 2000), and FACS analysis has indicated CB_2 receptor protein in the dorsal root; however, immunohistochemistry failed to detect CB_2 receptors on DRG neurones, implying the receptors may be present on non-neuronal cells such as microglia and fibroblasts ganglia (Ross et al., 2001). It is thought that CB_2 receptors are largely involved in the modulation of the immune system, including antigen processing by macrophages, helper T cell activation and the migratory response of certain immunocompetent cells (Howlett et al., 2002).

In addition to the cannabinoid receptors discussed above, evidence is accumulating that cannabinoids, and endocannabinoids in particular, are capable of acting via additional sites that remain yet to be cloned. Firstly, GPR55, an orphan GPCR that AstraZenaca plc and GlaxoSmithKline, Inc.

have independently demonstrated to be activated by endogenous, phyto and synthetic cannabinoids, as well as by lysophophatidylinositol (LPI) (Brown and Wise, 2001; Drmota et al., 2004). Secondly, evidence for a putative CB_2-like cannabinoid receptor has been obtained from experiments involving inhibition of electrically evoked contractions of the mouse vas deferens; and from others with palmitoylethanolamide (PEA), an endogenous lipid that does not bind CB_1 or CB_2 receptors yet exhibits antinociceptive activity in various animal pain models that can be blocked by the CB_2-selective inverse agonist SR144528 (Bisogno et al., 2001; Petitet et al., 1998; Wiley and Martin, 2002). Thirdly, a putative non-CB_1, non-CB_2 GPCR for AEA and WIN55212-2 has been reported from GTPγS binding in wild type and $CB_1^{-/-}$ mice (Breivogel et al., 1997; Di Marzo et al., 2000). Fourthly, experiments involving mouse and rat mesenteric arteries from double $CB_1^{-/-}/CB_2^{-/-}$ knockout animals, and also from studies of microglial cell migration, indicate a novel cannabinoid receptor that is activated by AEA and two CBD analogues, abnormal cannabidiol (Abn-CBD) and O-1602, neither of which bind appreciably to CB_1 or CB_2 receptors (Bukoski et al., 2002; Járai et al., 1999; Kunos et al., 2000; Offertáler et al., 2003; Showalter et al., 1996; Wagner et al., 1999; Walter et al., 2003). CBD and N-arachidonoyl-L-serine (ARA-S) are thought to behave as partial agonists at the Abn-CBD receptor, while another CBD analogue, O-1918 and SR141716A, act an antagonists (McHugh et al., 2008; Milman et al., 2006; Offertáler et al., 2003; Pertwee, 2004; Zhang et al., 2005). Fifthly, studies with isolated segments of rat mesenteric and hepatic arteries have yielded evidence for a novel cannabinoid receptor on capsaicin-sensitive perivascular sensory neurons that can be equipotently activated by Δ^9-THC and cannabinol (CBN; another nonpsychoactive phytocannabinoid) to induce calcitonin gene-related peptide (CGRP) release and the relaxation of phenylephrine-precontracted vessels (Zygmunt et al., 2002). Sixthly, evidence has been obtained that cannabinoids can interact with and modulate non-cannabinoid receptors via allosteric binding sites. The primary binding site on receptors recognized by an endogenous agonist is referred to as the orthosteric site, while a distinct site on a receptor protein that modulates the binding properties of the orthosteric site, via conformational change, is termed an allosteric site (Christopoulos and Kenakin, 2002). Therefore, an allosteric modulator is a compound that interacts with this secondary site to modulate the affinity of a distinct ligand for the orthosteric site and can result in positive or negative modulation (Lee and El-Fakahany, 1991). Evidence has been obtained that the 5-HT_3 receptor contains an allosteric site with which CP55940, WIN55212-2, and AEA can interact (Barann et al., 2002). Other reports indicate there may be allosteric sites for cannabinoids on other receptors, including M1 and M4 muscarinic receptors (Christopoulos and Wilson, 2001) and ionotropic AMPA, GLUA1, and GLUA3 receptors (Akinshola et al., 1999a,b).

III. Cannabinoids Modulate Cell Migration

As CB_1 and CB_2 receptors couple to pertussis toxin-sensitive $G_{i/o}$-proteins, it was theorized that these receptors may be involved in cell migration. Subsequent studies have indeed shown that $G_{i/o}$ protein-coupled receptors, including CB_1 and CB_2, modulate cell migration (Derocq et al., 2000; Song and Zhong, 2000; Walter et al., 2003). Song and Zhong found that the cannabinoid agonists HU-210, WIN55212-2, and AEA are able to induce chemotaxis and chemokinesis (stimulated random motion), via the CB_1 receptor in stably transfected HEK293 cells. The selective CB_1 antagonist SR141716A (Rimonabant) attenuates the migratory response in a concentration-dependent manner and pretreatment with pertussis toxin eliminates the cannabinoid-induced cell migration. While Walter et al. (2003) have demonstrated that 2-AG recruits microglial cells by engaging CB_2 and Abn-CBD receptors. SR141716A (30 nM), at a concentration that is sufficient to antagonize CB_1 receptors but not Abn-CBD receptors, has no effect on 2-AG-induced migration. However, CBN and CBD, both prevent the 2-AG-induced cell migration by antagonizing CB_2 and Abn-CBD receptors, respectively. Δ^9-THC and Abn-CBD induce a synergistic increase in microglial cell migration. Walter et al. state that concurrent activation of CB_2 and Abn-CBD receptors is necessary to trigger microglial cell migration.

Research to fully elucidate the exact functions of CB_1 and CB_2, as well as putative novel cannabinoid receptors, and their endogenous ligands with regard to the regulation of cellular motility is ongoing. It is already becoming increasingly clear that the effects cannabinoids have on cell migration are complex, as both stimulation and inhibition of chemotaxis have been reported (Franklin and Stella, 2003; Franklin et al., 2003; Joseph et al., 2004; Kishimoto et al., 2003, 2005; McHugh et al., 2008; Millar and Stella, 2008; Mo et al., 2004; Sacerdote et al., 2000; Song and Zhong, 2000; Oka et al., 2004; Vaccani et al., 2005; Walter et al., 2003). This bidirectionality is not particularly surprising given that both families of the structurally distinct chemokines and neurotransmitters have members that not only induce migration but exert inhibitory functions as well. With the above backdrop in mind, it is only relatively recently that studies have begun to characterize the role cannabinoids and their receptors play in the modulation of neutrophil migration. Of the migration studies conducted with regard to neutrophils to date, it has been found that certain endogenous, plant and synthetic cannabinoids were able to modulate either spontaneous or induced neutrophil migration, while others were not. The focus of this review concerns the effects that endogenous cannabinoids exert upon neutrophil migration.

IV. Endocannabinoid Effects on Basal Locomotion of Neutrophils

In 2004, Joseph et al. and Oka et al. reported, respectively, that the endogenous cannabinoids, AEA (40 nM) and 2-AG (10 nM–10 μM), have no stimulatory effect on the basal locomotion of human neutrophils. Kurihara et al. (2006) similarly found that 2-AG (300 nM) had no effect. McHugh et al. (2008) confirmed these findings for AEA and 2-AG over a concentration range of 0.01 nM–1 μM and also demonstrated that PEA does not stimulate migration above basal levels of spontaneous activity. In contrast, McHugh et al. found that virodhamine (O-arachidonoylethanolamine, that is, arachidonic acid linked to ethanolamine via an ester bond) acted as a weak chemoattractant at 100 nM, inducing a small but significant stimulation of human neutrophil migration.

V. Endocannabinoid Effects on Induced Migration of Neutrophils

With regard to induced migration Kurihara et al. reported that 2-AG (300 nM) inhibited both human neutrophils and HL-60 cells differentiated into neutrophil-like cells toward 100 nM N-formly-L-methionyl-L-leucyl-L-phenylalanine (fMLP), following a 2 min pretreatment. McHugh et al. have shown that AEA potently inhibits 1 μM fMLP- and 100 nM LTB$_4$-evoked chemotaxis of human neutrophils in a biphasic concentration-dependent manner; preincubation with AEA was necessary to observe any significant inhibitory effect. In other words, when neutrophils were simultaneously exposed to either fMLP or LTB$_4$ and AEA, AEA was unable to prevent induced migration. This is consistent with the observations of Joseph et al., who using a collagen-based three-dimensional migration assay and time-lapse videomicroscopy found that when neutrophils are exposed to fMLP and AEA concomitantly, AEA is unable to inhibit fMLP-induced migration. McHugh et al. also found that virodhamine and NADA inhibit fMLP-induced migration of human neutrophils in a concentration-dependent manner; like AEA, preincubation was necessary to observe any inhibitory effect with virodhamine. While, fMLP-induced migration remained unaffected by PEA, ARA-S, LPI, or 2-AG. The data regarding 2-AG is in direct contrast to that obtained by Kurihara et al., but something as simple as the difference in fMLP concentration used to induce neutrophil migration may allow us to resolve this discrepancy. Kurihara et al. administered a concentration of 100 nM fMLP to induce cell migration following pretreatment with 300 nM 2-AG, whereas McHugh *et al.* employed a 1 μM concentration of fMLP after preincubation with 0.01 nM–1 μM 2-AG. It is a reasonable argument that 1 μM fMLP may be sufficient to overcome any inhibition exerted by 1 μM 2-AG, resulting in a lack of any observed inhibitory activity. McHugh *et al.* go on to reveal an intriguing pharmacology underlying the inhibition of human neutrophil chemotaxis indicating that certain endocannabinoids and phytocannabinoids are potent inhibitors of human neutrophil migration; however, discussion of phyto- and synthetic cannabinoid effects on neutrophil motility observed by McHugh *et al.* and others are outside the scope of this article. Of the endogenous cannabinoids and related lipids that McHugh *et al.* tested, AEA, virodhamine and NADA are potent inhibitors of fMLP-induced migration with IC_{50} values of 0.14, 0.18, and 8.80 nM, respectively. Interestingly, AEA is roughly twofold more potent at inhibiting neutrophil migration induced by LTB_4 compared with fMLP, which implies that endocannabinoids may be able to inhibit migration stimulated by numerous chemotactic agents. Consistent with this speculation, Nilsson *et al.* (2006) tested the ability of several cannabinoids to inhibit tumor necrosis factor-α (TNF-α)-induced neutrophil migration across endothelial-like ECV304 cells. TNF-α is a potent chemoattractant for neutrophils and helps them to stick to endothelial cells as they initiate migration. Nilsson *et al.* found that the CB_1 agonist, WIN55212-2, could decrease the migration of neutrophils in a concentration-dependent manner; AM251 was able to partially attenuate the effect of 1 μM WIN55212-2.

VI. Cannabinoid Receptor Expression in Neutrophils

The next logical step in characterizing the pharmacology of endocannabinoids toward neutrophil migration is to resolve the question of which cannabinoid receptors mediate the inhibition observed.

CB_1 and CB_2 receptor expression in neutrophils has largely been implied from studies involving the quantification of mRNA transcripts. Although CB_1 receptors are predominantly expressed in the brain, CB_1 mRNA sequences have also been detected in testis, spleen, and leukocytes; with levels in B cells > NK cells > neutrophils > $CD8^+$ cells > monocytes > $CD4^+$ cells (Bouaboula *et al.*, 1993; Galiégue *et al.*, 1995; Kaminski *et al.*, 1992). The CB_2 subtype is present principally in the immune system where mRNA transcripts are 10–100-fold more abundant than that of CB_1 (Galiégue *et al.*, 1995; Kaminski *et al.*, 1992; Massi *et al.*, 1997). That said, discrepant evidence has been published regarding whether neutrophils actually express functional CB_1 or CB_2 receptors despite the detection of mRNA sequences encoding their protein. Derocq et al. (2000) conducted studies with wild type and differentiated HL-60 cells, which, as previously mentioned, are derivative of committed neutrophil progenitors of the granulocyte/monocyte lineage (Gallagher et al., 1979), they observed a dramatic decrease of CB_2 receptor mRNA transcripts during the course of DMSO-induced differentiation, implying a downregulation of the expressed CB_2 receptor. Deusch et al. (2003) were also unable to find any CB_2 receptor protein with antihuman CB_2 antibodies via chemiluminescent probes. In a similar vein, Oka et al. (2004) using RT-PCR reported that the amount of CB_2 receptor mRNA in neutrophils was negligible and that Western blot analysis with an anti-CB_2 receptor antibody failed to discern CB_2 receptor protein. More recently and in direct contrast to these findings, Kurihara et al. detected cell-surface CB_2 expression in both human neutrophils and differentiated neutrophil-like HL-60 cells using flow cytometry. In trying to resolve this conflict, it is worth noting that receptor antibodies are notoriously problematic in the cannabinoid field. Grimsey et al. (2008) conducted a systematic study that found commercially available CB_1 antibodies not only fail to detect proteins of the correct molecular weight but also recognize multiple proteins in brain tissues, rendering them ineffective for specific detection of cannabinoid receptors. It is entirely likely that the same is true of commercially available CB_2 receptor antibodies, which may be an underlying factor in those studies that failed to detect CB_2 expression. In other words, a lack of evidence is not the same as evidence of a lack. Another reasonable explanation arises from the knowledge that CB_2 receptor expression in immune cells can be exceptionally plastic and that the specific methods used to prepare the neutrophils may in themselves alter CB_2 expression levels (Miller and Stella, 2008). Given the flow cytometry results and the scale of pharmacological data provided by Kuirhara et al. and McHugh et al., which will be discussed shortly, the weight of evidence currently indicates that human neutrophils do indeed express functional CB_2 receptors, although support for CB_1 receptors remains sparse.

VII. Inhibition of Induced Migration: Which Receptors are Involved?

Kurihara et al. and McHugh et al. went on to try to elucidate the identity of the receptors the pertinent endocannabinoid ligands activate, by investigating their observed effects in the presence of established CB_1, CB_2, or TRPV1 antagonists. It is important to note at this point that CB_1- or CB_2-selective antagonists, such as SR141716A and SR144528, respectively, together with behaving as inverse agonists, can block non-CB_1 and non-CB_2 targets when administered at concentrations greater than the K_d values determined for their corresponding cannabinoid receptors, that is, at concentrations in the micromolar range (Rinaldi-Carmona et al., 1998; Walter et al., 2003). Kurihara et al. observed that pretreatment of human neutrophils and differentiated HL-60 cells with 1 μM SR144528, but not 1 μM AM251 (a CB_1-selective antagonist), attenuated the inhibitory effect produced by 2-AG; however, caution must be exercised in interpreting these results as plainly indicating CB_2 receptor involvement. It may well be that CB_2 receptors are responsible for transducing the inhibitory signal of 2-AG, but it is also remains a distinct possibility that putative CB_2-like cannabinoid receptors mediate the 2-AG effect either alone or in conjunction with CB_2. Given the implied scarcity of CB_1 expression in neutrophils, a concentration of 1 μM SR141716A was deliberately chosen by McHugh et al. to reduce its antagonism selectivity in the hope of detecting either CB_1 or a non-CB_1 SR141716A-sensitive receptor. While, 100 nM SR144528 was used in order to ensure its antagonism of CB_2 receptors alone. As AEA and NADA are both additionally capable of activating TRPV1 receptors (Ross et al., 2001; Smart et al., 2000; Zygmunt et al., 1999), and although Heiner et al. (2003) have reported that neutrophils do not express TRPV1 receptors, McHugh et al. investigated the inhibitory effect observed in the presence of 1 μM capsazepine (CPZ) to exclude TRPV1 as a receptor candidate.

McHugh et al. tested the ability of the two most potent endocannabinoid compounds in the presence of the antagonist concentrations specified above and found that the inhibition produced by virodhamine and AEA was significantly attenuated by SR141716A (1 μM); however, the apparent K_B was calculated to be 745.2 nM, which is considerably higher than that expected ($\sim$10 nM) for CB_1-related antagonism (MacLennan et al., 1998). Surprisingly in the presence of 100 nM SR144528, they observed the inhibition exerted by 100 nM virodhamine and AEA was significantly enhanced rather than attenuated. This enhancement was reproduced in the presence of another CB_2-selective antagonist, AM630 (100 nM), to an even greater degree. And lastly, the ability of 100 nM virodhamine and AEA to inhibit fMLP-induced neutrophil migration remained unaffected by 1 μM CPZ, excluding any involvement of TRPV1 receptors as expected. Likewise, SR141716A (1 μM), SR144528 (100 nM), CPZ (1 μM), and AM630 (100 nM) alone had no effect on neutrophil migration triggered by fMLP.

A. CB_1 cannabinoid receptors

The bulk of evidence gathered indicates that CB_1 receptors do not play any role in the cannabinoid-mediated modulation of spontaneous or induced neutrophil migration for the following reasons: there is scant evidence of functional CB_1 receptor expression; AEA and virodhamine display potent inhibitory behavior that is inconsistent with their pharmacology at CB_1

receptors, that is, AEA acts as a CB_1 partial agonist (Mackie et al., 1993; Sugiura et al., 2000), virodhamine as a CB_1 antagonist (Porter et al., 2002); 2-AG, which is a full agonist at CB_1 receptors (Pertwee and Ross, 2002), in the hands of McHugh et al., had no effect on neutrophil migration, while Kurihara et al. report an inhibitory effect for 2-AG that remains unaffected by the presence of 1 μM AM251; and although the inhibition of fMLP-induced migration observed by McHugh et al. was significantly attenuated by SR141716A, the apparent K_B value was inconsistent with a CB_1-mediated effect (MacLennan et al., 1998). Taken together, these results are indicative of a role for a non-CB_1, SR141716A-sensitive receptor mediating the effect of these cannabinoids.

B. CB_2 cannabinoid receptors

Kurihara et al. base their case for CB_2 receptors transducing the inhibitory signal exerted by 2-AG toward fMLP-induced migration solely upon the observation that this effect is attenuated in the presence of 1 μM SR144528. As noted earlier, it is not possible to confidently assert this interpretation of the data as SR144528 loses its CB_2-selectivity in the micromolar concentration range. McHugh et al. imply a role for functional CB_2 receptors in the modulation of neutrophil migration on the basis that two, structurally distinct, CB_2-selective antagonists, administered at appropriate concentrations, significantly enhanced the cannabinoid-mediated inhibition. There is considerable evidence that cannabinoid CB_1 and CB_2 receptors exist in a conformation that is precoupled to the G protein (Pertwee, 2005). Both SR144528 and AM630 are CB_2 receptor inverse agonists (Bouaboula et al., 1999; Ross et al., 1999), and as such they bind with a high affinity to precoupled CB_2 receptors. CB_2 receptors on human neutrophils may be constitutively activated, exerting high basal levels of CB_2 receptor-mediated signaling; thereby precluding inhibition mediated by certain other receptors. Furthermore, constitutively activation may underlie the lack of significant stimulation of neutrophil migration above basal levels by CB_2 agonists, such as 2-AG (Kurihara et al., 2006; McHugh et al., 2008). In line with the findings of McHugh et al., Lunn et al. (2006) have demonstrated that CB_2 receptor inverse agonists inhibit leukocyte migration both *in vivo* and *in vitro*, and the level of effect is proportional to the degree of inverse efficacy.

C. Novel cannabinoid receptors

Many aspects of the endocannabinoid system have yet to be fully elucidated, and as mentioned previously, a range of putative cannabinoid receptors that exert cannabimimetic activity has been described in the literature. By and large, the reported pharmacology of these receptors is insufficient to account for the array of ligands exerting modulatory activity on spontaneous or induced neutrophil migration observed by Kurihara *et al.* and McHugh *et al.* The exception being the non-CB_1, non-CB_2 pharamcological target named the Abn-CBD receptor, which is antagonized by SR141716A at concentrations considerably higher than those predicted from its CB_1 receptor affinity (Begg *et al.*, 2005). Evidence for the existence of Abn-CBD receptors initially emerged from studies in certain blood vessels from $CB_1^{-/-}$ mice (Jarái *et al.*, 1999). In addition, virodhamine and NADA induce a relaxation of mesenteric arteries, being more potent than either AEA or Abn-CBD (Begg *et al.*, 2005; Ho and Hiley, 2003; O'Sullivan *et al.*, 2006). Probably of greatest relevance are studies in microglial cells, which provide robust evidence for a role of Abn-CBD receptors in cell migration (Walter *et al.*, 2003). Thus, 2-AG triggers microglial cells by acting through CB_2 and Abn-CBD receptors. Stimulation of microglial cell migration by cannabinoids is antagonized by high but not low concentrations of SR141716A (Franklin and Stella, 2003). Certain parallels exist between the pharmacology that we observed in human neutrophils and the pharmacology of the Abn-CBD receptor. First, the effects of both AEA and virodhamine are sensitive to antagonism by 1 μM SR141716A, which is in line with the proposed affinity of the CB_1 receptor antagonist for the Abn-CBD receptor (Begg *et al.*, 2005). Second, agonist efficacy and potency closely match that previously obtained for the Abn-CBD receptor in blood vessels and microglial cells; of the endocannabinoids and related lipids, AEA, NADA, and virodhamine are active, the latter having the highest potency, while PEA is inactive. McHugh *et al.* also report the antagonism of endocannabinoid-induced inhibition by *N*-arachidonoyl-L-serine (ARA-S), a brain constituent previously shown to be an agonist at the Abn-CBD receptor (Milman *et al.*, 2006). They report that this endogenous compound attenuates the inhibition of human neutrophil migration by AEA, virodhamine, and Abn-CBD. Although 1 μM ARA-S abolishes the effect of Abn-CBD completely, it reduces the E_{max} value of virodhamine, implying that the effect of virodhamine may involve more than one receptor, only one of which is blocked by ARA-S. Alternatively, ARA-S may be acting allosterically on the target receptor; a reduction in E_{max} is characteristic of an allosteric inhibitor. In line with these findings, in a recent abstract Zhang *et al.* (2005) report that Abn-CBD inhibits angiogenesis, an effect that is antagonized by ARA-S. This raises two possibilities: either this endogenous lipid mediator is a partial agonist at the Abn-CBD receptor and thereby also acts as an antagonist in certain conditions; or the effects observed by McHugh *et al.* are due, at least in part, to activation of another, perhaps related target.

Recent publications (Johns *et al.*, 2007; Ryberg *et al.*, 2007) have emerged suggesting certain cannabinoid ligands interact with the orphan receptor GPR55 and that this, and possibly other orphan receptors, may account for the pharmacological and functional evidence for some novel cannabinoid receptors (Baker *et al.*, 2006; Pertwee, 2007). Given the possible relevance of orphan GPCRs with regard to migratory signaling, it is pertinent to discuss any potential involvement of this novel cannabinoid receptor in the modulation of neutrophil migration. GPR55 was cloned by Sawzdargo *et al.* (1999) and maps to chromosome 2q37; it is poorly related to other receptors, sharing closest identity with the orphan receptors, GPR35 (37%) and GPR92 (30%), and 13.5 and 14.4% homology with CB_1 and CB_2 receptors, respectively (Baker *et al.*, 2006). The expression profile of GPR55 has been reported to be lung > spleen > kidney > brain > heart (Brown and Wise, 2001). Additionally, it has been located in smaller mesenteric arteries of human colon tissue and is highly expressed in adipose (Brown *et al.*, 2005; Drmota *et al.*, 2004). Some controversy surrounds the pharmacology of GPR55; two pharmaceutical companies, AstraZeneca plc and GlaxoSmithKline Inc, have independently demonstrated that GPR55 can be activated by endogenous, phyto and synthetic cannabinoid ligands (Brown and Wise, 2001; Drmota *et al.*, 2004). However, the pharmacology of GPR55 reported by AstraZeneca and GlaxoSmithKline is somewhat conflicting; moreover, Oka *et al.* (2007) report that LPI, but not cannabinoid ligands, induces extracellular signal-related kinase (ERK) phosphorylation in GPR55-expressing cells.

Most pertinent to the discussion of endocannabinoid-induced inhibition of neutrophil migration is the suggestion that the Abn-CBD receptor is, in fact, GPR55; however, there are substantial differences in the reported pharmacology of these two targets, which argues against this being the case (Baker *et al.*, 2006; Mackie and Stella, 2006). Furthermore, although Abn-CBD has high affinity for GPR55, it retains its vasodilator effects in $GPR55^{-/-}$ mice, the implication being that this atypical cannabinoid activates more additional novel receptors (Hiley and Kaup, 2007; Johns *et al.*, 2007).

That said and with regard to human neutrophils, it is relevant that GPR55 is expressed in splenic tissue and that virodhamine, which was the highest efficacy and potency compound in the study by McHugh *et al.*, is also reported to be a high-efficacy agonist of GPR55 (Ryberg *et al.*, 2007). Kurihara *et al.* have demonstrated that their CB_2-like receptor plays a role in human neutrophil migration by modulating RhoA activation, which will be discussed shortly. McHugh *et al.* report that CBD potently inhibits, while LPI does not inhibit human neutrophil migration, which is in conflict with the reported pharmacology of GPR55 by Oka *et al.* (2007) and Whyte *et al.* (2008). Reports suggest that GPR55 and related orphan receptors are G_{13}-coupled and that they can activate the Rho pathway (Henstridge *et al.*, 2008; Ryberg *et al.*, 2007), which plays an important role in the regulation of myosin light chain phosphorylation and subsequent cytoskeletal-dependent locomotion (Buhl *et al.*, 1995; Kozasa *et al.*, 1998). Given such G protein-coupling, as shall be discussed shortly, agonists at GPR55 would be expected to stimulate neutrophil migration rather than inhibiting it. Currently, further work is required to establish the molecular identity of the Abn-CBD receptor.

VIII. Inhibitory Signal Transduction Mechanisms: Receptor Crosstalk

Interactions among GPCRs are complex (Snyderman and Uhing, 1992); processes termed "receptor crosstalk" may offer additional explanations for the impairment of induced neutrophil migration. Two such crosstalk mechanisms include heterologous desensitization and dimerization of receptors.

Heterologous desensitization occurs when the activation of one GPCR causes the phosphorylation of a distinct, yet co-localized, receptor, with a consequent decrease of its function. For example, thrombin receptor activation has been shown to cause phosphorylation of several receptors for chemoattractants, for example, IL-8, C5a and platelet activating factor (PAF) (Ali *et al.*, 1996). In terms of evidence for heterologous desensitization in neutrophils, Grimm *et al.* (1998) demonstrated that opioids acting through δ and μ receptor subtypes were capable of inhibiting subsequent migratory responses of neutrophils to IL-8. Additionally, Campbell *et al.* (1997) have found that fMLP inhibits IL-8-induced neutrophil migration; however, the reverse was not observed, indicating that heterologous desensitization of chemokine responses may be selective. Consistent with this interpretation, it has been shown that α_1-adrenergic receptors are unable to desensitize chemoattractant receptors (Didsbury *et al.*, 1991).

Receptor dimerization is a second crosstalk mechanism potentially able to explain the cannabinoid-evoked modulation of chemotactic receptor function in human neutrophils. Gomes *et al.* (2000) have demonstrated that μ and δ opioid receptors form heterodimers, which exhibit distinct ligand binding and signaling properties. They found that extremely low doses of subtype-selective drugs (both agonists and antagonists) were able to significantly enhance the potency and efficacy of the second subtype, and *vice versa*. Receptor dimerization of CB_2 receptors may explain the potent synergism observed for SR144528 and AM630. The data provided by McHugh *et al.* suggest that in human neutrophils, there is crosstalk/negative co-operativity between the cannabinoid precoupled CB_2 receptors and a novel receptor such that inhibition of precoupled CB_2 receptors enhances the inhibition observed in response to activators of this receptor. This is in line with the positive co-operativity observed between CB_2 and Abn-CBD receptors in microglial cells (Walter *et al.*, 2003). A realistic explanation for the apparent bidirectional co-operativity exhibited between these two receptors is a difference in underlying signal transduction depending on cell type. CB_2 and/or the novel non-CB_1, SR141716A-sensitive target may also cross-modulate with fMLP receptors in a similar manner to inhibit induced neutrophil migration. Consistent with this, Rahaman et al. (2006) reported that the IL-8 receptor, CXCR1 formed constitutive heterodimers with S1P4 receptors in human neutrophils. Exogenously applied S1P (sphingosine-1-phosphate) was found to reduce IL-8 induced neutrophil chemotaxis together with transcription of the IL-8 receptor; a complete understanding of the molecular basis of this inhibition has not yet been fully elucidated. Such reports in the literature of established chemotactic receptors undergoing heterologous desensitization or dimerization, support "receptor crosstalk" as another potential mechanism through which the cannabinoids may be acting.

IX. Inhibitory Signal Transduction Mechanisms: Disruption of the Actin Cytoskeleton

In the absence of a chemoattractant, neutrophils spontaneously polymerize actin filaments in order to extend pseudopodia, which probe the extracellular environment for directional cues. Upon stimulation with chemotactic ligands such as fMLP, neutrophils rapidly change from a spherical to an elongated shape with front/rear polarity and migrate within minutes; and, as stated earlier, the actin–myosin cytoskeleton is the ultimate structure responsible for the generation of locomotory force in migratory cells (Entschladen et al., 2005). The cellular actin cytoskeleton is a system of fibers critical to motility and comprised of well-ordered polymers, which are constructed from small protein subunits held together by noncovalent bonds. The fibers can be grouped into three categories, according to diameter size: microfilaments (7–9 nm), intermediate filaments (10 nm), and microtubules (24 nm) (Lodish et al., 2000). The assembly of actin filaments, and their subsequent formation into bundles and networks, is one of two mechanisms necessary in order for cells to migrate. The second consists of translational movement along actin cables generated by actin-activated ATPases known as myosins. Although several classes of myosin have been identified, myosin II is the best characterized in nonmuscle cells. Members of the myosin II family consist of two heavy chains (200 kDa) and two sets of light chains (16–20 kDa), and are widely distributed in eukaryotic cells. Myosin II function is controlled by phosphorylation of the regulating light chains via the Ca^{2+}/calmodulin (CaM)-dependent enzyme, myosin light chain kinase (MLCK). Phosphorylation of myosin light chains (MLC) by MLCK is a critical step in myosin function, as it promotes myosin

ATPase activity and polymerization of actin cables. This results in a fully functional actin–myosin motor unit, which generates the contractile force necessary for cell migration.

Two distinct signaling pathways, of consequence to the cytoskeleton and myosin motor units, are initiated by receptor activation. Once a GPCR is agonist-bound, its G protein disengages into two distinct subunits: a GTP-bound α-subunit and a βγ-complex, each of which signals via an independent route (Neer, 1995). Firstly, depending on the type of GPCR, stimulatory (G_s) or inhibitory (G_i) α-proteins are activated (see Fig. 13.1).

They increase or decrease, respectively, the activity of adenylyl cyclase enzymes, which in turn regulate the production of cAMP (Neer, 1995). cAMP activates protein kinase A (PKA), a serine/threonine kinase that can modulate multiple molecular targets, such as actin assembly, via action on Ena/VASP proteins and prolifin (Jockusch et al., 1995; Kwiatkowski et al., 2003); the sequestration of intracellular Ca^{2+}, via phospholamban (PLB) and the SERCA (Joseph et al., 2002); and the activity of the monomeric G protein Rho, via guanine dissociation inhibitors (GDi) (Ellerbroek et al., 2003; Narumiya, 1996; Somlyo and Somlyo, 2003). Rho regulates the Rho-kinase/Rho-dependent coiled-coil kinase (ROCK) (Riento and Ridley, 2003), which is a MLC phosphorylating enzyme that acts on the myosin II light chain. The intracellular Ca^{2+} concentration, modulated via PKA activity, as stated above, is important for the activation of CaM, which subsequently phosphorylates the MLCK; this enzyme also acts on the myosin II light chain.

The second pathway is initiated by the βγ-complex, which activates G protein-coupled receptor kinases (GRKs) (Pierce et al., 2002) (see Fig. 13.2).

GRKs engage src-protein tyrosine kinases (srcPTKs) via β-arrestin (Luttrel et al., 1999); an important substrate of the srcPTKs is phospholipase C (PLC). PLCγ catalyzes the breakdown of phosphatidyl-inositolbiphosphate (PIP2) into diacylglycerol (DAG) and inositol-1,4,5-triphosphate (IP3). IP3 opens Ca^{2+} channels located on intracellular stores, with the resultant CaM-dependent stimulation of myosin II light chains recently discussed above (Masur et al., 2001), while DAG activates protein kinase C (PKC). PKCα phosporylates actin-binding proteins, such as vinculin and the myristoylated alanine-rich C kinase substrate (MARCKS), and focal adhesion kinase (FAK) enzymes (Luo et al., 2003; Ziegler et al., 2002). FAK initiates a phosphorylation cascade by acting on MAP kinases, which phosphorylate MLCKs, which in turn phosphorylate the myosin II light chain. Lastly, PKC activates Rho, via guanine exchange factors (GEF) (Sagi et al., 2001), with consequent effects on Rho kinase/ROCK activity and myosin II light chains.

While a complete understanding of these transduction pathways remains yet to be achieved, from what has been elucidated thus far about the complex array of signaling cascades involved in migration, it is clear that modulation of

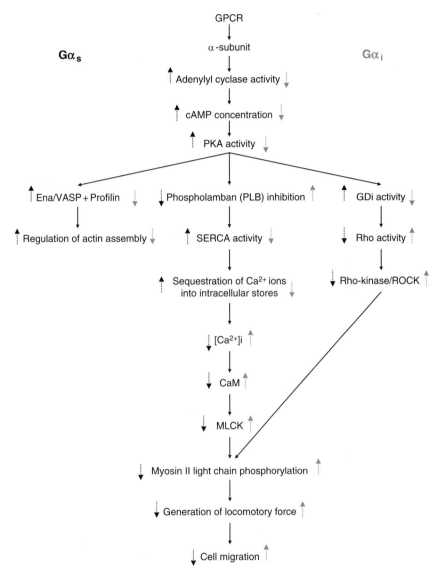

Figure 13.1 Signal transduction pathway for the modulation of cell migration initiated by GPCR G_α-subunits.

actin/myosin function is possible via multiple targets. Pivotal effectors include cAMP, PLC, and PKC, as amplification of the transduction signal is dependent upon their activity. In short, ligands that lead to reduced cAMP production will initiate a transduction cascade, terminating with stimulation of cell migration (Wright et al., 1990), while those that lead to increased cAMP production will

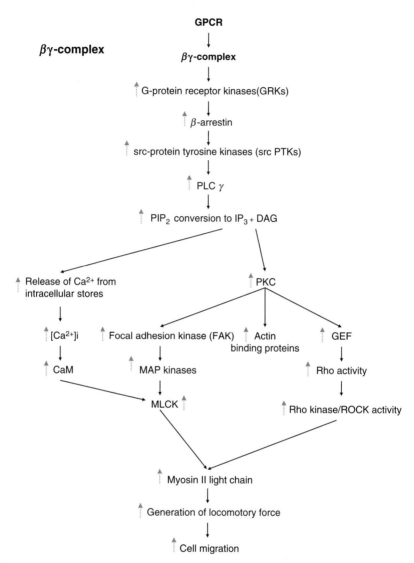

Figure 13.2 Signal transduction pathway for the modulation of cell migration initiated by GPCR $\beta\gamma$-complex.

cause subsequent inhibition of cell migration. Indeed, established neutrophil chemoattractants, such as fMLP, LTB$_4$, IL-8, and C5a, activate G$_i$-protein-coupled cell-surface receptors in order to exert their pro-migratory signal (Hwang *et al.*, 2004; Neptune and Bourne, 1997; Wenzel-Seifert *et al.*, 1998; Yokomizo *et al.*, 2000).

Among the numerous molecules implicated in cytoskeletal reorganization and cell migration, the Rho-GTPases (RhoA, rac1, rac2, and cdc42) feature in both the signaling cascades outlined above and play particularly important roles in achieving the migratory behavior associated with chemotaxis. As stated, migrating cells clearly possess a polarized morphology, that is, a front and a back. Exactly how this spatial arrangement is formulated at a molecular level is still unclear. In a cell that is meandering randomly, the front can easily give way and become passive as some other region of the cell forms a new front. While in chemotaxing cells, the stability of the front is enhanced and the cell seems purposively to advance toward its target. This polarity is reflected at a molecular level by a restriction of certain molecules to particular regions of the cell surface: thus RhoA and PTEN (phosphatase and tensin homologue) are found toward the rear and in forming the retracting tail (Bokoch, 2005; Van Haastert and Devreotes, 2004; Xu et al., 2003), whereas the phospholipid PIP3 together with activated rac and cdc42 are found at the front of the cell and in forming the lamellipodium, which is a cytoskeletal actin projection on the mobile edge of the cell. Lamellipodia are believed to be the actual motor that pulls the cell forward as well as acting as a steering device for cells during the process of chemotaxis (Alberts et al., 2002; Bokoch, 2005; Dong et al., 2005; Sánchez-Madrid and del Pozo, 1999; Xu et al., 2003).

Kurihara et al. examined both RhoA activity and front/rear polarization as part of their investigation into the effect of cannabinoids on the migration of human neutrophils. They state that more than 70% of the neutrophils examined responded to 100 nM fMLP, namely: RhoA activity was increased, the neutrophils polarized, developing a front/rear morphology, and began to migrate 5 min after stimulation. In contrast, neutrophils pretreated with 2-AG prior to stimulation with 100 nM fMLP displayed decreased RhoA activity, they failed to develop distinct front/rear polarity, and instead extended one or more pseudopods (temporary lamellipodium-like projections) in different directions. In line with these findings, Xu et al. (2003) reported that two inhibitors of RhoA pathways, Y27632 and a dominant-negative RhoA mutant (RhoA-T17), disrupt fMLP-induced cell polarization and induce formation of multiple pseudopods in neutrophil-like HL-60 cells. It seems plausible therefore that at least some of the endocannabinoids able to inhibit induced neutrophil migration may do so at the level of RhoA activity, disrupting normal actin/myosin function in response to chemotactic stimuli such as fMLP. Additionally, it is also possible to speculate that crosstalk between the CB_2 receptor and the novel target located in human neutrophils occurs at the level of Rho signaling.

X. Conclusion

Neutrophils are the earliest inflammatory cell to infiltrate tissue and dominate the acute inflammatory stage, playing an important role in early phagocytosis (Ainsworth et al., 1996; Hurley, 1983; Movat, 1985;). Neutrophils release various inflammatory mediators, as well as chemotactic factors specific for monocytes and other inflammatory cells (Ainsworth et al., 1996; Antony et al., 1985; Levine et al., 1986; Pereira et al., 1990; Territo et al., 1989). Thus, neutrophils have the potential to amplify the cellular inflammatory response and are pivotal in its early onset (Ainsworth et al., 1996). Although neutrophils are generally replaced by macrophages as the primary cell type present in tissue as the acute inflammatory response progresses, abnormal chronic inflammatory states are usually pleomorphic, involving a variety of cell types, including neutrophils (Hurley, 1983; Movat, 1985). Under pathological conditions, the pro-inflammatory actions of neutrophils contribute to the development of inflammatory pain (Bennett et al., 1998; Levine and Taiwo, 1994; Levine et al., 1984, 1985; Schuligoi, 1998), unstable angina pectoris (Wysocka et al., 1997), inflammatory bowel disease (Kane, 2001), atherosclerosis (Cross et al., 2005) and also play a critical role in the development of arthritic inflammation (Ward, 1997). In addition, there is a growing body of evidence suggesting that neutrophils make a crucial contribution to a number of autoimmune, autoinflammatory, and neoplastic disorders (Nathan, 2006).

It is clear from the volume of empirical evidence contained in the literature that cell-surface receptors are an essential component of the majority of pro-migratory responses (Arai et al., 1997; Hwang et al., 2004; Neptune and Bourne, 1997; Wenzel-Seifert et al., 1998; Yokomizo et al., 2000). To date, few investigations regarding inhibitors of cell migration have been reported (Armstrong, 1995; De la Fuente et al., 1998; Garrido et al., 1996; Harvath et al., 1991; Mitsui et al., 1997; Okamoto et al., 2000); they include a μ-opioid agonist, prostaglandin E, S1P and adenylyl cyclase-activating peptide, all of which act mainly on leukocytes, via cell-surface receptors. Together, the work of Kurihara et al. and McHugh et al. has revealed that certain endogenous cannabinoids and lipids are potent inhibitors of induced human neutrophil migration. McHugh et al. implicate a novel SR141716A-sensitive pharmacological target distinct from cannabinoid CB_1 and CB_2 receptors, which is antagonized by the endogenous compound N-arachidonoyl-L-serine; and that, in neutrophils, the CB_2 receptor exerts negative co-operativity upon this receptor. In addition, Kurihara et al. have demonstrated that, at least in the case of 2-AG, fMLP-induced RhoA activity is decreased following endocannabinoid pretreatment, disrupting the front/rear polarization necessary for neutrophils to engage in chemotaxis. From this review, the therapeutic potential of exploiting endocannabinoids as neutrophilic chemorepellants is plain to see.

REFERENCES

Ainsworth, T. M., Lynam, E. B., and Sklar, L. A. (1996). Neutrophil function in inflammation and infection. *In* "Cellular and Molecular Pathogenesis" (A. E. Sirica, Ed.), pp. 37–55. Lippincott-Raven, Philadelphia.

Akinshola, B. E., Chakrabarti, A., and Onaivi, E. S. (1999a). *In-vitro* and *in-vivo* action of cannabinoids. *Neurochem. Res.* **24**(10), 233–240.

Akinshola, B. E., Taylor, R. E., Ogunseitan, A. B., and Onaivi, E. S. (1999b). Anandamide inhibition of recombinant AMPA receptor subunits in Xenopus oocytes is increased by forskolin and 8-bromo-cyclic AMP. *Naunyn-Schmiedebergs Arch. Pharmacol.* **360**(3), 242–248.

Alberts, B., Johnson, A., Lewis, J., Raff, M., Roberts, K., and Walter, P. (2002). "Molecular Biology of the Cell" 4th edn. Garland Science, Taylor & Francis Group, New York, pp. 908, 931, 973–975.

Ali, H., Tomhave, E. D., Richardson, R. M., Haribabu, B., and Snyderman, R. (1996). Thrombin primes responsiveness of selective chemoattractant receptors at a site distal to G protein activation. *J. Biol. Chem.* **271**(6), 3200–3206.

Antony, V. B., Sahn, S. A., Antony, A. C., and Repine, J. E. (1985). Bacillus Calmette-Guerin-stimulated neutrophils release chemotaxis for monocytes in rabbit pleural spaces and *in vitro*. *J. Clin. Invest.* **76**(4), 1514–1521.

Arai, H., Tsou, C.-L., and Charo, I. F. (1997). Chemotaxis in a lymphocyte cell line transfected with C-C chemokine receptor 2B: Evidence that directed migration is mediated by betagamma dimers released by activation of G(alphai)-coupled receptors. *Proc. Nat. Acad. Sci. USA* **94**(26), 14495–14499.

Armstrong, R. A. (1995). Investigation of the inhibitory effects of PGE2 and selective EP agonists on chemotaxis of human neutrophils. *Br. J. Pharmacol.* **116**(7), 2903–2908.

Baker, D., Pryce, G., Davies, W. L., and Hiley, R. (2006). In silico patent searching reveals a new cannabinoid receptor. *Trends Pharamacol. Sci.* **27**, 1–4.

Barann, M., Molderings, G., Bruss, M., Bonisch, H., Urban, B. W., and Gothert, M. (2002). Direct inhibition by cannabinoids of human 5-HT3A receptors: Probable involvement of an allosteric modulatory site. *Br. J. Pharmacol.* **137**(5), 589–596.

Begg, M., Pacher, P., Batkai, S., Osei-Hyiaman, D., Offertaler, L., Fong, M. M., Liu, J., and Kunos, G. (2005). Evidence for novel cannabinoid receptors. *Pharmacol. Ther.* **106**(2), 133–145.

Bennett, G., al-Rashed, S., Hoult, J. R. S., and Brain, S. D. (1998). Nerve growth factor induced hyperalgesia in the rat hind paw is dependent on circulating neutrophils. *Pain* **77**(3), 315–322.

Bisogno, T., Hanus, L., De Petrocellis, L., Tchilibon, S., Ponde, D. E., Brandi, I., Moriello, A. S., Davis, J. B., Mechoulam, R., and Di Marzo, V. (2001). Molecular targets for cannabidiol and its synthetic analogues: Effect on vanilloid VR1 receptors and on the cellular uptake and enzymatic hydrolysis of anandamide. *Br. J. Pharmacol.* **134**(4), 845–852.

Bokoch, G. M. (2005). Regulation of innate immunity by Rho GTPases. *Trends Cell Biol.* **15**, 163–171.

Bouaboula, M., Rinaldi, M., Carayon, P., Carillon, C., Delpech, B., Shire, D., Le Fur, G., and Casellas, P. (1993). Cannabinoid-receptor expression in human leukocytes. *Eur. J. Biochem.* **214**(1), 173–180.

Bouaboula, M., Dussossoy, D., and Casellas, P. (1999). Regulation of peripheral cannabinoid receptor CB2 phosphorylation by the inverse agonist SR144528. Implications for receptor biological responses. *J. Biol. Chem.* **274**, 20397–20405.

Breivogel, C. S., Sim, L. J., and Childers, S. R. (1997). Regional differences in cannabinoid receptor/G-protein coupling in rat brain. *J. Pharmacol. Exp. Ther.* **282**(3), 1632–1642.

Brown, A. J., and Wise, A. (2001). Glaxosmithkline, Identification of modulators of GPR55 activity. International Patent #: WO0186305.

Brown, A. J., Ueno, S., Suen, K., Dowell, S. J., and Wise, A. (2005). *In* Molecular idenfication of GPR55 as a third G protein-coupled receptor responsive to cannabinoid ligands *Symp. Cannabinoids, Int. Cannabinoid Res. Soc.*, Burlington, Vermont, Abstract, p. 16.

Buhl, A. M., Johnson, N. L., Dhanasekaran, N., and Johnson, G. L. (1995). Galpha12 and Galpha13 stimulate Rho-dependent stress fiber formation and focal adhesion assembly. *J. Biol. Chem.* **270**(42), 24631–24634.

Bukoski, R. D., Batkai, S., Jarai, Z., Wang, Y., Offertaler, L., Jackson, W. F., and Kunos, G. (2002). CB1 receptor antagonist SR141716A inhibits Ca^{2+}-induced relaxation in CB1 receptor-deficient mice. *Hypertension* **39**(2), 251–257.

Campbell, J. J., Foxman, E. F., and Butcher, E. C. (1997). Chemoattractant receptor cross talk as a regulatory mechanism in leukocyte adhesion and migration. *Eur. J. Immunol.* **27**(10), 2571–2578.

Christopoulos, A., and Kenakin, T. (2002). G protein-coupled receptor allosterism and complexing. *Pharmacol. Rev.* **54**(2), 323–374.

Christopoulos, A., and Wilson, K. (2001). Interaction of anandamide with the M1 and M4 muscarinic acetylcholine receptors. *Brain Res.* **915**(1), 70–78.

Cross, A., Bakstad, D., Allen, J. C., Thomas, L., Moots, R. J., and Edwards, S. W. (2005). Neutrophil gene expression in rheumatoid arthritis. *Pathophysiology* **12**(3), 191–202.

De la Fuente, M., Carrasco, M., Del Rio, M., and Hernanz, A. (1998). Modulation of murine lymphocyte functions by sulfated cholecystokinin octapeptide. *Neuropeptides* **32**(3), 225–233.

Derocq, J.-M., Jbilo, O., Bouaboula, M., Segui, M., Clere, C., and Casellas, P. (2000). Genomic and functional changes induced by the activation of the peripheral cannabinoid receptor CB2 in the promyelocytic cells HL-60. Possible involvement of the CB2 receptor in cell differentiation. *J. Biol. Chem.* **275**(21), 15621–15628.

Deusch, E., Kraft, B., Nahlik, G., Weigl, L., Hohenegger, M., and Kress, H. G. (2003). No evidence for direct modulatory effects of Delta^9-tetrahydrocannabinol on human polymorphonuclear leukocytes. *J. Neuroimmunol.* **141**(1–2), 99–103.

Devane, W. A. (1992). Isolation and structure of a brain constituent that binds to the cannabinoid receptor. *Science* **252,** 1946–1949.

Didsbury, J. R., Uhing, R. J., Tomhave, E., Gerard, C., Gerard, N., and Snyderman, R. (1991). Receptor class desensitization of leukocyte chemoattractant receptors. *Proc. Natl. Acad. Sci. USA* **88**(24), 11564–11568.

Di Marzo, V., Bisogono, T., and De Petrocellis, L. (2000). Endocannabinoids: New targets for drug development. *Curr. Pharm. Des.* **6,** 1361–1380.

Dong, X., Mo, Z., Bokoch, G., Guo, C., Li, Z., and Wu, D. (2005). P-Rex1 is a primary Rac2 guanine nucleotide exchange factor in mouse neutrophils. *Curr. Biol.* **15,** 1874–1879.

Drmota, T., *et al.* (2004). AstraZenica, "Screening assays for cannabinoid-ligand type modulators of GPR55. International Patent #: WO2004074844.

Ellerbroek, S. M., Wennerberg, K., and Burridge, K. (2003). Serine phosphorylation negatively regulates RhoA *in vivo. J. Biol. Chem.* **278**(21), 19023–19031.

Entschladen, F., Drell, T. L. IV, Palm, D., Bastian, P., Potthoff, S., Zanker, K. S., and Lang, K. (2005). A comparative review on leukocyte and tumor cell migration with regard to the regulation by serpentine receptor ligands. *Signal Transduct.* **5**(1–2), 9–18.

Fezza, F., Bisogno, T., Minassi, A., Appendino, G., Mechoulam, R., and Di Marzo, V. (2002). Noladin ether, a putative novel endocannabinoid: Inactivation mechanisms and a sensitive method for its quantification in rat tissues. *FEBS Lett.* **513**(2–3), 294–298.

Franklin, A., and Stella, N. (2003). Arachidonylcyclopropylamide increases microglial cell migration through cannabinoid CB2 and abnormal-cannabidiol-sensitive receptors. *Eur. J. Pharmacol.* **474**(2–3), 195–198.

Franklin, A., Parmentier-Batteur, S., Walter, L., Greenberg, D. A., and Stella, N. (2003). Palmitoylethanolamide increases after focal cerebral ischemia and potentiates microglial cell motility. *J. Neurosci.* **23**(21), 7767–7775.

Galiégue, S., Mary, S., Marchand, J., Dussossoy, D., Carriere, D., Carayon, P., Bouaboula, M., Shire, D., Le Fur, G., and Casellas, P. (1995). Expression of central and peripheral cannabinoid receptors in human immune tissues and leukocyte subpopulations. *Eur. J. Biochem.* **232**(1), 54–61.

Gallagher, R., Collins, S., Trujillo, J., McCredie, K., Ahearn, M., Tsai, S., Metzgar, R., Aulakh, G., Ting, R., Ruscetti, F., and Gallo, R. (1979). Characterization of the continuous, differentiating myeloid cell line (HL-60) from a patient with acute promyelocytic leukemia. *Blood* **54**(3), 713–733.

Garrido, E., Delgado, M., Martinez, C., Gomariz, R. P., and De la Fuente, M. (1996). Pituitary adenylate cyclase-activating polypeptide (PACAP38) modulates lymphocyte and macrophage functions: Stimulation of adherence and opposite effect on mobility. *Neuropeptides* **30**(6), 583–595.

Gillitzer, R., and Goebeler, M. (2001). Chemokines in cutaneous wound healing. *J. Leukoc. Biol.* **69**(4), 513–521.

Gomes, I., Gupta, J. A., Trapaidze, N., Nagy, V., and Devi, L. A. (2000). Heterodimerization of μ and δ opioid receptors: A role in opiate synergy. *J. Neurosci.* **20**, RC110.

Grimm, M. C., Ben-Baruch, A., Taub, D. D., Howard, O. M. Z., Resau, J. H., Wang, J. M., Ali, H., Richardson, R., Snyderman, R., and Oppenheim, J. J. (1998). Opiates transdeactivate chemokine receptors: Delta and mu opiate receptor-mediated heterologous desensitization. *J. Exp. Med.* **188**(2), 317–325.

Grimsey, N. L., Goodfellow, C. E., Scotter, E. L., Dowie, M. J., Glass, M., and Graham, E. S. (2008). Specific detection of CB1 receptors: Cannabinoid CB1 receptor antibodies are not all created equal. *J. Neurosci. Methods* **171**, 78–86.

Hanus, L., Gopher, A., Almog, S., and Mechoulam, R. (1993). 2-Arachidonoyl glyceryl ether, an endogenous agonist of the cannabinoid CB1 receptor. *Proc. Natl. Acad. Sci. USA* **98**, 3032–3034.

Harvath, L., Robbins, J. D., Russell, A. A., and Seamon, K. B. (1991). cAMP and human neutrophil chemotaxis. Elevation of cAMP differentially affects chemotactic responsiveness. *J. Immunol.* **146**(1), 224–232.

Heiner, I., Eisfeld, J., and Luckhoff, A. (2003). Role and regulation of TRP channels in neutrophil granulocytes. *Cell Calcium* **33**, 533.

Henstridge, C., Balenga, N., Campbell, G., Waldhoer, M., and Irving, A. (2008). The GPR55 ligand, L-α-lysophosphatidylinositol, promotes RHO-dependent Ca^{2+} signaling. *18th Ann. Symp. ICRS*, Abstract, p. 14.

Hiley, C. R., and Kaup, S. S. (2007). GPR55 and the vascular receptors for cannabinoids. *Br. J. Pharmacol.* **152**, 559–561.

Ho, W.-S. V., and Hiley, C. R. (2003). Vasodilator actions of abnormal-cannabidiol in rat isolated small mesenteric artery. *Br. J. Pharmacol.* **138**(7), 1320–1332.

Howlett, A. C., Barth, F., Bonner, T. I., Cabral, G., Casellas, P., Devane, W. A., Felder, C. C., Herkenham, M., Mackie, K., Martin, B. R., Mechoulam, R., and Pertwee, R. G. (2002). International union of pharmacology. XXVII. Classification of cannabinoid receptors. *Pharmacol. Rev.* **54**(2), 161–202.

Hurley, J. V. (1983). Acute Inflammation, 2nd ed Churchill Livingston, Edinburgh.

Hwang, J.-I., Fraser, I. D. C., Choi, S., Qin, X.-F., and Simon, M. I. (2004). Analysis of C5a-mediated chemotaxis by lentiviral delivery of small interfering RNA. *Proc. Natl. Acad. Sci. USA* **101**(2), 488–493.

Imhof, B. A., and Dunon, D. (1997). Basic mechanism of leukocyte migration. *Hormone Metab. Res.* **29**(12), 614–621.
Jarái, Z., Wagner, J. A., Varga, K., Lake, K. D., Compton, D. R., Martin, B. R., Zimmer, A. M., Bonner, T. I., Buckley, N. E., Mezey, E., Razdan, R. K., Zimmer, A., et al. (1999). Cannabinoid-induced mesenteric vasodilation through an endothelial site distinct from CB1 or CB2 receptors. *Proc. Natl. Acad. Sci. USA* **96**(24), 14136–14141.
Jockusch, B. M., Bubeck, P., Giehl, K., Kroemker, M., Moschner, J., Rothkegel, M., Rudiger, M., Schluter, K., Stanke, G., and Winkler, J. (1995). The molecular architecture of focal adhesions. *Ann. Rev. Cell Dev. Biol.* **11**, 379–416.
Johns, D. G., Behm, D. J., Walker, D. A., Ao, Z., Shapland, E. M., Daniels, D. A., Riddick, M., Dowell, S., Staton, P. C., and Green, P. (2007). The novel endocannabinoid receptor GPR55 is activated by atypical cannabinoids but does not mediate their vasodilator effects. *Br. J. Pharmacol.* **152**, 825–831.
Joseph, J., Niggemann, B., Zaenker, K. S., and Entschladen, F. (2002). The neurotransmitter gamma-aminobutyric acid is an inhibitory regulator for the migration of SW 480 colon carcinoma cells. *Cancer Res.* **62**(22), 6467–6469.
Joseph, J., Niggemann, B., Zaenker, K. S., and Entschladen, F. (2004). Anandamide is an endogenous inhibitor for the migration of tumor cells and T lymphocytes. *Cancer Immunol. Immunother.* **53**(8), 723–728.
Kaminski, N. E., Abood, M. E., Kessler, F. K., Martin, B. R., and Schatz, A. R. (1992). Identification of a functionally relevant cannabinoid receptor on mouse spleen cells that is involved in cannabinoid-mediated immune modulation. *Mol. Pharmacol.* **42**(5), 736–742.
Kane, S. (2001). Caring for women with inflammatory bowel disease. *JGSM* **4**(1), 54–59.
Kayali, A. G., Van Gunst, K., Campbell, I. L., Stotland, A., Kritzik, M., Liu, G., Flodstrom-Tullberg, M., Zhang, Y.-Q., and Sarvetnick, N. (2003). The stromal cell-derived factor-1a/CXCR4 ligand-receptor axis is critical for progenitor survival and migration in the pancreas. *J. Cell Biol.* **163**(4), 859–869.
Kishimoto, S., Gokoh, M., Oka, S., Muramatsu, M., Kajiwara, T., Waku, K., and Sugiura, T. (2003). 2-Arachidonoylglycerol induces the migration of HL-60 cells differentiated into macrophage-like cells and human peripheral blood monocytes through the cannabinoid CB2 receptor-dependent mechanism. *J. Biol. Chem.* **278**(27), 24469–24475.
Kishimoto, S., Muramatsu, M., Gokoh, M., Oka, S., Waku, K., and Sugiura, T. (2005). Endogenous cannabinoid receptor ligand induces the migration of human natural killer cells. *J. Biochem.* **137**(2), 217–223.
Kozasa, T., Jiang, X., Hart, M. J., Sternweis, P. M., Singer, W. D., Gilman, A. G., Bollag, G., and Sternweis, P. C. (1998). p115 RhoGEF, a GTPase activating protein for Galpha12 and Galpha13. *Science* **280**(5372), 2109–2111.
Kucharzik, T., and Williams, I. R. (2003). Neutrophil migration across the intestinal epithelial barrier—Summary of *in vitro* data and description of a new transgenic mouse model with doxycycline-inducible interleukin-8 expression in intestinal epithelial cells. *Pathobiology* **70**(3), 143–149.
Kunos, G., Jarai, Z., Batkai, S., Goparaju, S. K., Ishac, E. J. N., Liu, J., Wang, L., and Wagner, J. A. (2000). Endocannabinoids as cardiovascular modulators. *Chem. Phys. Lipids* **108**(1–2), 159–168.
Kurihara, R., Tohyama, Y., Matsusaka, S., Naruse, H., Kinoshita, E., Tsujioka, T., Katsumata, Y., and Hirohei, Y. (2006). Effects of peripheral cannabinoid receptor ligands on motility and polarization in neutrophil-like HL60 cells and human neutrophils. *J. Biol. Chem.* **281**, 12908–12918.
Kwiatkowski, A. V., Gertler, F. B., and Loureiro, J. J. (2003). Function and regulation of Ena/VASP proteins. *Trends Cell Biol.* **13**(7), 386–392.

Lee, N. H., and El-Fakahany, E. E. (1991). Allosteric antagonists of the muscarinic acetylcholine receptor. *Biochem. Pharmacol.* **42**(2), 199–205.
Levine, J., and Taiwo, Y. (1994). Inflammatory pain. *In* "Textbook of Pain" (P. D. Wall and R. Melzack, Eds.), 3rd ed. Churchill Livingston, London.
Levine, J. D., Lau, W., Kwiat, G., and Goetzl, E. J. (1984). Leukotriene B4 produces hyperalgesia that is dependent on polymorphonuclear leukocytes. *Science* **225**(4663), 743–745.
Levine, J. D., Gooding, J., and Donatoni, P. (1985). The role of the polymorphonuclear leukocyte in hyperalgesia. *J. Neurosci.* **5**(11), 3025–3029.
Levine, J. D., Lam, D., and Taiwo, Y. O. (1986). Hyperalgesic properties of 15-lipoxygenase products of arachidonic acid. *Proc. Natl. Acad. Sci. USA* **83**(14), 5331–5334.
Lodish, H., Berk, A., Baltimore, D., and Darnell, J. (2000). Molecular Cell Biology. W. H. Freeman & Co Ltd., New York.
Lu, Q., Straiker, A., Lu, Q., and Maguire, G. (2000). Expression of CB2 cannabinoid receptor mRNA in adult rat retina. *Visual Neurosci.* **17**(1), 91–95.
Lunn, C. A., Fine, J. S., Rojas-Triana, A., Jackson, J. V., Fan, X. D., Kung, T. T., Gonsiorek, W., Schwarz, M. A., Lavey, B., and Kozlowski, J. A. (2006). A novel cannabinoid peripheral cannabinoid receptor-selective inverse agonist blocks leukocyte recruitment *in vivo*. *J. Pharmacol. Exp. Ther.* **316**, 780–788.
Luo, B., Prescott, S. M., and Topham, M. K. (2003). Protein kinase Cα phosphorylates and negatively regulates diacylglycerol kinase zeta. *J. Biol. Chem.* **278**(41), 39542–39547.
Luttrel, L. M., Ferguson, S. S. G., Daaka, Y., Miller, W. E., Maudsley, S., Della Rocca, G. J., Lin, F.-T., Kawakatsu, H., Owada, K., Luttrell, D. K., Caron, M. G., and Lefkowitz, R. J. (1999). Beta-arrestin-dependent formation of Beta2 adrenergic receptor-Src protein kinase complexes. *Science* **283**(5402), 655–661.
Mackie, K., and Stella, N. (2006). Cannabinoid receptors and endocannabinoids: Evidence for new players. *AAPS J.* **8**, E298–E306.
Mackie, K., Devane, W. A., and Hille, B. (1993). Anandamide, an endogenous cannabinoid, inhibits calcium currents as a partial agonist in N18 neuroblastoma cells. *Mol. Pharmacol.* **44**(3), 498–503.
MacLennan, S. J., Reynen, P. H., Kwan, J., and Bonhaus, D. W. (1998). Evidence for inverse agonism of SR141716A at human recombinant cannabinoid CB1 and CB2 receptors. *Br. J. Pharmacol.* **124**(4), 619–622.
Massi, P., Patrini, G., Rubino, T., Fuzio, D., and Parolaro, D. (1997). Changes in rat spleen cannabinoid receptors after chronic CP-55,940: An autoradiographic study. *Pharmacol. Biochem. Behav.* **58**(1), 73–78.
Masur, K., Niggemann, B., Zanker, K. S., and Entschladen, F. (2001). Norepinephrine-induced migration of SW 480 colon carcinoma cells is inhibited by beta-blockers. *Cancer Res.* **61**(7), 2866–2869.
Matsuda, L. A., Lolait, S. J., Brownstein, M. J., Young, A. C., and Bonner, T. I. (1990). Structure of a cannabinoid receptor and functional expression of the cloned cDNA. *Nature* **346**(6284), 561–564.
McHugh, D., Tanner, C., Mechoulam, R., Pertwee, R. G., and Ross, R. A. (2008). Inhibition of human neutrophil chemotaxis by endogenous cannabinoids and phytocannabinoids: Evidence for a site distinct from CB_1 and CB_2. *Mol. Pharmacol.* **73**(2), 441–450.
Mechoulam, R., Ben-Shabata, S., Hanua, L., Ligumskya, M., Kaminskib, N. E., Schatzd, A. R., Gopherc, A., Almogc, S., Martind, B. R., Comptond, D. R., Pertwee, R. G., Griffine, G., *et al.* (1995). Identification of an endogenous 2-monoglyceride, present in canine gut, that binds to cannabinoid receptors. *Biochem. Pharmacol.* **50**, 83–90.

Miller, A., and Stella, N. (2008). CB$_2$ receptor-mediated migration of immune cells: It can go either way. *Br. J. Pharmacol.* **153**, 299–308.

Milman, G., Maor, Y., Abu-Lafi, S., Horowitz, M., Gallily, R., Batkai, S., Mo, F.-M., Offertaler, L., Pacher, P., and Kunos, G. (2006). N-arachidonoyl-L-serine, an endogenous endocannabinoid-like brain constituent with vasodilatory properties. *Proc. Natl. Acad. Sci. USA* **103**, 2428–2433.

Mitsui, H., Takuwa, N., Kurokawa, K., Exton, J. H., and Takuwa, Y. (1997). Dependence of activated Galpha12-induced G1 to S phase cell cyclin progression on both Ras/mitogen-activated protein kinase and Ras/Rac1/Jun N-terminal kinase cascades in NIH3T3 fibroblasts. *J. Biol. Chem.* **272**(8), 4904–4910.

Mo, F. M., Offertaler, L., and Kunos, G. (2004). Atypical cannabinoid stimulates endothelial cell migration via a G i/Go-coupled receptor distinct from CB1, CB2 or EDG-1. *Eur. J. Pharmacol.* **489**(1–2), 21–27.

Moser, B., and Ebert, L. (2003). Lymphocyte traffic control by chemokines: Follicular B helper T cells. *Immunol. Lett.* **85**(2), 105–112.

Movat, H. Z. (1985). The Inflammatory Reaction. Elsevier, Amsterdam.

Munro, S., Thomas, K. L., and Abu-Shaar, M. (1993). Molecular characterization of a peripheral receptor for cannabinoids. *Nature* **365**(6441), 61–65.

Narumiya, S. (1996). The small GTPase Rho: Cellular functions and signal transduction. *J. Biochem.* **120**(2), 215–228.

Nathan, C. (2006). Neutrophils and immunity: Challenges and opportunities. *Nat. Rev. Immunol.* **6**, 173–182.

Neer, E. J. (1995). Heterotrimeric G proteins: Organizers of transmembrane signals. *Cell* **80**(2), 249–257.

Neptune, E. R., and Bourne, H. R. (1997). Receptors induce chemotaxis by releasing the betagamma subunit of G(i), not by activating G(q) or G(s). *Proc. Natl. Acad. Sci. USA* **94**(26), 14489–14494.

Nilsson, O., Fowler, C. J., and Jacobsson, S. O. (2006). The cannabinoid agonist WIN 55, 212-2 inhibits TNF-alpha-induced neutrophil transmigration across ECV304 cells. *Eur. J. Pharmacol.* **547**, 165–173.

Offertáler, L., Mo, F.-M., Batkai, S., Liu, J., Begg, M., Razdan, R. K., Martin, B. R., Bukoski, R. D., and Kunos, G. (2003). Selective ligands and cellular effectors of a G protein-coupled endothelial cannabinoid receptor. *Mol. Pharmacol.* **63**(3), 699–705.

Oka, S., Ikeda, S., Kishimoto, S., Gokoh, M., Yanagimoto, S., Waku, K., and Sugiura, T. (2004). 2-Arachidonoylglycerol, an endogenous cannabinoid receptor ligand, induces the migration of EoL-1 human eosinophilic leukemia cells and human peripheral blood eosinophils. *J. Leukoc. Biol.* **76**(5), 1002–1009.

Oka, S., Nakajima, K., Yamashita, A., Kishimoto, S., and Sugiura, T. (2007). Identification of GPR55 as a lysophosphatidylinositol receptor. *Biochem. Biophys. Res. Commun.* **362**, 928–934.

Okamoto, H., Takuwa, N., Yokomizo, T., Sugimoto, N., Sakurada, S., Shigematsu, H., and Takuwa, Y. (2000). Inhibitory regulation of Rac activation, membrane ruffling, and cell migration by the G protein-coupled sphingosine-1-phosphate receptor EDG5 but not EDG1 or EDG3. *Mol. Cell. Biol.* **20**(24), 9247–9261.

O'Sullivan, S. E., Kendall, D. A., and Randall, M. D. (2006). Characterisation of the vasorelaxant properties of the novel endocannabinoid N-arachidonoyl dopamine (NADA). *Br. J. Pharmacol.* **141**, 803–812.

Pereira, H. A., Shafer, W. M., Pohl, J., Martin, L. E., and Spitznagel, J. K. (1990). CAP37, a human neutrophil-derived chemotactic factor with monocyte specific activity. *J. Clin. Invest.* **85**(5), 1468–1476.

Pertwee, R. G. (1997). Pharmacology of cannabinoid CB1 and CB2 receptors. *Pharmacol. Ther.* **74**(2), 129–180.

Pertwee, R. G. (2001). Cannabinoid receptors and pain. *Prog. Neurobiol.* **63**(5), 569–611.
Pertwee, R. G. (2004). Novel pharmacological targets for cannabinoids. *Curr. Neuropharmacol.* **2**(1), 9–29.
Pertwee, G. R. (2005). The therapeutic potential of drugs that target cannabinoid receptors or modulate the tissue levels or actions of endocannabinoids. *AAPS J.* **7**(3), E625–E654.
Pertwee, R. G. (2007). GPR55: A new member of the cannabinoid clan? *Br. J. Pharmacol.* **152,** 984–986.
Pertwee, R. G., and Ross, R. A. (2002). Cannabinoid receptors and their ligands. *Prostaglandins Leukot. Essent. Fatty Acids* **66**(2–3), 101–121.
Petitet, F., Jeantaud, B., Reibaud, M., Imperato, A., and Dubroeucq, M. (1998). Complex pharmacology of natural cannabinoids: Evidence for partial agonist activity of Delta9-tetrahydrocannabinol and antagonist activity of cannabidiol on rat brain cannabinoid receptors. *Life Sci.* **63**(1), P1–P6.
Pierce, K. L., Premont, R. T., and Lefkowitz, R. J. (2002). Seven-transmembrane receptors. *Nat. Rev. Mol. Cell Biol.* **3**(9), 639–650.
Porter, A. C., Sauer, J.-M., Knierman, M. D., Becker, G. W., Berna, M. J., Bao, J., Nomikos, G. G., Carter, P., Bymaster, F. P., Leese, A. B., and Felder, C. C. (2002). Characterization of a novel endocannabinoid, virodhamine, with antagonist activity at the CB1 receptor. *J. Pharmacol. Exp. Ther.* **301**(3), 1020–1024.
Rahaman, M., Costello, R. W., Belmonte, K. E., Gendy, S., and Walsh, M. (2006). Neutrophil sphingosine-1-phosphate and lysophosphatidic acid receptors in pneumonia. *AJRCMB* **34,** 233–241.
Riento, K., and Ridley, A. J. (2003). Rocks: Multifunctional kinases in cell behaviour. *Nat. Rev. Mol. Cell Biol.* **4**(6), 446–456.
Rinaldi-Carmona, M., Barth, F., Millan, J., Derocq, J.-M., Casellas, P., Congy, C., Oustric, D., Sarran, M., Bouaboula, M., Calandra, B., Portier, M., Shire, D., *et al.* (1998). SR 144528, the first potent and selective antagonist of the CB2 cannabinoid receptor. *J. Pharmacol. Exp. Ther.* **284**(2), 644–650.
Ross, R. A., Brockie, H. C., Stevenson, L. A., Murphy, V. L., Templeton, F., Makriyannis, A., and Pertwee, R. G. (1999). Agonist-inverse agonist characterization at CB1 and CB2 cannabinoid receptors of L759633, L759656 and AM630. *Br. J. Pharmacol.* **126**(3), 665–672.
Ross, R. A., Gibson, T. M., Brockie, H. C., Leslie, M., Pashmi, G., Craib, S. J., Di Marzo, V., and Pertwee, R. G. (2001). Structure-activity relationship for the endogenous cannabinoid, anandamide, and certain of its analogues at vanilloid receptors in transfected cells and vas deferens. *Br. J. Pharmacol.* **132**(3), 631–640.
Ryberg, E., Larsson, N., Sjorgen, S., Hjorth, S., Hermansson, N.-O., Leonova, J., Elebring, T., Nilsson, K., Drmota, T., and Greasley, P. (2007). The orphan receptor GPR55 is a novel cannabinoid receptor. *Br. J. Pharmacol.* **152,** 1092–1101.
Sacerdote, P., Massi, P., Panerai, A. E., and Parolaro, D. (2000). *In vivo* and *in vitro* treatment with the synthetic cannabinoid CP55,940 decreases the *in vitro* migration of macrophages in the rat: Involvement of both CB1 and CB2 receptors. *J. Neuroimmunol.* **109**(2), 155–163.
Sagi, S. A., Seasholtz, T. M., Kobiashvili, M., Wilson, B. A., Toksoz, D., and Brown, J. H. (2001). Physical and functional interactions of Gaq with Rho and its exchange factors. *J. Biol. Chem.* **276**(18), 15445–15452.
Sallusto, F., Schaerli, P., Loetscher, P., Schaniel, C., Lenig, D., Mackay, C. R., Qin, S., and Lanzavecchia, A. (1998). Rapid and coordinated switch in chemokine receptor expression during dendritic cell maturation. *Eur. J. Immunol.* **28**(9), 2760–2769.
Sánchez-Madrid, F., and Del Pozo, M. A. (1999). Leukocyte polarization in cell migration and immune interactions. *EMBO J.* **18,** 501–511.

Sawzdargo, M., Nguyen, T., Lee, D. K., Lynch, K. R., Cheng, R., Heng, H. H. Q., George, S. R., and O'Dowd, B. F. (1999). Identification and cloning of three novel human G protein-coupled receptor genes GPR52, PsiGPR53 and GPR55: GPR55 is extensively expressed in human brain. *Mol. Brain Res.* **64**(2), 193–198.

Schuligoi, R. (1998). Effect of colchicine on nerve growth factor—Induced leukocyte accumulation and thermal hyperalgesia in the rat. *Naunyn Schmiedebergs Arch. Pharmacol.* **358**(2), 264–269.

Senior, R. M., Griffin, G. L., and Huang, J. S. (1983). Chemotactic activity of platelet alpha granule proteins for fibroblasts. *J. Cell Biol.* **96**(2), 382–385.

Showalter, V. M., Compton, D. R., Martin, B. R., and Abood, M. E. (1996). Evaluation of binding in a transfected cell line expressing a peripheral cannabinoid receptor (CB2): Identification of cannabinoid receptor subtype selective ligands. *J. Pharmacol. Exp. Ther.* **278**(3), 989–999.

Smart, D., Gunthorpe, M. J., Jerman, J. C., Nasir, S., Gray, J., Muir, A. I., Chambers, J. K., Randall, A. D., and Davis, J. B. (2000). The endogenous lipid anandamide is a full agonist at the human vanilloid receptor (hVR1). *Br. J. Pharmacol.* **129**(2), 227–230.

Snyderman, R., and Uhing, R. J. (1992). Chemoattractant stimulus-response coupling. *In* "Inflammation: Basic Principles and Clinical Correlates" (J. I. Gallin, I. M. Goldstein, and R. Snyderman, Eds.), 2nd ed., pp. 421–439. Raven Press Ltd., New York.

Somlyo, A. P., and Somlyo, A. V. (2003). Ca^{2+} sensitivity of smooth muscle and nonmuscle myosin II: Modulated by G proteins, kinases, and myosin phosphatase. *Physiol. Rev.* **83**(4), 1325–1358.

Song, Z.-H., and Zhong, M. (2000). CB1 cannabinoid receptor-mediated cell migration. *J. Pharmacol. Exp. Ther.* **294**(1), 204–209.

Stevens, A., and Lowe, J. S. (1996). Human Histology, 2nd ed. Mosby, London.

Sugiura, T., Kondo, S., Sukagawa, A., Nakane, S., Shinoda, A., Itoh, K., Yamashita, A., and Waku, K. (1995). 2-Arachidonoylglycerol: A possible endogenous cannabinoid receptor ligand in brain. *Biochem. Biophys. Res. Commun.* **215**(1), 89–97.

Sugiura, T., Kondo, S., Kishimoto, S., Miyashita, T., Nakane, S., Kodaka, T., Suhara, Y., Takayama, H., and Waku, K. (2000). Evidence that 2-arachidonoylglycerol but not N-palmitoylethanolamine or anandamide is the physiological ligand for the cannabinoid CB2 receptor. Comparison of the agonistic activities of various cannabinoid receptor ligands in HL-60 cells. *J. Biol. Chem.* **275**(1), 605–612.

Territo, M. C., Ganz, T., Selsted, M. E., and Lehrer, R. (1989). Monocyte-chemotactic activity of defensins from human neutrophils. *J. Clin. Invest.* **84**(6), 2017–2020.

Thiele, J., Varus, E., Wickenhauser, C., Kvasnicka, H. M., Metz, K. A., and Beelen, D. W. (2004). Regeneration of heart muscle tissue: Quantification of chimeric cardiomyocytes and endothelial cells following transplantation. *Histol. Histopathol.* **19**(1), 201–209.

Vaccani, A., Massi, P., Colombo, A., Rubino, T., and Parolaro, D. (2005). Cannabidiol inhibits human glioma cell migration through a cannabinoid receptor-independent mechanism. *Br. J. Pharmacol.* **144**(8), 1032–1036.

Van Haastert, P. J. M., and Devreotes, P. N. (2004). Chemotaxis: Signalling the way forward. *Nat. Rev. Mol. Cell Biol.* **5**, 626–634.

Wagner, J. A., Varga, K., Jarai, Z., and Kunos, G. (1999). Mesenteric vasodilation mediated by endothelial anandamide receptors. *Hypertension* **33**(Part II), 429–434.

Walker, J. M., Krey, J. F., Chu, C. J., and Huang, S. M. (2002). Endocannabinoids and related fatty acid derivatives in pain modulation. *Chem. Phys. Lipids* **121**(1–2), 159–172.

Walter, L., Franklin, A., Witting, A., Wade, C., Xie, Y., Kunos, G., Mackie, K., and Stella, N. (2003). Nonpsychotropic cannabinoid receptors regulate microglial cell migration. *J. Neurosci.* **23**(4), 1398–1405.

Ward, P. A. (1997). Neutrophils and adjuvant arthritis. *Clin. Exp. Immunol.* **107**(2), 225–226.

Wenzel-Seifert, K., Hurt, C. M., and Seifert, R. (1998). High constitutive activity of the human formyl peptide receptor. *J. Biol. Chem.* **273**(37), 24181–24189.

Whyte, L., Ryberg, E., Ridge, S., Mackie, K., Greasley, P., Rogers, M. J., and Ross, R. (2008). GPR55 is involved in the regulation of Osteoclast activity *in vitro* and bone mass *in vivo* 18th Ann. Symp. ICRS, Abstract p. 50.

Wiley, J. L., and Martin, B. R. (2002). Cannabinoid pharmacology: Implications for additional cannabinoid receptor subtypes. *Chem. Phys. Lipids* **121**(1–2), 57–63.

Wright, C. D., Kuipers, P. J., Kobylarz-Singer, D., Devall, L. J., Klinkefus, B. A., and Weishaar, R. E. (1990). Differential inhibition of human neutrophil functions: Role of cyclic AMP-specific, cyclic GMP-insensitive phosphodiesterases. *Biochem. Pharmacol.* **40**(4), 699–707.

Wysocka, J., Turowski, D., Lipska, A., and Milkolajczyk-Kwapisz, M. (1997). Participation of neutrophils in pathophysiology of unstable angina pectoris. *Pol. Merkur. Lekarski.* **3**(15), 145–148.

Xu, J., Wang, F., Van Keymeulen, A., Herzmark, P., Straight, A., Kelly, K., Takuwa, Y., Sugimoto, N., Mitchison, T., and Bourne, H. R. (2003). Divergent signals and cytoskeletal assemblies regulate self-organizing polarity in neutrophils. *Cell* **114**, 201–214.

Yokomizo, T., Masuda, K., Kato, K., Toda, A., Izumi, T., and Shimizu, T. (2000). Leukotriene B4 receptor: Cloning and intracellular signaling. *Am. J. Respir. Crit. Care Med.* **161**(2 II), S51–S55.

Zhang, X. F., Wang, J. F., Maor, Y., Kunos, G., and Groopman, J. E. (2005). Endogenous cannabinoid-like arachidonoyl serine induces angiogensis through novel pathways (Abstract). *Blood* **106**, 3690.

Ziegler, W. H., Tigges, U., Zieseniss, A., and Jockusch, B. M. (2002). A lipid-regulated docking site on vinculin for protein kinase C. *J. Biol. Chem.* **277**(9), 7396–7404.

Zygmunt, P. M., Petersson, J., Andersson, D. A., Chuang, H., Sorgard, M., Di Marzo, V., Julius, D., and Hogestatt, E. D. (1999). Vanilloid receptors on sensory nerves mediate the vasodilator action of anandamide. *Nature* **400**(6743), 452–457.

Zygmunt, P. M., Andersson, D. A., and Hogestatt, E. D. (2002). Delta^9-tetrahydrocannabinol and cannabinol activate capsaicin-sensitive sensory nerves via a CB1 and CB2 cannabinoid receptor-independent mechanism. *J. Neurosci.* **22**(11), 4720–4727.

CHAPTER FOURTEEN

CB1 Activity in Male Reproduction: Mammalian and Nonmammalian Animal Models

Riccardo Pierantoni,* Gilda Cobellis,* Rosaria Meccariello,[†] Giovanna Cacciola,* Rosanna Chianese,* Teresa Chioccarelli,* *and* Silvia Fasano*

Contents

I. Introduction	368
II. Receptor Properties	368
III. Brain–Pituitary Axis	371
IV. Testis	374
V. Excurrent Duct System	379
VI. Concluding Remarks	382
References	382

Abstract

The importance of the endocannabinoid system (ECBS) and its involvement in several physiological processes is still increasing. Since the isolation of the main active compound of *Cannabis sativa*, Δ^9-THC, several lines of research have evidenced the basic roles of this signaling system mainly considering its high conservation during evolution.

In this chapter the attention is focussed on the involvement of the ECBS in the control of male reproductive aspects at both central and local levels which are both considered from a comparative point of view.

* Dipartimento di Medicina Sperimentale, Seconda Università di Napoli, Via Costantinopoli, Napoli, Italy
[†] Dipartimento di Studi delle Istituzioni e dei Sistemi Territoriali, Università Parthenope, Napoli, Italy

I. INTRODUCTION

Biochemical basis of marihuana activity has remained unclear until 1990 when a cannabinoid receptor able to bind the Δ^9-tetrahydrocannabinol (Δ^9-THC), the psychoactive constituent of *Cannabis sativa*, was discovered (Matsuda *et al.*, 1990). Furthermore, the discovery of an endogenous ligand (endocannabinoid, eCB) (Devane *et al.*, 1992) clearly demonstrated the existence of a novel signaling system named endocannabinoid system (ECBS).

Nowadays, ECBS comprises eCBs, several endocannabinoid receptors (CBs), many enzymatic machineries responsible for eCB degradation and biosynthesis and putative lipid transporters. Endocannabinoids are lipid compounds, in general amides, esters, and ethers of long chain polyunsaturated fatty acid, isolated from brain, peripheral tissues, and reproductive fluids (Devane *et al.*, 1992; Schuel *et al.*, 2002a,b; Sugiura *et al.*, 1995). The best known are the *N*-arachidonoyl-ethanolamine (AEA, known as anandamide), the first eCB discovered in porcine brain (Devane *et al.*, 1992), and 2-arachidonoylglycerol (2-AG) (Sugiura *et al.*, 1995). Endocannabinoid family also is comprehensive of 2-arachidonoylglycerol ether (noladin ether), *O*-arachidonoyl-ethanolamine (virodhamine), *N*-oleoyl-ethanolamine (OEA), *N*-palmitoyl-ethanolamine (PEA), and other compounds with eCB-like properties (Padgett, 2005). Due to their lipid nature, eCBs should be able to cross membranous bilayers, but several authors report the existence of specific eCB carriers (EMT) transporting eCBs in a temperature and saturation dependent manner (Hillard and Jarrahian, 2000). Endocannabinoid uptake and degradation rate, the latter mainly due to the action of a fatty acid amide hydrolase (FAAH), are the main modulators of eCB activity in biological systems. A schematic representation of eCB components is depicted in Fig. 14.1.

II. RECEPTOR PROPERTIES

Classic CBs are CB1 and CB2, two seven-transmembrane spanning G-coupled receptors (Matsuda *et al.*, 1990; Munro *et al.*, 1993). While CB1 is by far the most abundant G-coupled receptor in the brain, for longtime CB2 expression has exclusively been linked to immune system. Nowadays, a widespread distribution at peripheral level, gonads included, for both CB1 and CB2 has been reported (Brown *et al.*, 2002; Galiegue *et al.*, 1995; Shire *et al.*, 1995); in addition, in central nervous system (CNS) CB2 expression has been detected both in perikarya/neuronal processes as well as in glial cells (Gong *et al.*, 2006). CB1 and CB2 are highly evolutionarily conserved.

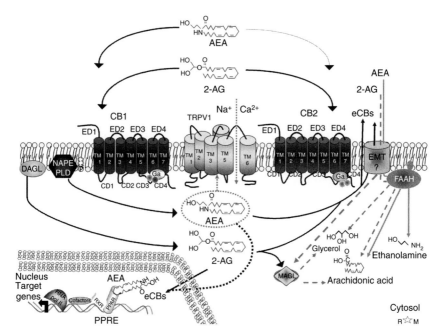

Figure 14.1 The endocannabinoid system (ECBS). AEA and 2-AG selectively bind CB1 and CB2, classical seven transmembrane spanning G-coupled receptors; AEA also binds TRPV1 at intracellular-binding sites; lastly, eCBs and their metabolites (mainly arachidonic acid) also bind the nuclear receptor PPAR family members, activating the transcription of genes involved in fatty acid metabolism or inflammatory response. Endocannabinoids are selectively released from membrane lipid precursors as a consequence of the action of specific enzymes (i.e., N-acylphosphatidylethanolamine (NAPE) specific phospholipase D (NAPE-PLD) for AEA and diacylglycerol lipase (DAGL) for 2-AG. A putative eCB transporter (EMT) is responsible for eCB release/uptake in a temperature and saturation dependent manner. Lastly, the fatty acid amide hydrolase (FAAH) is responsible for AEA degradation in arachidonic acid and ethanolamine; 2-AG is hydrolyzed in glycerol and arachidonic acid by FAAH and mainly by monoacylglycerol lipase (MAGL).

Studies correlate the first appearance of eCBs signaling to deuterostomian organisms (Chianese et al., 2008a,b; Elphick and Egertovà, 2001; Salzet and Stefano, 2002). Orthologs of human *cb1* and *cb2* genes (currently also named *cnr1* and *cnr2*) have been cloned in most vertebrates (Elphick and Egertovà, 2001), in the urochordate, the sea squirt *Ciona intestinalis* (Elphick et al., 2003) and in cephalochordate, the amphioxus, *Branchiostoma floridae* (Elphick, 2007); evidence for eCB activity has been reported in nonchordate deuterostomian invertebrates (Salzet and Stefano, 2002; Schuel and Burkman, 2005), but no receptor have yet been cloned. In fish, *Fugu rubripes* and *Danio rerio*, whose genome has undergone duplication event, two *cnr1* and *cnr2* paralog genes have been detected, respectively (Rodriguez-Martin *et al.*, 2007; Yamaguchi *et al.*, 1996). In humans, splicing derived forms of CB1 have been detected (Ryberg *et al.*, 2005; Shire *et al.*, 1995); in addition, several single nucleotide polymorphisms (SNPs), sometimes associated to diseases or substance abuse, have been reported for *cnr1, cnr2* (Chianese *et al.*, 2008a,b). Synonymous and nonsynonymous mutations have been observed comparing *cnr1* sequence obtained from genome and from testis and brain cDNA in the anuran amphibian, *Rana esculenta* (Meccariello *et al.*, 2007). Such mutations differently affect the mRNA folding predicted from genomic DNA and cDNA, suggesting they might represent a key control of RNA stability and turnover. Lastly, nonsynonymous mutations found in the coding region might also affect receptor functionality. Also in other vertebrates differences between CB1 cDNA sequences and the corresponding genome have been reported (Meccariello *et al.*, 2007) and posttranscriptional modifications of CB1 mRNA cannot be excluded.

Studies carried out on $CB1^{-/-}$, $CB2^{-/-}$, and CB1/CB2 double knockout (KO) mice reveal the existence of no CB1/CB2 mediated response to cannabinoids. In this respect, besides CB1 and CB2, the orphan G-coupled receptor GPR55 is currently accounted as the third CB (Lauckner *et al.*, 2008). Furthermore, AEA acts as an intracellular messenger amplifying Ca^{2+} influx via transient potential type 1 vanilloid receptor (TRPV1) channels (Van der Stelt *et al.*, 2005). Unlike CB1, CB2, and GPR55, TRPV1 is a six transmembrane spanning receptor, with cytosolic amino and carboxyl terminal regions. Fifth and sixth transmembrane domains form a ligand gated nonselective cationic channel activated by capsaicin, a component of red chilli pepper. TRPV1 is unable to bind 2-AG, but its cytosolic domain binds AEA (Van der Stelt and Di Marzo, 2004). However, it is not excluded that other orphan receptors might be included in the list (Brown *et al.*, 2007). Novel studies propose eCBs and their metabolites as ligands of peroxisome proliferator-activated receptor (PPAR) family of nuclear receptor transcription factors (O'Sullivan, 2007; Sun and Bennett, 2007), suggesting a CB1/CB2 independent effects of eCBs exerted at nuclear level. PPAR family members (PPARα, β, and δ) are widely expressed and differentially bind eCBs (OEA, PEA, AEA, 2-AG, noladin ether, and virodhamine), phytocannabinoid (Δ^9-THC), CB agonists (HU210, WIN55121-2) as well as cannabinoid metabolites (O'Sullivan, 2007, for review). PPARs heterodymerize with other receptors, mainly retinoid X receptor, bind to DNA sequences called PPAR response elements (PPREs) and induce transcriptional activation of genes involved in fatty acid lipid metabolisms, anti-inflammatory response, analgesia, neuroprotection, and vasorelaxation (O'Sullivan, 2007).

In peripheral systems, eCBs signaling controls blood pressure (hypotension), vasodilatation and platelet aggregation, affects immune and inflammatory response, inhibits peristalsis and intestinal motility, controls liver activity by affecting lipogenesis, and lastly, controls reproductive functions in both sexes (Klein *et al.*, 2003; Mendizabal *et al.*, 2003; Pagotto *et al.*, 2006; Sanger, 2007; Wang *et al.*, 2006). At central level, eCBs and congeners, via CBs, control pain, wake/sleep cycles, thermogenesis, food intake, induce psychomotor disorders, tremor and spasticity, catalepsy, impair working memory, and memory consolidation, inhibits long-term potentiation and glutamatergic transmission (Elphick and Egertovà, 2001; Fride, 2002; Pagotto *et al.*, 2006; Wang *et al.*, 2006). In CNS, often CBs are localized in same brain structures with possible differences in the distribution amid neuronal elements where they are localized, as it happens in the cerebellum, where CB1 has been localized in molecular and granular layers while CB2 has been found in Purkinje cells (Gong *et al.*, 2006). At central level, AEA is suggested to act as a retrograde messenger to inhibit neurotransmitter release (Elphick and Egertovà, 2001); it is released by postsynaptic neurons and binds CB1 located at presynaptic level; CB1 activation, via Gi/o proteins, inhibits adenylcyclase and modulates ionic current mainly activating K^+ channels and inhibiting N and P/Q Ca^{2+} channels blocking neurotransmitter exocytosis. As reported at striatal level, eCB retrograde signaling requires a regulated postsynaptic release step (Adermark and Lovinger, 2007); while CB1 has been suggested to have a presynaptic localization, CB2, at least in cerebellum, has a postsynaptic localization (Gong *et al.*, 2006). Furthermore, in several brain areas, CB1 and TRPV1 colocalize (Cristino *et al.*, 2006). In this respect, both CB1 and CB2 seem likely to work both independently or/and in cooperation in different neuronal population to regulate same physiological activities in CNS, while CB1 and TRPV1 colocalization (Cristino *et al.*, 2006) suggests a sticky relationship in their biological effects.

III. Brain–Pituitary Axis

CBs are also involved in neuroendocrine regulation of hormone secretion since cannabinoid exposure affects reproduction, lactation, food intake, and stress. Exposure to Δ^9-THC inhibits the release of gonadotrophin (LH), prolactin, growth hormone, thyroid-stimulating hormone, and stimulates the release of the stress hormone corticotrophin (for review Murphy *et al.*, 1998; Pagotto *et al.*, 2006). Effects of eCBs on endocrine axis involve several actors. For example, GH inhibition is mediated through somatostatin activation while AEA suppresses prolactin release in male rat through the activation of the tuberoinfundibular dopaminergic system (Pagotto *et al.*, 2006). Direct effect on thyroid gland has been reported (Porcella *et al.*, 2002); the colocalization of CB1 and corticotrophin-releasing hormone (CRH) mRNA, reported in hypothalamic neurons of paraventricular nucleus, and a CRH increase observed in CB1$^{-/-}$ mice suggest a direct involvement of CB1 in stress response (Cota et al., 2003).

Reproductive function is under control of gonadotrophin-releasing hormone (GnRH), pituitary gonadotrophins (FSH and LH), and sexual steroids. Hypothalamic GnRH, the decapeptide responsible for gonadotrophin discharge, is released in the median eminence, it reaches the adenohypophysis, through a closed portal system connecting the hypothalamus and the anterior pituitary, and stimulate gonadotroph cells to release FSH and LH in the general circulation. Gonadotrophins, in turn, induce the release of gonadal steroids. Both gonadotrophins and gonadal steroids exert a negative feedback on hypothalamus to inhibit both GnRH and LH/FSH release (short and long feedback loop); GnRH may also reduce its own synthesis at hypothalamic level (ultra short feedback). Both Δ^9-THC and AEA are responsible for the decrease of circulating LH and sexual steroids (Murphy et al., 1998; Wang et al., 2006; Wenger et al., 2001, for review). CB1 has been localized in pituitary of both mammals and nonmammalian vertebrates (Cesa et al., 2002; Wenger et al., 1999) and CB1 immunereactivity has been detected in lactotrophs and gonadotrophs (Cesa et al., 2002; Wenger et al., 1999). Such inhibitory activity of eCBs on LH is not exerted at pituitary level but directly at hypothalamic level, the brain region where GnRH secreting neurons reside and CB1 is expressed (Matsuda et al., 1993). Incubations of mid-basal hypothalamus, collected from male/ovariectomized female rats, with AEA are reported to inhibit GnRH discharge through CB1 activation in both sexes (Scorticati et al., 2004). Such an effect is dependent on the gonadal steroid environment since in ovariectomized oestrogen-primed rats AEA induces GnRH release and increases LH circulating levels (Scorticati et al., 2004). Several mechanisms might be responsible for the inhibition of GnRH release (for review see Murphy et al., 1998). In fact, eCB activity might activate neuronal systems (i.e., GABA, dopamine, CRF, opioid) able to inhibit the activity of GnRH secreting neurons; by contrast they might also repress the activity of neuronal systems (glutamate, norepinphrine) involved in the stimulation of GnRH secreting neurons (Murphy et al., 1998). In a recent paper we suggest that eCBs might be directly involved in the modulation of GnRH secreting neuron activity (Meccariello et al., 2008). We have used as experimental model the anuran amphibian, the frog *Rana esculenta*, a seasonal breeder, characterized by a laminar brain showing a discrete localization of GnRH secreting neurons (Di Matteo et al., 1996; Meccariello et al., 2004; Pierantoni et al., 2002). Two GnRH molecular forms, GnRH-I and GnRH-II (previously known as mammalian GnRH and chicken II GnRH), have been purified by HPLC-RIA, localized in the brain and recently cloned from male frog brain (Di Matteo et al., 1996; Fasano et al.,

1993; Meccariello et al., 2008; Meccariello et al., unpublished). Both GnRH-I and GnRH-II secreting neurons have been detected in the forebrain, mainly in the anterior preoptic area and median eminence, suggesting an involvement of both forms in gonadotrophin discharge (Pierantoni et al., 2002). Lastly, while in mammals GnRH pulse is fast and limited in time, in R. esculenta GnRH molecular forms accumulate in the brain in postreproductive period (summer) and are released in autumn and during the winter stasis (Di Matteo et al., 1996; Fasano et al., 1993; Meccariello et al., 2004) to sustain gonadotrophin discharge (Polzonetti-Magni et al., 1998). In the prosencephalon of male frogs, *GnRH-I* and *CB1* mRNA have opposite expression profiles during the annual sexual cycle (Chianese et al., 2008a,b; Meccariello et al., 2008). By contrast, in telencephalon, GnRH-II expression matches CB1 expression while in diencephalon, the CB1 expression increases in December, just between GnRH-II mRNA expression peaks (Meccariello et al., unpublished). Besides GnRH secreting hormone cell line (GT1), possessing a complete ECBS, in mammals, a scanty CB1 and GnRH-I mRNA colocalization has been detected; furthermore, the majority of hypothalamic GnRH secreting neurons are surrounded by neurons expressing CB1 (Gammon et al., 2005). Also in frog, the majority of CB1 immunoreactive neurons are surrounded by GnRH-I immunoreactive fibers but, in a subset of subventral hypothalamic neurons (20% of total GnRH-I secreting neurons), both GnRH-I and CB1 are clearly present in neuronal perikarya. Such a neuroanatomical relationship might have a functional relevance since, as in rat, *in vitro* incubation of male frog diencephalons with AEA clearly reduces GnRH-I mRNA synthesis (Meccariello et al., 2008); such an effect is mediated by CB1. In fact, administration of Rimonabant, a specific CB1 antagonist, prevents AEA effects (Rinaldi-Carmona et al., 1994). Results obtained using a GnRH-I analog (buserelin) administration show a reduction of GnRH-I mRNA expression, and, in the meantime, an increase of CB1 mRNA expression. This let us to hypothesize the existence of a crosstalk between GnRH-I secreting neurons and CB1: GnRH-I ultra short feedback might be exerted through the release of eCBs and CB1 activation (Fig. 14.2). Unlike GnRH-I, GnRH-II protein and mRNA increase in midbrain, hindbrain, and spinal cord since such a decapeptide exerts several other roles as neuromodulator/neurotransmitter. For example, GnRH-II is involved in the control of sexual behavior, a process which also involves CB1 (Pierantoni et al., 2002; Soderstrom et al., 2000). To date, GnRH-II expression in the posterior area has been found to be higher than in the forebrain, but no relationship with CB1 expression has emerged yet (Fig. 14.3). Nevertheless, in diencephalon, AEA, via CB1, decreases GnRH-II mRNA levels (Meccariello et al., unpublished), indicating that in the areas involved in the control of gonadotrophin discharge, both GnRH-I and GnRH-II might be regulated in the same manner.

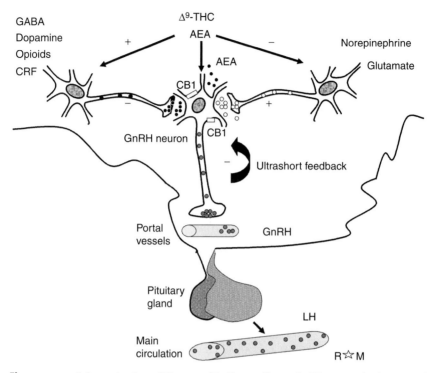

Figure 14.2 Schematization of direct and indirect effects of eCB system in the control of GnRH/LH release. See text for explanation.

IV. Testis

It is becoming evident that ECBS plays a key role in the development of testis and in the physiology of male reproductive tract. The evidence stems from old studies, involving exposure of male mice to exogenous cannabinoids (Dalterio et al., 1982), and recent studies, that have localized CB1 (Cobellis et al., 2006; Gye et al., 2005; Wenger et al., 2001) and CB2 (Maccarrone et al., 2003), as well as the use of CB1 KO mice (Wenger et al., 2001). Nevertheless, it is apparent that male CB1-KO mice develop both normal testis (Cacciola et al., 2008b) and reproductive tract (unpublished), but a deeper analysis of testicular developmental stages has shown a Leydig cell deficit (Cacciola et al., 2008a) that endorses the important role of eCBs in neonatal testicular development.

Several *in vivo* and *in vitro* studies show that cannabinoids affect spermatogenesis. Sperm concentration and sperm count, as well as serum LH levels decrease in marihuana smokers (Hembree et al., 1978; Vescovi et al., 1992).

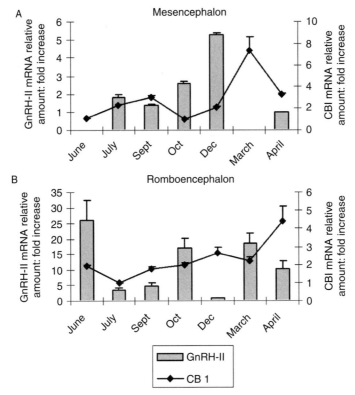

Figure 14.3 Expression analysis of *GnRH-II* and *CB1* from male frog midbrain (A) and hindbrain (B) during the annual sexual cycle in *R.esculenta*. Gene expression is normalized against the expression of the housekeeping gene *fp1*. Data are expressed as mean fold increase ± S.D.

Cannabinoid-treated rodents show abnormal sperm morphology (Zimmerman *et al.*, 1979). In rodents and men, chronic exposure to (or use of) cannabinoids decreases Leydig cell testosterone production (Dalterio *et al.*, 1983) and secretion (Kolodny *et al.*, 1974), depresses spermatogenesis (Dixit *et al.*, 1974), and reduces testes and accessory reproductive organs weight (Harclerode *et al.*, 1978; Wang *et al.*, 2006, for review). Direct intracerebroventricular THC administration decreases plasma LH levels and increases GnRH hypothalamic levels (Wenger *et al.*, 1987) suggesting, as stated in the above section, that the lowered GnRH concentration in the portal system is responsible for the suppressed LH levels (Pielecka and Moenter, 2006). Furthermore, it has been demonstrated that there is an association between AEA injection and LH/testosterone level reduction in CD1-wild-type (WT) mice (Wenger *et al.*, 2001). As a consequence, the adverse effects on spermatogenesis might be mediated by indirect eCB action on hypothalamus–pituitary axis (through endocrine or local control, see above sections) or directly on the testis (through tubular/interstitial cells). Indeed, additionally to hypothalamus (Cristino *et al.*, 2006) and pituitary (Cottone *et al.*, 2008; Pagotto *et al.*, 2001; Wenger *et al.*, 2001), CBs have been found in the testis of several vertebrates (Cacciola *et al.*, 2008b; Cobellis *et al.*, 2006; Maccarrone *et al.*, 2003), and invertebrates (McPartland *et al.*, 2006).

In vitro, cultured mouse immature Sertoli cells (4–16 dpp) express a functional CB2 but not CB1 (Maccarrone *et al.*, 2003). Discrepancy about protein localization has recently been discussed (Cacciola *et al.*, 2008a) because rat Sertoli cells express CB1 at 41 dpp. Probably, apart from species differences, Sertoli cells may express CB1 later during testicular development through a possible induction exerted by germ cell contact (Fig. 14.4). In Sertoli cells, CB2 inactivation by SR144528 increases AEA-induced apoptosis, suggesting that CB2 has a protective role against the cytotoxic effects of AEA (Maccarrone *et al.*, 2003).

CB1 and FAAH are present in mammalian (Cacciola *et al.*, 2008a; Gye *et al.*, 2005) and nonmammalian germ cells (Cacciola *et al.*, 2008a; Cobellis *et al.*, 2006). In mammalian germinal epithelium, apart specie-specific differences (see for discussion Cacciola *et al.*, 2008a), CB1 is present in spermatids (SPT) (Cacciola *et al.*, 2008a; Gye *et al.*, 2005) and spermatozoa

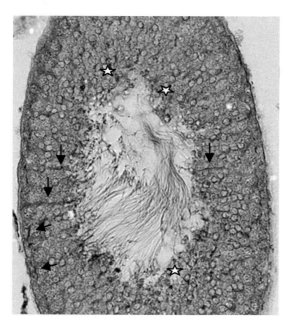

Figure 14.4 Section of adult rat testis immunostained for CB1. CB1 is evident in round SPT (stars) and Sertoli cells (arrows). Scale bars: 20 μm.

(SPZ) (Cacciola et al., 2008a; Gye et al., 2005; Maccarrone et al., 2005; Rossato et al., 2005). In rat, CB1 appears in round SPT, around the nucleus, contemporaneously with the acrosome formation. The signal is retained in elongating SPT, in the head, always close to the acrosome region, suggesting the involvement of CB1 in spermiogenesis (probably in acrosome and/or cellular shape configuration) and also confirming the activity in acrosome reaction (Schuel et al., 2002b). In nonmammalian testis, during the annual reproductive cycle of the frog, R. esculenta, CB1 and FAAH levels increase in September when spermiogenesis occurs (Cobellis et al., 2006). Thus, the immune-histochemical analysis shows weak CB1 and FAAH signals in spermatocytes (SPC) and a more intense signal in elongating SPT and SPZ(Cobellis et al., 2006). The localization of CB1 and FAAH in the same cell type hints the hypothesis that FAAH regulates the physiological eCB levels (endocannabinoid tone) to accurately drive SPT morphogenesis. Accordingly, marihuana smokers (Issidorides, 1978) and animals treated with THC show alterations of sperm morphology (Zimmerman et al., 1979) as well as CB1 KO male mice show defected spermiogenesis (unpublished). The effects of eCBs on spermiogenesis may be due to low testosterone levels described in both THC-treated (Kolodny et al., 1974) and CB1 KO male mice (Wenger et al., 2001). Indeed, androgens are critical for testis function (Tan et al., 2005). Androgen (ARKO) or LH receptor (LuRKO) KO male mice both show a block of spermatogenesis at round SPT stage (Yeh et al., 2002; Zhang et al., 2001). Androgen-binding protein (ABP) transgenic mice show apoptosis of pachytene SPC and round SPT as well as a decreased SPT number (Selva et al., 2000). In rat, low doses of estrogens reduce LH and testosterone levels as well as the number of SPT and SPZ (Eddy et al., 1996; Kalla, 1987). As a consequence, both withdrawal and deficit of LH or testosterone dramatically affect spermatogenesis progression. Unexpectedly, CB1 KO male mice are fertile (Ledent et al., 1999) and produce a canonical number of SPZ (unpublished). These results suggest that the observed low levels of testosterone are in a physiological range and, as a consequence, they are not responsible for the spermiogenesis deficit. Such findings further support the hypothesis that eCBs, probably produced by Sertoli (Maccarrone et al., 2003) or germ cells (Maccarrone et al., 2005), may control SPT differentiation by regulating eCB tone through FAAH. Features of FAAH-KO SPT might confirm and indicate more specifically the eCB activity in spermiogenesis.

As far as interstitial compartment concerns, in nonmammalian testis, CB1 is not expressed in the amphibian, R. esculenta. On the contrary, in mammalian testis, CB1 has been observed in mouse and rat Leydig cells (Cacciola et al., 2008a; Wenger et al., 2001). Recent results, during rat postnatal testicular development, confirm the involvement of eCBs in differentiation and/or proliferation events concerning adult Leydig cells (ALC). CB1 is present in Leydig cells during differentiation and maintenance of their

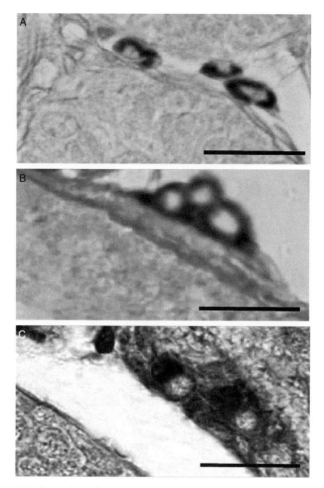

Figure 14.5 Testis sections from rats aged 14 (A), 31 (B), and 60 days (C) showing differentiating ALC immunostained for CB1. Scale bar: 20 μm.

fully developed phenotype (Fig. 14.5). Our preliminary observations support the idea that eCBs constitute a new family of lipidic signaling responsible for the regulation of progenitor proliferation and differentiation through CB1. Indeed, we have observed that mesenchymal/progenitor cells, at 14 and 31 dpp, respectively, differentiating in immature Leydig cells, express CB1. The receptor expression switches-off at 41 dpp, when synchronized immature ALC undergo a single massive mitotic division to produce fully developed ALC (Fig. 14.6). Consequentially, endocannabinoid/CB1 signaling may regulate positively Leydig cell lineage differentiation (from progenitor to immature cells) and negatively the proliferation of immature cells committed

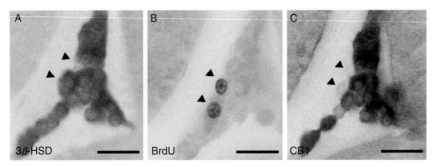

Figure 14.6 Consecutive sections of testis from a rat aged 41 days immunostained for 3β-HSD (A), BrdU (B), and CB1 (C). Note that proliferating Leydig cells (arrowhead) labeled by BrdU (B) are not detected by CB1 antiserum (C), but are unequivocally stained by 3β-HSD antiserum (A). Scale bar: 20 μm.

to become ALC. Accordingly, $CB1^{-/-}$ mice develop few ALC and this might explain the low levels of circulating and testicular testosterone *in vitro* secretion (Wenger *et al.*, 2001).

More recent studies prove that eCBs, produced *in vitro* by hypothalamic GnRH neurons (Gammon *et al.*, 2005; Meccariello *et al.*, 2008), reduce hypothalamic GnRH synthesis in organ culture experiments (Meccariello *et al.*, 2008), and LH secretion (Wenger *et al.*, 2001) through a CB1-dependent mechanism (see previous sections). Altogether, these data do not explain the low levels of LH in $CB1^{-/-}$ mice serum (Wenger *et al.*, 2002). Finally, results concerning the effect of AEA in decreasing GnRH mRNA (Meccariello *et al.*, 2008), LH and testosterone (both serum and intratesticular) levels (Wenger *et al.*, 2001), let to speculate about a higher concentration of AEA in CB1 KO male mice as compared with WT animals.

The presence of eCBs at brain–pituitary–gonadal levels (Pagotto *et al.*, 2006) suggests that a deeper knowledge of endocrine, paracrine and autocrine feedback mechanisms, involving endocannabinoid/CB signaling, is necessary to reconcile the above described results.

V. Excurrent Duct System

Once formed within the seminiferous tubules, the immotile SPZ are released into the luminal fluid and transported through the excurrent duct system, differentially organized according to the species (Amann *et al.*, 1976; Birkhead and Fletcher, 1998; Cobellis *et al.*, 2005). In amphibians, a simple excurrent duct system with several efferent ducts, leaving testis and entering only one external duct (properly indicated as spermatic duct) which confluences into the cloacae has been observed. In mammals, testicular SPZ are passively transported to the rete testis, which is a branched reservoir of the openings of the seminiferous tubules. From the rete testis, the transport of SPZ to the epididymis takes place via the efferent ducts (Jonte and Holstein, 1987; Saitoh *et al.*, 1990). The epithelium lining these ducts is columnar and it consists of two cell types, ciliated and nonciliated cells (Hermo *et al.*, 1992). Both cell types are able to perform endocytosis, and ciliated cells also sustain the movement of luminal fluid and sperms. Nonciliated cells are mainly responsible for the absorption of water and ions. The efferent ducts absorb most of the fluid discharged from the testis with SPZ, thus increasing epididymal sperm concentration (Clulow *et al.*, 1994). The epididymis can be divided in three parts, *caput* (the proximal part), *corpus*, and *cauda* (the distal part). In mammals, the transit of SPZ through the epididymis usually takes 10–13 days, whereas in humans the estimated transit time is 2–6 days (Amann and Howards, 1980; Hermo *et al.*, 1992; Johnson and Varner, 1988). The epididymis segment, where most SPZ attain their full fertilizing capacity, appears to be the *cauda*. Spermatozoa from that region are able to move progressively, typical feature of SPZ preceding fertilization, and to bind to zona-free hamster oocyte *in vitro* at a higher percentage than SPZ from *caput* and *corpus* (Turner, 1995; Yanagimachi, 1994). To attain the fertilizing capacity, sperm undergoes several maturational changes during the transit in the epididymis (Yanagimachi, 1994). These include, for instance, changes in plasma membrane lipids (membrane remodeling), proteins, and glycosylation, alterations in the outer acrosome membrane, morphological changes in acrosome, chromatin condensation, and cross-linking of nuclear protamines (Evenson *et al.*, 1989). The potential progressive motility of mammalian SPZ undergoes two modifications before fertilization: the first occurs in the epididymis, the second in the female reproductive tract (ipermotility), where SPZ show a frantic flagellar movement called hyperactivation (Yanagimachi, 1970). Hyperactivation is believed to enable SPZ to generate adequate forces to rend them free from the oviductal isthmus and to potentiate their movement through the cumulus matrix, or the oocyte zona pellucida (Suarez, 1996; Yanagimachi, 1994). In boar SPZ, CB1 activation by methanandamide, an AEA-stable analog, inhibits capacitation and acrosome reaction (Maccarrone *et al.*, 2005). Recently, it has been shown that eCBs are present in amphibian cloacae fluid (Cobellis *et al.*, 2006) and in human seminal plasma (Schuel *et al.*, 2002a), suggesting that they may influence important steps, including sperm maturation or function (Wang *et al.*, 2006). Consistently, a complete ECBS has been characterized in boar SPZ. These cells produce AEA and express TRPV1, FAAH, CB2, and CB1 (Maccarrone *et al.*, 2005). Accordingly, frog, mouse, and rat SPZ express both CB1 and FAAH (Cobellis et al., 2006). In rat immune-fluorescent analysis localizes CB1 in the head (along the head and close to the acrosome) and tail (Cacciola et al., in 2008a) of SPZ epididymis and confirms previous results obtained in humans (Rossato et al., 2005). Conversely, in ejaculated boar SPZ, CB1 is reported to be expressed in the postacrosome region (Maccarrone et al., 2005). Likely, such intracellular distribution might be species-specific and/or it might be affected by membrane remodeling mechanism occurring in male reproductive tract before ejaculation (Jones, 1998). The presence of CB1 in SPZ (both testicular and isolated) of all analyzed vertebrates, such as frog (Cobellis et al., 2006), mouse (Cobellis et al., 2006; Gye et al., 2005), rat (Cacciola et al., 2008a), boar (Maccarrone et al., 2005), and human (Rossato et al., 2005), strongly suggests an evolutionarily conserved role of eCBs in sperm physiology. *In vitro* studies show that AEA inhibits motility in human (Rossato et al., 2005) and frog SPZ (Cobellis et al., 2006). In particular, the inhibitory effect on frog SPZ is mediated by CB1 (as in humans), because it is counteracted by Rimonabant. Furthermore, AEA effect on SPZ motility is counteracted also by SPZ washing or dilution, suggesting that endocannabinoids control the number of motile SPZ, via CB1, keeping sperm motility quiescent until their release in aquatic environment ("dilution mechanism"). Since the ECBS has been demonstrated to be highly conserved in evolution (Matias et al., 2005), it is possible that in mammals it may operate into the epididymis to regulate SPZ potential motility acquisition. As previously described, SPZ acquire potential motility along the epididymis, traveling from the *caput* to the *cauda* (Cooper, 1998). In CB1 KO mice, instead, the percentage of motile SPZ significantly increases in *caput* reaching the *cauda* levels, suggesting that CB1 signaling controls the number of motile SPZ along the epididymis (lower in *caput* and higher in *cauda*) keeping quiescent sperm motility in the *caput* (Ricci et al., 2007). From the epididymis, SPZ enter a muscular tube, the *vas deferens* whose contractions, during a process called ejaculation, move SPZ into the ejaculatory duct, where they are mixed with secretions from the seminal vesicles. Mouse isolated *vas deferens* is a nerve-smooth muscle preparation that serves as a highly sensitive and quantitative functional *in vitro* bioassay for CB1 agonists. The bioassay of CB1 agonists shows that both AEA and cannabinoids are able to produce concentration-related decrease in the amplitude of electrically evoked contractions of the *vas deferens* (Pertwee et al., 2005). They act on naturally expressed prejunctional neuron CB1 receptors to inhibit release of the contractile neurotransmitters and noradrenaline, provoked by the electrical stimulation.

The above reported evidences support the suggested analogy on mechanisms occurring in sperm and neurons (Pierantoni et al., 2009; Schuel and Burkman, 2005) regulated by eCBs.

VI. Concluding Remarks

There are little doubts about the importance of ECBS in the control of several male reproductive aspects. Furthermore, it is particularly important to underline that, once again, the use of nonmammalian models has been useful to open new research avenues and to give insight into evolutionarily conserved mechanisms.

Therefore, the battleground is going to be, in the next future, to design new molecules able to influence both positively or negatively male reproduction. Finally, an interesting point to emerge from data here presented concerns the possibility of new therapeutic approaches.

REFERENCES

Adermark, L., and Lovinger, D. M. (2007). Retrograde endocannabinoid signaling at striatal synapses requires a regulated postsynaptic release step. *Proc. Natl. Acad. Sci. USA* **104**, 20564–20569.

Amann, R. P., and Howards, S. S. (1980). Daily spermatozoal production and epididymal spermatozoal reserves of the human male. *J. Urol.* **124**, 211–215.

Amann, R.P, Johnson, L., Thompson, D. L., and Pickett, B. W. (1976). Spermatozoal production, epididymal spermatozoal reserves and transit time of spermatozoa through the epididymis of the rhesus monkey. *Biol. Reprod.* **15**, 586–592.

Birkhead, T. R., and Fletcher, F. (1998). Sperm transport in the reproductive tract of female zebra finches (*Taeniopygia guttata*). *J. Reprod. Fert.* **114**, 141–145.

Brown, A. L. (2007). Novel cannabinoid receptors. *Br. J. Pharmacol.* **152**, 567–575.

Brown, S. M., Wager-Miller, J., and Mackie, K. (2002). Cloning and molecular characterization of the rat CB2 cannabinoid receptor. *Biochim. Biophys. Acta* **1576**, 255–264.

Cacciola, G., Chioccarelli, T., Mackie, K., Meccariello, R., Ledent, C., Fasano, S., Pierantoni, R., and Cobellis, G. (2008a). Expression of type-1 cannabinoid receptor during rat postnatal testicular development: Possible involvement in adult Leydig cell differentiation. *Biol. Reprod.* **79**, 758–765.

Cacciola, G., Chioccarelli, T., Ricci, G., Meccariello, R., Fasano, S., Pierantoni, R., and Cobellis, G. (2008b). The endocannabinoid system in vertebrate male reproduction: A comparative overview. *Mol. Cell. Endocrinol.* **286**, S24–S30.

Cesa, R., Guastalla, A., Cottone, E., Mackie, K., Beltramo, M., and Franzoni, M. F. (2002). Relationships between CB1 cannabinoid receptors and pituitary endocrine cells in *Xenopus laevis*: An immunohistochemical study. *Gen. Comp. Endocrinol.* **25**, 17–24.

Chianese, R., Cobellis, G., Pierantoni, R., Fasano, S., and Meccariello, R. (2008a). Non mammalian vertebrate models and the endocannabinoid system: Relationships with gonadotropin-releasing hormone. *Mol. Cell. Endocrinol.* **286**, S46–S51.

Chianese, R., Pierantoni, R., Cobellis, G., Fasano, S., and Meccariello, R. (2008b). Two GnRH mRNA molecular forms and GnRH receptors fluctuate in the frog, *Rana esculenta* brain and testis: A regulatory role exerted by endocannabinoids [submitted].

Clulow, J., Jones, R. C., and Hansen, L. A. (1994). Micropuncture and cannulation studies of fluid composition and transport in the ductuli efferentes testis of the rat: Comparisons with the homologous metanephric proximal tubule. *Exp. Physiol.* **79**, 915–928.

Cobellis, G., Cacciola, G., Scarpa, D., Meccariello, R., Chianese, R., Franzoni, M. F., Mackie, K., Pierantoni, R., and Fasano, S. (2006). Endocannabinoid system in frog and rodent testis: Type-1 cannabinoid receptor and fatty acid amide hydrolase activity in male germ cells. *Biol. Reprod.* **75,** 82–89.

Cobellis, G., Lombardi, M., Scarpa, D., Izzo, G., Fienga, G., Meccariello, R., Pierantoni, R., and Fasano, S. (2005). Fra1 activity in the frog, *Rana esculenta,* testis: A new potential role in sperm transport. *Biol. Reprod.* **72,** 1101–1108.

Cooper, T. G. (1998). Interactions between epididymal secretions and spermatozoa. *J. Reprod. Fert.* **53**(Suppl.), 119–136.

Cota, D., Marsicano, G., Tschöp, M., Grübler, Y., Flachskamm, C., Schubert, M, Auer, D., Yassouridis, A., Thöne-Reineke, C., Ortmann, S., Tomassoni, F., Cervino, C., et al. (2003). The endogenous cannabinoid system affects energy balance via central orexigenic drive and peripheral lipogenesis. *J. Clin. Invest.* **112,** 323–326.

Cottone, E., Guastalla, A., Mackie, K., and Franzoni, M. F. (2008). Endocannabinoids affect the reproductive functions in teleosts and amphibians. *Mol. Cell. Endocrinol.* **286,** S41–S45.

Cristino, L, de Petrocellis, L., Pryce, G., Baker, D., Guglielmotti, V., and Di Marzo, V. (2006). Immunohistochemical localization of cannabinoid type 1 and vanilloid transient receptor potential vanilloid type 1 receptors in the mouse brain. *Neuroscience* **139,** 1405–1415.

Dalterio, S., Badr, F., Bartke, A., and Mayfield, D. (1982). Cannabinoids in male mice: Effects on fertility and spermatogenesis. *Science* **216,** 315–316.

Dalterio, S. L., Bartke, A., and Mayfield, D. (1983). Cannabinoids stimulate and inhibit testosterone production *in vitro* and *in vivo*. *Life Sci.* **32,** 605–612.

Devane, W. A., Hanus, L., Breuer, A., Pertwee, R. G., Stevenson, L. A., Griffin, G., Gibson, D., Mandelbaum, A., Etinger, A., and Mechoulam, R. (1992). Isolation and structure of a brain constituent that binds to the cannabinoid receptor. *Science* **258,** 1946–1949.

Di Matteo, L., Vallarino, M., and Pierantoni, R. (1996). Localization of GnRH molecular forms in the brain, pituitary, and testis of the frog, *Rana esculenta.* *J. Exp. Zool.* **274,** 33–40.

Dixit, V. P., Sharma, V. N., and Lohiya, N. K. (1974). The effect of chronically administered cannabis extract on the testicular function of mice. *Eur. J. Pharmacol.* **26,** 111–114.

Eddy, E. M., Washburn, T. F., Bunch, D. O., Goulding, E. H., Gladen, B. C., Lubahn, D. B., and Korach, K. S. (1996). Targeted disruption of the estrogen receptor gene in male mice causes alteration of spermatogenesis and infertility. *Endocrinology* **137,** 4796–4805.

Elphick, M. R. (2007). BfCBR: A cannabinoid receptor ortholog in the cephalochordate *Branchiostoma floridae* (Amphioxus). *Gene* **399,** 65–71.

Elphick, M. R., and Egertovà, M. (2001). The neurobiology and evolution of cannabinoid signaling. *Philos. Trans. R. Soc. Lond. B Biol. Sci.* **356,** 381–408.

Elphick, M. R., Satou, Y., and Satoh, N. (2003). The invertebrate ancestry of endocannabinoid signaling: An orthologue of the vertebrate cannabinoid receptors in the urochordate *Ciona intestinalis*. *Gene* **302,** 95–101.

Evenson, D. P., Baer, R. K., and Jost, L. K. (1989). Flow cytometric analysis of rodent epididymal spermatozoal chromatin condensation and loss of free sulfhydryl groups. *Mol. Reprod. Dev.* **1,** 283–288.

Fasano, S., Goos, G. J., Janssen, C., and Pierantoni, R. (1993). Two GnRHs fluctuate in correlation with androgen levels in the male frog *Rana esculenta.* *J. Exp. Zool.* **266,** 277–283.

Fride, E. (2002). Endocannabinoids in the central nervous system. *Prostaglandins Leukot. Essent. Fatty Acids* **66,** 221–233.

Galiegue, S., Mary, S., Marchand, J., Dussossoy, D., Carriere, D., Carayon, P., Bouaboula, M., Shire, D., Le Fur, G., and Casellas, P. (1995). Expression of central and peripheral cannabinoid receptors in human immune tissues and leukocyte subpopulations. *Eur. J. Biochem.* **232,** 54–61.

Gammon, C. M., Freeman, G. M., Jr., Xie, W., Petersen, S. L., and Wetsel, W. C. (2005). Regulation of gonadotropin-releasing hormone secretion by cannabinoids. *Endocrinology* **46,** 4491–4499.

Gong, J. P., Onaivi, E.S, Ishiguro, H., Liu, Q. R., Tagliaferro, P. A., Brusco, A., and Uhl, G. R. (2006). Cannabinoid CB2 receptors: Immunohistochemical localization in rat brain. *Brain Res.* **1071,** 10–23.

Gye, M. C., Kang, H. H., and Kang, H. J. (2005). Expression of cannabinoid receptor 1 in mouse testes. *Arch. Androl* **51,** 247–255.

Harclerode, J., Nyquist, S. E., Nazar, B., and Lowe, D. (1978). Effects of cannabis on sex hormones and testicular enzymes of the rodent. *Adv. Biosci.* **22,** 395–405.

Hembree, W. C. I., Nahas, G. G., Zeidenberg, P., and Huang, H. F. (1978). Changes in human spermatozoa associated with high dose marihuana smoking. *Adv. Biosci.* **22,** 429–439.

Hermo, L., Barin, K., and Robaire, B. (1992). Structural differentiation of the epithelial cells of the testicular excurrent duct system of rats during postnatal development. *Anat. Rec.* **233,** 205–228.

Hillard, C. J., and Jarrahian, A. (2000). The movement of N-arachidonoylethanolamine (anandamide) across cellular membranes. *Chem. Phys. Lipids* **108,** 123–134.

Issidorides, M. R. (1978). Observations in chronic hashish users: Nuclear aberrations in blood and sperm and abnormal acrosome in spermatozoa. *Adv. Biosci.* 377–388.

Johnson, L., and Varner, D. D. (1988). Effect of daily spermatozoan production but not age on transit time of spermatozoa through the human epididymis. *Biol. Reprod.* **39,** 812–817.

Jones, R. (1998). Plasma membrane structure and remodelling during sperm maturation in the epididymis. *J. Reprod. Fert.* **53**(Suppl.), 73–84.

Jonte, G., and Holstein, A. F. (1987). On the morphology of the transitional zones from the rete testis into the ductuli efferentes and from the ductuli efferentes into the ductus epididymidis. Investigations on the human testis and epididymis. *Andrologia* **19,** 398–412.

Kalla, N. R. (1987). Demonstration of direct effect of estrogen on rat spermatogenesis. *Acta Eur. Fertil.* **18,** 293–302.

Klein, T. W., Newton, C., Larsen, K., Lu, L., Perkins, I., Nong, L., and Friedman, H. (2003). The cannabinoid system and immune modulation. *J. Leukoc. Biol.* **74,** 486–496.

Kolodny, R. C., Masters, W. H., Kolodner, R. M., and Toro, G. (1974). Depression of plasma testosterone levels after chronic intensive marihuana use. *N. Engl. J. Med.* **290,** 872–874.

Lauckner, J. E., Jensen, J.B, Chen, H. Y., Lu, H. C., Hille, B, and Mackie, K. (2008). GPR55 is a cannabinoid receptor that increases intracellular calcium and inhibits M current. *Proc. Natl. Acad. Sci. USA* **105,** 2699–2704.

Ledent, C., Valverde, O., Cossu, G., Petitet, F., Aubert, J. F., Beslot, F., Bohme, G. A., Imperato, A., Pedrazzini, T., Roques, B. P., Vassart, G., Fratta, W., *et al.* (1999). Unresponsiveness to cannabinoids and reduced addictive effects of opiates in CB1 receptor knockout mice. *Science* **283,** 401–404.

Maccarrone, M., Cecconi, S., Rossi, G., Battista, N., Pauselli, R., and Finazzi-Agro, A. (2003). Anandamide activity and degradation are regulated by early postnatal aging and follicle-stimulating hormone in mouse Sertoli cells. *Endocrinology* **144,** 20–28.

Maccarrone, M., Barboni, B., Paradisi, A., Bernabo, N., Gasperi, V., Pistilli, M. G., Fezza, F., Lucidi, P., and Mattioli, M. (2005). Characterization of the endocannabinoid system in boar spermatozoa and implications for sperm capacitation and acrosome reaction. *J. Cell Sci.* **118,** 4393–4404.

Matias, I., McPartland, J. M., and Di Marzo, V. (2005). Occurrence and possible biological role of the endocannabinoid system in the sea squirt *Ciona intestinalis*. *J. Neurochem.* **93**, 1141–1156.

Matsuda, L. A., Lolait, S. J., Brownstein, M. J., Young, A. C., and Bonner, T. I. (1990). Structure of a cannabinoid receptor and functional expression of the cloned cDNA. *Nature* **346**, 561–564.

Matsuda, L. A., Bonner, T. I., and Lolait, S. J. (1993). Localization of cannabinoid receptor mRNA in rat brain. *J. Comp. Neurol.* **327**, 535–550.

McPartland, J. M., Agraval, J., Gleeson, D., Heasman, K., and Glass, M. (2006). Cannabinoid receptors in invertebrates. *J. Evol. Biol.* **19**, 366–373.

Meccariello, R., Mathieu, M, Cobellis, G., Vallarino, M., Bruzzone, F., Fienga, G., Pierantoni, R., and Fasano, S. (2004). Jun localization in cytosolic and nuclear compartments in brain-pituitary system of the frog, *Rana esculenta*: An analysis carried out in parallel with GnRH molecular forms during the annual reproductive cycle. *Gen. Comp. Endocrinol.* **135**, 310–323.

Meccariello, R., Chianese, R, Cobellis, G., Pierantoni, R., and Fasano, S. (2007). Cloning of type 1 cannabinoid receptor in *Rana esculenta* reveals differences between genomic sequence and cDNA. *FEBS J.* **274**, 2909–2920.

Meccariello, R., Franzoni, M. F., Chianese, R., Cottone, E., Scarpa, D., Donna, D., Cobellis, G., Guastalla, A, Pierantoni, R., and Fasano, S. (2008). Interplay between the endocannabinoid system and GnRH-I in the forebrain of the anuran amphibian *Rana esculenta*. *Endocrinology* **149**, 2149–2158.

Mendizabal, V. E., and Adler-Graschinsky, E. (2003). Cannabinoid system as potential target for drug development in the treatment of cardiovascular disease. *Curr. Vasc. Pharmacol.* **1**, 301–313.

Munro, S., Thomas, K. L., and Abu-Shaar, M. (1993). Molecular characterization of a peripheral receptor for cannabinoids. *Nature* **365**, 61–65.

Murphy, L. L., Munoz, R. M., Adrian, B. A., and Villanua, M. A. (1998). Function of cannabinoid receptors in the neuroendocrine regulation of hormone secretion. *Neurobiol. Dis.* **5**, 432–446.

O'Sullivan, S. E. (2007). Cannabinoids go nuclear: Evidence for activation of peroxisome proliferator-activated receptors. *Br. J. Pharmacol.* **152**, 576–582.

Padgett, L. W. (2005). Recent developments in cannabinoid ligands. *Life Sci.* **77**, 1767–1798.

Pagotto, U., Marsicano, G., Fezza, F., Theodoropoulou, M., Grubler, Y., Stalla, J., Arzberger, T., Milone, A., Losa, M., Di Marzo, V., Lutz, B., and Stalla, G. K. (2001). Normal human pituitary gland and pituitary adenomas express cannabinoid receptor type 1 and synthesize endogenous cannabinoids: First evidence for a direct role of cannabinoids on hormone modulation at the human pituitary level. *J. Clin. Endocrinol. Metab.* **86**, 2687–2696.

Pagotto, U., Marsicano, G., Cota, D., Lutz, B., and Pasquali, R. (2006). The emerging role of the endocannabinoid system in endocrine regulation and energy balance. *Endocr. Rev.* **27**, 73–100.

Pertwee, R. G., Thomas, A., Stevenson, L. A., Maor, Y., and Mechoulam, R. (2005). Evidence that (-)-7-hydroxy-4'-dimethylheptyl-cannabidiol activates a non-CB(1), non-CB(2), non-TRPV1 target in the mouse vas deferens. *Neuropharmacology* **48**, 1139–1146.

Pielecka, J., and Moenter, S. M. (2006). Effect of steroid milieu on gonadotropin-releasing hormone-1 neuron firing pattern and luteinizing hormone levels in male mice. *Biol. Reprod.* **74**, 931–937.

Pierantoni, R., Cobellis, G., Meccariello, R., Cacciola, G., Chianese, R., Chioccarelli, T., and Fasano, S. (2009). Testicular GnRH activity, progression of spermatogenesis and sperm transport in vertebrates. *NY Acad. Sci.* **1163**, 279–291.

Pierantoni, R., Cobellis, G., Meccariello, R., and Fasano, S. (2002). Evolutionary aspects of cellular communication in the vertebrate hypothalamo-hypophysio-gonadal axis. *Int. Rev. Cytol.* **218**, 69–141.

Polzonetti-Magni, A. M., Mosconi, G., Carnevali, O., Yamamoto, K., Hanaoka, Y., and Kikuyama, S. (1998). Gonadotropins and reproductive function in the anuran amphibian, *Rana esculenta. Biol. Reprod.* **58**, 88–93.

Porcella, A., Marchese, G, Casu, M. A., Rocchitta, A., Lai, M. L., Gessa, G. L., and Pani, L. (2002). Evidence for functional CB1 cannabinoid receptor expressed in the rat thyroid. *Eur. J. Endocrinol.* **147**, 255–261.

Ricci, G., Cacciola, G., Altucci, L., Meccariello, R., Pierantoni, R., Fasano, S., and Cobellis, G. (2007). Endocannabinoid control of sperm motility: The role of epididymis. *Gen. Comp. Endocrinol.* **153**, 320–322.

Rinaldi-Carmona, M., Barth, F., Hèaulme, M., Shire, D., Calandra, B., Congy, C., Martinez, S., Maruani, J., Nèliat, G., Caput, D., Ferrara, P., and Sobrie, P. (1994). SR141716A, a potent and selective antagonist of the brain cannabinoid receptor. *FEBS Lett.* **350**, 240–244.

Rodriguez-Martin, I., Herrero-Turrion, M. J., Marron Fdez de Velasco, E, Gonzalez-Sarmiento, R., and Rodriguez, R.E (2007). Characterization of two duplicate zebrafish Cb2-like cannabinoid receptors. *Gene* **389**, 36–44.

Rossato, M., Ion, P. F., Ferigo, M., Clari, G., and Foresta, C. (2005). Human sperm express cannabinoid receptor Cb1, the activation of which inhibits motility, acrosome reaction, and mitochondrial function. *J. Clin. Endocrinol. Metab.* **90**, 984–991.

Ryberg, E., Vu, H. K., Larsson, N., Groblewski, T., Hjorth, S., Elebring, T., Sjogren, S., and Greasley, P. J. (2005). Identification and characterization of a novel splice variant of the human CB1 receptor. *FEBS Lett.* **57**, 259–264.

Saitoh, K., Terada, T., and Hatakeyama, S. (1990). A morphological study of the efferent ducts of the human epididymis. *Int. J. Androl.* **13**, 369–376.

Salzet, M., and Stefano, G. B. (2002). The endocannabinoid system in invertebrates. *Prostaglandins Leukot. Essent. Fatty Acids* **66**, 353–361.

Sanger, G. J. (2007). Endocannabinoids and the gastrointestinal tract: What are the key questions? *Br. J. Pharmacol.* **152**, 663–670.

Schuel, H., and Burkman, L. J. (2005). A tale of two cells: Endocannabinoid-signaling regulates functions of neurons and sperm. *Biol. Reprod.* **73**, 1078–1086.

Schuel, H., Burkman, L. J., Lippes, J., Crickard, K., Forester, E., Piomelli, D., and Giuffrida, A. (2002a). N-Acylethanolamines in human reproductive fluids. *Chem. Phys. Lipids* **121**, 211–227.

Schuel, H., Burkman, L. J., Lippes, J., Crickard, K., Mahony, M. C., Giuffrida, A., Picone, R. P., and Makriyannis, A. (2002b). Evidence that anandamide-signaling regulates human sperm functions required for fertilization. *Mol. Reprod. Dev.* **63**, 376–387.

Schuel, H., and Burkman, L. J. (2005). A tale of two cells: Endocannabinoid-signaling regulates functions of neurons and sperm. *Biol. Reprod.* **73**, 1078–1086.

Scorticati, C., Fernandez-Solari, J., De Laurentiis, A., Mohn, C., Prestifilippo, J. P., Lasaga, M., Seìlicovich, A., Billi, S., Franchi, A., McCann, S., and Rettori, V. (2004). The inhibitory effect of anandamide on luteinizing hormone-releasing hormone secretion is reversed by estrogen. *Proc. Natl. Acad. Sci. USA* **32**, 11891–11896.

Selva, D. M., Tirado, O. M., Toran, N., Suarez-Quian, C. A., Reventos, J., and Munell, F. (2000). Meiotic arrest and germ cell apoptosis in androgen-binding protein transgenic mice. *Endocrinology* **141**, 1168–1177.

Shire, D., Carillon, C., Kaghad, M., Calandra, B., Rinaldi-Carmona, M., Le Fur, G., Caput, D, and Ferrara, P. (1995). An amino-terminal variant of the central cannabinoid receptor resulting from alternative splicing. *J. Biol. Chem.* **270**, 3726–3731.

Soderstrom, K., Leid, M., Moore, F. L., and Murray, T. F. (2000). Behavioural, pharmacological, and molecular characterization of an amphibian cannabinoid receptor. *J. Neurochem.* **75,** 413–423.
Suarez, S. S. (1996). Hyperactivated motility in sperm. *J. Androl.* **17,** 331–335.
Sugiura, T., Kondo, S., Sukagawa, A., Nakane, S., Shinoda, A., Itoh, K., Yamashita, A., and Waku, K. (1995). 2-Arachidonoylglycerol: A possible endogenous cannabinoid receptor ligand in brain. *Biochem. Biophys. Res. Commun.* **215,** 89–97.
Sun, Y., and Bennett, A. (2007). Cannabinoids: A new group of agonists of PPARs. *PPAR Res.* **2007,** Article ID: 23513.
Tan, K. A., De, G. K., Atanassova, N., Walker, M., Sharpe, R. M., Saunders, P. T., Denolet, E., and Verhoeven, G. (2005). The role of androgens in Sertoli cell proliferation and functional maturation: Studies in mice with total or Sertoli cell-selective ablation of the androgen receptor. *Endocrinology* **146,** 2674–2683.
Turner, T. T. (1995). On the epididymis and its role in the development of the fertile ejaculate. *J. Androl.* **16,** 292–298.
Van der Stelt, M., and Di Marzo, V. (2004). Endovanilloids. Putative endogenous ligands of transient receptor potential vanilloid 1 channels. *Eur. J. Biochem.* **271,** 1827–1834.
Van der Stelt, M., Trevisani, M., Vellani, V., De Petrocellis, L., Schiano Moriello, A., Campi, B., McNaughton, P., Geppetti, P., and Di Marzo, V. (2005). Anandamide acts as an intracellular messenger amplifying Ca^{2+} influx via TRPV1 channels. *EMBO J.* **24,** 3026–3037.
Vescovi, P. P., Pedrazzoni, M., Nichelini, M., Maninetti, L., Bernardelli, F., and Passeri, M. (1992). Chronic effects of marihuana smoking on luteinizing hormone, follicle-stimulating hormone and prolactin levels in human males. *Drug Alcohol Depend.* **30,** 59–63.
Wang, H., Dey, S. K., and Maccarrone, M. (2006). Jekyll and Hyde: Two faces of cannabinoid signaling in male and female fertility. *Endocr. Rev.* **27,** 427–448.
Wenger, T., Rettori, V., Snyder, G. D., Dalterio, S., and McCann, S. M. (1987). Effects of delta-9-tetrahydrocannabinol on the hypothalamic-pituitary control of luteinizing hormone and follicle-stimulating hormone secretion in adult male rats. *Neuroendocrinology* **46,** 488–493.
Wenger, T., Fernández-Ruiz, J. J., and Ramos, J. A. (1999). Immunocytochemical demonstration of CB1 cannabinoid receptors in the anterior lobe of the pituitary gland. *J. Neuroendocrinol.* **11,** 873–878.
Wenger, T., Ledent, C., Csernus, V., and Gerendai, I. (2001). The central cannabinoid receptor inactivation suppresses endocrine reproductive functions. *Biochem. Biophys. Res. Commun.* **284,** 363–368.
Yamaguchi, F., Macrae, A. D., and Brenner, S. (1996). Molecular cloning of two cannabinoid type 1-like receptor genes from the puffer fish *Fugu rubripes*. *Genomics* **35,** 603–605.
Yanagimachi, R. (1970). The movement of golden hamster spermatozoa before and after capacitation. *J. Reprod. Fertil.* **23,** 193–196.
Yanagimachi, R. (1994). Fertility of mammalian spermatozoa: Its development and relativity. *Zygote* **2,** 371–372.
Yeh, S., Tsai, M. Y., Xu, Q., Mu, X. M., Lardy, H., Huang, K. E., Lin, H., Yeh, S. D., Altuwaijri, S., Zhou, X., Xing, L., Boyce, B. F., *et al.* (2002). Generation and characterization of androgen receptor knockout (ARKO) mice: An *in vivo* model for the study of androgen functions in selective tissues. *Proc. Natl. Acad. Sci. USA* **99,** 13498–13503.
Zhang, F. P., Poutanen, M., Wilbertz, J., and Huhtaniemi, I. (2001). Normal prenatal but arrested postnatal sexual development of luteinizing hormone receptor knockout (LuRKO) mice. *Mol. Endocrinol.* **15,** 172–183.
Zimmerman, A. M., Bruce, W. R., and Zimmerman, S. (1979). Effects of cannabinoids on sperm morphology. *Pharmacology* **18,** 143–148.

CHAPTER FIFTEEN

Anandamide and the Vanilloid Receptor (TRPV1)

Attila Tóth,[*] Peter M. Blumberg,[†] and Judit Boczán[‡]

Contents

I. Cannabinoid and Vanilloid Receptors		390
A. Endogenous cannabinoid/vanilloid receptor ligands		391
B. General overlap between the cannabinoid system and the vanilloid receptor		392
II. Biochemistry of Anandamide		392
A. Anandamide synthesis		392
B. Anandamide transport		394
C. Anandamide metabolism		395
III. Anandamide as Vanilloid Receptor (TRPV1) Ligand		396
A. Affinity, potency, and efficacy of anandamide on TRPV1		397
B. Regulation of TRPV1 responsiveness to anandamide		398
C. Anandamide activation of cannabinoid receptors affects TRPV1 responsiveness		399
D. TRPV1 activation regulates anandamide synthesis		401
E. Anandamide metabolism affects TRPV1 responses		402
F. Convergent physiological actions of anandamide and TRPV1 agonists do not necessarily represent direct effects on TRPV1 in both cases		403
G. Coactivation of the cannabinoid receptors and TRPV1 often complicates the distinction between these pathways		403
IV. Other Anandamide Receptors		404
V. Physiological Actions of Anandamide on TRPV1		404
A. Central nervous system (CNS)		404
B. Analgesia		406
C. Peripheral nervous system		407
D. Vascular effects		408

[*] Division of Clinical Physiology, Institute of Cardiology, University of Debrecen, Debrecen, Hungary
[†] Molecular Mechanisms of Tumor Promotion Section, Laboratory of Cancer Biology and Genetics, National Cancer Institute, National Institutes of Health, Bethesda, Maryland, USA
[‡] Department of Neurology, University of Debrecen, Debrecen, Hungary

E. Temperature regulation 409
F. Body weight/fat regulation 410
VI. Future Directions 410
References 411

Abstract

Arachidonylethanolamide (anandamide) was identified some 15 years ago as a brain constituent that binds to the cannabinoid receptor. After this seminal discovery, multiple new receptors for anandamide have been identified, including the vanilloid receptor (TRPV1), and anandamide is now frequently referred as an "endovanilloid." Characterization of the action of anandamide on TRPV1 revealed that (1) the potency and efficacy of anandamide on TRPV1 very much depend on the species and tissue, (2) anandamide responsiveness *in vivo* is significantly controlled by its local metabolism, (3) anandamide activation of cannabinoid receptors regulates TRPV1 responsiveness, (4) TRPV1 activation regulates anandamide synthesis, (5) anandamide metabolites affect TRPV1 responses, (6) the often observed convergent physiological actions of anandamide and TRPV1 agonists in neither case necessarily represent direct effects on TRPV1, and (7) coactivation of the cannabinoid receptors and TRPV1 often complicates the distinction between these pathways. These issues are reviewed here together with the potential implications for the pathophysiological and pharmacological regulation of inflammatory, respiratory, and cardiovascular disorders, as well as of appetite and fat metabolism.

I. Cannabinoid and Vanilloid Receptors

There are two cloned and well-characterized cannabinoid receptors: cannabinoid receptors 1 (Matsuda *et al.*, 1990) and 2 (Munro *et al.*, 1993). Both cannabinoid receptors are seven transmembrane domain containing receptors, belonging to the G-protein-coupled receptor family. Their activation predominantly leads to activation of heterotrimeric G-proteins of the $G_{i/o}$ family (Howlett *et al.*, 1986; Matsuda *et al.*, 1990). As a consequence, agonist binding to the cannabinoid receptors results in inhibition of adenylyl cyclase and in decreased protein kinase A (PKA) activity, among other responses. Interestingly, stimulation of cannabinoid receptor 1 with different agonists results in activation of different members of the $G_{i/o}$ family.

The vanilloid receptor 1 (TRPV1) is a polymodal integrator of painful stimuli (Caterina *et al.*, 2000; Tominaga *et al.*, 1998). It is a nonspecific cation channel with some selectivity for Ca^{2+} and is expressed predominantly in the sensory neurons (Tominaga *et al.*, 1998). Knockout models of TRPV1 provided evidence for its role in mediating various forms of pain and gave a boost to a robust pharmaceutical effort to develop a new class of pain killers antagonizing TRPV1. However, in recent years, TRPV1 expression has been detected in tissues other than sensory neurons, such as neurons and astrocytes of the central nervous system (CNS) (Cristino et al., 2006; Mezey et al., 2000; Tóth et al., 2005a), vascular smooth muscle cells (Kark et al., 2008; Yang et al., 2006), and urothelium (Birder et al., 2002). Moreover, there is accumulating evidence that these receptor populations have important physiological functions: TRPV1 located in the brain may regulate motor functions (Starowicz et al., 2008), TRPV1 in smooth muscle mediates vasoconstriction (Kark et al., 2008), and TRPV1 in the bladder urothelium is under clinical trials as a target to regulate bladder activity (Birder, 2007).

A. Endogenous cannabinoid/vanilloid receptor ligands

After identification of the cannabinoid receptors, intense effort was made to identify their endogenous ligands. Arachidonylethanolamide (anandamide) was first identified in porcine brain (Devane et al., 1992). It was followed by the identification of additional potential endogenous regulators. These include 2-arachidonyl glycerol (2-AG) (Mechoulam et al., 1995), 2-arachidonyl glyceryl ether (noladin ether) (Hanus et al., 2001), virodhamine (O-arachidonoyl ethanolamine) (Porter et al., 2002), and N-arachidonyl-dopamine (NADA) (Bisogno et al., 2000), among others (Alexander and Kendall, 2007). While anandamide and 2-AG are equipotent on cannabinoid receptors 1 and 2, NADA and noladin ether are rather selective for the cannabinoid receptor 1 and virodhamine is selective for the cannabinoid receptor 2. Interestingly, anandamide and virodhamine are both arachidonyl ethanolamides, but anandamide is an agonist and virodhamine is an antagonist of the cannabinoid receptors. Importantly, these endocannabinoids have potencies on the cannabinoid receptors in the nanomolar range.

It has been proposed that lipid-based molecules and, in particular, anandamide-related structures may have activity at TRPV1 (Beltramo and Piomelli, 1999; Di Marzo et al., 1998; Szolcsanyi and Jancso-Gabor, 1975), suggesting an interplay between the cannabinoid and vanilloid systems. The first evidence regarding the activity of the endocannabinoids at TRPV1 was the identification of anandamide as a TRPV1 ligand (Zygmunt et al., 1999). NADA was also found to be a full agonist at TRPV1 (Huang et al., 2002; Tóth et al., 2003). In contrast, noladin ether has partial activity (36.5 ± 3.2%) at 10 μM on exogenously expressed TRPV1 and possessed cannabinoid receptor independent vasodilatative effects (Duncan et al., 2004), 2-AG is probably inactive at TRPV1, and no data are available on the effect of virodhamine on TRPV1.

B. General overlap between the cannabinoid system and the vanilloid receptor

The possible interplay between anandamide-like structures and TRPV1 was investigated many years before the identification of the receptors responsible for cannabinoid or capsaicin binding. Ethanolamides with long alkyl chains were suggested as early as 1975 by Szolcsányi and Jancsó to be TRPV1 (called capsaicin receptor at this time) ligands (Szolcsanyi and Jancso-Gabor, 1975). Some 13 years later, the chemical and physiological similarities between anandamide and some synthetic agonists of vanilloid receptors such as olvanil (N-(3-methoxy-4-hydroxy)-benzyl-cis-9-octadecenoamide) prompted a detailed investigation of the effects of these compounds on TRPV1 and cannabinoid receptors by Di Marzo and coworkers. Their findings suggested that some of the analgesic actions of olvanil may be due to its interactions with the endogenous cannabinoid system (Di Marzo et al., 1998). In subsequent work, they investigated the effects of changing the length and degree of unsaturation of the fatty acyl chain of olvanil on its ability to inhibit facilitated transport of anandamide into cells as well as anandamide metabolism, on its binding to CB1 and CB2 cannabinoid receptors, and on its activation of TRPV1. They found that N-acyl-vanillyl-amides behave as "hybrid" activators of cannabinoid and vanilloid receptors (Melck et al., 1999). In the same year, Beltramo and Piomelli investigated the effects of olvanil on anandamide transport, motivated by the structural similarity of olvanil to the anandamide transport inhibitor N-(4-hydroxyphenyl)-arachidonylamide (AM404). They found that olvanil reduce anandamide clearance at concentrations similar to those needed for vanilloid receptor activation (Beltramo and Piomelli, 1999). All of these data suggested interplay between the cannabinoid system and the vanilloid receptor.

II. Biochemistry of Anandamide

A. Anandamide synthesis

Although anandamide may be produced as a simple condensation product of arachidonic acid and ethanolamine, its *in vivo* synthesis seems to occur from the hydrolysis of N-acylphosphatidylethanolamine (NAPE) (Di Marzo et al., 1994) by a specific phospholipase D (NAPE-PLD) (Okamoto et al., 2004). In addition, two other synthetic pathways were identified recently. One involves the sequential deacylation of NAPE by α,β-hydrolase 4 (Abhd4) and the subsequent cleavage of glycerophosphate to yield anandamide (Simon and Cravatt, 2006); and the other one proceeds through phospholipase C-mediated hydrolysis of NAPE to yield phosphoanandamide, which is then dephosphorylated by phosphatases, including the tyrosine phosphatase PTPN22 and the inositol 5′ phosphatase SHIP1, reviewed in Liu *et al.* (2008). It needs to be noted that all three pathways are metabolizing NAPE and that NAPE is only a minor membrane component; therefore, NAPE synthesis may well be the limiting step in anandamide synthesis (Fig. 15.1). NAPE synthesis occurs through the transfer of

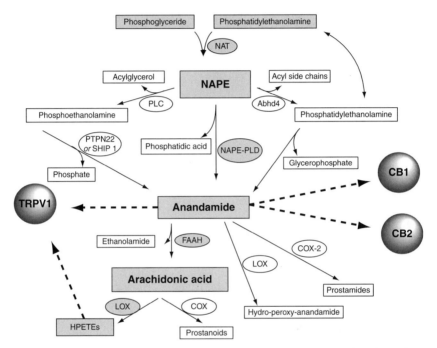

Figure 15.1 Anandamide metabolism. The rate-limiting step in anandamide synthesis is the Ca^{2+}-dependent synthesis of *N*-arachidonylethanolamide (NAPE) from phosphatidylethanolamine and phosphoglyceride catalyzed by *N*-acyltransferase (NAT). NAPE may be transformed to anandamide by three different pathways. The most prominent is catalyzed by a NAPE selective phospholipase D (NAPE-PLD). Alternatively, NAPE may be metabolized to phosphoethanolamide by phospholipase C (PLC) and subsequently to anandamide by tyrosine phosphatase PTPN2 or by inositol phosphatase SHIP1. In addition, NAPE may be transformed to phosphatidylethanolamine by α,β-hydrolase 4 (Abhd4) prior to the metabolism to anandamide. The synthesized anandamide can activate many receptors, including cannabinoid receptors (CB1 and CB2) and the vanilloid receptor 1 (TRPV1). Anandamide breakdown is predominantly catalyzed by the fatty acid amide hydrolase (FAAH) to produce arachidonic acid, which in turn leads to the formation of hydroperoxyeicosatetraenoic acids (HPETE) and leukotrienes (catalyzed by lipoxygenase, LOX) or alternatively to the formation of prostanoids (catalyzed by cyclooxygenase, COX). Important to note that some of the HPETEs are activators of TRPV1. Finally, anandamide itself may also be metabolized by LOX and COX enzymes, to yield hydroperoxy-anandamide and prostamides, respectively. Please note the different representations on the figures: small biomolecules (rectangular), enzymes (oval), and receptors (globe). In addition, the most important items are highlighted by gray background.

arachidonic acid from the sn-1 position of 1,2-sn-diarachidonylphosphatidylcholine to phosphatidylethanolamine catalyzed by a Ca^{2+}-dependent N-acyltransferase (NAT) (Reddy et al., 1983). Unfortunately, the cloning and characterization of NAT still needs to be done.

Although NAPE-PLD knockout mice showed decreased Ca^{2+}-dependent conversion of NAPEs to N-acylethanolamines, the anandamide level was not different from the wild-type animals in the brain, suggesting the importance of the additional pathways for the *in vivo* anandamide synthesis (Leung et al., 2006).

B. Anandamide transport

As a putative neurotransmitter (e.g., in the brain) anandamide has many unusual features. One of them is that it is not stored in synaptic vesicles. Instead it is synthesized upon stimulation and is immediately released. The most important mechanism for the inactivation of anandamide is (re)uptake and fast metabolism to arachidonic acid and ethanolamine.

Currently, there is a debate about the mechanism(s) by which anandamide crosses the cell membrane. One group of researchers interprets their findings in terms of a putative membrane transporter (anandamide membrane transporter, AMT), while another group of researchers believes that this hypothetical transporter actually reflects the behavior of the anandamide metabolizing enzyme fatty acid amide hydrolase (FAAH) as detailed below. This significance of this issue is that, if anandamide requires a transport protein to penetrate the cell membrane, then this transport protein may be of pharmacological importance in regulating anandamide effects *in vivo*.

The first relatively potent molecule able to inhibit the hypothetical AMT was olvanil (N-vanillyl-cis-9-octadecenoamide) (Di Marzo et al., 1998). It was also described that V_{max} of AMT was about an order of magnitude lower than that for the metabolizing enzyme FAAH and the AMT was activated twofold by nitric monoxide (Maccarrone et al., 1998). Soon, additional AMT inhibitors were reported (De Petrocellis et al., 2000; Melck et al., 1999; Piomelli et al., 1999) and it was postulated that the actions of anandamide at TRPV1 required facilitated transport and were limited by intracellular metabolism (De Petrocellis et al., 2001a). However, it has been shown that modulation of FAAH activity affects anandamide uptake (Day et al., 2001), suggesting that FAAH dependent metabolism creates and maintains an inward concentration gradient for anandamide contributing to anandamide uptake. Moreover, the earlier identified "AMT specific molecules" were ineffective at inhibiting short term anandamide uptake and were found to be inhibitors of FAAH. In addition, anandamide uptake was not saturable at short incubation times, in contrast to its metabolizing enzymes (Glaser et al., 2003). Additional data suggested that anandamide spontaneously diffuses into cells and its diffusion is significantly affected by FAAH activity (Kaczocha et al., 2006). These data suggested that anandamide uptake is mediated by the regulation of its concentration gradient. Later, additional specific inhibitors of AMT were synthesized (Ligresti et al., 2004; Lopez-Rodriguez et al., 2001, 2003a,b; Ortar et al., 2003) and, again, it was claimed that they acted through the inhibition of FAAH (Alexander and Cravatt, 2006; Dickason-Chesterfield et al., 2006; Kaczocha et al., 2006).

Independent of whether there is facilitated anandamide transport or not, it seems clear that FAAH inhibitors are able to increase the efficacy and potency of anandamide, suggesting a predominant role of metabolism as compared to transport. It means that intracellular anandamide concentration is limited; therefore binding of anandamide to intracellular sites may be regulated by AMT/FAAH modulation. Finally, if AMT or FAAH plays a role in the availability of anandamide following its synthesis, then anandamide release is probably regulated/initiated by local activation mechanisms (Adermark and Lovinger, 2007).

C. Anandamide metabolism

1. Anandamide breakdown by fatty acid amide hydrolase (FAAH)

The primary enzyme responsible for anandamide metabolism is the FAAH (Fig. 15.1), which was cloned in 1996 (Cravatt et al., 1996). FAAH knockout mice were also generated (Cravatt et al., 2001). FAAH knockout mice had attenuated anandamide degradation and exhibited an array of intense CB(1)-dependent behavioral responses, including hypomotility, analgesia, catalepsy, and hypothermia upon anandamide treatment. Moreover, a 15-fold higher concentration of anandamide was measured in the brain, and there was reduced pain sensation which was reversed by the CB1 antagonist SR141716A, suggesting a pivotal role of FAAH in regulating anandamide bioavailability. However, most of the classical cannabinoid responses (like hypomotility, lower body weight, hypothermia) were not induced by the elevated levels of the endogenous anandamide in the FAAH knockout mice. These latter data suggest that some cannabinoid responses cannot be evoked by the increased anandamide level; therefore these responses are not mediated by anandamide *in vivo* or, alternatively, they are controlled by localized anandamide synthesis [(FAAH inhibition alone elevates anandamide levels where significant steady state (unstimulated) synthesis occurs]).

2. Anandamide metabolism by cyclooxigenase-2 (COX-2)

Anandamide is metabolized not only by FAAH but also by human recombinant cyclooxygenase-2 (COX-2) (Fig. 15.1), while COX-1 is less effective (Yu et al., 1997). Cyclooxygenation of anandamide (Fig. 15.1) results in the formation of prostaglandin (PG)D_2-, PGE_2-, and PGF_2-ethanolamides, and there is evidence that PGH2-ethanolamide is formed as an intermediate in the production of PGF_2-EA in cultured cells treated with anandamide (Yang et al., 2005). The question remains regarding the importance of this pathway under in vivo conditions. It is important to note that prostaglandin production is usually determined by methods (ELISA) relying on the immunological recognition of a cyclooxygenated acyl side chain, which is the same in anandamide and in arachidonic acid metabolites. For example, amnion-derived cells responded to both interleukin-1 and anandamide stimulation by apparent PGE_2 production, but in the case of anandamide the cells produced PGE_2-EA rather than PGE_2 (Glass et al., 2005). Prostaglandin ethanolamides are able to evoke prostaglandin-like effects, having only two- to fivefold lower potencies than the respective protaglandins (Matias et al., 2004). Moreover, they may act on prostaglandin ethanolamide selective receptor(s) or on prostaglandin receptors with different pharmacology (Woodward et al., 2007).

III. Anandamide as Vanilloid Receptor (TRPV1) Ligand

It was a breakthrough for both the endocannabinoid and vanilloid fields when the endocannabinoid anandamide itself was shown to be an activator of TRPV1 in vitro and the vascular effects of anandamide were shown to be mediated by TRPV1 in vivo (Zygmunt et al., 1999). Following these seminal findings, the question of anandamide effectiveness on TRPV1 became the subject of heated debate. Zygmunt et al. (2000b) heralded a new era of vanilloid research by the identification of anandamide as the first endovanilloid, Szolcsányi (2000) expressed skepticism, while Smart and Jerman (2000) predicted that this question "will continue to challenge pharmacologists for years to come." Indeed, almost a decade later the role of anandamide as an endovanilloid is still under debate.

Some of the questions have been answered, however. It now seems clear that (1) anandamide metabolism plays an important role in the regulation of in vivo anandamide concentrations; inhibition of anandamide breakdown may result in high-enough anandamide concentrations to activate TRPV1, (2) TRPV1 sensitivity to anandamide is enhanced by "entourage" effects of other endogenous lipids or by phosphorylation of TRPV1, and (3) anandamide may be synthesized and released in significant quantities, potentially sufficient to activate TRPV1 (0.5–10 μM). In spite of its relatively low potency, anandamide may thus be present in sufficient quantities to activate TRPV1.

However, it has also became clear that (1) anandamide effects on TRPV1 are greatly dependent on the cell types and their environment.

Anandamide may activate or inactivate TRPV1 depending on the conditions; (2) selective activation of cannabinoid and TRPV1 receptors often lead to the same effects, which makes it hard to pinpoint the identity of the actual anandamide target which is engaged; (3) activation of cannabinoid receptors regulates heat, pH, and anandamide responsiveness of TRPV1; (4) on the other hand, activation of TRPV1 results in increased anandamide synthesis; (5) anandamide metabolites may also regulate TRPV1; (6) CB1 and TRPV1 are often expressed in the same cell; and (7) these interlocking relationships suggest that even knockout models of an individual element (e.g., CB1 or TRPV1) of these systems may not give decisive information about the physiological role of anandamide in the regulation of TRPV1 responsiveness.

A. Affinity, potency, and efficacy of anandamide on TRPV1

First of all, anandamide is able to inhibit the binding of tritiated resiniferatoxin to the vanilloid binding site of TRPV1 (De Petrocellis et al., 2001b; Ross et al., 2001; Tóth et al., 2003), suggesting a direct interaction between anandamide and TRPV1. Formally, however, this competition might result from allosteric regulation (in this case the anandamide and vanilloid binding sites would be different) or by the disruption of hypothetical interaction(s) of TRPV1 with proteins or lipids (in this case anandamide would disrupt an interaction necessary for resiniferatoxin binding). Indeed, there are data that capsaicin analogs may bind differently to the plasma membrane and to the intracellular membrane located TRPV1, suggesting a role of microenvironment in the interactions with vanilloids (Tóth et al., 2004).

The affinity of anandamide for TRPV1 binding seems to be similar (De Petrocellis et al., 2001b; Ross et al., 2001) or about fivefold weaker than that for capsaicin (Tóth et al., 2003). Since capsaicin is a well-characterized activator of TRPV1, anandamide may well be an endogenous ligand. However, the potency of anandamide [usually in the range of 0.5–10 μM, summarized in Ross (2003)] on TRPV1 in functional assays is significantly lower than that for capsaicin (10–100 nM), suggesting lower intrinsic efficiency. An important consequence of these data is that, to activate TRPV1, anandamide must be present in relatively high concentrations (at least 10 times higher than that necessary to activate cannabinoid receptor 1).

In general, efficacy of anandamide on TRPV1 appears to be tissue, expression system, and species dependent (Ross, 2003). Accordingly, anandamide is a partial activator of TRPV1 when the receptor expression is low, while it is a full agonist when receptor expression is high. In addition, the intrinsic efficiency of anandamide is apparently regulated.

B. Regulation of TRPV1 responsiveness to anandamide

In general, the *in vivo* activity of TRPV1 appears to be dependent on its sensitization state. There are numerous mechanisms identified which regulate TRPV1 activity. One of the most studied pathways is the phosphorylation of TRPV1, which affects its sensitivity to vanilliods and to anandamide (Fig. 15.2). TRPV1 is phosphorylated on serine and threonine residues by protein kinase C (PKC) (Bhave *et al.*, 2003; Premkumar and Ahern, 2000); PKA (Bhave *et al.*, 2002), protein kinase D (Wang *et al.*, 2004), and calcium-calmodulin dependent protein kinase II (Jung *et al.*, 2004). Protein phosphatase 2B (calcineurin) was found to be important for dephosphorylation, leading to receptor desensitization (Cortright and Szallasi, 2004; Docherty *et al.*, 1996; Lizanecz *et al.*, 2006; Mohapatra and Nau, 2005). In addition, phosphorylation of TRPV1 on tyrosine has also been reported (Jin *et al.*, 2004). A further regulatory mechanism for TRPV1 is through interaction with phosphatidylinositol-4,5-bisphosphate (Chuang *et al.*, 2001; Lukacs *et al.*, 2007). Finally, other lipids may also regulate TRPV1 (De Petrocellis *et al.*, 2001b, 2004; Smart *et al.*, 2002; Vandevoorde *et al.*, 2003).

Activation of PKC was found to sensitize TRPV1 to anandamide (Bhave *et al.*, 2003; Premkumar and Ahern, 2000), probably through phosphorylation of serines 502 and 800 (Bhave *et al.*, 2003) leading to increased gating probability of the channel (Vellani *et al.*, 2001).

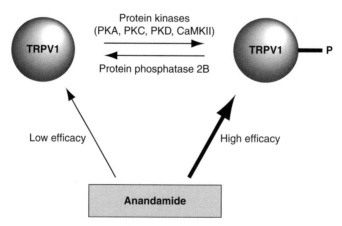

Figure 15.2 Sensitization of TRPV1 to anandamide. The vanilloid receptor 1 (TRPV1) may be phosphorylated by various kinases (e.g., by protein kinase C, PKC; by protein kinase A, PKA; and by calcium-calmodulin dependent protein kinase II, CaMKII) on critical serine/threonine residues. These phosphorylations sensitize TRPV1 to anandamide, resulting in an apparent increase in the efficacy of anandamide and other vanilloids on TRPV1. In contrast, phosphorylated TRPV1 may be dephosphorylated by the Ca^{2+}-dependent protein phosphatase 2B (calcineurin), resulting in decreased sensitivity and efficacy.

The potentiation of anandamide effects caused by PKC activation was accompanied by a significant increase in CGRP release induced by anandamide (Orliac *et al.*, 2007). PKA also sensitizes TRPV1 to anandamide, probably through the phosphorylation of serine 116 (Bhave *et al.*, 2002). A decisive role of PKC and protein phosphatase 2B (calcineurin) was found in the regulation of anandamide responsiveness of TRPV1 (Lizanecz *et al.*, 2006). These latter findings have suggested the following model. According to this model, TRPV1 is phosphorylated under resting conditions. Upon activation by anandamide, Ca^{2+} flows through the channel and activates the Ca^{2+}-dependent protein phosphatase 2B. The activated phosphatase then dephosphorylates the TRPV1, making it insensitive to anandamide and other vanilloids (acute desensitization) (Lizanecz *et al.*, 2006). Closure of the channel may result in restored (low) intracellular Ca^{2+} concentrations, inactivation of protein phosphatase 2B, and in the subsequent phosphorylation of TRPV1.

An additional regulatory pathway was suggested in the case of inflammation. Pretreatment with CCL3 (a proinflammatory chemokine) enhanced the response of DRG neurons to capsaicin or anandamide. This sensitization was inhibited by pertussis toxin, U73122, or chelerythrine chloride, inhibitors of Gi-protein, phospholipase C, and PKC, respectively (Zhang *et al.*, 2005).

The first lipid with an "entourage" effect on TRPV1 was the palmitoylethanolamide, which enhanced the effect of anandamide (De Petrocellis *et al.*, 2001b). N-morpholino- and N-diethyl-analogs of palmitoylethanolamide also increased the sensitivity of TRPV1 to activation by anandamide without affecting fatty acid amidohydrolase activity (Vandevoorde *et al.*, 2003). In addition, N-palmitoyl- and N-stearoyl-dopamine may also play an "entourage" role (De Petrocellis *et al.*, 2004).

In conclusion, the host cell expression system modulates the regulation of TRPV1 receptor activity and suggests that anandamide activation of native human TRPV1 receptors *in vivo* is dependent on cell-specific regulatory factors/pathways (Bianchi *et al.*, 2006). Because TRPV1 is now thought to be present in numerous cell types other than sensory neurons and these various cell types will have different levels of expression of different regulatory factors for TRPV1, the clear expectation is that TRPV1 behavior and pharmacology will be different in these different contexts.

C. Anandamide activation of cannabinoid receptors affects TRPV1 responsiveness

There are many potential mechanisms by which cannabinoid receptor activation may regulate TRPV1 responsiveness (Fig. 15.3). The best characterized ones include (1) phospholipase C (PLC)-mediated metabolism of

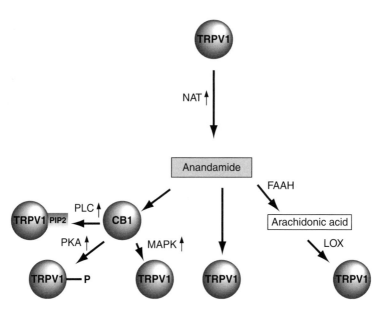

Figure 15.3 Anandamide-mediated activation of TRPV1. There is a complex crosstalk between cannabinoid receptor 1 (CB1) and the vanilloid receptor 1 (TRPV1). First, activation of TRPV1 results in an increase of intracellular Ca^{2+} concentrations, which may activate the Ca^{2+}-dependent N-acyltransferase (NAT). NAT controls the rate-limiting step in anandamide synthesis, consequently, TRPV1 activation may result in increased anandamide formation. On the other hand, anandamide may activate TRPV1 either directly or indirectly. In the case of direct activation, anandamide binds to and activates TRPV1 as mentioned in Fig. 15.2. Direct activation apparently occurs when anandamide concentration is high (concentration is in the low micromolar range). On the other hand, a low concentration of anandamide may selectively activate cannabinoid receptor 1 (CB1), which in turn may activate three independent pathways: (1) activation of phospholipase C (PLC) resulting in increased synthesis of phosphatidylinositol 4,5-bisphosphate (PIP2), which regulates TRPV1 activity; (2) activation of protein kinase A (PKA) which phosphorylates and sensitizes TRPV1 to anandamide and also to other vanilloids; and (3) activation of mitogen activated protein kinases (MAPK, in particular ERK, c-Jun, JNK, and p38) again sensitizing TRPV1. Finally, anandamide breakdown catalyzed by fatty acid amide hydrolase (FAAH) produces arachidonic acid, which is transformed to hydroperoxyeicosatetraenoic acids by lipoxygenase (LOX). These latter molecules are activators of TRPV1.

inhibitory phosphatidylinositol 4,5-bisphosphate (PIP2); (2) activation of MAP kinases such as ERK, c-Jun, JNK, and p38; and (3) activation/inhibition of PKA.

The stimulatory role of bradykinin and NGF on TRPV1 activity has been attributed to the activation of PLC and the subsequent decrease of membrane PIP2 level (Chuang et al., 2001). The release of diacylglycerol from PIP2 may activate PKC resulting in the sensitization of TRPV1 to other stimuli, like vanilloids, anandamide itself, or heat. In addition, PIP2 itself emerged as an endogenous regulator of TRPV1. Although originally it has been reported that

PIP2 inhibits (Chuang et al., 2001) TRPV1, later works suggested that PIP2 may rather activate TRPV1 (Lukacs et al., 2007) which effect is probably mediated by the TRPV1 regulatory protein pirt (Kim et al., 2008). These controversial findings may be explained by differences in the availability of TRPV1 regulatory proteins: when PLC stimulation sensitizes TRPV1 the effect is probably mediated by the activation of PKC, while in other cases PLC-mediated depletion of PIP2 may inhibit TRPV1 through pirt.

CB1 receptor stimulation results in JNK, p38 MAP kinase (Bron et al., 2003), and ERK (Bouaboula et al., 1995) activation. These MAP kinases were also implicated in the stimulatory role of NGF on the TRPV1 expression/responsiveness (Bron et al., 2003; Ji et al., 2002).

In contrast to these effects of the cannabinoid receptor on TRPV1 expression/availability, PKA-mediated direct phosphorylation of TRPV1 in response to cannabinoid receptor stimulation was implicated in the regulation of TRPV1 sensitivity (Bhave et al., 2002; De Petrocellis et al., 2001c). The most prominent and best characterized effect of cannabinoid receptor stimulation is the activation of heterotrimeric G-proteins of the $G_{i/o}$ family (Howlett et al., 1986; Matsuda et al., 1990), which in turn inhibits adenylyl cyclase and results in decreased PKA activity. However, the functional effects of CB1 receptor stimulation on TRPV1 responsiveness remain controversial. Stimulation of CB1 in cells heterologously expressing both CB1 and TRPV1 resulted in an elevated TRPV1-mediated increase in intracellular Ca^{2+} concentrations, while CB1 receptor activation had an inhibitory effect when the cells were prestimulated with forskolin (Hermann et al., 2003). This observation highlights a dual role of the cannabinoid receptors in the regulation of TRPV1 functions: at low endogenous PKA activity they stimulate TRPV1 responses, while at elevated endogenous PKA activity they inhibit the PKA-mediated increase in TRPV1 responses.

In conclusion, the G-protein-coupled cannabinoid receptors interfere both with the expression/availability and also with the sensitization/desensitization of TRPV1 *in vivo*. This may impact all cells where cannabinoid receptor(s) and TRPV1 are coexpressed. The effect of cannabinoid receptor stimulation on TRPV1 is determined by the physiological circumstances, ranging from inhibition to activation. Finally, cannabinoid receptor stimulation not only regulates TRPV1 responsiveness to anandamide but also regulates its sensitivity to other still not identified endogenous ligands, to heat and to acid. Therefore, anandamide activation of endogenous TRPV1 may be an indirect, regulatory/modulatory effect.

D. TRPV1 activation regulates anandamide synthesis

As pointed out above, anandamide synthesis is limited by a Ca^{2+}-dependent (NAT) (Reddy et al., 1983), responsible for the synthesis of the anandamide precursor NAPE. Elevation of intracellular Ca^{2+} concentrations results in increased NAPE synthesis (Cadas et al., 1996). Moreover, activation of TPRV1 in capsaicin-sensitive primary sensory neurons (Ahluwalia et al., 2003b) or in TRPV1 expressing HEK cells (Di Marzo et al., 2001) evokes anandamide production and release. Taking into account that TRPV1 stimulation increases intracellular Ca^{2+} concentrations, it is possible that activation of TRPV1 mediates anandamide synthesis. Moreover, the newly formed anandamide may further activate TRPV1 to further increase the intracellular Ca^{2+} (Van der Stelt et al., 2005). Nevertheless, this Ca^{2+}-dependent NAT has not been cloned yet and other NAPE producing, Ca^{2+} independent enzyme have also been identified recently (Jin et al., 2007).

Taken together, TRPV1 may have a modulatory role in anandamide formation in vivo, which can provide a positive feedback mechanism in the physiological regulation of TRPV1 responses.

E. Anandamide metabolism affects TRPV1 responses

As has been mentioned, anandamide binding sites for both TRPV1 and cannabinoid receptors appear to be intracellular and the anandamide concentration is determined by its uptake (passive or AMT facilitated) and metabolism (catalyzed by FAAH). These factors are of particular importance in the stimulation of TRPV1, which requires substantial local (intracellular) concentrations of anandamide (0.5–10 μM). Indeed, inhibition of FAAH activity significantly increases intracellular anandamide concentrations and also anandamide efficacy at TRPV1.

In particular, blockade of anandamide cellular efflux/metabolism led to enhanced TRPV1-mediated currents upon stimulation of metabotropic receptors. These data suggested that metabotropic receptors are coupled to the phospholipase C/inositol 1,4,5-triphosphate pathway and then signal to TRPV1 channels (Van der Stelt et al., 2005). This behavior is not limited to dorsal root ganglion cells. In the case of endothelial cells, the presence of FAAH limits the vasodilator action of anandamide in rat small mesenteric arteries (Ho and Randall, 2007).

Anandamide breakdown not only limits anandamide availability but also generates arachidonic acid and ethanolamine. Importantly, arachidonic acid formed by anandamide breakdown may be the precursor of a set of molecules, in particular, lipoxygenase (LOX) products (Hwang et al., 2000), which may activate TRPV1 (Fig. 15.1). Indeed, the anandamide-mediated contractile action in the isolated bronchus of the guinea pig was inhibited by LOX inhibitors suggesting that the anandamide effects may be due, at least in part, to lipoxygenase metabolites of this fatty acid amide that are vanilloid receptor agonists (Craib et al., 2001).

In conclusion, anandamide metabolism has dual effects on TRPV1. On the one hand, intracellular anandamide concentration is a function of its metabolism, since the efficiency of the transport mechanisms (passive or facilitated by AMT) is not much higher than the efficiency of the enzyme(s) responsible for its breakdown (FAAH). This suggests a therapeutic rationale for FAAH inhibitors in the modulation of cannabinoid/TRPV1 responses *in vivo*. On the other hand, the FAAH-mediated breakdown of anandamide produces arachidonic acid in high concentrations, which may activate a local arachidonic acid pathway (Fig. 15.1). One of the enzymes of particular significance is LOX, which may metabolize arachidonic acid to TRPV1 activating ligands like 12-HPETE. It suggests that anandamide released from one cell may diffuse to another, where it is internalized and transformed to lipoxygenase products via arachidonic acid, leading to TRPV1 activation. It is important to note that this hypothetical pathway does not require any cannabinoid receptors. Moreover, the synthesized lipoxygenase products may be released to activate neighboring TRPV1 expressing cells.

F. Convergent physiological actions of anandamide and TRPV1 agonists do not necessarily represent direct effects on TRPV1 in both cases

Similar physiological effects of anandamide and TRPV1 agonists do not necessarily prove that anandamide is an endovanilloid. As mentioned above, anandamide may sensitize TRPV1 to other endogenous regulators (including yet unidentified ligands or body temperature) or alternatively anandamide may be transformed to other TRPV1 activating substances. In these cases the anandamide-evoked responses may be specifically antagonized by TRPV1 inhibitors, even though anandamide has no direct effect on TRPV1.

G. Coactivation of the cannabinoid receptors and TRPV1 often complicates the distinction between these pathways

As has been mentioned, cannabinoid receptor activation regulates TRPV1 activity and, conversely, TRPV1 stimulation regulates anandamide synthesis. These findings suggest that there is a complex interplay between cannabinoid and vanilloid systems *in vivo*, which complicates the evaluation of the actual effects of endocannabinoids on TRPV1. For example, while anandamide seems to be a ligand for TRPV1 *in vitro*, it does not necessarily directly activate TRPV1 under *in vivo* conditions.

Knockout models are only an imperfect tool to dissect such questions since deletion of a single gene may result in the activation of compensatory mechanisms. For example, while TRPV1 knockout mice have normal thermoregulation (Caterina *et al.*, 2000), acute treatment with various TRPV1 antagonists results in hyperthermia (Gavva *et al.*, 2007, 2008).

IV. Other Anandamide Receptors

An important observation was that anandamide induces cell death in numerous cell lines. This effect was independent of cannabinoid or vanilloid receptors but was antagonized by methyl-β cyclodextrin (MCD), a compound which depletes membrane cholesterol. In addition, MCD also blocked anandamide-induced superoxide generation, phosphatidylserine exposure on the cell surface and p38 MAPK/JNK activation. Thus, anandamide may play a novel role in the modulation of responses associated with membrane lipid rafts (Sarker and Maruyama, 2003). Further direct targets for anandamide were also identified in some detail. These include Na^+ channels (inhibition, $K_d = 5$–40 μM) (Kim et al., 2005a), T-type Ca^{2+} channels (inhibition, $K_d = 0.3$–4 μM) (Chemin et al., 2001), peroxisome proliferator activated receptor γ (activation, EC_{50} ~5 μM) (Bouaboula et al., 2005) and α (Sun et al., 2007), and glycine receptors (inhibition, $EC_{50} < 1$ μM) (Lozovaya et al., 2005).

V. Physiological Actions of Anandamide on TRPV1

There is a vast literature for both TRPV1- and anandamide-mediated physiological effects reviewed separately in this book. Here in this chapter we concentrate on those findings for which anandamide effects were mediated by TRPV1.

A. Central nervous system (CNS)

TRPV1 was first identified as a sensory neuron specific integrator of painful stimuli (Tominaga et al., 1998). However, TRPV1 mRNA and protein have also been detected in the rat brain (Mezey et al., 2000). TRPV1 was expressed in neurons of various regions of the rat brain, both post- and presynaptically (Tóth et al., 2005a). Moreover, TRPV1 expression was not restricted to neurons, but was also present in pericytes and astrocytes (Tóth et al., 2005a). In addition, coexpression of CB1 and TRPV1 was found in the brain and the earlier findings were confirmed by knockout animals (Cristino et al., 2006). These results suggested a possible role for TRPV1 in mediating anandamide effects in the CNS.

Evidence is accumulating that anandamide functionally activates TRPV1 in the CNS. The first results suggesting functional vanilloid receptor-mediated anandamide effects were in the hippocampus where anandamide increased paired-pulse depression of population spikes, which was mimicked by TRPV1 activators and inhibited by TRPV1 inhibitors (Al Hayani et al., 2001). In the same year, Di Marzo et al. (2001) suggested that vanilloid receptor stimulation led to anandamide formation, probably contributing to the decreased motor behavior upon TRPV1 stimulation. Indeed, application of a high dose of anandamide (30 μM) activated presynaptic TRPV1 and enhanced the release of glutamate within the superficial medullary dorsal horn (Jennings et al., 2003). It was suggested that a tonic facilitation of glutamate release in the substantia nigra pars compacta may be maintained by stimulation of TRPV1 by endovanilloids, including anandamide, probably contributing to the motor and cognitive functions involving the dopaminergic system (Marinelli et al., 2003). The administration of the endocannabinoid anandamide to rats produced motor depression in the open-field test, presumably related to a decrease in nigrostriatal dopaminergic activity. These effects were completely reversed by the vanilloid-like receptor antagonist capsazepine, implicating TRPV1 in the hypokinetic effects of anandamide (de Lago et al., 2004). Inhibition of anandamide uptake/metabolism also attenuated spontaneous hyperlocomotion in dopamine transporter knockout mice, which was significantly reversed by coadministration of the TRPV1 antagonist capsazepine but not by the selective CB1 antagonist AM251. In addition, TRPV1 binding was increased in the striatum, while CB1 receptor binding was unaffected (Tzavara et al., 2006).

Anandamide not only affected hippocampal and dopaminergic neurons but also induced apoptosis which may be mediated via aberrantly expressed TRPV1 in glioma cell lines, since both CB1 and CB2 stimulation were partially protective (Contassot et al., 2004). Similarly, stimulation of TRPV1 and/or CB1 receptors-mediated cell death of dopaminergic neurons, which may contribute to neurodegeneration in response to endogenous anandamide (Kim et al., 2005b). Inhibition of CB1 receptors by the CB1 receptor antagonist rimonabant evokes neuroprotective effects against EEG flattening, memory impairment and CA1 hippocampal neuronal loss, and these responses were antagonized by the TRPV1 antagonist capsazepine. These findings suggested that VR1 vanilloid receptors are involved in rimonabant's neuroprotection (Pegorini et al., 2006).

At the level of the spinal cord, a low dose of anandamide reduced the frequency of miniature excitatory postsynaptic currents and electrical stimulation-evoked or capsaicin-induced excitatory postsynaptic currents in substantia gelatinosa cells in the spinal cord without any effect on their amplitude, acting on presynaptic CB1. In contrast, anandamide at higher concentrations increased the frequency of miniature excitatory postsynaptic currents recorded from substantia gelatinosa neurons, acting on TRPV1 (Morisset et al., 2001). A similar dual mechanism was suggested in the case of anandamide regulation of neurotransmitter release from dorsal root ganglia. TRPV1 stimulation with high doses of anandamide excited central terminals of capsaicin-sensitive DRG neurons, contrasting with the CB1 receptor-mediated inhibitory action of low concentrations of anandamide (Tognetto et al., 2001).

Taken together, TRPV1 activation of anandamide may play a significant role in the hippocampus (cognitive functions), in dopaminergic neurons (motor functions), and in the spinal cord (pain transmission), where cannabinoid actions are well characterized. In addition, apoptotic effects of anandamide may also be mediated by TRPV1. These findings not only suggest that TRPV1 may be a therapeutic target within the CNS but also that therapeutic regulation of peripheral TRPV1 activity may cause side effects in the CNS.

B. Analgesia

One of the most pharmaceutically explored areas of cannabinoid as well as vanilloid research is analgesia. Both systems are primary targets for development of new classes of painkillers, useful in conditions where other drugs are ineffective (e.g., neuropathic pain) or problematic because of side effects.

The role of TRPV1 in the mediation of anandamide responses in particular has been explored in some detail. Low concentrations of anandamide usually evoke analgesia by activating CB1 and by decreasing TRPV1 responsiveness, as showed by Millns et al. (2001). Thus, topically administered cannabinoid receptor ligand HU210 significantly reduced the perception of pain following the administration of capsaicin on human skin (Rukwied et al., 2003).

On the other hand, high doses of anandamide may excite sensory neurons through TRPV1 stimulation, evoking pain. This was detected in the rat knee joint; where anandamide-induced hyperalgesia was abolished by the TRPV1 antagonist capsazepine. (Gauldie et al., 2001). The dual regulation of TRPV1 responsiveness by anandamide was also shown in the regulation of CGRP release from capsaicin-sensitive primary sensory neurons in vivo, and it was suggested that anandamide could be one of the molecules responsible for the development of inflammatory heat hyperalgesia (Ahluwalia et al., 2003a). Indeed, inhibition of TRPV1 by capsazepine or application of inhibitors of PKA, PKC, or phospholipase C antagonized the anandamide-evoked cobalt-uptake by cultured rat primary sensory neurons both in the presence and absence of bradykinin and prostaglandin E2, suggesting that increased anandamide concentrations might rather evoke rather than decrease inflammatory heat hyperalgesia (Tahim et al., 2005).

In summary, low concentrations of anandamide have analgesic effects through the activation of CB1 and subsequent inhibition of TRPV1. In contrast, under inflammatory conditions, where TRPV1 sensitivity to anandamide is elevated, anandamide may activate TRPV1, evoking pain.

Similarly, painful effects of TRPV1 may also dominate in case of high anandamide concentrations. Nevertheless, it should be noted that the relative responsiveness of cannabinoid and TRPV1 receptors is sensitively regulated in a microenvironment dependent manner.

C. Peripheral nervous system

Sensory neurons not only mediate central perception of pain but also local sensory-efferent effects (Szolcsanyi, 1983), which may be of substantial relevance.

In particular, anandamide induces a modest contractile response in guinea pig isolated bronchus that is dependent upon the activation of vanilloid receptors on airway sensory nerves (Tucker et al., 2001). Intravenous bolus injection of anandamide evoked a TRPV1 dependent stimulatory effect on pulmonary C-fiber terminals (Lin and Lee, 2002). Anandamide (2-AG) stimulates not only pulmonary neurons but also intestinal primary sensory neurons via TRPV1 to release SP, probably resulting in enteritis (Mcvey et al., 2003). Similar to the situation in the spinal cord, in the bladder anandamide at lower concentrations inhibited (TRPV1 mediated) neuropeptide release in a CB1 dependent manner, while higher concentrations increased neuropeptide release. Moreover, intravesical instillation of anandamide increased c-fos expression in the spinal cord, which was reduced by capsazepine or by resiniferatoxin pretreatment. These results suggest that anandamide, through activation of TRPV1, contributes to the development of hyperreflexia and hyperalgesia during cystitis (Dinis et al., 2004). Experiments with TRPV1 knockout mice revealed an obligatory role for TRPV1 in the vagal C-fiber activation by capsaicin and anandamide (Kollarik and Undem, 2004).

It is important to note that anandamide effects on TRPV1 are different from those of the archetypical TRPV1 activators capsaicin or resiniferatoxin. For example, their potencies, efficacies, kinetics of action, and desensitization are all different in heterologous expression systems (Tóth et al., 2005b). Likewise, anandamide triggers an influx of calcium through TRPV1 but no intracellular store depletion in dorsal root ganglion cells. It also facilitates the heat responsiveness of TRPV1 in a calcium-independent manner (Fischbach et al., 2007).

Moreover, as mentioned above, some of the physiological effects of anandamide may be indirectly mediated, for example, anandamide-evoked depolarization of guinea pig vagus nerve; the mechanism was the activation of TRPV1 in response to the generation of lipoxygenase products (Kagaya et al., 2002). It also needs to be noted that activation of capsaicin-sensitive primary sensory neurons evokes anandamide production and release and that anandamide might be a key endogenous regulator of the excitability of these neurons (Ahluwalia et al., 2003b).

In conclusion, a similar pattern of *in vivo* activity can be seen for the sensory-effector functions as for the analgesic pathway. In addition, anandamide activation of TRPV1 may result in local anandamide synthesis, providing a plausible positive feedback mechanism to increase intracellular Ca^{2+} elevations and sensory-effector functions *in vivo*.

D. Vascular effects

Stimulation of cannabinoid and vanilloid receptors leads to vascular effects. These are mediated by various receptors and cell types, including cannabinoid receptors in endothelial cells and in sensory neurons and vanilloid receptors in sensory neurons and in vascular smooth muscle. Vasodilation is generally seen, but vasoconstriction is observed in some cases.

Anandamide induced vasodilation by activating vanilloid receptors on perivascular sensory nerves and causing release of CGRP (Zygmunt *et al.*, 1999). Anandamide induced reflex bradycardia and hypotension in anesthetized rats through TRPV1 activation. In contrast, the prolonged hypotension following this reflex is mediated by CB1 (Malinowska *et al.*, 2001). There is evidence that both anandamide and the TRPV1 agonist capsaicin evoke cardiovascular reflexes when injected intra-arterially into the rat hindlimb, probably mediated via TRPV1 located on sensory nerve endings within the hindlimb vasculature (Smith and McQueen, 2001). Anandamide can also activate TRPV1 in the trigeminovascular system, causing vasodilation independently of CB1 (Akerman *et al.*, 2004). Interestingly, anandamide responsiveness may be different in different regions of the vasculature. In mesenteric small resistance vessels, vasorelaxation upon anandamide treatment occurred through stimulation of vanilloid receptors, CB1 receptors, and an endothelial receptor coupled to EDHF release. By contrast, in the larger mesenteric artery, vasorelaxation was almost entirely due to stimulation of vanilloid receptors and CB1 receptors and was endothelium-independent (O'Sullivan *et al.*, 2004). Vasodilatative vascular TRPV1 may also be regulated in a gender specific manner. Anandamide caused greater reductions of the contractile-induced responses to noradrenaline in mesenteric vascular beds in female than in male rats. This greater relaxant response in females was the result of enhanced TRPV1 function (Peroni *et al.*, 2004). Dermal application of anandamide significantly increased microcirculatory flow through the activation of TRPV1 in the human skin (Movahed *et al.*, 2005).

An emerging topic is whether TRPV1 may contribute to the long-term regulation of blood pressure and, if yes, what role anandamide may have in this regulation. A high salt containing diet evoked upregulation of mesenteric TRPV1 expression and increased sensitivity to anandamide; the diet-induced enhancement of the depressor effect of anandamide was prevented only when both TRPV1 and CB1 receptors were blocked, suggesting a contribution of TRPV1 (Wang et al., 2005).

As mentioned above in detail, anandamide metabolism may produce vasoactive substances which may contribute to the vascular actions. Anandamide increased pulmonary arterial pressure via COX-2 metabolites following enzymatic degradation by FAAH into arachidonic acid products (Wahn et al., 2005). Moreover, the relaxant effects of anandamide in sheep coronary arteries were mediated in part via the conversion of anandamide to a vasodilatory prostanoid (Grainger and Boachie-Ansah, 2001).

The above-mentioned data suggested a uniform vasodilatatory action upon TRPV1 activation. However, recent data suggest functional, vasoconstrictive TRPV1 expression in vascular smooth muscle cells (Kark et al., 2008; Lizanecz et al., 2006). In particular, anandamide evoked TRPV1-mediated vasoconstriction has been shown to be regulated by phosphorylation (Lizanecz et al., 2006).

The contribution of TRPV1 to anandamide-evoked cardiovascular responses was also investigated in TRPV1 knockout mice. The results indicated that the predominant cardiovascular depressor response to anandamide was mediated through cannabinoid receptors. The role of TRPV1 was limited to the transient activation of the Bezold-Jarisch reflex by very high initial plasma concentrations of anandamide (Pacher et al., 2004).

Taken together, there is substantial evidence that anandamide has a potential to act on TRPV1 to evoke vasoactive effects. However, the physiological importance of the interaction is not clear. Many reports suggest TRPV1 dependent effects, while TRPV1 knockout data do not. It is also clear that anandamide may be metabolized locally, and these metabolites may have vasoactive effects, which may be dependent (e.g., lipoxygenase products of anandamide) or independent (e.g., prostaglandin ethanolamide) of TRPV1. Finally, TRPV1 responsiveness is most probably affected by the local environment, leading to substantial change in its anandamide activation which may be only be studied in animal or disease models. In addition, there is only limited information on the role of anandamide synthesized *in vivo*.

E. Temperature regulation

TRPV1-mediated temperature regulation is controversial. In particular, while TRPV1 knockout mice have normal thermoregulation (Caterina et al., 2000), acute treatment with various TRPV1 antagonists results in hyperthermia (Gavva et al., 2007, 2008), suggesting complex compensatory/regulatory mechanisms. This may be an important side effect of TRPV1 antagonists developed as analgesic drugs. It is suggested that anandamide uptake/breakdown inhibition by AM404 affects the tonic activation of TRPV1 which plays an important role in the physiological thermoregulation (Rawls *et al.*, 2006). Unfortunately, AM 404 may also affect TRPV1 activity by directly activating/desensitizing it (Zygmunt *et al.*, 2000a).

Nevertheless, activation of both cannabinoid and vanilloid receptors result in hypothermia and this feature may be a core issue in the pharmaceutical development of new drugs acting on both targets. It is expected that the physiological regulation of TRPV1 dependent body temperature regulation and the role of anandamide will be investigated in detail.

F. Body weight/fat regulation

Another example for the physiological convergence between cannabinoid and vanilloid systems is the regulation of body weight. The CB1 selective inhibitor rimonabant is a promising drug to reduce body weight. Since this drug apparently does not discriminate between CB1 receptors mediating specific responses (e.g., satiety and emotional responses) its main side effect is depression, resulting from the inhibition of cannabinoid receptors in the CNS, as it is discussed in detail in this issue. On the other hand, pharmacological inhibition of TRPV1, as well as knockout models revealed a similar effect of TRPV1 inhibition when animals were fed on a high-fat diet (Motter and Ahern, 2008; Zhang *et al.*, 2007). Although the molecular mechanisms are not clear, adipose tissue may be affected, based on the fact that it has functional TRPV1 and cannabinoid receptors as well as the biosynthetic machinery to produce anandamide (Spoto *et al.*, 2006).

VI. Future Directions

In the last decade, anandamide has emerged as a multifunctional lipid mediator of various stimuli. Some of its effects are most probably mediated by TRPV1. However, anandamide activation of TRPV1 probably occurs in conditions when TRPV1 is sensitized.

Anandamide metabolism is an emerging hot topic. Anandamide has been proven to be synthesized in significant quantities. It is intriguing that anandamide may function as a releasable hormone-like molecule. For example, activation of anandamide synthesis and subsequent release in one cell may induce the synthesis of prostaglandin ethanolamides or arachidonic acid metabolites in a remote cell. Moreover, lipoxygenase-mediated metabolism of anandamide may contribute to TRPV1 activation.

It has also been found that anandamide may influence signaling elements located in lipid rafts. It is possible that a lipid raft regulatory role of anandamide will be revealed.

Finally, anandamide is often considered as an endovanilloid and the vanilloid binding site of TRPV1 (where anandamide binds) is also usually considered as the physiological ligand binding site of this receptor. However, some species, such as rabbit and chicken, have no functional vanilloid binding site. Since TRPV1-related physiological functions (pain, thermoregulation, etc.) in these species do not seem to be drastically affected, it is expected that the real physiological regulator of this channel is still remains to be revealed.

REFERENCES

Adermark, L., and Lovinger, D. M. (2007). Retrograde endocannabinoid signaling at striatal synapses requires a regulated postsynaptic release step. *Proc. Natl. Acad. Sci. USA* **104**(51), 20564–20569.

Ahluwalia, J., Urban, L., Bevan, S., and Nagy, I. (2003a). Anandamide regulates neuropeptide release from capsaicin-sensitive primary sensory neurons by activating both the cannabinoid 1 receptor and the vanilloid receptor 1 *in vitro*. *Eur. J. Neurosci.* **17**(12), 2611–2618.

Ahluwalia, J., Yaqoob, M., Urban, L., Bevan, S., and Nagy, I. (2003b). Activation of capsaicin-sensitive primary sensory neurones induces anandamide production and release. *J. Neurochem.* **84**(3), 585–591.

Akerman, S., Kaube, H., and Goadsby, P. J. (2004). Anandamide acts as a vasodilator of dural blood vessels *in vivo* by activating TRPV1 receptors. *Br. J. Pharmacol.* **142**(8), 1354–1360.

Alexander, J. P., and Cravatt, B. F. (2006). The putative endocannabinoid transport blocker LY2183240 is a potent inhibitor of FAAH and several other brain serine hydrolases. *J. Am. Chem. Soc.* **128**(30), 9699–9704.

Alexander, S. P., and Kendall, D. A. (2007). The complications of promiscuity: Endocannabinoid action and metabolism. *Br. J. Pharmacol.* **152**(5), 602–623.

Al Hayani, A., Wease, K. N., Ross, R. A., Pertwee, R. G., and Davies, S. N. (2001). The endogenous cannabinoid anandamide activates vanilloid receptors in the rat hippocampal slice. *Neuropharmacology* **41**(8), 1000–1005.

Beltramo, M., and Piomelli, D. (1999). Anandamide transport inhibition by the vanilloid agonist olvanil. *Eur. J. Pharmacol.* **364**(1), 75–78.

Bhave, G., Zhu, W. G., Wang, H. B., Brasier, D. J., Oxford, G. S., and Gereau, R. W. (2002). cAMP-dependent protein kinase regulates desensitization of the capsaicin receptor (VR1) by direct phosphorylation. *Neuron* **35**(4), 721–731.

Bhave, G., Hu, H. J., Glauner, K. S., Zhu, W., Wang, H., Brasier, D. J., Oxford, G. S., and Gereau, R. W. (2003). Protein kinase C phosphorylation sensitizes but does not activate the capsaicin receptor transient receptor potential vanilloid 1 (TRPV1). *Proc. Natl. Acad. Sci. USA* **100**(21), 12480–12485.

Bianchi, B. R., Lee, C. H., Jarvis, M. F., El Kouhen, R., Moreland, R. B., Faltynek, C. R., and Puttfarcken, P. S. (2006). Modulation of human TRPV1 receptor activity by extracellular protons and host cell expression system. *Eur. J. Pharmacol.* **537**(1–3), 20–30.

Birder, L. A. (2007). TRPs in bladder diseases. *Biochim. Biophys. Acta* **1772**(8), 879–884.

Birder, L. A., Nakamura, Y., Kiss, S., Nealen, M. L., Barrick, S., Kanai, A. J., Wang, E., Ruiz, G., De Groat, W. C., Apodaca, G., Watkins, S., and Caterina, M. J. (2002). Altered urinary bladder function in mice lacking the vanilloid receptor TRPV1. *Nat. Neurosci.* **5**(9), 856–860.

Bisogno, T., Melck, D., Bobrov, M. Y., Gretskaya, N. M., Bezuglov, V. V., De Petrocellis, L., and Di Marzo, V. (2000). N-acyl-dopamines: Novel synthetic CB (1) cannabinoid-receptor ligands and inhibitors of anandamide inactivation with cannabimimetic activity *in vitro* and *in vivo*. *Biochem. J.* **351**(Pt 3), 817–824.

Bouaboula, M., Poinot-Chazel, C., Bourrie, B., Canat, X., Calandra, B., Rinaldi-Carmona, M., Le Fur, G., and Casellas, P. (1995). Activation of mitogen-activated protein kinases by stimulation of the central cannabinoid receptor CB1. *Biochem. J.* **312** (Pt 2), 637–641.

Bouaboula, M., Hilairet, S., Marchand, J., Fajas, L., Le Fur, G., and Casellas, P. (2005). Anandamide induced PPARgamma transcriptional activation and 3T3-L1 preadipocyte differentiation. *Eur. J. Pharmacol.* **517**(3), 174–181.

Bron, R., Klesse, L. J., Shah, K., Parada, L. F., and Winter, J. (2003). Activation of Ras is necessary and sufficient for upregulation of vanilloid receptor type 1 in sensory neurons by neurotrophic factors. *Mol. Cell Neurosci.* **22**(1), 118–132.

Cadas, H., Gaillet, S., Beltramo, M., Venance, L., and Piomelli, D. (1996). Biosynthesis of an endogenous cannabinoid precursor in neurons and its control by calcium and cAMP. *J. Neurosci.* **16**(12), 3934–3942.

Caterina, M. J., Leffler, A., Malmberg, A. B., Martin, W. J., Trafton, J., Petersen-Zeitz, K. R., Koltzenburg, M., Basbaum, A. I., and Julius, D. (2000). Impaired nociception and pain sensation in mice lacking the capsaicin receptor. *Science* **288**(5464), 306–313.

Chemin, J., Monteil, A., Perez-Reyes, E., Nargeot, J., and Lory, P. (2001). Direct inhibition of T-type calcium channels by the endogenous cannabinoid anandamide. *EMBO J.* **20**(24), 7033–7040.

Chuang, H. H., Prescott, E. D., Kong, H., Shields, S., Jordt, S. E., Basbaum, A. I., Chao, M. V., and Julius, D. (2001). Bradykinin and nerve growth factor release the capsaicin receptor from PtdIns(4,5)P2-mediated inhibition. *Nature* **411**(6840), 957–962.

Contassot, E., Wilmotte, R., Tenan, M., Belkouch, M. C., Schnuriger, V., De Tribolet, N., Burkhardt, K., and Dietrich, P. Y. (2004). Arachidonylethanolamide induces apoptosis of human glioma cells through vanilloid receptor-1. *J. Neuropathol. Exp. Neurol.* **63**(9), 956–963.

Cortright, D. N., and Szallasi, A. (2004). Biochemical pharmacology of the vanilloid receptor TRPV1. An update. *Eur. J. Biochem.* **271**(10), 1814–1819.

Craib, S. J., Ellington, H. C., Pertwee, R. G., and Ross, R. A. (2001). A possible role of lipoxygenase in the activation of vanilloid receptors by anandamide in the guinea-pig bronchus. *Br. J. Pharmacol.* **134**(1), 30–37.

Cravatt, B. F., Giang, D. K., Mayfield, S. P., Boger, D. L., Lerner, R. A., and Gilula, N. B. (1996). Molecular characterization of an enzyme that degrades neuromodulatory fatty-acid amides. *Nature* **384**(6604), 83–87.

Cravatt, B. F., Demarest, K., Patricelli, M. P., Bracey, M. H., Giang, D. K., Martin, B. R., and Lichtman, A. H. (2001). Supersensitivity to anandamide and enhanced endogenous cannabinoid signaling in mice lacking fatty acid amide hydrolase. *Proc. Natl. Acad. Sci. USA* **98**(16), 9371–9376.

Cristino, L., De Petrocellis, L., Pryce, G., Baker, D., Guglielmotti, V., and Di Marzo, V. (2006). Immunohistochemical localization of cannabinoid type 1 and vanilloid transient receptor potential vanilloid type 1 receptors in the mouse brain. *Neuroscience* **139**(4), 1405–1415.

Day, T. A., Rakhshan, F., Deutsch, D. G., and Barker, E. L. (2001). Role of fatty acid amide hydrolase in the transport of the endogenous cannabinoid anandamide. *Mol. Pharmacol.* **59**(6), 1369–1375.

de Lago, E., de Miguel, R., Lastres-Becker, I., Ramos, J. A., and Fernandez-Ruiz, J. (2004). Involvement of vanilloid-like receptors in the effects of anandamide on motor behavior

and nigrostriatal dopaminergic activity: In vivo and in vitro evidence. *Brain Res.* **1007** (1–2), 152–159.
De Petrocellis, L., Bisogno, T., Davis, J. B., Pertwee, R. G., and Di Marzo, V. (2000). Overlap between the ligand recognition properties of the anandamide transporter and the VR1 vanilloid receptor: Inhibitors of anandamide uptake with negligible capsaicin-like activity. *FEBS Lett.* **483**(1), 52–56.
De Petrocellis, L., Davis, J. B., and Di Marzo, V. (2001a). Palmitoylethanolamide enhances anandamide stimulation of human vanilloid VR1 receptors. *FEBS Lett.* **506**(3), 253–256.
De Petrocellis, L., Bisogno, T., Maccarrone, M., Davis, J. B., Finazzi-Agro, A., and Di Marzo, V. (2001b). The activity of anandamide at vanilloid VR1 receptors requires facilitated transport across the cell membrane and is limited by intracellular metabolism. *J. Biol. Chem.* **276**(16), 12856–12863.
De Petrocellis, L., Harrison, S., Bisogno, T., Tognetto, M., Brandi, I., Smith, G. D., *et al.* (2001c). The vanilloid receptor (VR1)-mediated effects of anandamide are potently enhanced by the cAMP-dependent protein kinase. *J. Neurochem.* **77**(6), 1660–1663.
De Petrocellis, L., Chu, C. J., Moriello, A. S., Kellner, J. C., Walker, J. M., and Di Marzo, V. (2004). Actions of two naturally occurring saturated N-acyldopamines on transient receptor potential vanilloid 1 (TRPV1) channels. *Br. J. Pharmacol.* **143**(2), 251–256.
Devane, W. A., Hanus, L., Breuer, A., Pertwee, R. G., Stevenson, L. A., Griffin, G., Gibson, D., Mandelbaum, A., Etinger, A., and Mechoulam, R. (1992). Isolation and structure of a brain constituent that binds to the cannabinoid receptor. *Science* **258**(5090), 1946–1949.
Dickason-Chesterfield, A. K., Kidd, S. R., Moore, S. A., Schaus, J. M., Liu, B., Nomikos, G. G., and Felder, C. C. (2006). Pharmacological characterization of endocannabinoid transport and fatty acid amide hydrolase inhibitors. *Cell Mol. Neurobiol.* **26**(4–6), 407–423.
Di Marzo, V., Bisogno, T., Melck, D., Ross, R., Brockie, H., Stevenson, L., Pertwee, R., and De Petrocellis, L. (1998). Interactions between synthetic vanilloids and the endogenous cannabinoid system. *FEBS Lett.* **436**(3), 449–454.
Di Marzo, V., Lastres-Becker, I., Bisogno, T., De Petrocellis, L., Milone, A., Davis, J. B., and Fernandez-Ruiz, J. J. (2001). Hypolocomotor effects in rats of capsaicin and two long chain capsaicin homologues. *Eur. J. Pharmacol.* **420**(2–3), 123–131.
Di Marzo, V., Fontana, A., Cadas, H., Schinelli, S., Cimino, G., Schwartz, J. C., and Piomelli, D. (1994). Formation and inactivation of endogenous cannabinoid anandamide in central neurons. *Nature* **372**(6507), 686–691.
Dinis, P., Charrua, A., Avelino, A., Yaqoob, M., Bevan, S., Nagy, I., and Cruz, F. (2004). Anandamide-evoked activation of vanilloid receptor 1 contributes to the development of bladder hyperreflexia and nociceptive transmission to spinal dorsal horn neurons in cystitis. *J. Neurosci.* **24**(50), 11253–11263.
Docherty, R. J., Yeats, J. C., Bevan, S., and Boddeke, H. W. (1996). Inhibition of calcineurin inhibits the desensitization of capsaicin-evoked currents in cultured dorsal root ganglion neurones from adult rats. *Pflugers Arch.* **431**(6), 828–837.
Duncan, M., Millns, P., Smart, D., Wright, J. E., Kendall, D. A., and Ralevic, V. (2004). Noladin ether, a putative endocannabinoid, attenuates sensory neurotransmission in the rat isolated mesenteric arterial bed via a non-CB1/CB2 G(i/o) linked receptor. *Br. J. Pharmacol.* **142**(3), 509–518.
Fischbach, T., Greffrath, W., Nawrath, H., and Treede, R. D. (2007). Effects of anandamide and noxious heat on intracellular calcium concentration in nociceptive DRG neurons of rats. *J. Neurophysiol.* **98**(2), 929–938.
Gauldie, S. D., McQueen, D. S., Pertwee, R., and Chessell, I. P. (2001). Anandamide activates peripheral nociceptors in normal and arthritic rat knee joints. *Br. J. Pharmacol.* **132**(3), 617–621.

Gavva, N. R., Bannon, A. W., Surapaneni, S., Hovland, D. N. Jr., Lehto, S. G., Gore, A., Juan, T., Deng, H., Han, B., Klionsky, L., Kuang, R., and Le, A. (2007). The vanilloid receptor TRPV1 is tonically activated in vivo and involved in body temperature regulation. *J. Neurosci.* **27**(13), 3366–3374.

Gavva, N. R., Treanor, J. J., Garami, A., Fang, L., Surapaneni, S., Akrami, A., Alvarez, F., Bak, A., Darling, M., Gore, A., Jang, G. R., and Kesslak, J. P. (2008). Pharmacological blockade of the vanilloid receptor TRPV1 elicits marked hyperthermia in humans. *Pain* **136**(1–2), 202–210.

Glaser, S. T., Abumrad, N. A., Fatade, F., Kaczocha, M., Studholme, K. M., and Deutsch, D. G. (2003). Evidence against the presence of an anandamide transporter. *Proc. Natl. Acad. Sci. USA* **100**(7), 4269–4274.

Glass, M., Hong, J., Sato, T. A., and Mitchell, M. D. (2005). Misidentification of prostamides as prostaglandins. *J. Lipid Res.* **46**(7), 1364–1368.

Grainger, J., and Boachie-Ansah, G. (2001). Anandamide-induced relaxation of sheep coronary arteries: The role of the vascular endothelium, arachidonic acid metabolites and potassium channels. *Br. J. Pharmacol.* **134**(5), 1003–1012.

Hanus, L., Abu-Lafi, S., Fride, E., Breuer, A., Vogel, Z., Shalev, D. E., Kustanovich, I., and Mechoulam, R. (2001). 2-Arachidonyl glyceryl ether, an endogenous agonist of the cannabinoid CB1 receptor. *Proc. Natl. Acad. Sci. USA* **98**(7), 3662–3665.

Hermann, H., De Petrocellis, L., Bisogno, T., Moriello, A. S., Lutz, B., and Di Marzo, V. (2003). Dual effect of cannabinoid CB1 receptor stimulation on a vanilloid VR1 receptor-mediated response. *Cell Mol. Life Sci.* **60**(3), 607–616.

Howlett, A. C., Qualy, J. M., and Khachatrian, L. L. (1986). Involvement of Gi in the inhibition of adenylate cyclase by cannabimimetic drugs. *Mol. Pharmacol.* **29**(3), 307–313.

Ho, W. S. V., and Randall, M. D. (2007). Endothelium-dependent metabolism by endocannabinoid hydrolases and cyclooxygenases limits vasorelaxation to anandamide and 2-arachidonoylglycerol. *Br. J. Pharmacol.* **150**(5), 641–651.

Huang, S. M., Bisogno, T., Trevisani, M., Al Hayani, A., De Petrocellis, L., Fezza, F., Tognetto, M., Petros, T. J., Krey, J. F., Chu, C. J., Miller, J. D., and Davies, S. N. (2002). An endogenous capsaicin-like substance with high potency at recombinant and native vanilloid VR1 receptors. *Proc. Natl. Acad. Sci. USA* **99**(12), 8400–8405.

Hwang, S. W., Cho, H., Kwak, J., Lee, S. Y., Kang, C. J., Jung, J., Cho, S., Min, K. H., Suh, Y. G., Kim, D., and Oh, U. (2000). Direct activation of capsaicin receptors by products of lipoxygenases: Endogenous capsaicin-like substances. *Proc. Natl. Acad. Sci. USA* **97**(11), 6155–6160.

Jennings, E. A., Vaughan, C. W., Roberts, L. A., and Christie, M. J. (2003). The actions of anandamide on rat superficial medullary dorsal horn neurons *in vitro*. *J. Physiol. Lond.* **548**(1), 121–129.

Jin, X., Morsy, N., Winston, J., Pasricha, P. J., Garrett, K., and Akbarali, H. I. (2004). Modulation of TRPV1 by nonreceptor tyrosine kinase, c-Src kinase. *Am. J. Physiol. Cell Physiol.* **287**(2), C558–C563.

Jin, X. H., Okamoto, Y., Morishita, J., Tsuboi, K., Tonai, T., and Ueda, N. (2007). Discovery and characterization of a Ca^{2+}-independent phosphatidylethanolamine N-acyltransferase generating the anandamide precursor and its congeners. *J. Biol. Chem.* **282**(6), 3614–3623.

Ji, R. R., Samad, T. A., Jin, S. X., Schmoll, R., and Woolf, C. J. (2002). p38 MAPK activation by NGF in primary sensory neurons after inflammation increases TRPV1 levels and maintains heat hyperalgesia. *Neuron* **36**(1), 57–68.

Jung, J., Shin, J. S., Lee, S. Y., Hwang, S. W., Koo, J., Cho, H., and Oh, U. (2004). Phosphorylation of vanilloid receptor 1 by Ca^{2+}/calmodulin-dependent kinase II regulates its vanilloid binding. *J. Biol. Chem.* **279**(8), 7048–7054.

Kaczocha, M., Hermann, A., Glaser, S. T., Bojesen, I. N., and Deutsch, D. G. (2006). Anandamide uptake is consistent with rate-limited diffusion and is regulated by the degree of its hydrolysis by fatty acid amide hydrolase. *J. Biol. Chem.* **281**(14), 9066–9075.

Kagaya, M., Lamb, J., Robbins, J., Page, C. P., and Spina, D. (2002). Characterization of the anandamide induced depolarization of guinea-pig isolated vagus nerve. *Br. J. Pharmacol.* **137**(1), 39–48.

Kark, T., Bagi, Z., Lizanecz, E., Pasztor, E. T., Erdei, N., Czikora, A., Papp, Z., Edes, I., Pórszász, R., and Tóth, A. (2008). Tissue-specific regulation of microvascular diameter: Opposite functional roles of neuronal and smooth muscle located vanilloid receptor-1. *Mol. Pharmacol.* **73**(5), 1405–1412.

Kim, H. I., Kim, T. H., Shin, Y. K., Lee, C. S., Park, M., and Song, J. H. (2005a). Anandamide suppression of Na+ currents in rat dorsal root ganglion neurons. *Brain Res.* **1062**(1–2), 39–47.

Kim, S. R., Lee, D. Y., Chung, E. S., Oh, U. T., Kim, S. U., and Jin, B. K. (2005b). Transient receptor potential vanilloid subtype 1 mediates cell death of mesencephalic dopaminergic neurons *in vivo* and *in vitro*. *J. Neurosci.* **25**(3), 662–671.

Kim, A. Y., Tang, Z., Liu, Q., Patel, K. N., Maag, D., Geng, Y., and Dong, X. (2008). Pirt, a phosphoinositide-binding protein, functions as a regulatory subunit of TRPV1. *Cell* **133**(3), 475–485.

Kollarik, M., and Undem, B. J. (2004). Activation of bronchopulmonary vagal afferent nerves with bradykinin, acid and vanilloid receptor agonists in wild-type and TRPV1$^{-/-}$ mice. *J. Physiol. Lond.* **555**(1), 115–123.

Leung, D., Saghatelian, A., Simon, G. M., and Cravatt, B. F. (2006). Inactivation of N-acyl phosphatidylethanolamine phospholipase D reveals multiple mechanisms for the biosynthesis of endocannabinoids. *Biochemistry* **45**(15), 4720–4726.

Ligresti, A., Morera, E., Van der Stelt, M., Monory, K., Lutz, B., Ortar, G., and Di Marzo, V. (2004). Further evidence for the existence of a specific process for the membrane transport of anandamide. *Biochem. J.* **380**, 265–272.

Lin, Y. S., and Lee, L. Y. (2002). Stimulation of pulmonary vagal C-fibres by anandamide in anaesthetized rats: Role of vanilloid type 1 receptors. *J. Physiol. Lond.* **539**(3), 947–955.

Liu, J., Wang, L., Harvey-White, J., Huang, B. X., Kim, H. Y., Luquet, S., Palmiter, R. D., Krystal, G., Rai, R., Mahadeven, A., Razdan, R. K., and Kunos, G. (2008). Multiple pathways involved in the biosynthesis of anandamide. *Neuropharmacology* **54**(1), 1–7.

Lizanecz, E., Bagi, Z., Pasztor, E. T., Papp, Z., Edes, I., Kedei, N., Blumberg, P. M., and Tóth, A. (2006). Phosphorylation-dependent desensitization by anandamide of vanilloid receptor-1 (TRPV1) function in rat skeletal muscle arterioles and in Chinese hamster ovary cells expressing TRPV1. *Mol. Pharmacol.* **69**(3), 1015–1023.

Lopez-Rodriguez, M. L., Viso, A., Ortega-Gutierrez, S., Lastres-Becker, I., Gonzalez, S., Fernandez-Ruiz, J., and Ramos, J. A. (2001). Design, synthesis and biological evaluation of novel arachidonic acid derivatives as highly potent and selective endocannabinoid transporter inhibitors. *J. Med. Chem.* **44**(26), 4505–4508.

Lopez-Rodriguez, M. L., Viso, A., Ortega-Gutierrez, S., Fowler, C. J., Tiger, G., de Lago, E., *et al.* (2003a). Design, synthesis, and biological evaluation of new inhibitors of the endocannabinoid uptake: Comparison with effects on fatty acid amidohydrolase. *J. Med. Chem.* **46**(8), 1512–1522.

Lopez-Rodriguez, M. L., Viso, A., Ortega-Gutierrez, S., Fowler, C. J., Tiger, G., de Lago, E., Fernández-Ruiz, J., and Ramos, J. A. (2003b). Design, synthesis and biological evaluation of new endocannabinoid transporter inhibitors. *Eur. J. Med. Chem.* **38**(4), 403–412.

Lozovaya, N., Yatsenko, N., Beketov, A., Tsintsadze, T., and Burnashev, N. (2005). Glycine receptors in CNS neurons as a target for nonretrograde action of cannabinoids. *J. Neurosci.* **25**(33), 7499–7506.

Lukacs, V., Thyagarajan, B., Varnai, P., Balla, A., Balla, T., and Rohacs, T. (2007). Dual regulation of TRPV1 by phosphoinositides. *J. Neurosci.* **27**(26), 7070–7080.

Maccarrone, M., van der, S. M., Rossi, A., Veldink, G. A., Vliegenthart, J. F., and Agro, A. F. (1998). Anandamide hydrolysis by human cells in culture and brain. *J. Biol. Chem.* **273**(48), 32332–32339.

Malinowska, B., Kwolek, G., and Gothert, M. (2001). Anandamide and methanandamide induce both vanilloid VR1-and cannabinoid CB1 receptor-mediated changes in heart rate and blood pressure in anaesthetized rats. *Naunyn-Schmiedebergs Arch. Pharmacol.* **364**(6), 562–569.

Marinelli, S., Di Marzo, V., Berretta, N., Matias, I., Maccarrone, M., Bernardi, G., and Mercuri, N. B. (2003). Presynaptic facilitation of glutamatergic synapses to dopaminergic neurons of the rat substantia nigra by endogenous stimulation of vanilloid receptors. *J. Neurosci.* **23**(8), 3136–3144.

Matias, I., Chen, J., De Petrocellis, L., Bisogno, T., Ligresti, A., Fezza, F., Krauss, A. H., Shi, L., Protzman, C. E., Li, C., Liang, Y., and Nieves, A. L. (2004). Prostaglandin ethanolamides (prostamides): *In vitro* pharmacology and metabolism. *J. Pharmacol. Exp. Ther.* **309**(2), 745–757.

Matsuda, L. A., Lolait, S. J., Brownstein, M. J., Young, A. C., and Bonner, T. I. (1990). Structure of a cannabinoid receptor and functional expression of the cloned cDNA. *Nature* **346**(6284), 561–564.

Mcvey, D. C., Schmid, P. C., Schmid, H. H. O., and Vigna, S. R. (2003). Endocannabinoids induce ileitis in rats via the capsaicin receptor (VR1). *J. Pharmacol. Exp. Ther.* **304**(2), 713–722.

Mechoulam, R., Ben Shabat, S., Hanus, L., Ligumsky, M., Kaminski, N. E., Schatz, A. R., Gopherc, A., Almogc, S., Martind, B. R., Comptond, D. R., Pertweee, R. G., and Griffine, G. (1995). Identification of an endogenous 2-monoglyceride, present in canine gut, that binds to cannabinoid receptors. *Biochem. Pharmacol.* **50**(1), 83–90.

Melck, D., Bisogno, T., De Petrocellis, L., Chuang, H. H., Julius, D., Bifulco, M., and Di Marzo, V. (1999). Unsaturated long-chain N-acyl-vanillyl-amides (N-AVAMs): Vanilloid receptor ligands that inhibit anandamide-facilitated transport and bind to CB1 cannabinoid receptors. *Biochem. Biophys. Res. Commun.* **262**(1), 275–284.

Mezey, E., Toth, Z. E., Cortright, D. N., Arzubi, M. K., Krause, J. E., Elde, R., Guo, A., Blumberg, P. M., and Szallasi, A. (2000). Distribution of mRNA for vanilloid receptor subtype 1 (VR1), and VR1-like immunoreactivity, in the central nervous system of the rat and human. *Proc. Natl. Acad. Sci. USA* **97**(7), 3655–3660.

Millns, P. J., Chapman, V., and Kendall, D. A. (2001). Cannabinoid inhibition of the capsaicin-induced calcium response in rat dorsal root ganglion neurones. *Br. J. Pharmacol.* **132**(5), 969–971.

Mohapatra, D. P., and Nau, C. (2005). Regulation of Ca2+-dependent desensitization in the vanilloid receptor TRPV1 by calcineurin and cAMP-dependent protein kinase. *J. Biol. Chem.* **280**(14), 13424–13432.

Morisset, V., Ahluwalia, J., Nagy, I., and Urban, L. (2001). Possible mechanisms of cannabinoid-induced antinociception in the spinal cord. *Eur. J. Pharmacol.* **429**(1–3), 93–100.

Motter, A. L., and Ahern, G. P. (2008). TRPV1-null mice are protected from diet-induced obesity. *FEBS Lett.* **582**(15), 2257–2262.

Movahed, P., Evilevitch, V., Andersson, T. L. G., Jonsson, B. A. G., Wollmer, P., Zygmunt, P. M., and Högestätt, E. D. (2005). Vascular effects of anandamide and N-acylvanillylamines in the human forearm and skin microcirculation. *Br. J. Pharmacol.* **146**(2), 171–179.

Munro, S., Thomas, K. L., and Abu-Shaar, M. (1993). Molecular characterization of a peripheral receptor for cannabinoids. *Nature* **365**(6441), 61–65.

O'Sullivan, S. E., Kendall, D. A., and Randall, M. D. (2004). Heterogeneity in the mechanisms of vasorelaxation to anandamide in resistance and conduit rat mesenteric arteries. *Br. J. Pharmacol.* **142**(3), 435–442.

Okamoto, Y., Morishita, J., Tsuboi, K., Tonai, T., and Ueda, N. (2004). Molecular characterization of a phospholipase D generating anandamide and its congeners. *J. Biol. Chem.* **279**(7), 5298–5305.

Orliac, M. L., Peroni, R. N., Abramoff, T., Neuman, I., Podesta, E. J., and Adler-Graschinsky, E. (2007). Increases in vanilloid TRPV1 receptor protein and CGRP content during endotoxemia in rats. *Eur. J. Pharmacol.* **566**(1–3), 145–152.

Ortar, G., Ligresti, A., De Petrocellis, L., Morera, E., and Di Marzo, V. (2003). Novel selective and metabolically stable inhibitors of anandamide cellular uptake. *Biochem. Pharmacol.* **65**(9), 1473–1481.

Pacher, P., Batkai, S., and Kunos, G. (2004). Haemodynamic profile and responsiveness to anandamide of TRPV1 receptor knock-out mice. *J. Physiol. Lond.* **558**(2), 647–657.

Pegorini, S., Zani, A., Braida, D., Guerini-Rocco, C., and Sala, M. (2006). Vanilloid VR1 receptor is involved in rimonabant-induced neuroprotection. *Br. J. Pharmacol.* **147**(5), 552–559.

Peroni, R. N., Orliac, M. L., Becu-Villalobos, D., Huidobro-Toro, J. P., Adler-Graschinsky, E., and Celuch, S. M. (2004). Sex-linked differences in the vasorelaxant effects of anandamide in vascular mesenteric beds: Role of oestrogens. *Eur. J. Pharmacol.* **493**(1–3), 151–160.

Piomelli, D., Beltramo, M., Glasnapp, S., Lin, S. Y., Goutopoulos, A., Xie, X. Q., and Makriyannis, A. (1999). Structural determinants for recognition and translocation by the anandamide transporter. *Proc. Natl. Acad. Sci. USA* **96**(10), 5802–5807.

Porter, A. C., Sauer, J. M., Knierman, M. D., Becker, G. W., Berna, M. J., Bao, J., Nomikos, G. G., Carter, P., Bymaster, F. P., Leese, A. B., and Felder, C. C. (2002). Characterization of a novel endocannabinoid, virodhamine, with antagonist activity at the CB1 receptor. *J. Pharmacol. Exp. Ther.* **301**(3), 1020–1024.

Premkumar, L. S., and Ahern, G. P. (2000). Induction of vanilloid receptor channel activity by protein kinase C. *Nature* **408**(6815), 985–990.

Rawls, S. M., Ding, Z., and Cowan, A. (2006). Role of TRPV1 and cannabinoid CB1 receptors in AM 404-evoked hypothermia in rats. *Pharmacol. Biochem. Behav.* **83**(4), 508–516.

Reddy, P. V., Natarajan, V., Schmid, P. C., and Schmid, H. H. (1983). N-Acylation of dog heart ethanolamine phospholipids by transacylase activity. *Biochim. Biophys. Acta* **750**(3), 472–480.

Ross, R. A. (2003). Anandamide and vanilloid TRPV1 receptors. *Br. J. Pharmacol.* **140**(5), 790–801.

Ross, R. A., Gibson, T. M., Brockie, H. C., Leslie, M., Pashmi, G., Craib, S. J., Di Marzo, V., and Pertwee, R. G. (2001). Structure-activity relationship for the endogenous cannabinoid, anandamide, and certain of its analogues at vanilloid receptors in transfected cells and vas deferens. *Br. J. Pharmacol.* **132**(3), 631–640.

Rukwied, R., Watkinson, A., McGlone, F., and Dvorak, M. (2003). Cannabinoid agonists attenuate capsaicin-induced responses in human skin. *Pain* **102**(3), 283–288.

Sarker, K. P., and Maruyama, I. (2003). Anandamide induces cell death independently of cannabinoid receptors or vanilloid receptor 1: Possible involvement of lipid rafts. *Cell Mol. Life Sci.* **60**(6), 1200–1208.

Simon, G. M., and Cravatt, B. F. (2006). Endocannabinoid biosynthesis proceeding through glycerophospho-N-acyl ethanolamine and a role for alpha/beta-hydrolase 4 in this pathway. *J. Biol. Chem.* **281**(36), 26465–26472.

Smart, D., and Jerman, J. C. (2000). Anandamide: An endogenous activator of the vanilloid receptor. *Trends Pharmacol. Sci.* **21**(4), 134.

Smart, D., Jonsson, K. O., Vandevoorde, S., Lambert, D. M., and Fowler, C. J. (2002). "Entourage" effects of N-acyl ethanolamines at human vanilloid receptors. Comparison of effects upon anandamide-induced vanilloid receptor activation and upon anandamide metabolism. *Br. J. Pharmacol.* **136**(3), 452–458.

Smith, P. J. W., and McQueen, D. S. (2001). Anandamide induces cardiovascular and respiratory reflexes via vasosensory nerves in the anaesthetized rat. *Br. J. Pharmacol.* **134**(3), 655–663.

Spoto, B., Fezza, F., Parlongo, G., Battista, N., Sgro, E., Gasperi, V., Zoccali, C., and Maccarrone, M. (2006). Human adipose tissue binds and metabolizes the endocannabinoids anandamide and 2-arachidonoylglycerol. *Biochimie* **88**(12), 1889–1897.

Starowicz, K., Cristino, L., and Di, M. V. (2008). TRPV1 receptors in the central nervous system: Potential for previously unforeseen therapeutic applications. *Curr. Pharm. Des.* **14**(1), 42–54.

Sun, Y., Alexander, S. P., Garle, M. J., Gibson, C. L., Hewitt, K., Murphy, S. P., Kendall, D. A., and Bennett, A. J. (2007). Cannabinoid activation of PPAR alpha; a novel neuroprotective mechanism. *Br. J. Pharmacol.* **152**(5), 734–743.

Szolcsanyi, J. (1983). Tetrodotoxin-resistant non-cholinergic neurogenic contraction evoked by capsaicinoids and piperine on the guinea-pig trachea. *Neurosci. Lett* **42**(1), 83–88.

Szolcsanyi, J. (2000). Anandamide and the question of its functional role for activation of capsaicin receptors. *Trends Pharmacol. Sci.* **21**(6), 203–204.

Szolcsanyi, J., and Jancso-Gabor, A. (1975). Sensory effects of capsaicin congeners. 1. Relationship between chemical-structure and pain-producing potency of pungent agents. *Arzneimittel-Forschung/Drug Res.* **25**(12), 1877–1881.

Tahim, A. S., Santha, P., and Nagy, I. (2005). Inflammatory mediators convert anandamide into a potent activator of the vanilloid type 1 transient receptor potential receptor in nociceptive primary sensory neurons. *Neuroscience* **136**(2), 539–548.

Tognetto, M., Amadesi, S., Harrison, S., Creminon, C., Trevisani, M., Carreras, M., Matera, M., Geppetti, P., and Bianchi, A. (2001). Anandamide excites central terminals of dorsal root ganglion neurons via vanilloid receptor-1 activation. *J. Neurosci.* **21**(4), 1104–1109.

Tominaga, M., Caterina, M. J., Malmberg, A. B., Rosen, T. A., Gilbert, H., Skinner, K., Raumann, B. E., Basbaum, A. I., and Julius, D. (1998). The cloned capsaicin receptor integrates multiple pain-producing stimuli. *Neuron* **21**(3), 531–543.

Tóth, A., Kedei, N., Wang, Y., and Blumberg, P. M. (2003). Arachidonyl dopamine as a ligand for the vanilloid receptor VR1 of the rat. *Life Sci.* **73**(4), 487–498.

Tóth, A., Blumberg, P. M., Chen, Z., and Kozikowski, A. P. (2004). Design of a high-affinity competitive antagonist of the vanilloid receptor selective for the calcium entry-linked receptor population. *Mol. Pharmacol.* **65**(2), 282–291.

Tóth, A., Boczan, J., Kedei, N., Lizanecz, E., Bagi, Z., Papp, Z., Edes, I., Csiba, L., and Blumberg, P. M. (2005a). Expression and distribution of vanilloid receptor 1 (TRPV1) in the adult rat brain. *Mol. Brain Res.* **135**(1–2), 162–168.

Tóth, A., Wang, Y., Kedei, N., Tran, R., Pearce, L. V., Kang, S. U., Jin, M. K., Choi, H. K., Lee, J., and Blumberg, P. M. (2005b). Different vanilloid agonists cause different patterns of calcium response in CHO cells heterologously expressing rat TRPV1. *Life Sci.* **76**(25), 2921–2932.

Tucker, R. C., Kagaya, M., Page, C. P., and Spina, D. (2001). The endogenous cannabinoid agonist, anandamide stimulates sensory nerves in guinea-pig airways. *Br. J. Pharmacol.* **132**(5), 1127–1135.

Tzavara, E. T., Li, D. L., Moutsimilli, L., Bisogno, T., Di Marzo, V., Phebus, L. A., Nomikos, G. G., and Giros, B. (2006). Endocannabinoids activate transient receptor potential vanilloid 1 receptors to reduce hyperdopaminergia-related hyperactivity: Therapeutic implications. *Biol. Psychiatry* **59**(6), 508–515.

Van der Stelt, M., Trevisani, M., Vellani, V., De Petrocellis, L., Moriello, A. S., Campi, B., McNaughton, P., Geppetti, P., and Di Marzo, V. (2005). Anandamide acts as an intracellular messenger amplifying Ca^{2+} influx via TRPV1 channels. *EMBO J.* **24**(17), 3026–3037.

Vandevoorde, S., Lambert, D. M., Smart, D., Jonsson, K. O., and Fowler, C. J. (2003). N-morpholino- and N-diethyl-analogues of palmitoylethanolamide increase the sensitivity of transfected human vanilloid receptors to activation by anandamide without affecting fatty acid amidohydrolase activity. *Bioorg. Med. Chem.* **11**(6), 817–825.

Vellani, V., Mapplebeck, S., Moriondo, A., Davis, J. B., and McNaughton, P. A. (2001). Protein kinase C activation potentiates gating of the vanilloid receptor VR1 by capsaicin, protons, heat and anandamide. *J. Physiol. Lond.* **534**(3), 813–825.

Wahn, H., Wolf, J., Kram, F., Frantz, S., and Wagner, J. A. (2005). The endocannabinoid arachidonyl ethanolamide (anandamide) increases pulmonary arterial pressure via cyclooxygenase-2 products in isolated rabbit lungs. *Am. J. Physiol. Heart Circ. Physiol.* **289**(6), H2491–H2496.

Wang, Y., Kedei, N., Wang, M., Wang, Q. J., Huppler, A. R., Tóth, A., Tran, R., and Blumberg, P. M. (2004). Interaction between protein kinase Cmu and the vanilloid receptor type 1. *J. Biol. Chem.* **279**(51), 53674–53682.

Wang, Y. P., Kaminski, N. E., and Wang, D. H. (2005). VR1-mediated depressor effects during high-salt intake—Role of anandamide. *Hypertension* **46**(4), 986–991.

Woodward, D. F., Krauss, A. H., Wang, J. W., Protzman, C. E., Nieves, A. L., Liang, Y., Donde, Y., Burk, R. M., Landsverk, K., and Struble, C. (2007). Identification of an antagonist that selectively blocks the activity of prostamides (prostaglandin-ethanolamides) in the feline iris. *Br. J. Pharmacol.* **150**(3), 342–352.

Yang, W., Ni, J., Woodward, D. F., Tang-Liu, D. D., and Ling, K. H. (2005). Enzymatic formation of prostamide F2alpha from anandamide involves a newly identified intermediate metabolite, prostamide H2. *J. Lipid Res.* **46**(12), 2745–2751.

Yang, X. R., Lin, M. J., McIntosh, L. S., and Sham, J. S. (2006). Functional expression of transient receptor potential melastatin- and vanilloid-related channels in pulmonary arterial and aortic smooth muscle. *Am. J. Physiol. Lung Cell Mol. Physiol.* **290**(6), L1267–L1276.

Yu, M., Ives, D., and Ramesha, C. S. (1997). Synthesis of prostaglandin E2 ethanolamide from anandamide by cyclooxygenase-2. *J. Biol. Chem.* **272**(34), 21181–21186.

Zhang, N., Inan, S., Cowan, A., Sun, R. H., Wang, J. M., Rogers, T. J., Caterina, M., and Oppenheim, J. J. (2005). A proinflammatory chemokine, CCL3, sensitizes the heat- and capsaicin-gated ion channel TRPV1. *Proc. Natl. Acad. Sci. USA* **102**(12), 4536–4541.

Zhang, L. L., Yan, L. D., Ma, L. Q., Luo, Z. D., Cao, T. B., Zhong, J., Yan, Z. C., Wang, L. J., Zhao, Z. G., Zhu, S. J., Schrader, M., and Thilo, F. (2007). Activation of transient receptor potential vanilloid type-1 channel prevents adipogenesis and obesity. *Circ. Res.* **100**(7), 1063–1070.

Zygmunt, P. M., Petersson, J., Andersson, D. A., Chuang, H. H., Sorgard, M., Di Marzo, V., Julius, D., and Högestätt, E. D. (1999). Vanilloid receptors on sensory nerves mediate the vasodilator action of anandamide. *Nature* **400**(6743), 452–457.

Zygmunt, P. M., Chuang, H. H., Movahed, P., Julius, D., and Hogestatt, E. D. (2000a). The anandamide transport inhibitor AM404 activates vanilloid receptors. *Eur. J. Pharmacol.* **396**(1), 39–42.

Zygmunt, P. M., Julius, D., Di Marzo, V., and Hogestatt, E. D. (2000b). Anandamide—The other side of the coin. *Trends Pharmacol. Sci.* **21**(2), 43–44.

CHAPTER SIXTEEN

Endocannabinoid System and Fear Conditioning

Leonardo B. M. Resstel,* Fabrício A. Moreira,[†] *and* Francisco S. Guimarães*

Contents

I. Introduction	421
II. Fear Conditioning	423
III. Influence of Endocannabinoids on Fear Conditioning	426
IV. Brain Regions in which Endocannabinoids may Modulate Fear Conditioning	429
V. Conclusion	433
References	433

Abstract

The endocannabinoid system has been proposed to modulate neuronal functions involved in distinct types of defensive reactions, possibly counteracting the harmful consequences of stressful stimuli. However, the precise brain sites for this action remain to be further explored. This chapter summarizes the data about the role of the endocannabinoid system in the processing of conditioned fear as well as the potential neural subtract for its actions.

I. Introduction

In the last few years several pieces of evidence have emerged suggesting that endocannabinoids may modulate innate as well as conditioned fear. The term endocannabinoid refers to a group of neurotransmitters that are the endogenous counterparts of Δ^9-tetrahydrocannabinol (Δ^9-THC), the compound that accounts for most of the effects induced by the herb *Cannabis sativa* (Howlett *et al.*, 2002). In the central nervous system

---

* Department of Pharmacology, School of Medicine of Ribeirão Preto, University of São Paulo, Ribeirão Preto, São Paulo, Brazil
[†] Department of Pharmacology, Institute of Biological Sciences, Federal University of Minas Gerais, Belo Horizonte, Minas Gerais, Brazil

endocannabinoid and Δ^9-THC act primarily on a metabotropic receptor referred to as CB1 (Devane et al., 1988). The first endocannabinoid identified was arachidonoyl ethanolamide, also named anandamide, after the Sanskrit word "ananda," for "bliss" (Devane et al., 1992). Although the majority of studies have focussed on anandamide, another endocannabinoid, 2-arachidonoyl glycerol, is present in even higher concentrations in the brain (Mechoulam et al., 1995). Moreover, other arachidonic acid derivates have been proposed as endocannabinoids, such as arachidonoyl dopamine, virodhamine, and noladin ether (Pacher et al., 2006).

In addition to CB1, another cannabinoid receptor has been characterized, called CB2, whose expression and functions in the brain remain a matter of controversy (Munro et al., 1993). CB1 receptors, on the other hand, seem to play an important role in several physiological and pathological processes, such as such as memory, motor modulation, appetite control, analgesia, and defensive responses (Pacher et al., 2006). CB1 expression is considerably high in brain regions responsible for fear and anxiety, namely the medial prefrontal cortex (MPFC), hippocampus, amygdala, hypothalamus, and periaqueductal gray (PAG) (Herkenham et al., 1991; Tsou et al., 1998). Actually, this is the most densely expressed receptor in the brain.

A particular feature of CB1 receptor is its location in presynaptic rather than in postsynaptic neurons (Egertova et al., 1998). This is in line with the proposal that endocannabinoids may be produced in postsynaptic neurons and act as retrograde neurotransmitters (Wilson and Nicoll, 2001). The functions of CB1 receptors have been investigated by selective pharmacological agents, such as the agonist arachidonoyl chloroethilamide and the antagonists/inverse agonists rimonabant and AM251. By binding to CB1 receptors, cannabinoids can activate G-proteins (subtype Gi) that inhibit adenylate cyclase activity reducing the synthesis of the intracellular messenger cAMP. Furthermore, they can also interfere with ion conductance, inhibiting calcium and increasing potassium currents (Howlett et al., 2002). The overt effect is inhibition of neural activity and neurotransmitter release.

The effects mediated by endocannabinoids are usually limited and short-lasting due to their quick removal from the synaptic cleft by a two-step process. First, they are internalized in neurons, crossing the membrane by mechanisms that remain unclear. Two prevailing hypotheses, not mutually exclusive, are that they may either passively diffuse down a concentration gradient or be up-taken by a specific transporter (Giuffrida et al., 2001). Inside the neurons, anandamide and 2-AG are hydrolyzed by fatty acid amide hydrolase (FAAH) and monoacyl glycerol lipase (MGL), respectively (McKinney and Cravatt, 2005). Pharmacological tools such as AM404, a proposed anandamide uptake-inhibitor, and URB597, a selective FAAH blocker, have been developed to study the role of these mechanisms in endocannabinoids effects.

Some cannabinoids may also bind to other receptors, including the transient receptor potential vanilloid type 1 (TRPV1), the peroxisome-proliferator-activated receptor and the G-protein-coupled receptor GPR55 (Brown, 2007). In addition, an allosteric site in the CB1 receptor has been identified (Price *et al.*, 2005), but its physiological function is unknown.

II. Fear Conditioning

Central states of fear organize behavioral and physiological responses to perceived threat and danger that are fundamental to adaptation and survival (LeDoux, 1996). Either real or potentially threatening stimuli are able to activate defensive systems that coordinate avoidance or escape reactions (Fendt and Fanselow, 1999). In animal models, these threats may be innately recognized or learned (Blanchard and Blanchard, 1972; Dielenberg *et al.*, 2001). Thus, rats faced with a predator odor or a stimulus that predicts potential injury will freeze, a behavioral response characterized by complete movement cessation except for those associated with respiration. This state is associated with increased cardiovascular activity, temperature alterations, anti-nociception, and increased release of several hormones (Resstel *et al.*, 2006a,b, 2008a,b,c; Sullivan *et al.*, 2004; Vianna *et al.*, 2008). Moreover, chronic fear responses that continue well past their usefulness have been related to the development of pathological human states such as anxiety and depression (Maren *et al.*, 1997). The persistence of fear memories is thought to play a major role on the morbidity associated with posttraumatic stress disorder, panic disorder, and specific and social phobia (Cannistraro and Rauch, 2003).

Pavlovian classical fear conditioning is an experimental procedure in which an aversive unconditioned stimulus (US), normally an electrical footshock, is presented in association with a neutral conditioned stimulus (CS) (Fanselow, 1980; Fendt and Fanselow, 1999; Resstel *et al.*, 2008a). Fear conditioning is a rapidly acquired and long-lasting form of learning. It is, therefore, an attractive model for studying the neural basis of learning and memory, specifically, emotional learning and memory (Bouton, 2004; LeDoux, 1992; Phillips and LeDoux, 1992; Quirk and Mueller, 2008; Schafe *et al.*, 2005; Sotres-Bayon *et al.*, 2006).

Normally, the fear-evoking or -conditioning stimulus is a discrete stimulus such as a light or tone (Fanselow, 1980; Fendt and Fanselow, 1999; Maren *et al.*, 1997). However, stimuli present in the environment where a US is presented but not explicitly paired with the US in a temporally specific manner may also acquire aversive properties and thereby elicit conditioned emotional responses (Beck and Fibiger, 1995; Blanchard and

Blanchard, 1972; Fanselow, 1980; Fendt and Fanselow, 1999; Resstel *et al.*, 2006b). Thus, contextual conditioned fear is the fear evoked by reexposure to an environment that has previously been paired with an aversive or unpleasant stimulus. This is observed, for example, when a rat is reexposed to a footshock chamber in which it has previously received electric footshocks (Resstel *et al.*, 2006b; Sullivan *et al.*, 2004; Vianna *et al.*, 2008). Therefore, after repeated pairings, the CS or context is able to evoke conditioned emotional responses (fear) which can be measured by freezing behavior, changes in autonomic activity (increase of heart rate and arterial blood pressure), and facilitation of reflex action such as a startle (Resstel *et al.*, 2006b; Sullivan *et al.*, 2004; Vianna *et al.*, 2008). Qualitatively, the behavior- and autonomics-conditioned responses are similar in conditioned fear to a context or to a discrete stimulus, although there is a marked difference in the duration of the two.

Conditioned fear has been implicated in the development of anxiety disorders after stressful experiences (Fendt and Fanselow, 1999; LeDoux, 2000; Pare *et al.*, 2004; Resstel *et al.*, 2006b). Moreover, the fear-conditioned responses are reduced by anxiolytic drugs such as diazepam, further suggesting its relationship to anxiety-like behavior (Beck and Fibiger, 1995; Fanselow and Helmstetter, 1988; Malkani and Rosen, 2000; Resstel *et al.*, 2006b).

Experimental manipulations aimed at studying the neural circuitry mediating fear responses may be performed at three different situations. First, interferences immediately before training procedure may disrupt acquisition of conditioned fear. Second, interference immediately after training may affect the consolidation of conditioned fear. Finally, manipulations made immediately before the testing procedure affect the expression of the conditioned fear response (Frankland and Bontempi, 2005; Suzuki *et al.*, 2004).

Several forebrain structures have been related to fear conditioning, including the amygdala, hippocampus, and MPFC. The amygdala is thought to play an essential role in the acquisition and expression of conditioned fear. Lesions or pharmacological inactivation of the basolateral amygdala (BLA) or lateral amygdala impairs fear conditioning (Bishop, 2007; Fendt and Fanselow, 1999; Maren and Quirk, 2004). Associative learning between the conditioned and unconditioned stimuli is proposed to occur in the lateral nucleus. This nucleus projects to the central nucleus both directly and indirectly via projections to the BLA (Pare and Smith, 1998; Pitkanen *et al.*, 1997). The central nucleus of the amygdala has extensive projections to several nuclei in the midbrain and brainstem that orchestrate the behavioral, autonomic, and endocrine responses to threat and danger, such as the ventrolateral PAG, lateral hypothalamus, paraventricular nucleus, parabrachial nucleus, prefrontal cortex, and locus coeruleus (Davis, 1992; Rosen, 2004).

Neurons in the amygdala seem to be engaged by conditioning to both discrete and contextual stimuli (Beck and Fibiger, 1995; Phillips and LeDoux, 1992; Schafe et al., 2005; Sullivan et al., 2004). The amygdala supports the association of the unified contextual representation (acquired in the hippocampus) with the shock US in a way similar to which it supports tone–shock associations (Fanselow and Kim, 1994; Maren and Fanselow, 1995; Maren, 1996; Maren et al., 1996a,b). Local inhibition of dorsal hippocampus (DH) neurotransmission causes deficits in acquisition but not in expression of contextual fear, measured as freezing in the training apparatus (Resstel et al., 2008c). However, the cardiovascular changes were reduced in both cases. Thus, in addition to a crucial role on the consolidation of contextual fear, the DH may also play an important role on the cardiovascular responses generated by this model. In contrast, fear conditioning to discrete CS, such as a tone, is typically not affected by hippocampal formation lesions (Kim and Fanselow, 1992; Phillips and LeDoux, 1992, 1994).

Another forebrain region associated with learning and memory (Kesner et al., 1996; Ragozzino et al., 1998) that has been related to fear conditioning is the MPFC. Neurons located in this region show increased activity during the expression and acquisition of Pavlovian fear associations (Baeg et al., 2001; Laviolette et al., 2005; Sotres-Bayon et al., 2006) and the MPFC is proposed to play a crucial role in interfacing processes involved in fear responses originated in limbic and neocortical regions (Conde et al., 1995; Sesack et al., 1989; Takagishi and Chiba, 1991).

The MPFC receives information about context and electrical shock associations through afferents from the amygdala and the hippocampus (Jay and Witter, 1991). Moreover, similar to the hippocampus and amygdala, lesions of the MPFC decrease conditioned fear behavioral and cardiovascular responses (Anagnostaras et al., 1999; Galeno et al., 1984; Gentile et al., 1986; Iwata et al., 1986; Kim and Fanselow, 1992; LeDoux et al., 1988; Maren and Fanselow, 1997; Maren et al., 1997).

In addition, mechanisms involving the prefrontal cortex seem to be crucial to the process of fear extinction and active inhibition of previously conditioned fear responses (Bishop, 2007; Milad and Quirk, 2002). It has been increasingly recognized that extinction of conditioned fear does not entail the original CS–US association being eradicated, but rather it being overshadowed by a stronger association between the CS and the nonoccurrence of the US (Myers and Davis, 2007). Several reports have indicated that MPFC neurons, specifically those from the infralimbic cortex (Milad and Quirk, 2002; Milad et al., 2004), are important mediators of the extinction of conditioned fear associations (Garcia et al., 1999). These neurons are proposed to activate GABAergic neurons within the basolateral complex of the amygdala or the nearby intercalated cells. These cells, in turn, could inhibit output neurons in the central nucleus of the amygdala (Bishop, 2007; Grace and Rosenkranz, 2002; Quirk et al., 2003; Sotres-Bayon et al., 2004).

III. INFLUENCE OF ENDOCANNABINOIDS ON FEAR CONDITIONING

Studies employing systemic administration of CB1 agonists or antagonists have produced conflicting results on the expression of fear conditioning (see Table 16.1). Most studies investigating contextual fear conditioning indicate that endocannabinoids decrease conditioned emotional responses (Table 16.1). This result, however, contrasts with those obtained by Arenos et al. (2006) and Mikics et al. (2006) using mice and rats, respectively (Table 16.2). In these studies AM-251, a CB1 receptor antagonist, reduced the expression of contextual conditioned fear (Arenos et al., 2006; Mikics et al., 2006). Moreover, this response was increased by the CB1 agonist WIN-55,212-2 (Arenos et al., 2006; Mikics et al., 2006). In opposition to these results, studies using rimonabant (SR141716A), another CB1 receptor antagonist, produced opposite effects than those observed with AM-251 (see Table 16.1). These apparently contradictory data could partially be explained by specific pharmacological properties of the antagonists employed in each study. It has been suggested that rimonabant and AM251 have distinct pharmacological profiles, differently affecting GABAergic or glutamatergic neurons (Haller et al., 2007). The reasons for this remain unknown, but may involve binding to receptors other than CB1 such as GPR55, a still unknown cannabinoid receptor (Brown, 2007). In addition, the use of different species (rats versus mice) could also have accounted for some discrepancies. In this regard Haller et al. (2007) have suggested that cannabinoids may interfere with the balance between GABA and glutamate in a species-dependent fashion, leading to anxiolytic- and anxiogenic-like effects in mice and rats, respectively. Finally, the diversity of experimental protocols is another potential explanation. In the study by Arenos et al. (2006), no changes in the expression of contextual fear were observed when the conditioning session was performed without presentation of a discrete CS (tone).

Results regarding the effects of cannabinoids in fear-conditioned responses elicited by discrete cues are sparser. Again, whereas AM-251 increased conditioned responses in rats (Arenos et al., 2006), rimonabant inhibited fear extinctions but did not interfere with acquisition and consolidation of these responses in mice (Marsicano et al., 2002). Further complicating the picture, one study reported that CB1 knockout mice may have reduced fear conditioning responses to CS cues or context, when compared to wild-type control animals (Mikics et al., 2006). On the other hand Marsicano et al. (2002) found that the phenotype has a selective impairment in extinction to a conditioned tone, without changes in acquisition or consolidation.

Table 16.1 Systemic effects of drugs that affect endocannabinoid-mediated neurotransmission in fear conditioning

Drug	Mechanism	Dose (mg/kg)	Species	Conditioning	Time of injection	Effect	Reference
AM-251	CB1 antagonist	0.3–3	Mice	Context	Pretest	↓ CFR	Mikics et al. (2006)
AM-251	CB1 antagonist	3	Rats	Context	Pretraining or pretest	↓ CFR	Arenos et al. (2006)
AM-251	CB1 antagonist	3	Rats	CS discrete cue	Pretest	↑ CFR	Arenos et al. (2006)
SR141716A	CB1 antagonist	3	Mice	CS discrete cue	Pretraining or pretest	↓ Extinction (pretest only)	Marsicano et al. (2002)
SR141716A	CB1 antagonist	1–10	Mice	Context	Pretraining or pretest	↑ CFR (pretest only)	Suzuki et al. (2004)
SR141716A	CB1 antagonist	1	Rats	Context	Pretest	↑ CFR	Finn et al. (2004)
SR141716A	CB1 antagonist	0.2–2	Rats	Context	Pretest	↓ Extinction	Pamplona et al. (2006)
SR141716A	CB1 antagonist	0.15–5	Rats	CS discrete cue	Pretest	↓ Extinction	Chhatwal et al. (2005)
WIN-55,212-2	CB1 agonist	0.3–3	Mice	Context	Pretest	↑ CFR	Mikics et al. (2006)
WIN 55,212-2	CB1 agonist	0.25–2.5	Rats	Context	Pretest	↑ Extinction	Pamplona et al. (2006)
AM404	Inhibitor of cannabinoid reuptake	2 and 10	Rats	Context	Pretest	↑ Extinction	Chhatwal et al. (2005)

Table 16.2 Results of experiments testing the behavioral of knockout mice in fear conditioning

Species/control	Conditioning	Effect	Reference
CB1 KO/ C57BL/ 6JOlaHsd mice	CS discrete cue	↓ Extinction	Marsicano et al. (2002)
CB1 KO/CD1 mice	Context	↓ CFR	Mikics et al. (2006)

A recent study described that the administration of AM251 prior to fear-conditioning training enhances the contextual fear response 24 h later, suggesting that endocannabinoids play a role in the acquisition of fear memory (Suzuki et al., 2004). Moreover, WIN 55,212–2, a CB1 receptor agonist, facilitated extinction of contextual fear conditioning in rats (Pamplona and Takahashi, 2006; Pamplona et al., 2006) whereas rimonabant disrupted it (Pamplona et al., 2006). In fear-conditioning tasks using auditory stimuli, fear extinction can be observed by repeated nonreinforced tone presentations, which decreases the responsiveness to the tone in the stimulus–response pathways because of habituation-like processes (Kamprath and Wotjak, 2004; McSweeney and Swindell, 2002). Tone presentation during an extinction trial triggered an increase in the levels of endocannabinoids in selected brain regions of wild-type mice (Marsicano et al., 2002). Moreover, genetic ablation or pharmacological blockade of CB1 receptors impairs the extinction of fear memories (Chhatwal et al., 2005; Marsicano et al., 2002; Suzuki et al., 2004) and spatial memories acquired under stressful conditions (Varvel and Lichtman, 2002; Varvel et al., 2005). Conversely, the treatment of rats with an inhibitor of cannabinoid reuptake, AM404, enhanced extinction (Chhatwal et al., 2005). In addition, animals that had received AM404 during extinction training exhibited less fear reinstatement.

Taken together, the results so far seems to indicate that extinction is the process that most critically depends on the endocannabinoid system (Chhatwal et al., 2005; Marsicano et al., 2002; Suzuki et al., 2004), even though the mechanism by which CB1 mediates fear extinction remains unclear. In any case, this is in line with the evidence that CB1 receptors facilitate systems that promote adaptation to innate stressful events (Hill and Gorzalka, 2005; Viveros et al., 2005). This has potential implications for the interaction of these processes, in particular because tone presentations after fear-conditioning or -sensitization procedures can be regarded as psychological stressors (Korte, 2001). Long-term fear extinction, in contrast, might well have common mechanisms with adaptation to stressors. In this context, it was described that the endocannabinoid system mediates habituation to repeated restraint stress (Patel et al., 2005). Pharmacological antagonism of CB1 receptors during the fifth restraint episode resulted in a reversal of the habituation-like reduction in Fos expression in the MPFC (Patel et al., 2005). Interestingly, the MPFC is known to play a important role in extinction of conditioned fear (Pare et al., 2004; Quirk et al., 2003) and therefore may be involved in the endocannabinoid-mediated long-term habituation. Thus, this information provides sources for understanding the role of CB1 receptors in short- and long-term adaptation to aversive situations and emphasizes the importance of CB1-mediated habituation in extinction of acquired fear.

IV. Brain Regions in which Endocannabinoids may Modulate Fear Conditioning

As mentioned before, CB1 cannabinoid receptors are widely distributed in the central nervous system, mainly in hippocampus, MPFC, amygdala, basal ganglia, and PAG (Herkenham et al., 1991; Tsou et al., 1998). Moreover, they are predominantly localized on axon terminals of GABAergic and glutamatergic neurons, and their activation decrease both GABA and glutamate release in the hippocampus, amygdala, and other regions of the brain (Freund et al., 2003). Modulation of these neurotransmitters could be involved in the regulation of emotional behavior by endocannabinoids. In addition, anandamide might also influence the release of corticotropin-releasing factor, cholecystokinin-octapeptide, acetylcholine, and noradrenaline (Beinfeld and Connolly, 2001; Davies et al., 2002; Weidenfeld et al., 1994). Studies had shown that the endocannabinoid transport inhibitor AM404 selectively increases levels of anandamide, but not 2-AG, in the rat MPFC, hippocampus, and thalamus, three brain regions which are intimately involved in the regulation of stress and conditioned fear responses (Cahill and McGaugh 1998; Fendt and Fanselow, 1999; Nestler et al., 2002). This drug evokes significant anxiolytic-like effects, which are prevented by rimonabant, suggesting the involvement of tonic AEA production in anxiety regulation (Bortolato et al., 2006; Lafenetre et al., 2007). These results agree with several pieces of evidence suggesting a role for AEA in the modulation of emotional responses to stress. For example, electrical footshocks have been shown to increase anandamide levels in the midbrain, while in mice physical restraint decreases these levels in the amygdala (Hohmann et al., 2005; Patel et al., 2004). Moreover, pharmacological blockade or genetic depletion of CB1 receptors increases normal reactions to acute stress, probably by disabling an endocannabinoid-mediated modulation of these reactions (Haller et al., 2004;

Navarro *et al.*, 1997; Uriguen *et al.*, 2004). Finally, the FAAH inhibitor URB597 enhances stress-coping behaviors, an effect that was prevented by rimonabant (Gobbi *et al.*, 2005; Kathuria *et al.*, 2003).

Memory impairment evoked by intrahippocampal administration of CB1 agonist such as Δ^9-THC, CP55,940, and AEA was observed in aversive tasks, suggesting that some degree of emotion (aversiveness) would be required in order to recruit the CB1-sensitive response (Egashira *et al.*, 2002; Lichtman *et al.*, 1995).

CB1 receptor is specially prominent in the hippocampus (Ameri, 1999; Hampson and Deadwyler, 1999). Consistently, cannabinoids acting upon CB1 receptors have been shown to inhibit the release of glutamate, GABA, acetylcholine, and noradrenaline in hippocampal preparations (Davies *et al.*, 2002). As described earlier, contextual fear conditioning, like spatial learning, is strong dependent on the hippocampus (Anagnostaras *et al.*, 1999; Barrientos *et al.*, 2002; Knowlton and Fanselow, 1998; Matus-Amat *et al.*, 2004; Resstel *et al.*, 2008c). Bilateral infusion of AM251 into the DH of rats facilitated the reconsolidation of contextual fear-conditioning memory. In contrast, local infusion of anandamide blocked memory reconsolidation, an effect that was dependent on CB1 receptors. These results support a CB1-mediated role of the hippocampal endocannabinoid system in the modulation of the memory reconsolidation. Moreover, AM251 blocked memory extinction whereas the administration of anandamide facilitated it (de Oliveira Alvares *et al.*, 2008). These results suggest a possible role of the hippocampal endocannabinoid system as a switching mechanism deciding which processes will take place, either maintaining the original memory (reconsolidation) or promoting a new learning (extinction) (Table 16.3).

Systemic activation or blockade of cannabinoid CB1 receptors modulates emotional associative learning and memory formation in neurons located in the MPFC that receive functional input from the BLA (Laviolette and Grace, 2006). Thus, CB1 receptors within the amygdala–MPFC circuit can potently modulate emotional associative learning processes during both the acquisition and expression of emotionally salient conditioned fear associations. Recent results of our laboratory have demonstrated that CB1 receptors within the MPFC strongly modulate the expression of behavior and autonomic responses associated to contextual fear conditioning (unpublished data). Administration of either anandamide or AM404 before reexposition to an aversively conditioned context reduces the expression of contextual fear. Moreover, the effects were inhibited by AM251. The latter was able to increase the freezing behavior by itself, supporting a tonic endocannabinoid role in this region. However, the protocol induced strong behavioral and autonomic responses, which could have caused a ceiling effect. Previous studies have indicated that interference with MPFC-glutamate-mediated neurotransmission modify contextual fear responses (Resstel *et al.*, 2008b). A CB1-mediated inhibition

Table 16.3 Intracerebral effects of drugs that affect endocannabinoid-mediated neurotransmission in fear conditioning

Drug	Mechanism	Brain area/dose	Species	Conditioning	Time of injection	Effect	Reference
AM251	CB1 antagonist	DH/−0.27 and 5.5 ng	Rats	Context	After conditioning test	↑ Reconsolidation	de Oliveira Alvares et al. (2008)
AEA	CB1 agonist	DH/−0.17 ng	Rats	context	After conditioning test	↓ Reconsolidation	de Oliveira Alvares et al. (2008)
AM251	CB1 antagonist	MPFC/100 pmol	Rats	Context	Pretraining	↑ CFR	Resstel et al. data not published
AEA	CB1 agonist	MPFC/5 pmol	Rats	Context	Pretraining	↓ CFR	Resstel et al. (unpublished data)
AM404	Inhibitor of cannabinoid reuptake	MPFC/−50 pmol	Rats	Context	Pretraining	↓ CFR	Resstel et al. (unpublished data)
AM251	CB1 antagonist	dlPAG/100–300 pmol	Rats	Context	Pretraining	No effect	Resstel et al. (2008d)
AEA	CB1 agonist	dlPAG/5 pmol	Rats	Context	Pretraining	↓ CFR	Resstel et al. (2008d)
AM404	Inhibitor of cannabinoid reuptake	dlPAG/−50 pmol	Rats	Context	Pretraining	↓ CFR	Resstel et al. (2008d)

of glutamate release, therefore, could be a potential mechanism of these effects.

CB1 receptors are expressed in the dorsolateral column of the PAG (dlPAG) and modulate defensive responses (Egertova *et al.*, 2003; Herkenham *et al.*, 1991; Lisboa *et al.*, 2008; Moreira *et al.*, 2007). Moreover, conditioned fear to context activates the dlPAG (Carrive *et al.*, 1997, 2000). Studies from our laboratory showed that intra-dlPAG injection of AEA or AM404 immediately before the test session decreases conditioned fear responses, an effect blocked by AM251 (Resstel *et al.*, 2008d). This observation reinforces the idea of an important role of the dlPAG on the expression of conditioned fear responses. It also indicates that local endocannabinoid-mediated neurotransmission can modulate these responses and could be involved in anxiety modulation (Bortolato *et al.*, 2006; Chhatwal *et al.*, 2005; Kathuria *et al.*, 2003; Patel and Hillard, 2006).

CS tone presentation during extinction increases endogenous levels of anandamide in the amygdala (Kamprath *et al.*, 2006; Marsicano *et al.*, 2002). It has been proposed that endocannabinoids facilitate extinction of aversive memories through their selective inhibitory effects on local inhibitory networks in this region. CB1 mRNA is densely expressed within the rat BLA, a region implicated in the extinction of conditioned fear (Chhatwal *et al.*, 2005). Acute CB1 receptor antagonist administration evokes anxiety responses either alone or after long-term exposure to cannabinoids (Navarro *et al.*, 1997), which reduces corticotropin-releasing hormone level in the central nucleus of amygdala (Rodriguez de Fonseca *et al.*, 1997). The central nucleus is the major amygdala output to the autonomic and endocrine centers of the brain and mediates stress and fear responses to aversive stimuli, which often correlates with elevated corticotropin-releasing hormone level (Davis, 2000). Therefore, the apparently lower density of CB1 receptors in the central nucleus of amygdala, in contrast with the high density in the basolateral complex, may seem to be surprising. However, the aversive or appetitive nature of a stimulus is processed in part by the basolateral complex, and afferent inputs from these nuclei to the central nucleus constitute an important pathway in the induction of different kinds of emotional responses (Everitt *et al.*, 2000).

As described above, recent anatomical and physiological findings have revealed that GABAergic neurons of the so-called intercalated nuclei may serve as an important intermediate station in this pathway by generating feed forward inhibition of the central nucleus after activation of the BLA (Pare and Smith, 1993; Royer *et al.*, 1999). Consistently, tone presentation during extinction trials resulted in elevated levels of endo-cannabinoids in the basolateral amygdala complex (Kamprath *et al.*, 2006; Marsicano *et al.*, 2002). In the BLA endocannabinoids and CB1 were crucially involved in long-term depression of GABA-mediated inhibitory currents (Katona *et al.*, 2001). Therefore, by reducing the inhibitory tone on BLA pyramidal cells, cannabinoids may indirectly enhance the activity of GABAergic cell population in the intercalated nuclei and thereby inhibit neuronal activity in the central nucleus. Thus, the inhibition of GABA release from axon terminals of local GABAergic interneurons in the BLA by presynaptic CB1 receptors may constitute an important neurobiological substrate of cannabinoid-induced emotional responses (Pare et al., 2004). Despite these pieces of evidence linking the amygdaloid CB1-mediated neurotransmission to fear conditioning, no study so far has investigated the effects of direct intra-amygdala administration of endocannabinoid-related drugs on this process.

V. Conclusion

The results reviewed above clearly indicate that endocannabinoid play an important modulatory role in fear conditioning. They also stressed several contradictory data regarding their specific role. Future experiments should address these contradictions by, for example, testing in the same study using context and cue-specific conditioned stimuli AM251 and rimonabant and different animal species. Moreover, experiments using direct injections of cannabinoid-related drugs into structures related to fear conditioning, particularly the amygdaloid complex, would also help to unveil the role of endocannabinoid in this process.

However, despite the controversies, the results reviewed here corroborate the proposal that endocannabinoid act on-demand as a protective mechanism, being released after a stressful stimulus to inhibit its consequences. Thus, drugs that potentiate endocannabinoid may be potentially useful in psychiatric conditions such as posttraumatic stress disorder. Considering the relevance of fear conditioning paradigms, understanding how the endocannabinoid system interfere with acquisition, consolidation, expression, or extinction of aversive responses could offer new opportunities for more specific pharmacological approaches.

REFERENCES

Ameri, A. (1999). The effects of cannabinoids on the brain. *Prog. Neurobiol.* **58,** 315–348.
Anagnostaras, S. G., Maren, S., and Fanselow, M. S. (1999). Temporally graded retrograde amnesia of contextual fear after hippocampal damage in rats: Within-subjects examination. *J. Neurosci.* **19,** 1106–1114.
Arenos, J. D., Musty, R. E., and Bucci, D. J. (2006). Blockade of cannabinoid CB1 receptors alters contextual learning and memory. *Eur. J. Pharmacol.* **539,** 177–183.
Baeg, E. H., Kim, Y. B., Jang, J., Kim, H. T., Mook-Jung, I., and Jung, M. W. (2001). Fast spiking and regular spiking neural correlates of fear conditioning in the medial prefrontal cortex of the rat. *Cereb. Cortex* **11,** 441–451.

Barrientos, R. M., O'Reilly, R. C., and Rudy, J. W. (2002). Memory for context is impaired by injecting anisomycin into dorsal hippocampus following context exploration. *Behav. Brain Res.* **134,** 299–306.

Beck, C. H., and Fibiger, H. C. (1995). Conditioned fear-induced changes in behavior and in the expression of the immediate early gene c-fos: With and without diazepam pretreatment. *J. Neurosci.* **15,** 709–720.

Beinfeld, M. C., and Connolly, K. (2001). Activation of CB1 cannabinoid receptors in rat hippocampal slices inhibits potassium-evoked cholecystokinin release, a possible mechanism contributing to the spatial memory defects produced by cannabinoids. *Neurosci. Lett.* **301,** 69–71.

Bishop, S. J. (2007). Neurocognitive mechanisms of anxiety: An integrative account. *Trends Cogn. Sci.* **11,** 307–316.

Blanchard, D. C., and Blanchard, R. J. (1972). Innate and conditioned reactions to threat in rats with amygdaloid lesions. *J. Comp. Physiol. Psychol.* **81,** 281–290.

Bortolato, M., Campolongo, P., Mangieri, R. A., Scattoni, M. L., Frau, R., Trezza, V., La Rana, G., Russo, R., Calignano, A., Gessa, G. L., Cuomo, V., and Piomelli, D. (2006). Anxiolytic-like properties of the anandamide transport inhibitor AM404. *Neuropsychopharmacology* **31,** 2652–2659.

Bouton, M. E. (2004). Context and behavioral processes in extinction. *Learn. Mem.* **11,** 485–494.

Brown, A. J. (2007). Novel cannabinoid receptors. *Br. J. Pharmacol.* **152,** 567–575.

Cahill, L., and McGaugh, J. L. (1998). Mechanisms of emotional arousal and lasting declarative memory. *Trends Neurosci.* **21,** 294–299.

Cannistraro, P. A., and Rauch, S. L. (2003). Neural circuitry of anxiety: Evidence from structural and functional neuroimaging studies. *Psychopharmacol. Bull.* **37,** 8–25.

Carrive, P., Leung, P., Harris, J., and Paxinos, G. (1997). Conditioned fear to context is associated with increased Fos expression in the caudal ventrolateral region of the midbrain periaqueductal gray. *Neuroscience* **78,** 165–177.

Carrive, P., Lee, J., and Su, A. (2000). Lidocaine blockade of amygdala output in fear-conditioned rats reduces Fos expression in the ventrolateral periaqueductal gray. *Neuroscience* **95,** 1071–1080.

Chhatwal, J. P., Davis, M., Maguschak, K. A., and Ressler, K. J. (2005). Enhancing cannabinoid neurotransmission augments the extinction of conditioned fear. *Neuropsychopharmacology* **30,** 516–524.

Conde, F., Maire-Lepoivre, E., Audinat, E., and Crepel, F. (1995). Afferent connections of the medial frontal cortex of the rat. II. Cortical and subcortical afferents. *J. Comp. Neurol.* **352,** 567–593.

Davies, S. N., Pertwee, R. G., and Riedel, G. (2002). Functions of cannabinoid receptors in the hippocampus. *Neuropharmacology* **42,** 993–1007.

Davis, M. (1992). The role of the amygdala in fear and anxiety. *Annu. Rev. Neurosci.* **15,** 353–375.

Davis, M. (2000). The role of amygdala in conditioned and unconditioned fear and anxiety. *In* "The Amygdala" (J. P. Aggleton, Ed.), 2nd edn., pp. 213–287. Oxford University Press, Oxford.

de Oliveira Alvares, L., Pasqualini Genro, B., Diehl, F., Molina, V. A., and Quillfeldt, J. A. (2008). Opposite action of hippocampal CB1 receptors in memory reconsolidation and extinction. *Neuroscience* **154,** 1648–1655.

Devane, W. A., Dysarz, F. A. 3rd, Johnson, M. R., Melvin, L. S., and Howlett, A. C. (1988). Determination and characterization of a cannabinoid receptor in rat brain. *Mol. Pharmacol.* **34,** 605–613.

Devane, W. A., Hanus, L., Breuer, A., Pertwee, R. G., Stevenson, L. A., Griffin, G., Gibson, D., Mandelbaum, A., Etinger, A., and Mechoulam, R. (1992). Isolation and

structure of a brain constituent that binds to the cannabinoid receptor. *Science* **258,** 1946–1949.
Dielenberg, R. A., Carrive, P., and McGregor, I. S. (2001). The cardiovascular and behavioral response to cat odor in rats: Unconditioned and conditioned effects. *Brain Res.* **897,** 228–237.
Egashira, N., Mishima, K., Iwasaki, K., and Fujiwara, M. (2002). Intracerebral microinjections of delta 9-tetrahydrocannabinol: Search for the impairment of spatial memory in the eight-arm radial maze in rats. *Brain Res.* **952,** 239–245.
Egertova, M., Giang, D. K., Cravatt, B. F., and Elphick, M. R. (1998). A new perspective on cannabinoid signalling: Complementary localization of fatty acid amide hydrolase and the CB1 receptor in rat brain. *Proc. Biol. Sci.* **265,** 2081–2085.
Egertova, M., Cravatt, B. F., and Elphick, M. R. (2003). Comparative analysis of fatty acid amide hydrolase and cb(1) cannabinoid receptor expression in the mouse brain: Evidence of a widespread role for fatty acid amide hydrolase in regulation of endocannabinoid signaling. *Neuroscience* **119,** 481–496.
Everitt, B., Cardinal, R., Hall, J., Parkinson, J., and Robbins, T. (2000). Differential involvement of amygdala subsystems in appetitive conditioning and drug addiction. *In* "The Amygdala: A Functional Analysis" (J. P. Aggleton, Ed.), 2nd edn., pp. 353–390. Oxford University Press, Oxford.
Fanselow, M. S. (1980). Conditioned and unconditional components of post-shock freezing. *Pavlov J. Biol. Sci.* **15,** 177–182.
Fanselow, M. S., and Helmstetter, F. J. (1988). Conditional analgesia, defensive freezing, and benzodiazepines. *Behav. Neurosci.* **102,** 233–243.
Fanselow, M. S., and Kim, J. J. (1994). Acquisition of contextual Pavlovian fear conditioning is blocked by application of an NMDA receptor antagonist D,L-2-amino-5-phosphonovaleric acid to the basolateral amygdala. *Behav. Neurosci.* **108,** 210–212.
Fendt, M., and Fanselow, M. S. (1999). The neuroanatomical and neurochemical basis of conditioned fear. *Neurosci. Biobehav. Rev.* **23,** 743–760.
Finn, D. P., Beckett, S. R., Richardson, D., Kendall, D. A., Marsden, C. A., and Chapman, V. (2004). Evidence for differential modulation of conditioned aversion and fear-conditioned analgesia by CB1 receptors. *Eur. J. Neurosci.* **20,** 848–852.
Frankland, P. W., and Bontempi, B. (2005). The organization of recent and remote memories. *Nat. Rev. Neurosci.* **6,** 119–130.
Freund, T. F., Katona, I., and Piomelli, D. (2003). Role of endogenous cannabinoids in synaptic signaling. *Physiol. Rev.* **83,** 1017–1066.
Galeno, T. M., Van Hoesen, G. W., and Brody, M. J. (1984). Central amygdaloid nucleus lesion attenuates exaggerated hemodynamic responses to noise stress in the spontaneously hypertensive rat. *Brain Res.* **291,** 249–259.
Garcia, R., Vouimba, R. M., Baudry, M., and Thompson, R. F. (1999). The amygdala modulates prefrontal cortex activity relative to conditioned fear. *Nature* **402,** 294–296.
Gentile, C. G., Jarrell, T. W., Teich, A., McCabe, P. M., and Schneiderman, N. (1986). The role of amygdaloid central nucleus in the retention of differential pavlovian conditioning of bradycardia in rabbits. *Behav. Brain Res.* **20,** 263–273.
Giuffrida, A., Beltramo, M., and Piomelli, D. (2001). Mechanisms of endocannabinoid inactivation: Biochemistry and pharmacology. *J. Pharmacol. Exp. Ther.* **298,** 7–14.
Gobbi, G., Bambico, F. R., Mangieri, R., Bortolato, M., Campolongo, P., Solinas, M., Cassano, T., Morgese, M. G., Debonnel, G., Duranti, A., Tontini, A., Tarzia, G., *et al.* (2005). Antidepressant-like activity and modulation of brain monoaminergic transmission by blockade of anandamide hydrolysis. *Proc. Natl. Acad. Sci. USA* **102,** 18620–18625.
Grace, A. A., and Rosenkranz, J. A. (2002). Regulation of conditioned responses of basolateral amygdala neurons. *Physiol. Behav.* **77,** 489–493.

Haller, J., Varga, B., Ledent, C., and Freund, T. F. (2004). CB1 cannabinoid receptors mediate anxiolytic effects: Convergent genetic and pharmacological evidence with CB1-specific agents. *Behav. Pharmacol.* **15,** 299–304.

Haller, J., Matyas, F., Soproni, K., Varga, B., Barsy, B., Nemeth, B., Mikics, E., Freund, T. F., and Hajos, N. (2007). Correlated species differences in the effects of cannabinoid ligands on anxiety and on GABAergic and glutamatergic synaptic transmission. *Eur. J. Neurosci.* **25,** 2445–2456.

Hampson, R. E., and Deadwyler, S. A. (1999). Cannabinoids, hippocampal function and memory. *Life Sci.* **65,** 715–723.

Herkenham, M., Lynn, A. B., Johnson, M. R., Melvin, L. S., de Costa, B. R., and Rice, K. C. (1991). Characterization and localization of cannabinoid receptors in rat brain: A quantitative in vitro autoradiographic study. *J. Neurosci.* **11,** 563–583.

Hill, M. N., and Gorzalka, B. B. (2005). Pharmacological enhancement of cannabinoid CB1 receptor activity elicits an antidepressant-like response in the rat forced swim test. *Eur. Neuropsychopharmacol.* **15,** 593–599.

Hohmann, A. G., Suplita, R. L., Bolton, N. M., Neely, M. H., Fegley, D., Mangieri, R., Krey, J. F., Walker, J. M., Holmes, P. V., Crystal, J. D., Duranti, A., Tontini, A., *et al.* (2005). An endocannabinoid mechanism for stress-induced analgesia. *Nature* **435,** 1108–1112.

Howlett, A. C., Barth, F., Bonner, T. I., Cabral, G., Casellas, P., Devane, W. A., Felder, C. C., Herkenham, M., Mackie, K., Martin, B. R., Mechoulam, R., and Pertwee, R. G. (2002). International Union of Pharmacology. XXVII. Classification of cannabinoid receptors. *Pharmacol. Rev.* **54,** 161–202.

Iwata, J., LeDoux, J. E., Meeley, M. P., Arneric, S, and Reis, S (1986). Intrinsic neurons in the amygdaloid field projected to by the medial geniculate body mediate emotional responses conditioned to acoustic stimuli. *Brain Res.* **383,** 195–214.

Jay, T. M., and Witter, M. P. (1991). Distribution of hippocampal CA1 and subicular efferents in the prefrontal cortex of the rat studied by means of anterograde transport of Phaseolus vulgaris-leucoagglutinin. *J. Comp. Neurol.* **313,** 574–586.

Kamprath, K., and Wotjak, C. T. (2004). Nonassociative learning processes determine expression and extinction of conditioned fear in mice. *Learn. Mem.* **11,** 770–786.

Kamprath, K., Marsicano, G., Tang, J., Monory, K., Bisogno, T., Di Marzo, V., Lutz, B., and Wotjak, C. T. (2006). Cannabinoid CB1 receptor mediates fear extinction via habituation-like processes. *J. Neurosci.* **26,** 6677–6686.

Kathuria, S., Gaetani, S., Fegley, D., Valino, F., Duranti, A., Tontini, A., Mor, M., Tarzia, G., La Rana, G., Calignano, A., Giustino, A., Tattoli, M., *et al.* (2003). Modulation of anxiety through blockade of anandamide hydrolysis. *Nat. Med.* **9,** 76–81.

Katona, I., Rancz, E. A., Acsady, L., Ledent, C., Mackie, K., Hajos, N., and Freund, T. F. (2001). Distribution of CB1 cannabinoid receptors in the amygdala and their role in the control of GABAergic transmission. *J. Neurosci.* **21,** 9506–9518.

Kesner, R. P., Hunt, M. E., Williams, J. M., and Long, J. M. (1996). Prefrontal cortex and working memory for spatial response, spatial location, and visual object information in the rat. *Cereb. Cortex* **6,** 311–318.

Kim, J. J., and Fanselow, M. S. (1992). Modality-specific retrograde amnesia of fear. *Science* **256,** 675–677.

Knowlton, B. J., and Fanselow, M. S. (1998). The hippocampus, consolidation and on-line memory. *Curr. Opin. Neurobiol.* **8,** 293–296.

Korte, S. M. (2001). Corticosteroids in relation to fear, anxiety and psychopathology. *Neurosci. Biobehav. Rev.* **25,** 117–142.

Lafenetre, P., Chaouloff, F., and Marsicano, G. (2007). The endocannabinoid system in the processing of anxiety and fear and how CB1 receptors may modulate fear extinction. *Pharmacol. Res.* **56,** 367–381.

Laviolette, S. R., and Grace, A. A. (2006). Cannabinoids potentiate emotional learning plasticity in neurons of the medial prefrontal cortex through basolateral amygdala inputs. *J. Neurosci.* **26,** 6458–6468.

Laviolette, S. R., Lipski, W. J., and Grace, A. A. (2005). A subpopulation of neurons in the medial prefrontal cortex encodes emotional learning with burst and frequency codes through a dopamine D4 receptor-dependent basolateral amygdala input. *J. Neurosci.* **25,** 6066–6075.

LeDoux, J. E. (1992). Brain mechanisms of emotion and emotional learning. *Curr. Opin. Neurobiol.* **2,** 191–197.

LeDoux, J. (1996). Emotional networks and motor control: A fearful view. *Prog. Brain Res.* **107,** 437–446.

LeDoux, J. E. (2000). Emotion circuits in the brain. *Annu. Rev. Neurosci.* **23,** 155–184.

LeDoux, J. E., Iwata, J., Cicchetti, P., and Reis, D. J. (1988). Different projections of the central amygdaloid nucleus mediate autonomic and behavioral correlates of conditioned fear. *J. Neurosci.* **8,** 2517–2529.

Lichtman, A. H., Dimen, K. R., and Martin, B. R. (1995). Systemic or intrahippocampal cannabinoid administration impairs spatial memory in rats. *Psychopharmacology (Berl.)* **119,** 282–290.

Lisboa, S. F., Resstel, L. B., Aguiar, D. C., and Guimaraes, F. S. (2008). Activation of cannabinoid CB(1) receptors in the dorsolateral periaqueductal gray induces anxiolytic effects in rats submitted to the Vogel conflict test. *Eur. J. Pharmacol.* **593,** 73–78.

Malkani, S., and Rosen, J. B. (2000). Induction of NGFI-B mRNA following contextual fear conditioning and its blockade by diazepam. *Brain Res. Mol. Brain Res.* **80,** 153–165.

Maren, S. (1996). Synaptic transmission and plasticity in the amygdala. An emerging physiology of fear conditioning circuits. *Mol. Neurobiol.* **13,** 1–22.

Maren, S., and Fanselow, M. S. (1995). Synaptic plasticity in the basolateral amygdala induced by hippocampal formation stimulation *in vivo*. *J. Neurosci.* **15,** 7548–7564.

Maren, S., and Fanselow, M. S. (1997). Electrolytic lesions of the fimbria/fornix, dorsal hippocampus, or entorhinal cortex produce anterograde deficits in contextual fear conditioning in rats. *Neurobiol. Learn. Mem.* **67,** 142–149.

Maren, S., and Quirk, G. J. (2004). Neuronal signalling of fear memory. *Nat. Rev. Neurosci.* **5,** 844–852.

Maren, S., Aharonov, G., and Fanselow, M. S. (1996a). Retrograde abolition of conditional fear after excitotoxic lesions in the basolateral amygdala of rats: Absence of a temporal gradient. *Behav. Neurosci.* **110,** 718–726.

Maren, S., Aharonov, G., Stote, D. L., and Fanselow, M. S. (1996b). N-methyl-D-aspartate receptors in the basolateral amygdala are required for both acquisition and expression of conditional fear in rats. *Behav. Neurosci.* **110,** 1365–1374.

Maren, S., Aharonov, G., and Fanselow, M. S. (1997). Neurotoxic lesions of the dorsal hippocampus and Pavlovian fear conditioning in rats. *Behav. Brain Res.* **88,** 261–274.

Marsicano, G., Wotjak, C. T., Azad, S. C., Bisogno, T., Rammes, G., Cascio, M. G., Hermann, H., Tang, J., Hofmann, C., Zieglgansberger, W., Di Marzo, V., and Lutz, B. (2002). The endogenous cannabinoid system controls extinction of aversive memories. *Nature* **418,** 530–534.

Matus-Amat, P., Higgins, E. A., Barrientos, R. M., and Rudy, J. W. (2004). The role of the dorsal hippocampus in the acquisition and retrieval of context memory representations. *J. Neurosci.* **24,** 2431–2439.

McKinney, M. K., and Cravatt, B. F. (2005). Structure and function of fatty acid amide hydrolase. *Annu. Rev. Biochem.* **74,** 411–432.

McSweeney, F. K., and Swindell, S. (2002). Common processes may contribute to extinction and habituation. *J. Gen. Psychol.* **129,** 364–400.

Mechoulam, R., Ben-Shabat, S., Hanus, L., Ligumsky, M., Kaminski, N. E., Schatz, A. R., Gopher, A., Almog, S., Martin, B. R., Compton, D. R., Pertweee, R. G., Griffine, G., et al. (1995). Identification of an endogenous 2-monoglyceride, present in canine gut, that binds to cannabinoid receptors. *Biochem. Pharmacol.* **50,** 83–90.

Mikics, E., Dombi, T., Barsvari, B., Varga, B., Ledent, C., Freund, T. F., and Haller, J. (2006). The effects of cannabinoids on contextual conditioned fear in CB1 knockout and CD1 mice. *Behav. Pharmacol.* **17,** 223–230.

Milad, M. R., and Quirk, G. J. (2002). Neurons in medial prefrontal cortex signal memory for fear extinction. *Nature* **420,** 70–74.

Milad, M. R., Vidal-Gonzalez, I., and Quirk, G. J. (2004). Electrical stimulation of medial prefrontal cortex reduces conditioned fear in a temporally specific manner. *Behav. Neurosci.* **118,** 389–394.

Moreira, F. A., Aguiar, D. C., and Guimaraes, F. S. (2007). Anxiolytic-like effect of cannabinoids injected into the rat dorsolateral periaqueductal gray. *Neuropharmacology* **52,** 958–965.

Munro, S., Thomas, K. L., and Abu-Shaar, M. (1993). Molecular characterization of a peripheral receptor for cannabinoids. *Nature* **365,** 61–65.

Myers, K. M., and Davis, M. (2007). Mechanisms of fear extinction. *Mol. Psychiatry* **12,** 120–150.

Navarro, M., Hernandez, E., Munoz, R. M., del Arco, I., Villanua, M. A., Carrera, M. R., and Rodriguez de Fonseca, F. (1997). Acute administration of the CB1 cannabinoid receptor antagonist SR 141716A induces anxiety-like responses in the rat. *Neuroreport* **8,** 491–496.

Nestler, E. J., Barrot, M., DiLeone, R. J., Eisch, A. J., Gold, S. J., and Monteggia, L. M. (2002). Neurobiology of depression. *Neuron* **34,** 13–25.

Pacher, P., Batkai, S., and Kunos, G. (2006). The endocannabinoid system as an emerging target of pharmacotherapy. *Pharmacol. Rev.* **58,** 389–462.

Pamplona, F. A., and Takahashi, R. N. (2006). WIN 55212–2 impairs contextual fear conditioning through the activation of CB1 cannabinoid receptors. *Neurosci. Lett.* **397,** 88–92.

Pamplona, F. A., Prediger, R. D., Pandolfo, P., and Takahashi, R. N. (2006). The cannabinoid receptor agonist WIN 55,212–2 facilitates the extinction of contextual fear memory and spatial memory in rats. *Psychopharmacology (Berlin)* **188,** 641–649.

Pare, D., and Smith, Y. (1993). The intercalated cell masses project to the central and medial nuclei of the amygdala in cats. *Neuroscience* **57,** 1077–1090.

Pare, D., and Smith, Y. (1998). Intrinsic circuitry of the amygdaloid complex: Common principles of organization in rats and cats. *Trends. Neurosci.* **21,** 240–241.

Pare, D., Quirk, G. J., and Ledoux, J. E. (2004). New vistas on amygdala networks in conditioned fear. *J. Neurophysiol.* **92,** 1–9.

Patel, S., and Hillard, C. J. (2006). Pharmacological evaluation of cannabinoid receptor ligands in a mouse model of anxiety: Further evidence for an anxiolytic role for endogenous cannabinoid signaling. *J. Pharmacol. Exp. Ther.* **318,** 304–311.

Patel, S., Roelke, C. T., Rademacher, D. J., Cullinan, W. E., and Hillard, C. J. (2004). Endocannabinoid signaling negatively modulates stress-induced activation of the hypothalamic-pituitary-adrenal axis. *Endocrinology* **145,** 5431–5438.

Patel, S., Roelke, C. T., Rademacher, D. J., and Hillard, C. J. (2005). Inhibition of restraint stress-induced neural and behavioural activation by endogenous cannabinoid signalling. *Eur. J. Neurosci.* **21,** 1057–1069.

Phillips, R. G., and LeDoux, J. E. (1992). Differential contribution of amygdala and hippocampus to cued and contextual fear conditioning. *Behav. Neurosci.* **106,** 274–285.

Phillips, R. G., and LeDoux, J. E. (1994). Lesions of the dorsal hippocampal formation interfere with background but not foreground contextual fear conditioning. *Learn. Mem.* **1,** 34–44.

Pitkanen, A., Savander, V., and LeDoux, J. E. (1997). Organization of intra-amygdaloid circuitries in the rat: An emerging framework for understanding functions of the amygdala. *Trends Neurosci.* **20,** 517–523.

Price, M. R., Baillie, G. L., Thomas, A., Stevenson, L. A., Easson, M., Goodwin, R., McLean, A., McIntosh, L., Goodwin, G., Walker, G., Westwood, P., Marrs, J., et al. (2005). Allosteric modulation of the cannabinoid CB1 receptor. *Mol. Pharmacol.* **68,** 1484–1495.

Quirk, G. J., and Mueller, D. (2008). Neural mechanisms of extinction learning and retrieval. *Neuropsychopharmacology* **33,** 56–72.

Quirk, G. J., Likhtik, E., Pelletier, J. G., and Pare, D. (2003). Stimulation of medial prefrontal cortex decreases the responsiveness of central amygdala output neurons. *J. Neurosci.* **23,** 8800–8807.

Ragozzino, M. E., Adams, S., and Kesner, R. P. (1998). Differential involvement of the dorsal anterior cingulate and prelimbic-infralimbic areas of the rodent prefrontal cortex in spatial working memory. *Behav. Neurosci.* **112,** 293–303.

Resstel, L. B., Joca, S. R., Guimaraes, F. G., and Correa, F. M. (2006a). Involvement of medial prefrontal cortex neurons in behavioral and cardiovascular responses to contextual fear conditioning. *Neuroscience* **143,** 377–385.

Resstel, L. B., Joca, S. R., Moreira, F. A., Correa, F. M., and Guimaraes, F. S. (2006b). Effects of cannabidiol and diazepam on behavioral and cardiovascular responses induced by contextual conditioned fear in rats. *Behav. Brain Res.* **172,** 294–298.

Resstel, L. B., Alves, F. H., Reis, D. G., Crestani, C. C., Correa, F. M., and Guimaraes, F. S. (2008a). Anxiolytic-like effects induced by acute reversible inactivation of the bed nucleus of stria terminalis. *Neuroscience* **154,** 869–876.

Resstel, L. B., Correa, F. M., and Guimaraes, F. S. (2008b). The expression of contextual fear conditioning involves activation of an NMDA receptor-nitric oxide pathway in the medial prefrontal cortex. *Cereb. Cortex* **18,** 2027–2035.

Resstel, L. B., Joca, S. R., Correa, F. M., and Guimaraes, F. S. (2008c). Effects of reversible inactivation of the dorsal hippocampus on the behavioral and cardiovascular responses to an aversive conditioned context. *Behav. Pharmacol.* **19,** 137–144.

Resstel, L. B., Lisboa, S. F., Aguiar, D. C., Correa, F. M., and Guimaraes, F. S. (2008d). Activation of CB1 cannabinoid receptors in the dorsolateral periaqueductal gray reduces the expression of contextual fear conditioning in rats. *Psychopharmacology (Berlin)* **198,** 405–411.

Rodriguez de Fonseca, F., Carrera, M. R., Navarro, M., Koob, G. F., and Weiss, F. (1997). Activation of corticotropin-releasing factor in the limbic system during cannabinoid withdrawal. *Science* **276,** 2050–2054.

Rosen, J. B. (2004). The neurobiology of conditioned and unconditioned fear: A neurobehavioral system analysis of the amygdala. *Behav. Cogn. Neurosci. Rev.* **3,** 23–41.

Royer, S., Martina, M., and Pare, D. (1999). An inhibitory interface gates impulse traffic between the input and output stations of the amygdala. *J. Neurosci.* **19,** 10575–10583.

Schafe, G. E., Bauer, E. P., Rosis, S., Farb, C. R., Rodrigues, S. M., and LeDoux, J. E. (2005). Memory consolidation of Pavlovian fear conditioning requires nitric oxide signaling in the lateral amygdala. *Eur. J. Neurosci.* **22,** 201–211.

Sesack, S. R., Deutch, A. Y., Roth, R. H., and Bunney, B. S. (1989). Topographical organization of the efferent projections of the medial prefrontal cortex in the rat: An anterograde tract-tracing study with Phaseolus vulgaris leucoagglutinin. *J. Comp. Neurol.* **290,** 213–242.

Sotres-Bayon, F., Bush, D. E., and LeDoux, J. E. (2004). Emotional perseveration: An update on prefrontal-amygdala interactions in fear extinction. *Learn. Mem.* **11,** 525–535.

Sotres-Bayon, F., Cain, C. K., and LeDoux, J. E. (2006). Brain mechanisms of fear extinction: Historical perspectives on the contribution of prefrontal cortex. *Biol. Psychiatry* **60,** 329–336.

Sullivan, G. M., Apergis, J., Bush, D. E., Johnson, L. R., Hou, M., and Ledoux, J. E. (2004). Lesions in the bed nucleus of the stria terminalis disrupt corticosterone and freezing responses elicited by a contextual but not by a specific cue-conditioned fear stimulus. *Neuroscience* **128,** 7–14.

Suzuki, A., Josselyn, S. A., Frankland, P. W., Masushige, S., Silva, A. J., and Kida, S. (2004). Memory reconsolidation and extinction have distinct temporal and biochemical signatures. *J. Neurosci.* **24,** 4787–4795.

Takagishi, M., and Chiba, T. (1991). Efferent projections of the infralimbic (area 25) region of the medial prefrontal cortex in the rat: An anterograde tracer PHA-L study. *Brain Res.* **566,** 26–39.

Tsou, K., Brown, S., Sanudo-Pena, M. C., Mackie, K., and Walker, J. M. (1998). Immunohistochemical distribution of cannabinoid CB1 receptors in the rat central nervous system. *Neuroscience* **83,** 393–411.

Uriguen, L., Perez-Rial, S., Ledent, C., Palomo, T., and Manzanares, J. (2004). Impaired action of anxiolytic drugs in mice deficient in cannabinoid CB1 receptors. *Neuropharmacology* **46,** 966–973.

Varvel, S. A., and Lichtman, A. H. (2002). Evaluation of CB1 receptor knockout mice in the Morris water maze. *J. Pharmacol. Exp. Ther.* **301,** 915–924.

Varvel, S. A., Anum, E. A., and Lichtman, A. H. (2005). Disruption of CB(1) receptor signaling impairs extinction of spatial memory in mice. *Psychopharmacology (Berlin)* **179,** 863–872.

Vianna, D. M., Allen, C., and Carrive, P. (2008). Cardiovascular and behavioral responses to conditioned fear after medullary raphe neuronal blockade. *Neuroscience* **153,** 1344–1353.

Viveros, M. P., Marco, E. M., and File, S. E. (2005). Endocannabinoid system and stress and anxiety responses. *Pharmacol. Biochem. Behav.* **81,** 331–342.

Weidenfeld, J., Feldman, S., and Mechoulam, R. (1994). Effect of the brain constituent anandamide, a cannabinoid receptor agonist, on the hypothalamo-pituitary-adrenal axis in the rat. *Neuroendocrinology* **59,** 110–112.

Wilson, R. I., and Nicoll, R. A. (2001). Endogenous cannabinoids mediate retrograde signalling at hippocampal synapses. *Nature* **410,** 588–592.

CHAPTER SEVENTEEN

Regulation of Gene Transcription and Keratinocyte Differentiation by Anandamide

Nicoletta Pasquariello,* Sergio Oddi,*,† Marinella Malaponti,†
and Mauro Maccarrone*,†

Contents

I. Introduction	442
II. Epidermis	445
A. Stages of epidermal differentiation	446
III. Transcriptional Control of Skin Differentiation	453
IV. Endocannabinoid System in Epidermis	455
V. Modulation of the Endocannabinoid System in Differentiating Keratinocytes	456
VI. Repression of Gene Transcription by Anandamide	457
VII. Conclusions	459
Acknowledgments	460
References	461

Abstract

Anandamide (AEA) is a member of an endogenous class of lipid mediators, known as endocannabinoids, which are involved in various biological processes. In particular, AEA regulates cell growth, differentiation, and death. Accumulating evidence demonstrates that AEA controls also epidermal differentiation, one of the best characterized mechanisms of cell specialization. Indeed, the epidermis is a keratinized multistratified epithelium that functions as a barrier to protect the organism from dehydration, mechanical trauma, and microbial insults. Its function is established during embryogenesis and is maintained during the whole life span of the organism, through a complex and tightly controlled program, termed epidermal terminal differentiation

---

* Department of Biomedical Sciences, University of Teramo, Teramo, Italy
† European Center for Brain Research (CERC)/Santa Lucia Foundation, Rome, Italy

(or cornification). Whereas the morphological changes that occur during cornification have been extensively studied, the molecular mechanisms that underlie this process remain poorly understood.

In this chapter, we summarize current knowledge about the molecular regulation of proliferation and terminal differentiation in mammalian epidermis. In this context, we show that endocannabinoids are finely regulated by, and can interfere with, the differentiation program. In addition, we review the role of AEA in the control of cornification, and show that it occurs by maintaining a transcriptional repression of gene expression through increased DNA methylation.

I. Introduction

Endocanabinoids (eCBs) represent a growing family of lipid signaling mediators found in the brain and peripheral tissues. These bioactive lipids comprise amides, esters, and ethers of long-chain polyunsaturated fatty acids. Among these, anandamide (N-arachidonoylethanolamine, AEA) and 2-arachidonoylglycerol (2-AG) are the major endogenous agonists of the cannabinoid receptors, thus mimicking some effects of Δ^9-tetrahydrocannabinol (THC), the psychoactive principle of *Cannabis sativa* preparations (Howlett, 2005). Type-1 (CB1R) and type-2 (CB2R) cannabinoid receptors belong to the rhodopsin family of the G-protein-coupled receptors, particularly those of the $G_{i/o}$ family (Howlett, 2002). The binding of the eCBs to these receptors triggers several signaling pathways, including the inhibition of adenylyl cyclase, the regulation of ionic currents (e.g., inhibition of voltage-gated L, N, and P/Q-type Ca^{2+} channels, activation of K^+ channels), the activation of focal adhesion kinase, of mitogen-activated protein kinase (MAPK), and of cytosolic phospholipase A_2, as well as the activation (CB1R) or the inhibition (CB2R) of nitric oxide synthase (Bari *et al.*, 2006; Di Marzo, 2008; Piomelli, 2003). Furthermore, ample evidence suggests that additional non-CB1/CB2 receptors may contribute to the behavioral, vascular, and immunological actions of eCBs, like a purported CB3 (GPR55) receptor (Lauckner *et al.*, 2008; Sawzdargo *et al.*, 1999), and the transient receptor potential vanilloid 1, TRPV1 (Starowicz *et al.*, 2007).

AEA is synthesized and released "on demand" from membrane glycerophospholipids of neurons and peripheral cells, through a two-step pathway. The first step consists in the transfer of the *sn*-acyl group of phospholipids onto the primary amine of phosphatidylethanolamine, to originate an N-arachidonoylphosphatidylethanolamine (NArPE) precursor. This reaction is catalyzed by a calcium-dependent *trans*-acylase (N-acyltransferase, NAT) (Sugiura *et al.*, 2002), although it can be catalyzed also by a recently discovered calcium-independent NAT (Jin et al., 2007). The second step is the hydrolysis of NArPE to produce AEA and phosphatidic acid, catalyzed by a specific N-acylphosphatidylethanolamine (NAPE)-hydrolyzing phospholipase D (NAPE-PLD) (Tsuboi et al., 2005). In addition, a parallel pathway exists through which AEA can be generated, consisting of a two-step process composed of a phospholipase C-catalyzed cleavage of NArPE, that yields phospho-AEA, followed by dephosphorylation to AEA by phosphatases (Liu et al., 2006). A third pathway for AEA biosynthesis has emerged in brains from *nape-pld* knockout mice that are still able to synthesize AEA by the sequential action of a fluorophosphonate-sensitive serine hydrolase and a metal-dependent phosphodiesterase (Simon and Cravatt, 2006).

The biological actions of AEA depend on its life span in the extracellular space, which in turn depends on cellular uptake and intracellular degradation. The existence of a protein responsible for the transport of AEA across the plasma membrane is currently under debate (Glaser et al., 2005). In fact, there are several lines of evidence that speak in favor of a selective AEA membrane transporter (AMT), yet the molecular identity of this putative entity remains elusive. At any rate, once taken up by the cell, AEA is degraded by fatty acid amide hydrolase (FAAH) (N-arachidonoylethanolamine amidohydrolase), to release arachidonic acid and ethanolamine (McKinney and Cravatt, 2005). Additionally, it should be mentioned that AEA is also a good substrate for a recently discovered N-acylethanolamine-hydrolyzing acid amidase (NAAA) (Tsuboi et al., 2005). Altogether AEA and its congeners, their metabolic enzymes, purported transporters and molecular targets form the "endocannabinoid system (ECS)." The ECS elements present in human keratinocytes and other skin cells are summarized in Table 17.1.

In the last decade, a large body of evidence has been accumulated to support the role of AEA as a lipid messenger, which elicits a variety of biological actions both in the central nervous system (CNS) and in the periphery (Bari et al., 2006; Klein, 2005; Piomelli, 2003). In addition, a robust therapeutic potential for the modulation of AEA signaling has been shown in an ever-growing number of pathological conditions, including chronic pain, arthritis, anxiety, glaucoma, eating disorders, and infertility (Di Marzo, 2008; Maccarrone et al., 2007; Piomelli, 2005; Wang et al., 2006).

Furthermore, AEA has emerged as a key mediator of cell growth, differentiation, and death (Gutzmán, 2003; Maccarrone, 2006), also through the modulation of gene expression (Pasquariello et al., 2008). In particular, it has been shown that AEA is involved in skin differentiation (Kariniemi et al., 1984; Maccarrone et al., 2003a), and in fact a fully functional ECS has been described in human keratinocytes (Maccarrone et al., 2003a). It should be stressed that epidermis is an ideal target tissue for

Table 17.1 Elements of the ECS demonstrated in human keratinocytes and other skin cells, and effect of differentiation (see text for details)

ECS member	Description	Cell/Tissue	Change upon differentiation
AEA	Prototype of fatty acid amides	HaCaT cells, NHEK cells, human skin	↓
NAT, NAPE-PLD	Enzymes responsible for AEA biosynthesis	HaCaT cells, NHEK cells, human skin	=
FAAH	Hydrolytic enzyme responsible for AEA degradation	HaCaT cells, NHEK cells, human skin	↑
CB1R	Type-1 cannabinoid receptor	Keratinocytes of spinous and granular layers, epithelial cells of hair follicles, differentiated sebocytes, HaCaT cells, NHEK cells, HPV-16 cells	=
CB2R	Type-2 cannabinoid receptor	Basal keratinocytes, infundibular hair follicle cells, root sheath and bulge cells, undifferentiated sebaceous cells, HPV-16 cells, sebocytes	N.D.
TRPV1	Vanilloid receptor	Human skin, hair follicle keratinocytes	N.D.
AMT	AEA membrane transporter	HaCaT cells, NHEK cells, human skin	↑

N.D., not determined.

differentiation studies, because both progenitor and differentiating cells can be identified and selected, by means of their distinctive molecular and morphological features.

The ECS has an important role in controlling the balance between proliferation and differentiation, as reported in several cellular models (Bellocchio *et al.*, 2008; Fride, 2008; Trezza *et al.*, 2008; Watson *et al.*, 2008; Wolf and Ullrich, 2008). For instance, in the CNS it has been demonstrated that eCBs regulate neuritogenesis, axonal growth, and synaptogenesis in differentiated neural cells (Galve-Roperth et al., 2006; Rueda et al., 2002). Interestingly, ECS is expressed in neural progenitor cells, where eCBs control differentiation (Aguado et al., 2007). eCBs have also been found to modulate signaling pathways that are more directly involved in the control of cell fate (Guzmàn, 2003; Maccarrone, 2006). Indeed, they stimulate p42/p44 MAPK, the extracellular signal-regulated kinase (ERK) (Bouaboula, 1996), the stress-activated kinases JUN amino-terminal kinase (JNK), and p38 MAPK. eCBs can stimulate also the AKT pathway (Gomez del Pulsar et al., 2000; Vivanco and Sawyers, 2002) that includes molecules able to inhibit specific transcription factors (TFs) implicated in growth control, differentiation, and apoptosis (Guzmàn, 2003).

Much alike the CNS, also epidermal development is finely regulated by a correct balance between proliferation, differentiation and death of totipotent progenitors. In particular, we have recently reported that AEA blocks differentiation of human keratinocytes by a CB1R-dependent mechanism that activates protein kinase C (PKC), leading to a decrease of typical differentiative markers like cornified envelope (CE), and transglutaminase (TG) activity (Maccarrone et al., 2003a). Taken together, these data suggest that eCBs are involved in both neural and epidermal differentiation, even though the underlying molecular mechanisms remain to be clarified.

II. Epidermis

Placed at the interface between the organism and the environment, the epidermis functions as a stable, water-proof barrier adapted to withstand a variety of physical, chemical, and biological insults. The epidermis also has immunologic functions and provides some protection of the skin from ultraviolet light via the pigment system. These functions are brought about by the particular histological organization of the epidermis, a multistratified squamous epithelium that is generated by the keratinocytes through a tightly regulated differentiation process, called epidermal terminal differentiation or cornification (Fuchs, 1990). The innermost layer of cuboidal cells—the basal layer—represents the proliferative compartment for generation and continuous renewal of the tissue. On detachment from the basement membrane, newborn keratinocytes migrate outward through successive stages of differentiation (represented by the spinous and granular layers) towards the terminal differentiation on the surface (the cornified layer); here they are transformed in flattened, enucleated squames, termed corneocytes. The latter cells are sloughed from the surface, and continually replaced by inner cells (Fig. 17.1).

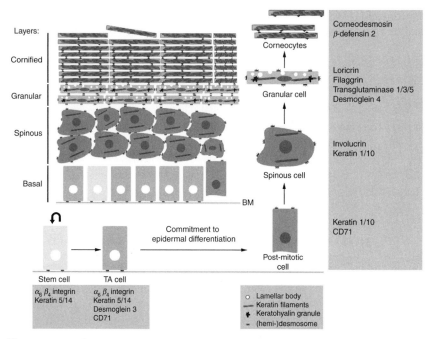

Figure 17.1 Schematic representation of the morphological and biochemical changes during keratinocyte differentiation. Epidermal stem cells reside in the basal layer of the epidermis, and can undergo self-renewal, thereby giving rise to another stem cell; alternatively, they can divide to give rise to TA cells, which are rapidly proliferating cells that ensure skin homeostasis by new epithelial cells. After a limited number of mitotic divisions, new epidermal basal cells exit the cell cycle and commit to differentiation as postmitotic cells. The latter cells detach from the basement membrane (BM) and start the process of keratinocyte cornification. As cells differentiate, they move up to form the different layers of the epidermis (spinous, granular, and cornified layers). Stem cells and TA cells can be distinguished from terminally differentiated cells on the basis of protein expression (bottom left box). Each stage of the differentiation programme (i.e., each epidermal layer) can also be distinguished according to the proteins that are expressed (box on the right side). A legend to the symbols used is reported in the bottom right box.

A. Stages of epidermal differentiation

Basal layer. The epidermis is a highly dynamic, self-renewing tissue, in which the corneocytes lost from the surface are continuously replaced by the proliferation and the subsequent differentiation of new epithelial cells. Like in most other tissues—the epidermal homeostasis is believed to be ensured by a population of resident somatic stem cells—that persist all through the life of an individual (Watt, 1998). These cells reside in the proliferative basal compartment, are relatively undifferentiated, retain a high capacity for self-renewal throughout their life span, have a large proliferative potential, and are rarely cycling (Blanpain *et al.*, 2004; Taylor *et al.*, 2000). A few divisions of stem cells give rise to another stem cell and one transit amplifying (TA) cell. Compared to stem cells, TA cells have limited proliferative potential, and most likely they have limited differentiation potential as well (Dunnwald *et al.*, 2001; Jones *et al.*, 1995; Potten and Booth, 2002; Potten and Morris, 1988). However, they divide rapidly, providing cells that are committed to terminal differentiation, called postmitotic cells. These three subpopulations of the basal layer may be distinguished on the basis of their different locations, mitotic potential, and adhesive properties (Fig. 17.1 and Table 17.2). The cytoskeleton of the undifferentiated keratinocytes of the basal layer is mainly formed by a relatively dispersed, but extensive, network of keratins, a subclass of epithelial-specific intermediate filaments that constitute the major structural proteins of the epidermis. These filaments are made of polymers of the coiled-coil dimer of keratin 5 (K5) and 14 (K14), which are specifically expressed by basal keratinocytes (Nelson and Sun, 1983). Keratinocytes in the basal proliferative compartment of the epidermis can be also distinguished from their suprabasal postmitotic neighbors by their specific pattern of intercellular junctions, including desmosomes and adherens junctions (Table 17.3).

Spinous layer. After a limited number of mitotic divisions, new epidermal basal cells exit the cell cycle and commit to differentiation. The first consequence of the execution of this program is that attachment proteins, like integrins, are downregulated, and postmitotic cells disassemble their cell–matrix contacts to move upward in the spinous layer. Commitment to terminally differentiate is also marked by suppression of the expression of K5 and K14, along with synthesis of a new pair of keratins, K1 and K10 (Fuchs and Green, 1980; Moll *et al.*, 1982); as a result, cytoskeletal filaments are formed, that aggregate into thin bundles termed tonofilaments (Eichner *et al.*, 1986). Moreover, desmosome expression increases as keratinocytes mature in the spinous stage, thus giving them a "spiny" appearance under the light microscope. Finally, at the end of their maturation, spinous cells make glutamine and lysine-rich envelope proteins, such as involucrin, which are deposited on the inner surface of the plasma membrane of each cell (Rice and Green, 1979).

Granular layer. The spinous layer is followed by the granular layer, characterized by the granules that appear in the cells at this stage of the epidermal differentiation. There are two types of granules: keratohyalin granules and lamellar bodies. Keratohyalin granules are mainly formed by profilaggrin, a large, insoluble, highly phosphorylated precursor of filaggrin that is a protein involved in the bundling of tonofilaments into larger, microfibril cables (Rothnagel and Steinert, 1990). Lamellar bodies are secretory organelles containing phospholipids, cholesterol, and glucosylceramides, as well as several specialized proteins (Table 17.2), which are extruded into the extracellular environment at the apical portion of the

Table 17.2 Specific markers of keratinocyte differentiation

Protein	Description/Function	Cell compartment
CD71	Transferrin receptor	Postmitotic/ TA cell suprabasal
Corneodesmosin	It coats the external face of desmosomes	Granular
Desmocollin 1; desmoglein 1,4	Desmosomal components	Granular
Desmocollin 3; desmoglein 3		Basal
Filaggrin	Histidine-rich protein involved in the bundling of keratin filaments	Granular
Integrins $\beta_1/\alpha_6\beta_4$	They mediate cell–cell or cell–basement membrane adhesion, forming adherens junctions or hemidesmosomes, respectively	Stem cell TA cell
Involucrin	Major components of the cornified envelope	Spinous/ granular
Loricrin		Granular
SPRs		
Keratin 1/10	Epithelial-specific intermediate filaments	Suprabasal
Keratin 5/14		Basal
SCCE	Lytic enzymes stored in the lamellar bodies, and necessary for the barrier function of the cornified layer	Spinous/ granular
Secretory phospolipase A_2 sphingomyelinase		
Steroid sulfatase		
β-glucocerebrosidase		
Sulfhydryl oxidase	It catalyzes disulfide bond formation between cornified envelope structural proteins	Granular
Transglutaminases 1,3,5	Ca^{2+}-dependent enzymes that catalyze cross-links between cornified envelope structural proteins	Granular
β-Defensin 2	Antimicrobial peptide	Granular

TA, transit amplifying; SPRs, small proline-rich proteins; SCCE, stratum corneum chymotryptic enzyme.

Table 17.3 Transcription factors (TFs) involved in epidermal differentiation

Transcription factors	Role in epidermal differentiation	References
P63	• Regulation by p63 involves an intricate interplay between various p63 isoforms. • Data suggest a balancing action, that pits antiproliferative and prodifferentiating isoforms that drive epidermal cells toward cornification, against proliferation-promoting isoforms that maintain the regenerative potential of the stem cell compartment.	Koster et al. (2004)
$p21^{WAF/Cip1}$	• ΔNp63 binds directly to $p21^{WAF/Cip1}$ promoter and inhibits its transcription, thus favoring cell proliferation. • $p21^{WAF/Cip1}$ is a direct target of Notch signaling.	Nguyen et al. (2006), Rangarajan et al. (2001), and Westfall et al. (2003)
Retinoblastoma (Rb)	• Reporter assays indicate that pRb may be acting through E2F, whose role in promoting proliferation and differentiation in the epidermis has been studied thoroughly. • Members of the retinoblastoma family control cell cycle progression and differentiation in several tissues, and may act as upstream inducers of p63 expression.	Classon and Dyson (2001), Lipinski and Jacks (1999), and Weinberg (1995)
Notch	• Notch signaling may be important for the initial growth arrest signals that allow the cell to enter a differentiation program. • Notch is involved also in the cornifcation process. • ΔNp63α and Notch signaling synergistically induce K1 expression in mature keratinocytes.	Nickoloff et al. (2002) and Nguyen et al. (2006)

(continued)

Table 17.3 (continued)

Transcription factors	Role in epidermal differentiation	References
Peroxisome proliferators-activated receptor (PPAR)	• PPAR activators accelerate differentiation and cornification in fetal and adult epidermis. • Notch may function upstream of PPARs to induce terminal differentiation.	Hanley et al. (1998, 1999), Kömüves et al. (2000), and Kim et al. (2006)
Nuclear factor-kappa B (NF-κB)	• NF-κB is known to play important roles in growth arrest and morphogenesis in the developing and renewing epidermis. • In normal epidermis, NF-κB proteins were found to exist in the cytoplasm of basal cells and then to localize in the nuclei of suprabasal cells, suggesting a role for NF-κB in the switch from proliferation to growth arrest.	(Seitz et al. 1998)
Activator protein 1 (AP-1 B)	• AP-1 is involved in the expression of cornified envelope precursors, including loricrin, involucrin, as well as TG1. • The maintenance of K14 expression in mature epidermis requires not only ΔNp63, but also the cooperative function of AP-1 and AP-2.	Di Sepio et al. (1995), Liew and Yamanishi (1992), Romano et al. (2007), Sinha et al. (2000), Takahashi and Iizuka (1993), Welter and Eckert (1995), and Yamada et al. (1994)

c-Myc	• Increasing expression of c-Myc in cultured keratinocytes promotes terminal differentiation and causes a progressive reduction in growth. • c-Myc drives stem cells into transit-amplifying cells, thereby initiating the differentiation pathway.	Bull et al. (2001)
Sp1	• In human epidermal keratinocytes, Sp1 is a differentiation-specific activator and a downstream target of E2F-mediated suppression of the differentiation-specific marker TG1.	Huang et al. (2008)
E2F	• E2F family members play a crucial role in the control of the cell cycle and of the action of tumor suppressor proteins. • The dimerization domain determines interaction with the differentiation-regulated transcription factor proteins.	Ruiz et al. (2003)

The gray boxes of the table list some of the TFs able to recognize specific responsive elements in the *faah* gene promoter region.

outermost granular cells. These proteins contribute to barrier homeostasis, desquamation, CE formation, and antimicrobial defense of the epidermis (Grayson et al., 1985; Oren et al., 2003; Sondell et al., 1994). Also at the granular stage, CE (a lipoproteinaceous wrapping) replaces the plasma membrane that is crucial for the barrier function in all stratified squamous epithelia. The CE is an extremely insoluble structure, ~15 nm thick, and consists of a layer of cross-linked proteins coated with covalently bound lipids (Swartzendruber et al., 1987). The protein envelope (~10 nm thick) is thought to contribute to the biomechanical properties of the CE, as a result of cross-linking of specialized structural proteins by both disulfide bonds and ε-(γ-glutamyl)lysine isopeptide bonds, generated by sulfhydryl oxidases and TGs, respectively (Hohl, 1990; Polakowska et al., 1991; Schmidt et al., 1988). Three distinct TGs are known to be present in the epidermis, TG1, TG3, and TG5, which are coordinately involved in various stages of CE assembly (Candi et al., 2005). Some of the structural proteins that are cross-linked by TGs during the protein scaffold formation are in a temporal order of deposition: desmosomal plaque proteins, involucrin, keratin intermediate filaments, loricrin, and small proline-rich proteins. At the upper granular layer, the protein scaffold is coated by the lipid envelope which is a lipid monolayer (~5 nm thick), mainly consisting of ω-hydroxyceramides and fatty acids attached and cross-linked by TG5 and TG1 on the already cross-linked proteins, and then exposed on the outside of the membrane (Candi et al., 2005; Stewart and Downing, 2001; Wertz, 1992). This monomolecular layer of lipids creates cohesion between the cornified cells and the surrounding intercellular lipids, and is essential for alignment of these lipids into lamellae (Swartzendruber et al., 1987; Wertz et al., 1989).

Cornified layer. As cells reach the outermost cornified layer, they enter an apoptotic-like destructive phase, becoming metabolically inert. At this point, they extrude lipid bilayers and lose cytoplasmic organelles, including the nucleus. The barrier function of the skin mainly resides in this layer that could be modeled as a brick wall, where the individual corneocytes represent the bricks and the layers of intercellular dispersed lipids constitute the mortar (Fig. 17.1). At the end of the differentiation program, these cellular bricks consist of flattened, dead cells made of tightly bundled keratin filaments encased within the CE. The cohesiveness of this architecture is ensured by the lipid layers extruded by lamellar bodies, as well as through specialized junctional structures, the corneodesmosomes, which are a corneocyte-specific variant of desmosomes (Jonca et al., 2002). Corneodesmolysis, which is catalyzed by proteolytic enzymes present within the intercellular space, must be tightly controlled and confined at the outermost superficial corneocytes for proper desquamation (Deraison et al., 2007; Jonca et al., 2002; Menon et al., 1992) (Fig. 17.1).

III. Transcriptional Control of Skin Differentiation

The specific function of epidermis is established during embryogenesis and is the result of a complex and precisely coordinated stratification program. The morphological changes that are a hallmark of epidermal stratification are associated with modifications in the expression of differentiation marker genes.

In humans, multiple mechanisms can trigger the cellular differentiation program. One of the major system of cellular fate control involves DNA methylation, an epigenetic mechanism able to modify gene expression and cell functioning, without changing DNA sequence. DNA methylation results from the activity of a family of DNA methyltransferase (DNMT) enzymes, which catalyze the addition of a methyl group to cytosine residues at CpG dinucleotides (Bird, 1996). Therefore, human genome contains small stretches, up to a few kilobases in length, that are rich in CpG dinucleotides, and are termed CpG islands. Commonly, in human DNA the CpG dinucleotides are highly methylated, in contrast with the CpG dinucleotides in these islands that are typically methylation-free in adult tissues. Consequently, this pattern of DNA methylation is stably inherited from one cell generation to the other (Bird, 1996). The genome consists of ~30,000 CpG islands, 50–60% of which are within the promoter region of genes (Costello and Plass, 2001). The potential functional importance of CpG islands was revealed by studies demonstrating that their methylation within gene promoters is associated with transcriptional repression of gene expression (Bird, 1996) (Fig. 17.2).

The precise role of methylation of eukaryotic DNA is yet unclear, even though it was originally suggested that methylated DNA would be transcriptionally less active. This is true for certain genes, such as *MyoD* (Jones *et al.*, 1990; Kay *et al.*, 1993), a master control gene regulating the differentiation of muscle cells. Indeed, the under-methylation of *MyoD* is responsible for the conversion of fibroblasts to myoblasts. On the other hand, demethylation changes gene structure, and thereby enables transcription by making DNA more accessible to TFs.

Interestingly, emerging evidence has revealed the role of several TFs in integrating upstream signals to execute specification and differentiation of epidermal cells. In fact, TFs bring signals from several pathways in a coordinate mode to allow downstream differentiation events (Dai and Segre, 2004). Table 17.3 summarizes TFs that have been shown to be engaged in the regulation of both keratinocyte proliferation and differentiation. Remarkably, an unprecedented *in silico* analysis of the *faah* promoter region by the Clustal 2.0.3 multiple sequence alignment program has

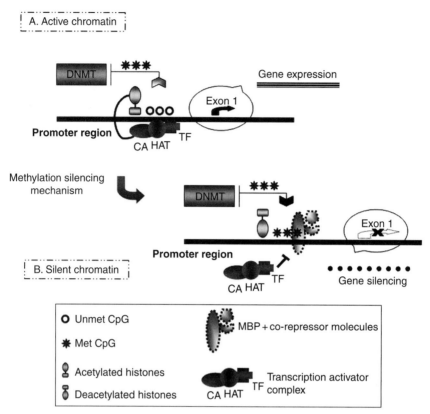

Figure 17.2 (A) Scheme of a transcriptionally active promoter that contains a CpG island. DNMTs do not get in contact with the promoter region that is protected by acetylated histones and by a flanking methylated zone near the first exon (not shown). (B) Transcriptional repression associated with hypermethylation of a gene promoter, and deacetylation of histones. DNMT, DNA methyltransferase; CA, coactivator molecules; HAT, histone acetyltransferase; TF, transcription factor.

revealed that some of these TFs (gray boxes in Table 17.3) could bind to specific responsive elements within the *faah* gene, further supporting the involvement of ECS in keratinocytes cornification.

Among the TFs that could regulate FAAH, NF-κB seems of particular interest, because it is known to act as a key transcription factor in several cellular events. In normal epidermis, NF-κB proteins were found to exist in the cytoplasm of basal cells, and then to localize in the nuclei of suprabasal cells, suggesting a role for this TF in the switch from proliferation to growth arrest (Seitz *et al.*, 1998).

Another family of TFs believed to regulate the balance between epidermal proliferation and differentiation includes *Hox* genes, which function as critical regulators of axial patterning (Krumlauf, 1994; Manak and Scott,

1994) and organogenesis (Hombria and Lovegrove, 2003). Hox proteins may act early in epidermal development as transcriptional activators of cellular proliferation and at least a few *Hox* genes are regulated by consensus binding sites for the (activator protein 1 and 2) AP-1 and AP-2 families of transcriptional regulators. The AP-1 transcriptional complex is involved in a paracrine loop whereby interleukin-1 in keratinocytes induces c-Jun in fibroblasts to produce keratinocyte growth factor, which leads keratinocytes to promote cell growth and differentiation (Maas-Szabowski *et al.*, 2001; Szabowski *et al.*, 2000). AP-1 is also involved in the expression of CE precursors, including loricrin (Di Sepio *et al.*, 1995) and involucrin (Takahashi and Iizuka, 1993; Welter and Eckert, 1995), as well as TG1 (Liew and Yamanishi, 1992; Yamada *et al.*, 1994). Moreover, the *faah* gene contains c-Myc responsive elements, as well as Sp1 and E2F sites (Table 17.3). c-Myc drives stem cells into TA cells initiating the differentiation process *in vitro*, whereas Sp1 is a differentiation activator and a target of E2F in the TG1 suppression pathway.

IV. ENDOCANNABINOID SYSTEM IN EPIDERMIS

Growing evidence has been accumulated to suggest a role for the ECS in the control of epidermis physiology (Berdyshev *et al.*, 2001; Casanova *et al.*, 2003; Maccarrone *et al.*, 2003a; Oddi *et al.*, 2005; Pasquariello *et al.*, 2008; Stander *et al.*, 2005). In line with this, it has been proposed that cannabinoid receptors present in epidermal cells might even act as cutaneous nociceptors (Ibrahim *et al.*, 2005). Immunohistochemical investigations of the precise localization of CB1R and CB2R in sections of human skin revealed intense immunoreactivity for both receptors in epidermal keratinocytes, in the epithelial cells of hair follicles and in sebocytes (Casanova *et al.*, 2003; Ibrahim *et al.*, 2005; Stander *et al.*, 2005). In particular, CB1R was observed in keratinocytes of the spinous and granular layers. Additionally, the differentiated cells of the infundibulum and the inner hair root sheet were positively stained for CB1R, while in sebaceous glands CB1R was found in the central differentiated cells (Stander *et al.*, 2005). On the other hand, conflicting results have been reported on the histological localization of CB2R. In particular, some investigators found that CB2R immunolabeling is strictly associated with that of CB1R (Casanova *et al.*, 2003; Ibrahim *et al.*, 2005), whereas others found major CB2R immunoreactivity only in basal keratinocytes of the epidermis, in undifferentiated cells of the infundibulum, in the outer hair root sheet and in the bulge, which is where the hair follicle stem cells have been shown to reside (Stander *et al.*, 2005). The latter authors also identified CB2R in peripheral undifferentiated cells of the sebaceous glands, and more recently an independent study has shown this receptor subtype also in human sebocytes (Dobrosi et al., 2008). Taken together, these data suggest that further investigations are needed to establish the actual expression of CB2R in the epidermis.

The presence of the components of the ECS has been also investigated in both immortalized and normal keratinocytes (Table 17.1). From these studies, it has been demonstrated that cultured keratinocytes have the biochemical machinery to bind and metabolize AEA. For instance, it has been found that mouse epidermal cells can synthesize AEA in response to UV irradiation (Berdyshev et al., 2000), and that normal human keratinocytes, beside expressing high levels of CB1R, contain also significant levels of FAAH (Maccarrone et al., 2003a; Oddi et al., 2005). Remarkably, it has also been demonstrated that exogenous AEA is able to inhibit CE formation in human keratinocytes through a CB1R-dependent mechanism, further supporting an involvement of the ECS in the control of epidermal cell growth and differentiation (Maccarrone et al., 2003a). On the other hand, studies using either $CB1R^{-/-}$ or $CB2R^{-/-}$ knockout mice did not report gross abnormalities in the epidermis (Buckley et al., 2000; Buckley, 2008; Valverde et al., 2005), thereby suggesting that the genetic ablation of cannabinoid receptors can be compensate for during skin development.

V. Modulation of the Endocannabinoid System in Differentiating Keratinocytes

The differentiation process has been shown to modulate ECS elements in human keratinocytes (Maccarrone et al., 2003a). Indeed, treatment of HaCaT cells with 12-O-tetradecanoylphorbol 13-acetate (TPA) plus calcium, typical inducers of keratinocyte differentiation (Maccarrone et al., 2003a, and references therein), led to a time-dependent increase in the activity of AMT and FAAH, that reached a statistically significant increase (160–180% of the control values) after 24 h, and a maximum (210–280%) after 5 days. Western blot analysis of HaCaT cell extracts showed that specific anti-FAAH antibodies recognized a single immunoreactive band of the molecular size expected for FAAH, the intensity of which increased time-dependently upon treatment with TPA plus calcium (Maccarrone et al., 2003a). Further densitometric analysis indicated that FAAH protein content in 5-day-treated cells increased up to 260% with respect to the vehicle-treated cells. On the other hand, *faah* gene transcription was not affected by TPA plus calcium, neither was the activity of NAPE-PLD, nor CB1R binding (Maccarrone et al., 2003a). Yet, treatment of HaCaT cells with TPA plus calcium time-dependently decreased the endogenous levels of AEA, reaching statistical significance after 24 h and a minimum (25% of the control value) after 5 days (Maccarrone et al., 2003a). Therefore, the levels of AEA inversely correlated with those of the AEA-degrading agents AMT and FAAH. The effects of keratinocyte differentiation on ECS elements are summarized in Table 17.1. To further generalize the analysis of the effect of keratinocyte differentiation on ECS, normal human epidermal keratinocytes (NHEKs) were also investigated (Maccarrone et al., 2003a). These cells showed a full ECS, that is, CB1R, AMT, FAAH, and NAPE-PLD (Table 17.1). When NHEK cells were induced to differentiate under the same experimental conditions as HaCaT cells, it was found that treatment with TPA plus calcium modulated the ECS, and reduced AEA content in NHEK cells, in a way quite similar to that observed in HaCaT cells (Table 17.1). Indeed, the activity of AMT and FAAH increased time-dependently, reaching 200% and 250% of the controls in 5-day-treated cells, whereas CB1R and NAPE-PLD were not significantly affected, and AEA content was reduced down to 25% of controls (Maccarrone et al., 2003a). It line with these findings, treatment of HaCaT cells with TPA plus calcium led also to a 400% increase in CE formation that was paralleled by increased activity of TG1 (330% of the controls) and PKC (250%). As expected, administration of AEA to HaCaT cells dose-dependently reduced CE formation, TG1 activity and PKC activity induced by TPA plus calcium, according to a CB1R-dependent mechanism (Maccarrone et al., 2003a).

Taken together, these studies suggest that AEA has an antidifferentiating effect, which may be more generally involved in epidermal development. To further characterize the molecular mechanisms responsible for this effect, we investigated the expression of epidermal differentiation-related genes after AEA treatment.

VI. Repression of Gene Transcription by Anandamide

The first evidence that AEA can mediate transcriptional effects associated with skin development appeared only recently (Maccarrone et al., 2003a). Human HaCaT keratinocytes were transiently transfected in vitro with a vector containing the chloramphenicol-acetyl transferase (cat) gene under the control of the loricrin promoter, in which a wild-type or a mutated AP-1 responsive sites were included (Rossi et al., 1998; Welter and Eckert, 1995). Functional AP-1 and AP-2 binding sites have been demonstrated in the promoters of many epidermal structural proteins (Table 17.3), and indeed mice with specific epidermal ablation of AP-1 show defects in epidermal growth factor signaling and altered differentiation (Li et al., 2003; Zenz et al., 2003). Exogenous AEA inhibited CAT activity only in the wild-type loricrin promoter, and this effect was completely abolished by a selective CB1R antagonist like SR141716; these data suggest a CB1R-depending effect of

AEA that engages an intact AP-1 element (Maccarrone et al., 2003a). As mentioned above, also *faah* promoter exhibits AP-1 responsive sites, suggesting an involvement of this gene in epidermal differentiation (Table 17.3). The upregulation of FAAH (and AMT) in differentiated keratinocytes, and the subsequent reduction of the endogenous tone of AEA, seem to be in keeping with the effects of exogenous AEA on keratinocytes: inhibition of CE formation and reduction of TG1 and PKC activities that are both responsive to AP-1 (Maccarrone et al., 2003a).

In this line, AEA has been shown to inhibit keratinocyte differentiation by reducing the expression of genes known to be upregulated during cornification, such as *K1, K10, involucrin*, and *TG5* (Pasquariello et al., 2008). Generally, gene activation requires the opening of chromatin structure, and demethylation of specific genomic promoter regions. Thus, it is commonly accepted that specific chromatin modifications distinguish proliferating from differentiated cells in a wide range of tissues (Orford et al., 2008; Shivapurkar et al., 2008). Moreover, many histone modifications, including acetylation, phosphorylation, ubiquitination, sumoylation, and methylation, are also known to regulate chromatin structure and gene expression (Jenuwein and Allis, 2001; Lund and van Lohuizen, 2004). Yet, the most commonly reported variation between differentiated and undifferentiated cells is an overall DNA methylation (Ehrlich et al., 1982; Lyon et al., 1987). In particular, DNA of differentiated keratinocytes has been shown to contain less 5-methylcytosine than DNA of undifferentiated keratinocytes (Veres et al., 1989). Agents known to inhibit DNA methylation, like 5-azacytidine (5AC), and histone deacetylation, like sodium butyrate (NaB), are also able to inhibit cell growth and to promote keratinocyte differentiation (Rosl et al., 1988; Schmidt et al., 1989; Staiano-Coico et al., 1989). We have reported that inhibition of DNA methylation in the presence of 5AC prevents the effect of AEA on the expression of *K10* gene, and more in general its effect on the differentiation process, suggesting that inhibition of differentiation by AEA occurs through changes in chromatin methylation patterns (Pasquariello et al., 2008). A role for methylation in the regulation of epidermal cornification is not totally unexpected, because an inverse correlation between DNA methylation and the expression of differentiating genes has been already shown in human keratinocytes (Elder and Zhao, 2002; Engelkamp et al., 1993). We have also demonstrated that AEA induces DNA methylation of keratinocyte-differentiating genes by increasing DNMT activity, and that this effect depends on CB1 receptors (Pasquariello et al., 2008). It seems noteworthy that CB1R activation by AEA and congeners triggers two major signaling pathways, that engage p38 and p42/p44 MAPK (Bouaboula et al., 1995; Liu et al., 2000). The involvement of these kinases in the arrest of differentiation due to AEA was demonstrated by using selective inhibitors, able to fully restore K10 expression and to abolish DNMT activation in AEA-treated keratinocytes (Pasquariello et al., 2008).

AEA and Gene Expression

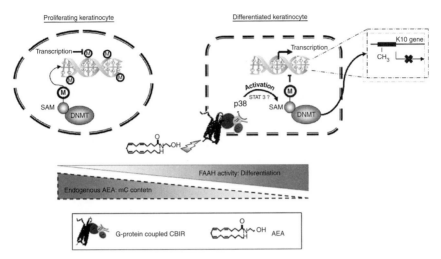

Figure 17.3 Effect of AEA on keratinocyte differentiation. The left part of the scheme represents a typical proliferating keratinocyte, in which it is possible to observe a high genomic methylation level and elevated endogenous tone of AEA. A differentiated epidermal cell (on the right) is known to be less methylated, also because of the partial inactivation of DNA methyltransferase (DNMT). AEA reactivates the latter enzyme in a CB1R-dependent manner, possibly through p38/STAT3 signaling; this leads to an increase of methyl cytosine (mC) content and, ultimately, to repression of gene expression (see the box on the right for the state of *K10* expression after AEA treatment). SAM, *S*-adenosylmethionine.

Additionally, execution of the p38 pathway requires the activation of specific TFs, such as signal transducers and activators of transcription (i.e., STAT3), the responsive site of which has been found also in the human *faah* promoter (Maccarrone *et al.*, 2003b). Against this background, it is tempting to speculate that AEA might induce the upregulation of DNMTs via a p38/STAT3 signaling pathway, and that the subsequent methylation of specific DNA regions can decrease gene expression. This hypothesis is schematically depicted in Fig. 17.3.

Overall, these data suggest a novel biological activity of AEA as a transcriptional repressor, an action that might explain a number of effects of eCBs on cell proliferation and differentiation.

VII. Conclusions

Epidermal differentiation provides an excellent model to analyze the state and the variability of gene expression in response to specific events. The ECS is fully functional in keratinocytes, and it has been shown to play a role in several aspects of skin physiology, although the underlying molecular details are as yet unclear. For instance, CB1R and TRPV1 are involved in itch (Paus et al., 2006), and in line with this a protective role for the ECS has been recently shown in contact allergy of the skin (Karsak et al., 2007). Moreover, in human sebocytes eCBs enhance lipid synthesis and apoptosis through a CB2R-dependent signaling, and through engagement of peroxisome proliferator-activated receptors (Dobrosi et al., 2008). In this context, the transcriptional activity of AEA might represent a common mechanism to explain at once several cellular effects of this compound and its congeners. Indeed, we believe that the importance of AEA in regulating gene expression goes beyond its role in keratinocyte differentiation. In fact, regulation of DNA methylation is a fundamental epigenetic modification of the genome that is involved in regulating a large number of cellular processes, including: embryonic development, transcription, chromatin structure, X chromosome inactivation, genomic imprinting, and chromosome stability. The importance of DNA methylation is also demonstrated by the growing number of diseases that occur when methylation is not properly established or maintained in cells (Robertson, 2005). Among many other diseases, a role for altered methylation has been established in cancer. Cancer cells are usually hypomethylated and loss of genomic methylation is usually an early event in cancer development, that also correlates with disease severity and metastatic potential (Robertson, 2005). Genes involved in apoptosis, cell cycle regulation, DNA repair, cell signaling, and transcription have been shown to be silenced by hypermethylation. There is therefore a growing interest in developing ways of pharmacologically reversing methylation abnormalities. The observations reviewed above might open the avenue to the potential exploitation of ECS-oriented drugs to enhance DNA methylation in those human pathologies associated with an hypomethylation state. In this context, it seems noteworthy that stimulation of CB1R has been shown to inhibit in vivo Ras oncogene-dependent tumor growth and metastasis (Portella et al., 2003). Moreover, it has been recently demonstrated that also melanoma cells express CB1 receptors (Blázquez et al., 2006; Casanova et al., 2003), and this observation may contribute to design new therapeutic strategies for the management of such a widespread skin cancer. In conclusion, the finding that CB1R activation by AEA triggers DNA methylation in human keratinocytes could be relevant for the development of novel pharmacological treatments, able to reduce skin cancer and allergic inflammation through the promotion of epigenetic modifications.

ACKNOWLEDGMENTS

This work was supported by Ministero dell'Università e della Ricerca (COFIN 2006), and by Fondazione TERCAS (Research Program 2005) to M.M.

REFERENCES

Aguado, T., Romero, E., Monory, K., Palazuelos, J., Sendtner, M., Marsicano, G., Lutz, B., Guzman, M., and Ismael, G. (2007). The CB1 cannabinoid receptor mediates excitotoxicity-induced neural progenitor proliferation and neurogenesis. *J. Biol. Chem.* **282,** 23892–23898.

Bari, M., Battista, N., Fezza, F., Gasperi, V., and Maccarrone, M. (2006). New insights into endocannabinoid degradation and its therapeutic potential. *Mini Rev. Med. Chem.* **6,** 257–268.

Bellocchio, L., Cervino, C., Vicennati, V., Pasquali, R., and Fagotto, U. (2008). Cannabinoid type 1 receptor: Another arrow in the adipocytes' bow. *J. Neuroendocrinol.* **1,** 130–138.

Berdyshev, E. V., Schmid, P. C., Dong, Z., and Schmid, H. H. (2000). Stress-induced generation of N-acylethanolamines in mouse epidermal JB6 P + cells. *Biochem. J.* **346,** 369–374.

Berdyshev, E. V., Schmid, P. C., Krebsbach, R. J., Hillard, C. J., Huang, C., Chen, N., Dong, Z., and Schmid, H. H. (2001). Cannabinoid-receptor-independent cell signalling by N-acylethanolamines. *Biochem. J.* **360,** 67–75.

Bird, A. P. (1996). The relationship of DNA methylation to cancer. *Cancer Surv.* **28,** 87–101.

Blanpain, C., Lowry, W. E., Geoghegan, A., Polak, L., and Fuchs, E. (2004). Self-renewal, multipotency, and the existence of two cell populations within an epithelial stem cell niche. *Cell* **118,** 635–648.

Blázquez, C., Carracedo, A., Barrado, L., Real, P. J., Fernández-Luna, J. L., Velasco, G., Malumbres, M., and Guzmán, M. (2006). Cannabinoid receptors as novel targets for the treatment of melanoma. *FASEB J.* **20,** 2633–2635.

Bouaboula, M. (1996). Signaling pathway associated with stimulation of CB2 peripheral cannabinoid receptor. Involvement of both mitogen-activated protein kinase and induction of Krox-24 expression. *Eur J. Biochem.* **237,** 704–711.

Bouaboula, M., Poinot-Chazel, C., Bourrié, B., Canat, X., Calandra, B., Rinaldi-Carmona, M., Le Fur, G., and Casellas, P. (1995). Activation of mitogen-activated protein kinases by stimulation of the central cannabinoid receptor CB1. *Biochem. J.* **312,** 637–641.

Buckley, N. E. (2008). The peripheral cannabinoid receptor knockout mice: An update. *Br. J. Pharmacol.* **153,** 309–318.

Buckley, N. E., McCoy, K. L., Mezey, E., Bonner, T., Zimmer, A., Felder, C. C., Glass, M., and Zimmer, A. (2000). Immunomodulation by cannabinoids is absent in mice deficient for the cannabinoid CB(2) receptor. *Eur. J. Pharmacol.* **396,** 141–149.

Bull, J. J., Muller-Rover, S., Patel, S. V., Chronnell, C. M., McKay, I. A., and Philpott, M. P. (2001). Contrasting localization of c-Myc with other Myc superfamily transcription factors in the human hair follicle and during the hair growth cycle. *J. Invest. Dermatol.* **116,** 617–622.

Candi, E., Schmidt, R., and Melino, G. (2005). The cornified envelope: A model of cell death in the skin. *Nat. Rev. Mol. Cell Biol.* **6,** 328–340.

Casanova, M. L., Blázquez, C., Martínez-Palacio, J., Villanueva, C., Fernández-Aceñero, M. J., Huffman, J. W., Jorcano, J. L., and Guzmán, M. (2003). Inhibition of skin tumor growth and angiogenesis *in vivo* by activation of cannabinoid receptors. *J. Clin. Invest.* **111,** 43–50.

Classon, M., and Dyson, N. (2001). p107 and p130: Versatile proteins with interesting pockets. *Exp. Cell Res.* **264,** 135–147.

Costello, J. F., and Plass, C. (2001). Methylation matters. *J. Med. Genet.* **38,** 285–303.

Dai, X., and Segre, J. A. (2004). Transcriptional control of epidermal specification and differentiation. *Curr. Opin. Genet. Dev.* **14,** 485–491.

Deraison, C., Bonnart, C., Lopez, F., Besson, C., Robinson, R., Jayakumar, A., Wagberg, F., Brattsand, M., Hachem, J. P., Leonardsson, G., and Hovnanian, A. (2007). LEKTI fragments specifically inhibit KLK5, KLK7, and KLK14 and control desquamation through a pH-dependent interaction. *Mol. Biol. Cell* **18,** 3607–3619.

Di Marzo, V. (2008). Targeting the endocannabinoid system: To enhance or reduce? *Nat. Rev. Drug Discov.* **7,** 438–455.

Di Sepio, D., Jones, A., Longley, M. A., Bundman, D., Rothnagel, J. A., and Roop, D. R. (1995). The proximal promoter of the mouse loricrin gene contains a functional AP-1 element and directs keratinocyte-specific but not differentiation-specific expression. *J. Biol. Chem.* **270,** 10792–10799.

Dobrosi, N., Tóth, B. I., Nagy, G., Dózsa, A., Géczy, T., Nagy, L., Zouboulis, C. C., Paus, R., Kovács, L., and Bíró, T. (2008). Endocannabinoids enhance lipid synthesis and apoptosis of human sebocytes via cannabinoid receptor-2-mediated signaling. *FASEB J.* **10,** 3685–3695.

Dunnwald, M., Tomanek-Chalkley, A., Alexandrunas, D., Fishbaugh, J., and Bickenbach, J. R. (2001). Isolating a pure population of epidermal stem cells for use in tissue engineering. *Exp. Dermatol.* **10,** 45–54.

Ehrlich, M., Gama-Sosa, M. A., Huang, L. H., Midgett, R. H., Kuo, K. C., McCune, R. A., and Gehrke, C. (1982). Amount and distribution of 5-methylcytosine in human DNA from different types of tissues of cells. *Nucleic Acids Res.* **10,** 2709–2721.

Eichner, R., Sun, T. T., and Aebi, U. (1986). The role of keratin subfamilies and keratin pairs in the formation of human epidermal intermediate filaments. *J. Cell. Biol.* **102,** 1767–1777.

Elder, J. T., and Zhao, X. (2002). Evidence for local control of gene expression in the epidermal differentiation complex. *Exp. Dermatol.* **11,** 406–412.

Engelkamp, D., Schafer, B. W., Mattei, M. G., Erne, P., and Heizmann, C. W. (1993). Six S100 genes are clustered on human chromosome 1q21: Identification of two genes coding for the two previously unreported calcium-binding proteins S100D and S100E. *Proc. Natl. Acad. Sci. USA* **90,** 6547–6551.

Fride, E. (2008). Multiple roles for the endocannabinoid system during the earliest stages of life: Pre- and postnatal development. *J. Neuroendocrinol.* **20,** 75–81.

Fuchs, E. (1990). Epidermal differentiation: The bare essentials. *J. Cell. Biol.* **111,** 2807–2814.

Fuchs, E., and Green, H. (1980). Changes in keratin gene expression during terminal differentiation of the keratinocyte. *Cell* **19,** 1033–1042.

Galve-Roperth, I., Aguado, T., Rueda, D., Velasco, G., and Guzman, M. (2006). Endocannabinoids: A new family of lipid mediators involved in the regulation of neural cell development. *Curr. Pharm. Des.* **12,** 2319–2325.

Glaser, S. T., Kaczocha, M., and Deutsch, D. G. (2005). Anandamide transport: A critical review. *Life Sci.* **77,** 1584–1604.

Gomez del Pulsar, T., Velasco, G., and Guzman, M. (2000). The CB1 cannabinoid receptor is coupled to the activation of protein kinase B/Akt. *Biochem. J.* **347,** 369–373.

Grayson, S., Johnson-Winegar, A. G., Wintroub, B. U., Isseroff, R. R., Epstein, E. H. Jr., and Elias, P. M. (1985). Lamellar body-enriched fractions from neonatal mice: Preparative techniques and partial characterization. *J. Invest. Dermatol.* **85,** 289–294.

Guzmán, M. (2003). Cannabinoids: Potential anticancer agents. *Nat. Rev. Cancer* **10,** 745–755.

Hanley, K., Jiang, Y., He, S. S., Friedman, M., Elias, P. M., Bikle, D. D., Williams, M. L., and Feingold, K. R. (1998). Keratinocyte differentiation is stimulated by activators of the nuclear hormone receptor PPARalpha. *J. Invest. Dermatol.* **110,** 368–375.

Hanley, K., Kömüves, L. G., Bass, N. M., He, S. S., Jiang, Y., Crumrine, D., Appel, R., Friedman, M., Bettencourt, J., Min, K., Elias, P. M., Williams, M. L., *et al.* (1999). Fetal epidermal differentiation and barrier development *in vivo* is accelerated by nuclear hormone receptor activators. *J. Invest. Dermatol.* **113,** 788–795.

Hohl, D. (1990). Cornified cell envelope. *Dermatologica* **180,** 201–211.
Hombria, J. C., and Lovegrove, B. (2003). Beyond homeosis—HOX function in morphogenesis and organogenesis. *Differentiation* **71,** 461–476.
Howlett, A. C. (2002). The cannabinoid receptors. *Prostaglandins Other Lipid Mediat.* **68–69,** 619–631.
Howlett, A. C. (2005). A short guide to the nomenclature of seven-transmembrane spanning receptors for lipid mediators. *Life Sci.* **77,** 1522–1530.
Huang, C. S., Ho, W. L., Lee, W. S., Shew, M. T., Wang, Y. J., Tu, S. H., Chen, R. J., Chu, J. S., Chen, L. C., Lee, C. H., Tseng, H., Ho, Y. S., *et al.* (2008). Sp1-regulated p27/Kip 1 gene expression is involved in terbinafine-induced A431 cancer cell differentiation: An *in vitro* and *in vivo* study. *Biochem. Pharmacol.* **75,** 1783–1796.
Ibrahim, M. M., Porreca, F., Lai, J., Albrecht, P. J., Rice, F. L., Khodorova, A., Davar, G., Makriyannis, A., Vanderah, T. W., Mata, H. P., and Malan, T. P. Jr. (2005). CB2 cannabinoid receptor activation produces antinociception by stimulating peripheral release of endogenous opioids. *Proc. Natl. Acad. Sci. USA* **102,** 3093–3098.
Jenuwein, T., and Allis, C. D. (2001). Translating the histone code. *Science* **293,** 1074–1080.
Jin, X. H., Okamoto, Y., Morishita, J., Tsuboi, K., Tonai, T., and Ueda, N. (2007). Discovery and characterization of a Ca^{2+}-independent phosphatidylethanolamine N-acyltransferase generating the anandamide precursor and its congeners. *J. Biol. Chem.* **282,** 3614–3623.
Jonca, N., Guerrin, M., Hadjiolova, K., Caubet, C., Gallinaro, H., Simon, M., and Serre, G. (2002). Corneodesmosin, a component of epidermal corneocyte desmosomes, displays homophilic adhesive properties. *J. Biol. Chem.* **277,** 5024–5029.
Jones, G. E., Murphy, S. J., and Watt, D. J. (1990). Segregation of the myogenic cell lineage in mouse muscle development. *J. Cell Sci.* **97,** 659–667.
Jones, P. H., Harper, S., and Watt, F. M. (1995). Stem cell patterning and fate in human epidermis. *Cell* **80,** 83–93.
Kariniemi, A. L., Forsman, L. M., Wahlström, T., and Andersson, L. C. (1984). Expression of differentiation antigens in benign sweat gland tumours. *Br. J. Dermatol.* **111,** 175–182.
Karsak, M., Gaffal, E., Date, R., Wang-Eckhardt, L., Rehnelt, J., Petrosino, S., Starowicz, K., Steuder, R., Schlicker, E., Cravatt, B., Mechoulam, R., Buettner, R., *et al.* (2007). Attenuation of allergic contact dermatitis through the endocannabinoid system. *Science* **316,** 1494–1497.
Kay, P. H., Marlow, S. A., Mitchell, C. A., and Papadimitriou, J. M. (1993). Studies on the evolution and function of different forms of the mouse myogenic gene Myo-D1 and upstream flanking region. *Gene* **124,** 215–222.
Kim, D. J., Bility, M. T., Billin, A. N., Willson, T. M., Gonzalez, F. J., and Peters, J. M. (2006). PPARbeta/delta selectively induces differentiation and inhibits cell proliferation. *Cell Death Differ.* **13,** 53–60.
Klein, T. W. (2005). Cannabinoid-based drugs as anti-inflammatory therapeutics. *Nat. Rev. Immunol.* **5,** 400–411.
Kömüves, L. G., Hanley, K., Lefebvre, A. M., Man, M. Q., Ng, D. C., Bikle, D. D., Williams, M. L., Elias, P. M., Auwerx, J., and Feingold, K. R. (2000). Stimulation of PPARalpha promotes epidermal keratinocyte differentiation *in vivo*. *J. Invest. Dermatol.* **115,** 353–360.
Koster, M. I., Kim, S., Mills, A. A., DeMayo, F. J., and Roop, D. R. (2004). p63 is the molecular switch for initiation of an epithelial stratification program. *Genes Dev.* **18,** 126–131.
Krumlauf, R. (1994). Hox genes in vertebrate development. *Cell* **78,** 191–201.
Lauckner, J. E., Jensen, J. B., Chen, H. Y., Lu, H. C., Hille, B., and Mackie, K. (2008). GPR55 is a cannabinoid receptor that increases intracellular calcium and inhibits M current. *Proc. Natl. Acad. Sci. USA* **105,** 2699–2704.

Liew, F. M., and Yamanishi, K. (1992). Regulation of transglutaminase 1 gene expression by 12-O-tetradecanoylphorbol-13-acetate, dexamethasone, and retinoic acid in cultured human keratinocytes. *Exp. Cell Res.* **202,** 310–315.

Li, G., Gustafson-Brown, C., Hanks, S. K., Nason, K., Arbeit, J. M., Pogliano, K., Wisdom, R. M., and Johnson, R. S. (2003). c-Jun is essential for organization of the epidermal leading edge. *Dev. Cell* **4,** 865–877.

Lipinski, M. M., and Jacks, T. (1999). The retinoblastoma gene family in differentiation and development. *Oncogene* **18,** 7873–7882.

Liu, J., Gao, B., Mirshahi, F., Sanyal, A. J., Khanolkar, A. D., Makriyannis, A., and Kunos, G. (2000). Functional CB1 cannabinoid receptors in human vascular endothelial cells. *Biochem. J.* **346,** 835–840.

Liu, J., Wang, L., Harvey-White, J., Osei-Hyiaman, D., Razdan, R., Gong, Q., Chan, A. C., Zhou, Z., Huang, B. X., Kim, H. Y., and Kunos, G. (2006). A biosynthetic pathway for anandamide. *Proc. Natl. Acad. Sci. USA* **103,** 13345–13350.

Lund, A. H., and van Lohuizen, M. (2004). Epigenetics and cancer. *Genes Dev.* **18,** 2315–2335.

Lyon, S. B., Buonocore, L., and Miller, M. (1987). Naturally occurring methylation inhibitor: DNA hypomethylation and hemoglobin synthesis in human K562 cells. *Mol. Cell. Biol.* **7,** 1759–1763.

Maas-Szabowski, N., Szabowski, A., Stark, H. J., Andrecht, S., Kolbus, A., Schorpp-Kistner, M., Angel, P., and Fusenig, N. E. (2001). Organotypic cocultures with genetically modified mouse fibroblasts as a tool to dissect molecular mechanisms regulating keratinocytes growth and differentiation. *J. Invest. Dermatol.* **116,** 816–820.

Maccarrone, M. (2006). Fatty acid amide hydrolase: A potential target for next generation therapeutics. *Curr. Pharm. Des.* **12,** 759–772.

Maccarrone, M., Di Rienzo, M., Battista, N., Gasperi, V., Guerrieri, P., Rossi, A., and Finazzi-Agrò, A. (2003a). The endocannabinoid system in human keratinocytes. Evidence that anandamide inhibits epidermal differentiation through CB1 receptor-dependent inhibition of protein kinase C, activation protein-1, and transglutaminase. *J. Biol. Chem.* **278,** 33896–33903.

Maccarrone, M., Bari, M., Di Rienzo, M., Finazzi-Agrò, A., and Rossi, A. (2003b). Progesterone activates fatty acid amide hydrolase (FAAH) promoter in human T lymphocytes through the transcription factor Ikaros. Evidence for a synergistic effect of leptin. *J. Biol. Chem.* **278,** 32726–32732.

Maccarrone, M., Battista, N., and Centonze, D. (2007). The endocannabinoid pathway in Huntington's disease: A comparison with other neurodegenerative diseases. *Prog. Neurobiol.* **81,** 349–379.

Manak, J. R., and Scott, M. P. (1994). A class act: Conservation of homeodomain protein functions. *Dev. Suppl.* 61–77.

McKinney, M. K., and Cravatt, B. F. (2005). Structure and function of fatty acid amide hydrolase. *Annu. Rev. Biochem.* **74,** 411–432.

Menon, G. K., Ghadially, R., Williams, M. L., and Elias, P. M. (1992). Lamellar bodies as delivery systems of hydrolytic enzymes: Implications for normal and abnormal desquamation. *Br. J. Dermatol.* **126,** 337–345.

Moll, R., Franke, W. W., Volc-Platzer, B., and Krepler, R. (1982). Different keratin polypeptides in epidermis and other epithelia of human skin: A specific cytokeratin of molecular weight 46,000 in epithelia of the pilosebaceous tract and basal cell epitheliomas. *J. Cell Biol.* **95,** 285–295.

Nelson, W. G., and Sun, T. T. (1983). The 50- and 58-kdalton keratin classes as molecular markers for stratified squamous epithelia: Cell culture studies. *J. Cell Biol.* **97,** 244–251.

Nguyen, B. C., Lefort, K., Mandinova, A., Antonimi, D., Devgan, V., Della Gatta, G., Koster, M. I., Zhang, Z., Wang, J., Tommasi di Vignano, A., Kitajewski, J.,

Chionino, G., et al. (2006). Cross-regulation between Notch and p63 in keratinocyte commitment to differentiation. *Genes Dev.* **20,** 1028–1042.

Nickoloff, B. J., Qin, J. Z., Chaturvedi, V., Denning, M. F., Bonish, B., and Miele, L. (2002). Jagged-1 mediated activation of notch signaling induces complete maturation of human keratinocytes through NF-κB and PPARγ. *Cell Death Differ.* **9,** 842–855.

Oddi, S., Bari, M., Battista, N., Barsacchi, D., Cozzani, I., and Maccarrone, M. (2005). Confocal microscopy and biochemical analysis reveal spatial and functional separation between anandamide uptake and hydrolysis in human keratinocytes. *Cell Mol. Life Sci.* **62,** 386–395.

Oren, A., Ganz, T., Liu, L., and Meerloo, T. (2003). In human epidermis, beta-defensin 2 is packaged in lamellar bodies. *Exp. Mol. Pathol.* **74,** 180–182.

Orford, K., Kharchenko, P., Lai, W., Dao, M. C., Worhunsky, D. J., Ferro, A., Janzen, V., Park, P. J., and Scadden, D. T. (2008). Differential H3K4 methylation identifies developmentally poised hematopoietic genes. *Dev. Cell* **14,** 798–809.

Pasquariello, N., Paradisi, A., Barcaroli, D., and Maccarrone, M. (2008). Anandamide regulates keratinocyte differentiation by inducing DNA methylation in a CB1 receptor-dependent manner. *J. Biol. Chem.* **283,** 6005–6012.

Paus, R., Schmelz, M., Bíró, T., and Steinhoff, M. (2006). Frontiers in pruritus research: Scratching the brain for more effective itch therapy. *J. Clin. Invest.* **116,** 1174–1186.

Piomelli, D. (2003). The molecular logic of endocannabinoid signalling. *Nat. Rev. Neurosci.* **4,** 873–884.

Piomelli, D. (2005). The endocannabinoid system: A drug discovery perspective. *Curr. Opin. Investig. Drugs* **6,** 672–679.

Polakowska, R. R., Eddy, R. L., Shows, T. B., and Goldsmith, L. A. (1991). Epidermal type I transglutaminase (TGM1) is assigned to human chromosome 14. *Cytogenet. Cell Genet.* **56,** 105–107.

Portella, G., Laezza, C., Laccetti, P., De Petrocellis, L., Di Marzo, V., and Bifulco, M. (2003). Inhibitory effects of cannabinoid CB1 receptor stimulation on tumor growth and metastatic spreading: Actions on signals involved in angiogenesis and metastasis. *FASEB J.* **17,** 1771–1773.

Potten, C. S., and Booth, C. (2002). Keratinocyte stem cells: A commentary. *J. Invest. Dermatol.* **119,** 888–899.

Potten, C. S., and Morris, R. J. (1988). Epithelial stem cells *in vivo*. *J. Cell Sci.* **10**(Suppl.), 45–62.

Rangarajan, A, Talora, C, Okuyama, R, Nicolas, M., Mammuccari, C., Oh, H., Aster, J. C., Krishna, S., Metzger, D., Chambon, P., Miele, L., Aguet, M., et al. (2001). Notch signaling is a direct determinant of keratinocyte growth arrest and entry into differentiation. *EMBO J.* **20,** 3427–3436.

Rice, R. H., and Green, H. (1979). Presence in human epidermal cells of a soluble protein precursor of the cross-linked envelope: Activation of the cross-linking by calcium ions. *Cell* **18,** 681–694.

Robertson, K. D. (2005). DNA methylation and human disease. *Nat. Rev. Genet.* **6,** 597–610.

Romano, R. A., Birkaya, B., and Sinha, S. (2007). A functional enhancer of Keratin14 is a direct transcriptional target of ΔNp63. *J. Invest. Dermatol.* **127,** 1175–1186.

Rosl, F., Durst, M., and Zur Hausen, H. (1988). Selective suppression of human papillomavirus transcription in non-tumorigenic cells by 5-azacytidine. *EMBO J.* **7,** 1321–1328.

Rossi, A., Jang, S. I., Ceci, R., Steinert, P. M., and Markova, N. G. (1998). Effect of AP1 transcription factors on the regulation of transcription in normal human epidermal keratinocytes. *J. Invest. Dermatol.* **110,** 34–40.

Rothnagel, J. A., and Steinert, P. M. (1990). The structure of the gene for mouse filaggrin and a comparison of the repeating units. *J. Biol. Chem.* **265,** 1862–1865.

Rueda, D., Navarro, B., Martinez-Serrano, A., Guzman, M., and Galve-Roperth, I. (2002). The endocannabinoid anandamide inhibits neuronal progenitor cell differentiation through attenuation of the Rap1/B-Raf/ERK pathway. *J. Biol. Chem.* **277,** 46645–46650.

Ruiz, S., Segrelles, C., Bravo, A., Santos, M., Perez, P., Leis, H., Jorcano, J. L., and Paramio, J.M (2003). Abnormal epidermal differentiation and impaired epithelial-mesenchymal tissue interactions in mice lacking the retinoblastoma relatives p107 and p130. *Development* **130,** 2341–2353.

Sawzdargo, M., Nguyen, T., Lee, D. K., Lynch, K. R., Cheng, R., Heng, H. H., George, S. R., and O'Dowd, B. F. (1999). Identification and cloning of three novel human G protein-coupled receptor genes GPR52, PsiGPR53 and GPR55: GPR55 is extensively expressed in human brain. *Brain Res. Mol. Brain Res.* **64,** 193–198.

Schmidt, R., Michel, S., Shroot, B., and Reichert, U. (1988). Plasma membrane transglutaminase and cytosolic transglutaminase form distinct envelope-like structures in transformed human keratinocytes. *FEBS Lett.* **229,** 193–196.

Schmidt, R., Cathelineau, C., Cavey, M. T., Dionisius, V., Michel, S., Shroot, B., and Reichert, U. (1989). Sodium butyrate selectively antagonizes the inhibitory effect of retinoids on cornified envelope formation in cultured human keratinocytes. *J. Cell Physiol.* **140,** 281–287.

Seitz, C. S., Lin, Q., Deng, H., and Kavary, P. A. (1998). Alterations in NF-kappaB function in transgenic epithelial tissue demonstrate a growth inhibitory role for NF-kappaB. *Proc. Natl. Acad. Sci. USA* **95,** 2307–2312.

Shivapurkar, N., Stastny, V., Xie, Y., Prinsen, C., Frenkel, E., Czerniak, B., Thunnissen, F. B., Minna, J. D., and Gazdar, A. F. (2008). Differential methylation of a short CpG-rich sequence within exon 1 of TCF21 gene: A promising cancer biomarker assay. *Cancer Epidemiol. Biomarkers Prev.* **17,** 995–1000.

Simon, G. M., and Cravatt, B. F. (2006). Endocannabinoid biosynthesis proceeding through glycerophospho-N-acyl ethanolamine and a role for alpha/beta-hydrolase 4 in this pathway. *J. Biol. Chem.* **281,** 26465–26472.

Sinha, S., Degenstein, L., Copenhaver, C., and Fuchs, E. (2000). Defining the regulatory factors required for epidermal gene expression. *Mol. Cell. Biol.* **20,** 2543–2555.

Sondell, B., Thornell, L. E., Stigbrand, T., and Egelrud, T. (1994). Immunolocalization of stratum corneum chymotryptic enzyme in human skin and oral epithelium with monoclonal antibodies: Evidence of a proteinase specifically expressed in keratinizing squamous epithelia. *J. Histochem. Cytochem.* **42,** 459–465.

Staiano-Coico, L., Helm, R. E., McMahon, C. K., Pagan-Charry, I., LaBruna, A., Piraino, V., and Higgins, P. (1989). Sodium-N-butyrate induces cytoskeletal rearrangements and formation of cornified envelopes in cultured adult human keratinocytes. *Cell Tissue Kinet.* **22,** 361–375.

Stander, S., Schmelz, M., Metze, D., Luger, T., and Rukwied, R. (2005). Distribution of cannabinoid receptor 1 (CB1) and 2 (CB2) on sensory nerve fibers and adnexal structures in human skin. *J. Dermatol. Sci.* **38,** 177–188.

Starowicz, K., Maione, S., Cristino, L., Palazzo, E., Marabese, I., Rossi, F., de Novellis, V., and Di Marzo, V. (2007). Tonic endovanilloid facilitation of glutamate release in brainstem descending antinociceptive pathways. *J. Neurosci.* **27,** 13739–13749.

Stewart, M. E., and Downing, D. T. (2001). The omega-hydroxyceramides of pig epidermis are attached to corneocytes solely through omega-hydroxyl groups. *J. Lipid Res.* **42,** 1105–1110.

Sugiura, T., Kobayashi, Y., Oka, S., and Waku, K. (2002). Biosynthesis and degradation of anandamide and 2-arachidonoylglycerol and their possible physiological significance. *Prostaglandins Leukot. Essent. Fatty Acids* **66,** 173–192.

Swartzendruber, D. C., Wertz, P. W., Madison, K. C., and Downing, D. T. (1987). Evidence that the corneocyte has a chemically bound lipid envelope. *J. Invest. Dermatol.* **88,** 709–713.

Szabowski, A., Maas-Szabowski, N., Andrecht, S., Kolbus, A., Schorpp-Kistner, M., Fusenig, N. E., and Angel, P. (2000). c-Jun and JunB antagonistically control cytokine-regulated mesenchymal-epidermal interaction in skin. *Cell* **103,** 745–755.

Takahashi, H., and Iizuka, H. (1993). Analysis of the 5'-upstream promoter region of human involucrin gene: Activation by 12-O-tetradecanoylphorbol-13- acetate. *J. Invest. Dermatol.* **100,** 10–15.

Taylor, G., Lehrer, M. S., Jensen, P. J., Sun, T. T., and Lavker, R. M. (2000). Involvement of follicular stem cells in forming not only the follicle but also the epidermis. *Cell* **102,** 451–461.

Trezza, V., Cuomo, V., and Vanderschuren, L. J. (2008). Cannabis and the developing brain: Insights from behavior. *Eur. J. Pharmacol.* **585,** 441–452.

Tsuboi, K., Sun, Y. X., Okamoto, Y., Araki, N., Tonai, T., and Ueda, N. (2005). Molecular characterization of N-acylethanolamine-hydrolyzing acid amidase, a novel member of the choloylglycine hydrolase family with structural and functional similarity to acid ceramidase. *J. Biol. Chem.* **280,** 11082–11092.

Valverde, O., Karsak, M., and Zimmer, A. (2005). Analysis of the endocannabinoid system by using CB1 cannabinoid receptor knockout mice. *Handb. Exp. Pharmacol.* **168,** 117–145.

Veres, D. A., Wilkins, L., Coble, D. W., and Lyon, S. B. (1989). DNA methylation and differentiation of human keratinocytes. *J. Invest. Dermatol.* **93,** 687–690.

Vivanco, I., and Sawyers, C. L. (2002). The phosphatidylinositol 3-kinase AKT pathway in human cancer. *Nat. Rev. Cancer* **2,** 489–501.

Wang, H., Dey, S. K., and Maccarrone, M. (2006). Jekyll and hyde: Two faces of cannabinoid signaling in male and female fertility. *Endocr. Rev.* **5,** 427–448.

Watson, S., Chambers, D., Hobbs, C., Doherty, P., and Graham, A. (2008). The endocannabinoid receptor, CB1, is required for normal axonal growth and fasciculation. *Mol. Cell Neurosci.* **38,** 89–97.

Watt, F. M. (1998). Epidermal stem cells: Markers, patterning and the control of stem cell fate. *Philos. Trans. R. Soc. Lond. B Biol. Sci.* **353,** 831–837.

Weinberg, R. A. (1995). The retinoblastoma protein and cell cycle control. *Cell* **81,** 323–330.

Welter, J. F., and Eckert, R. L. (1995). Differential expression of the fos and jun family members c-fos, fosB, Fra-1, Fra-2, c-jun, junB and junD during human epidermal keratinocyte differentiation. *Oncogene* **11,** 2681–2687.

Wertz, P. W. (1992). Epidermal lipids. *Semin. Dermatol.* **11,** 106–113.

Wertz, P. W., Swartzendruber, D. C., Kitko, D. J., Madison, K. C., and Downing, D. T. (1989). The role of the corneocyte lipid envelopes in cohesion of the stratum corneum. *J. Invest. Dermatol.* **93,** 169–172.

Westfall, M. D., Mays, D. J., Sniezek, J. C., and Pietenpol, J. A. (2003). The Delta Np63 alpha phosphoprotein binds the p21 and 14–3–3 sigma promoters *in vivo* and has transcriptional repressor activity that is reduced by Hay-Wells syndrome-derived mutations. *Mol. Cell Biol.* **23,** 2264–2276.

Wolf, S. A., and Ullrich, O. (2008). Endocannabinoids and the brain immune system: New neurones at the horizon? *J. Neuroendocrinol.* **1,** 15–19.

Yamada, K., Yamanishi, K., Kakizuka, A., Kibe, Y., Doi, H., and Yasuno, H. (1994). Transcriptional regulation of human transglutaminase1 gene by signaling systems of protein kinase C, RAR/RXR and Jun/Fos in keratinocytes. *Biochem. Mol. Biol. Int.* **34,** 827–836.

Zenz, R., Scheuch, H., Martin, P., Frank, C., Eferl, R., Kenner, L., Sibilia, M., and Wagner, E. F. (2003). c-Jun regulates eyelid closure and skin tumor development through EGFR signaling. *Dev. Cell* **4,** 879–889.

CHAPTER EIGHTEEN

Changes in the Endocannabinoid System May Give Insight into new and Effective Treatments for Cancer

Gianfranco Alpini*,‡,§ *and* Sharon DeMorrow*,†

Contents

I. Introduction	470
II. Changes in the Endocannabinoid System in Cancer	471
III. Antiproliferative Effects of Anandamide	473
A. Receptor mediated effects	473
B. Receptor-independent effects	475
IV. Effects of AEA on Migration, Invasion, and Angiogenesis	476
V. Targeting Degradation Enzymes of Cannabinoids as an Anticancer Therapy	479
VI. Tumor Promoting Effects of Anandamide	480
VII. Conclusions	480
Acknowledgments	481
References	481

Abstract

The endocannabinoid system comprises specific cannabinoid receptors such as Cb1 and Cb2, the endogenous ligands (anandamide and 2-arachidonyl glycerol among others) and the proteins responsible for their synthesis and degradation. This system has become the focus of research in recent years because of its potential therapeutic value several disease states. The following review describes our current knowledge of the changes that occur in the endocannabinoid system during carcinogenesis and then focuses on the effects of anandamide on various aspects of the carcinogenic process such as growth, migration, and angiogenesis in tumors from various origins.

---

* Department of Medicine, Texas A&M Health Science Center, College of Medicine, Temple, Texas, USA
† Division of Research and Education, Scott & White Hospital, Temple, Texas, USA
‡ Systems Biology and Translational Medicine, Texas A&M Health Science Center, College of Medicine, Temple, Texas, USA
§ Division of Research, Central Texas Veterans Health Care System, Temple, Texas, USA

Vitamins and Hormones, Volume 81   © 2009 Elsevier Inc.
ISSN 0083-6729, DOI: 10.1016/S0083-6729(09)81018-2   All rights reserved.

I. INTRODUCTION

Marijuana and its derivatives have been used in medicine for centuries, however, it was not until the isolation of the psychoactive component of Cannabis sativa (Δ^9-tetrahydrocannabinol; Δ^9-THC) and the subsequent discovery of the endogenous cannabinoid signaling system that research into the therapeutic value of this system reemerged. Ongoing research is determining that regulation of the endocannabinoid system may be effective in the treatment of pain (Calignano *et al.*, 1998; Manzanares *et al.*, 1999), glaucoma (Voth and Schwartz, 1997), and neurodegenerative disorders such as Parkinson's disease (Piomelli *et al.*, 2000) and multiple sclerosis (Baker *et al.*, 2000). In addition, cannabinoids might be effective antitumoral agents because of their ability to inhibit the growth of various types of cancer cell lines in culture (De Petrocellis *et al.*, 1998; Ruiz *et al.*, 1999; Sanchez *et al.*, 1998, 2001) and in laboratory animals (Galve-Roperh *et al.*, 2000).

The endogenous cannabinoid system (endocannabinoid) consists of the cannabinoid receptors, their endogenous ligands (endocannabinoids) and the proteins for their synthesis and inactivation (Bisogno *et al.*, 2005). The cannabinoid receptors are seven-transmembrane-domain proteins coupled to $G_{i/o}$ type of G-proteins (Bisogno *et al.*, 2005). To date, there are two definitive cannabinoid receptors, Cb1 and Cb2, as well as a putative involvement of the vanilloid receptor VR1. More recently, the orphan receptor GPR55 was shown to function as a novel cannabinoid receptor (Ryberg *et al.*, 2007). Cb1 receptors are found predominantly in the central nervous system, but also in most peripheral tissues including immune cells, the reproductive system, the gastrointestinal tract and the lungs (Devane *et al.*, 1988; Matsuda *et al.*, 1990; Munro *et al.*, 1993). Cb2 receptors are found predominantly in the immune system, that is, in tonsils, spleen, macrophages, and lymphocytes (Devane *et al.*, 1988; Matsuda *et al.*, 1990; Munro *et al.*, 1993).

To date, many endocannabinoids have been identified with varying affinities for the receptors and all of which are lipid molecules. Anandamide (AEA) was the first endogenous ligand to be identified (Devane *et al.*, 1988), which acts as a partial Cb1 agonist and weak Cb2 agonist. It has also been shown to activate the GPR55 (Ryberg *et al.*, 2007). While the physiological roles of many of the other ligands have not yet been fully clarified, AEA has been implicated in a wide variety of physiological and pathological processes.

Currently, there are two biosynthesis pathways for AEA. The first is via the remodeling of an existing membrane phosphoglyceride, that is, the calcium-dependent *N*-transacylation of phosphotidylethanolamine with arachidonic acid to form *N*-arachidonyl-phosphatidyl-ethanolamine, which is then hydrolyzed to AEA (Bisogno *et al.*, 2005; Di Marzo and Deutsch, 1998). The enzyme responsible for the catalysis of this pathway is phospholipase D (Bisogno *et al.*, 2005). The second pathway is via the *de novo* synthesis of AEA from arachidonic acid and ethanolamine by the enzyme anandamide amidohydrolase catalyzing the reverse reaction from high levels of ethanolamine (Di Marzo and Deutsch, 1998). After synthesis, AEA is rapidly inactivated via a tightly controlled series of events involving sequestration by cells and enzymatic hydrolysis. The mechanism of AEA uptake is largely unknown with data suggesting that it is via passive diffusion and others suggesting that it is through the presence of an active transporter (Glaser *et al.*, 2005). Regardless of the mechanism, this uptake is a rapid event with a half-life of approximately 2.5 min (Di Marzo and Deutsch, 1998). After uptake, AEA is hydrolyzed and degraded by the enzyme anandamide amidohydrolase (also called fatty acid amide hydrolase or FAAH) (Di Marzo and Deutsch, 1998).

II. Changes in the Endocannabinoid System in Cancer

Evidence suggests that the endocannabinoid system may be dysregulated in a number of cancers (summarized in Table 18.1). Indeed, both AEA and 2-arachidonyl glycerol (2-AG) have been shown to be increased in human colorectal adenomatous polyps and carcinomas compared to normal colorectal mucosa (Ligresti *et al.*, 2003), suggesting that these endocannabinoids increase when passing from normal to transformed mucosa. No consistent differences were observed in the expression levels of Cb1, Cb2, or FAAH as assessed by both RT-PCR and immunoblotting between normal and colorectal cancer tissue (Ligresti *et al.*, 2003). Similarly, AEA levels were enhanced by 17-fold in glioblastomas whereas meningiomas were characterized by a massively enhanced level of 2-AG (Petersen *et al.*, 2005). Coupled with these changes was a 60% reduction in the enzyme activities of the AEA degradation enzymes, *N*-acylphosphotidylethanolamine-hydrolyzing phospholipase D and fatty acid amide hydrolase in the glioblastoma tissue and an enhanced *in vitro* conversion of phosphotidyl choline to monoacyl glycerol in the meningioma tissue (Petersen *et al.*, 2005). Similarly, the levels of AEA and 2-AG were found to be increased in human pituitary adenomas compared to normal pituitary gland (Pagotto *et al.*, 2001). Moreover, endocannabinoid content in the different pituitary adenomas correlated with the presence of Cb1, being elevated in the tumoral samples positive for Cb1 and lower in the samples in which no or low levels of Cb1 were found (Pagotto *et al.*, 2001). In another study, the levels of AEA were found to differ depending upon the source of the tumor

Table 18.1 Changes in the endocannabinoid system in various types of human cancer

Cancer type	In vivo/ in vitro	Changes in the endocannabinoid system	References
Colorectal	In vivo	AEA ↑, 2-AG ↑	Ligresti et al. (2003)
Glioblastomas	In vivo	AEA ↑, FAAH ↓, Phopholipase D ↓, VR1 ↑	Contassot et al. (2004b) and Petersen et al. (2005)
Meningioma	In vivo	2-AG ↑	Petersen et al. (2005)
Pituitary adenomas		AEA ↑, 2-AG ↑, Cb1 ↑	Pagotto et al. (2001)
Prostate carcinoma	In vivo	AEA ↑, Cb1 ↑, Cb2 ↑	Schmid et al. (2002) and Sarfaraz et al. (2005)
Endometrial sarcoma	In vivo	AEA ↑	Schmid et al. (2002)
Thigh histiocytoma	In vivo	AEA ↑	Schmid et al. (2002)
Stomach carcinoma	In vivo	AEA ↓	Schmid et al. (2002)
Mantle cell lymphoma	In vivo	Cb1 ↑, Cb2 ↑	Ek et al. (2002) and Islam et al. (2003)
Acute myeloid leukemia	In vitro	Cb2 ↑	Alberich Jorda et al. (2004)
Breast cancer	In vivo	Cb1 ↓, Cb2 ↑	Caffarel et al. (2006)

(Schmid et al., 2002). For example, AEA was increased in prostate carcinoma, endometrial sarcoma, and thigh histiocytoma, with no change in ileum lymphoma, and bladder carcinoma and a significant decrease in stomach carcinoma (Schmid et al., 2002).

Furthermore, the cannabinoid receptor system has also been shown to be altered during the carcinogenic process. Indeed, in Mantle cell lymphoma and prostate cancer cells, both Cb1 and Cb2 are upregulated compared to nonmalignant tissue (Ek et al., 2002; Islam et al., 2003; Sarfaraz et al., 2005), whereas in acute myeloid leukemia, only Cb2 is upregulated (Alberich Jorda et al., 2004). A differential effect on receptor expression is shown in breast cancer, with a marked suppression of Cb1 and an increased Cb2 expression (Caffarel et al., 2006). In another study using human brain tissue, the expression of VR1 was aberrantly expressed in glioma brain tumors compared to nonmalignant brain tissue sampled from epileptic patients undergoing surgery (Contassot et al., 2004b).

Dysregulation of the endocannabinoid system still needs to be examined in a number of other cancer types and the implications of a dysregulated endocannabinoid system on malignant transformation and tumor progression remains to be addressed.

III. ANTIPROLIFERATIVE EFFECTS OF ANANDAMIDE

There is a large volume of data indicating antiproliferative effects of AEA on various cancer types via a number or receptor-dependent and receptor-independent mechanisms. A schematic diagram of the proposed mechanisms is depicted in Fig. 18.1.

A. Receptor mediated effects

Cb1: AEA has been shown to decrease the proliferation of breast cancer cells *in vitro* by decreasing the levels of the long form of the prolactin receptor (De Petrocellis *et al.*, 1998) and inhibiting the nerve growth factor (NGF)-mediated proliferation by decreasing the expression of the *trk* NGF receptor (Melck *et al.*, 2000). This was shown to be via a Cb1-dependent mechanism and involves the activation of the cAMP/protein kinase A pathway and subsequent activation of the mitogen-activated protein kinase (MAPK) pathway (Melck *et al.*, 1999). Furthermore, these growth suppressing effects of AEA are not due to toxicity or apoptosis, but by cell cycle arrest in the S phase which was correlated with an activation of Chk1 (an S phase checkpoint kinase), and a suppression of Cdk2 activity (Laezza *et al.*, 2006).

Similar antiproliferative and apoptotic effects of AEA were found in human prostate cancer cell lines (Mimeault *et al.*, 2003), which were also via the Cb1-dependent downregulation of growth factor signaling (Mimeault *et al.*, 2003). Specifically, AEA decreased the expression of the epidermal growth factor receptor (EGFR) in several prostate cancer cell lines which was associated with a G_1 arrest and cell death (Mimeault *et al.*, 2003). Furthermore, this effect could be partially blocked by acidic ceramide inhibitors indicating that AEA might be induced via the cellular ceramide production (Mimeault *et al.*, 2003).

Cb2: In contrast, the antiproliferative actions of AEA on glioma and asytocytoma cells are via the activation of the Cb2 receptor (Sanchez *et al.*, 2001). Treatment of mice inoculated with human glioma tumor cells or human astrocytoma cells with the specific Cb2 agonist, JWH-133 completely blocked the tumor growth *in vivo* (Sanchez *et al.*, 2001), and was, once again, associated with increased *de novo* ceramide synthesis *in vitro* (Sanchez *et al.*, 2001). Furthermore, targeting the Cb2 receptor has been suggested as a therapy to treat malignant lymphoblastic diseases

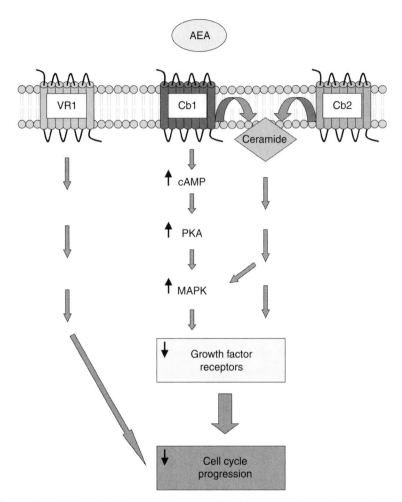

Figure 18.1 Schematic diagram of the cannabinoid receptor-dependent mechanisms whereby AEA leads to tumor growth suppression and/or cell death. AEA may act via the Cb1 receptor to activate the cAMP/PKA/MAPK pathway or to increase the production of ceramide. These effects ultimately result in a decrease in the expression of various growth factor receptors and decrease cell cycle progression. Alternatively, AEA may activate either Cb2 or VR1 to elicit a similar response, although the mechanism by which this occurs is not clear.

(McKallip et al., 2002). In these types of cancers, AEA induced apoptosis *in vitro* and *in vivo* and it is suggested that because the Cb2 agonists lack psychotropic effects, targeting of the Cb2 receptor would be preferential to targeting the Cb1 receptor in cancers of immune origin (McKallip et al., 2002). Another study demonstrated that interleukin-12 treatment and/or overexpression in thyroid carcinoma cells leads to an increase in Cb2 expression and that either overexpression of interleukin-12 or Cb2 resulted in tumor regression and increased sensitivity to chemotherapy (Shi et al., 2008). Taken together, these data indicate that targeting Cb2 may be of therapeutic value in certain tumor types and warrants further investigation.

VR1: The actions of AEA on VR1 were shown to be antiproliferative in glioma cells (Contassot et al., 2004b) and in uterine cervix cancer cells (Contassot et al., 2004a). In both of these cell types, the VR1 was aberrantly expressed in the tumor cells and tissue compared to their nonmalignant counterparts (Contassot et al., 2004a,b). Furthermore, stimulation of Cb1 or Cb2 was protective against the VR1-mediated antiproliferative effects of AEA (Contassot et al., 2004a,b).

Combination of receptors: In addition to the above-mentioned findings, studies have shown an antiproliferative/growth suppressive effect of AEA that could not be attributable to just one receptor. Using rat C6 glioma cells, AEA was shown to have growth suppressing effects that were time and dose-dependent (Jacobsson et al., 2001). These effects could be partially blocked by the Cb1, Cb2, and VR1 antagonists alone, but was completely attenuated when the three receptor antagonists were added in combination (Jacobsson et al., 2001). In addition, in osteosarcoma cells, AEA induced apoptosis by increasing intracellular calcium levels, activation of p38 MAPK and subsequent activation of caspase 3 (Hsu et al., 2007). Unfortunately, the authors do not address the issue of receptor-dependence in this study, therefore for the purpose of this review we will assume potential involvement of all three receptors as none can be ruled out.

B. Receptor-independent effects

Cannabinoid receptor-independent actions of AEA have been described in several tumor cell types (DeMorrow et al., 2007; Hinz et al., 2004a,b; Patsos et al., 2005; Ramer et al., 2001). Furthermore, most of these effects are via the actions of AEA at the cell membrane (DeMorrow et al., 2007; Hinz et al., 2004a,b; Ramer et al., 2001). Treatment of human neuroglioma cells with the stable analogue of AEA (Met-AEA) resulted in the induction of apoptosis (Hinz et al., 2004a,b; Ramer et al., 2001). This effect could not be blocked by the coadministration of antagonists of the Cb1, Cb2, or VR1 receptors, nor the $G_{i/o}$ protein inactivator pertussis toxin (Ramer et al., 2001). Coupled with the induction of apoptosis by Met-AEA, there was an increased synthesis of ceramide, expression of cyclooxygenase-2 (Cox-2) and subsequent prostaglandin E2 synthesis via a mechanism involving p38 and p42/44 MAPK activation (Ramer et al., 2001). Specific Cox-2 inhibitors, such as celecoxib and diclofenac, or the specific silencing of Cox-2 expression with small-interfering RNA blocked the Met-AEA-induced apoptosis (Hinz et al., 2004b). Furthermore, the lipid raft disruptor methyl-beta-cyclodextrin blocked the Met-AEA-induced effects of ceramide synthesis, phosphorylation of p38 and p42/44 MAPK, expression of Cox-2, and subsequent prostaglandin E2 synthesis (Hinz et al., 2004a). Together, these data suggest that Met-AEA, via lipid raft-mediated events, induces apoptosis of neuroglioma cells. A similar effect of AEA was observed in colorectal cancer cells (Patsos et al., 2005). Treatment of these cells with AEA increased the expression of Cox-2 and induced a subsequent cell death pathway (Patsos et al., 2005). Interestingly, inhibition of FAAH potentiated this cell death, suggesting that AEA-induced cell death was mediated via the metabolism of AEA by Cox-2 rather than through the classical AEA degradation pathway (Patsos et al., 2005).

More recently, we have shown that AEA can induce growth-suppressive/pro-apoptotic effects in cholangiocarcinoma cells which could not be blocked by any Cb1, Cb2, or VR1 antagonists nor the $G_{i/o}$ protein inactivator pertussis toxin (DeMorrow et al., 2007). This is via the stabilization of the lipid raft structures within the plasma membrane, the increased production of ceramide, and the subsequent recruitment of the death receptor complex components into the lipid raft structures (DeMorrow et al., 2007). Interestingly, the other more prevalent endocannabinoid, 2-AG, had growth-promoting effects, which were shown to be via the complete disruption of the lipid raft structure (DeMorrow et al., 2007), an event which has previously been shown to result in growth-promoting effects in other cell types (Lambert et al., 2006; Mathay et al., 2008).

These receptor-independent mechanisms are summarized in Fig. 18.2. These so-called "receptor-independent" effects of AEA must be taken with a note of caution. Most of these studies were performed prior to the identification of GPR55 as a putative cannabinoid receptor. Therefore, the possibility that AEA is exerting its effects through GPR55 or some other, as yet unidentified cannabinoid receptor cannot be ruled out.

IV. Effects of AEA on Migration, Invasion, and Angiogenesis

The acquisition of metastatic abilities by cancer cells often leads to clinically incurable disease. Metastasis consists of a series of sequential steps including detachment of cells from the primary tumor, survival of the cells in circulation, arrest in a new organ, initiation and maintenance of growth in the new tissue, and vascularization of the metastatic tumor (Fidler, 2002). AEA has been shown to have a regulatory role at each of these stages in the metastatic process (Fig. 18.3). Firstly, AEA was shown to be an endogenous inhibitor of migration of both colon carcinoma cells and nonmalignant T lymphocytes in vitro (Joseph et al., 2004). There was a differential mechanism involved in the regulation of the tumor versus

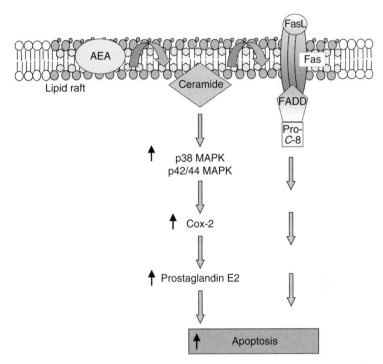

Figure 18.2 Schematic diagram of the cannabinoid receptor-independent mechanisms whereby AEA induces apoptosis. AEA stabilizes the lipid raft microdomains at the plasma membrane and increases ceramide production. This then has an effect on Cox-2 expression and prostaglandin E2 production with a subsequent increase in apoptosis. Alternatively, AEA-induced ceramide production facilitates the Fas/FasL death receptor complex within the lipid raft structure, which ultimately results in increased apoptosis.

immune cells, that is, tumor cell migration could be stimulated by specific agonists for Cb1 receptor, whereas the immune cell migration was inhibited by a Cb2-dependent mechanism (Joseph et al., 2004). Furthermore, in an *in vivo* model of metastatic spreading using breast cancer cell lines, the AEA analogue, met-AEA significantly reduced the number and dimension of metastatic nodes, an effect that was inhibited by specific Cb1 receptor antagonists (Grimaldi et al., 2006). Molecular changes in the organization and distribution of cytoskeleton proteins are necessary for focal adhesion; cell motility and cell invasion were then assessed. No changes in expression of any of the integrins were detected after Met-AEA treatment, however, there was a marked decrease in the phosphorylation of focal adhesion kinase and src, both of which are normally localized to the focal adhesions and are involved in the metastatic formation and development (Grimaldi et al., 2006).

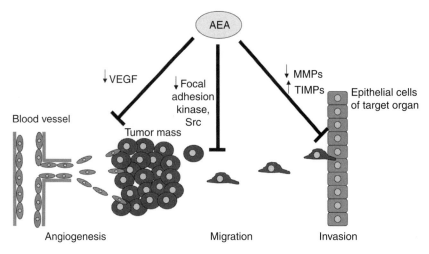

Figure 18.3 AEA inhibits other aspects of tumorigenesis such as angiogenesis, tumor cell migration, and tumor invasion. AEA inhibits angiogenesis via a decrease in VEGF expression, whereas the decrease in migration is thought to be via a decrease in the activation of focal adhesion kinases and src kinase, both of which are thought to be involved in cell migration and metastasis. Lastly, AEA inhibits tumor cell invasion by decreasing the expression of proteins responsible for breaking down the extracellular matrix of the target organ, such as MMPs and increasing the expression of the tissue inhibitors of MMPs.

In order for a migrating cancer cell to then invade another organ, the existing extracellular matrix components (e.g., collagens and proteoglycans) must be broken down and hence the rigid architecture of the target organ must be compromised. Matrix metalloproteinases (MMPs) are emerging as a family of enzymes that exerts important functions during tumor invasion (Curran and Murray, 2000; Stamenkovic, 2000). Tissue inhibitors of MMPs (TIMPs), and in particular TIMP-1, have also been shown to inhibit the proteolytic activity of MMPs and suppress vascular tumor growth and angiogenesis in xenograft animal models (Zacchigna *et al.*, 2004). Furthermore, there appears to be a correlation between high cancer invasiveness and decreased TIMP-1 expression (Chan *et al.*, 2005; Khokha *et al.*, 1989). Recently, the effects of AEA on MMP and TIMP expression were evaluated in various cancer cell types. Using a cervical cancer cell line, Met-AEA as well as Δ^9-THC, inhibited the invasive properties of these cells via the increased expression of TIMP-1 (Ramer and Hinz, 2008). The effects of the cannabinoids on invasion and TIMP-1 expression were inhibited by the pretreatment of the cells with Cb1 and Cb2 antagonists as well as specific inhibitors of the p38 and p42/44 MAPkinases (Ramer and Hinz, 2008). This effect of cannabinoids on TIMP-1 expression was mimicked by treatment of glioma cells with Δ^9-THC (Blazquez *et al.*, 2008a).

Conversely, in glioma cells, cannabinoid treatment selectively decreased the expression of MMP-2 via a Cb2-dependent mechanism and requiring the synthesis of ceramide (Blazquez *et al.*, 2008b). By manipulating the levels of MMP-2 expression by siRNA and cDNA overexpression, the authors were able to show that the decrease in MMP-2 expression was critical for the cannabinoid-mediated inhibition of cell invasion (Blazquez *et al.*, 2008b).

The last aspect of the metastatic process that is regulated by cannabinoids is angiogenesis. Met-AEA inhibited the basic fibroblast growth factor-stimulated endothelial cell proliferation and induced apoptosis in a Cb1-dependent manner (Pisanti *et al.*, 2007). Furthermore, Met-AEA was able to inhibit bidimensional capillary-like tube formation and tumor-induced angiogenesis in a three-dimensional model of endothelial and thyroid tumor cell spheroid cocultures (Pisanti *et al.*, 2007). In support of this, we have shown that AEA treatment of an *in vivo* xenograft model of cholangiocarcinoma also markedly inhibits the expression of members of the vascular endothelial growth factor family (DeMorrow *et al.*, 2008), which are key regulators of both normal and abnormal angiogenesis (Ferrara, 2005; Ferrara and Kerbel, 2005).

Together, these data suggest that cannabinoids, and in particular anandamide, may be key endogenous inhibitors of various stages in metastatic processes, including migration, invasion, and angiogenesis. This further supports the notion that drugs directed at regulating the endocannabinoid system may prove to be valuable tools in the fight against various cancers.

V. Targeting Degradation Enzymes of Cannabinoids as an Anticancer Therapy

As mentioned previously, the degradation of endocannabinoids is an active and rapid process. Therefore, blocking the degradation pathway may enhance the antiproliferative effects of AEA and have beneficial effects in cancer treatment. Indeed, treatment of human breast cancer cells *in vitro* with palmitoylethanolamide enhances the antiproliferative effects of AEA (Di Marzo *et al.*, 2001). This agent was shown to reduce the expression of FAAH up to 30–40% thereby allowing the accumulation of AEA and increasing its antiproliferative effects (Di Marzo *et al.*, 2001). In addition, treatment of athymic mice with thyroid tumor xenografts with VDM-11 (a selective inhibitor of endocannabinoid cellular reuptake) or arachidonyl-serotonin (a selective blocker of endocannabinoid hydrolysis) increased the intratumoral levels of anandamide and significantly decreased tumor volume (Bifulco *et al.*, 2004). The antiproliferative actions of these agents could be only partly inhibited by the pretreatment of the Cb1 receptor antagonist, suggesting that endocannabinoids tonically control tumor growth *in vivo* by both Cb1-mediated and non-Cb1-mediated mechanisms (Bifulco *et al.*, 2004). Regardless of the molecular mechanism by which anandamide and other endocannbinoids regulate tumor growth, inhibitors of their inactivation might be useful for the development of novel anticancer drugs (Bifulco *et al.*, 2004).

VI. Tumor Promoting Effects of Anandamide

The evidence supporting growth-promoting effect of AEA in tumors is pallid in comparison to the antitumoral effects described above. There is a greater volume of research indicating that the structurally similar, plant-derived cannabinoid, Δ^9-THC stimulates growth in a number of cancer cell lines via Cb1 and Cb2 receptor-independent mechanisms (McKallip *et al.*, 2005; Takeda *et al.*, 2008). However, several cannabinoids, including AEA, have been shown to accelerate proliferation via the transactivation of the EGFR in a TACE/ADAM17 metalloprotease-dependent manner (Hart *et al.*, 2004). This effect was observed in several cell lines from various origins including lung cancer, squamous cell carcinoma, bladder carcinoma, glioblastoma, astrocytoma, and kidney cancer (Hart *et al.*, 2004). The cannabinoid-induced activation of the EGFR leads to the subsequent phosphorylation and activation of the adaptor protein Src homology 2 domain-containing (shc), and downstream activation of the ERK1/2 and Akt/PKB pathways (Hart *et al.*, 2004). Thus, the cross-communication of cannabinoid receptors and EGFR may provide an explanation as to how cannabinoids may stimulate cancer cell proliferation (Hart *et al.*, 2004).

In addition, the stable analogue of AEA, methanandamide, has a mitogenic effect on an androgen-dependent prostate cancer cell line that could be blocked by antagonists for either the Cb1 or Cb2 receptors as well as by the PI-3kinase inhibitor (Sanchez *et al.*, 2003). The downstream consequence of activation of the endocannabinoid system was an increase in the expression of the androgen receptor, which is directly linked to the growth of these cells (Sanchez *et al.*, 2003).

VII. Conclusions

In conclusion, the endocannabinoid system exerts a myriad of effects on tumor cell growth, progression, angiogenesis, and migration. With a notable few exceptions, targeting the endocannabinoid system with agents that activate cannabinoid receptors or increase the endogenous levels of

AEA may prove to have therapeutic benefit in the treatment of various cancers. Further studies into the downstream consequences of AEA treatment are required and may illuminate other potential therapeutic targets.

ACKNOWLEDGMENTS

This work was supported by an NIH K01 grant award (DK078532) to Dr DeMorrow and by the Dr. Nicholas C. Hightower Centennial Chair of Gastroenterology from Scott & White, the VA Research Scholar Award, a VA Merit Award and the NIH grant DK58411 and DK062975 to Dr. Alpini.

We acknowledge Glen Cryer of the Scott & White Hospital, Grants Administration Office for his assistance with proof reading and Bryan Moss of the Graphics Services, Biomedia Communications, Scott & White Hospital for his assistance with the figures.

REFERENCES

Alberich Jorda, M., Rayman, N., Tas, M., Verbakel, S. E., Battista, N., van Lom, K., Lowenberg, B., Maccarrone, M., and Delwel, R. (2004). The peripheral cannabinoid receptor Cb2, frequently expressed on AML blasts, either induces a neutrophilic differentiation block or confers abnormal migration properties in a ligand-dependent manner. *Blood* **104,** 526–534.

Baker, D., Pryce, G., Croxford, J. L., Brown, P., Pertwee, R. G., Huffman, J. W., and Layward, L. (2000). Cannabinoids control spasticity and tremor in a multiple sclerosis model. *Nature* **404,** 84–87.

Bifulco, M., Laezza, C., Valenti, M., Ligresti, A., Portella, G., and Di Marzo, V. (2004). A new strategy to block tumor growth by inhibiting endocannabinoid inactivation. *FASEB J.* **18,** 1606–1608.

Bisogno, T., Ligresti, A., and Di Marzo, V. (2005). The endocannabinoid signalling system: Biochemical aspects. *Pharmacol. Biochem. Behav.* **81,** 224–238.

Blazquez, C., Carracedo, A., Salazar, M., Lorente, M., Egia, A., Gonzalez-Feria, L., Haro, A., Velasco, G., and Guzman, M. (2008a). Down-regulation of tissue inhibitor of metalloproteinases-1 in gliomas: A new marker of cannabinoid antitumoral activity? *Neuropharmacology* **54,** 235–243.

Blazquez, C., Salazar, M., Carracedo, A., Lorente, M., Egia, A., Gonzalez-Feria, L., Haro, A., Velasco, G., and Guzman, M. (2008b). Cannabinoids inhibit glioma cell invasion by down-regulating matrix metalloproteinase-2 expression. *Cancer Res.* **68,** 1945–1952.

Caffarel, M. M., Sarrio, D., Palacios, J., Guzman, M., and Sanchez, C. (2006). Delta9-tetrahydrocannabinol inhibits cell cycle progression in human breast cancer cells through Cdc2 regulation. *Cancer Res.* **66,** 6615–6621.

Calignano, A., La Rana, G., Giuffrida, A., and Piomelli, D. (1998). Control of pain initiation by endogenous cannabinoids. *Nature* **394,** 277–281.

Chan, V. Y., Chan, M. W., Leung, W. K., Leung, P. S., Sung, J. J., and Chan, F. K. (2005). Intestinal trefoil factor promotes invasion in non-tumorigenic Rat-2 fibroblast cell. *Regul. Pept.* **127,** 87–94.

Contassot, E., Tenan, M., Schnuriger, V., Pelte, M. F., and Dietrich, P. Y. (2004a). Arachidonyl ethanolamide induces apoptosis of uterine cervix cancer cells via aberrantly expressed vanilloid receptor-1. *Gynecol. Oncol.* **93,** 182–188.

Contassot, E., Wilmotte, R., Tenan, M., Belkouch, M. C., Schnuriger, V., de Tribolet, N., Burkhardt, K., and Dietrich, P. Y. (2004b). Arachidonylethanolamide induces apoptosis of human glioma cells through vanilloid receptor-1. *J. Neuropathol. Exp. Neurol.* **63,** 956–963.

Curran, S., and Murray, G. I. (2000). Matrix metalloproteinases: Molecular aspects of their roles in tumour invasion and metastasis. *Eur. J. Cancer* **36,** 1621–1630.

DeMorrow, S., Glaser, S., Francis, H., Venter, J., Vaculin, B., Vaculin, S., and Alpini, G. (2007). Opposing actions of endocannabinoids on cholangiocarcinoma growth: Recruitment of Fas and Fas ligand to lipid rafts. *J. Biol. Chem.* **282,** 13098–13113.

DeMorrow, S., Francis, H., Gaudio, E., Venter, J., Franchitto, A., Kopriva, S., Onori, P., Mancinelli, R., Frampton, G., Coufal, M., Mitchell, B., Vaculin, B., and Alpini, G. (2008). The endocannabinoid anandamide inhibits cholangiocarcinoma growth via activation of the noncanonical Wnt signaling pathway. *Am. J. Physiol. Gastrointest Liver Physiol.* **295,** G1150–G1158.

De Petrocellis, L., Melck, D., Palmisano, A., Bisogno, T., Laezza, C., Bifulco, M., and Di Marzo, V. (1998). The endogenous cannabinoid anandamide inhibits human breast cancer cell proliferation. *Proc. Natl. Acad. Sci. USA* **95,** 8375–8380.

Devane, W. A., Dysarz, F. A. III., Johnson, M. R., Melvin, L. S., and Howlett, A. C. (1988). Determination and characterization of a cannabinoid receptor in rat brain. *Mol. Pharmacol.* **34,** 605–613.

Di Marzo, V., and Deutsch, D. G. (1998). Biochemistry of the endogenous ligands of cannabinoid receptors. *Neurobiol. Dis.* **5,** 386–404.

Di Marzo, V., Melck, D., Orlando, P., Bisogno, T., Zagoory, O., Bifulco, M., Vogel, Z., and De Petrocellis, L. (2001). Palmitoylethanolamide inhibits the expression of fatty acid amide hydrolase and enhances the anti-proliferative effect of anandamide in human breast cancer cells. *Biochem. J.* **358,** 249–255.

Ek, S., Hogerkorp, C. M., Dictor, M., Ehinger, M., and Borrebaeck, C. A. (2002). Mantle cell lymphomas express a distinct genetic signature affecting lymphocyte trafficking and growth regulation as compared with subpopulations of normal human B cells. *Cancer Res.* **62,** 4398–4405.

Ferrara, N. (2005). VEGF as a therapeutic target in cancer. *Oncology* **69**(Suppl. 3), 11–16.

Ferrara, N., and Kerbel, R. S. (2005). Angiogenesis as a therapeutic target. *Nature* **438,** 967–974.

Fidler, I. J. (2002). Critical determinants of metastasis. *Semin. Cancer Biol.* **12,** 89–96.

Galve-Roperh, I., Sanchez, C., Cortes, M. L., del Pulgar, T. G., Izquierdo, M., and Guzman, M. (2000). Anti-tumoral action of cannabinoids: Involvement of sustained ceramide accumulation and extracellular signal-regulated kinase activation. *Nat. Med.* **6,** 313–319.

Glaser, S. T., Kaczocha, M., and Deutsch, D. G. (2005). Anandamide transport: A critical review. *Life Sci.* **77,** 1584–1604.

Grimaldi, C., Pisanti, S., Laezza, C., Malfitano, A. M., Santoro, A., Vitale, M., Caruso, M. G., Notarnicola, M., Iacuzzo, I., Portella, G., Di Marzo, V., and Bifulco, M. (2006). Anandamide inhibits adhesion and migration of breast cancer cells. *Exp. Cell Res.* **312,** 363–373.

Hart, S., Fischer, O. M., and Ullrich, A. (2004). Cannabinoids induce cancer cell proliferation via tumor necrosis factor alpha-converting enzyme (TACE/ADAM17)-mediated transactivation of the epidermal growth factor receptor. *Cancer Res.* **64,** 1943–1950.

Hinz, B., Ramer, R., Eichele, K., Weinzierl, U., and Brune, K. (2004a). R(+)-methanandamide-induced cyclooxygenase-2 expression in H4 human neuroglioma cells: Possible involvement of membrane lipid rafts. *Biochem. Biophys. Res. Commun.* **324,** 621–626.

Hinz, B., Ramer, R., Eichele, K., Weinzierl, U., and Brune, K. (2004b). Up-regulation of cyclooxygenase-2 expression is involved in R(+)-methanandamide-induced apoptotic death of human neuroglioma cells. *Mol. Pharmacol.* **66,** 1643–1651.

Hsu, S. S., Huang, C. J., Cheng, H. H., Chou, C. T., Lee, H. Y., Wang, J. L., Chen, I. S., Liu, S. I., Lu, Y. C., Chang, H. T., Huang, J. K., Chen, J. S., et al. (2007). Anandamide-induced Ca2+ elevation leading to p38 MAPK phosphorylation and subsequent cell death via apoptosis in human osteosarcoma cells. *Toxicology* **231,** 21–29.

Islam, T. C., Asplund, A. C., Lindvall, J. M., Nygren, L., Liden, J., Kimby, E., Christensson, B., Smith, C. I., and Sander, B. (2003). High level of cannabinoid receptor 1, absence of regulator of G protein signalling 13 and differential expression of Cyclin D1 in mantle cell lymphoma. *Leukemia* **17,** 1880–1890.

Jacobsson, S. O., Wallin, T., and Fowler, C. J. (2001). Inhibition of rat C6 glioma cell proliferation by endogenous and synthetic cannabinoids. Relative involvement of cannabinoid and vanilloid receptors. *J. Pharmacol. Exp. Ther.* **299,** 951–959.

Joseph, J., Niggemann, B., Zaenker, K. S., and Entschladen, F. (2004). Anandamide is an endogenous inhibitor for the migration of tumor cells and T lymphocytes. *Cancer Immunol. Immunother.* **53,** 723–728.

Khokha, R., Waterhouse, P., Yagel, S., Lala, P. K., Overall, C. M., Norton, G., and Denhardt, D. T. (1989). Antisense RNA-induced reduction in murine TIMP levels confers oncogenicity on Swiss 3T3 cells. *Science* **243,** 947–950.

Laezza, C., Pisanti, S., Crescenzi, E., and Bifulco, M. (2006). Anandamide inhibits Cdk2 and activates Chk1 leading to cell cycle arrest in human breast cancer cells. *FEBS Lett.* **580,** 6076–6082.

Lambert, S., Vind-Kezunovic, D., Karvinen, S., and Gniadecki, R. (2006). Ligand-independent activation of the EGFR by lipid raft disruption. *J. Invest. Dermatol.* **126,** 954–962.

Ligresti, A., Bisogno, T., Matias, I., De Petrocellis, L., Cascio, M. G., Cosenza, V., D'Argenio, G., Scaglione, G., Bifulco, M., Sorrentini, I., and Di Marzo, V. (2003). Possible endocannabinoid control of colorectal cancer growth. *Gastroenterology* **125,** 677–687.

Manzanares, J., Corchero, J., Romero, J., Fernandez-Ruiz, J. J., Ramos, J. A., and Fuentes, J. A. (1999). Pharmacological and biochemical interactions between opioids and cannabinoids. *Trends Pharmacol. Sci.* **20,** 287–294.

Mathay, C., Giltaire, S., Minner, F., Bera, E., Herin, M., and Poumay, Y. (2008). Heparin-binding EGF-like growth factor is induced by disruption of lipid rafts and oxidative stress in keratinocytes and participates in the epidermal response to cutaneous wounds. *J. Invest. Dermatol.* **128,** 717–727.

Matsuda, L. A., Lolait, S. J., Brownstein, M. J., Young, A. C., and Bonner, T. I. (1990). Structure of a cannabinoid receptor and functional expression of the cloned cDNA. *Nature* **346,** 561–564.

McKallip, R. J., Lombard, C., Fisher, M., Martin, B. R., Ryu, S., Grant, S., Nagarkatti, P. S., and Nagarkatti, M. (2002). Targeting CB2 cannabinoid receptors as a novel therapy to treat malignant lymphoblastic disease. *Blood* **100,** 627–634.

McKallip, R. J., Nagarkatti, M., and Nagarkatti, P. S. (2005). Delta-9-tetrahydrocannabinol enhances breast cancer growth and metastasis by suppression of the antitumor immune response. *J. Immunol.* **174,** 3281–3289.

Melck, D., Rueda, D., Galve-Roperh, I., De Petrocellis, L., Guzman, M., and Di Marzo, V. (1999). Involvement of the cAMP/protein kinase A pathway and of mitogen-activated protein kinase in the anti-proliferative effects of anandamide in human breast cancer cells. *FEBS Lett.* **463,** 235–240.

Melck, D., De Petrocellis, L., Orlando, P., Bisogno, T., Laezza, C., Bifulco, M., and Di Marzo, V. (2000). Suppression of nerve growth factor Trk receptors and prolactin receptors by endocannabinoids leads to inhibition of human breast and prostate cancer cell proliferation. *Endocrinology* **141,** 118–126.

Mimeault, M., Pommery, N., Wattez, N., Bailly, C., and Henichart, J. P. (2003). Anti-proliferative and apoptotic effects of anandamide in human prostatic cancer cell lines:

Implication of epidermal growth factor receptor down-regulation and ceramide production. *Prostate* **56,** 1–12.

Munro, S., Thomas, K. L., and Abu-Shaar, M. (1993). Molecular characterization of a peripheral receptor for cannabinoids. *Nature* **365,** 61–65.

Pagotto, U., Marsicano, G., Fezza, F., Theodoropoulou, M., Grubler, Y., Stalla, J., Arzberger, T., Milone, A., Losa, M., Di Marzo, V., Lutz, B., and Stalla, G. K. (2001). Normal human pituitary gland and pituitary adenomas express cannabinoid receptor type 1 and synthesize endogenous cannabinoids: First evidence for a direct role of cannabinoids on hormone modulation at the human pituitary level. *J. Clin. Endocrinol. Metab.* **86,** 2687–2696.

Patsos, H. A., Hicks, D. J., Dobson, R. R., Greenhough, A., Woodman, N., Lane, J. D., Williams, A. C., and Paraskeva, C. (2005). The endogenous cannabinoid, anandamide, induces cell death in colorectal carcinoma cells: A possible role for cyclooxygenase 2. *Gut* **54,** 1741–1750.

Petersen, G., Moesgaard, B., Schmid, P. C., Schmid, H. H., Broholm, H., Kosteljanetz, M., and Hansen, H. S. (2005). Endocannabinoid metabolism in human glioblastomas and meningiomas compared to human non-tumour brain tissue. *J. Neurochem.* **93,** 299–309.

Piomelli, D., Giuffrida, A., Calignano, A., and Rodriguez de Fonseca, F. (2000). The endocannabinoid system as a target for therapeutic drugs. *Trends Pharmacol. Sci.* **21,** 218–224.

Pisanti, S., Borselli, C., Oliviero, O., Laezza, C., Gazzerro, P., and Bifulco, M. (2007). Antiangiogenic activity of the endocannabinoid anandamide: Correlation to its tumor-suppressor efficacy. *J. Cell Physiol.* **211,** 495–503.

Ramer, R., and Hinz, B. (2008). Inhibition of cancer cell invasion by cannabinoids via increased expression of tissue inhibitor of matrix metalloproteinases-1. *J. Natl. Cancer Inst.* **100,** 59–69.

Ramer, R., Brune, K., Pahl, A., and Hinz, B. (2001). R(+)-methanandamide induces cyclooxygenase-2 expression in human neuroglioma cells via a non-cannabinoid receptor-mediated mechanism. *Biochem. Biophys. Res. Commun.* **286,** 1144–1152.

Ruiz, L., Miguel, A., and Diaz-Laviada, I. (1999). Delta9-tetrahydrocannabinol induces apoptosis in human prostate PC-3 cells via a receptor-independent mechanism. *FEBS Lett.* **458,** 400–404.

Ryberg, E., Larsson, N., Sjogren, S., Hjorth, S., Hermansson, N. O., Leonova, J., Elebring, T., Nilsson, K., Drmota, T., and Greasley, P. J. (2007). The orphan receptor GPR55 is a novel cannabinoid receptor. *Br. J. Pharmacol.* **152,** 1092–1101.

Sanchez, C., Galve-Roperh, I., Canova, C., Brachet, P., and Guzman, M. (1998). Delta9-tetrahydrocannabinol induces apoptosis in C6 glioma cells. *FEBS Lett.* **436,** 6–10.

Sanchez, C., de Ceballos, M. L., del Pulgar, T. G., Rueda, D., Corbacho, C., Velasco, G., Galve-Roperh, I., Huffman, J. W., Ramon y Cajal, S., and Guzman, M. (2001). Inhibition of glioma growth *in vivo* by selective activation of the CB(2) cannabinoid receptor. *Cancer Res.* **61,** 5784–5789.

Sanchez, M. G., Sanchez, A. M., Ruiz-Llorente, L., and Diaz-Laviada, I. (2003). Enhancement of androgen receptor expression induced by (R)-methanandamide in prostate LNCaP cells. *FEBS Lett.* **555,** 561–566.

Sarfaraz, S., Afaq, F., Adhami, V. M., and Mukhtar, H. (2005). Cannabinoid receptor as a novel target for the treatment of prostate cancer. *Cancer Res.* **65,** 1635–1641.

Schmid, P. C., Wold, L. E., Krebsbach, R. J., Berdyshev, E. V., and Schmid, H. H. (2002). Anandamide and other N-acylethanolamines in human tumors. *Lipids* **37,** 907–912.

Shi, Y., Zou, M., Baitei, E. Y., Alzahrani, A. S., Parhar, R. S., Al-Makhalafi, Z., and Al-Mohanna, F. A. (2008). Cannabinoid 2 receptor induction by IL-12 and its potential as a therapeutic target for the treatment of anaplastic thyroid carcinoma. *Cancer Gene Ther.* **15,** 101–107.

Stamenkovic, I. (2000). Matrix metalloproteinases in tumor invasion and metastasis. *Semin. Cancer Biol.* **10,** 415–433.

Takeda, S., Yamaori, S., Motoya, E., Matsunaga, T., Kimura, T., Yamamoto, I., and Watanabe, K. (2008). Delta(9)-Tetrahydrocannabinol enhances MCF-7 cell proliferation via cannabinoid receptor-independent signaling. *Toxicology* **245,** 141–146.

Voth, E. A., and Schwartz, R. H. (1997). Medicinal applications of delta-9-tetrahydrocannabinol and marijuana. *Ann. Intern. Med.* **126,** 791–798.

Zacchigna, S., Zentilin, L., Morini, M., Dell'Eva, R., Noonan, D. M., Albini, A., and Giacca, M. (2004). AAV-mediated gene transfer of tissue inhibitor of metalloproteinases-1 inhibits vascular tumor growth and angiogenesis *in vivo*. *Cancer Gene Ther.* **11,** 73–80.

CHAPTER NINETEEN

USE OF CANNABINOIDS AS A NOVEL THERAPEUTIC MODALITY AGAINST AUTOIMMUNE HEPATITIS

Rupal Pandey,* Venkatesh L. Hegde,* Narendra P. Singh,*
Lorne Hofseth,[†] Uday Singh,* Swapan Ray,* Mitzi Nagarkatti,*
and Prakash S. Nagarkatti*

Contents

I. Introduction	488
II. The Endogenous Cannabinoid System	489
III. The Biosynthesis of Endocannabinoids	490
IV. Endocannabinoid System is Autoprotective	490
A. The endocannabinoid system and pathophysiology of liver diseases	490
B. Anti-inflammatory effects of endocannabinoids	492
V. Autoimmune Hepatitis	492
VI. Treatment Drawbacks	494
VII. Cannabinoid/Endocannabinoid System in Hepatitis	494
A. Controversies on the beneficial and deleterious roles of cannabinoid receptors in the regulation of liver disease	498
VIII. Conclusions and Future Directions	499
Acknowledgments	500
References	500

Abstract

Autoimmune hepatitis is a severe immune mediated chronic liver disease with a prevalence range between 50 and 200 cases per million in Western Europe and North America and mortality rates of up to 80% in untreated patients. The induction of CB1 and CB2 cannabinoid receptors during liver injury and the potential involvement of endocannabinoids in the regulation of this process have sparked significant interest in further evaluating the role of cannabinoid

---

* Department of Pathology, Microbiology, and Immunology, University of South Carolina School of Medicine, Columbia, South Carolina, USA
[†] Department of Pharmaceutical and Biomedical Sciences, College of Pharmacy, University of South Carolina, Columbia, South Carolina, USA

systems during hepatic disease. Cannabinoids have been shown to possess significant immunosuppressive and anti-inflammatory properties. Cannabinoid abuse has been shown to exacerbate liver fibrogenesis in patients with chronic hepatitis C infection involving CB_1 receptor. Nonetheless, CB_2 receptor activation may play a protective role during chronic liver diseases. Thus, differential targeting of cannabinoid receptors may provide novel therapeutic modality against autoimmune hepatitis. In this review, we summarize current knowledge on the role of endocannabinoids and exocannabinoids in the regulation of autoimmune hepatitis.

I. INTRODUCTION

Medicinal properties of *Cannabis sativa*, Marijuana have been explored for centuries. Interest in the potential medicinal use of cannabinoids grew following the isolation and characterization of delta-9-tetrahydrocannabinol (THC) as the major psychoactive component in marijuana (Gaoni and Mechoulam, 1971) and subsequently the discovery of two cannabinoid receptors, CB1 and CB2 (Matsuda *et al.*, 1990; Munro *et al.*, 1993). Cannabinoids have been used as potential therapeutic agents in alleviating such complications as intraocular pressure in glaucoma, cachexia, nausea, and pain (Watson *et al.*, 2000). Nonetheless, the clinical use of Marijuana cannabinoids has been controversial due to their psychotropic effects and potential abuse. To date, over 60 other unique phytocannabinoids have been identified from the marijuana plant, generating renewed interest in its medicinal potential.

It is well established now that cannabis acts primarily on mammalian tissue through at least two specific cannabinoid receptor subtypes which are both G protein-coupled membrane receptors. Both the receptors share little sequence homology, only 44% at the protein level or 68% in the transmembrane domains which are thought to contain the binding sites for cannabinoids (Howlett, 2002; Lutz, 2002). The CB1 receptor is found in abundance in the brain areas which control motor activity, memory, cognition, emotions, endocrine functions, and sensory perception. It is also expressed in peripheral nerve terminals and other sites such as eye, spleen, vascular endothelium, adipose tissue, gut, liver, and testis. Its activation results in central and many peripheral effects of cannabinoids by modulating the neurotransmitter release. The CB2 receptor is primarily expressed on subsets on immune cells and several leukocyte lines of hematopoietic systems (macrophages, both B and T lymphocytes), secondary lymphoid tissues such as spleen, tonsils, Peyer's patches, Lymphatic ganglia, microglia and hepatic myofibroblastic cells (Gong *et al.*, 2006; Munro *et al.*, 1993). The precise function of CB2 receptors on immune cells is not clear. However, studies from our laboratory demonstrated for the first time that activation of CB2 can induce apoptosis in immune cells, including T-cells and dendritic cells (McKallip et al., 2002b). Other studies have shown that activation of CB2 may alter the cytokine profile and promote the differentiation of Th2 cells (Roth et al., 2002). Intracellular CB1 and CB2 dependent signaling pathways include $G_{i/o}$-dependent inhibition of adenylyl cyclase, PI3-kinase, MAP kinase and FAK pathways, and modulation of calcium and potassium channels. The discovery of expression of cannabinoid receptors in vertebrate body was followed by rapid discovery of naturally occurring endogenous ligands for these receptors. It is, however, noteworthy that endocannabinoids may also bind to vanilloid receptors and peroxisome-proliferators activated receptors or can also exhibit receptor—independent properties due to their high lipophilicity (Batkai et al., 2001; Maccarrone and Finazzi-Agro, 2003).

II. THE ENDOGENOUS CANNABINOID SYSTEM

The discovery of cannabinoid receptors occurring naturally throughout the vertebrate body and the availability of highly selective and potent canabimimetics led to identification of a naturally occurring lipid signaling system termed as the endocannabinoid system. This comprises of a family of lipid transmitters serving as natural ligands for the cannabinoid receptors. The first endocannabinoid discovered was named as anandamide (AEA) from the Sanskrit "internal bliss" (Devane et al., 1992). AEA is an amide of arachidoic acid and ethanolamine. It binds to brain cannabinoid receptor with high affinity and mimics the behavioral actions of THC when injected into rodents. Subsequently, 3 years later, another ligand named 2-arachidinoiylethanolamide (2-AG) was discovered independently by Mechoulam et al. (1995) and Sugiura et al. (1995), which was found to be in much higher concentration in serum and brain than anandamide. This triggered an exponential growth of studies describing the endocannabinoid synthesis, their release (Di Marzo et al.,1994), transport (Beltramo et al., 1997) and their degradation (Cravatt et al., 1996) constituting the new ubiquitous "endogenous cannabinoid system."

The main pharmacological function of the endocannabinoid system is in neuromodulation controlling motor functions, cognition, emotional responses, homeostasis, and motivation. However, in the periphery, this system is an important modulator of autonomic nervous system, the immune system and microcirculation.

III. The Biosynthesis of Endocannabinoids

Endocannabinoids are not stored in vesicles or cells like neurotransmitters; rather their biosynthesis takes place on demand from lipid precursors in the cytoplasmic membrane through enzyme activation in response to elevations of intracellular calcium. They are produced by receptor-stimulated cleavage of membrane lipid precursors and then released from cells. Anandamide is the product of arachydonate and phosphoethanolamide. Arachydonate is released from cytoplasmic membrane through enzymatic cleavage of N-arachydonoyl phophatidiethanolamide by a Ca^{++} dependent N-acyl transferase (Schmid et al., 1983). 2-AG is produced by hydrolysis of a diacygliceride containing arachdonate in glycerol position2 released by a specific diacylglycerol lipase (Lichtman et al., 2002). The endocannabinoids are then removed from the site by cellular uptake processes such as simple diffusion, through membrane associated binding proteins or by a transmembrane carrier protein. Inside the tissue, their metabolism is catalyzed by fatty acid amide hydrolase (FAAH) (Kozak and Marnett, 2002; Lichtman et al., 2002) which is located on cytosolic surfaces of SER cisternae and mitochondria. In some tissues, however, endocannabinoids also suffer an oxidative catabolism through lipoxygenases, cyclooxygenase-2 and cytochrome P450 (Kozak and Marnett, 2002), and palmitoylethanolamide-preferring acid amidase (PAA) (Goparaju et al., 1999).

IV. Endocannabinoid System is Autoprotective

It certain disorders such as multiple sclerosis, cancer, intestinal disorder, cardiovascular disorder, pain, Parkinson's disease and excitotoxicity, the tissue concentration of endocannabinoids, cannabinoid receptor density and the cannabinoid receptor coupling efficiency increases resulting in reduction of symptoms of these disorders. The endocannabinoid system has been shown to be involved in various physiological processes like lipogenesis, inflammation, food intake, and nociception (Di Marzo et al., 2004).

A. The endocannabinoid system and pathophysiology of liver diseases

There is increasing evidence suggesting the role of endocannabinoids in the pathophysiology of liver diseases such as in hepatic injury and as a mediator of complications of cirrhosis. The normal adult liver hepatocytes and non-parenchymal cells have been found to produce low levels of endocannabinoids (Julien et al., 2005; Teixeira-Clerc et al., 2006). This can be attributed to presence of high hepatocellular expression of FAAH responsible for AEA degradation (Cravatt et al., 1996). Hepatic and serum levels of AEA increase during acute hepatitis and fatty liver disease (Batkai et al., 2007; Biswas et al., 2003). In fatty liver, this increase in AEA was attributed to a decrease in AEA degradation by FAAH.

Although embryonic liver has been shown to express CB2 receptor mRNA, adult liver hepatocytes and endothelial cells display a faint physiological level of expression of CB1 receptors. However, both CB1 and CB2 receptors are upregulated in early stages of liver injury (Batkai et al., 2007; Biecker et al., 2004; Julien et al., 2005; Schwabe and Siegmund, 2005; Teixeira-Clerc et al., 2006). CB1 receptors have been found to be upregulated in the vascular endothelium and in myofibroblasts located in fibrotic bands of cirrhotic livers in humans and rodents (Teixeira-Clerc et al., 2006). CB2 receptors are also expressed in myofibroblasts and in inflammatory cells and biliary epithelial cells (Julien et al., 2005). Thus, induction of endocannabinoids and their receptors are potentially two mechanisms rendering liver responsive to endocannabinoids in course of liver fibrogenesis.

Nonetheless, many recent studies indicate strongly the increased upregulation of the endocannabinoid system during liver diseases affecting hepatocyte injury, inflammation, fibrogenesis, hepatic encephalopathy, cirrhotic cardiomyopathy, and portal hypertension.

Liver injury is associated with Kupffer cell activation, release of proinflammatory cytokines and generation of reactive oxygen species leading to infiltration of the liver by activated polymorphonuclear leukocytes. In a recent study, hepatic ischemia-reperfusion (I/R) injury was shown to be associated with dramatic induction of hepatic expression of anadamide and 2-AG directly related to extent of liver damage (Batkai et al., 2007). The data also points out the protective role of CB2 receptor activation in inflammatory response related with chronic liver diseases such as viral hepatitis and alcoholic or nonalcoholic fatty liver diseases.

Similarly CB1 and CB2 receptors are also markedly upregulated in cirrhotic human liver samples demonstrating the impact of endocannabinoid in liver fibrogenesis. Elevated circulating levels of anadamide and of hepatic 2-AG in cirrhosis and liver fibrosis, respectively, have been consistently reported (Batkai et al., 2001). The role of CB2 receptors in mediating antifibrogenic response has been corroborated using CB2 knockout mice which when exposed to CCl4 showed enhanced liver fibrosis and increased liver fibrogenic cell accumulation as compared to their wild type counterparts (Julien et al., 2005). The role of CB2 was corroborated in vitro showing that CB2 receptor activation induces apoptosis in cultured liver fibrogenic cells by activation of oxidative stress (Julien et al., 2005). However, CB1 receptors were found to signal profibrotic response (Teixeira-Clerc et al., 2006) depicting that CB1 and

CB2 receptors exert opposite effects on liver fibrosis and further suggesting that endocannabinoid system regulates both pro- and antifibrogenic responses in the liver.

In an analysis carried out on patients with chronic hepatitis C, it was observed that cannabis used over the span of the disease was an independent predictor of fibrosis severity (Hezode et al., 2005) indicating that cannabinoids may exacerbate liver fibrogenesis and thus CB1 antagonists may play as antifibrosing molecules. It should be noted that CB2 is expressed at higher density on monocytes and the immunosuppression exerted through CB2 may have an effect in patients with hepatitis C because such patients require an intact immune component to keep hepatitis in check. Thus, chronic marijuana consumption may promote fibrogenesis via CB2 mediated suppression of antiviral immunity (Schwabe and Siegmund, 2005).

Studies have also been carried out to evaluate the role of endocannabinoids as mediators of vascular and cardiac abnormalities in cirrhosis. Reports indicate that endocannabinoids trigger vasorelaxing effects and an upregulated CB1-dependent cannabinoid tone causes enhanced mesenteric vasodialation leading to portal hypertension (Batkai et al., 2001; Ros et al., 2002).

B. Anti-inflammatory effects of endocannabinoids

Hepatic inflammation is linked to hepatic fibrosis in models of fibrogenesis. Kupffer cells are crucial and promote activation of hepatic stellate cells (HSCs). Activated T lymphocytes and neutrophils also contribute to inflammatory microenvironment leading to HSC activation and fibrogenesis. Kupffer cells are shown to express high CB2 mRNA which is known to mediate anti-inflammatory effect by suppression of TNF-α and IFN-γ and stimulate anti-inflammatory cytokines such as IL-10 and also inhibit macrophage migration at sites of inflammation leading to anti-fibrogenic effects of endocannabinoids in the liver. The selective CB2 receptor agonists JWH133 and HU-308 have also been shown to decrease TNF-α, ICAM-1, and VCAM-1 expression in human liver sinusoidal endothelial cells (HLSECs) expressing CB2 receptors and thus decreased adhesion of human neutrophils to HLSECs, thus depicting a role for CB2 receptor in endothelial cell activation and endothelial-inflammatory cell interactions (Rajesh et al., 2008).

V. Autoimmune Hepatitis

Autoimmune liver disease (ALD) includes a spectrum of diseases which comprises both cholestatic and hepatitic forms: autoimmune hepatitis (AIH), primary biliary cirrhosis (PBC), primary sclerosing cholangitis (PSC) and the so-called "overlap" syndromes where hepatic and cholestatic damage coexist. All these diseases are characterized by an extremely high heterogeneity of presentation, varying from asymptomatic, acute (as in a subset of AIH) or chronic (with a specific symptoms such as fatigue and myalgia in AIH or fatigue and pruritus in PBC and PSC). The incidence and prevalence of the forms of ALDs remain to be defined worldwide, although information is available for distinct ethnic populations (Berdal et al., 1998). The prevalence of ALD ranges from 13 to 20/100,000 population and the incidence from 1 to 2/100,000 in a Norwegian population (Boberg et al., 1998).

AIH comprises a relatively diverse group of liver diseases associated with autoantibody formation which are thought to occur as a result of an uncontrolled, self-directed inflammatory attack on hepatocytes or bile ducts. The immune injury commonly leads to rapidly progressive liver disease, progressive fibrosis, and ultimately cirrhosis. AIH occurs mainly in women and is characterized by elevated serum transaminase activity, hypergammaglobulinemia, circulating positive organ and nonorgan autoantibodies, and morphologic changes of interface hepatitis on liver biopsy. The prevalence of AIH is estimated to range between 50 and 200 cases per million in North America and accounts for 5–9% of liver transplantation in United States. Mortality rates of up to 80% have been reported in untreated patients who presented with greater than fivefold elevations in serum aspartate (AST) or alanine (ALT) aminotransferase activities (Soloway et al., 1972). The cause of ALD is not known, but factors that influence the development of clinical disease include genetic predisposition, prior liver injury caused by viruses, environmental exposure to toxins, and possibly, ongoing infectious factors, including viruses. Immune reactions against host liver antigens are believed to be the major pathogenic mechanisms.

Infectious causes of ALD have been proposed, with prior viral hepatitis A, B, or C, cytomegalovirus (CMV), paramyxovirus, and most recently retrovirus infections, suggested as possibilities (Manns and Obermayer-Straub, 1997; Mason et al., 1998). The development of recurrent AIH, PBC, and PSC in liver transplant recipients provides strong support for an infectious origin for of these disorders. If environmental agents are found to be the cause or a component of disease activity and progression, advances in antibiotic and antiviral therapies may also delay or prevent progressive liver disease.

The standard therapy for AIH involves treatment with prednisone or prednisolone (40/60 mg/d) alone or a lower steroid dose (20–30 mg/d) in combination with azathioprine (1 mg/kg/d) (Murray-Lyon et al., 1973). Corticosteroids and azathioprine also represent the standard therapy for AIH; both compliance and treatment outcome can be monitored by transaminase serum levels; a slow tapering of the immunosuppressive therapy is always recommended to avoid the relapse of the disease, frequently observed in cases of too fast tapering. Corticosteroids bind to corticosteroid receptor elements in the promoter regions of numerous steroid responsive genes thereby affecting transcription. They act rapidly on the immune system mainly by affecting the cytokine production and inhibition of T lymphocyte activation (Bellary et al., 1995). Azathioprine acts as immunosuppressive drug by blocking the maturation of lymphocyte precursors and requires 3 months or more to be fully effective.

VI. Treatment Drawbacks

The common side effects of corticosteroids include fluid retention, obesity, diabetes mellitus, osteoporosis, psychiatric disturbances, cataracts (Beswick et al., 1985; Biagini et al., 1991). Azathioprine treatments can cause nausea, pancreatitis, hepatotoxicity, and bone marrow suppression (Beswick et al., 1985). Additionally, about 15% patients are unresponsive to this standard therapy. Liver transplantation remains the only therapeutical approach for the end stage of liver disease with an actual survival rate over 60% after 10 years.

VII. Cannabinoid/Endocannabinoid System in Hepatitis

Cannabinoids which act as cannabinoid receptor agonists, including THC, have shown promising results in many models of inflammation (Hegde et al., 2008; Pertwee, 2002; Quartilho et al., 2003). Cannabinoid compounds have been shown to possess significant immunosuppressive and anti-inflammatory properties (Croxford and Yamamura, 2005). The histological picture of AIH with its striking infiltrate of lymphocytes, plasma cells, and macrophages was the first to suggest an autoaggressive cellular immune attack in the pathogenesis of AIH. Whatever is the initial trigger, this massive recruitment of activated inflammatory cells causes damage. The T lymphocyes have been identified as the predominant cells in immunohistochemical studies.

Recently our lab has reported that both exogenous and endogenous cannabinoids can attenuate concanavalin A-induced acute hepatitis (a well-established model for viral or AIH in which liver injury is T-cell-mediated). TNF-α, IFN-γ, and IL-2 play crucial roles in Con-induced hepatitis (Ganter et al., 1995; Tagawa et al., 1997). We demonstrated that a single intraperitoneal injection of THC (a CB1 and CB2 receptor agonist) and anandamide (an endocannabinoid, partial agonist of CB1 and CB2 receptors) can ameliorate ConA-induced hepatitis via a negative inflammatory cytokine regulation (IL-2, TNF-α, IFN-γ, IL-1α, IL-1β, IL-5, IL-6, IL-10, IL-17, GM-CSF, G-CSF, KC, MIP-1α, and RANTES) in mice and by upregulating forkhead helix transcription factor p3(Foxp3)$^+$ regulatory T-cells (Hegde et al., 2008).

An impairment of immunoregulatory mechanisms, which would enable the autoimmune response to develop, has been repeatedly reported in the setting of both human and experimental autoimmunity. In early studies it was shown that patients with AIH have low levels of circulating T-cells (Lobo-Yeo et al., 1987) and impaired suppressor cell function. Recent experimental evidence confirms an impairment of the immunoregulatory function in AIH. Thus, among recently defined T-cell subsets, with potential immunosuppressive function, CD4$^+$ T-cells constitutively expressing the interleukin two receptor α chain (CD25) (T-regulatory cells, T-regs) have emerged as a dominant immunoregulatory lymphocyte (Shevach et al., 2003). In addition to CD25, which is also present on activated T-cells, T-regs express a number of additional markers such as the glucocorticoid-induced tumor necrosis factor receptor, CD62L, the cytotoxic T lymphocyte associated protein-4 (CTLA-4) and the forkhead/winged helix transcription factor FOXP3, the expression of which is closely associated with the acquisition of regulatory properties. In patients with AIH, T-regs are defective both in number and function compared to normal controls and these abnormalities relate to the stage of disease, being more evident at diagnosis than during drug induced remission (Longhi et al., 2005, 2006). The percentage of T-regs inversely correlates with markers of disease severity, such as levels of antibodies to antisoluble liver antigen (Ma et al., 2002), and anti-LKM-1 autoantibody titers, suggesting that a reduction in T-regs favors the serological manifestations of ALD. If loss of immunoregulation was central to the pathogenesis of ALDe, treatment should concentrate on restoring T-regs ability to expand, with consequent increase in their number and function. This is at least partially achieved by standard immunosuppression, since T-regs numbers do increase during remission (Longhi et al., 2005). With reference to the reports stating that T regulatory cells play a crucial role in natural tolerance in ConA-induced hepatitis (Erhardt et al., 2007), we observed that THC treatment significantly enhanced T regulatory cells over vehicle controls which may be considered as a critical factor to reduce the severity of hepatitis. Decreased IL-6 levels observed in ConA$^+$ THC-injected mice may be contributing to suppression of effector T-cell function by T-regs. We found that THC treatment in ConA-injected mice triggered a significant increase in the number of CD4$^+$ CD25$^+$ T-regs in the liver that was responsible for decreased hepatitis (Hegde et al., 2008). We also analyzed Foxp3 mRNA expression by reverse transcription-PCR in splenocyte and liver cells which were obtained 4 h after ConA challenge *in vivo*. We found that ConA$^+$ THC treated mice had higher levels of Foxp3 mRNA in the liver and spleen. Mice injected with

ConA and pretreated with anti-CD25 antibodies (which effectively depleted $CD25^+$ cells) showed higher serum AST levels compared with mice pretreated with control IgG suggesting that regulatory T-cells are required for THC-mediated suppression of hepatitis. It is important to note that $CD25^+$ cell-depleted mice showed increased AST levels after ConA injection when compared to naive mice. One possible explanation for this would be that depletion of T-regs resulted in uncontrolled hepatotoxicity upon polyclonal activation of T-cells and other inflammatory cells by ConA. This supports the recent observation that T-regs mediate tolerance in ConA-induced hepatitis model (Erhardt et al., 2007). Previous studies have shown that THC suppresses cytokine secretion by its direct action on lymphocytes (Blanchard et al., 1986; Klein et al., 2000). In the current model, THC may be acting directly on T-cells and also by enhancing T-reg function in the presence of ConA to suppress cytokine production and disease attenuation. Suppression of M-CSF, G-CSF, RANTES, KC, and MIP-1α can be considered important to decrease chemotaxis and activation of macrophages and eosinophils which is associated with ConA-induced hepatitis (Louis et al., 2002; Bonder et al., 2004).

Circulating T-cells specific for liver autoantigens is found in normal subjects. However, in AIH, their frequency is at least 10-fold higher (Wen et al., 2001). This finding suggests that the pool of autoreactive T-cells undergoes a significant expansion in patients with AIH and hence may be involved in the initiation and perpetuation of the immune attack to the liver. Previous studies showed that activated T- and NK T-cells are increased in liver after ConA challenge. However, these cells rapidly undergo apoptosis, which may be crucial for mice to recover from ConA-induced hepatitis (Russell, 1995). We have noted previously that THC induces apoptosis in thymic and splenic T-cells (McKallip et al., 2002b). Thus, we also investigated the frequency of hepatic T lymphocytes undergoing apoptosis which we found was increased in ConA-injected mice upon THC treatment (Hegde et al., 2008). Nonetheless, absolute numbers of hepatic T-cells in THC-treated hepatitis-induced mice were significantly higher. This may result from induction of T-regs, as well as, possible increased migration of T-cells into the liver. Interestingly, we also noted that IL-2 was not decreased upon THC treatment, and it may be contributing to proliferation of T-cells. However, T-cells were functionally suppressed by THC as indicated by decreased serum cytokine levels and liver injury. This could be due to direct suppressive effect of THC on activated T-cells as well as an increase in $CD4^+$ $Foxp3^+$ T-regs. IL-2 is an essential growth and survival factor for T-regs, and it is required for their function (Furtado et al., 2002). Sufficient levels of IL-2 observed in ConA-injected mice treated with THC or anandamide could contribute to increased T-reg function, resulting in significant suppression of inflammatory cytokines.

Because THC acts through both CB1 and CB2 receptors, we also tested CB1/CB2 mixed agonists (CP55,940 and WIN55212) and found that these compounds were also effective to suppress ConA-induced hepatitis (Hegde et al., 2008). Interestingly, we found that CB1 or CB2 activation alone had no anti-inflammatory effect on hepatitis. However, cannabinoids that bind to both CB1 and CB2 receptors (THC, CP55,940, WIN55212, and anandamide) effectively attenuated hepatitis. That CB1/CB2 mixed agonists could suppress the pathogenesis but not the coadministered CB1 and CB2 agonists indicated that both cannabinoid receptors need to be activated simultaneously to produce the observed effect and that different pharmacokinetics of the two coadministered agonists may not allow this to happen. Signaling through both receptors is important because blocking either CB1 or CB2 could reverse the effect of THC. We also noted that activation through CB1 *per se* worsened the effect of ConA, which is in agreement with previous studies showing that CB1 may contribute toward liver inflammatory disorders (Teixeira-Clerc et al., 2006). However, activation of both CB1 and CB2 not only prevented the worsening effect of CB1 stimulation alone, but also resulted in a strong counteraction of inflammation. The need for simultaneous activation of CB1 and CB2 to see the beneficial effect is consistent with some previous observations (Van Sickle et al., 2005). However, these results are contradictory to previous understanding that anti-inflammatory properties of cannabinoids are mainly mediated by CB2. The finding that hepatocytes (Batkai et al., 2007; Biswas et al., 2003) and dendritic cells (Do et al., 2004; Matias et al., 2002) express CB1 and CB2 receptors may explain the contribution of both receptors toward protection observed in this study. To further confirm the role of both cannabinoid receptors, we blocked either of these receptors by pretreating mice with CB1 (AM251) or CB2 (SR144258) antagonists before injecting ConA+ THC and observed that blocking of either receptor was sufficient to reverse the effects of THC (Hegde et al., 2008). These data demonstrated that immunomodulatory function of THC in ConA-induced hepatitis is mediated by signaling through CB1 and CB2 receptors, and blocking either of these receptors was sufficient to reverse the effects of THC.

Similar to our findings with THC, exogenously administered anandamide in the hepatitis model also caused a decrease in hepatic injury which correlated with decreased AST and ALT levels, and inflammatory cytokines. We also observed that 12 h after anandamide treatment, in ConA injected mice, there was a significant decrease in inflammatory cytokines TNF-α, IL-1β, IL-6, IL-9, and IL-17 and in chemokines KC, eotaxin and monocyte chemoattractant protein-1. Blocking of CB1 or CB2 receptors reversed the anandamide-mediated suppression of hepatitis thereby indicating that anandamide was acting in a CB1 and CB2 dependent manner. Anandamide levels are elevated in the absence of FAAH activity

(Cravatt et al., 1996). Mice which lacked FAAH enzyme (responsible for hydrolysis of anandamide) were found to be resistant to ConA-induced hepatitis and showed less severe liver tissue damage and less leukocyte infiltration (Hegde et al., 2008). Similarly, increased endogenous anandamide levels caused by administering mice with FAAH inhibitors MAFP or URB532 also decreased hepatic injury with significant decrease in AST levels upon ConA challenge. These findings are exciting and suggest the potential manipulation of endocannabinoids, including the use of FAAH inhibitors in the treatment of AIH.

A. Controversies on the beneficial and deleterious roles of cannabinoid receptors in the regulation of liver disease

As seen from the discussion above, the role of endocannabinoid system in hepatitis is somewhat controversial. On one hand, human studies have revealed that chronic marijuana use is detrimental to the liver (Hezode et al., 2005; Ishida et al., 2008). It is known to increase the risk for progression of liver fibrosis, in patients with hepatitis C infection. These studies contradicted the experimental studies in rodents that CB2 activation may protect from fibrosis (Julien et al., 2005). However, subsequent studies from the same group suggested that activation of CB1 receptors promotes progression of fibrosis and therefore CB1 antagonists may be useful for the treatment of liver fibrosis (Teixeira-Clerc et al., 2006). In contrast, studies from our laboratory seem to suggest that cannabinoids including THC seem to protect hepatic cell injury in a rodent model of hepatitis triggered by T-cell mitogen, ConA. Our findings are consistent with another report which demonstrated that a synthetic nonpsychoactive cannabinoid derivative (PRS-211,092) was found to attenuate ConA-induced hepatitis, although it should be noted that the effect was noted to be independent of cannabinoid receptors. These discrepancies can be explained in many ways: First, ConA-induced rodent model of hepatitis, although mimics human viral and AIH in many aspects, is an acute model of hepatitis in which activation of T-cells with the release of cytokines including TNF-α, IFN-γ, IL-6, and IL-1 seems to play a crucial role. In addition to T-cells, neutrophils, δT, NK T-cells, and Kupffer cells have been implicated in ConA-induced hepatitis and similar involvement of immune cells has also been noted in AIH. Thus, cannabinoids that are anti-inflammatory are likely to provide beneficial effects. In contrast, chronic liver injury may include additional nonimmune mechanisms. Likewise, in human chronic hepatitis C infection, a sustained immune response against the virus may be crucial for regulating the infection (Yu et al., 2008) and smoking marijuana which is know to suppress the immune response is likely to promote chronic infection and liver fibrosis. Secondly, while in an acute model, cannabinoid receptors expressed on immune cells may play a crucial role, in the chronic disease, the cannabinoid receptors expressed on hepatocytes may also regulate the disease process as seen from their upregulation (Julien et al., 2005; Schwabe and Siegmund, 2005). It is interesting to note that even in the acute murine model of ConA-induced hepatitis, we noted that use of CB1 select agonists at higher doses worsened the effect of ConA (Hegde et al., 2008). Thirdly, in human studies using marijuana smoke, the patients are potentially exposed to a large number of other compounds besides THC; thus the net effect may be different from the use of THC as a single compound used in animal studies. Of particular interest is cannabidiol which has been shown be induce apoptosis and cause immunosuppression (Straus, 2000).

VIII. Conclusions and Future Directions

Taken together, our data suggest that exogenous cannabinoids such as THC, upon binding to CB1 and CB2 receptors on immune cells, mediate anti-inflammatory effects through multiple pathways including induction of apoptosis in effector T-cells (Do et al., 2004; Hegde et al., 2008; Jia et al., 2006; Lombard et al., 2005; McKallip et al., 2002a,b), upregulation of T-reg function and suppression of inflammatory cytokines (Hegde et al., 2008), thereby preventing ConA-induced activated T-cell-mediated liver injury. The observation that the anandamide treatment ameliorates ConA-induced hepatitis, together with FAAH deficiency or inhibition leading to increased resistance to the disease, strongly suggests that the endocannabinoid system serves to attenuate the inflammatory response in acute model of ConA-induced hepatitis. These findings raise the promising potential of developing novel pharmacological treatments for T-cell-mediated liver diseases. Due to the controversial nature of potential deleterious effects of marijuana smoke on liver fibrosis as seen from human studies versus the beneficial effects of THC and endocannabinoids as seen in acute rodent models of hepatitis, clearly, additional studies are necessary to address the precise nature and role of cannabinoid receptors and their ligands in acute and chronic hepatitis. Better animal models that mimic chronic hepatitis need to be developed and the effect of cannabinoids addressed. Whether activation of cannabinoid receptors on immune cells found in the liver versus hepatocytes and endothelial cells, leads to differential effects needs to be investigated. While some studies, including those from our lab, have shown that CB2 activation alone is sufficient to cause immunosuppression and trigger an anti-inflammatory response (Lombard et al., 2005, 2007), in some disease models, activation of both CB receptors seems to be necessary to trigger immunosuppression and attenuation of the disease (Hegde et al., 2008). Thus, additional studies should be pursued that delineate the signaling events that follow activation of individual versus combined CB receptors on immune cells. Moreover, the potential manipulation of endocannabinoids through use of FAAH inhibitors to protect the liver from immune-mediated attacks, and the unique ability of CB receptor agonists to trigger regulatory T-cells (Hegde et al., 2008), deserves further attention.

ACKNOWLEDGMENTS

This work was supported in part by NIH grants R01AI053703, R01ES09098, R01 AI058300, R01DA016545, R01HL058641 and P01AT00396.

REFERENCES

Batkai, S., Jarai, Z., Wagner, J. A., Goparaju, S. K., Varga, K., Liu, J., Wang, L., Mirshahi, F., Khanolkar, A. D., Makriyannis, A., Urbaschek, R., Garcia, N. Jr., et al. (2001). Endocannabinoids acting at vascular CB1 receptors mediate the vasodilated state in advanced liver cirrhosis. *Nat. Med.* **7,** 827–832.

Batkai, S., Osei-Hyiaman, D., Pan, H., El-Assal, O., Rajesh, M., Mukhopadhyay, P., Hong, F., Harvey-White, J., Jafri, A., Hasko, G., Huffman, J. W., Gao, B., et al. (2007). Cannabinoid-2 receptor mediates protection against hepatic ischemia/reperfusion injury. *FASEB J.* **21,** 1788–1800.

Bellary, S., Schiano, T., Hartman, G., and Black, M. (1995). Chronic hepatitis with combined features of autoimmune chronic hepatitis and chronic hepatitis C: Favorable response to prednisone and azathioprine. *Ann. Intern. Med.* **123,** 32–34.

Beltramo, M., Stella, N., Calignano, A., Lin, S. Y., Makriyannis, A., and Piomelli, D. (1997). Functional role of high-affinity anandamide transport, as revealed by selective inhibition. *Science* **277,** 1094–1097.

Berdal, J. E., Ebbesen, J., and Rydning, A. (1998). Incidence and prevalence of autoimmune liver diseases. *Tidsskr Nor Laegeforen* **118,** 4517–4519.

Beswick, D. R., Klatskin, G., and Boyer, J. L. (1985). Asymptomatic primary biliary cirrhosis. A progress report on long-term follow-up and natural history. *Gastroenterology* **89,** 267–271.

Biagini, M. R., McCormick, P. A., Guardascione, M., Surrenti, C., and Burroughs, A. K. (1991). Prognosis in primary biliary cirrhosis. A review. *Ital. J. Gastroenterol.* **23,** 222–226.

Biecker, E., Sagesser, H., and Reichen, J. (2004). Vasodilator mRNA levels are increased in the livers of portal hypertensive NO–synthase 3-deficient mice. *Eur. J. Clin. Invest.* **34,** 283–289.

Biswas, K. K., Sarker, K. P., Abeyama, K., Kawahara, K., Iino, S., Otsubo, Y., Saigo, K., Izumi, H., Hashiguchi, T., Yamakuchi, M., Yamaji, K., Endo, R., et al. (2003). Membrane cholesterol but not putative receptors mediates anandamide-induced hepatocyte apoptosis. *Hepatology* **38,** 1167–1177.

Blanchard, D. K., Newton, C., Klein, T. W., Stewart, W. E. 2nd, and Friedman, H. (1986). In vitro and in vivo suppressive effects of delta-9-tetrahydrocannabinol on interferon production by murine spleen cells. *Int. J. Immunopharmacol.* **8,** 819–824.

Boberg, K. M., Aadland, E., Jahnsen, J., Raknerud, N., Stiris, M., and Bell, H. (1998). Incidence and prevalence of primary biliary cirrhosis, primary sclerosing cholangitis, and autoimmune hepatitis in a Norwegian population. *Scand. J. Gastroenterol.* **33,** 99–103.

Bonder, C. S., Ajuebor, M. N., Zbytnuik, L. D., Kubes, P., and Swain, M. G. (2004). Essential role of neutrophil recruitment to the liver concanavalin A-induced hepatitis. *J. Immunol.* **172,** 45–53.

Cravatt, B. F., Giang, D. K., Mayfield, S. P., Boger, D. L., Lerner, R. A., and Gilula, N. B. (1996). Molecular characterization of an enzyme that degrades neuromodulatory fatty-acid amides. *Nature* **384,** 83–87.

Coxford, J. L., and Yamamura, T. (2005). Cannabinoids and the immune system: Potential for the treatment of inflammatory diseases? *J. Neuroimmunol.* **166,** 3–18.

Devane, W. A., Hanus, L., Breuer, A., Pertwee, R. G., Stevenson, L. A., Griffin, G., Gibson, D., Mandelbaum, A., Etinger, A., and Mechoulam, R. (1992). Isolation and structure of a brain constituent that binds to the cannabinoid receptor. *Science* **258,** 1946–1949.

Di Marzo, V., *et al.* (1994). Formation and inactivation of endogenous cannabinoid anandamide in central neurons. *Nature* **372,** 686–691.

Di Marzo, V., Bifulco, M., and De Petrocellis, L. (2004). The endocannabinoid system and its therapeutic exploitation. *Nat. Rev. Drug Discov.* **3,** 771–784.

Do, Y., McKallip, R. J., Nagarkatti, M., and Nagarkatti, P. S. (2004). Activation through cannabinoid receptors 1 and 2 on dendritic cells triggers NF-kappaB-dependent apoptosis: Novel role for endogenous and exogenous cannabinoids in immunoregulation. *J. Immunol.* **173,** 2373–2382.

Erhardt, A., Biburger, M., Papadopoulos, T., and Tiegs, G. (2007). IL-10, regulatory T cells, and Kupffer cells mediate tolerance in concanavalin A-induced liver injury in mice. *Hepatology* **45,** 475–485.

Furtado, G. C., Curotto de Lafaille, M. A., Kutchukhidze, N., and Lafaille, J. J. (2002). Interleukin 2 signaling is required for CD4(+) regulatory T cell function. *J. Exp. Med.* **196,** 851–857.

Gantner, F., Leist, M., Lohse, A. W., Germann, P. G., and Tiegs, G. (1995). Concanavalin-A induced T-Cell-mediated hepatic injury in mice: The role of tumor necrosis factor. *Hepatology* **21,** 190–198.

Gaoni, Y., and Mechoulam, R. (1971). The isolation and structure of delta-1-tetrahydrocannabinol and other neutral cannabinoids from hashish. *J. Am. Chem. Soc.* **93,** 217–224.

Gong, J. P., Onaivi, E. S., Ishiguro, H., Liu, Q. R., Tagliaferro, P. A., Brusco, A., and Uhl, G. R. (2006). Cannabinoid CB2 receptors: Immunohistochemical localization in rat brain. *Brain Res.* **1071,** 10–23.

Goparaju, S. K., Ueda, N., Taniguchi, K., and Yamamoto, S. (1999). Enzymes of porcine brain hydrolyzing 2-arachidonoylglycerol, an endogenous ligand of cannabinoid receptors. *Biochem. Pharmacol.* **57,** 417–423.

Hegde, V. L., Hegde, S., Cravatt, B. F., Hofseth, L. J., Nagarkatti, M., and Nagarkatti, P. S. (2008). Attenuation of experimental autoimmune hepatitis by exogenous and endogenous cannabinoids: Involvement of regulatory T cells. *Mol. Pharmacol.* **74,** 20–33.

Hezode, C., Roudot-Thoraval, F., Nguyen, S., Grenard, P., Julien, B., Zafrani, E. S., Pawlotsky, J. M., Dhumeaux, D., Lotersztajn, S., and Mallat, A. (2005). Daily cannabis smoking as a risk factor for progression of fibrosis in chronic hepatitis C. *Hepatology* **42,** 63–71.

Howlett, A. C. (2002). The cannabinoid receptors. *Prostaglandins Other Lipid. Mediat.* **68–69,** 619–631.

Ishida, J. H., Peters, M. G., Jin, C., Louie, K., Tan, V., Bacchetti, P., and Terrault, N. A. (2008). Influence of cannabis use on severity of hepatitis C disease. *Clin. Gastroenterol. Hepatol.* **6,** 69–75.

Jia, W., Hegde, V. L., Singh, N. P., Sisco, D., Grant, S., Nagarkatti, M., and Nagarkatti, P. S. (2006). Delta9-tetrahydrocannabinol-induced apoptosis in Jurkat leukemia T cells is regulated by translocation of Bad to mitochondria. *Mol. Cancer Res.* **4,** 549–562.

Julien, B., Grenard, P., Teixeira-Clerc, F., Van Nhieu, J. T., Li, L., Karsak, M., Zimmer, A., Mallat, A., and Lotersztajn, S. (2005). Antifibrogenic role of the cannabinoid receptor CB2 in the liver. *Gastroenterology* **128,** 742–755.

Klein, T. W., Newton, C. A., Nakachi, N., and Friedman, H. (2000). Delta 9-tetrahydrocannabinol treatment suppresses immunity and early IFN-gamma, IL-12, and IL-12 receptor beta 2 responses to Legionella pneumophila infection. *J. Immunol.* **164,** 6461–6466.

Kozak, K. R., and Marnett, L. J. (2002). Oxidative metabolism of endocannabinoids. *Prostaglandins Leukot. Essent. Fatty Acids* **66,** 211–220.

Lichtman, A. H., Hawkins, E. G., Griffin, G., and Cravatt, B. F. (2002). Pharmacological activity of fatty acid amides is regulated, but not mediated, by fatty acid amide hydrolase in vivo. *J. Pharmacol. Exp. Ther.* **302,** 73–79.

Lobo-Yeo, A., Alviggi, L., Mieli-Vergani, G., Portmann, B., Mowat, A. P., and Vergani, D. (1987). Preferential activation of helper/inducer T lymphocytes in autoimmune chronic active hepatitis. *Clin. Exp. Immunol.* **67,** 95–104.

Lombard, C., Nagarkatti, M., and Nagarkatti, P. S. (2005). Targeting cannabinoid receptors to treat leukemia: Role of cross-talk between extrinsic and intrinsic pathways in Delta9-tetrahydrocannabinol (THC)-induced apoptosis of Jurkat cells. *Leuk. Res.* **29,** 915–922.

Lombard, C., Nagarkatti, M., and Nagarkatti, P. (2007). CB2 cannabinoid receptor agonist, JWH-015, triggers apoptosis in immune cells: Potential role for CB2-selective ligands as immunosuppressive agents. *Clin. Immunol.* **122,** 259–270.

Longhi, M. S., Ma, Y., Mitry, R. R., Bogdanos, D. P., Heneghan, M., Cheeseman, P., Mieli-Vergani, G., and Vergani, D. (2005). Effect of CD4+ CD25+ regulatory T-cells on CD8 T-cell function in patients with autoimmune hepatitis. *J. Autoimmun.* **25,** 63–71.

Longhi, M. S., Hussain, M. J., Mitry, R. R., Arora, S. K., Mieli-Vergani, G., Vergani, D., and Ma, Y. (2006). Functional study of CD4+CD25+ regulatory T cells in health and autoimmune hepatitis. *J. Immunol.* **176,** 4484–4491.

Louis, H., et al. (2002). Critical role of interleukin 5 and eosinophils in concanavalin A-induced hepatitis in mice. *Gastroenterology* **122,** 2001–2010.

Lutz, B. (2002). Molecular biology of cannabinoid receptors. *Prostaglandins Leukot. Essent. Fatty Acids* **66,** 123–142.

Ma, Y., Okamoto, M., Thomas, M. G., Bogdanos, D. P., Lopes, A. R., Portmann, B., Underhill, J., Durr, R., Mieli-Vergani, G., and Vergani, D. (2002). Antibodies to conformational epitopes of soluble liver antigen define a severe form of autoimmune liver disease. *Hepatology* **35,** 658–664.

Maccarrone, M., and Finazzi-Agro, A. (2003). The endocannabinoid system, anandamide and the regulation of mammalian cell apoptosis. *Cell Death Differ.* **10,** 946–955.

Manns, M. P., and Obermayer-Straub, P. (1997). Viral induction of autoimmunity: Mechanisms and examples in hepatology. *J. Viral Hepat.* 4(Suppl. 2), 42–47.

Mason, A. L., Xu, L., Guo, L., Munoz, S., Jaspan, J. B., Bryer-Ash, M., Cao, Y., Sander, D. M., Shoenfeld, Y., Ahmed, A., Van de Water, J., Gershwin, M. E., Gershwin, M. E., et al. (1998). Detection of retroviral antibodies in primary biliary cirrhosis and other idiopathic biliary disorders. *Lancet* **351,** 1620–1624.

Matias, I., Pochard, P., Orlando, P., Salzet, M., Pestel, J., and Di Marzo, V. (2002). Presence and regulation of the endocannabinoid system in human dendritic cells. *Eur. J. Biochem.* **269,** 3771–3778.

Matsuda, L. A., Lolait, S. J., Brownstein, M. J., Young, A. C., and Bonner, T. I. (1990). Structure of a cannabinoid receptor and functional expression of the cloned cDNA. *Nature* **346,** 561–564.

McKallip, R. J., Lombard, C., Fisher, M., Martin, B. R., Ryu, S., Grant, S., Nagarkatti, P. S., and Nagarkatti, M. (2002a). Targeting CB2 cannabinoid receptors as a novel therapy to treat malignant lymphoblastic disease. *Blood* **100,** 627–634.

McKallip, R. J., Lombard, C., Martin, B. R., Nagarkatti, M., and Nagarkatti, P. S. (2002b). Delta(9)-tetrahydrocannabinol-induced apoptosis in the thymus and spleen as a mechanism of immunosuppression in vitro and in vivo. *J. Pharmacol. Exp. Ther.* **302,** 451–465.

Mechoulam, R., et al. (1995). Identification of an endogenous 2-monoglyceride, present in canine gut, that binds to cannabinoid receptors. *Biochem. Pharmacol.* **50,** 83–90.

Munro, S., Thomas, K. L., and Abu-Shaar, M. (1993). Molecular characterization of a peripheral receptor for cannabinoids. *Nature* **365,** 61–65.

Murray-Lyon, I. M., Stern, R. B., and Williams, R. (1973). Controlled trial of prednisone and azathioprine in active chronic hepatitis. *Lancet* **1,** 735–737.

Pertwee, R. G. (2002). Cannabinoids and multiple sclerosis. *Pharmacol. Ther.* **95,** 165–174.

Quartilho, A., et al. (2003). Inhibition of inflammatory hyperalgesia by activation of peripheral CB2 cannabinoid receptors. *Anesthesiology* **99,** 955–960.

Rajesh, M., Mukhopadhyay, P., Hasko, G., Huffman, J. W., Mackie, K., and Pacher, P. (2008). CB2 cannabinoid receptor agonists attenuate TNF-alpha-induced human vascular smooth muscle cell proliferation and migration. *Br. J. Pharmacol.* **153,** 347–357.

Ros, J., Claria, J., To-Figueras, J., Planaguma, A., Cejudo-Martin, P., Fernandez-Varo, G., Martin-Ruiz, R., Arroyo, V., Rivera, F., Rodes, J., and Jimenez, W. (2002). Endogenous cannabinoids: A new system involved in the homeostasis of arterial pressure in experimental cirrhosis in the rat. *Gastroenterology* **122,** 85–93.

Roth, M. D., Baldwin, G. C., and Tashkin, D. P. (2002). Effects of delta-9-tetrahydrocannabinol on human immune function and host defense. *Chem. Phys. Lipids* **121,** 229–239.

Russell, J. H. (1995). Activation-induced death of mature T cells in the regulation of immune responses. *Curr. Opin. Immunol.* **7,** 382–388.

Schmid, P. C., Reddy, P. V., Natarajan, V., and Schmid, H. H. (1983). Metabolism of N-acylethanolamine phospholipids by a mammalian phosphodiesterase of the phospholipase D type. *J. Biol. Chem.* **258,** 9302–9306.

Schwabe, R. F., and Siegmund, S. V. (2005). Potential role of CB2 receptors in Cannabis smokers with chronic hepatitis C. *Hepatology* **42,** 975–976author reply 976–977.

Shevach, E. M., Piccirillo, C. A., Thornton, A. M., and McHugh, R. S. (2003). Control of T cell activation by CD4+CD25+ suppressor T cells. *Novartis Found Symp.* **252,** 24–36discussion 36–44, 106–14..

Soloway, R. D., Summerskill, W. H., Baggenstoss, A. H., Geall, M. G., Gitnick, G. L., Elveback, I. R., and Schoenfield, L. J. (1972). Clinical, biochemical, and histological remission of severe chronic active liver disease: A controlled study of treatments and early prognosis. *Gastroenterology* **63,** 820–833.

Straus, S. E. (2000). Immunoactive cannabinoids: Therapeutic prospects for marijuana constituents. *Proc. Natl. Acad. Sci. USA* **97,** 9363–9364.

Sugiura, T., et al. (1995). 2-Arachidonoylglycerol: A possible endogenous cannabinoid receptor ligand in brain. *Biochem. Biophys. Res. Commun.* **215,** 89–97.

Tagawa, Y., Sekikawa, K., and Iwakura, Y. (1997). Suppression of concanavalin A-induced hepatitis in IFN-gamma(-/-) mice, but not in TNF-alpha(-/-) mice: Role for IFN-gamma in activating apoptosis of hepatocytes. *J. Immunol.* **159,** 1418–1428.

Teixeira-Clerc, F., Julien, B., Grenard, P., Tran Van Nhieu, J., Deveaux, V., Li, L., Serriere-Lanneau, V., Ledent, C., Mallat, A., and Lotersztajn, S. (2006). CB1 cannabinoid receptor antagonism: A new strategy for the treatment of liver fibrosis. *Nat. Med.* **12,** 671–676.

Van Sickle, M. D., Duncan, M., Kingsley, P. J., Mouihate, A., Urbani, P., Mackie, K., Stella, N., Makriyannis, A., Piomelli, D., Davison, J. S., Marnett, L. J., Di Marzo, V., *et al.* (2005). Identification and functional characterization of brainstem cannabinoid CB2 receptors. *Science* **310,** 329–332.

Watson, S. J., Benson, J. A. Jr., and Joy, J. E. (2000). Marijuana and medicine: Assessing the science base: A summary of the 1999 Institute of Medicine report. *Arch. Gen. Psychiatry* **57,** 547–552.

Wen, L., Ma, Y., Bogdanos, D. P., Wong, F. S., Demaine, A., Mieli-Vergani, G., and Vergani, D. (2001). Pediatric autoimmune liver diseases: The molecular basis of humoral and cellular immunity. *Curr. Mol. Med.* **1,** 379–389.

Yu, J., Wu, C. W., Chu, E. S., Hui, A. Y., Cheng, A. S., Go, M. Y., Ching, A. K., Chui, Y. L., Chan, H. L., and Sung, J. J. (2008). Elucidation of the role of COX-2 in liver fibrogenesis using transgenic mice. *Biochem. Biophys. Res. Commun.* **372,** 571–577.

Index

A

AA-CoA acyltransferase, 267
AA-containing endocannabinoids, 291–292
Abh4-mediated pathway, 18
Abn-CBD receptor, 348
Acetylcholine, 27, 128, 163, 429
   nicotinic, 317
   receptors, 87, 95
Acetyl coenzyme-A carboxylase (ACC1), 162
Acomplia, 163
Actin cytoskeleton, 351
Activin receptor-like kinase 3 (Alk3), 238
N-Acyl amides, 193–194, 198
Acyl CoA synthetase, 35
N-Acyl conjugates, 57
N-Acylethanolamides, 57, 271
N-Acylethanolaminehydrolyzing acid amidase (NAAA), 443
N-Acylethanolamines, 2, 394
N-Acyl glycines, 195–197
   biological activity, 200
   members characterization, 198–200
   molecular and fragmentation ions, 200
N-Acylphosphatidylethanolamine (NAPE), 3, 392, 443
   alternative pathways for NAEs, 15–18
   synthesis, 393–394
N-Acyl-phosphatidylethanolamine-selective phosphodiesterase, 209, 240, 242, 392–394
N-Acylphosphatidylethanolamine–selective phospholipase D, 142
N-Acyl taurines, 194
N-Acyltransferase (NAT), 3–4, 209, 240, 442
   Ca-NAT, 4–6
   iNAT, 6–9
N-Acylyamino acids, 57
Adenosine A1 receptor agonist, 115
Adenylyl cyclase, 27, 112, 163, 233, 236, 276, 352, 390, 401, 442, 489
β_2-Adrenoceptors, 124
Adrenocorticotropic hormone (ACTH), 265
AEA degradation, 28–29
AEA inhibition, 322–323
AEA-mediated cell death, glioma cells, 176
AEA signaling, 27–28
   implications of pharmacologically altered, 44–45
AEA synthesis, 26–27
AEA transport, 31–32
AEA metabolites, lipid rafts and fate, 29–31
   cellular AEA accumulation, models, 37–44
      facilitated processes, 39–40, 42–44
      passive diffusion model, 37–39
   characteristics, 36–37
   fatty acid transporters, 32–35
   lipid transfer proteins, 35–36
AEA uptake inhibitors, 41
Alanine (ALT) aminotransferase activities, 493
Allosteric modulator, 341
Alzheimer's disease, 212
AM-251, CB_1 receptor antagonist, 115, 126, 426, 428
γ-Aminobutyric acid, 163
Ammonium ion, 62
AMT/FAAH modulation, 395
Analgesia, 56
Anandamide, 57, 160, 271–273, 391, 422, 470–471, 489–490
   antiproliferative effects, 473
      receptor-independent effects, 475–476
      receptor mediated effects, 473–475
   biological actions, 443
   biological inactivation, 209
   biosynthetic and metabolic pathways, 211
   cannabinoid system and vanilloid receptor, 387
   (see also Vanilloid receptor 1)
      analgesia and, 406–407
      body weight regulation, 410
      in CNS, 404–406
      interplay between, evidence, 392
      in peripheral nervous system, 407–408
      temperature regulation and, 409–410
      vascular effects, 408–409
   cellular uptake mechanism of, 209
   de novo biosynthesis, 210
   effect on keratinocyte differentiation, 457–459
   and future research, 410–411
   hydrolysis, 81
   inhibitory effect, on 5-HT_3 receptors, 321–324
   levels in glioblastomas, 471
   as lipid messenger, 443
   metabolism, 393
      and COX-2 action, 395–396
      FAAH role, 395
   in microglia, 216
   and multiple sclerosis
      inhibitors, of AEA uptake and, 219–220

Anandamide (cont.)
    as neuroprotective agent, 222–223
    upregulation under, 220–221
  as neuroimmune signal, 212
    as cytokine network modulator, 214–215
    as immune function modulator, 212–214
    in cytokines production, 215–219, 223
  receptors, 404
  role in
    metastatic process, 476–479
    skin differentiation, 443–444
  synthesis, 392–394, 442–443
  transport, 394–395
  as TRPV1 ligand, 396–397
    affinity and efficacy, on TRPV1, 397
    anandamide-mediated activation, 399–401
    anandamide metabolism, effects, 402–403
    complex interplay between, 403
    no direct effect on, 403
    role, 401–402
    sensitization, 398–399
  tumor promoting effects, 480
Anandamide amidohydrolase. See Fatty acid amide hydrolase
Anandamide membrane transporter (AMT), 142, 394, 443
Androgen-binding protein (ABP), 237
Androgens, 236, 244
Angiogenesis, 476
  role of cannabinoids, 479
Angiotensin II, 124
Antinociception, 56
Antitumor properties, of cannabinoid receptor agonists, 177
Apoptosis, 67
  alpha-glycyrrhetinic acid and gap junction blocker, 69
  bystander killing, 69
  caspase activation 9, 68
  gap junction communication, 69, 70
  neuronal depolarization and, 69
  release of cytochrome c from, 67
  role for oleamide signaling in, 69
AP-1 transcriptional complex, 455
Arachidonic acid (AA), 29, 236, 265–266
  AA-containing endocannabinoids, 271–274
  biochemical identities, 265
  cascade, 266–267
  role in CNS, 265
N-Arachidonoyl alanine, 193
N-Arachidonoyl dopamine, 194
N-Arachidonoyl ethanolamine, 2, 113, 192, 208, 239. See also AEA; Anandamide
N-Arachidonoyl GABA, 193
2-Arachidonoylglycerol (2-AG), 113, 160, 192, 208
  agonist for CB1R and CB2R and, 240

biosynthetic and metabolic pathways for, 211
in maternal milk, 149
N-Arachidonoyl glycine (NAGly), 193–194
  biological activity, 194
  biosynthesis, 194–195
  FAAH metabolism, 203
N-Arachidonoyl-L-serine (ARA-S), 341, 348
N-Arachidonoyl-lyso-PE, 17
N-Arachidonoylphenolamine (AM40), 280
N-Arachidonoyl-phosphatidylethanolamine, 17–18, 209, 240, 272, 442–443
N-Arachidonoyl serine, 193
N-Arachidonyl-dopamine (NADA), 339, 391
N-Arachidonylethanolamide. See AEA; Anandamide
2-Arachidonylglycerol, 26, 273–274, 339, 368, 391, 471, 489–490
2-Arachidonyl glyceryl ether, 391
N-Arachidoylation, 209
Autoimmune hepatitis (AIH), 492–494
Autoimmune liver disease (ALD), 492–493
Azathioprine, 494
600-Azidohex-200-yne-cannabidiol (O-2654), 168

B

Bombesin, 290
Bone morphogenetic protein 4 (BMP4), 237
Brain-derived 2-AG, 150
Brain derived neurotrophic factor (BDNF), 222
Brain–pituitary axis, 371–374
Branchiostoma floridae, 369

C

Ca^{2+}-activated K^+ channels, 126
Ca^{2+}-dependent phospholipase D activity, 27
Ca^{2+}-dependent PLA2, 123
Ca^{2+} influx, 236
Calcitonin gene-related peptide (CGRP), 114, 341
Calcium channels, 85–86
Calmodulin-dependent protein kinases (CaM kinases), 236
cAMP-induced transcription, 234
cAMP/protein kinase A pathway, 473
cAMP-responsive element (CRE), 233
cAMP-responsive element modulator (CREM), 233
Cancer
  cells, antiproliferative effects of AEA in, 473, 479
  dysregulation of endocannabinoid system in, 471–473
  role of CB1 in, 473
  role of CB2 in, 473–475
  VR1, role, 475
Cannabidiol (CBD), 340–341

Index

Cannabinoid CB_1 receptor, 81–82, 162. *See also* Cannabinoids; CB_1 receptors
  activity, in newborn, 148
  antagonists (*see also* CB_1 antagonists, in treatment of obesity)
    as anticancer drugs, 180
    exhibiting anorexigenic effect, 180
    psychiatric side effects, 179–180
  blockade at birth, 152
  density (Bmax), 151
Cannabinoid receptor 2, 89, 113. *See also* Cannabinoids; CB_2 receptors
  antagonists, 115, 177
  signaling pathways, 89–90
Cannabinoids, 486
  additional sites for activation, 340–341
  classification
    endocannabinoids, 339–340
    phytocannabinoids, 340
    synthetic cannabinoids, 340
  definition, 339
  inhibition of 5-HT_3 receptors, 318, 321–324
  modulation of Gly receptor function by, 324–328
  receptor agonist
    antitumor properties of, 177
    HU210, 145
    WIN 55,212–2, 145
  receptors, 390, 470, 486, 491
    antibodies, 345
    CB_1 receptors, 340, 422, 473, 486
    CB_2 receptors, 340, 422, 486–489
    expression in neutrophils, 344–345
    properties, 368–371
  role in cell migration, 342
  role in hepatitis, 494–498
    controversies on, 498–499
Cannabinol (CBN), 341
Cannabis sativa, 80, 112, 160, 208, 340, 486
Capsaicin, 114
Capsazepine (CPZ), 346
Cardiovascular diseases (CVD), 160
Catecholamines, 289
Caveolae/CRM domains, 29
Caveolae-related endocytic uptake, 31
Caveolin-rich membrane (CRM) domains, 29
CB_1 antagonists, in obesity treatment, 169
  rimonabant as antiobesity drug, 169–172
  safety and adverse effects, 174–176
    ADAGIO study, 176
    FDA assessement, 175
    RIO studies, 174–176
    STRADIVARIUS trial, 175
  use of taranabant, 172
  in vitro and in animal models, 172–174
CB1 KO mice, 374
$CB1^{-/-}$ mice, 115, 149
CB_1 receptors. *See also* Cannabinoids
  activating $G\alpha$ subunits, 113
  antagonists, 115, 164–167
    LY320135, 115
    SR141716, 143, 148
    VCHSR1, 148
  on blastocytes, 141
  in cancer therapy, 180
  in development of emotional processing, 146
  expression, in pathophysiological situations, 160
  G proteins activation, 113
  knockout mice, 115, 142
  in neural development, 147
  and neutrophil migration, 346–347
  null mice, 115–116
  presence in developing brain, 147
  and preterm birth, 144–146
  signaling, 27–28
  stimulating adenylyl cyclase activity in, 113
CB_2 receptors. *See also* Cannabinoids
  apoptosis through ceramide signaling, 113
  and neutrophil migration, 347
  signaling, 27–28
CCAAT enhancer-binding protein C/EBP, 234
Cell migration
  cannabinoids modulating, 342
  initiated by GPCR $\beta\gamma$-complex, 354
  initiated by GPCR G_α-subunits., 353
Cellular motility, 338
Central nervous system (CNS), 80, 161, 264
Central stimulation, of SAS, 290
Ceramide-induced apoptosis, 113
Cerebrospinal fluid, 56
Chemokines, 214
Chemotaxis, 335. *See also* Neutrophils
CHO cells, 123
Cholesterol, 29
Ciona intestinalis, 116
CNR1 gene, 179
Cocaine-and amphetamine-related transcript (CART), 161
Compound SR141716, 163
Concanavalin A-induced acute hepatitis, 494
Corneocytes, 445
Corneodesmolysis, 452
Corneodesmosomes, 452
Cornification. *See* Epidermal terminal differentiation
Cornified envelope (CE), 452
Corticosteroids, for treatment of AIH, 494–495
Corticosterone (CCS), 144
Corticotropinreleasing hormone (CRH), 265
CP55940, 116
CpG islands, 453
C-reactive protein, 160
CREB-binding proteins (CBP/p300), 233
CRH, and appetite, 290
CRH-mediated suppression of food intake, 290

Cyclic adenosine monophosphate (cAMP), 80
  regulation, 83
Cyclooxygenase-2 (COX-2), 212, 391.
  See also Glucocorticoids (GCs)
  anandamide metabolism by, 395–396
  dependent SAS activation, 290
  endocannabinoids metabolization by, 278–282
  inhibitor, 289
  knockout mice, 285
Cyclooxygenases (COXs), 267
  activity, 283
  blockers, 286
  inhibitors, 292
  isozymes, 283
Cys-loop ligand-gated ion channels (LGICs), 317
  CB receptor-independent effects of cannabinoids and AA on, 319–320
  structure and function of, 317–318
Cytochrome c, 67
  catalyzing formation of oleoylglycine and, 64–66
  effect of hydrogen peroxide on, 64
  as oleamide synthase, 62
  roles in programmed cell death, 68
Cytochrome P450, 267
Cytokine IL-1β, 292

D

Delta-9-tetrahydrocannabinol (¤9THC), 80
2-Deoxyglucose, 162
Deuterium-labeled NAGly, 193
Diacylglycerol (DAG), 241
Diarylimidazolyl oxadiazole, 168
DNA methylation, 453
DNA methyltransferase (DNMT), 453
Docosahexaenoic acid (DHA), 329–330
N-Docosahexaenoyl glycine, 198
Double-O-deacylation of NAPEs, 17

E

eCB. See Endocannabinoid
Eicosanoids, 160
Endocannabinoid, 2, 26, 113, 335–337, 364, 387, 417–419, 438–441, 466. See also Anandamide; Endocannabinoid system
  and AKT pathway, 445
  biosynthesis of, 490
  in bovine and human milk, 148
  control of GnRH/LH release, 372–374
  effects on
    basal locomotion of neutrophils, 343
    endocrine axis, 371–372
    induced migration of neutrophils, 343–344
    and excurrent duct system, 379–381
    fear conditioning, influence on, 426–429
      brain regions involved, 429–433
    inhibition of nACh receptors by, 328–329
  metabolization by, 278–282
  and receptors in developing nervous system, 147
  role in
    CB receptor-independent effect, 316–317
    synaptic plasticity, 316
  signaling
    in peripheral systems, 370–371
    in regulation of physiological processes, 209
    use in anticancer therapy, 479–480
Endocannabinoid 2-arachidonoylglycerol (2-AG), 115, 142
Endocannabinoid membrane transporter (EMT), 241
Endocannabinoid reuptake inhibitor OMDM-1, 143
Endocannabinoid system, 140, 160, 368–369, 443, 489
  anti-inflammatory effects, 492
  changes in cancers, 471–473
  control of energy balance, 161–162
  developmental manipulation, on offspring, 150–152
  in differentiating keratinocytes, 456–457
  in early gestation, 143
    CB1 receptors and preterm birth, 144–146
    preimplantation embryo and implantation, 141–144
  elements in human keratinocytes and, 444
  in neural development, 146–147
  for oviductal transport, 141
  and pathophysiology of liver diseases, 490–492
  possessing immunomodulatory activity, 212, 214
  postnatal development, 147–150
  role in development of testis, 374–379
Endogenous cannabinoids. See Endocannabinoid
Endogenous CB receptor lipid ligand, 26
Endogenous signaling lipids, with cannabimimetic activity, 192–193
Energy-independent condensation, 2
Epidermal differentiation, 446–447, 452
  transcription factors (TFs), 449–451
Epidermal growth factor (EGF), 237
Epidermal growth factor receptor (EGFR), 473, 480
Epidermal terminal differentiation, 445
  markers of keratinocyte differentiation, 448
  morphological and biochemical changes, 446
  stages, 446
    basal layer, 446–447
    cornified layer, 452
    granular layer, 447, 452
    spinous layer, 447
  transcriptional control, 453–455
  transcription factors (TFs) involved, 449–451
Epidermis, 445
  endocannabinoid system in, 455–456

Epoxins, 265
Epoxyeicosatrienoic acid (EET), 267
17β-Estradiol (E2), 236
Estrogens, 236
Ethanolamine, 29
Ets variant gene 5 (ETV5), 238
Extracellular signal-regulated kinase (ERK), 445
　cascade, 113

F

FAAH. *See* Fatty acid amide hydrolase
faah genes, 246
　transcription factors, 247–253
FAAH-like compounds, 142
FAT/CD36 expression, 34
FATP/VLACS motif, 34
Fatty acid amide hydrolase, 2, 28–30,
　81, 142, 209, 219, 368, 394–395, 443,
　471, 490
　activity of AMT and, 457
　AEA, substrate for, 241
　as integrator of fertility signals, 244–246
　knockout mice, 395
　leptin stimulatory effect, 291
　in pregnancy outcome, 142
　regulation, by FSH, 246
Fatty acid amide messengers, 56–57
Fatty acid translocase (FAT/CD36), 32–34
Fatty acid transport protein (FATP), 32, 34–35
Fatty acyl CoA synthetase, 35
Fear conditioning, 423–425
　behavioral of knockout mice in, 428
　forebrain structures related
　　amygdala, 424–425
　　hippocampus, 425
　　MPFC, 425
　influence of endocannabinoids, 426–429
　and effects of drugs, 427
Fetal 2-AG levels, 147
Fetal ECS development, maternal stress-induced
　changes in, 144
Fibroblast growth factor (FGF), 237
Focal adhesion kinase (FAK), 352
Follicle-stimulating hormone (FSH), 232.
　See also Sertoli cells
　activation and inhibition of MAPK
　　pathway, 236
　effect on AEA-induced apoptosis and, 243
　on ERK kinase activation, 236
　induces rapid Ca^{2+} influx, 236
　interaction with FSHR, 236
　PLA_2 in mechanism of action, 236
　responsive genes, 234
　signal transduction pathways activated by, 233
　transcription factors mediating, 235
N-Formly-L-methionyl-L-leucyl-L-phenylalanine
　(fMLP), 343
FSH-induced aromatase (ARO), 235

FSH receptor (FORKO), 232
FSH receptor (FSHR), 232

G

$GABA_A$ receptor
　function, 329
　modulation of function by fatty acids, 329–330
$GABA_C$ receptors, 330
Gap junctions, in programmed cell death, 68
$G\alpha_{q/11}$-PKC pathway-coupled receptor, 275
$G_\alpha 13$ signaling pathway, 124
GC-induced suppression of excitation
　(GSE), 274
Gdnf null allele, 238
Germ stem cells (GSCs), 238
Ghrelin, 275–276
Glial cell line-derived neurotrophic factor
　(GDNF), 237–238
Glial cells, 212
Glucagon-like peptide-1 (GLP-1), 289
Glucocorticoids (GCs), 264–265
　and COX_2, crosstalk between
　　neuroinflammation and neuroprotection,
　　282–285
　　synaptic plasticity and learning processes,
　　285–288
　inhibition of AA release from, 269–271
　nongenomic GC-induced activation of,
　　274–278
Glutamatergic principal neurons, 210
Glycerol-3-phosphate, 17
Glycerophosphodiesterases (GDEs), 17
Glycerophospholipid, 2
Glycine receptors, 95–96, 324
　cannabinoid modulation of, 325–328
Gonadotropin-releasing hormone (GnRH),
　232, 372
GP-NAE phosphodiesterase, 17
GP-N-arachidonoylethanolamine, 17
G protein-coupled receptor kinases
　(GRKs), 352
G protein-coupled receptors (GPCRs), 80, 112,
　163, 192, 339
GPR55 receptor (orphan G-coupled receptor),
　97–98, 117, 340–341, 348–350, 370, 470
　CB_1 receptors, interactions with, 126–128
　cellular signaling pathways, 123
　　Ca^{2+} signals, 124–126
　　$G_\alpha 12$ and $G_\alpha 13$, 123–124
　endogenous ligands for, 121
　　LPI and cannabinoids, biased agonism,
　　122–123
　　LPI, as signaling molecule, 122
　　lysophosphatidylinositol, 121–122
　patent reports, 118–119
　pharmacology, 119–121
GPR119 receptor, 98
GTPγS binding studies, 116

H

HEK293 cells, 339
  chemotaxis, 342
Hepatic inflammation, 492
Hepatic stellate cells (HSCs), 492
Heterologous desensitization, 350
Hippocampus, 286
Histamine, 290
Hox genes, 454–455
5-HT$_{3A}$/CB1-expressing neurons, 318
5-HT$_3$ receptor, 94–95, 318, 341
  inhibition by cannabinoids, 318, 321–324
HU210, cannabinoid agonists, 115
  hyporesponsiveness, 146
  promoting neurogenesis, 147
Human chorionic gonadotropin (hCG), 232
Human liver sinusoidal endothelial cells (HLSECs), 492
α/β-Hydrolase 4 (Abh4), 17
Hydroxyeicosatetraenoic acids (HETEs), 267
Hyperalgesia, 163
Hypothalamic–pituitary–adrenal (HPA) axis, 265, 274, 293
Hypothermia, 410

I

IB secretory PLA2 (sPLA2-IB), 16
IGF-binding proteins (IGF-BPs), 239
IL-1β expression in macrophages, 291
IL-12 gene expression, 215
Indomethacin, 289
Inositol 5'-phosphatase, 18
Insulin-like growth factor-I (IGF-I), 237, 239
Involucrin, 447

J

c-Jun N-terminal kinase (JNK), 124

K

Keratinocytes, 445
  differentiation, 446, 448
Keratohyalin granules, 447
Kit ligand (KL), 237–239
Kupffer cells, 492

L

Lactate dehydrogenase, 236
Lamellar bodies, 447, 452
Lamellipodia, 355
Leptin, 143, 275, 291
  defective *ob/ob* mice, 143
  receptor in orexigenic neurons, 161
Leptinergic signaling, 161

Leukocyte migration, and host immunity, 338–339
Leukotrienes, 265
LH-21
  oleoylethanolamide (OEA) efficacy enhancement, 174
  poor penetration into brain, 168
LH surge, 232
Ligand specificities, for GPR55, 129
N-Linoleoyl glycine, 198
Lipase, 17
Lipid signaling discoveries, 192
Lipocortin-1, 269
Lipopolysaccharides (LPS), 266
Lipoxins, 265
Lipoxygenases (LOXs), 267
Liver diseases, 490
Locomotion, 56
  of neutrophils, endocannabinoid effects, 343
Long-chain fatty acid amides. *See* Oleamide
LPS-induced anorexia, 291
Luteinizing hormone (LH), 232
Lysophosphatidylinositol (LPI), 121–122, 274, 341

M

MAG lipase, 291
  inhibitor, 290
Marijuana, 152, 290, 470, 486, 498
Mass spectrometry, 275
Matrix metalloproteinases (MMPs), 478
Medial prefrontal cortex (MPFC), 422, 425, 429
Melanocortin (MC$_1$–MC$_5$ receptors), 117
α-Melanocyte-stimulating hormone (MSH), 162
Melittin effect, 270
Met-AEA treatment, 477
Metastasis, 476
Met-F-AEA, effect on tumor growth, 177
Methyl-β cyclodextrin (MCD), 404
N-Methyl-D-aspartate-sensitive (NMDA) receptors, 284
MIR16, 17
Mitogen-activated protein kinase (MAPK), 113
  cascade, 124
  pathway, in cancer, 473
  phosphorylation, 162
Mitogen-activated protein kinase phosphatase-1 (MKP-1), 215
Monoacylglycerol lipase (MAGL), 143, 241
Monoacylglycerol (MAG), 279
Mullerian-inhibiting substance (MIS), 237
Multiple sclerosis, 212
MyoD, 453
Myosin light chain kinase (MLCK), 351
Myosins, 351

N

nACh receptors
  function, 328
  inhibition by endocannabinoids, 328–329
NAE-hydrolyzing acid amidase (NAAA), 2
NAPE-hydrolyzing phospholipase
  D, 3, 27, 272, 443
  function, 12–14
  independent pathways, 4, 17
  knockout mice, 394
  structure, 9–12
  tissue distribution, 14–15
NAPE-phospholipase D, 221
NAPE-PLD. See N-Acyl-phosphatidylethanolamine-selective phosphodiesterase; NAPE-hydrolyzing phospholipase D
nape-pld knockout mice, 443
NAPE-PLD$^{-/-}$ mice, 15, 17
NArPE. See N-Arachidonoyl-phosphatidylethanolamine
Neomycin, 18
Neurochemical and homeostatic switch, in central coordination, 293
Neuroglioma cells, apoptosis of, 475–476
Neuronal plasticity, 87–89
Neuropeptides urocortin, 289
Neuropeptide Y (NPY), 161
Neutral CB_1 receptor antagonists, 168
Neutrophils, 338
  cannabinoid receptor expression in, 344–345
  and cell motility, 338–339
  endocannabinoid effects, 343–344
  heterologous desensitization, 350
  inhibitory signal transduction mechanisms, 351–355
    actin–myosin motor unit, 351–352
    initiated by GPCR $\beta\gamma$-complex, 353
    initiated by GPCR G_α-subunits, 354
  receptor dimerization in, 350–351
  receptors involved, in inhibition of induced migration, 345–346
    CB_1/CB_2, 346–347
    novel cannabinoid receptors, 347–350
NF-κB proteins, 454
Nicotinic acetylcholine receptors, 95
Nitric oxide (NO), 266
NMDA receptors, 96
Nodose ganglion neurons (NGN), 321
Noladin, 115
Non-CB1/non-CB2 receptor-mediated responses, 115
Noradrenaline, 163, 289, 408
Normal human epidermal keratinocytes (NHEKs), 457
Nuclear signaling pathways, 83–85

O

O-1602 activated GTPγS binding, 119, 123
O-1918, analogue of abnormal cannabidiol, 115
obese gene, 244
Oleamide, 57
  biologic actions, 58–59
  biosynthesis, 59–60 (*see also* Apoptosis)
    by peptidylglycine alpha-amidating monooxygenase, 60–62
    protocols, need to be optimized for, 70
  natural occurrence, 57–58
  in programmed cell death, 68
Oleamide synthesome, 66–67
Oleoyl-CoA, 62
N-Oleoyl dopamine, 194
Oleoylethanolamide (OEA), 61, 66, 174
N-Oleoylethanolamine, 2
N-Oleoyl glycine (OlGly), 198–200
Oligozoospermia, 244
Olvanil (N-(3-methoxy-4-hydroxy)-benzyl-*cis*-9-octadecenoamide), 392, 394
μ Opioid agonist, 115
Orphan receptor GPR55. See GPR55 receptor
Otenabant, 179
Overlap syndromes, 493

P

PalGly metabolism, 197–198
N-Palmitoyl dopamine, 194
Palmitoylethanolamide (PEA), 341, 399, 479
Palmitoylethanolamide-preferring acid amidase (PAA), 490
N-Palmitoylethanolamine, 2, 17
N-Palmitoyl glycine
  biological activity, 195–197
  biosynthesis, 197
N-Palmitoyl-lyso-PE, 16–17
N-Palmitoyl-PE, 17
Paracetamol, 280
Parkinson's disease, 212
PAR-1 thrombin receptors, 123
Pavlovian classical fear conditioning. See Fear conditioning
Peripheral antagonist, of CB1 receptor, 168
Peroxisome proliferator-activated receptor (PPAR), 96, 212, 370
Pertussis toxin (PTX), 112
Phosphatase inhibitor, 18
Phosphatidylinositol 3'-kinase (PI3K), 113, 162
  signaling pathway, 234
Phosphodiesterase (PDE), 17, 234
Phospholipase A_2 (PLA$_2$), 236
Phospholipase C (PLC), 241
Phospholipase (PL) D-type hydrolysis, 3
Photoaffinity labeling, 113, 124
Phylogenetic tree, for *CiCBR*, 117
PI3K/PKB signaling pathway, 235

PI-PLCβ isozymes, 277
PLA$_1$/A$_2$ activity to hydrolyze N-palmitoyl-PE, 16
PLA$_2$ inhibitors, 290
Plasma membrane fatty acid binding protein (FABPpm), 32–33
Plasmenylethanolamine, 29
PLC inhibitor, 18
PLC-like enzyme-dependent conversion, 18
PLC-mediated pathway, 18
PLD-type hydrolyzing activities, 16
Polymorphic CB1 expression and treatments, 179
Positron emission tomography, 172
Postmitotic cells, 447
Potassium channels, 86–87
Prednisone, 493
Prenatal stress, 144–145
p115 Rho guanine nucleotide exchange factor (p115Rho-GEF), 124
Primary biliary cirrhosis (PBC), 492
Primary sclerosing cholangitis (PSC), 492
Prostaglandin biosynthesis, 267
Prostaglandin D2 (PGD2), 266
Prostaglandin E2 (PGE2), 266
Prostaglandin ethanolamides, 396
Prostanoid DP$_1$ and DP$_2$ receptors, 117
Prostanoids, 289–290, 294
Prostate cancer cells, antiproliferative effects of AEA in, 473
Protein kinase, 113
Protein kinase A (PKA), 233, 352, 390
p38/STAT3 signaling pathway, 459
PTX sensitivity and cannabinoid signaling, 112
PVN neurons, 292
P2Y$_2$ purinoceptors, 123
Pyrazole derivative, 168

R

RAW macrophage cells, 282–283
Receptor
  crosstalk
    heterologous desensitization, 350
    receptor dimerization, 350–351
  dimerization, 350–351
  for lysophosphatidic acid, 117
  properties, 368–371
RhoA activity, 355
Rho-GTPases, 355
Rho-kinase/Rho-dependent coiled-coil kinase (ROCK), 352
Rimonabant, 143, 163, 342
  anticancer and anti-inflammatory effects, 177, 179
  attenuating antitumor effects, 176
  clinical efficacy, 169–172
  decreasing levels of proinflammatory cytokines, 178

exertsing immunomodulatory and anti-inflammatory effects, 178
in LDLR$^{-/-}$mice, 178
potential antitumor action, 177
treated patients, 171

S

Schizophrenia-like symptoms, 144
Serotonin receptors, 318
Sertoli cells, 233–234, 376
  activities and biological relevance, 237–239
  cAMP-induced genes in, 234
  ECS in, 241–243
  paracrine growth factors and action, 240
  regulation of FAAH by FSH in, 243–244
Sertoli–Sertoli adherent junction, 236
Serum and glucocorticoid-induced kinase 3 (SGK), 234
Serum aspartate (AST), 493
Skin differentiation, transcriptional control, 453–455
27 (SLC27) of proteins, 34
Sodium orthovanadate, 18
Spermatogenesis, 237
  cannabinoids affect on, 374–375
Sphingomyelin, 29
Sphingosine-1-phosphate, 117
SR144528, CB$_2$ antagonist, 115
Src-protein tyrosine kinases (srcPTKs), 352
START protein, 35–36
N-Stearoyl dopamine, 194
N-Stearoyl glycine, 198
Sulfonamide analogue of D8-THC, 168
Superoxide dismutase, 66
Sympathoadrenomedullary system (SAS), 288

T

Tansacylation–phosphodiesterase pathway, 2–3
Taranabant, 172
TASK-1 channel, 94
TATA box, 234
Teratozoospermia, 244
Testis, and ECBS role, 374–379
Testosterone, 244
12-O-Tetradecanoylphorbol 13-acetate (TPA), 456
Δ^9-Tetrahydrocannabinol (THC), 112, 141, 148, 150–151, 160, 163, 192, 208, 244, 272, 326, 340, 486
Tetrazole–biarylpyrazole analogues, 168
Theiler's murine encephalomyelitis virus-induced demyelinating disease (TMEV-IDD), 216, 219
T helper 2 (Th2) cytokines, 245
Thermoregulation, 56
Thiadiazole, 168
Thromboxane A(2) synthase, 289

Index

Thyroid-stimulating hormone (TSH), 232
Tissue inhibitors of MMPs (TIMPs), 478
Tonofilaments, 447
Transcription factor
   mediating FSH response, 235
   NFAT, 125
Transforming growth factor-β (TGFβ), 237
Transglutaminase (TG), 445, 452
Transient receptor potential (TRP) family of proteins, 90–91
Transient receptor potential vanilloid type-1, 27, 90–91
   activation, 401–402
   desensitization with capsaicin, 114
   knockout mice, 407, 409
   in nervous system, 91
   physiological actions, of anandamide on, 404–410
   as physiological target of AEA, 239
   receptor signaling, 28
   regulation of, 398
   in vasodilation and bronchoconstriction, 92
Transit amplifying (TA) cell, 447
Transmembrane domains (TMDs), 117
1,2,4-Triazole, 168
TRPV1. *See* Transient receptor potential vanilloid type-1
Tumor necrosis factor-α (TNF-α), 344
Type 2 diabetes, 160
Tyrphostins AG18 and AG561, 122

V

Vanilloid receptor 1, 366, 386. *See also* Anandamide; Arachidonylethanolamide; Transient receptor potential vanilloid type-1
   expression of, 390–391
   functions, 391
Vasodilation, 408–409
Vasodilator-associated protein (VASP) pathway, 124
Vasopressin, 265, 274, 290
Very long-chain acyl-CoA synthetase (VLACS), 34
Virodhamine (O-arachidonoylethanolamine), 343–344, 391
VLACS protein, 35
Voltage-gated Ca^{2+} channels, 92–93
Voltage-gated potassium channels, 94
Voltage-gated sodium channels, 93

W

WIN55,212–2, cannabinoid agonists, 115–116

X

X sPLA2, 17

Z

Zymosan, 281